Frontiers in Earth Sciences

Series Editors: J.P. Brun, O. Oncken, H. Weissert, W.-C. Dullo

Victor A. Melezhik *Editor-in-Chief*
Anthony R. Prave
Eero J. Hanski
Anthony E. Fallick
Aivo Lepland
Lee R. Kump
Harald Strauss *Editors*

Reading the Archive of Earth's Oxygenation

Volume 1: The Palaeoproterozoic of Fennoscandia as Context for the Fennoscandian Arctic Russia - Drilling Early Earth Project

Springer

Editor-in-Chief
Victor A. Melezhik
Geological Survey of Norway,
Trondheim

Centre of Excellence in Geobiology
University of Bergen
Norway

Editors
Anthony R. Prave
Department of Earth Science
University of St Andrews
United Kingdom

Eero J. Hanski
Department of Geosciences
University of Oulu
Oulu
Finland

Anthony E. Fallick
Environmental Research Centre
Scottish Universities
East Kilbride
United Kingdom

Aivo Lepland
Geological Survey of Norway
Trondheim
Norway

Lee R. Kump
Department of Geosciences
Pennsylvania State University
Pennsylvania
USA

Harald Strauss
Institut für Geologie
Westfälische Wilhelms-Univ. Münster
Münster
Germany

ISBN 978-3-662-52201-1 ISBN 978-3-642-29682-6 (eBook)
DOI:10.1007/978-3-642-29682-6
Springer Heidelberg New York Dordrecht London

Printed on acid-free paper

Springer is part of Springer Science+Business Media (www.springer.com)

Dedication

The editors respectfully dedicate this three-volume treatise to Dr. Alexander Predovsky of the Geological Institute of the Russian Academy of Sciences in Apatity. He is one of the earliest explorers of the Precambrian geology in Russian Fennoscandia, and his half century of active work on the geochemistry of sedimentary and igneous rocks provided important foundations for the current understanding of Palaeoproterozoic stratigraphy, geochemistry of sedimentary and volcanic processes and ore formation in the region.

Contributors to Three-Volume Treatise Reading the Archive of Earth's Oxygenation, Volume 1: The Palaeoproterozoic of Fennoscandia as Context for the Fennoscandian Arctic Russia - Drilling Early Earth Project

Anthony E. Fallick Scottish Universities Environmental Research Centre, Rankine Avenue, East Kilbride, Glasgow G75 0QF, Scotland, UK

Eero J. Hanski Department of Geosciences, University of Oulu, P.O. Box 3000, 90014 Oulu, Finland

Lee R. Kump Department of Geosciences, Pennsylvanian State University, 503 Deike Building, University Park, PA 16870, USA

Pavel V. Medvedev Institute of Geology, Karelian Research Centre, Russian Academy of Sciences, Pushkinskaya 11, 185910 Petrozavodsk, Russia

Victor A. Melezhik Geological Survey of Norway, Postboks 6315 Sluppen, NO-7491 Trondheim, Norway/Centre for Geobiology, University of Bergen, Allegaten 41, Bergen N-5007, Norway

Anthony R. Prave Department of Earth Science, University of St. Andrews, St Andrews KY16 9AL, Scotland, UK

Sergey A. Svetov Institute of Geology, Karelian Research Centre, Russian Academy of Sciences, Pushkinskaya 11, 185910 Petrozavodsk, Russia

Reviewers for Three-Volume Treatise: Reading the Archive of Earth's Oxygenation

Prof. Ariel Anbar
Arizona State University, Tempe, AZ, USA
Dr. Victor Balagansky
Geological Institute, Apatity, Russia
Prof. Mark Barley
University of Western Australia, Crawley, Australia
Prof. Jochen Brocks
The Australian National University, Canberra, Australia
Prof. Louis Derry
Cornell University, Ithaca, NY, USA
Prof. Anton Eisenhauer
GEOMAR, Kiel, Germany
Prof. Anthony E. Fallick
Scottish Universities Environmental Research Centre, East Kilbride, Scotland, UK
Prof. David Fike
Washington University, St. Louis, USA
Prof. Karl Föllmi
University of Lausanne, Lausanne, Switzerland
Prof. Robert Frei
University of Copenhagen, Copenhagen, Denmark
Dr. Dieter Garbe-Schönberg
Christian-Albrechts-Universität, Kiel, Germany
Prof. Eero Hanski
University of Oulu, Oulu, Finland
Prof. Raimo Lahtinen
Geological Survey of Finland, Espoo, Finland
Prof. Michail Mints
Geological Institute, Moscow, Russia
Prof. David Mossman
Mount Allison University, Sackville, NB, Canada
Prof. Richard Ojakangas
University of Minnesota, Duluth, USA
Prof. John Parnell
University of Aberdeen, Aberdeen, Scotland, UK
Prof. Adina Paytan
University of California, Santa Cruz, USA
Dr. Bernhard Peucker-Ehrenbrink
Woods Hole Oceanographic Institution, Woods Hole, MA, USA

Prof. Igor Puchtel
University of Maryland, College Park, MD, USA
Prof. Robert Raiswell
University of Leeds, Leeds, UK
Prof. Robert Riding
University of Tennessee, USA/
Cardiff University, Wales, UK
Dr. Nathaniel Sheldon
University of Michigan, Ann Arbor, MI, USA
Dr. Graham Shields-Zhou
University College, London, UK
Dr. Craig Storey
University of Portsmouth, Portsmouth, UK
Prof. Kari Strand
Thule Institute, University of Oulu, Oulu, Finland
Prof. Frances Westall
Centre de Biophysique Moléculaire, CNRS, Orléans, France

Acknowledgements

The idea of making an atlas with comprehensive descriptions and illustrations of the Palaeoproterozoic rocks from the Fennoscandian Shield was initiated in 2009 during a workshop held in Trondheim, Norway, under the auspices of the International Continental Scientific Drilling Program (ICDP). Starting from this workshop, a plan was developed and finalised. Chris Bendall, Senior Editor for Springer, is acknowledged for encouragement and editorial supervision of the project.

The three-volume set has three major underpinnings. The first is many years of research in Precambrian geology of the Fennoscandian Shield by many workers, and we acknowledge particularly the support of the Geological Survey of Norway; the University of Oulu, Finland; and the Institute of Geology, Petrozavodsk, Russia.

The second is the unique core material obtained during the drilling operations by the Fennoscandian Arctic Russia – Drilling Early Earth Project (FAR-DEEP). The drilling operations were largely supported by the ICDP and by additional funding from several other agencies and institutions. We are grateful for the financial support to the Norwegian Research Council (NFR), the German Research Council (DFG), the National Science Foundation (NSF), the NASA Astrobiology Institutes, the Geological Survey of Norway (NGU) and the Centre of Excellence in Geobiology, the University of Bergen, Norway. The core archive and associated analytical work were supported by NGU, the Scottish Universities Environmental Research Centre (SUERC) and by the Pennsylvanian State University.

The third is a multidisciplinary approach to investigate complicated geological processes. This was provided by the international scientific community and we acknowledge the support of many universities in Scandinavia, Europe and the USA.

Many individuals helped in the preparation of the drilling operations and offered logistical support. We are sincerely grateful to Anatoly Borisov (Kola Geological Information and Laboratory Centre) for providing geological assistance for precise positioning of drillholes in unexposed, boggy and forested terrains in the Imandra/Varzuga Greenstone Belt. Stanislav Sokolov (Kola Mining Metallurgical Company) is acknowledged for logistic support and geological guidance in the Pechenga Greenstone Belt. Logistical organisation of the drilling operations and core transport across national borders by the State Company Mineral, St. Petersburg, Russia, is appreciated. The Finnish company, SMOY, performed the drilling.

Many organisations and people have provided rock samples, photographs, SEM images and permission to use figures. Thanks are due to the Geological Museum of the Department of Geosciences, University of Oulu; Geological Museum of the Geological Institute, Kola Science Centre, Apatity; Geological Survey of Finland; and to the following people: Wlady Altermann, Lawrence Aspler, Alex Brasier, Ronald Conze, Alenka Črne, Kathleen Grey, Jens Gutzmer, Eero Hanski, Emmanuelle Javaux, Yrjö Kähkönen, Vadim Kamenetsky, Reino Kesola, Andrew Knoll, Kauko Laajoki, Reijo Lampela, Aivo Lepland, Kevin Lepot, Zhen-Yu Luo, Vladimir Makarikhin, Tuomo Manninen, Jukka Marmo, Nicola McLoughlin, Pavel Medvedev, Victor Melezhik, Satu Mertanen, Tapani Mutanen, Lutz Nasdala,

Richard W. Ojakangas, Domenic Papineau, Petri Peltonen, Vesa Perttunen, Vladimir Pozhilenko, Anthony Prave, Igor Puchtel, Jorma Räsänen, Pentti Rastas, Raimo Ristimäki, Alexander Romashkin, Dmitry Rychanchik, Ronny Schoenberg, Evgeny Sharkov, Igor Sokolov, Hubert Staudigel, Kari Strand, Sergey Svetov, Vladimir Voloshin, Frances Westall, Grant Young, Valery Zlobin and Bouke Zwaan.

Unpublished geochemical data were kindly provided by the Geological Survey of Finland, Zhorzh Fedotov, Valery Smol'kin and Peter Skuf'in. Permission to use published material was kindly given by the Royal Society of Edinburgh, Elsevier, John Wiley and Sons and Blackwell Publishing Ltd.

The book preparation was supported by the NFR (grant 191530/V30 to Victor Melezhik), Natural Environment Research Council (grant NE/G00398X/1 to Anthony Fallick, Anthony Prave and Daniel Condon), DFG (grants Str281/29, 32, 35 to Harald Strauss), NASA (grant NNA09DA76A to Lee Kump), NSF (grant EAR 0704984 to Lee Kump) and the Academy of Finland (grant 116845 to Eero Hanski).

To be embedded in the family of science always requires sacrifices such as time lost in family contact. We wish to extend our gratitude to our families for patience, understanding and constant encouragement.

Finally and most importantly, the editors wish to thank those colleagues and students who will use and read these books or some parts of them. We hope that this will encourage them to reach a more complete understanding of those processes that played an important role in the irreversible modification of Earth's surface environments and in shaping the face of our emerging aerobic planet. We would also like to thank those scientists who will use the offered advantage of rich illustrative material linked to the core collection to undertake new research projects.

Preface to Volume 1

Earth's present-day environments are the outcome of a 4.5-billion-year period of evolution reflecting the interaction of global-scale geological and biological processes. Punctuating that evolution were several extraordinary events and episodes that perturbed the entire Earth system and led to the creation of new environmental conditions, sometimes even to fundamental changes in how planet Earth operated. One of the earliest and arguably the greatest of these events was a substantial increase (orders of magnitude) in the atmospheric oxygen abundance, sometimes referred to as the Great Oxidation Event. Given our present knowledge, this oxygenation of the terrestrial atmosphere and the surface ocean, during the Palaeoproterozoic Era between 2.4 and 2.0 billion years ago, irreversibly changed the course of Earth's evolution. Understanding why and how it happened and what its consequences were are among the most challenging problems in Earth sciences.

The three-volume treatise entitled "Reading the Archive of Earth's Oxygenation" (1) provides a comprehensive review of the Palaeoproterozoic Eon with an emphasis on the Fennoscandian Shield geology; (2) serves as an initial report of the preliminary analysis of one of the finest lithological and geochemical archives of early Palaeoproterozoic Earth history, created under the auspices of the International Continental Scientific Drilling Programme (ICDP); (3) synthesises the current state of our understanding of aspects of early Palaeoproterozoic events coincident with and likely related to Earth's progressive oxygenation with an emphasis on still-unresolved problems that are ripe for and to be addressed by future research. Combining this information in three coherent volumes offers an unprecedented cohesive and comprehensive elucidation of the Great Oxidation Event and related global upheavals that eventually led to the emergence of the modern aerobic Earth System.

The format of these books centres on high-quality photo-documentation of Fennoscandian Arctic Russia – Drilling Early Earth Project (FAR-DEEP) cores and natural exposures of the Palaeoproterozoic rocks of the Fennoscandian Shield. The photos are linked to geochemical data sets, summary figures and maps, and time-slice reconstructions of basinal and palaeoenvironmental settings that document the response of the Earth system to the Great Oxidation Event. The emphasis on a thorough, well-illustrated characterisation of rocks reflects the importance of sedimentary and volcanic structures that form a basis for interpreting ancient depositional environments, and chemical, physical and biological processes operating on Earth's surface. Most of the structural features are sufficiently complex as to challenge the description by other than a visual representation, and high-quality photographs are themselves a primary resource for presenting essential information. Although nothing can replace the wealth of information that a geologist can obtain from examining an outcrop first hand, the utility of photographs offers the next best source of data for assessing and evaluating palaeoenvironmental reconstructions. This three-volume treatise will, thus, act as an information source and guide to other researchers and help them identify and interpret such features elsewhere, and will serve as an illustrated guidebook to the Precambrian for geology students.

Finally, the three-volume treatise provides a link to the FAR-DEEP core collection archived at the Geological Survey of Norway. These drillcores are a unique resource that can be used to help solve the outstanding problems in understanding the causes and consequences of the multiple processes associated with the progressive oxygenation of terrestrial environments. It is anticipated that the well-archived core will provide the geological foundation for future research aimed at testing and generating new ideas about the Palaeoproterozoic Earth. The three-volume treatise will be of interest to researchers involved directly in studying this hallmark period in Earth history, as well as professionals and students interested in Earth System evolution in general.

Volume 1: "The Palaeoproterozoic of Fennoscandia as Context for the Fennoscandian Arctic Russia – Drilling Earth Project" describes the implementation of the FAR-DEEP drilling project in Arctic Russia. It summarises the knowledge of more than 50 years of largely Russian-led fieldwork, information hitherto virtually unavailable in the West, and provides geological description of drilling areas with an exhaustive illustration of rocks by high-quality, representative photographs. The volume offers a comprehensive review and rich photo-illustration of palaeotectonic, palaeogeographic and magmatic evolution of the Fennoscandian Shield in the early Palaeoproterozoic and links the evolution of the shield to the emergence of an aerobic Earth system. The volume unfolds the event-based Fennoscandian chronostratigraphy and discusses the chronology of the Palaeoproterozoic global events as the basis for a new subdivision of Palaeoproterozoic time.

Welcome to the illustrative journey through one of the most exciting periods of planet Earth!

Contents to Volume 1

Contents to Volume 2

Contents to Volume 3

Part I

Palaeoproterozoic Earth

1.1 Tectonic Evolution and Major Global Earth-Surface Palaeoenvironmental Events in the Palaeoproterozoic

V.A. Melezhik, L.R. Kump, E.J. Hanski, A.E. Fallick, and A.R. Prave

Many, if not all, of the long-term fluctuations in geological processes operating on Earth's surface are tectonically driven and related to the interplay of plate tectonics and deep mantle dynamics resulting in supercontinental cycles and (super) plume events (Condie et al. 2001; Condie 2004). These processes include the amalgamation, dispersal, collision and geographic position of major land-masses which dictate volcanic and hydrothermal activities, changes in sea level and the global patterns of ocean circulation, thermal isolation of continents, climate change, rate of continental weathering and its influence on seawater composition, and atmospheric oxygen budget via control of burial and recycling of carbon and sulphur. Further, all of these are reflected in biological processes. However, well-documented and well-constrained examples of this conceptual model have been developed and tested largely on Phanerozoic rocks (Valentine and Moores 1970; Fischer 1984; Marshall et al. 1988; Hardebeck and Anderson 1996; Berner 2006; Rampino 2010). Although there have been a number of attempts to apply such concepts to "Deep Time", in particular, the Palaeoproterozoic (Nance et al. 1986; Windley 1993; Lindsay and Brasier 2002; Condie et al. 2009), testing and verification of the models is challenging. The existence of continental masses, their palaeogeography and sizes in the late Archaean-early Palaeoproterozoic remain hypothetical and robust plate reconstructions are hampered by the small number of reliable palaeomagnetic data (Evans and Pisarevsky 2008).

In this review we will focus on those global environmental events that can be documented by physical evidence (i.e. features preserved in rocks) or well-established geochemical proxies (e.g. S and C isotope systems). Hence, this is not a comprehensive review of tectonically driven processes operating on Earth's surface during the Palaeoproterozoic; such details can be found in Reddy and Evans (2009). We limit this overview to the time interval, which closely preceded, was associated with, and postdated the emergence of an aerobic Earth system between 2.5 and 2.0 Ga (e.g. Melezhik et al. 2005a). We will focus on the series of environmental upheavals whose understanding has great potential to address a key question: what caused the irreversible oxygenation of the terrestrial atmosphere (the answers to which currently remain highly uncertain; Kump 2008)?

1.1.1 Late Archaean Tectonic Restructuring

The pattern of radiometric ages for Palaeoproterozoic rocks exhibits two prominent peaks at 2.7–2.6 and 1.9–1.8 Ga (Gastil 1960; Worsley et al. 1984; Nance et al. 1986, 1988; Condie 1995, 1998, 2000; Campbell and Allen 2008; Condie et al. 2009). Both episodes are seen best in North America (Hoffman 1988, 1989) and have also been documented in the Fennoscandian Shield (e.g. Hanski et al. 2001; Nironen 2005; Lahtinen et al. 2005), though the main Archaean crustal growth episode seems to have occurred slightly earlier, at c. 2.8 Ga (e.g. Bibikova et al. 2005; Käpyaho et al. 2006; Mikkola et al. 2011). The major global growth peak of juvenile crust at 2.7 Ga has been suggested to be linked with a catastrophic mantle overturn event in a transitional period when mantle dynamics changed between layered mantle convection to whole-mantle convection favouring modern-day plate tectonic processes (Arndt 2004; Nelson 2004).

Palaeogeography

Archaean palaeotectonic reconstructions are still in their infancy (Sorjonen-Ward and Luukkonen 2005). Bleeker (2003) suggested that the global Archaean record includes c. 35 large cratonic fragments and several smaller slivers.

V.A. Melezhik (✉)
Geological Survey of Norway, Postboks 6315 Sluppen, NO-7491 Trondheim, Norway

Centre for Geobiology, University of Bergen, Allegaten 41, N-5007 Bergen, Norway
e-mail: victor.melezhik@ngu.no

V.A. Melezhik et al. (eds.), *Reading the Archive of Earth's Oxygenation*, Volume 1: The Palaeoproterozoic of Fennoscandia as Context for the Fennoscandian Arctic Russia - Drilling Early Earth Project,
DOI 10.1007/978-3-642-29682-6_1, © Springer-Verlag Berlin Heidelberg 2013

Three very different late Archaean global plate reconstructions have been proposed based on palaeomagnetic data with acknowledged limitations (Mertanen and Pesonen 2005; Buchan et al. 2000; Halls et al. 2008; Evans and Pisarevsky 2008). Bleeker (2003) proposed two: one reconstruction includes all the cratons as a single supercraton named Kenorland, while the other proposes three separate supercontinents termed Vaalbara, Superia and Slavia, distinguished by slightly different cratonisation ages of 2.9, 2.7, and 2.6 Ga, respectively. The third of these reconstructions suggests the existence of two supercontinents, "Kenorland" and "Zimvaalbara" (for detailed discussion, see Reddy and Evans 2009). Kenorland includes the Laurentian, Fennoscandian and Siberian shields, whereas Zimvaalbara arguably comprises Zimbabwe, Kaapval, Pilbara, and perhaps the São Francisco and Indian cratons (Williams et al. 1991; Aspler and Chiarenzelli 1998).

Regardless of the exact number of supercontinents, geological evidence indicates that the late Archaean – early Palaeoproterozoic was a time of protracted break-ups, driven by inferred mantle plumes and associated intraplate rifting. Break-up of Zimvaalbara started at 2.65 Ga, Kenorland at 2.45 Ga and the final fragmentation of these supercontinents was during the 2.42–2.0 Ga time interval (Aspler and Chiarenzelli 1998). The initial break-up of Kenorland started at a low palaeolatitude (0–20 °; Christie et al. 1975) and was associated with the formation of a large continental flood basalt province including giant radiating dyke swarms and layered gabbro intrusions (Heaman 1997; Vogel et al. 1998). Attenuation of Zimvaalbara involved crust- and mantle-driven mafic magmatism at 2.47–2.43 Ma (Jones et al. 1975; Hamilton 1977) followed by rifting. In Australia, a 2.45 Ga mantle plume was also emplaced equatorially (05 °; Evans 2003), generating a large igneous province. Vast deposition of the 2.48–2.46 Ga banded iron formations in fringing seas, (Figs. 1.1 and 1.2a) coeval with initial rifting in Kaapvaal and Pilbara (Barley et al. 1997; Nelson et al. 1999; Martin et al. 1998; Pickard 2003) differentiates Zimvaalbara Paleoproterozoic events from those of Kenorland. This difference has been ascribed to contrasting palaeotectonic settings involving convergent (Kenorland) and rifted (Kaapvaal and Pilbara) margins of the large late Archaean cratons (Barley et al. 2005). In contrast, the latest significant BIF deposition in Kenorland (Fig. 1.2b) is dated at c. 2750 Ma (e.g. Bibikova 1989).

1.1.2 Newly Born Supercontinent(s) Through 2.4–c. 2.3 Ga Time Interval

Despite apparent differences in various palaeogeographic reconstructions, major cratons share several common features through the Archaean-Palaeoproterozoic transition: low-latitudinal and high-standing position, mantle plume generated gabbro plutons (Fig. 1.2c, d) and intraplate rifting, formation of large igneous provinces and widespread magmatic shut- or slowdown. These features were apparently palaeotectonic and palaeogeographic requirements for a series of unprecedented palaeoenvironmental upheavals that began around 2.4 Ga.

Global Magmatic Shut- or Slow-Down

The global distribution of U-Pb ages of subduction-associated granitoids and ages of detrital zircons suggest a widespread reduction in terrestrial magmatic activity and a significant lull in volcanism and crustal production during the c. 2.45–2.2 Ga time interval (Condie et al. 2009). It is also suggested that this was the period when the global subduction system shutdown since most terrestrial volcanism is related to plate tectonics. The mechanisms by which the global subduction system could completely shutdown were advanced by O'Neil et al. (2007) and Silver and Behn (2008).

Condie et al. (2009) provided a plausible causative link between magmatic stagnation and several global events. It is suggested that a drastic decrease of Fe^{2+} delivery to the oceans via submarine volcanism could shut down deposition of banded iron formations. This, combined with a drop in the rate of CO_2 venting from the mantle, could reduce the sink for oxygen thus leading to a cooler, more oxidising atmosphere capable of supporting global glaciation. It is important to point out that such a link offers a complementary "geotectonic" component, seemingly missing from the widely accepted explanations in which major modifications related to atmospheric composition and climatic changes are driven either geochemically or biogeochemically (Pavlov et al. 2000; Kasting 2004, 2005; Kopp et al. 2005).

Condie et al. (2009) also emphasise that the shutdown of the global subduction system should cool oceanic lithosphere, which would decrease ocean ridge volume, thus leading to deeper oceanic basins and a lower sea level (Moucha et al. 2008). Inevitable results include extensive and deep erosion of continental shelves, widespread unconformities, and the incompleteness of craton-margin sedimentary successions spanning the 2.45–2.2 Ga time-interval. This, combined with the scarcity of magmatic rocks suitable for dating, hampers the chronological reconstruction of events and the causative relationships between various factors operating during this enigmatic part of Earth's history.

Precise dating of sedimentation, perhaps by the Re-Os technique, and tracking of continental erosion rate by $^{87}Sr/^{86}Sr$ ratio in marine carbonates, represent major, and challenging, components of future research. Interestingly, the

Fennoscandian Shield is marked by an intensive intermediate-mafic volcanism dated at 2.43 Ga, and perhaps even younger (see Chaps. 3.2, 3.3, and 3.4), thus covering at least part of the period otherwise marked by stagnant magmatic activity. Sedimentary formations of the same age contain marine carbonate rocks that could be targeted for chemostratigraphy.

Global Glaciation and Oxygenation of Atmosphere

Radical modification of the Earth's surface environments may have started with the seemingly rapid onset of wide-spread glaciation (Fig. 1.2e, f; Young 1970; Visser 1971; Miall 1983; Marmo and Ojakangas 1984; Martin 1999; Young et al. 2001) from otherwise climatically stable conditions in the early Palaeoproterozoic (for details see Chap. 7.2).

With a partial overlap in time, this wide-spread glaciation was accompanied by the disappearance of mass-independent fractionation of sulphur isotopes (MIF-S) (Guo et al. 2009; for details see Chap. 7.1), a significant rise in atmospheric oxygen concentration (the "Great Oxidation Event"; Holland 2006), and the onset of a short-term global (however, for opposite view see Frauenstein et al. 2009), positive $\delta^{13}C$ anomaly in sedimentary carbonates reported from the Duitschland Formation in the Transvaal Supergroup of South Africa (Bekker et al. 2001). The MIF-S disappearance occurs between the first and the second (out of two) glacial units and can be constrained between 2.48 and 2.32 Ga (Hannah et al. 2004). A somewhat similar pattern is reported by Papineau et al. (2007) from within the Huronian Supergroup, Ontario, Canada, where the MIF-S signal was lost after the second (out of three) glacial unit. Similar to the Transvaal Supergroup, radiometric age constraint for the onset of the glaciation and the disappearance of MIF-S remains imprecise (between 2.45 and 2.3–2.2 Ga), as does the onset of an oxidising atmosphere.

Hence, it is evident that the Palaeoproterozoic started with several extraordinary events associated with radical changes in Earth surface environments and composition of the atmosphere. The question remains: what were the driving forces for such rapid and unprecedented modification of Earth surface conditions after several hundreds of million years of relative stability?

Several hypotheses advanced for explanation of these early Palaeoproterozoic events invoke the onset of oxygenic bacterial photosynthesis and the collapse of a methane-rich atmosphere; however, all suggested hypotheses have conflicting cause-and-effect relationships (Pavlov and Kasting 2002; Kopp et al. 2005; Zahnle et al. 2006; Konhauser et al. 2009). But, did the tectonically-driven palaeogeographic restructuring of major land masses and related changes in magmatic and metamorphic systems play any role in the geochemical and climatic modification of the Earth system (Catling and Claire 2005; Kump and Barley 2007; Condie et al. 2009)?

With tectonics as the major factor in mind, Barley et al. (2005) addressed the cause-and-effect relationships between tectonics and early Palaeoproterozoic environmental changes. Mantle plume breakouts were suggested as factors limiting ocean productivity, the rate of photosynthesis, and maintaining chemically reduced conditions throughout the Archaean. They also proposed that the establishment of large, thick and stable continental land masses (i.e. supercontinents, e.g. Kenorland) and a change in the style of plate tectonics at c. 2.4 Ga led to an increase in ocean productivity and photosynthesis rates. This, complemented by a decreased influx of chemically reduced gases produced by subaerial volcanism (Kump and Barley 2007), resulted in a reduced oxygen sink, the removal of methane greenhouse conditions and the build-up of an oxic atmosphere. The approach is definitely attractive and worthy of development; however, the resultant outcome appears to be convoluted due, mainly, to the incomplete geological record and poorly understood history of Palaeoproterozoic supercontinents and supercratons (Reddy and Evans 2009).

The most crucial factor hampering our understanding of the chronological sequence of events and their causative relationships during the 2.4–2.3 Ga time interval is the limited number of precise radiometric ages that may constrain onsets of different processes operating on Earth at that time. This concerns the important question: which of the three glacial episodes recorded in Canada correlate with the two reported in Africa and one known in Scandinavia? Which were global in nature? The incompleteness of the geological record, combined with a limited number of well-studied sections, causes a serious bias in geological and geochemical interpretations.

The disappearance of MIF-S is recorded in only two sections (in S. Africa and Canada) and so far cannot provide robust information on internal structure of such process. Was it an abrupt, single, irreversible process? Or perhaps it was marked by several reversals that signify a transitional period from anoxic to oxic conditions. Moreover, there is a high potential that some basal Palaeoproterozoic beds can still contain older detrital sulphides retaining MIF-S. This seems to be the case for some sulphides associated with the first glacial deposits above the BIFs in the Hamersley Group. They carry an MIF-S signal, however, the dating by a Re-Os technique shows that they are significantly older than the inferred age of the glaciations; a significant amount of sulphides in Archaean shales in Western Australia also have older detrital sulphides (Barley, personal communication). Consequently Re-Os dating of sulphides should be employed more widely in the study of sulphur isotopes during the Archaean-Palaeoproterozoic transitional time. This will help to confirm when the MIF stopped and how long it took for oxidative weathering to be established and stop older detrital sulphides retaining MIF-S being eroded, transported and deposited in younger sediments. This should also assist in unraveling the cause and effect of tectonics and the early Palaeoproterozoic environmental upheavals which continues to be a challenge.

1.1.3 Rifted Supercontinent(s) Through c. 2.3–2.1 Ga

The world database on igneous radiometric ages, including those from the Fennoscandian Shield (Hanski et al. 2001), suggests a rather stable tectonic regime with minimal or no collisional/accretional processes throughout the c. 2.4 to 2.2 Ga period (Condie et al. 2009). The response to this prolonged phase of high-standing, stable continents was the apparent formation of deeply weathered surfaces (e.g. Marmo 1992) and widespread unconformities (Condie et al. 2009; also see Chap. 7.9.2). Intraplate rifting resumed at c. 2.2 Ga and resulted in erosion of the deeply weathered rocks (Fig. 1.2g) and deposition of chemically mature arenites (Fig. 1.2h). This was accompanied by carbonate sedimentation in coeval, wide-spread, shallow-water platforms and epeiric seas.

Formation of Giant Manganese Deposits and World-Wide "Red Beds"

The depositional age of the Hotazel Formation containing a giant Mn deposit (Figs. 1.1 and 1.2i–j) remains poorly constrained between 2.4 and 2.2 Ga but very likely close to 2.4–2.3 Ga as, the banded iron formation rests immediately on Makganyene glacial beds. Manganese-rich strata of the Hotazel Formation in South Africa contain c. four billion tons of Mn within the Kalahari manganese field; this is by far the world's largest land-based Mn deposit (Gutzmer et al. 1997). Mn silicate (braunite) finely intergrown with Mn-rich carbonates (kutnahorite) form thick units within the magnetite-haematite type of banded iron formation, all of which record deposition on continental shelfs (Klein et al. 1987).

Deposition of the Kalahari manganese field represents a unique oxidation and Mn extraction event in the world oceans. Deposition requires large quantities of O_2, thus suggesting that the oxygenic photosynthetic apparatus (photosystem II) must have evolved before the formation of the Kalahari manganese deposit (Kirschvink et al. 2000). The oxidation and Mn extraction from seawater is also documented in the Fennoscandian Shield, though at a more modest scale (Fig. 1.2k).

In the Fennoscandian Shield, thin beds of manganite-cemented arenites occur within a shallow-water shelf dolostone-siliciclastic succession of the Tri-Ostrova Formation in the Ust'Ponoj Greenstone Belt (Melezhik and Predovsky 1984). Here, the manganite deposits are associated with ^{13}C-rich carbonate rocks of the Lomagundi-Jatuli type (Melezhik and Fallick 1996), hence represent a later and much more modest occurrence.

The post-2.2 Ga period was the time when Palaeoproterozoic "red beds" (rocks stained red by iron oxide, Fig. 1.2l) became widespread (Fig. 1.1). Red beds, together with lithified ancient sols that accumulated iron during weathering, reflect the presence of oxygen in subaerial environments.

One of the Greatest Perturbations of the Global Carbon Cycle: The Lomagundi-Jatuli Isotopic Event

Approximately contemporaneous with the world-wide appearance of red beds and formation of manganese deposits is the first global record of sedimentary carbonates unusually enriched in ^{13}C. Although accumulated in a variety of depositional settings, the ^{13}C-rich carbonate rocks appear to be mainly shallow-water, red coloured stromatolitic dolostones with abundant evaporitic features (Fig. 1.2h, m–o; reviewed in Melezhik et al. 1999a). This unprecedented perturbation of the carbon cycle was first recognised globally by Baker and Fallick (1989a, b), and is known as the Lomagundi-Jatuli isotopic excursion (e.g. Melezhik et al. 2005a). The event apparently lasted over 140 million years, and its duration was best constrained on the Fennoscandian Shield (Karhu and Holland 1996; Karhu 2005; Melezhik et al. 2007; Martin et al. 2010).

The causes and internal structure of this long-lasting isotopic perturbation of the global carbon cycle, as well as its exact chronological relationship with other roughly coeval global environmental changes, remain a matter of continuous debate. Several conflicting models have been advanced to explain the Lomagundi-Jatuli isotopic event starting with enhanced burial of organic carbon (Bakker and Fallick 1989a, b; Karhu and Holland 1996), methane cycling (Yudovich et al. 1991; Hayes and Waldbauer 2006), and a redox-stratified ocean (Aharon 2005; Bekker et al. 2008).

Melezhik et al. (1999a, 2005b) and Frauenstein et al. (2009) have argued that there has been considerable basinal modification (amplification) of the global $\delta^{13}C$ excursion, giving the impression that the global carbon cycle perturbation was much larger than it actually was. It also remains unresolved what role tectonics/plate-tectonics played in the onset, maintenance for over 140 million years, and the demise of the unprecedented global carbon cycle perturbation. It is evident that the excursion was preceded by the c. 2.4–2.2 Ga magmatic shutdown (Condie et al. 2009), and the major part of the isotopic event coincided with the palaeogeography of high-standing rifted continents, extensive chemical weathering, development of shallow-water carbonate platforms and epeiric seas, and flourishing stromatolite-forming cyanobacteria (e.g. Melezhik et al. 1999a).

The termination is contemporaneous with the final break-up of the Kenorland supercontinent and formation of deep-water siliciclastic seas (Bekker and Eriksson 2003; Bekker et al. 2003; Wanke and Melezhik 2005). Other details on the Lomagundi-Jatuli isotopic event are presented in Chap. 7.3; however, determining the global timing and regional influences for one of the most profound carbon isotopic excursions in Earth's history represent major challenges for future research.

Upper Mantle Oxidising Event

Towards the end of the Lomagundi-Jatuli event, a vast volume of highly oxidised lavas (Fig. 1.2p) was extruded across the Fennoscandian Shield (Chaps. 3.4 and 7.4). The eruption has been precisely dated at 2.06 Ga in the Pechenga and Imandra/Varzuga greenstone belts (Melezhik et al. 2007; Martin et al. 2010). These lavas form units up to 2 km in thickness in several greenstone belts, over an area of 5,000 km^2. If this considerable volume of lavas of large regional extent is not the result of surface oxidation, then they may represent the preserved remnants of high fO_2 eruptions, and reflect original upper mantle conditions modified through Archaean subduction of highly oxidised, oceanic slabs. These rocks are apparently too young to be a "smoking gun" for mantle redox evolution and the major cause for the Great Oxidation Event. Nevertheless they highlight the heterogeneity of upper mantle redox state. The reduction in oxygen demand associated with the eruption of these 2.06 Ga high fO_2 lavas might still have played an important role in sustaining oxic surface conditions, including the final accumulation of oxygen in the atmosphere during the course of the Great Oxidation Event.

Radical Change of Seawater Sulphate Reservoir

Towards the end of the positive $\delta^{13}C$ isotopic excursion many continents record formation of Ca-sulphates, which occur as ubiquitous pseudomorphs in various marine sedimentary rocks (Fig. 1.2q–s; for details see Chap. 7.5). The sulphate-bearing strata formed in a variety of depositional settings ranging from lacustrine and marine intracratonic basins through passive margin and back-arc environments (El Tabakh et al. 1999; Melezhik et al. 2001, 2005b; Pope and Grotzinger 2003; Evans 2006; Schröder et al. 2008). Oceanic sulphate abundance remains unknown, but a sizeable sulphate reservoir suggested as early as 2.1–2.2 Ga (e.g. Melezhik et al. 2005b) would contrast with the view that, prior to the Mesoproterozoic, gypsum precipitation was inhibited by a small marine sulphate reservoir and higher marine carbonate saturation (Grotzinger 1989; Kah et al. 2004). The irregular, cyclic, secular variations of geochemical parameters known for Phanerozoic terrestrial hydrosphere and atmosphere systems (Budyko et al. 1985; Veizer 2005) should be, in principle, applicable to the Palaeoproterozoic and may provide a means for reconciling the two conflicting views. However, recently the small marine sulphate reservoir hypothesis has been challenged by the discovery of several tens of meters of massive anhydrite beds in the Onega Basin (Fig. 1.2t). The anhydrite strata are associated with the ^{13}C-rich dolostones of the Tulomozero Formation and represent the Lomagundi-Jatuli isotopic excursion, thus falling within the 2.22–2.06 Ga time-interval. This new discovery (Morozov et al. 2010) may cause us to rethink our interpretations of Palaeoproterozoic oceanic sulphate chemistry.

1.1.4 Dispersed Kenorland Supercontinent Through 2.1–1.9 Ga

Several hundred millions of years of tectonic stability punctuated with repeated intraplate rifting eventually resulted in supercontinent fragmentation at around 2.1–2.0 Ga. This is best documented in the Wyoming craton and Fennoscandian Shield (Bekker and Eriksson 2003; Daly et al. 2006; Lahtinen et al. 2005). How such a radical palaeogeographic reconfiguration and change in tectonic regime of this large land mass might have influenced weathering and burial cycles remains to be studied; however, one of the apparent consequences was the termination of the Lomagundi-Jatuli positive isotopic excursion of carbonate carbon, the greatest in Earth's history. Whether this was a coincidental or a causative relationship is the task for future research.

Enhanced Gglobal Accumulation of Organic Matter: The Shunga Event, and Giant Iron-Ore Deposits

The aftermath of the Lomagundi-Jatuli excursion was marked by another apparently global-scale event, an unprecedented accumulation of organic matter known as the Shunga Event (Figs. 1.1 and 1.2u; Melezhik et al. 1999b, 2004). Although documented on several continents, it is best represented in basins across the Fennoscandian Shield and in the Francevillian Series in Gabon (Chap. 7.6). The initial stage of "black shale" accumulation may partially overlap with the end of the Lomagundi-Jatuli excursion in Gabon (supplementary material in El Albani et al. 2010); this, though, is an exception and such shales are unlikely to be a coeval and complementary reduced carbon reservoir compensating for the 140 Ma-long Lomagundi-Jatuli positive isotopic excursion of $\delta^{13}C_{carb}$. Moreover, the continuous accumulation of vast volume of "black shales" in post-2.0 Ma time period, reaching several hundreds of meters in thickness, and containing up to 10–30 wt.% C, in numerous basins world-wide (Salop 1982; Melezhik et al. 1988; Condie et al. 2001), has not been reflected in isotopic composition of contemporaneous sedimentary carbonates (Karhu 1993; Melezhik et al. 2007).

The duration and the cause of enhanced accumulation of organic matter at c. 2.0 Ga have been only cursorily investigated and addressed in the published literature (e.g. Melezhik et al. 1999b; Mossman et al. 2005). Available radiometric dates from the Pechenga Greenstone Belt suggest that the event may have lasted less than 50 Ma (Hanski 1992; Melezhik et al. 2007).

This period of time is also known for the generation of supergiant oil deposits, perhaps for the first time in Earth's history in such volume and scale (Mossman et al. 2005; Melezhik et al. 2009). One of these petrified oil fields is preserved in the Onega Basin in the southeastern Fennoscandian Shield (Fig. 1.2v–x; for details see Chap. 7.6). This oil field possesses rich information with a great potential to elucidate petroleum generation and migration in the Palaeoproterozoic (Melezhik et al. 2009).

Why did the oldest known significant accumulation of organic-carbon-rich sediments and generation of large-scale petroleum deposits occur at around 2.0 Ga, if the fundamental features of the biologic carbon cycle were established by 3500 Ma (Schidlowski et al. 1975; Hayes et al. 1983; Grassineau et al. 2002)? Does this accumulation of organic-carbon-rich sediments reflect an episode of enhanced biological productivity in a nutrient-enriched ocean? Or, perhaps such 2.0 Ga occurrences were the result of biased preservation or are the result of modest productivity, but enhanced preservation or even a preservation bias? All these questions constitute a programme for investigating this fascinating period of the Palaeoproterozoic (Melezhik et al. 1999b; Papineau 2010).

The c. 2.1–1.9 Ga time interval is also characterised by formation of giant supergene iron-ore deposits. Recently obtained new precise Pb/Pb ages on baddeleyite provided 2.05–2.0 Ga constraint on ore genesis in the Hamersley province in Western Australia. Müller et al. (2005) demonstrated that hypogene iron-ore mineralisation (shortly postdating the emplacement of c. 2008 Ma dikes) was oxidised, reworked and enriched by supergene processes during c. 2.0 Ga uplift and continental rifting. Müller et al. (2005) also suggested that these rift-bound giant supergene iron-ores were coeval with another class of giant iron-oxide accumulations (i.e., iron oxide (rare earth element–Cu-Au-U) deposits, Barton and Johnson (1996)) occurring in a similar continental rift setting. They suggested further that several world-class iron-ore deposits in India, South Africa, Brazil, and Ukraine exhibit remarkably similar enrichment processes and relative timing relationships close to 2.0 Ga (Beukes et al. 2003; Dalstra et al. 2003; Müller et al. 2005). In Western Australia, similar to the Fennoscandian, Canadian shields and many other cratons, the period between 2.1 and 2.0 Ga was characterised by continental sedimentation and mafic magmatism in extensional settings within rift basins that were linked to the breakup of a Paleoproterozoic supercontinent (e.g. Aspler and Chiarenzelli 1998). Supercontinent breakup may have caused the accumulation of hypogene iron ores followed by their first exposure and supergene oxidation in rift basins worldwide Müller et al. (2005).

The Earliest Phosphorites: A Radical Change in the Phosphorous Cycle

The 2.0 Ga enhanced accumulation of organic matter also corresponds in time with the formation of the oldest phosphorites (reviewed in Papineau 2010; Fig. 1.1). The phosphate accumulations occur in different forms (Chap. 7.7). The most remarkable c. 2.0 Ga occurrence is the Jhamarkotra stromatolitic phosphorites (Fig. 1.2y) containing up to 35 wt.% P_2O_5 and reaching 25 m in thickness in the Lower Aravalli Group in Rajasthan, India (Banerjee 1971; Chauhan 1979). Similar age phosphorites, though in redeposited, clastic form (Fig. 1.2z), occur in the c. 2.0 Ga Pilgujärvi Sedimentary Formation of the Pechenga Greenstone Belt (Bekasova and Dudkin 1982). Concretions enriched in phosphorus (Fig. 1.2aa, ab) have been documented in the Il'mozero Sedimentary Formation in the Imandra/Varzuga Greenstone Belt (Melezhik 1992). Phosphorites occur also as discrete bands in a c. 2.0 Ga iron formation in the Central Lapland Greenstone Belt (Gehör 1994). Other phosphorite occurrences are summarised in (Chap. 7.7).

The appearance of phosphorites in various forms in sedimentary successions at 2.0 Ga reflects a critical change in the Precambrian phosphorous cycle. The most common pathway of precipitation involves sedimentation of organic matter that carries biologically concentrated phosphate (Knudsen and Gunter 2002). Most of the reported 2.0 Ga phosphorites are commonly associated with sulphides and organic matter, and thus are consistent with such a mechanism. However, high bioproductivity and enhanced concentration of organic matter (e.g. Papineau 2010) is probably not the only factor controlling the generation of Palaeoproterozoic phosphorites; several older C_{org}-rich formations are known (Hayes et al. 1983), but contain no phosphorites. Thus, it is possible that 2.0 Ga sediments record a major change in the diagenetic mineralisation of organic matter, which seemingly was coeval with accumulation of the first phosphorites. One of the possible factors preventing the diffusion of P out of the sediment could be the establishment of a ferric oxide "trap" at the sediment-water interface overlain by oxic water. Other factors, which controlled the formation of the oldest known phosphorites and phosphate concretions remain unresolved.

Establishment of an Aerobic Pathway in Recycling of Organic Matter

The scarcity of diagenetic carbonate concretions in the Archaean and in the early Palaeoproterozoic sedimentary formations older than c. 2.0 Ga was acknowledged long ago (e.g. Pettijohn 1940; Melezhik 1992), and is evident from a later extensive carbonate concretion bibliography listing 700 articles (Dietrich 1997). Importantly, a compilation of published global data shows that $\delta^{13}C$ values of both primary and diagenetic carbonates of pre-c. 2.0 Ga rocks mostly cluster near 0 ± 3‰, which are not characteristic of microbially recycled organic matter (Fallick et al. 2008). However, carbonates associated temporally and spatially with banded iron formations (BIF) represent an exception (summarised in Fallick et al. 2011).

The first known appearance, and then worldwide development, of microbially meditated, ^{13}C-depleted, diagenetic carbonate concretions outside a "BIF-environment" are documented in published literature at around 2.0 Ga (Winter and Knauth 1992; Melezhik et al. 1998, 1999b; Fallick et al. 2008). In c. post-2.0 Ga sedimentary successions, the concretions are varied and abundant (Fig. 1.2ac) and associated with other diagenetic products, such as phosphate nodules and redeposited bedded phosphorites, all of which are seemingly absent from older rocks. Thus, it appears that the first abundant occurrence of isotopically light diagenetic carbonate concretions signifies an important hallmark in biospheric evolution apparently linked to the emergence of 'modern-style' recycling of organic matter (Melezhik et al. 2005a). It has been suggested (Fallick et al. 2008) that prior to c. 2.0 Ga organic matter was recycled and remineralised predominantly in the anoxic water column and near the sediment/water interface allowing isotopically distinctive CO_2 and CH_4 to readily escape to the atmosphere, thus leaving no isotopic traces in diageneticaly formed carbonates. Fallick et al. (2008) also suggested that around 2.0 Ga, the water column became oxic enough to be toxic for the anaerobic microbial recyclers, forcing them to retreat deep into the substrate. As the result, redox gradients developed in the sedimentary column and abundant carbonate concretions formed incorporating ^{13}C-depleted by-products derived from remineralised organic carbon, similar to what is observed in the Phanerozoic world. This hypothesis represents another subject for future research and tests. Another question to be answered is why diagenetic carbonates with distinct isotopic fingerprints of methanogenesis and methanotrophy (e.g. ^{13}C-rich) have not been confidently recorded in the early Palaeoproterozoic sedimentary formations? Or, perhaps their absence is the result of biased sampling? Hayes (1994) documents organic matter in 2.7 Ga sedimentary rocks that they attributed to methanotrophy. If this were widespread, we would expect ^{13}C-rich diagenetic carbonates to have been preserved. Moreover, younger Neoarchaean rocks do not display such ^{13}-C depleted organic matter. Was this a false start of the aerobic biosphere (whether the oxidant used in methanotropy was sulphate or oxygen)? Detailed and systematic mineralogical, petrographic, geochemical and in situ isotopic studies of diagenetic carbonates of this age may provide the focus necessary to examine this problem.

1.1.5 Summary

A brief review of some of the major modifications of the Earth surface environments emphasises that the Archaean-Palaeoproterozoic transition was a major step towards emergence of an aerobic Earth system. What remains unresolved is precise absolute chronology of all these major events. At present, even a relative chronological sequence can only be established with a certain amount of approximation. This, together with the limited palaeomagnetic database and unreliable palaeogeographic reconstructions, hampers our understanding of causative effect of palaeotectonics and Earth's surface alterations, as well as the cause-and-effect relationships between various global environmental changes.

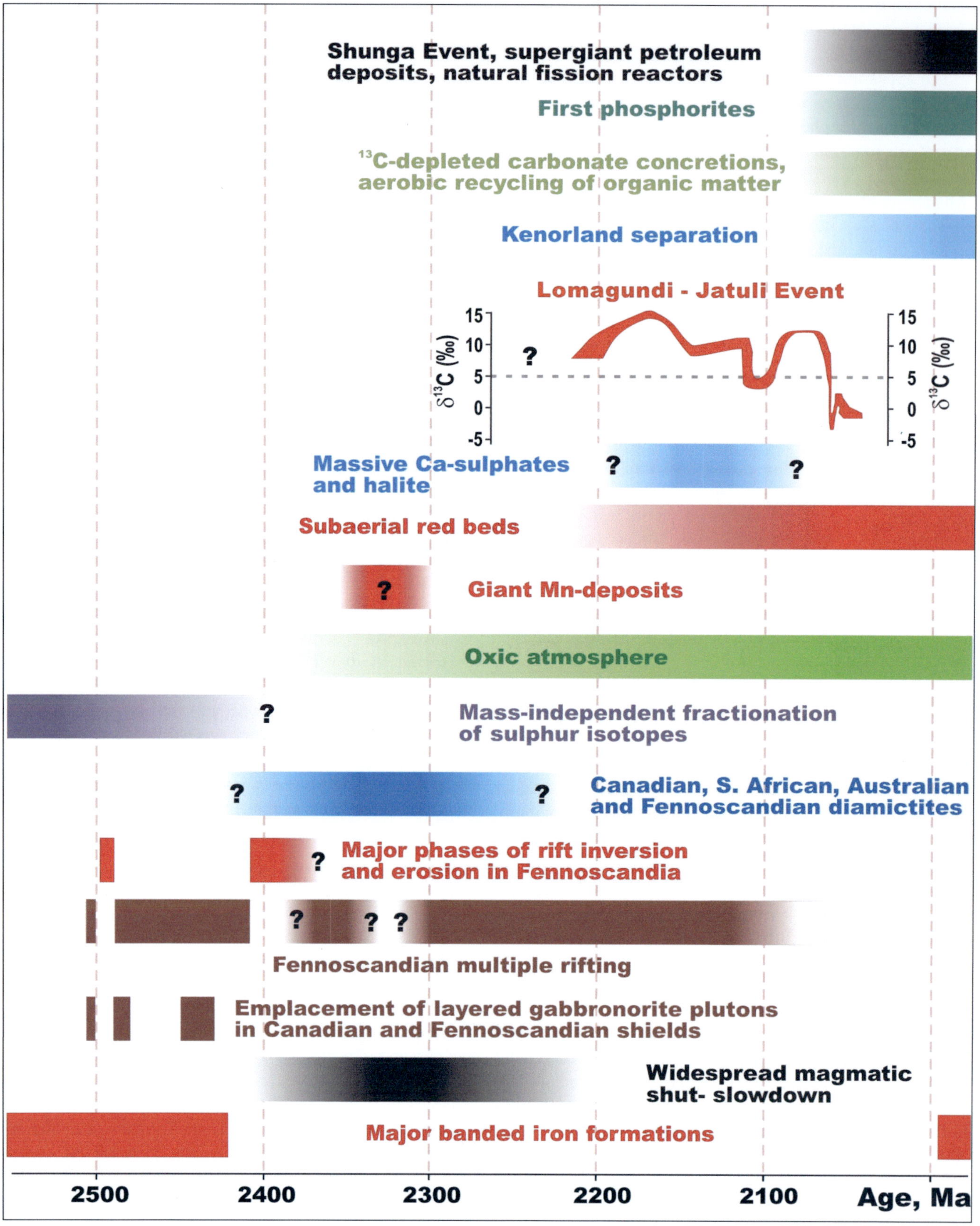

Fig. 1.1 Major global palaeoenvironmental and tectonic events during the Early Palaeoproterozoic (Data sources are presented in Chaps. 3.2, 3.3, 3.4, 7.1, 7.2, 7.3, 7.4, 7.5, 7.6, and 7.7, and in Müller et al. (2005), Melezhik et al. (2007) and Condie et al. (2009))

Fig. 1.2 Selected rock images illustrating major global events through the Archaean-Palaeoproterozoic transitional time interval. *Major Archaean banded iron formations, seafloor deposition*: (**a**) Late Archaean, 2.5 Ga, banded iron formation from the Hamersley Group in Western Australia. (**b**) Archaean, c. 2.7 Ga banded iron formation from Børnevatn in northern Norway, Fennoscandian Shield. *Mantle - plume-generated early Palaeoproterozoic gabbro-norite intrusions, and incipient rifting in the Fennoscandian Shield*: (**c**) Magmatic layering in 2.44 Ga gabbro-norite intrusion from the Peräpohja Schist belt in central Finland, Fennoscandian Shield; hammer head is 15 cm long. (**d**) Layering in mottled anorthosite from the 2.44 Ga Koitilainen intrusion in Finnish Lapland, Fennoscandian Shield; match box length is 5 cm. *Huronian glaciation, advanced intraplate rifting*: (**e**) Post-2.44 Ga "varved" siltstone with dropstone (*arrowed*) from the Polisarka Sedimentary Formation in the Imandra/Varzuga Greenstone Belt, central Kola Peninsula, Fennoscandian Shield

Fig. 1.2 (continued) (**f**) Diamictite from the Gowganda Formation in the Huronian Supergroup, Ontario, Canada. *Deeply weathered rocks, advanced intraplate rifting:* (**g**) Quartz-kyanite schist (metamorphosed kaolin soil) at Hokkalampi in North Karelia Schist Belt, the Fennoscandian Shield (for details see Marmo 1992). (**h**) Rippled and cross-bedded, chemically mature, quartz arenite from c. 2.1 Ga Kuetsjärvi Sedimentary Formation in Pechenga Greenstone Belt, NW Kola Peninsula, NE Fennoscandia

Fig. 1.2 (continued) *First land-based, sedimentary, manganese deposits:* (**i**) View of open pit at Mamatwan Mine in the Kalahari manganese field, the worlds largest manganese mine; manganese ore (*black*) overlain by Late Mesozoic/Cenozoic calcretised fluvial and lacustrine sediments. (**j**) Finely laminated, unmetamorphosed, braunite (Mn carbonate ore) lutite from the southern portion of the Kalahari manganese field. (**k**) Manganite bed from the Tri-Ostrova Formation in the Ust'Ponoj Greenstone Belt, eastern Kola Peninsula in NE Fennoscandia. W*ide-spread subaerial "red beds", large epeiric seas*: (**l**) Red, desiccation cracked surface of a red mudstone bed from the Segozero Formation, central Karelia, eastern Fennoscandia

Fig. 1.2 (continued) *Global perturbation of carbon reservoir,* ^{13}*C-rich sedimentary carbonates (Lomagundi-Jatuli Event), wide-spread shallow-water carbonate platforms and epeiric seas*: (**m**) Intertidal, ^{13}C-rich, stromatolitic dolostone from the Rantamaa Formation in the Peräpohja Schist Belt, central Finland. (**n**) Lacustrine, ^{13}C-rich, pink, dolarenite from the Kuetsjärvi Sedimentary Formation in the Pechenga Greenstone Belt, NW Kola Peninsula in NE Fennoscandia. (**o**) Massive, ^{13}C-rich dolostone with weathered-out Ca-sulphates from the Tulomozero Formation in the Onega basin, southern Karelia in NE Fennoscandia

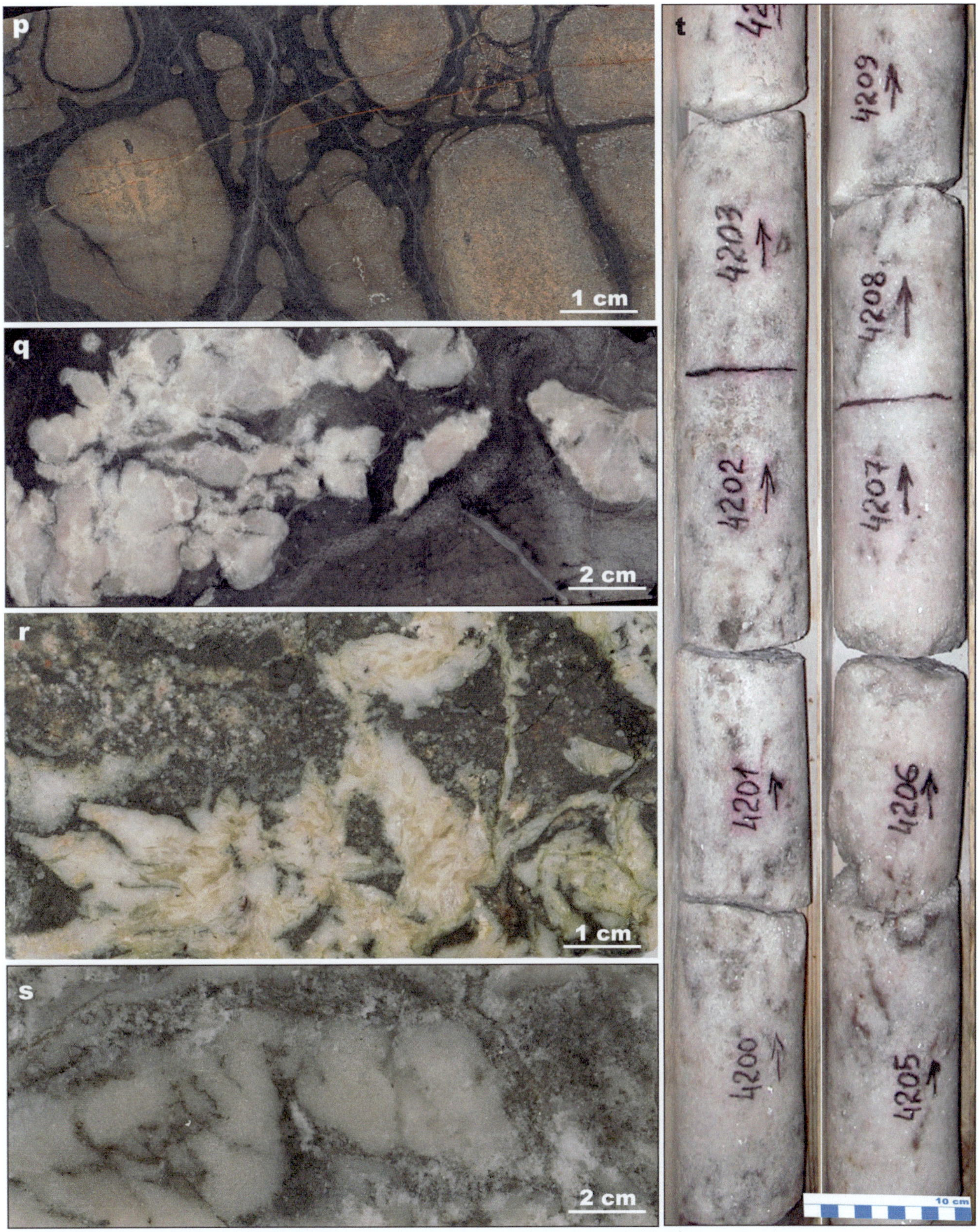

Fig. 1.2 (continued) *Apparent upper mantle oxidising event*: (**p**) Highly oxidised, subaerially erupted lava breccia with haematite-magnetite cement (*black*) from the Kuetsjärvi Volcanic Formation in the Pechenga Greenstone Belt, NW Kola Peninsula in NE Fennoscandia. *Abundant marine Ca-sulphates, wide-spread shallow-water carbonate platforms and epeiric seas*: (**q**) Quartz-pseudomorphed sulphate nodules with relics of anhydrite. (**r**) Gypsum rosettes replaced by quartz and dolomite. (**s**) Dolomite-pseudomorphed, enterolithic layer of Ca-sulphate with relics of anhydrite. (**t**) Massive anhydrite. Images of sulphates are from the Tulomozero Formation in the Onega basin, southern Karelia in SE Fennoscandia

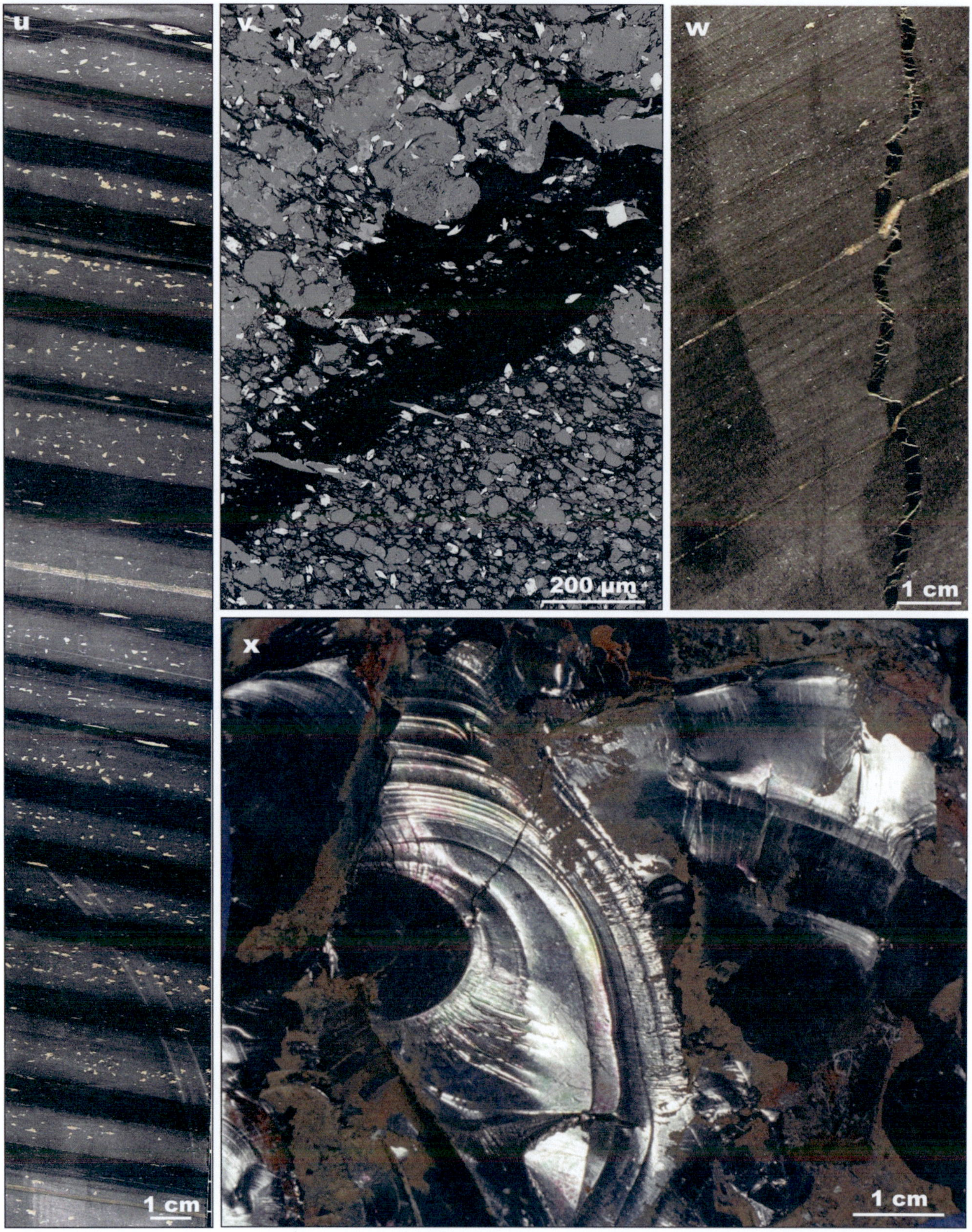

Fig. 1.2 (continued) *World-wide enhanced accumulation of organic carbon (Shunga Event), giant petroleum deposits*: (**u**) Rhythmically bedded, C_{org}-rich, turbiditic siltstone-mudstone. (**v**) Back-scatter electron microprobe image of graded greywacke beds with pyrobitumen (black) trapped between layers. (**w**) Pyrobitumen vein cross-cutting laminated greywacke. (**x**) Pyrobitumen from oil trap in shale-dolostone succession. All images are from the Zaonega Formation in the Onega basin, south Karelia in SE Fennoscandia

Fig. 1.2 (continued) *First phosphorites*: (**y**) Phosphatic stromatolites from the Jhamarkotra Formation in the Lower Aravalli Group, Rajasthan, India. (**z**) Gritstone with redeposited phosphorites (*black* clasts) and sulphides (*dark yellow*) from the Pilgujärvi Sedimentary Formation in the Pechenga Greenstone Belt, NW Kola Peninsula in Fennoscandia. (**aa, ab**) Non-zoned and zoned phosphatic diagenetic concretions in turbiditic greywackes of the Il'mozero Sedimentary Formation in the Imandra/Varzuga Greenstone Belt, central Kola Peninsula in eastern Fennoscandia. *Aerobic recycling of organic matter* (**ac**) ^{13}C-depleted ankeritic concretions (*brown*) in lacustrine turbiditic greywacke from the Kondopoga Formation in the Onega basin, south Karelia, SE Fennoscandia

(Photographs courtesy of Vesa Perttunen (**c** and **m**), Tapani Mutanen (**d**), Kauko Laajoki (**h**), Jens Gutzmer (**i** and **j**), Dmitry Rychanchik (**t**) and Domenic Papineau (**y**). Other photographs by Aivo Lepland (**a** and **ab**), Eero Hanski (**g**) and Victor Melezhik (**b**, **c**, **h**, **i**, **l**, **n–s**, **u–x**, **z**, **aa** and **ac**))

References

Aharon P (2005) Redox stratification and anoxia of the early Precambrian oceans: implications for carbon isotope excursions and oxidation events. Precambrian Res 137:207–222

Arndt NT (2004) Crustal growth rates. In: Eriksson PG, Altermann W, Nelson DR, Mueller WU, Catuneanu O (eds) The Precambrian Earth: tempos and events, 12th edn, Developments in Precambrian. Elsevier, Amsterdam, pp 155–158

Aspler LB, Chiarenzelli JR (1998) Two Neoarchean supercontinents? Evidence from the Paleoproterozoic. Sediment Geol 120:75–104

Baker AJ, Fallick AE (1989a) Heavy carbon in two-billion-year-old marbles from Lofoten-Vesterålen, Norway: implications for the Precambrian carbon cycle. Geochim Cosmochim Acta 53:1111–1115

Baker AJ, Fallick AE (1989b) Evidence from Lewisian limestone for isotopically heavy carbon in two-thousand-million-year-old sea water. Nature 337:352–354

Banerjee DM (1971) Precambrian stromatolitic phosphorites of Udaipur, Rajasthan, India. Geol Soc Am Bull 82:2319–2329

Barley ME, Pickard AL, Sylvester PL (1997) Emplacement of a large igneous province as a possible cause of banded iron formation 2.45 billion years ago. Nature 385:55–58

Barley ME, Bekker A, Krapez B (2005) Late Archean to Early Paleoproterozoic global tectonics, environmental change and the rise of atmospheric oxygen. Earth Planet Sci Lett 238:156–171

Barton MD, Johnson DA (1996) An evaporitic-source model for igneous related Feoxide (REE-Cu-Au-U) mineralization. Geology 24:259–262

Bekasova NB, Dudkin OB (1982) Composition and nature of concretionary phosphorites from the early Precambrian of Pechenga (Kola Peninsula). Lithol Miner Resour 16:625–630 (in Russian)

Bekker A, Eriksson KA (2003) A Paleoproterozoic drowned carbonate platform on the southeastern margin of the Wyoming Craton: a record of the Kenorland breakup. Precambrian Res 120:327–364

Bekker A, Kaufman AJ, Karhu JA, Beukes NJ, Swart QD, Coetzee LL, Eriksson KA (2001) Chemostratigraphy of the Paleoproterozoic Duitschland Formation, South Africa: implications for coupled climate change and carbon cycling. Am J Sci 301:261–285

Bekker A, Karhu JA, Eriksson KA, Kaufman AJ (2003) Chemostratigraphy of Paleoproterozoic carbonate successions of the Wyoming Craton: tectonic forcing of biogeochemical change? Precambrian Res 120:279–325

Bekker A, Holmden C, Beukes NJ, Kenig F, Eglinton B, Patterson WP (2008) Fractionation of inorganic and organic carbon during the Lomagundi (2.22–2.2 Ga) carbon isotope excursion. Earth Planet Sci Lett 271:278–291

Berner RA (2006) GEOCARBSULF: a combined model for Phanerozoic atmospheric O_2 and CO_2. Geochim Cosmochim Acta 70:5653–5664

Beukes NJ, Gutzmer J, Mukhopadhyay J (2003) The geology and genesis of high-grade hematite iron ore deposits, Institution of Mining and Metallurgy. Trans Sect B Appl Earth Sci 112:18–25

Bibikova EV (1989) Uranium-lead geochronology of early stages of development of ancient shields. Nauka (Science), Moscow, p 180 (in)

Bibikova EV, Samsonov AV, Petrova AU, Kirnozova TI (2005) The Archean geochronology of Western Karelia. Stratigraph Geol Correlat 13:459–475

Bleeker W (2003) The late Archaean record: a puzzle in c. 35 pieces. Lithos 71:99–134

Buchan KL, Mertanen S, Park RG, Pesonen LJ, Elming S-Å, Abrahamsen N, Bylund G (2000) Comparing drift of Laurentia and Baltica in the Proterozoic: the importance of key palaeomagnetic poles. Tectonophysics 319:167–198

Budyko MI, Ronov AB, Yanshin AL (1985) History of the atmosphere. Gidrometeooizdat, Leningrad, p 208 (in Russian)

Campbell IH, Allen CM (2008) Formation of supercontinents linked to increases in atmospheric oxygen. Nat Geosci 1:554–558

Catling DC, Claire MW (2005) How Earth's atmosphere evolved to an oxic state: a status report. Earth Planet Sci Lett 237:1–20

Chauhan DS (1979) Phosphate-bearing stromatolites of the Precambrian Aravalli phosphorite deposits of the Udaipur region, their environmental significance and genesis of phosphorite. Precambrian Res 8:95–126

Christie KW, Davidson A, Fahrig WE (1975) The paleomagnetism of the Kaminak dikes – No evidence of significant Hudsonian plate motion. Can J Earth Sci 12:2048–2064

Condie KC (1995) Episodic ages of greenstones: a key to mantle dynamics? Geophys Res Lett 22:2215–2218

Condie KC (1998) Episodic continental growth and supercontinents; a mantle avalanche connection? Earth Planet Sci Lett 163:97–108

Condie KC (2000) Episodic continental growth models: afterthoughts and extensions. Tectonophysics 322:153–162

Condie KC (2004) Precambrian superplume events. In: Eriksson PG, Altermann W, Nelson DR, Mueller WU, Catuneanu O (eds) The Precambrian Earth: tempos and events, vol 12, Developments in Precambrian. Elsevier, Amsterdam, pp 163–172

Condie KC, Des Marais DJ, Abbott D (2001) Precambrian superplumes and supercontinents: a record in black shales, carbon isotopes, and paleoclimates? Precambrian Res 106:239–260

Condie KC, O'Neill C, Aster RC (2009) Evidence and implications for a widespread magmatic shutdown for 250 My on Earth. Earth Planet Sci Lett 282:294–298

Dalstra H, Harding T, Riggs T, Taylor D (2003) Banded iron formation hosted high grade hematite deposits, a coherent group? Institution of Mining and Metallurgy. Trans Sect B Appl Earth Sci 112:68–72

Daly JS, Balagansky VV, Timmerman MJ, Whitehouse MJ (2006) The Lapland-Kola orogen: Palaeoproterozoic collision and accretion of the northern Fennoscandian lithosphere. In: Gee DG, Stephenson RA (eds) European lithosphere dynamics, 32nd edn. Geological Society Memoir, London, pp 579–598

Dietrich RV (1997) Carbonate concretions: a bibliography, p 64. http://www.cst.cmich.edu/users/dieter1rv/concretions/

El Albani A, Bengtson S, Canfield DE, Bekker A, Macchiarelli R, Arnaud Mazurier A, Hammarlund EU, Boulvais P, Dupuy J-J, Fontaine C, Fürsich FT, Gauthier-Lafaye F, Janvier P, Javaux E, Ossa FO, Pierson-Wickmann A-C, Riboulleau A, Sardini P, Vachard D, Whitehouse M, Meunier A (2010) Large colonial organisms with coordinated growth in oxygenated environments 2.1 Gyr ago. Nature 466:100–104

El Tabakh M, Grey C, Pirajno F, Schreiber BC (1999) Pseudomorphs after evaporitic minerals interbedded with 2.2 Ga stromatolites of the Yerrida basin, Western Australia: origin and significance. Geology 27:871–874

Evans DAD (2003) A fundamental Precambrian-Phanerozoic shift in Earth's glacial style? Tectonophysics 375:353–385

Evans DAD (2006) Proterozoic low orbital obliquity and axial-dipolar geomagnetic field from evaporite palaeolatitudes. Nature 444:51–55

Evans DAD, Pisarevsky SA (2008) Plate tectonics on early Earth? Weighing the paleomagnetic evidence. In: Condie KC, Pease V (eds) When did plate tectonics begin on planet Earth? Geological Society of America, Special Paper 440: 249–263

Fallick AE, Melezhik VA, Simonson B (2008) The ancient anoxic biosphere was not as we know it. In: Dobretsov N, Kolchanov N, Rozanov A, Zavarzin G (eds) Biosphere origin and evolution. Springer, London, pp 169–188

Fallick AE, Melezhik VA, Simonson B (2011) On Proterozoic ecosystems and the carbon isotopic composition of carbonates associated with banded iron formations. In: Neves L et al (eds) Modelacao de sistemas geologicos. Univeridade de Coimbra, Portugal, pp 57–71

Fischer AG (1984) The two Phanerozoic supercycles. In: Berggren WA, Van Couvering JA (eds) Catastrophes and Earth history. Princeton University Press, Princeton, pp 129–148

Frauenstein F, Veizer J, Beukes N, Van Niekerk HS, Coetzee LL (2009) Transvaal supergroup carbonates: implications for Paleoproterozoic $\delta^{18}O$ and $\delta^{13}C$ records. Precambrian Res 175:149–160

Gastil RG (1960) The distribution of mineral dates in time and space. Am J Sci 258:1–35

Gehör S (1994) REE distribution in the phosphorite bands within the Paleoproterozoic Tuomivaara and Pahtavaara ironformations, central and northern Finland. Geol Surv Finl Spec Pap 19:71–83

Grassineau NV, Nisbet EG, Fowler CMR, Bickle MJ, Lowry D, Chapman HJ, Mattey DP, Abell P, Young J, Martin A (2002) Stable isotopes in the Archaean Belingwe Belt, Zimbabwe; evidence for a diverse microbial mat ecology. In: Fowler CMR, Ebinger CJ, Hawkesworth C (eds) The early Earth; physical, chemical and biological development, 199th edn. Geological Society Special Publication, London, pp 309–328

Grotzinger JP (1989) Facies and evolution of Precambrian carbonate depositional systems; emergence of the modern platform archetype. In: Crevello PD, Wilson JJ, Sarg JF, Read F (eds) Controls on carbonate platform and basin development, vol 44. Special Publication of Society of Economic Paleontologists and Mineralogists, Tulsa, pp 79–106

Guo Q, Strauss H, Kaufman AJ, Schröder S, Gutzmer J, Wing B, Baker MA, Bekker A, Jin Q, Kim S-T, Farquhar J (2009) Reconstructing Earth's surface oxidation across the Archean-Proterozoic transition. Geology 37:399–402

Gutzmer J, Beukes NJ, Yeh HW (1997) Fault-controlled metasomatic alteration of Early Proterozoic sedimentary manganese ore at Mamatwan Mine, Kalahari manganese field, South Africa. S Afr J Geol 100:53–71

Halls HC, Davis DW, Stott GM, Ernst RE, Hamilton MA (2008) The Paleoproterozoic marathon large igneous province: new evidence for a 2.1 Ga long-lived mantle plume event along the southern margin of the North American Superior Province. Precambrian Res 162:327–353

Hamilton J (1977) Sr isotope and trace element studies of the Great Dyke and Bushveld mafic phase and their relation to early Proterozoic magma genesis in southern Africa. J Petrol 18:24–52

Hannah JL, Bekker A, Stein HJ, Markey RJ, Holland HD (2004) Primitive Os and 2316 Ma age for marine shale: implications for Paleoproterozoic glacial events and the rise of atmospheric oxygen. Earth Planet Sci Lett 225:43–52

Hanski EJ (1992) Petrology of the Pechenga ferropicrites and cogenetic, Ni-bearing gabbro-wehrlite intrusions, Kola Peninsula, Russia. Bull Geol Surv Finl 367:192

Hanski E, Huhma H, Vaasjoki M (2001) Geochronology of northern Finland: a summary and discussion. In: Vaasjoki M (ed) Radiometric age determination from Finnish Lapland and their bearing on the timing of Precambrian volcano-sedimentary sequences, 33rd edn. Geological Survey of Finland Special Paper, Espoo, pp 255–279

Hardebeck J, Anderson DL (1996) Eustasy as a test of a Cretaceous superplume hypothesis. Earth Planet Sci Lett 137:101–108

Hayes JM (1994) Global methanotrophy at the Archean-Proterozoic transition, In: Bengtson S (ed) Early life on Earth: nobel symposium 84, Columbia University Press, New York, pp 220–236

Hayes JM, Waldbauer JR (2006) The carbon cycle and associated redox processes through time. Philos Trans R Soc B 361:931–950

Hayes JM, Kaplan IR, Wedeking KW (1983) Precambrian organic geochemistry, preservation of the record. In: Schopf JW (ed) Earth's earliest biosphere: its origin and evolution. Princeton University Press, Princeton, pp 93–135

Heaman LM (1997) Global mafic volcanism at 2.45 Ga: remnants of an ancient large igneous province? Geology 25:299–302

Hoffman PF (1988) United plates of America, the birth of a craton: early Proterozoic assembly and growth of Laurentia. Ann Rev Earth Planet Sci 16:543–603

Hoffman PF (1989) Speculations on Laurentia's first gigayear (2.0 to 1.0 Ga). Geology 17:135–138

Holland HD (2006) The oxygenation of the atmosphere and oceans. Philos Trans R Soc B 361:903–915

Jones DL, Robertson DM, McFadden PL (1975) A palaeomagnetic study of Precambrian dyke swarms associated with the Great Dyke of Rhodesia. Trans Geol Soc S Afr 78:57–65

Kah LC, Lyons TW, Frank TD (2004) Low marine sulphate and protracted oxygenation of the Proterozoic biosphere. Nature 431:834–838

Käpyaho A, Mänttäri I, Huhma H (2006) Growth of Archaean crust in the Kuhmo district, eastern Finland: U–Pb and Sm–Nd isotope constraints on plutonic rocks. Precambrian Res 146:95–119

Karhu JA (1993) Palaeoproterozoic evolution of the carbon isotope ratios of sedimentary carbonates in the Fennoscandian Shield. Geol Surv Finl Bull 371:1–87

Karhu JA (2005) Paleoproterozoic carbon isotope excursion, In: Lehtinen M, Nurmi PA, Rämö O (eds) Precambrian Geology of Finland—Key to the Evolution of the Fennoscandian Shield, Elsevier, Amsterdam, p. 669–680.

Karhu JA, Holland HD (1996) Carbon isotopes and the rise of atmospheric oxygen. Geology 24:867–879

Kasting JE (2004) When methane made climate. Sci Am 291:78–85

Kasting JE (2005) Methane and climate during the Precambrian era. Precambrian Res 137:119–129

Kirschvink JL, Gaidos EJ, Bertani LE, Beukes NJ, Gutzmer J, Maepa LN, Steinberg RE (2000) Paleoproterozoic snowball Earth: extreme climatic and geochemical global change and its biological consequences. Proc Natl Acad Sci 97:1400–1405

Klein C, Beukes NJ, Schopf JW (1987) Filamentous microfossils in the early Proterozoic Transvaal supergroup: their morphology, significance and paleoenvironmental setting. Precambrian Res 36:81–94

Knudsen AC, Gunter ME (2002) Sedimentary phosphates – an example: phosphoria formation, southeastern Idaho, U.S.A., In: Kohn MJ, Rakovan J, Hughes J (eds) Rev Mineral Geochemist: Phosphate: geochemical, geobiological, and materials importance 48: 363–389

Konhauser KO, Pecoits E, Lalonde SV, Papineau D, Nisbet EG, Barley ME, Arndt NT, Zahnle K, Kamber BS (2009) Oceanic nickel depletion and a methanogen famine before the great oxidation event. Nature 458:750–753

Kopp RE, Kirschvink JL, Hilburn IA, Nash CZ (2005) The Paleoproterozoic snowball Earth: a climate disaster triggered by the evolution of oxygenic photosynthesis. Proc Natl Acad Sci 102:11131–11136

Kump LR (2008) The rise of atmospheric oxygen. Nature 451:277–278

Kump LR, Barley ME (2007) Increased subaerial volcanism and the rise of atmospheric oxygen 2.5 billion years ago. Nature 448:1033–1036

Lahtinen R, Korja A, Nironen M (2005) Paleoproterozoic tectonic evolution. In: Lehtinen M, Nurmi PA, Rämö OT (eds) Precambrian geology of Finland – key to the evolution of the Fennoscandian Shield. Elsevier, Amsterdam, pp 669–680

Lindsay JF, Brasier MD (2002) Did global tectonics drive early biosphere evolution? Precambrian Res 114:1–34

Marmo JS (1992) The lower Proterozoic Hokkalampi paleosol in North Karelia, eastern Finland. In: Golubic S, Kimberley MM, Trudinger PA, Schidlowski M (eds) Early organic evolution: implications for energy and mineral resources. Springer, Berlin/Heidelberg, pp 41–66

Marmo JS, Ojakangas RW (1984) Lower Proterozoic glaciogenic deposits, eastern Finland. Geol Soc Am Bull 98:1055–1062

Marshall H, Walker J, Kuhn W (1988) Longterm climate change and the geochemical cycle of carbon. J Geophys Res 93:791–801

Martin DMcB (1999) Depositional setting and implications of Paleoproterozoic glaciomarine sedimentation in the Hamersley Province, Western Australia. Geol Soc Am Bull 111:189–203

Martin DMcB, Clendenin CW, Krapez B, McNaughton NJ (1998) Tectonic and geochronological constraints on late Archaean and Palaeoproterozoic stratigraphic correlation within and between the Kaapvaal and Pilbara Cratons. J Geol Soc Lond 155:311–322

Martin AP, Condon DJ, Prave AR, Melezhik VA, Fallick A (2010) Constraining the termination of the Lomagundi-Jatuli positive isotope excursion in the Imandra-Varzuga segment (Kola Peninsula, Russia) of the North Transfennoscandian Greenstone Belt by high-precision ID-TIMS, AGU, San Francisco, 13–17 Dec 2010

Melezhik VA (1992) Early Proterozoic sedimentary and rock-forming basins of the Baltic Shield. Nauka (Science), Leningrad, p 256 (in Russian)

Melezhik VA, Fallick AE (1996) A widespread positive $\delta^{13}C_{carb}$ anomaly at around 2.33–2.06 Ga on the Fennoscandian Shield: a paradox? Terra Nova 8:141–157

Melezhik VA, Predovsky AA (1984) The first discovery of Precambrian manganite-bearing rocks in the Kola Peninsula and their metallogenetic significance. Commun USSR Acad Sci 274 (2):392–394

Melezhik VA, Basalaev AA, Predovsky AA, Balabonin HL, Bolotov VI, Pavlova MA, Gavrilenko BV, Abzalov MZ (1988) Carbonaceous rocks of the early stages of Earth evolution. Nauka (Science), Leningrad, p 200 (in Russian)

Melezhik VA, Grinenko LN, Fallick AE (1998) 2000 Ma sulphide concretions from the 'Productive' Formation of the Pechenga Greenstone Belt, NW Russia: genetic history based on morphological and isotopic evidence. Chem Geol 148:61–94

Melezhik VA, Fallick AE, Medvedev PV, Makarikhin VV (1999a) Extreme $^{13}C_{carb}$ enrichment in ca. 2.0 Ga magnesite-stromatolite-dolomite-'red beds' association in a global context: a case for the world-wide signal enhanced by a local environment. Earth-Sci Rev 48:71–120

Melezhik VA, Fallick AE, Filippov MM, Larsen O (1999b) Karelian shungite – an indication of 2000 Ma-year-old metamorphosed oil-shale and generation of petroleum: geology, lithology and geochemistry. Earth-Sci Rev 47:11–40

Melezhik VA, Fallick AE, Medvedev PV, Makarikhin V (2001) Palaeoproterozoic magnesite: lithological and isotopic evidence for playa/sabkha environments. Sedimentology 48:379–397

Melezhik VA, Filippov MM, Romashkin AE (2004) A giant Palaeoproterozoic deposit of shungite in NW Russia: genesis and practical applications. Ore Geol Rev 24:135–154

Melezhik VA, Fallick AE, Hanski E, Kump L, Lepland A, Prave A, Strauss H (2005a) Emergence of the aerobic biosphere during the Archean-Proterozoic transition: challenges for future research. GSA Today 15:4–11

Melezhik VA, Fallick AE, Rychanchik DV, Kuznetsov AB (2005b) Palaeoproterozoic evaporites in Fennoscandia: implications for seawater sulphate, $\delta^{13}C$ excursions and the rise of atmospheric oxygen. Terra Nova 17:141–148

Melezhik VA, Huhma H, Condon DJ, Fallick AE, Whitehouse MJ (2007) Temporal constraints on the Paleoproterozoic Lomagundi-Jatuli carbon isotopic event. Geology 35:655–658

Melezhik VA, Fallick AE, Filippov MM, Lepland A, Rychanchik DV, Deines JE, Medvedev PV, Romashkin AE, Strauss H (2009) Petroleum surface oil seeps from Palaeoproterozoic petrified giant oilfield. Terra Nova 21:119–126

Mertanen S, Pesonen LJ (2005) Drift history of the shield. In: Lehtinen M, Nurmi PA, Rämö OT (eds) Precambrian geology of Finland – key to the evolution of the Fennoscandian Shield. Elsevier, Amsterdam, pp 645–668

Miall AD (1983) Glaciomarine sedimentation in the Gowganda Formation (Huronian), northern Ontario. J Sediment Res 53:477–491

Mikkola P, Huhma H, Heilimo E, Whitehouse M (2011) Archean crustal evolution of the Suomussalmi district as part of the Kianta Complex, Karelia: constraints from geochemistry and isotopes of granitoids. Lithos 125:287–307

Morozov AF, Hakhaev BN, Petrov OV, Gorbachev VI, Tarkhanov GB, Tsvetkov LD, Erinchek YuM, Akhmedov AM, Krupenik VA, Sveshnikova KYu (2010) Rock-salts in Palaeoproterozoic strata of the Onega depression of Karelia (based on data from the Onega parametric drillhole). Trans Russ Acad Sci 435(2):230–233

Mossman DJ, Gauthier-Lafaye F, Jackson SE (2005) Black shales, organic matter, ore genesis and hydrocarbon generation in the Paleoproterozoic Franceville Series, Gabon. Precambrian Res 137:253–272

Moucha R, Forte AM, Mitrova JX, Rowley DB, Quere S, Simmons NA, Grand SP (2008) Dynamic topography and long-term sea-level variations: there is no such thing as a stable continental platform. Earth Planet Sci Lett 271:101–108

Müller SG, Krapež B, Barley ME, Barley ME, Fletcher IR (2005) Giant iron-ore deposits of the Hamersley province related to the breakup of Paleoproterozoic Australia: new insights from in situ SHRIMP dating of baddeleyite from mafic intrusions. Geology 33:577–580

Nance RD, Worsley TR, Moody JB (1986) Post-Archean biogeochemical cycles and long-term episodicity in tectonic processes. Geology 14:514–518

Nance R, Worsley T, Moody J (1988) The supercontinent cycle. Sci Am 259:72–79

Nelson DR (2004) Episodic crustal growth during catastrophic global-scale mantle overturn events. In: Eriksson PG, Altermann W, Nelson DR, Mueller WU, Catuneanu O (eds) The Precambrian Earth: tempos and events, vol 12, Developments in Precambrian. Elsevier, Amsterdam, pp 180–183

Nelson DR, Trendall AF, Altermann W (1999) Chronological correlations between the Pilbara and Kaapvaal cratons. Precambrian Res 97:165–189

Nironen M (2005) Proterozoic orogenic granitoid rocks. In: Lehtinen M, Nurmi PA, Rämö OT (eds) Precambrian geology of Finland – key to the evolution of the Fennoscandian Shield. Elsevier, Amsterdam, pp 443–480

O'Neill C, Lenardic A, Moresi L, Torsvik TH, Lee C-TA (2007) Episodic Precambrian subduction. Earth Planet Sci Lett 262:552–562

Papineau D (2010) Global biogeochemical changes at both ends of the Proterozoic: insights from phosphorites. Astobiology 10:1–17

Papineau D, Mojzsis SJ, Schmitt AK (2007) Multiple sulfur isotopes from Paleoproterozoic Huronian interglacial sediments and the rise of atmospheric oxygen. Earth Planet Sci Lett 255:188–212

Pavlov AA, Kasting JF (2002) Mass-independent fractionation of sulfur isotopes in Archean sediments: strong evidence for an anoxic Archean atmosphere. Astrobiology 2:27–41

Pavlov AA, Kasting JF, Brown LL (2000) Greenhouse warming by CH_4 in the atmosphere of early Earth. J Geophys Res 105:11981–11990

Pettijohn FJ (1940) Archean metaconcretions of Thunder Lake, Ontario. Geol Soc Am Bull 51:1841–1850

Pickard AL (2003) SHRIMP U-Pb zircon ages for the Palaeoproterozoic Kurumn Iron Formation, Northern Cape Province, South Africa: evidence for simultaneous BIF deposition on Kaapvaal and Pilbara cratons. Precambrian Res 125:275–315

Pope MC, Grotzinger JP (2003) Paleoproterozoic Stark Formation, Athapuscow Basin, Northwest Canada: record of cratonic-scale salinity crisis. J Sediment Res 73:280–295

Rampino MR (2010) Volcanism, climatic change, and the geological record. Sedimentat Volcan Sett 1:9–18

Reddy SM, Evans DAD (2009) Palaeoproterozoic supercontinents and global evolution: correlations from core to atmosphere. In: Reddy SM, Mazumdr R, Evans DAD, Collins AS (eds) Palaeoproterozoic supercontinents and global evolution, 323rd edn. Geological Society, London, Special Publication, London, pp 165–198

Salop LI (1982) Geological evolution of the Earth in the Precambrian. Nedra, Leningrad, p 343 (in Russian)

Schidlowski M, Eichmann R, Junge CE (1975) Precambrian sedimentary carbonates: carbon and oxygen isotope geochemistry and implications for the terrestrial oxygen budget. Precambrian Res 2:1–69

Schneiderhan EA, Gutzmer J, Strauss H, Mezger K, Beukes NJ (2006) The chemostratigraphy of a Paleoproterozoic MnF-BIF succession – the Voëlwater Subgroup of the Transvaal Supergroup in Griqualand West South Africa. S Afr J Geol 109:63–80

Schröder S, Bekker A, Beukes NJ, Strauss H, van Niekerk HS (2008) Rise in seawater sulphate concentration associated with the Paleoproterozoic positive carbon isotope excursion: evidence from sulphate evaporates in the 2.2–2.1 Gyr shallow-marine Lucknow Formation, South Africa. Terra Nova 20:108–117

Silver PG, Behn MD (2008) Intermittent plate tectonics. Science 319:85–88

Sorjonen-Ward P, Luukkonen EJ (2005) Archean rocks In: Lehtinen M, Nurmi PA, Rämö OT (eds) Precambrian geology of Finland – key to the evolution of the Fennoscandian Shield, Elsevier, Amsterdam, pp 19–99

Valentine JW, Moores EM (1970) Plate-tectonic regulation of faunal diversity and sea level: a model. Nature 228:657–659

Veizer J (2005) Celestial climate driver: a perspective from four billion years of the carbon cycle. Geoscience Canada 32:3–28

Visser JNJ (1971) The deposition of the Griquatown glacial member in the Transvaal Supergroup. Trans Geol Soc S Afr 74:187–199

Vogel DC, Vuollo JI, Alapieti TT, James RS (1998) Tectonic, stratigraphic, and geochemical comparison between ca. 2500–2440 Ma mafic igneous events in the Canadian and Fennoscandian Shields. Precambrian Res 92:89–116

Wanke A, Melezhik VA (2005) Palaeoproterozoic sedimentation and stromatolite growth in an advanced intracontinental rift associated with the marine realm: a record of the Neoarchaean continent breakup? Precambrian Res 140:1–35

Williams H, Hoffman PF, Lewry JF, Monger JWH, Rivers T (1991) Anatomy of North America: thematic geologic portraits of the continent. Tectonophysics 187:117–134

Windley BF (1993) Uniformitarianism today: plate tectonics is the key to the past. J Geol Soc Lond 1:7–19

Winter BL, Knauth LP (1992) Stable isotope geochemistry of early Proterozoic carbonate concretions in the Animikie Group of the Lake Superior region; evidence for anaerobic bacterial processes. Precambrian Res 54:131–151

Worsley TR, Nance D, Moody JB (1984) Global tectonics and eustasy for the past 2 billion years. Marine Geol 58:373–400

Young GM (1970) An extensive early Proterozoic glaciation in North America. Palaeogeogr Palaeoclimatol Palaeoecol 7:85–100

Young GM, Long DGF, Fedo CH, Nesbitt HW (2001) The Paleoproterozoic Huronian Basin: product of a Wilson cycle accompanied by glaciation and meteorite impact. Sediment Geol 141–142:233–254

Yudovich YE, Makarikhin VV, Medvedev PV, Sukhanov NV (1991) Carbon isotope anomalies in carbonates of the Karelian Complex. Geochemist Int 28:56–62

Zahnle K, Claire M, Catling D (2006) The loss of mass-independent fractionation in sulfur due to a Palaeoproterozoic collapse of atmospheric methane. Geobiology 4:271–283

Part II

The Fennoscandian Arctic Russia: Drilling Early Earth Project (FAR-DEEP)

2.1 The International Continental Scientific Drilling Program

Victor A. Melezhik

The International Continental Scientific Drilling Program (ICDP) was founded in 1996. Its main goal is to enable geoscientific research by obtaining continuous rock sections through key geological formations which are not accessible by any other means but crucial for understanding Earth's surface processes and their evolution, and providing direct information on processes operating at depth. ICDP has overseen 15 years of successful coordination of multinational efforts in continental scientific drilling (Harms et al. 2007). Drilling operations are expensive and a series of robust criteria must be met for projects to obtain financial and operational support from the ICDP. Among these criteria, two of the most important are global significance and international impact, which dictate that projects must address geological problems of global significance, utilising world-class geological sites and establishing broad international cooperation by pooling resources from the best possible scientific teams. Other criteria include societal needs, need-for-drilling and depth-to-cost, which are defined explicitly by the ICDP in order to assess and concentrate on topics of high international priorities (Harms and Emmermann 2007).

Over its first decade, the thematic activities of ICDP-coordinated projects were focused on issues relating to palaeoclimate, natural hazards (earthquake, volcanism, impact events), geodynamics and alternative energy resources. In 2005, at a 10-year-anniversary forum, the ICDP expanded its scope and scientific focus to include research of the biosphere and evolution of the Earth through critical time intervals.

V.A. Melezhik (✉)
Geological Survey of Norway, Postboks 6315 Sluppen, NO-7491 Trondheim, Norway

Centre for Geobiology, University of Bergen, Allegaten 41, N-5007 Bergen, Norway
e-mail: victor.melezhik@ngu.no

2.1.1 Fennoscandian Arctic Russia–Drilling Early Earth Project (FAR-DEEP)

The Fennoscandian Arctic Russia–Drilling Early Earth Project (FAR-DEEP) was one of the responses to the ICDP's new initiatives (Melezhik et al. 2010). FAR-DEEP's main scientific objectives centre around one of the most intriguing periods in Earth history when terrestrial environments experienced a revolutionary transition from being largely anoxic during Earth's first 2100 Ma to irreversibly oxic; the environmental hallmark (Cloud 1987a, b) now often referred to as the Great Oxidation Event (e.g. Holland 2006).

Motivation and Goals of Drilling Project

The formation of Earth as a solid body at c. 4500 Ma (Manhes et al. 1980; Dalrymple 2001) was followed by a c. 2000-Ma-long period of environmentally stable conditions. From 2500 to 2000 Ma the Earth System experienced a series of fundamental upheavals (Fig. 2.1). This hallmark period, known as the Late Archaean-Early Palaeoproterozoic transition (APT), represents one of the most critical periods in Earth's history, as it reflects the emergence of an aerobic Earth System and a series of interrelated global events detailed in Chap. 1.1. Most important of the Early Palaeoproterozoic events were the establishment of an oxygenated atmosphere between 2450 and 2320 Ma (Bekker et al. 2004; Hannah et al. 2004) and the emergence of an aerobic biosphere. The remaining 1500 Ma of the Proterozoic exhibit clear evidence that the surface environments on Earth operated under oxic conditions, pretty much as they do today. Even though macroscopic multicellular organisms and plants did not evolve until much later, most biogeochemical recycling, in the oceans and on land, was dependent on highly energetic aerobic pathways. However, some workers suggest that deep marine basins or perhaps even the global deep ocean itself remained anoxic and perhaps even became euxinic (rich in H_2S) throughout much of the

V.A. Melezhik et al. (eds.), *Reading the Archive of Earth's Oxygenation*, Volume 1: The Palaeoproterozoic of Fennoscandia as Context for the Fennoscandian Arctic Russia - Drilling Early Earth Project,
DOI 10.1007/978-3-642-29682-6_2, © Springer-Verlag Berlin Heidelberg 2013

ensuing Proterozoic time slice (Canfield 1998; Anbar and Knoll 2002; Poulton et al. 2004).

Understanding of the causative relationships between the series of the APT global palaeoenvironmental events, (Fig. 2.1) and the combined processes that led to the irreversible alteration of Earth's surface environments, represents one of the most challenging fundamental problems in the geosciences (e.g. Melezhik et al. 2005). Essential to this is the need for a continuous rock record spanning the APT supported by studies that integrate the various proxy datasets that document the various processes operating at this time. However, obtaining a continuous rock record and, consequently, an enhanced understanding of Earth System behaviour and evolution during the APT period is hindered by three major shortcomings: (1) limitation of exposures of key stratigraphic intervals; (2) samples compromised by anthropogenic contamination; and (3) modification by recent oxidation and weathering. These can be overcome by implementing scientific drilling that targets key geological formations containing a record of the 2500–2000 Ma global events. Such a drilling project should have three major goals: (*1*) establish a well-characterised, well-dated and well-archived succession of rocks for the period of 2500–2000 Ma; (*2*) document the changes in the biosphere and the geosphere associated with the rise in atmospheric oxygen; and (*3*) develop a self-consistent model to explain the genesis and timing of the establishment of the aerobic Earth System.

Scientific Background for Implementation of the Drilling Project

There are five areas where significant Palaeoproterozoic successions occur that could be used to study the 2500–2000 Ma interval; northern and southern America, South Africa, Western Australia and Fennoscandia (Fig. 2.2). Consequently, any of these areas could have been potential targets for implementing an ICDP drilling project. Fennoscandia was chosen as a result of discussions initiated during the 2004 Nordic Academy for Advanced Studies field course held in NW Russia. Then, an international group of scientists from several European countries, the USA and Australia deliberated in the field and reached the conclusion that the eastern part of the Fennoscandian Shield contains the finest archive of the hallmark events that typify the APT. This was elaborated further during subsequent field studies and an ICDP Workshop during 2004–2005 and a decision was reached to undertake a multidisciplinary international research programme and scientific drilling project on the APT. Target intervals and drilling sites were identified and overall objectives defined.

The Fennoscandian Shield is a composite craton composed of late Archaean granite-greenstone belts and several extensively developed Palaeoproterozoic greenstone and orogenic belts (Fig. 2.3). Well-preserved igneous and sedimentary rocks record a long period of rifting and opening of a series of oceans and break-up of an Archaean continent followed by several orogenic episodes (see Chap. 3.4). The Palaeoproterozoic successions reach a cumulative thickness of over 20,000 m and provide a rich information of nearly 700 Ma of Earth history. The eastern (Russian) sector of the shield area was selected for scientific drilling (Fig. 2.3) because it is characterised by exceptionally well preserved and lithologically diverse rocks 2500–1900 Ma in age. They are considered to contain a fairly complete record of the hallmark events of the APT (Melezhik et al. 2005), including c. 2500 Ma global rifting (Chap. 3.4), the global Huronian-age glaciation (Chap. 7.2), an apparent upper-mantle oxidising trend and related redox change of volcanic rocks (Chap. 7.4), the oldest record of subaerial hydrothermal processes (Chap. 7.9.4), a rise in atmospheric oxygen (Chaps. 7.1 and 8), the protracted and large-magnitude carbon isotope excursion (Chap. 7.3), changes in the sulphur cycle and a substantial increase in the seawater sulphate reservoir (Chap. 7.5), changes in phosphorus cycle and accumulation of earliest sedimentary phosphates (Chap. 7.7), a radical modification in recycling of organic matter (Chap. 8), an unprecedented accumulation of organic-matter-rich sediments and the oldest known significant generation of petroleum (Chap. 7.6).

Implementation of the Scientific Drilling Project

The ICDP FAR-DEEP implemented in Fennoscandia represents a novel type of project among the ones typical of the ICDP in that it aimed to obtain geological material by drilling numerous holes in several areas (Fig. 2.3). Collectively, these provided a representative geological record of the most important global events occurring through the APT. The drilling operations were largely co-sponsored by the ICDP, with additional funding received from DFG (German Research Council), NFR (Norwegian Research Council), NASA Astrobiology Institute and the US NSF (National Science Foundation), the Centre for Geobiology of the University of Bergen, and the Geological Survey of Norway.

The drilling operations were carried out from late May through late October in 2007 by the Finnish operator SMOY. The Russian State Company "Mineral", based in St. Petersburg, Russia, provided all logistical support. On-site core documentation was supported by the Institute of Geology, the Karelian Science Centre, Petrozavodsk, Russia. Fifteen drillholes were drilled, ranging from 92 to 503 m in depth, totaling 3,650 m of recovered core. In order to accomplish the proposed program, the drilling rig, the field camp, all the equipment, and on-site operating scientists travelled together over a total distance of 5,000 km. The large distances and remoteness of the drilling sites with limited infrastructure required a considerable logistical effort (Fig. 2.4).

The ICDP FAR-DEEP with its Deep Time perspective has opened a new window for multidisciplinary research, with aims of obtaining new knowledge of the processes that operated on Earth's surface and interior at the dawn of the emergence of an aerobic world. The chosen approach was a success, resulting in collection of a rich archive of geological material that is detailed in **Part 6**.

An ambitious and successful FAR-DEEP research programme (http://far-deep.icdp-online.org) requires a considerable international effort for its accomplishment. Currently 13 countries have applied for research grants from various funding agencies. Five Ph.D. students from Finland, Germany and Norway are involved in the FAR-DEEP research program. Four postdoc projects in Finland, Great Britain and Norway have been financed by national funding agencies. A large group of scientists from the USA has received support from the National Science Foundation and NASA. Several research groups from Belgium, Czech Republic, Estonia and Russia have been also successful in obtaining research grants. The FAR-DEEP core is stored in Trondheim, at the Geological Survey of Norway, and is available for coordinated international multidisciplinary research.

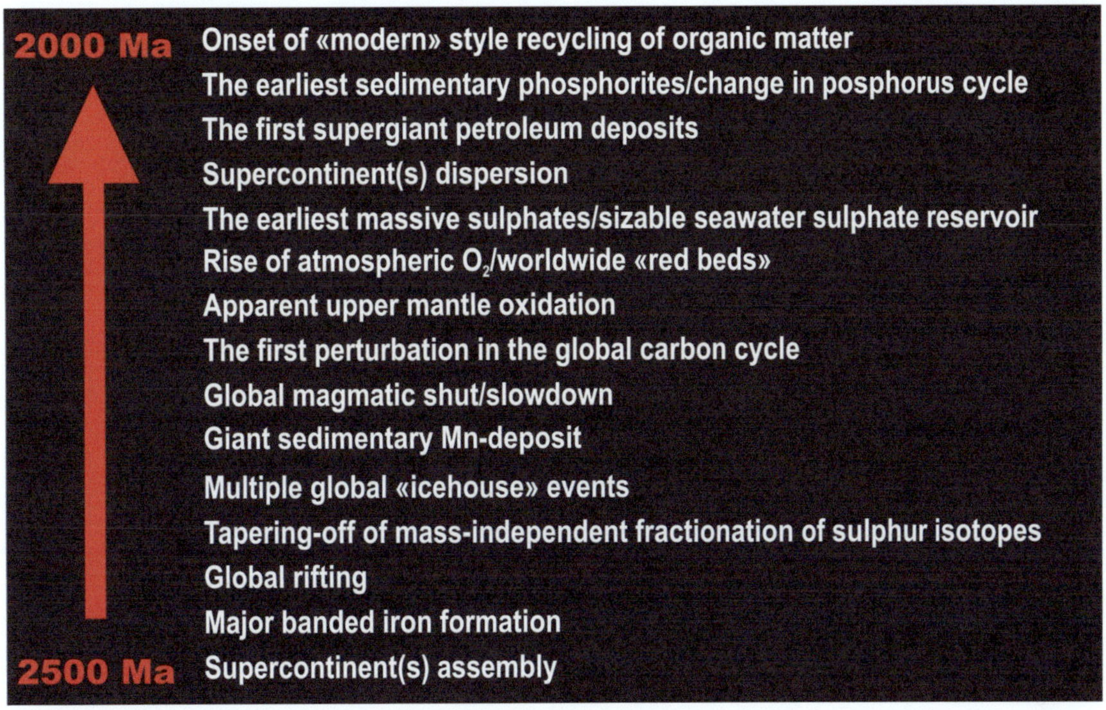

Fig. 2.1 A series of global events associated with the transition from an anoxic to oxic world: a "middle-age Earth crisis"?

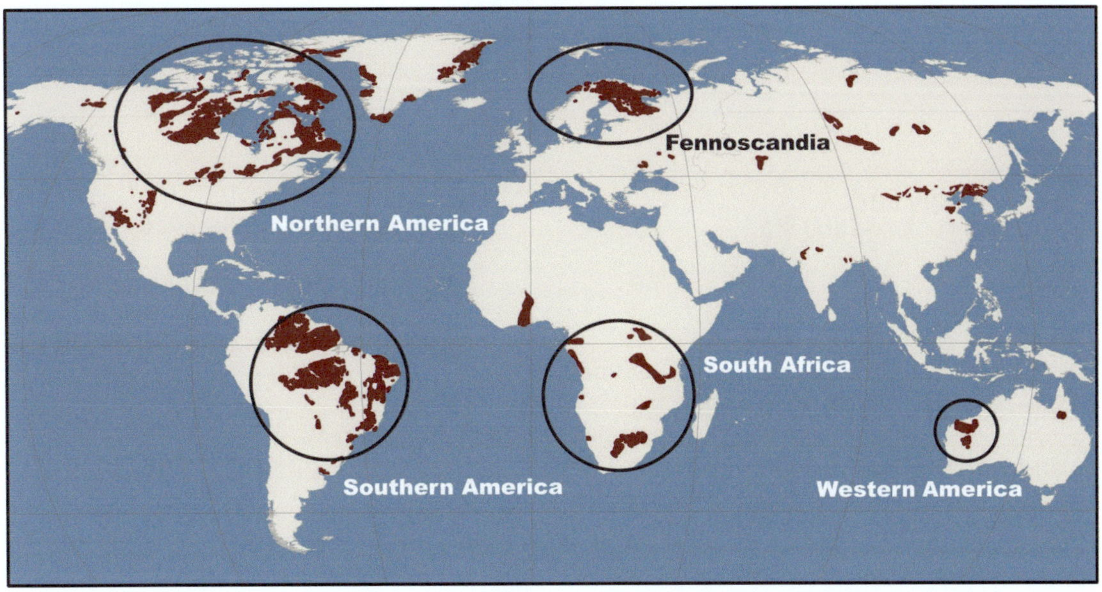

Fig. 2.2 Distribution of Palaeoproterozoic rocks (*red*) with circled areas highlighting the most continuous and thickly developed volcano-sedimentary successions sections (Map compiled by Aivo Lepland)

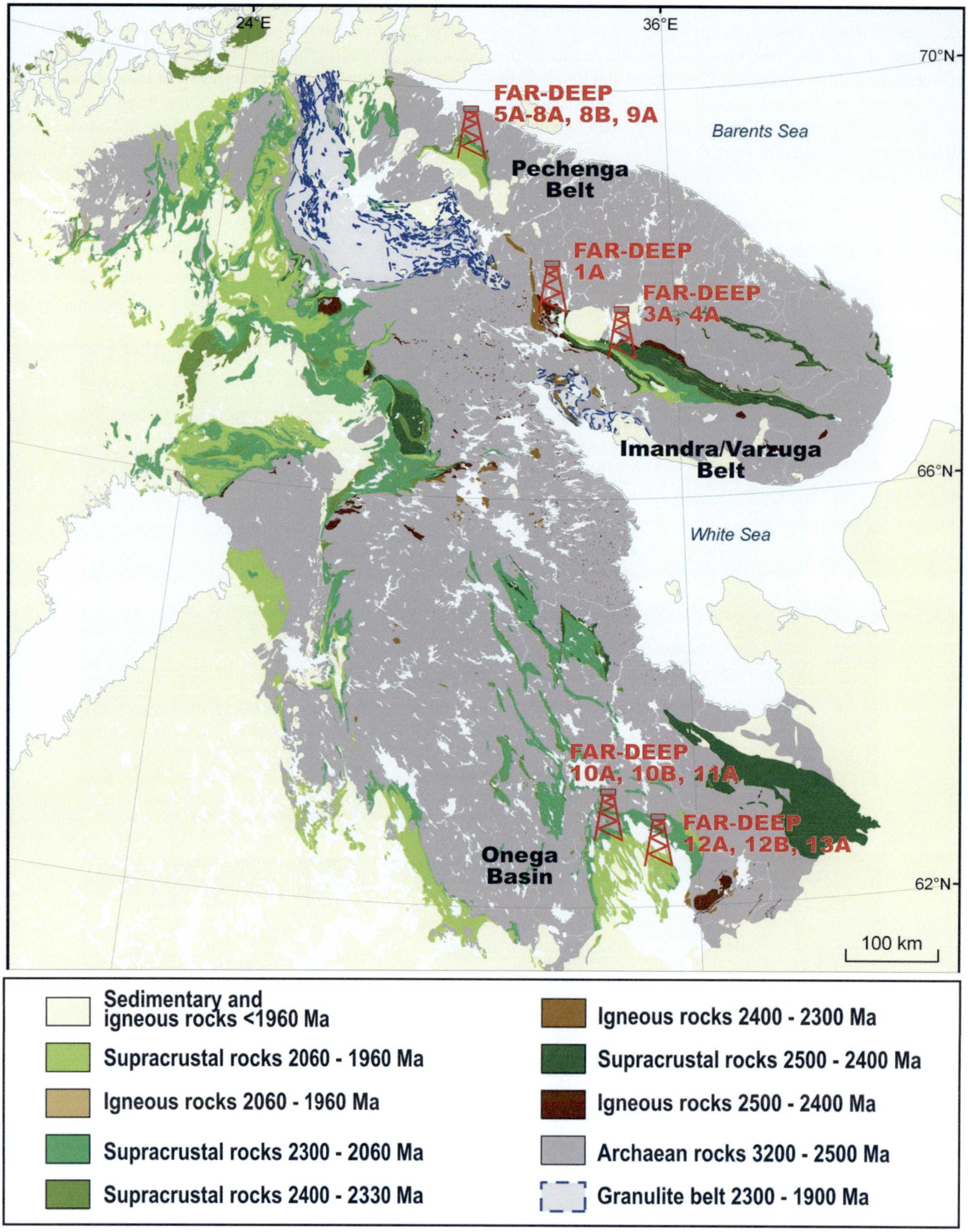

Fig. 2.3 Geological map of the eastern Fennoscandian Shield showing FAR-DEEP drilling sites and drillhole numbers (Geological map is modified by Aivo Lepland from Koistinen et al. (2001))

Fig. 2.4 Scientific drilling implemented in the eastern part of the Fennoscandian Shield. *The Pechenga Greenstone Belt*: (**a**) Peat bog near site 7A. (**b**) Military-terrain vehicle assists the mobile drilling rig to negotiate a swamp on its way to site 6A. (**c**) A field camp at site 8A and 8B. (**d**) Loading core in to terrain vehicle at the site 8B. *The Imandra/Varzuga Greenstone Belt*: (**e**) . . . even don't have a dream to step on it. . . (**f**) On the way to the site of Huronian-age glacial deposits. (**g**) Breakdown on the way to sites 3A and 4A

References

Anbar AD, Knoll AH (2002) Proterozoic ocean chemistry and evolution: a bioinorganic bridge? Science 297:1137–1142

Bekker A, Holland HD, Wang PL, Rumble D, Stein HJ, Hannah JL, Coetzee LL, Beukes NJ (2004) Dating the rise of atmospheric oxygen. Nature 427:117–120

Canfield DE (1998) A new model for Proterozoic ocean chemistry. Nature 396:450–453

Cloud P (1987a) Oasis in space: Earth history from the beginning. W.W. Norton and Co, New York, p 515

Cloud P (1987b) Trends, transitions, and events in Cryptozoic history and their calibration: apropos recommendations by the subcommission on Precambrian Stratigraphy. Precambrian Res 37:257–264

Dalrymple GB (2001) The age of the Earth in the twentieth century: a problem (mostly) solved. Spec Pub, Geol Soc Lond 190:205–221

Hannah JL, Bekker A, Stein HJ, Markey RJ, Holland HD (2004) Primitive Os and 2316 Ma age for marine shale: implications for Paleoproterozoic glacial events and the rise of atmospheric oxygen. Earth Planet Sci Lett 225:43–52

Harms U, Emmermann R (2007) History and status of the international continental scientific drilling program. In: Harms U, Koeberl Ch, Zoback MD (eds) Continental scientific drilling: a decade of progress, and challenges for the future. Springer, Berlin/Heidelberg, pp 1–52

Harms U, Koeberl Ch, Zoback MD (eds) (2007) Continental scientific drilling: a decade of progress, and challenges for the future. Springer, Berlin/Heidelberg, p 366

Holland HD (2006) The oxygenation of the atmosphere and oceans. Philos Trans R Soc B 361:903–915

Koistinen T, Stephens MB, Bogatchev V, Nordgulen Ø, Wennerström M, Korhonen J, (comps.) (2001) Geological map of the Fennoscandian Shield, scale 1:2 000 000, Espoo/Trondheim/Uppsala/Moscow

Manhes G, Allègrea CJ, Dupréa B, Hamelin B (1980) Lead isotope study of basic-ultrabasic layered complexes: speculations about the age of the earth and primitive mantle characteristics. Earth Planet Sci Lett 47:370–382

Melezhik VA, Fallick AE, Hanski E, Kump L, Lepland A, Prave A, Strauss H (2005) Emergence of the aerobic biosphere during the Archean-Proterozoic transition: challenges for future research. Geol Soc Am Today 15:4–11

Melezhik VA, Lepland A, Romashkin A, Rychanchik D, Mesli M, Finne TE, Conze R, the FAR-DEEP Scientists (2010) The great oxidation event recorded in Paleoproterozoic rocks from Fennoscandia. Sci Drill 9:23–29

Poulton SW, Fralick PW, Canfield DE (2004) The transition to a sulphidic ocean 1.84 billion years ago. Nature 431:173–177

Part III

Fennoscandia: The First 500 Million Years of the Palaeoproterozoic

3.1 The Early Palaeoproterozoic of Fennoscandia: Geological and Tectonic Settings

V.A. Melezhik and E.J. Hanski

The Fennoscandian Shield forms part of a large crustal segment sandwiched between the Scandinanvian Caledonian Orogenic Belt and the cover sequences of the Russian platform (Gorbatschev and Bogdanova 1993). Our main area of interest is the eastern part of the shield, which contains a long record of Archaean and Palaeoproterozoic evolution that, excluding minor remnants of c. 3500 Ga gneisses (Mutanen and Huhma 2003), started at c. 3200 Ma (Fig. 3.1, Koistinen et al. 2001). Five main geological subdivisions can be identified, from oldest to youngest: (1) late Archaean granitoid-gneiss basement with relicts of greenstone belts (Fig. 3.2); (2) early Palaeoproterozoic (2500–2000 Ma) volcanic and sedimentary complexes deposited on the Archaean sialic basement (Fig. 3.3); (3) early Palaeoproterozoic (c. 2100–1930 Ma) high-grade sedimentary and volcanic terranes with a major juvenile component (Lapland and Umba granulite belts, e.g. Daly et al. 2006; Fig. 3.3); (4) early Palaeoproterozoic ophiolite complexes forming a discontinuous chain from Outokumpu to central Finnish Lapland (Nuttio) (Fig. 3.3; c. 2015–1950 Ma; Hanski and Huhma 2005; Peltonen 2005); and (5) middle Palaeoproterozoic (1920–1880 Ma, Kähkönen 2005) sedimentary and volcanic rocks accreted together due to microcontinent-island arc collisions during the Svecofennian orogenies (Lahtinen et al. 2009) (Fig. 3.4).

The 2500–2000 Ma volcanic and sedimentary rocks are the main focus of this contribution (Figs. 3.1 and 3.3). The Archaean basement rocks comprise most of the north-eastern part of the Shield (they are not exposed in central and south-western areas) and form three segments, each with its associated cover of Palaeoproterozoic sedimentary and volcanic rocks. These are the Kola, Karelian (cf., Gaál and Gorbatchev 1987), and Norrbotten (Lahtinen et al. 2005) cratons (Fig. 3.2) and it is uncertain whether or not they were once assembled into a single Archaean plate. Some researchers regard the northern part of the Kola craton as a separate tectonic unit called the Murmansk craton (e.g. Daly et al. 2006). The Archaean rocks in the Belomorian Mobile Belt have been affected by strong crustal reworking (Bibikova et al. 2001) and there are no known occurrences of Palaeoproterozoic supracrustal rocks (however, see Bindeman et al. 2010).

Within the Fennoscandian Shield the early Palaeoproterozoic rocks fringe many of the Archaean complexes and Gaál and Gorbatchev (1987) have recognised several major tectonic units of which the Kola province, the Karelian province (including the Belomorian mobile belt) and Svecofennian domain are relevant for our case (Fig. 3.3). An integrated study involving petrological, geochronological and geophysical data applied to geological and tectonic modelling of the Fennoscandian Palaeoproterozoic history has complemented the work of Gaál and Gorbatchev (1987) and provided a more detailed and comprehensive tectonic subdivision (presented and summarised in Lahtinen et al. 2005). The position of the boundary between the Kola and Karelian provinces remains unresolved. Lahtinen et al. (2005) define it as the Palaeoproterozoic Pechenga and Imandra/Varzuga Greenstone Belts (suture zone of Berthelsen and Marker 1986, marked by SW-dipping electrical conductors) whereas others place it within the Lapland and Umba Granulite Belt (Melezhik and Sturt 1994).

In the **Kola province**, the 2500–2000 Ma rocks form a c. 1,000-km-long but discontinuous southeast-northwest trending belt from Russia through Finland to Norway. Each component is individually named (Imandra/Varzuga, Pechenga, Pasvik, Opukasjärvi and Polmak greenstone belts) and combined form the North Transfennoscandian Greenstone Belt (Fig. 3.1). They exhibit similar sedimentary and volcanic successions comprising volcanic rocks of tholeiitic, calc-alkaline, alkaline and komatiitic series ranging

V.A. Melezhik (✉)
Geological Survey of Norway, Postboks 6315 Sluppen, Trondheim NO-7491, Norway

Centre for Geobiology, University of Bergen, Allegaten 41, N-5007 Bergen, Norway
e-mail: victor.melezhik@ngu.no

V.A. Melezhik et al. (eds.), *Reading the Archive of Earth's Oxygenation*, Volume 1: The Palaeoproterozoic of Fennoscandia as Context for the Fennoscandian Arctic Russia - Drilling Early Earth Project,
DOI 10.1007/978-3-642-29682-6_3,

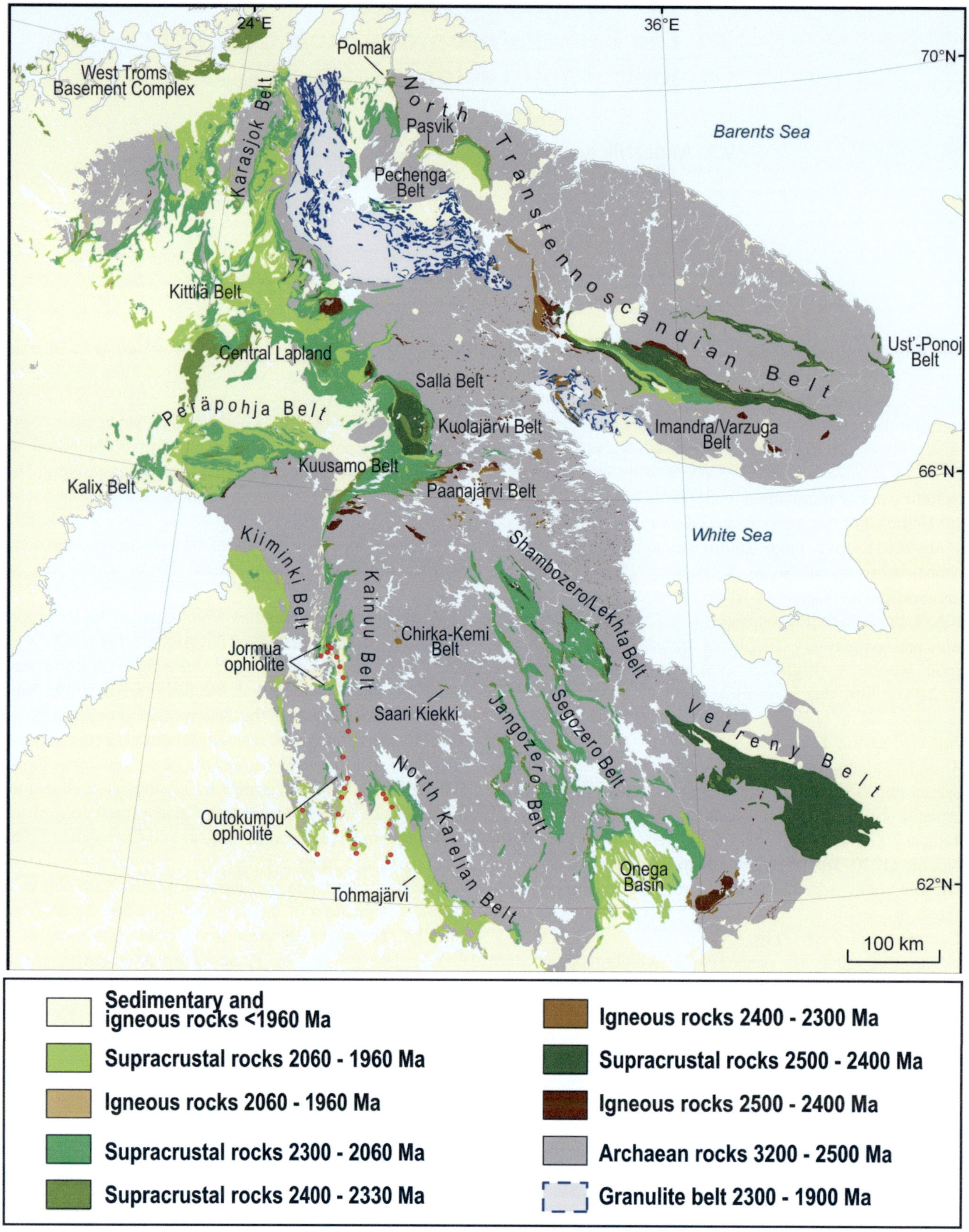

Fig. 3.1 Geological map of the eastern part of the Fennoscandian Shield emphasising the early Palaeoproterozoic rocks (The map is based on Koistinen et al. (2001), and modified by Aivo Lepland)

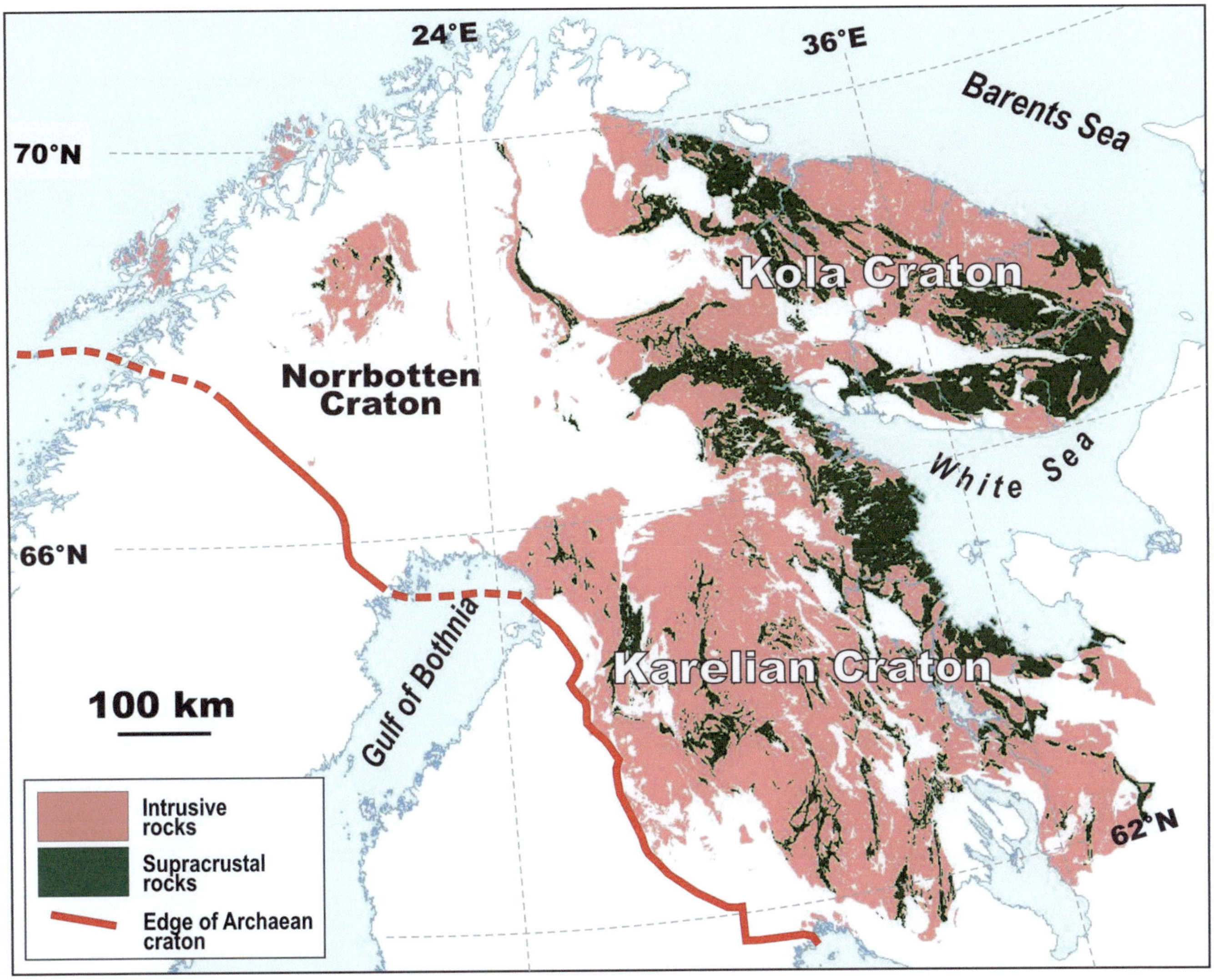

Fig. 3.2 Geological map of the Fennoscandian Shield emphasising the Archaean rocks (Based on Koistinen et al. 2001); the craton's position and names are based on Lahtinen et al. (2005)

from ultramafic to felsic and a much subordinate suite of diverse chemical and clastic sedimentary rocks. These have been interpreted as recording a complicated evolutionary history of multiple intracratonic rifting episodes leading eventually to drifting and formation of a short lived ocean (e.g. Melezhik and Sturt 1994). The Lapland and Umba Granulite belts, and the areas adjacent to them on either side of the Kola-Karelian boundary, comprise several Palaeoproterozoic oceanic-crust terranes accreted between c. 2100 and 1930 Ma (Meriläinen 1976; Daly et al. 2006).

The main geological feature of the **Karelian province** is a series of northwest-southeast trending belts that contain deposits recording intraplate rifting starting around 2450 Ma, followed by younger Palaeoproterozoic rocks accumulated in platform and marginal marine basins. Orthoquartzite-carbonate-pelite suites with varying abundances of intervening mafic volcanic rocks developed in this stage (e.g. Lahtinen et al. 2005). In addition, the Finnish part of the province possesses several allochthonous rock assemblages closely resembling ophiolites: the c. 1960 Ma Outokumpu mantle peridotites, the c. 1950 Ma Jormua sheeted dykes, gabbros, plagiogranites, lavas and hyaloclastites, and the c. 2015 Ma chromite-bearing peridotite and mafic lavas at Nuttio (details presented in Hanski and Huhma 2005; Peltonen 2005, and Peltonen et al. 2008). The southeastern part of the Karelian province is dominated by a carbonate platform and much reduced volume of volcanic infill (mainly mafic rocks) relative to other parts of the province.

The **Svecofennian domain** in the southern and western parts of the shield records significant growth of continental crust between c. 2000 and 1800 Ma ago. Since Hietanen's (1975) plate tectonic interpretations, the generation of the Svecofennian rocks has generally been related to subduction zone processes. However, the geological history of the area has turned out to be much more complicated than that of a single subduction zone. In the Finnish part of the Svecofennian

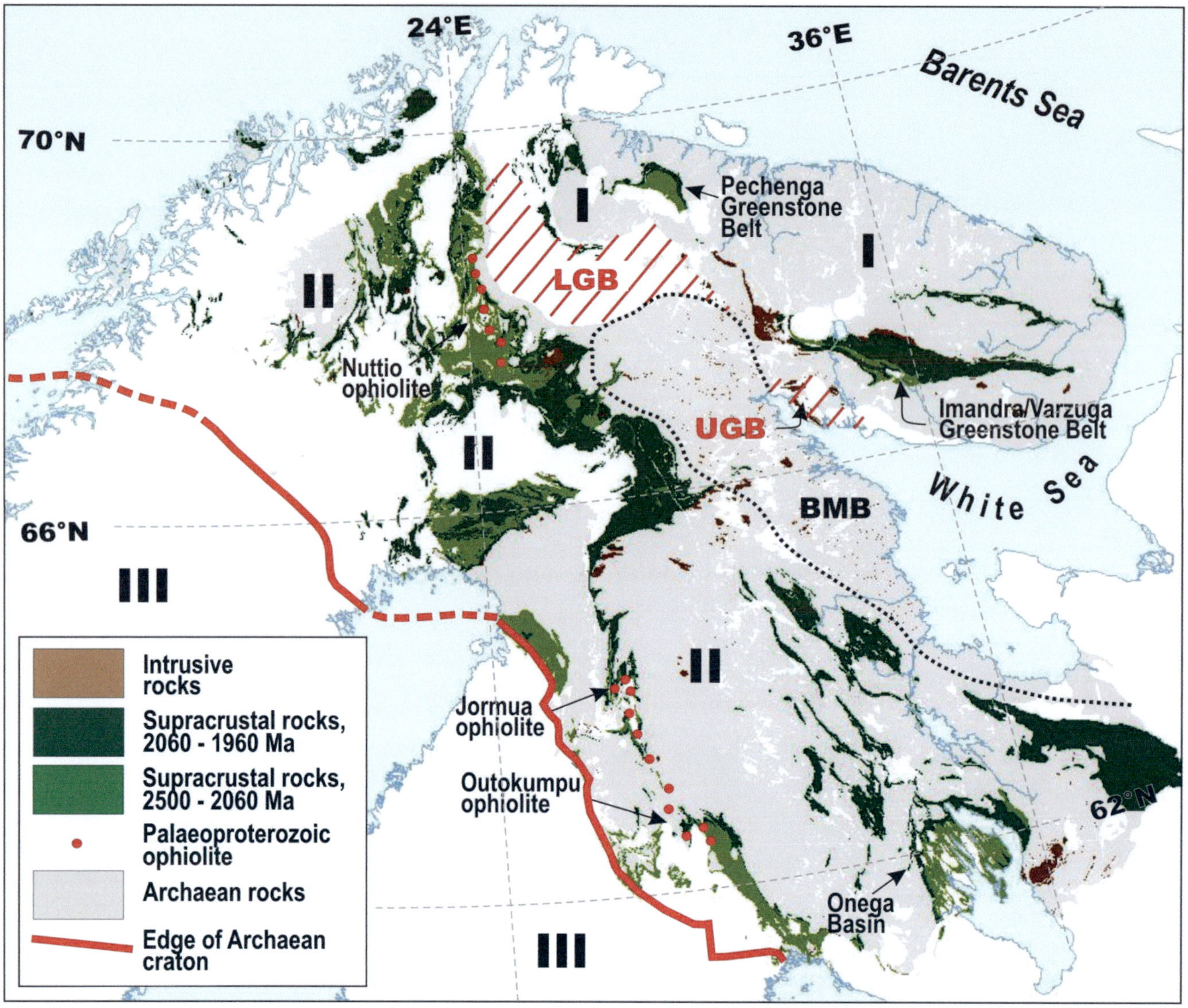

Fig. 3.3 Geological map of the Fennoscandian Shield emphasising the 2505–1960 Ma Palaeoproterozoic rocks (Based on Koistinen et al. 2001). ***I*** Kola province, ***II*** Karelian province, ***III*** Svecofennian domain, ***LGB*** Lapland Granulite Belt, ***UGB*** Umba Granulite Belt, ***BMB*** Belomorian Mobile Belt. The Pechenga and Imandra/Varzuga Greenstone Belt, and the Onega Basin are FAR-DEEP drilling sites; the tectonic provinces are based on Gaál and Gorbatschev (1987)

domain, Korsman et al. (1997) distinguished three separate arc complexes: (1) the Primitive arc complex of central Finland juxtaposed along the Archaean craton margin and the Accretionary arc complexes of (2) central and western Finland and (3) southern Finland (Fig. 3.4). These terranes were called the Savo belt, central Svecofennia and southern Svecofennia, respectively, by Kähkönen (2005). Characteristic rock types of the complexes include turbiditic greywackes and mudrocks, and calc-alkaline volcanic rocks ranging from low-K to shoshonitic basalts to rhyolites representing volcanic arc magmatism and associated accretionary prism sediments. MORB-, EMORB- and WPB-type pillow lavas are present locally. Carbonate formations are abundant in southern Finland (Kähkönen 2005).

The oldest volcanic and related plutonic rocks are found in the Savo belt, having ages in the range of c. 1930–1920 Ma, and in the Knaften area, Sweden, were granitoids have been dated at c. 1950 Ma. The evolved island arc volcanic rocks and associated sedimentary rocks in western and southern Finland are younger, recording ages of c. 1910–1890 Ma (Kähkönen 2005). With the exception of Mesoproterozoic rapakivi granites in central Sweden (Andersson et al. 2002), isotopic studies have not revealed any significant Archaean component in the source of igneous rocks of the Svecofennian domain (Vaasjoki et al. 2005). However, these studies have indicated that an older Palaeoproterozoic, c. 2100–2000 Ma source component is needed (Lahtinen and Huhma 1997). The isotopic results together with supportive geophysical

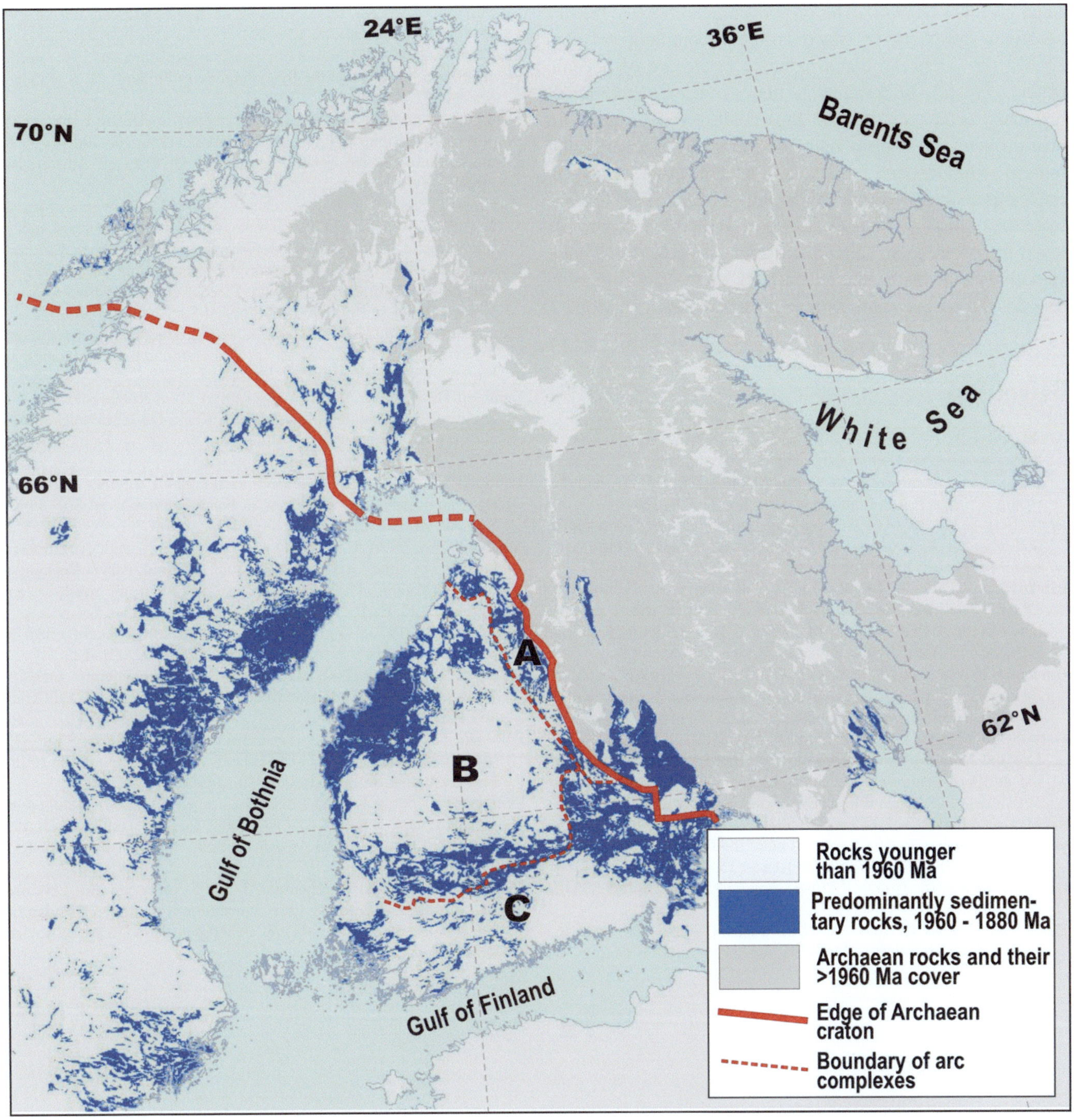

Fig. 3.4 Geological map of the Fennoscandian Shield emphasizing the 1960–1880 Ma Palaeoproterozoic rocks (Based on Koistinen et al. 2001). **A** = The Primitive arc complex of central Finland (Savo belt), **B** = The Accretionary arc complex of central and western Finland, **C** = The Accretionary arc complex of southern and western Finland. The edge of the Archaean craton and the boundaries of arc complexes in Finland after Vaasjoki et al. (2005)

evidence led Lahtinen et al. (2005) to suggest the presence, though without surface exposure, of several microcontinents (Keitele, Bothnia, Bergslagen) in the Svecofennian domain that were amalgamated with the island arc complexes.

Lahtinen et al. (2005, 2009) have published accounts on the tectonic evolution of the Shield and the reader is referred to these articles for detailed information. It suffices here to list the main Svecofennian orogenic stages as summarized by Lahtinen et al. (2005): (1) microcontinent accretion stage (1920–1870 Ma); (2) continental extension stage (1860–1840 Ma); (3) continent-continent collision stage (1840–1790 Ma); and (4) orogenic collapse and stabiliszation stage (1790–1770 Ma). The prolonged accretionary history involving collisions of island arc-microcontinent collages against the Archaean craton resulted in closure of oceanic basins, crustal thickening in the Archaean-Proterozoic boundary zone (currently 65 km), widespread orogenic to post-orogenic plutonism and overthrusting of continental margin sediments and fragments of oceanic crust. Archaean crust was reactivated (melted) within a distance of 150 km from the collisional zone (Vaasjoki et al. 2005) and isotopic resetting of minerals and disturbance of palaeomagnetic signatures due to the thermal effects of the Svecofennian orogenies took place still further inland within the craton.

References

Andersson UB, Neymark LA, Billström K (2002) Petrogenesis of Mesoproterozoic (Subjotnian) rapakivi complexes of central Sweden. Implications from U-Pb zircon ages, Nd, Sr and Pb isotopes. Trans R Soc Edinburgh Earth Sci 92:201–228

Berthelsen A, Marker M (1986) Tectonics of the Kola collision suture and adjacent Archean and Early Proterozoic terrains in the north-eastern region of the Baltic Shield. Tectonophysics 126:31–55

Bibikova E, Skiöld T, Bogdanova S, Gorbatschev R, Slabunov A (2001) Titanite-rutile thermochronology across the boundary between the Archaean Craton in Karelia and the Belomorian mobile Belt, eastern Baltic Shield. Precambrian Res 105:315–330

Bindeman IN, Schmitt AK, Evans DAD (2010) Limits of hydrosphere-lithosphere interaction: origin of the lowest-known $\delta^{18}O$ silicate rock on Earth in the Paleoproterozoic Karelian rift. Geology 38:631–634

Daly JS, Balagansky VV, Timmerman MJ, Whitehouse MJ (2006) The Lapland-Kola orogen: Palaeoproterozoic collision and accretion of the northern Fennoscandian lithosphere. In: Gee DG, Stephenson RA (eds) European lithosphere dynamics, Geological Society Memoir London: vol 32. pp 579–598

Gaál G, Gorbatchev R (1987) An outline of the Precambrian evolution of the Baltic Shield. Precambrian Res 64:15–32

Gorbatschev R, Bogdanova S (1993) Frontiers in the Baltic Shield. Precambrian Res 64:3–21

Hanski E, Huhma H (2005) Central Lapland greenstone belt. In: Lehtinen M, Nurmi PA, Rämö OT (eds) Precambrian geology of Finland – key to the evolution of the Fennoscandian Shield. Elsevier, Amsterdam, pp 139–194

Hietanen A (1975) Generation of potassium-poor magmas in the northern Sierra Nevada and the Svecofennian in Finland. J Res US Geol Surv 3:631–645

Kähkönen Y (2005) Svecofennian supracrustal rocks. In: Lehtinen M, Nurmi PA, Rämö OT (eds) Precambrian Geol Finland – key to the evolution of the Fennoscandian Shield. Elsevier, Amsterdam, pp 343–406

Koistinen T, Stephens MB, Bogatchev V, Nordgulen Ø, Wennesträm M, Korhonen, J (Comps.) (2001) Geological map of the Fennoscandian Shield, scale 1:2 000 000. Espoo/Trondheim/Upsala/Moscow

Korsman K, Koistinen T, Kohonen J, Wennerström M, Ekdahl E, Honkamo M, Idman H, Pekkala Y (eds) (1997) Bedrock map of Finland, scale 1: 1 000 000. Geological Survey of Finland, Espoo

Lahtinen R, Huhma H (1997) Isotopic and geochemical constraints on the evolution of the 1.93–1.79 Ga Svecofennian crust and mantle in Finland. Precambrian Res 82:13–34

Lahtinen R, Korja A, Nironen M (2005) Paleoproterozoic tectonic evolution. In: Lehtinen M, Nurmi PA, Rämö OT (eds) Precambrian geology of Finland – key to the evolution of the Fennoscandian Shield. Elsevier, Amsterdam, pp 669–680

Lahtinen R, Korja A, Nironen M, Heikkinen P (2009) Palaeoproterozoic accretionary processes in Fennoscandia. Geol Soc, Lond, Spe Pub 318:237–256

Melezhik VA, Sturt BA (1994) General geology and evolutionary history of the early Proterozoic Polmak-Pasvik-Pechenga-Imandra/Varzuga-Ust'Ponoy Greenstone Belt in the north-eastern Baltic Shield. Earth-Sci Rev 36:205–241

Meriläinen K (1976) The granulite complex and adjacent rocks in Lapland, northern Finland. Geol Surv Finl Bull 282:1–129

Mutanen T, Huhma H (2003) The 3.5 Ga Siurua trondhjemite gneiss in the Archean Pudasjärvi granulite belt, northern Finland. Bull Geol Soc Finl 75:51–68

Peltonen P (2005) Ophiolites. In: Lehtinen M, Nurmi PA, Rämö OT (eds) Precambrian geology of Finland – key to the evolution of the Fennoscandian Shield. Elsevier, Amsterdam, pp 237–278

Peltonen P, Kontinen A, Huhma H, Kuronen U (2008) Outokumpu revisited: new mineral deposit model for the mantle peridotite-associated Cu-Co-Zn-Ni-Ag-Au sulphide deposits. Ore Geol Rev 33:559–617

Vaasjoki M, Korsman K, Koistinen T (2005) Overview. In: Lehtinen M, Nurmi PA, Rämö OT (eds) Precambrian geology of Finland – key to the evolution of the Fennoscandian Shield. Elsevier, Amsterdam, pp 1–18

3.2 Litho- and Chronostratigraphy of the Palaeoproterozoic Karelian Formations

E.J. Hanski and V.A. Melezhik

3.2.1 Lithostratigraphic Framework

In Phanerozoic successions, the definition and interbasinal correlation of stratigraphic units is based on palaeontological evidence, generally supported by abundant high-precision radiometric dates. In contrast, Precambrian metasedimentary successions contain few fossils useful for correlation purposes and commonly lack precise geochronological control. This has hampered the development of robust chronostratigraphic frameworks for such successions, and restricted reconstruction of their geological histories. It has also led to the proliferation of many local stratigraphic nomenclatures rife with numerous formal and informal names.

The Finnish and Russian parts of the Fennoscandian Shield are notable exceptions in that efforts have been made to define and apply a common stratigraphic nomenclature for the entire region (see Fig. 3.5 as an example). This concerns the Palaeoproterozoic (c. 2500–1900 Ma) formations that have been traditionally termed Karelian and comprise widespread sedimentary and volcanic successions deposited or extruded on the Archaean basement in the eastern and northern parts of the Shield before experiencing the Svecofennian orogeny at c. 1900 Ma. Generally, and particularly away from the craton margins, the Karelian supracrustal formations have not undergone complex folding and overthrusting events during the orogeny and therefore geologists have been able to construct stratigraphic successions in different basins (schist and greenstone belts[1]) of the Fennoscandian Shield, covering an age range of several hundreds of millions of years. Stratigraphers have subdivided the Karelian rocks into several major units, including from oldest to youngest (adjectival forms in parenthesis): Sumi (Sumian), Sariola (Sariolian), Jatuli (Jatulian), Ludicovi (Ludicovian), and Kaleva (Kalevian); the latter is also called Livi (Livvian) in Russia. These names are derived from Karelian folklore and ethnographic names with some of them having been introduced almost a century ago (see Simonen 1986). The above-mentioned subdivisions have been widely used for correlation of sedimentary strata across the eastern part of the Fennoscandian Shield, both informally and formally in Finland (see Hanski 2001; Ojakangas et al. 2001) and formally in Russia (e.g. Semikhatov et al. 1991; Negrutza et al. 2000). In the formal nomenclature, the lithostratigraphic group (e.g. Jatuli Group) and superhorizon terms (e.g. Sumi Superhorizon) have been applied in the stratigraphic subdivisions in Finland and Russia, respectively.

Each subdivision was originally viewed as having its own unique suite of distinguishing lithological characteristics: the Sumian rocks contained immature clastic rocks associated with felsic volcanic rocks; the Sariolian rocks were marked by abundant polymictic conglomerates; the Jatulian rocks were recognised by the association of "red beds", dolostones (generally stromatolitic) and mature arenites, and quartz pebble conglomerates; the Ludicovian rocks were typified by "black shales"; and the Kalevian rocks were defined by thick successions of turbiditic greywackes. The Sumi through Kaleva framework has become deeply ingrained in the geological literature and, although laudable in reducing stratigraphic nomenclature (Fig. 3.5), an inadvertent outcome was an over-reliance on using sedimentological and volcanological features as a criterion for establishing interregional correlations (e.g. Sokolov 1987). In some instances this approach has proved useful, but in many others, it led to incorrect stratigraphic correlations, as is now known from radiometric datings (for an overview, see Melezhik and Sturt 1998; Hanski 2001).

Another complication derived from different connotations assigned to such names, is that they have been variably used

[1] In this account the names of the Palaeoproterozoic greenstone belts and schist belts are expressed in a short form, for example, the Peräpohja Belt instead of the Peräpohja Schist Belt.

E.J. Hanski (✉)
Department of Geosciences, University of Oulu, P.O. Box 3000, 90014 Oulu, Finland
e-mail: eero.hanski@oulu.fi

V.A. Melezhik et al. (eds.), *Reading the Archive of Earth's Oxygenation*, Volume 1: The Palaeoproterozoic of Fennoscandia as Context for the Fennoscandian Arctic Russia - Drilling Early Earth Project,
DOI 10.1007/978-3-642-29682-6_4, © Springer-Verlag Berlin Heidelberg 2013

in a lithostratigraphic, chronostratigraphic, or tectonic sense (Laajoki 1986). In addition, although the aim has been to develop unified nomenclature for the whole Shield, this has not been completely successful due to the fact that geotectonic conditions have been different in the intracratonic setting of the Russian part of the Shield as compared to the cratonic margin portions in the Finnish part, resulting in accumulation of rock successions with different physical appearances, particularly in post-Jatulian time. As the result, those sequences have been mainly assigned to Ludicovi on the Russian side of the border where they include large amounts of continental volcanic rocks, while on the Finnish side the post-Jatulian supracrustal rocks have been considered part of Kaleva, being dominated by deep-water greywackes and their high-grade metamorphic equivalents (mica schists and mica gneisses). The different tectonic histories are also manifested by the presence of allochthonous Kaleva units with ophiolitic fragments (upper Kaleva) in the west, whereas such rocks are absent in the east. Moreover, Karelian lithostratigraphic units may be only partially preserved and hence incompletely represented in the geological record. It is quite common that not only Sumian but also Sariolian and Jatulian successions show deposition directly upon the Archaean basement (e.g. Pekkarinen 1979; Negrutsa et al. 1986; Gaskelberg et al. 1986; Laajoki 1991). Even the Kalevian sediments may lie directly on Archaean gneisses (Honkamo 1985). Such features and lateral thickness variations of the individual stratigraphic units might have also been caused by original uneven palaeotopography, as well as syndepositional block movements, as demonstrated by the distribution of several Palaeoproterozoic mafic dyke swarms in the Archaean basement (Vuollo and Huhma 2005).

During the past two decades, individual local stratigraphic sections have been increasingly subdivided and named applying strict, formal, international lithostratigraphic rules (e.g. Hedberg 1976; for a recent Finnish guide, see Strand et al. 2010) in which geographical names are combined with lithostratigraphical rank terms. This has, though, led to the appearance of a plethora of local stratigraphic names, whose correlation is cumbersome and difficult to manage. Thus there is still a need for using informal stratigraphic terms when defining specific successions or types of sedimentary and volcanic rocks, which may then help to describe the geological evolution of each depositional basin and construct regional correlations. It should be noted that there have been previous attempts to redefine the traditional framework. One attempt used the concept of "tectofacies" (Laajoki 1988, 1990), in which lithologies were linked to inferred tectonic settings: Sumi-Sariola with rift tectofacies; Kainuu with narrow sea/inland basin tectofacies; Jatuli with open-sea tectofacies; and the threefold Kaleva tectofacies having lower rift-related, middle narrow-sea and upper open-sea components. However, this scheme has never gained popularity and has been largely ignored by Scandinavian geologists.

3.2.2 Event Stratigraphy and Age Constraints

We propose the use of radiometrically constrained global events (see Chap. 1) as a basis for chronostratigraphic subdivision of the early Palaeoproterozoic where it is possible. For the Fennoscandian Shield, we retain the traditional nomenclature, including Sumi, Sariola, Jatuli, Ludicovi, and Kaleva and their adjectival derivatives, but place it within the global-event-based chronostratigraphy described below (Fig. 3.6). When applying the terms in a chronostratigraphical sense, they should be compatible with an appropriate unit of the International Stratigraphic Chart. Accordingly, the duration of the Fennoscandian units is comparable with that of the "System". We stress that additional geochronological constraints are needed before this framework can be adopted formally or applied elsewhere.

The Palaeoproterozoic history of the Fennoscandian Shield commenced with the Sumian system rocks. These record the emplacement of large layered gabbronorite plutons, followed by a period of uplift and erosion, and subsequent sedimentation (see Chap. 3.4). This is best seen in the Imandra/Varzuga Belt where the basal conglomerate of the Kuksha Sedimentary Formation lies nonconformably on a gabbronorite intrusion dated at 2504.4 $\pm$ 1.5 Ma (Amelin et al. 1995) (Fig. 4.3 in Chap. 4.1). We use this age constraint to define the lower boundary of the Sumian system at 2505 Ma.

The next subdivision, the Sariolian system, includes Huronian-age glacial deposits and associated volcanic rocks. Determining the age of the base of this system is hindered by a phase of post-Sumian uplift and erosion of unknown duration. Nevertheless, the age of the Sumi-Sariola boundary must post-date the youngest Sumian volcanic rocks dated at 2442.2 $\pm$ 1.7 Ma (Amelin et al. 1995) and also the 2430 $\pm$ 4 Ma layered gabbronorite intrusions in the Peräpohja Belt (Perttunen and Vaasjoki 2001), which are overlain unconformably by Sariolian conglomerates. Hence we tentatively suggest that the base of the Sariolian system is 2430 Ma in age.

The lower boundary of the Jatulian system and the top of the Sariolian system is defined by the onset of the major phase of the Lomagundi-Jatuli Isotopic Event (for details see Chap. 7.3). The initiation of this event in the Fennoscandian Shield remains chronologically poorly constrained, but it predates the emplacement of the c. 2220 Ma mafic-ultramafic sills (Hanski et al. 2010). In South Africa, the main deposition of ^{13}C-rich carbonates postdates 2313 $\pm$ 7 Ma (Guo et al. 2009; Hannah et al. 2004) and thus we assign the age of 2300 Ma for the base of the Jatulian system. This is in agreement with previous propositions by Melezhik et al. (1997) and Meriläinen (1980). The termination of the Lomagundi-Jatuli Event is constrained at 2050 ± 8 Ma in the Peräpohja Belt (Karhu 2005) and more precisely at 2058 ± 2 Ma in the Pechenga Belt (Melezhik et al. 2007). Thus, the boundary between the Jatulian and Ludicovian systems is placed at 2060 Ma.

The boundary between the Ludicovian system and the youngest Palaeoproterozoic unit under consideration, the Kalevian system, remains poorly constrained. Lithologically it is marked by a rather rapid transition from organic-rich rocks (the Shunga Event, Fig. 3.6) associated with mafic-ultramafic lavas to a thick sequence of turbiditic greywackes with only locally developed mafic volcanic rocks. In eastern Finland, older Karelian units are cut by c. 1970 Ma diabase dykes (Vuollo et al. 1992), but there is no geochronological evidence for the existence of these dykes within Kalevian metasediments. Thus, we tentatively place the Ludicovi-Kaleva boundary at 1960 Ma. The upper limit of the Kalevian system is set at 1900 Ma, as it must be older than the age of the Svecofennian synorogenic plutonism that took place at c. 1890–1870 Ma (Shul'diner et al. 2000; Nironen 2005).

Figure 3.6 summarises the suggested chronostratigraphic subdivisions of the early Palaeoproterozoic of the Fennoscandian Shield and demonstrates their assigment to global palaeoenvironmental events. In the following discussion we define the chronostratigraphic units, accompanied with photo-documentation of the characteristic rock types.

3.2.3 The Sumian System, 2505–2430 Ma

The term Sumi has been used mostly by Russian geologists who applied it to the oldest Palaeoproterozoic rocks in the Imandra/Varzuga, Salla, Paanajärvi, Shambozero/Lekhta, and Vetreny belts (Fig. 3.1). Recently it has also been adopted for the Onega Basin in central Russian Karelia (Golubev et al. 2002). The term has been poorly defined, and its content and boundaries have been subject to continual changes (e.g. Heiskanen 1987a). In general, the Sumi has been recognised as a unit dominated by volcanogenic rocks containing polymictic volcaniclastic breccias and other volcaniclastic rocks genetically related to either interlayered rhyolites and dacites (e.g. Heiskanen 1987a; Ojakangas et al. 2001) or to the 2440–2430 Ma layered gabbronorite intrusions (Fig. 3.7a) (Alapieti 1982; Melezhik and Sturt 1994).

In the Kola Province, the Sumian succession is represented by the Strel'na Group in the Imandra/Varzuga Belt (Fig. 3.1) and consists of as much as a 5,500-m-thick pile of volcanic rocks with subordinate sedimentary rocks totalling 1,500 m in thickness. The Sumian succession starts with the Purnach Formation, having a tectonic contact with the basement, followed by the Kuksha Sedimentary Formation, which erosively overlies the Purnach Formation and various Archaean rocks. The Purnach Formation is more than 2,000 m thick and comprises amphibolite-facies volcanic rocks interbedded with four 100- to 200-m-thick greywacke units exhibiting well-preserved sedimentary features such as planar lamination and cross-bedding. The Kuksha Sedimentary Formation comprises similar greywacke and also arkosic sandstone (Fig. 3.7b) with minor small-pebble conglomerates and shows a variable thickness reaching more than 400 m in the eastern part of the belt. In the western part of the belt, polymict conglomerate and breccia (Fig. 3.7c) display a well-developed nonconformity on top of palaeo-weathered Archaean granites, gneisses, and layered gabbronorite intrusions (Figs. 4.5 and 4.6 in Chap. 4.1). The overlying Kuksha Volcanic Formation comprises massive, amygdaloidal, and rarer pillowed mafic lavas (Fig. 3.7d–f). These volcanites are overlain (the contact is marked by a palaeo-weathered surface) by the Seidorechka Sedimentary Formation, which consists of a thick (many tens of metres) succession of cross-bedded and flaser- to lenticular-bedded sandstone-shale cycles (Fig. 3.7h–k) followed by a unit of massive to bedded dolarenites and dolorudites (Fig. 3.7l–n) overlain unconformably by cross-bedded quartzites (Fig. 3.7o–r), shales, and marls (Fig. 3.7s–u). These rocks show mass-independent fractionation of sulphur isotopes (Fig. 3.6; Reuschel et al. 2009). The uppermost Sumian unit is the 2,800-m-thick Seidorechka Volcanic Formation composed of basaltic andesites (80 % of volume), dacites and rhyolites (Fig. 3.7v–x), and subordinate basalts with komatiitic affinities (Fedotov 1985).

In the Russian part of the Karelian craton (north-eastern Karelia; Fig. 3.1), a vast volume of Sumian rocks occurs in the Shombozero, Lekhta, and Vetreny belts (Fig. 3.7y–z). In the latter, a thick accumulation (4,000–8,000 m) of volcano-sedimentary rocks occurs in an area that is 15–85 km wide and more than 250 km long (Kulikov 1988). This includes a c. 500-m-thick unit of arkosic sandstones and quartzites (Toksha suite) and a c. 1,000-m-thick suite of andesitic basalts, basalts, and rare komatiitic basalts (Kirich suite) followed by a succession of tholeiitic basalts, dolostones (Kozhozero suite, c. 400 m), and conglomerates (Kalgachikha suite, c. 400 m). The upper part is composed of c. 1,500 m of turbiditic greywackes of the Vilenga suite overlain by c. 4,000 m of exceptionally well-preserved komatiitic basalts of the Vetreny Poyas suite (Figs. 3.7z and 3.32) (Puchtel et al. 1996, 1997; Kulikov et al. 2010). Kulikov et al. (2005) speculated that, of the six above-mentioned suites, only the Kalgachikha, Vilenga, and Vetreny Poyas rocks formed within a c. 2.5–2.4 Ga rift setting. However, no data supporting this assertion have been presented, and thus, similarly to the Imandra/Varzuga Belt, we tentatively include all six formations of the Vetreny Belt into the Sumian system until better radiometric constraints become available.

Sumian rocks are also known in the Paanajärvi Belt in northern Russian Karelia (Heiskanen 1987a) and in the northwestern rim of the Onega Basin in southern Russian Karelia (Heiskanen 1987a; Golubev et al. 2002). In the latter area, several laterally limited occurrences of basal clastic units sit unconformably on Archaean gneisses or granites. The composite section includes a 150-m-thick unit of quartzitic and arkosic sandstone and siltstone at the base (Fig. 3.7aa) followed by a c. 1,500-m-thick multiple-flow package of massive and pillowed, porphyritic, variolitic, and amygdaloidal andesitic basalts (Fig. 3.7ab). No age constraints exist for these rocks, and their tentative placement in the Sumian system is based on major and trace element geochemical data (e.g. Golubev et al. 2002).

In Finland, the Sumian system rocks are locally abundant in eastern Finnish Lapland, particularly in the Salla Belt (Fig. 3.1), from where they can be followed to central Finnish Lapland. In the Salla area, they are dominated by subaerial intermediate lava flows, but a continuous, cogenetic suite from basaltic andesite lavas to dacitic and rhyolitic lavas, ash-flow tuffs, and ignimbrites can be found (Fig. 3.7ac–ad; Manninen and Huhma 2001; Hanski and Huhma 2005). Further east, in the Kuolajärvi Belt (Fig. 3.1), the Sumian system consists of komatiitic lavas, shales and limestones that are thrust westward onto Ludicovian and Jatulian rocks. Sumian intermediate volcanic rocks also occur in central Finnish Lapland where they have acted as roof rocks for the Koitelainen and Akanvaara mafic layered intrusions.

Volcanic rocks, which comprise the bulk of the Sumian system, range widely in the mode of occurrence and chemical composition. They occur as massive, spinifex-textured, variolitic, amygdaloidal, and pillowed lava flows to tuff beds (Fig. 3.7d–g, v–z, ab–ad). In the Finnish part of the Karelian Province, the volcanic rocks vary from intermediate to felsic and have a mixed, tholeiitic to calc-alkaline affinity (Salla Belt, Manninen 1991), whereas in the Onega Basin, in central Russian Karelia, they show a calc-alkaline affinity (Golubev et al. 2002). In the Vetreny Belt along the northeastern margin of Karelia, they occur as andesitic basalts, basalts, and thick komatiitic lava flows (Puchtel et al. 1996, 1997), and in the Imandra/Varzuga Belt of the Kola Province, they are both komatiitic and tholeiitic (Fedotov 1985; Melezhik and Sturt 1994; Vrevskii et al. 2010).

Figures 3.9 and 3.10 summarise U-Pb geochronological data obtained from Sumian igneous rocks. Precise radiometric ages have not been easy to acquire due to the low content of Zr and/or the fine-grain size of zircon in these rocks. The most precise ages are the previously mentioned 2442.2 ± 1.7 Ma determined by Amelin et al. (1995) for a felsic subvolcanic body belonging to the Seidorechka Formation (Imandra/Varzuga Belt) and the age 2437 ± 3 Ma, which Puchtel et al. (1997) obtained for a Kirich suite dacite (Vetreny Belt) (Fig. 3.9a). The Sm-Nd, Re-Os and Pb-Pb systematics of the voluminous Vetreny Poyas suite komatiitic basalts are compatible with the age obtained from the Kirich suite dacite (Puchtel et al. 1997, 2001). Levchenkov et al. (1994) published ages of 2442.8 ± 4.8 Ma and 2448 ± 15 Ma for quartz porphyries from the Lekhta area. Manninen et al. (2001) dated intermediate volcanic rocks of the Salla Group hosting the Sumian-age Koitelainen layered intrusion in central Finnish Lapland, and obtained a U-Pb zircon age of 2438 ± 11 Ma. This matches the age of 2438 ± 8 Ma measured by Räsänen and Huhma (2001) for correlative felsic volcanic rocks c. 15 km to the east of the Koitelainen intrusion (Fig. 3.9a).

As revealed by Figs. 3.9 and 3.10, a great number of rock samples from Sumian mafic layered intrusions have been dated using the U-Pb system and conventional NTIMS analyses of zircon ± baddeleyite separates. Figure 3.9a, which shows data from the Kola and Karelian cratons, suggests that the Sumian intrusive magmatism can be divided broadly into two stages. The older one at c. 2500 Ma is represented by three layered intrusion complexes (Mt. Generalskaya intrusion and Fedorovo-Pansky and Monchegorsk Complexes) occurring on the northern flank of the Pechenga and Imandra/Varzuga belts in the Kola Peninsula, with the oldest nominal age being 2505.1 ± 1.6 Ma measured for the Mt. Generalskaya intrusion at Pechenga (Amelin et al. 1995). The younger period of magmatism took place at c. 2430–2450 Ma, during which the rest of the Sumian layered intrusions were emplaced in various parts of the Archaean cratons. The mafic layered intrusions of the younger age group are thus indistinguishable in age from the dated Sumian felsic to intermediate volcanic and subvolcanic rocks (see Fig. 3.9a). A-type granitic stocks and dykes with ages of c. 2440–2430 Ma have been documented recently from eastern Finland and adjacent parts of Russia, some of them in close association with layered intrusions (Lauri and Mänttäri 2002; Mikkola et al. 2010). They represent rare examples of felsic plutonism during early Palaeoproterozoic time (cf. Fig. 3.8), though such occurrences are potentially more common than previously thought (Mikkola et al. 2010).

Previous workers thought that the supracrustal rocks of the Purnach and Kuksha formations were cut by c. 2500 Ma layered intrusions (Zagorodny et al. 1982; Amelin et al. 1995), but more recent studies have shown that the basal conglomerates of the Kuksha Sedimentary Formation (see Chap. 4.1) or volcanic rocks of the Kuksha Volcanic Formation (Chashchin et al. 2008) lie unconformably on the Monchegorsk Layered Complex and are therefore younger than c. 2500 Ma. Further, neither the Strel'na Group nor other volcanic formations elsewhere in the Fennoscandian Shield have yielded ages approaching 2500 Ma. Thus, only the Purnach Formation in the Imandra/Varzuga Belt may contain potential candidates of pre-2505 Ma supracrustal rocks. Other candidates are the four lowermost formations in the Vetreny Belt whose depositional ages remain unknown.

In Figure 3.9a, the older magmatic phase is indicated by three intrusive massifs from the Kola Peninsula, showing apparently precise ages determined by Amelin et al. (1995), and the Olenogorsk olivine-bearing gabbronorite dyke (2495 ± 13 Ma; Smol'kin et al. 2004) that cuts Archaean banded iron formations. Amelin et al. (1995) also obtained equally precise data from the second magmatic phase and inferred that the older and younger magmatic events record two, relatively short-lived episodes of Palaeoproterozoic rifting separated by c. 60 Ma. According to this interpretation, the rifting initiated at different times in different parts of the Fennoscandian Shield with the older phase being limited to the Kola Peninsula. The only exceptions outside the Kola Peninsula are the Avdeevskiy and Shalskiy gabbronorite dykes close to the Burakovsky layered intrusion in southern Russian Karelia, which were dated by Bleeker et al. (2008) at c. 2504 Ma. The latter dates are not shown in Fig. 3.9a because no errors were given by those workers. It needs to be added that Chistyakov and Sharkov (2008) believe that the Avdeevskiy dyke is coeval with the 2.45 Ga Burakovka intrusion based on its Sm-Nd age of 2436 ± 46 Ma.

Later, Bayanova and her coworkers obtained a large range of ages for the same three intrusive complexes representing the older phase of Amelin (Fig. 3.10). Data from the Fedotovo-Pansky Complex show that several different dates have been measured using different rock types (gabbronorites, pegmatitic gabbronorites, anorthosites) from various stratigraphic levels of the same intrusive body. These results have been interpreted as recording a prolonged igneous activity fed from a long-lived mantle plume over a time span of 50 ± 20 Ma (Bayanova et al. 1999; Serov et al. 2007; Mitrofanov and Bayanova 1999). For the Imandra lopolith, which represents an intrusion of the younger age group in the Kola Peninsula, Bayanova et al. (2009) also obtained a large range of ages and distinguished three magmatic pulses between c. 2440 and 2395 Ma. Nominal dates obtained by the two mentioned scientific groups (and laboratories) on zircon separates from the same Sumian-age sample may differ by c. 10 Ma (see Mitrofanov and Bayanova 1999), but this cannot explain the large total spread of the plotted values in Fig. 3.10. A further complication is that even dates older than c. 2500 Ma have been produced. The oldest measured until now is 2585 ± 10 Ma (not shown in figures) from the Mt. Generalskaya intrusion, which Bayanova et al. (1999) considered to have been influenced by xenocrystic zircon grains. Abnormally old dates (2526 ± 6 Ma and 2521 ± 8 Ma) were also published by Nitkina (2006) from the Fedorovo-Pansky Complex (see Fig. 3.10).

The current paradigm for the relatively short duration of mantle plume head magmatism and the timing of cooling and crystallisation of basaltic magma chambers are at odds with the inferences for long-lived mantle plume activity. For example, the thermal modelling by Cawthorn and Walraven (1998) suggests that the emplacement of the c. 2060 Ma Bushveld Complex took place within c. 0.075 Ma and the whole complex was solidified within 0.2 Ma. A more detailed discussion of the dating results and their potential interpretations is beyond the scope of this chapter. The above, though, emphasises that more work is needed, preferably using in situ microbeam dating techniques on individual zircon grains. Nevertheless, the discussed problems do not impact on the proposed age boundaries of the Sumian system advocated in this chapter, especially the Sumian-Sariolian boundary, which is based on the relatively robust age for the cessation of the younger intrusive phase by 2430 Ma.

Figure 3.9b shows separately a collection of age determinations performed on Palaeoproterozoic igneous rocks from the Belomorian Mobile Belt, the reworked late Archaean terrane between the Kola and Karelian cratons that has a complex deformational and high-pressure metamorphic history. The dates of mafic intrusions and dykes (lherzolite-gabbronorite complex) are consistent with the younger of the above mentioned two Sumian magmatic phases, though spread across 50 Ma from c. 2460 to 2410 Ma (e.g. Sharkov et al. 2004; Krivolutskaya et al. 2010). The interpretation of these ages is not straightforward, as it has been presented that the rocks were metamorphosed as early as at 2440–2420 Ma (Balagansky et al. 2001). Based on studies of mafic dyke swarms, of which two may be regarded as Sumian in age, Stepanova and Stepanov (2010) suggested recently that the Belomorian Belt acted in the early Palaeoproterozoic as stable continental crust, which was affected by rifting similarly to the adjacent Karelian craton and could have been part of it but was later strongly reworked beyond recognition during Lapland-Kola collision events.

When examined across the entire Fennoscandian Shield, the Sumian system does not have a unique lithological character, as had been previously thought (e.g., Heiskanen 1987a). The system is marked by a lithological variety (Fig. 3.7), which is dependent upon the local palaeotectonic and depositional settings. However, despite their variable lithology and petrology, the Sumian system rocks share two important features: (*1*) the volcanic rocks are coeval and may even be partly comagmatic with the 2440–2430 Ma layered gabbronorite intrusions (Alapieti et al. 1990; Melezhik and Sturt 1994; Hanski et al. 2001b; Chashchin et al. 2008; Kulikov et al. 2010); and (*2*) isotope geochemistry of the various sedimentary units in the upper part of the Sumian system show mass-independent fractionation of sulphur isotopes (Fig. 3.6; Reuschel et al. 2009).

3.2.4 The Sariolian System, 2430–2300 Ma

Already almost a century ago, Eskola (1919) reported findings of tillite-like sedimentary rocks lying unconformably on older gneisses in the Onega region, Russian Karelia and linked the formation of these polymict, immature conglomerates and associated arkosic sandstones to an ancient glaciation event. He showed that they are older than the Jatulian rocks and named them later Sariolian deposits (Eskola 1925). A long time passed before the corresponding glaciogenic rocks were confidently identified and described in Finland. In the 1980s, Marmo and Ojakangas (1984) and Marmo (1986) reported a rock association with sedimentological features typical for glaciogenic deposits; these included the Urkkavaara Formation (North Karelia Belt, eastern Finland) diamictites and varved shales with outsized scattered polymict clasts (dropstones and lonestones, Fig. 3.11a–b) (see Chap. 7.2). Later Strand and Laajoki (1993) ascribed the deposition of the Honkala Formation at the eastern margin of the Kainuu Belt to glaciomarine sedimentation. Negrutsa and Negrutsa (1981) reinvestigated the tilloids from the Onega region, and Ojakangas (1988) confirmed their glaciogenic origin. Sedimentary rocks of glacial origin (diamictites and varved shales with dropstones) are also known from the Imandra/Varzuga (Polisarka Sedimentary Formation, Fig. 3.11c) and Shombozero (Pajozero Formation, Fig. 3.11d) belts.

The Sariolian sedimentary rocks, which may reach 600 m in thickness, rest either on the Sumian igneous rocks or Archaean basement complex (Fig. 3.11e–g). The former include partly eroded, c. 2505 and 2430–2440 Ma layered gabbronorite intrusions in the Pechenga and Peräpohja belts, thereby constraining the age of the base of the Sariolian system. The Sariolian deposits are thus temporal counterparts of the Huronian-age glacial deposits in North America (cf. Ojakangas 1985) and South Africa (Hannah et al. 2004) (for details, see Chap. 7.2). Apart from the glaciogenic deposits, typical features of the Sariolian sedimentary formations are their immature textures and compositions and a large lateral variation of their thicknesses. It was already inferred by Eskola (1919) that the deposition of the sediments and their thickness variations were controlled by uneven palaeotopography. In contrast, the overlying Jatulian sediments were deposited on a peneplained surface. Although the distinguishing sedimentary lithofacies of the Sariolian system is non-marine (immature, polymict conglomerate, gritstone and greywacke; Fig. 3.11h–o), marine marl, and limestone are also found in the Sariolian-age Polisarka Sedimentary Formation in the Imandra/Varzuga Belt (see Chap. 6.1.2).

In addition to sedimentary rocks, the Sariolian system contains locally thick accumulations (up to 1,000 m) of volcanic rocks (Heiskanen 1987b; Ojakangas et al. 2001). They were erupted either directly on Archaean gneisses as in the northernmost part of the Kainuu Belt (Strand 1988), or on older Sariolian sedimentary rocks in the Pechenga and Imandra/Varzuga belts (see Chap. 6.1.2 and 6.2.1). The volcanic rocks commonly consist of subaerially erupted, amygdaloidal basalts, and andesites and, in some areas, more magnesian komatiitic rocks. A general feature is a strong interaction of magma with the underlying Archaean basement as witnessed by the capture of crustal xenoliths by magmas during their ascent. Räsänen et al. (1989) reported a detailed description of a 250-m-thick section of volcanic rocks from central Finnish Lapland, belonging to the Onkamo Group. The drilled section between the Archaean basement and Jatulian quartzites consists of crustally contaminated, fragmental komatiites in the lower part and highly-amygdaloidal andesites and Cr-rich basalts (Fig. 3.11p) in the upper part. Supracrustal rocks of the Onkamo Group are also found discontinuously developed elsewhere in central Finnish Lapland (Hanski and Huhma 2005). These include siliceous high-Mg basalts and komatiitic basalts in the Salla area (Fig. 3.1) at the Finnish-Russian border where they are separated from the underlying Sumian rhyolitic rocks of the Salla Group (former Salla Formation; Manninen 1991) by an Al-rich palaeoweathering crust. In the Imandra/Varzuga Belt, the Sariolian-age Polisarka Volcanic Formation contains variolitic and spinifex-textured komatiitic pillow lavas and lava breccias (Fig. 3.11q–r).

The Sariolian volcanic rocks have not been reliably dated, and there are no direct age determinations constraining the depositional age of the sedimentary rocks. However, they are older than the c. 2220 Ma generation of layered sills and younger than the c. 2440–2430 Ma gabbro-noritic layered intrusions. The Runkaus Formation basalts that overlie the Sariolian basal conglomerates in the Peräpohja Belt (Fig. 3.1) yield an imprecise Sm-Nd age of 2330 $\pm$ 180 Ma (Huhma et al. 1990), and there is increasing isotopic (U-Pb, Sm-Nd) evidence for mafic magmatism at c. 2300–2320 Ma among the dyke swarms found in the Archaean basement (Vuollo and Huhma 2005). On the other hand, Kullerud et al. (2006) reported an age of 2403 $\pm$ 3 Ma for gabbronorite dykes in the West Troms Basement Complex, northern Norway, but it is presently not known whether cogenetic or coeval volcanic rocks exist in the Shield.

Without age constraints, the komatiitic rocks of the Sumian and Sariolian systems are indistinguishable. Thus the most distinctive feature of the Sariolian system is the Huronian-age glaciogenic deposits (Fig. 3.11). In addition, measured samples do not record mass-independent fractionation of sulphur isotopes in the Sariolian system sedimentary rocks of the Fennoscandian Shield (Fig. 3.6; Papineau et al. 2005).

3.2.5 The Jatulian System, 2300–2060 Ma

The Jatulian rocks variably overlie older Archaean, Sumian or Sariolian rocks. They record the positive isotopic excursion of carbon in sedimentary carbonates (Lomagundi-Jatuli Event, Fig. 3.6) and a time of strong chemical weathering as evidenced by regoliths composed of Al-rich shales and kyanite-bearing, quartzitic schists at the base of the Jatulian deposits (Fig. 3.12a, b) or the common occurrence of quartz-pebble conglomerates, and quartz arenites (Fig. 3.12c–i) (Negrutza 1984; Sokolov 1987; Marmo 1992). Chemical weathering beneath the Jatulian deposits appears to be a common phenomenon over the entire Shield. Marmo (1992) has published a detailed account of a palaeosol (80-m-thick Hokkalampi palaeosol) in North Karelia that was formed on Sariolian glacial deposits. A similar paleosol (Hallavaara) developed on Sariolian rocks has been found in the northern part of the Kainuu Belt (Strand 1988).

The Jatulian system spans ~240 m.y., and its thickness varies up to 2,000 m; for detailed, formation-level descriptions, readers are referred to Pekkarinen (1979), Sokolov (1980, 1987), Kontinen (1986), Negrutsa et al. (1986), Laajoki (1991), Kohonen and Marmo (1992), and Strand (2005). The system exhibits great lithological variability and is typified by mafic volcanic rocks, chemically mature arenites, quartz conglomerates, "red beds", and stromatolitic carbonate rocks. Relative to the other stratigraphic units, this system displays the greatest lithological variability (Figs. 3.12, 3.13, 3.14, 3.15, 3.16, 3.17, and 3.18). In Finnish and Russian Karelia, sedimentary rocks dominate over mafic lavas. The opposite relationship characterises the Jatuli in the Kola Province where sedimentary rocks form only 5–10 % of the total thickness, as exemplified by the Kuetsjärvi formations in the Pechenga Belt and the Umba formations in the Imandra/Varzuga Belt.

Many lithologies are red in colour due to haematite staining, hence application of the term "red beds" (Fig. 3.13). Haematite occurs in the oldest rocks of the Jatulian system as part of the weathering crust developed on underlying Sariolian rocks. It is also present in some quartz-pebble conglomerates, quartzose sandstones, and dolostones, and haematite-rich rocks occur as metre-scale layers in the uppermost Jatulian quartzites (Kiihtelysvaara area; Pekkarinen 1979). Haematite-rich, red and black mudstones can be abundant and commonly show desiccation cracks (Fig. 3.13a, c, d and 3.14). Chromium-bearing mica, fuchsite, is a characteristic mineral of some sericite quartzites in the Ust'-Ponoy Belt (the eastern coast of the Kola Peninsula) (Melezhik 1992) and in Finnish Lapland and the Kuusamo area, staining the rocks *green* (Figs. 3.12h, i and 3.15c). Cross-bedded and rippled, mature arenites are widespread throughout the eastern part of the Shield (Fig. 3.15). Carbonate rocks mainly occur in the Karelian Province as stromatolitic, clastic, and chemically precipitated dolostones exhibiting variable colour and textures (Fig. 3.16). In the Kola Province, marine carbonate successions in the Imandra/Varzuga Belt comprise fine-grained (micritic) dolostones (Fig. 3.16e), whereas lacustrine carbonate rocks in the Pechenga Belt are represented by stromatolitic and redeposited carbonates (Fig. 3.16f–g). The most remarkable feature of the Jatuli is the presence of carbonate rocks enriched in ^{13}C (e.g. Karhu 1993; Karhu and Holland 1996) and abundant Ca-sulphates partially or completely pseudomorphed by dolomite and quartz (Fig. 3.17) (Melezhik et al. 2005).

The Jatulian system volcanic rocks range in composition from tholeiitic to alkaline and are mostly mafic. In the Kuusamo and Peräpohja belts in Finland, they occur as piles of mafic amygdaloidal lava flows (Fig. 3.18a), though formations composed solely of mafic tuffs (Fig. 3.18b) or rarer pillow lavas (Fig. 3.18d) are also present. The thickest volcanic piles are made up by flood basalt-like accumulations up to 500 m thick (e.g. Perttunen 1985). Felsic ash-flow tuffs associated with pillowed and massive mafic lavas have been documented from the small Siilinjärvi Belt, Central Finland (Lukkarinen 2008). In the Russian central Karelian Province, the Jatulian system volcanic rocks are basaltic lavas and flow breccias (Fig. 3.18c). In the Pechenga and Imandra/Varzuga belts of the Kola Province, the volcanic rocks are more variable in composition (Predovsky et al. 1974; Melezhik and Sturt 1994; Smol'kin and Skuf'in 1995) and consist of subaerially erupted rhyolitic, dacitic, and andesitic lava-breccias and fluidal amygdaloidal lavas, basalts with columnar jointing, and tuffs (Fig. 3.18e–g).

Using unconformities and related conglomerate beds, or intervening volcanogenic formations as markers, several attempts have been made to subdivide the Jatulian successions into smaller subunits, such as the Lower, Middle and Upper Jatuli in Finland (Meriläinen 1980; Silvennoinen et al. 1980; Silvennoinen 1985) or the four 'Horizons' in Russian Karelia (e.g. Negrutsa et al. 1986; Sokolov 1987). Major breaks in sedimentation have been observed, for example, in the Kainuu Belt where Laajoki (1988, 2005) employed the so-called Nenäkangas conglomerates as a criterion for dividing the succession into two tectofacies (Kainuu and Jatuli). These subdivisions, though, are only applicable locally and have restricted utility for correlations across the Fennoscandian Shield.

Igneous rocks provide important age constraints for the Jatulian system. Differentiated mafic-ultramafic sills cut the lowermost Jatulian sedimentary rocks, including ^{13}C-rich carbonates in places (i.e., the Sericite Schist Formation in the Kuusamo Belt; Silvennoinen 1991; Karhu 1993), across the Karelian Province in Finland. This magmatism took place at c. 2220 Ma (Hanski et al. 2010) and provides a minimum age for the start of the post-Sariolian sedimentation. Mafic dykes with an age of c. 2100 Ma are known to cross-cut the basement and Jatulian rocks in eastern Finland (Vuollo and Huhma 2005). They have been precisely dated in the Kiihtelysvaara area, North Karelia Belt where they are regarded as feeder conduits for lava interbeds occurring in the middle part of the Jatulian sedimentary sequence. Two dykes sampled at three locations yielded very similar U-Pb zircon ages between 2113 ± 4 Ma and 2115 ± 5 Ma (Pekkarinen and Lukkarinen 1991). Within error, these are the same as the ages obtained for mafic volcanic rocks from the Tohmajärvi area, also in the North Karelia Belt (2105 ± 15 Ma; Huhma 1986). They are also consistent with the Sm-Nd isochron age of 2090 ± 70 Ma measured for the Jouttiaapa Formation basalts from the Peräpohja Belt (Huhma et al. 1990). Karhu et al. (2007) determined a SHRIMP zircon age of 2106 ± 8 Ma for a mafic tuff layer (Hirsimaa Formation) occurring between two ^{13}C-rich dolomite formations in the Peräpohja Belt. The upper age limit for the Lomagundi-Jatuli isotopic excursion, as well as for the Jatulian system, has been precisely constrained by dating several volcanic units separating ^{13}C-rich carbonates and those with δ^{13}C near zero. The felsic ash-flow tuffs at Siilinjärvi mentioned above are dated precisely at 2062 ± 2 Ma (Lukkarinen 2008), and Melezhik et al. (2007) obtained an equally precise age of 2058 ± 2 Ma for the Kuetsjärvi Volcanic Formation in the Pechenga Belt. Martin et al. (2010) have determined recently a comparative age of c. 2052 Ma for the correlative Umba Volcanic Formation in the Imandra-Varzuga Belt. A felsic porphyry in the Väystäjä Formation in the Peräpohja Belt has been dated at 2050 ± 8 Ma (Karhu 2005) and is associated with carbonate rocks showing δ^{13}C near zero. Although this formation appears to be allochthonous and its correlation with other rocks in the Peräpohja Belt remains unclear (Perttunen and Hanski 2003), it provides, together with the other c. 2060 Ma dates, a solid time constraint for the upper boundary of the Jatulian system.

The variability of the Jatulian rocks can cause that they are confused with older or younger rocks, particularly in isolated and/or structurally complicated exposures. Hence, the most robust criterion for identifying the Jatuli system is the enrichment of sedimentary carbonates in ^{13}C associated with the Lomagundi-Jatuli positive excursion (Fig. 3.6).

3.2.6 The Ludicovian System, 2060–1960 Ma

The Ludicovian system records unprecedented accumulation of C_{org}-rich rocks, the Earth's earliest known phosphorites, and first abundant formation of ^{13}C-depleted diagenetic carbonate concretions (Fig. 3.6). The Ludicovian rocks rest unconformably (see Chap. 6.3.2) and, in places, with a weathering crust (see Chap. 6.2.6) on Jatulian rocks and locally the pre-Jatulian basement (Galdobina 1987a). In Finland, the term Ludicovi per se has generally not been used, but instead the lithostratigraphically equivalent rocks have been assigned to the Marine Jatuli of Väyrynen (1933) (Fig. 3.5). Nevertheless, all these rocks, including the traditionally defined Ludicovi superhorizon, are typified by "black rocks" (i.e. any rocks enriched in organic carbon; Fig. 3.19a) commonly associated with widespread and thick accumulations of tholeiitic pillow lavas (Fig. 3.19b).

In the southern Russian Karelia, the Ludicovian rocks consist of two units; the older Zaonega Formation and younger Suisari Formation (horizons in the Russian literature) (Golubev et al. 1984; Galdobina and Melezhik 1986; Galdobina 1987a). The sedimentary rocks of the Zaonega Formation are exceptionally rich in organic carbon (Fig. 3.19a, c–e) and contain a significant volume of chert and carbonate rocks, both sedimentary and diagenetic in origin, as well as ^{13}C-depleted carbonate concretions (Fig. 3.19f) (Fallick et al. 2008). The 600-m-thick Suisari Formation comprises tholeiitic mafic tuffs, amygdaloidal, variolitic, and pillowed lavas ranging in composition from picritic to basaltic (Fig. 3.19g), all interbedded with minor "black shales". In the Finnish part of Karelia, the rocks that are stratigraphically equivalent to the Ludicovi system (Marine Jatulian) include "black shales", phyllites, and carbonate rocks with mafic lavas and tuffs; these are well studied in the Kiihtelysvaara area (Petäikkö Formation) by Pekkarinen (1979). In the the Peräpohja Belt, Ludicovian rocks consist of mafic pillow lavas, "black shales", and minor felsic tuffs (Fig. 6.19h, i, Väystäjä Formation) studied and dated by Perttunen and Vaasjoki (2001).

In Central Finnish Lapland (Fig. 3.1), the Savukoski Group can be regarded as Ludicovian based on its age, stratigraphic position and lithology (Lehtonen et al. 1998; Hanski and Huhma 2005). It lies conformably on a Jatulian-age quartzite-carbonate sequence (Sodankylä Group) and is composed of a phyllite-tuffite-"black schist" assemblage with minor dolostone and chert in its lower part and pyroclastic and pillowed komatiites and picrites and related basaltic rocks in its upper part (Fig. 3.19j–m) (Hanski and Huhma 2005). Komatiitic rocks can be followed northward to northern Norway in the Karasjok Belt (Barnes and Often 1990). Further southeast, in the Kuusamo and Kuolajärvi belts (Fig. 3.1), the Ludicovian system was described by Melezhik et al. (1988) and Silvennoinen (1992) and consists of C_{org}-rich shales, limestones, dolostones and amphibole schists. The c. 2000 Ma Kittilä Group, occupying a large part of the Central Lapland Belt, also belongs to the Ludicovian system, even though it is regarded as a major allochthonous unit (Hanski and Huhma 2005). It consists of a thick pile of massive, amygdaloidal, and pillowed basaltic lavas, shales, carbonate rocks, and banded iron formations and also contains bodies of refractory ultramafic rocks interpreted as fragments of dismembered ophiolites (Hanski 1997).

In the Pechenga Belt, the Ludicovian system encompasses the Kolosjoki (also spelled Kolasjoki in the literature translated from Russian into English) and Pilgujärvi formations. The former includes red, volcaniclastic greywackes, arkosic sandstones, stromatolitic dolostones, jaspers, and "black shales" overlain by a 1,500-m-thick, monotonous sequence of tholeiitic pillow basalts, subordinate flow breccias (Fig. 3.19n–s), and "black shales", all intruded by numerous gabbro sills. The overlying Pilgujärvi Sedimentary Formation attains a thickness of more than 1,000 m and is characterized by rhythmically bedded, turbiditic greywackes rich in organic matter and iron sulphides (as disseminated grains, concretions, and layers) and also contains redeposited phosphorites (Bekasova and Dudkin 1982) and carbonate concretions depleted in ^{13}C (Melezhik et al. 1998) (Fig. 3.17t–x). The greywacke succession is overlain by the >3,000-m-thick Pilgujärvi Volcanic Formation containing tholeiitic pillowed and massive basalts with minor ferropicrites and felsic tuffs (Fig. 3.17y–ab) (Zagorodny et al. 1964; Predovsky et al. 1974; Hanski 1992; Smol'kin and Skuf'in 1995).

In the Imandra/Varzuga Belt, the Ludicovian system includes the Il'mozero and and Solenozero formations. The former is composed of volcaniclastic greywackes with phosphate and ^{13}C-depleted carbonate concretions (Melezhik and Predovsky 1978; Melezhik and Fallick 1996; Melezhik 1992), stromatolitic dolostones (see Chap. 4.1 Imandra/Varzuga Greenstone Belt), cherts, and "black shales" overlain by a 500-m-thick, monotonous sequence of massive and amygdaloidal basaltic and dacitic andesites (Melezhik et al. 1982; Fedotov 1985). The Solenozero Formation of an unknown thickness contains intensively folded and sheared, rhythmically bedded turbiditic greywackes rich in organic matter and iron sulphides (Smol'kin and Dain 1985).

Radiometric age constraints of the Ludicovian system have been defined in different geographic areas. The minimum age for the pelitic metasediments of the Savukoski Group in Finnish Lapland is constrained by intrusive rocks that cut these sedimentary rocks, many of which have been dated using the U-Pb system (Mutanen and Huhma 2001; Rastas et al. 2001): Keivitsa ore-bearing mafic-ultramafic intrusion, 2058 ± 4 Ma and 2054 ± 5 Ma; a felsic porphyry, 2048 ± 5 Ma; mafic dykes from 2060 ± 8 to 2048 ± 8 Ma and from 2007 ± 8 to 1995 ± 8 Ma. The primitive volcanic rocks with komatiitic to picritic affinities in the upper part of the Savukoski Group have been dated at 2056 ± 25 Ma (Sm-Nd method, Hanski et al. 2001a). These data thus indicate that the sedimentary and volcanic rocks of the Savukoski Group are equal to or older than c. 2060–2050 Ma in age.

In the Pechenga Belt, the Kolosjoki Sedimentary Formation is younger than 2058 ± 2 Ma, which is the age of the underlying Kuetsjärvi lavas Melezhik et al. (2007). Intrusions and lavas of the ferropicritic magmatism within the Pilgujärvi Sedimentary and Volcanic Formations have been dated by several methods approximately at 1980–1990 Ma (Hanski 1992), with the most precise constraints being the U-Pb ages of 1987 ± 5 Ma and 1980 ± 10 Ma measured, respectively, on zircon and baddeleyite from the Pilgujärvi Ni-Cu sulphide ore-bearing intrusion (Skuf'in and Bayanova 2006). In the Onega region, Puchtel et al. (1998, 1999) have obtained age data for Ludicovian rocks. Sm-Nd, Pb-Pb and Re-Os isochrons date a differentiated sill intruding Suisari Formation basalts at 1988 ± 34, 1985 ± 57 and 1969 ± 18 Ma, respectively. These ages have been considered to be representative of the eruption time of the Ludicovian basalts, corresponding well to the dates obtained from the Pechenga Belt. Shul'diner et al. (2000) determined a U-Pb zircon age of 1963 ± 19 Ma for mafic volcanic rocks of the Sortavala Group from the Lake Ladoga region. These rocks are regarded as correlative with the Ludicovian volcanic rocks of the Onega region. However, the studied sample has a heterogeneous zircon population, and hence there is some uncertainty in the age, in addition to the relatively poor precision. A period of wide-spread tholeiitic basic magmatism at c. 1970 Ma is also recorded by one of the mafic dyke swarm generations observed in the Archaean basement (Vuollo et al. 1992; Vuollo and Huhma 2005).

The allochthonous Kittilä Group in central Finnish Lapland seems to be slightly older than the Ludicovian rocks in the Pechenga and Onega regions. Rastas et al. (2001) published U-Pb zircon ages for several felsic dykes and tuffs that are considered to be coeval with the associated mafic pillow lavas. These ages range from 2012 ± 3 to 2017 ± 8 Ma giving a combined age of 2015 ± 2 Ma (Hanski and Huhma 2005). A Sm-Nd isochron age of 1990 ± 34 Ma obtained for basalts is equal within error (Hanski and Huhma 2005). Recently a new age determination has been performed on zircon from the gabbroic part of a gabbro-mafic dyke assemblage from Kittilä, interpreted as representing an incipient stage of sheeted dyke complex development. This age, 2009 ± 3 Ma (H. Huhma, pers. comm. 2010), confirms that the upper part of the Kittilä Group (Vesmajärvi Formation) is c. 2010 Ma in age.

For many decades, Russian geologists have considered the boundary between the Jatulian "red beds" and Ludicovian "black shales" as a prominent and reliable lithostratigraphic boundary (Sokolov 1980; Galdobina 1987a; Melezhik and Predovsky 1982). However, this seemingly straightforward lithological criterion is now known to be incorrect. For example, in the Kola Province, the Kolosjoki Sedimentary Formation in the Pechenga Belt consists of red volcaniclastic greywackes, cherts, jaspers, and dolostones, including those with stromatolitic structures (Fig. 3.19n–q). This assemblage was thought to typify the former Jatuli Superhorizon, to which this formation was originally assigned (Fig. 3.5; e.g. Melezhik and Predovsky 1982; Smol'kin, 1992, 2005; Mokrousov et al. 1995). However, the 2058 ± 2 Ma age constraint for the end of the Lomagundi-Jatuli isotopic excursion in the Pechenga Belt, in combination with carbon isotope chemostratigraphy (Melezhik et al. 2007), makes the Kolosjoki Sedimentary Formation younger than the age interval for the Jatulian system (2300–2060 Ma). Consequently, the long-held criterion of a "red bed"-"black shale" boundary between the Jatulian and Ludicovian rocks can no longer be applied. This is not surprising given that, by 2100–2000 Ma, the Kola and Karelia cratons were separated by the incipient Kola Ocean (for details, see Chap. 3.4) and contemporaneous sedimentary basins on those cratons evolved differently, accumulating dissimilar rocks. Thus, age constraints are the main reliable criterion for distinguishing Ludicovian system rocks. Another would be phosphate-bearing strata, as well as ^{13}C-depleted carbonate concretions; these reflect the onset of aerobic recycling of organic matter in the sedimentary column, which has not been reported from older rocks (Fallick et al. 2008).

3.2.7 The Kalevian System, 1960–1900 Ma

The Kalevian system contains abundant pelitic metasediments and turbiditic greywackes (Fig. 3.20a) and developed on south-western (Finland) and southern (Russia) margins of the Karelian craton; its estimated thickness is between 3,000 and 5,000 m. It is present in the Peräpohja, Kiiminki, Kainuu and North Karelia belts in Finland and the latter's south-eastern continuation into the Lake Ladoga 560 region, as well as the Lake Onega area in southern Russian Karelia, where the Kalevian rocks were termed Livvian (Galdobina 1987b). The system does not appear to be associated with any well-defined global palaeoenvironmental events (Fig. 3.6).

The contact relationship between the Kalevian rocks and the older Jatulian or Ludicovian rocks has been the focus of many studies resulting in recognition of three kinds of relationships. Evidence for an erosional contact, in the form of basal polymictic conglomerates or sedimentary breccias, comes from North Karelia (Väyrynen 1933; Pekkarinen 1979; Kohonen 1995) and Kainuu (Väyrynen 1933; Kontinen 1986; Gehör and Havola 1988). A major unconformity is also indicated by the occurrence of suites of mafic dykes: they typically transect Jatulian quartzites but are absent within the Kalevian successions (Vuollo et al. 1992; Vuollo and Huhma 2005). It is worth of mentioning again that the Kalevian metasediments may also lie directly on the Archaean basement, as in the Kiiminki Belt where the Palaeoproterozoic sequence starts with coarse polymictic conglomerates on Archaean gneisses, grading upwards into turbiditic metagreywackes with dolostone and mafic pillow lavas (Fig. 3.20b–f) (Honkamo 1985). In the southern part of the Karelian province (Ladoga area), the base of the group cuts down section by as much as 1,300 m onto Ludicovian and Jatulian rocks over a distance of 100 km (Galdobina 1987b). Here, the basal unit is a polymict conglomerate with a thickness of up to 120 m.

Observations of a gradational contact with underlying rocks have also been reported (e.g. Nykänen 1971). At the base of the Kalevian sequence in the Peräpohja Belt, the lithofacies of the sedimentary rocks appears to change gradually, though rapidly, from shallow-water quartzite-dolomite association to deeper-water shales (Fig. 3.20g) (Perttunen 1985). Although quartzite-pebble conglomerates have been reported from the Kalevian rocks in the Peräpohja Belt (at Taivalkoski, Fig. 3.20h–i), the clasts are derived from the underlying Jatulian Kivalo Group and the conglomerates have been interpreted as part of a coarse-clastic turbiditic fan (Perttunen, op. cit).

The third contact type is a thrust zone. This separates the lowermost Kalevian rocks from the older formations in many places, particularly in North Karelia, as well as in the Pechenga and Imandra/Varzuga belts. This is due to the Svecofennian arc-collage/continent collision from the south-west at c. 1900 Ma, during which the Kalevian rocks were complexly folded and thrust over the platformal sedimentary cover of the basement or rifted continental margin (Kohonen 1995; Melezhik and Sturt 1994).

In Finland, Kontinen (1987) divided the Kalevian rocks into the autochthonous-parautochthonous Lower Kaleva and dominantly allochthonous Upper Kaleva. Ward (1987) applied the same style of informal division to the North Karelia Belt , but it was shown later that the two units differ in their lithology, age, and provenance (Lahtinen et al. 2010). In North Karelia, the Lower Kaleva grades south-westward without a clear boundary into the migmatitic Savo foldbelt of the Svecofennian domain. A detailed study of different lithologies has revealed four lithological assemblages that are prevalent in different, NE-SW trending segments of the Kalevian zone (Kohonen 1995) and are marked by a variable suite of metamorphosed siltstones, mudstones carbonaceous mudstones, quartz arenites, and greywackes indicating different depositional environments. Turbiditic sequences typify the Lower Kalevian rocks in different schist belts (Fig. 3.20j) (Kohonen 1995; Honkamo 1985; Gehör and Havola 1988) and, in the Kiiminki Belt, there is a significant component of mafic volcanic rocks occurring mainly as pillow lavas (Honkamo 1985) (Fig. 3.20f). Banded-iron formations with minor phosphorite bands occur in association with dolomites, phyllites, and black schists in the Kainuu Belt and were assigned to the Marine Jatuli (i.e. Ludicovian) by Laajoki and Saikkonen (1977), but have later been inferred as Kalevian (Gehör and Havola 1988; Laajoki 1991). Until the geochronological constraints are obtained, the chronostratigraphic position of these rocks remains unresolved.

Lithologically the Upper Kaleva is rather monotonous containing thick-bedded greywackes with minor intercalations of black schists. According to Lahtinen et al. (2010), a typical feature is the absence of volcanic rocks within the pelitic metasediments. An important observation is that the occurrence of tectonically emplaced slivers of ophiolitic rocks is confined to the Upper Kaleva both in Kainuu (Jormua ophiolite complex) and North Karelia (Outokumpu ophiolite complex) (Kontinen 1987; Peltonen et al. 2008). It has been suggested that at least the most distal parts of the Upper Kaleva were deposited on a newly formed oceanic crust.

The Jormua ophiolite comprises mantle tectonites, originally from subcontinental lithospheric mantle, and sheeted dykes, gabbros, plagiogranites, pillow lavas, and hyaloclastites (Fig. 3.20k–l). The formation of the oceanic crust took place at 1953 ± 2 Ma, as shown by the precise U-Pb zircon data from a gabbroic feeder dyke (Peltonen et al. 1998; Peltonen 2005). In the Outokumpu area, ophiolitic rocks consist mostly of serpentinite massifs, representing oceanic mantle peridotites, with some of them hosting major Cu-Co-Zn deposits (Peltonen 2005; Peltonen et al. 2008). In addition, there are minor gabbros and basaltic rocks. Gabbroic bodies cutting serpentinite have been dated at 1972 ± 18 Ma (Huhma 1986) and 1959 ± 5 Ma (Peltonen et al. 2008).

In Russian Karelia, the Kalevian rocks (termed Livvian; e.g. Galdobina 1987b) occur mainly as volcaniclastic, turbiditic, greywackes with minor conglomerates in the Onega area (Fig. 3.20m–s), and high-grade metamorphic schists, quartzites, and amphibolites in the Ladoga area (Fig. 3.20t). In both areas, ^{13}C-depleted carbonate concretions are common in greywackes and schists (Fig. 3.20p, u).

In the Kola Province, Kalevian system rocks are largely undated. They apparently constitute sedimentary-volcanic successions within southern parts of the Pechenga and Imandra/Varzuga belts. In the Pechenga Belt, they comprise tectonically imbricated units, several km in thickness, thrust from southwest upon Ludicovian tholeiitic pillow lavas. Lithologies vary from volcaniclastic andesitic and picritic turbiditic greywackes and shales to andesitic, dacitic, and rhyolitic lavas (Zagorodny et al. 1964; Predovsky et al. 1974; Melezhik and Sturt 1994). In the Imandra/Varzuga Belt, apparent Kalevian rocks are represented by arkosic and greywacke sandstones, and rhyolites and rhyodacites with minor basalts and "black shales" (e.g. Skuf'in et al. 2006).

In Russian Karelia, the emplacement of a granodiorite that intrudes the Ludicovian-Kalevian boundary has been dated at 1922 ± 11 Ma (U-Pb zircon; Matrenichev et al. 2006). This provides a minimum age for the Kalevian deposition in the northern lake Ladoga area. In-situ dating of detrital zircons recovered from the Lower Kaleva in Finland has provided dominantly Archaean ages, but Nd isotopic characteristics suggest that the rocks have a detrital component from Jatulian mafic lavas and dykes (Lahtinen et al. 2010). In contrast to the Lower Kaleva, only c. 25 % of the zircon populations in the Upper Kaleva are Archaean in age with the rest falling mainly in the approximate range of 2160 to 1920 Ma. The youngest zircon grains give maximum sedimentation ages of c. 1920 to 1950 Ma for the studied Upper Kaleva samples (Claesson et al. 1993; Lahtinen et al. 2010). The absolute minimum age of deposition is confined by an intrusive granite with an age of c. 1870 Ma (Huhma 1986), while the inferred timing of the tectonic emplacement of the Outokumpu ophiolitic complex would provide an older upper age limit of c. 1900 Ma (Peltonen et al. 2008). Based on the available zircon, Nd isotopic and geochemical data, Lahtinen et al. (2010) suggested that the main source area for the Upper Kaleva sediments was the Lapland-Kola orogen located now in the northern part of the Shield.

In the Kola Province, apparent Kalevian sedimentary-volcanic associations of the Imandra/Varzuga Belt are intruded by plagiomicrocline granite and subvolcanic dacite dated by the U-Pb technique at 1940 ± 5 and 1907 ± 18 Ma, respectively (Skuf'in et al. 2006). The latter date provides an upper age limit for deposition of the succession. In the southern Pechenga Belt, the emplacement of two large granodiorite plutons has been constrained at 1950 ± 9.5 and 1939 ± 7 Ma (U-Pb, Vetrin et al. 2008). The emplacement of the older (Kaskeljavr) pluton was into the tectonically imbricated Kalevian sedimentary-volcanic succession (Kozlova and Balagansky 1995).

The currently available detrital zircon data do not provide precise age limits for the initiation of the Kalevian sedimentation. A lower age limit can be obtained from the youngest dated underlying, pre-Kalevian volcanic rocks, which in Finland are c. 2060 Ma Jatulian felsic volcanic rocks. In Russian Karelia, Ludicovian volcanic rocks are c. 1970–1980 Ma in age allowing to place the lower boundary of the Kaleva at c. 1980 Ma (Ojakangas et al. 2001). Assuming that the contact between the Lower Kaleva and underlying Jatulian and Ludicovian (Marine Jatuli) rocks in North Karelia is an unconformity, as implied by the termination of the c. 1970 Ma mafic dykes at this surface (Vuollo et al. 1992), then the lower age limit of the Kalevian rocks can be implied to be c. 1960 Ma.

The Kalevian system volcanic and sedimentary rocks may be easily confused with those of the Ludicovian system. There are no robust and/or unequivocal lithological or geochemical criteria for distinguishing these rocks; thus the only reliable constraint is a precise depositional age.

3.2.8 Post-Karelian, Postorogenic Supracrustal Rocks

Although outside the time span considered herein, some discussion is worthwhile regarding the youngest Palaeoproterozoic supracrustal rocks that have been found in northern Finland, Norway and Sweden as they form an integral part of the supracrustal belts in this part of the Shield. The sedimentary association comprises cross-bedded arkosic arenites and quartzites (Fig. 3.21a), polymict conglomerates with greywacke interbeds, and siltstones with desiccation cracks. *Red-brown* or *purple* tints, caused by haematite pigment, are a characteristic feature of the polymict conglomerates (Fig. 3.21b, c). This sedimentary assemblage and minor intermediate volcanic rocks, lying discordantly on older Karelian metasedimentary and metavolcanic rocks, have been assigned to the Kumpu and Lainio Groups in northern Finland (Lehtonen et al. 1998; Hanski and Huhma 2005). The sedimentary rocks of these units contain clasts and detrital zircon derived from c. 1880 Ma felsic plutonic rocks, which implies that their deposition postdated the synorogenic magmatism. This time constraint precludes incorporation of the Kumpu and Lainio Groups into the Karelian time slice and negates previous interpretations that these rocks were correlative with Jatulian or Kalevian rocks (see Hanski 2001). Instead, potential counterparts are found among other conglomerates containing fragments of synorogenic granitoids in the Svecofennian domain. The rocks may also be correlative with the Vepsian red sandstones and quartzites in the western Lake Onega region, southern Russian Karelia (Fig. 3.21d) (Hanski et al. 2001c). There, Vepsian rocks sit erosively on sedimentary and volcanic rocks of the Ludicovian and Kalevian systems, and further to the south in the Lake Ladoga area, high-grade rocks of the Ladoga Group (Galdobina 1987c). The Vepsian sucsession is subdivided into a c. 400-m-thick lower unit (Petrozavodsk Formation) and a c. 1,000-m-thick upper unit (Shoksha Formation). The Petrozavodsk Formation comprises grey cross-bedded quartz-feldspar sandstones with lenses of polymict conglomerate whose clasts include basalts, C_{org}-rich tuffites and siltstones, and cherts. Some sandstone beds are rich in heavy minerals, whereas others are enriched in organic material redeposited from the Ludicovian system rocks (Galdobina 1987c). The upper part of the formation is marked by 10–20 m thick basaltic flows. The Shoksha Fromation is mainly composed of red, horizontally and cross-bedded quartz-sandstones deposited in subaerial environments. Minor conglomerates contain clasts derived mainly from the underlying Petrozavodsk Formation sandstones.

3.2.9 Global Implications of the Event-Based Fennoscandian Chronostratigraphy

In 1989 the Subcommission on Precambrian Stratigraphy (SPS) agreed formally on a chronometric subdivision and nomenclature of Precambrian time, and in 1990 the International Union of Geological Sciences (IUGS) ratified it for international use (Plumb 1991). The major goal of this collective effort was to establish a new time-scale nomenclature to facilitate international scientific communication (Plumb and James 1986). The new time-scale was based on absolute ages, without reference to particular rocks or type sections, with the Palaeoproterozoic-Archaean boundary placed at 2500 Ma and the Palaeoproterozoic subdivided into four periods (systems): Siderian (2500–2300 Ma), Rhyacian (2300–2050 Ma), Orosirian (2050–1800 Ma) and Statherian (1800–1600 Ma) (Fig. 3.6). This proposed nomenclature has not become widely adopted because many researchers favour the stratigraphic approach promoted by Cloud (1987b) rather than a geochronometrical approach. A purely number-based time-scale seemingly clashes with the basic principle of stratigraphy where subdivisions are based on the rocks and use biostratigraphic or other broadly recognisable events (Nisbet 1982; Preiss 1982; Trendall 1984; Eriksson et al. 2004).

With our Scandinavian attempt in linking the subdivision of Palaeoproterozoic time with global palaeoenvironmental events recorded in real rock sections, we resume the old discussion on whether "...stratigraphy or geochronometry shall take priority in the definition of transitions and boundaries in Earth history" (Cloud 1987a, p. 256). Whether or not the suggested chronostratigraphic subdivisions of the Fennoscandian Shield could form a base for a global subdivision of the early Palaeoproterozoic in general remains to be seen. We advocate that, in principle, at least several global events contained in the rock record, with some of them having been already precisely dated and others having a great potential to be better dated, may serve as the starting point in the creation of a common approach (Fig. 3.6). These include: (*1*) the initiation of world-wide rifting of the late Archaean supercontinent(s) at c. 2505 Ma recorded in emplacement of large gabbro complexes and formation of flood-basalt provinces on the Canadian and Fennoscandian Shields; (*2*) a radical change in atmospheric redox state, tapering-off of the mass-independent fractionation of sulphur isotopes at c. 2430 Ma as recorded in the Canadian Shield and South Africa; (*3*) the onset of the Huronian global ice age at c. 2400 Ma recorded in North America, Fennoscandia, South Africa, and Australia; (*4*) the onset and termination of the Lomagundi-Jatuli isotope excursion

(c. 2320–2060 Ma) recorded in several sedimentary successions worldwide, and perhaps (*5*) the onset of aerobic recycling of organic matter (^{13}C-depleted carbonate concretions) associated with the formation of the Earth's earlies phosphorites and unprecedented accumulation of organic matter at c. 2060 Ma (i.e. Shunga event), best recorded in Fennoscandia and India.

The first priority for building the Palaeoproterozoic stratigraphic nomenclature assigned to broadly recognisable global events is to obtain new, reliable precise ages, along with comprehensive understanding of rock geochemistry and rock depositional environments worldwide; this cannot be made by using data derived solely from Fennoscandia or any other individual craton due to the inherent incompleteness of the rock record on any given continent. Although 700 million years of early Palaeoproterozoic history have been recorded on the Fennoscandian Shield, the 20-km-thick sedimentary-volcanic succession would yield an average accumulation rate of c. 30 m per million years. Such rates cannot be considered to represent a complete geological record. Several prominent unconformities dividing the Fennoscandian stratigraphic units remain undated and thereby compromise stratigraphic completeness. However, the common effort involving successions from different continents should overcome such problems. Compilations presented in several chapters of this book demonstrate that a considerable achievement has already been made by collaborative international efforts in obtaining geochronological constraints of several global events recorded in rocks; many of these events were outlined as early as in the 1980s by Cloud (1987a), and many have remain unchanged ever since.

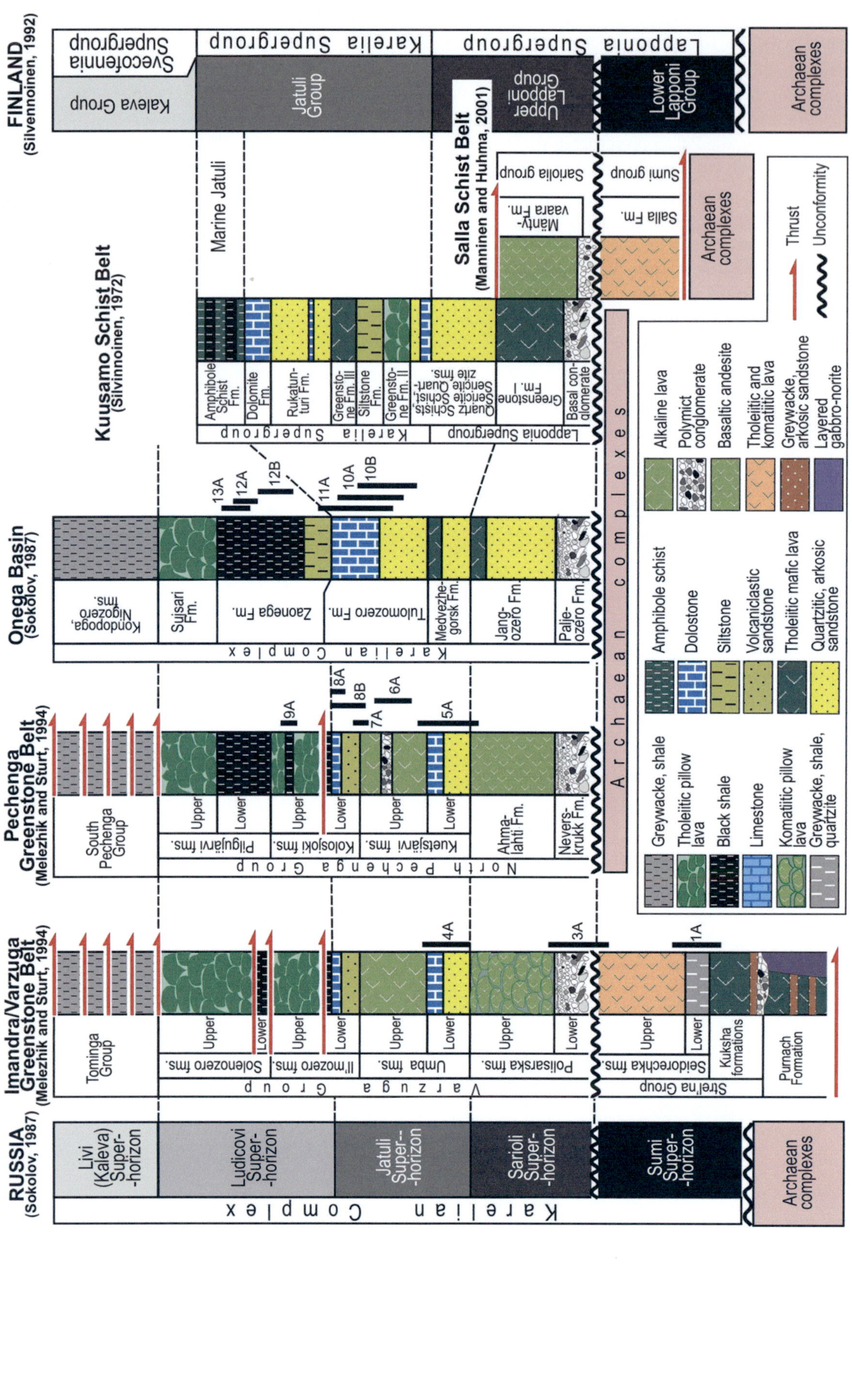

Fig. 3.5 An example of correlations across the Fennoscandian Shield in the 1980s and 1990s by using the traditional lithostratigraphic subdivisions

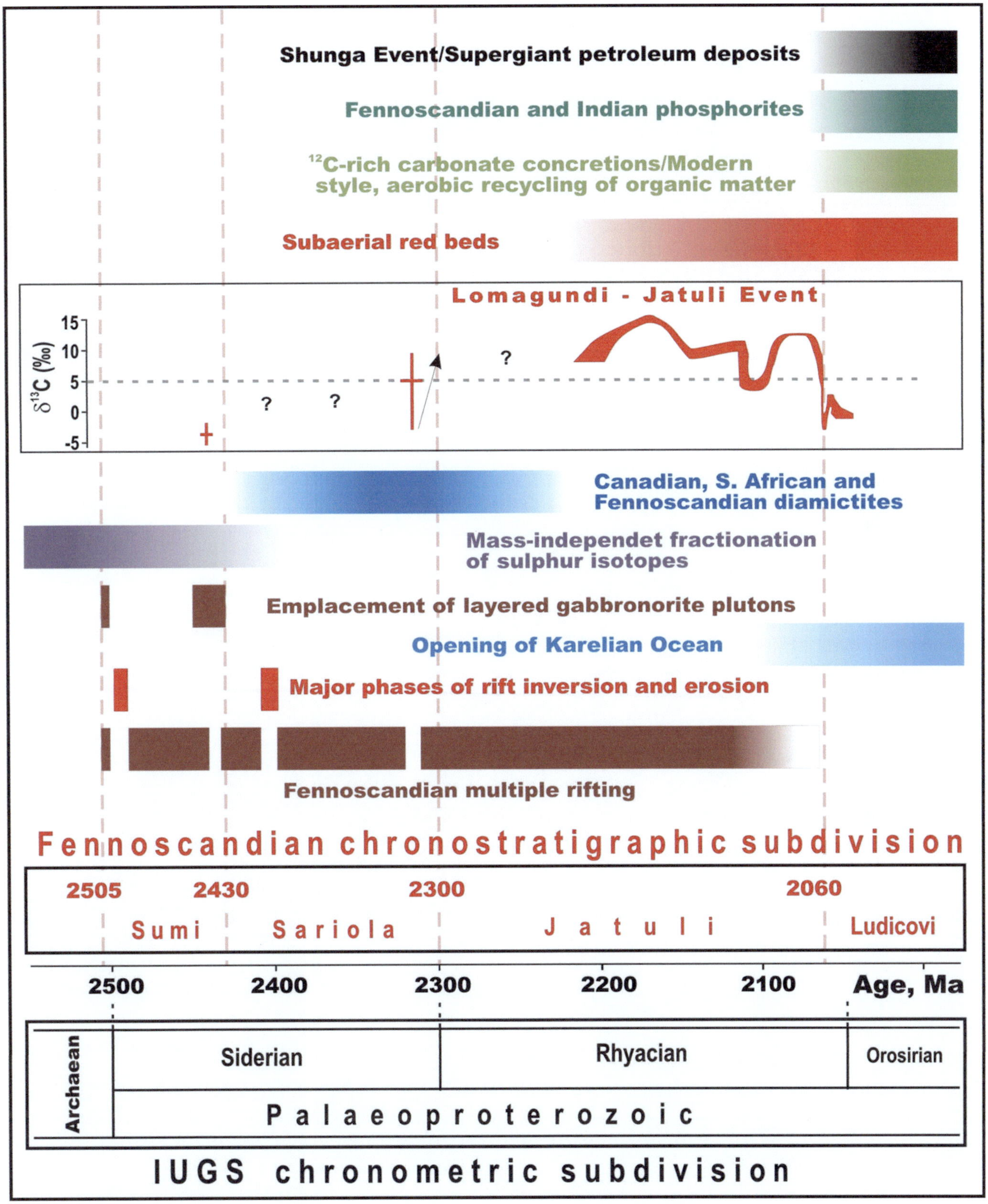

Fig. 3.6 Global environmental events in the early Palaeoproterozoic applied to the chronostratigraphic subdivisions of 2.5–2.0 Ga volcano-sedimentary formations of the Fennoscandian Shield

Fig. 3.7 Selected major sedimentary and igneous rock types representing the 2505–2430 Ma Sumian system. *Peräpohja Belt*: (**a**) Magmatic layering in a 2440 Ma gabbronorite intrusion at the base of Peräpohja sedimentary-volcanic succession; hammer head is 15 cm long. *Kuksha Sedimentary Formation, western Imandra/Varzuga Belt*: (**b**) Carbonate-cemented arkosic sandstone comprising cross-bedded and horizontal, parallel-bedded units in the Kuksha Sedimentary Formation in the western part of the Imandra/Varzuga Belt; lens cap is 6 cm in diameter, (**c**) Polymict conglomerate constituting the basal part of the Kuksha Sedimentary Formation; clasts are amphibolite and gabbro-anorthosite, whereas matrix is composed of gabbro particles; both the matrix and the clasts are derived from the underlying 2505 Ma gabbronorite intrusion; drillhole 2147, depth 120 m, core diameter is 6 cm

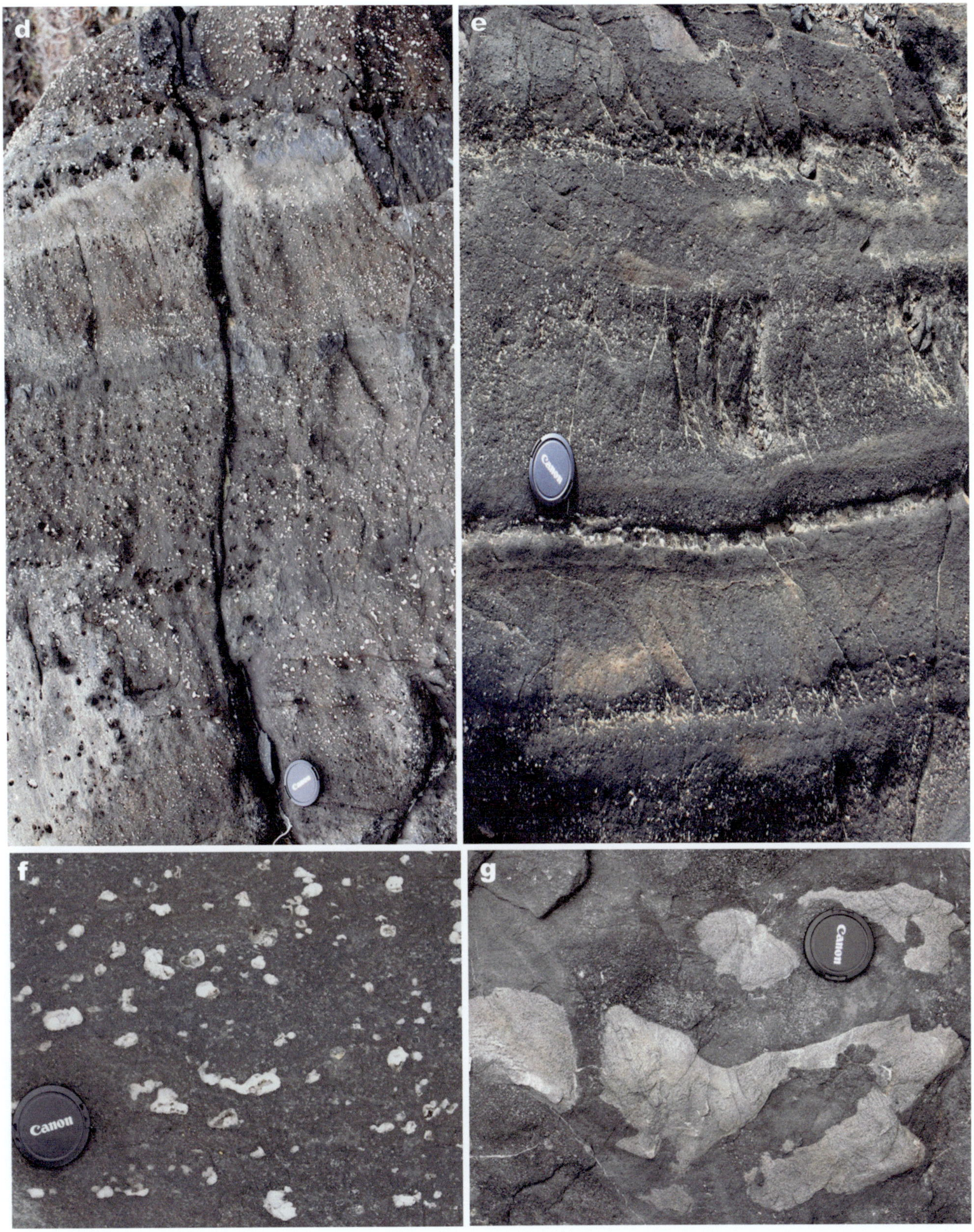

Fig. 3.7 (continued) *Kuksha Volcanic Formation, western Imandra/Varzuga Belt*: (**d**) A series of thin lava-flows composed of amygdaloidal basalts. (**e**) Several thin lava-flows comprised of massive basalts with amygdale-rich tops. (**f**) Mafic lava flow rich in quartz-filled amygdales. (**g**) Mafic lava flow containing numerous xenoliths of arkosic sandstone. Lens cap for scale in **b** and **d–g** is 6 cm in diameter

Fig. 3.7 (continued) *Seidorechka Sedimentary Formation, western Imandra/Varzuga Belt:* (**h**) Alternating wavy- to flaser-bedded sandstone-mudstone layers; many sandstone layers exhibit ripple cross-lamination and mudstone drapes; basal part of the formation. (**i**) Lenticular- to flaser-bedded rippled sandstone-mudstone couplets; basal part of the formation. (**j**) Flaser-bedded sandstone-mudstone unit; basal part of the formation. (**k**) Flaser-, wavy- and lenticular-bedded sandstone-mudstone injected by sandstone dyke; base of the formation

Fig. 3.7 (continued) *Seidorechka Sedimentary Formation, western Imandra/Varzuga Belt*: (**l**) Dolostone bed with in situ brecciated base resting on hummocky surface of calcareous shale with rounded dolostone fragments; middle part of the formation. (**m**) Thin, massive dolostone bed, which is discontinuously developed along strike due to partial or complete erosion prior to the deposition of overlying strata; middle part of the formation. (**n**) Siliceous doloarenite-matrix breccia with unsorted micritic dolostone clasts in the middle part of the formation

Fig. 3.7 (continued) *Seidorechka Sedimentary Formation, western Imandra/Varzuga Belt*: (**o**) Shale bed (*1*) overlain by breccia with unsorted clasts of clayey dolostones and a dolomite-cemented sandstone matrix, followed by (*2*) quartz-sandstone-matrix breccia with platy clasts of micritic dolostones caped by (*3*) mature, massive, quartz sandstone; middle part of the formation. (**p**) Sandstone-matrix conglomerate supported by tan and dark grey, unsorted fragments of dolostone followed by sandstone-supported, platy, imbricated clasts of dolostone, which pass gradually into mature, cross-bedded quartz sandstone; middle part of the formation. (**q**) Massive, silica-cemented, quartz sandstone erosively overlain by dolomite-cemented (brown-stained on weathered surface), cross-bedded, quartz sandstone; *middle* part of the formation. (**r**) Quarry wall exhibiting sets of alternating cross- and ripple-bedded quartzite beds in the middle of the formation; note unidirectional cross bedding

Fig. 3.7 (continued) *Seidorechka Sedimentary Formation, western Imandra/Varzuga Belt*: (**s**) Parallel-bedded dolostone with thin shale laminae overlain by laminated marl in the upper part of the formation. (**t**) Tectonically modified interbedded limestone (*white*), dolostone (*pale grey*) and shale (*black* and *brown*) in the *upper* part of the formation. (**u**) Light brown, sandy, crudely stratified marl containing grit-size, rounded dolostone fragments (*white* and *pale grey*) grading into laminated, silty dolarenite followed by fine-grained, laminated dolarenite; coin diameter is 2 cm

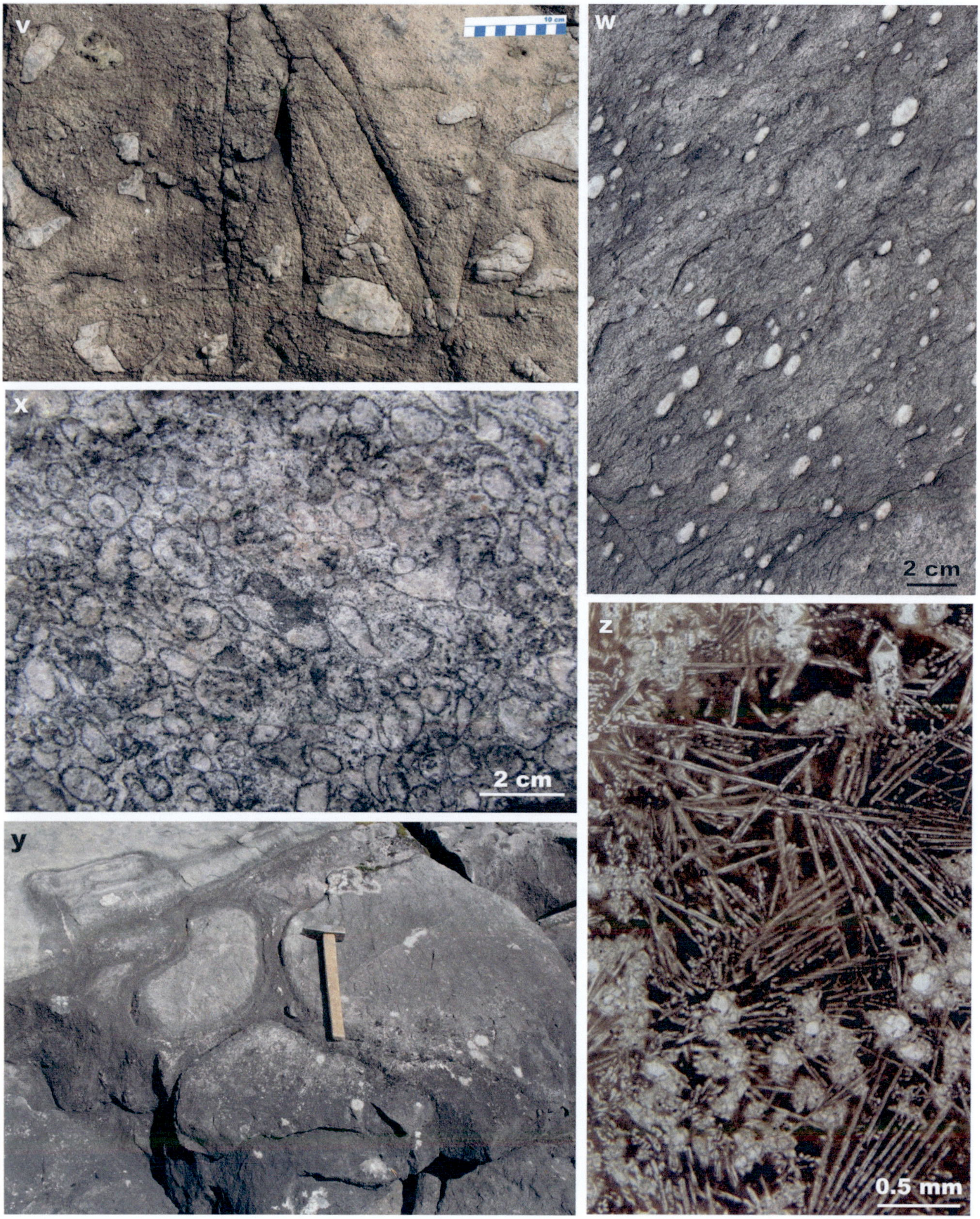

Fig. 3.7 (continued) *Seidorechka Volcanic Formation, western Imandra/Varzuga Belt*: (**v**) Dacitic lava with large quartz-filled amygdales. (**w**) Andesitic lava with sub-spherical, quartz-filled amygdales. (**x**) Accretionary lapilli in quartz-feldspar-phyric rhyolitic tuff. *Tunguda Formation, Lekhta Belt*: (**y**) Pillowed basaltic andesite. *Vetreny Belt*: (**z**) Photomicrograph of pyroxene spinifex-textured komatiitic lava; transmitted, polarised light

Fig. 3.7 (continued) *Central Karelia*: (**aa**) Low-angle cross-bedded quartzite; length of hammer head is 15 cm. (**ab**) Poligonally jointed, komatiitic pillow lava; pen length is 13 cm. *Rookkiaapa Formation, Central Lapland Belt*: (**ac**) Pumice fragments in felsic tuff. *Salla Group, Salla Belt*: (**ad**) Flat-bottomed segregation vesicle in andesitic lava; coin diameter is 2 cm (Photographs courtesy of Vesa Perttunen (**a**), Vladimir Pozhilenko (**x**), Valery Zlobin (**y**), Evgeny Sharkov (**z**), Sergey Svetov (**aa, ab**) and Tuomo Manninen (**ac**, **ad**). The rest of the photographs by Victor Melezhik)

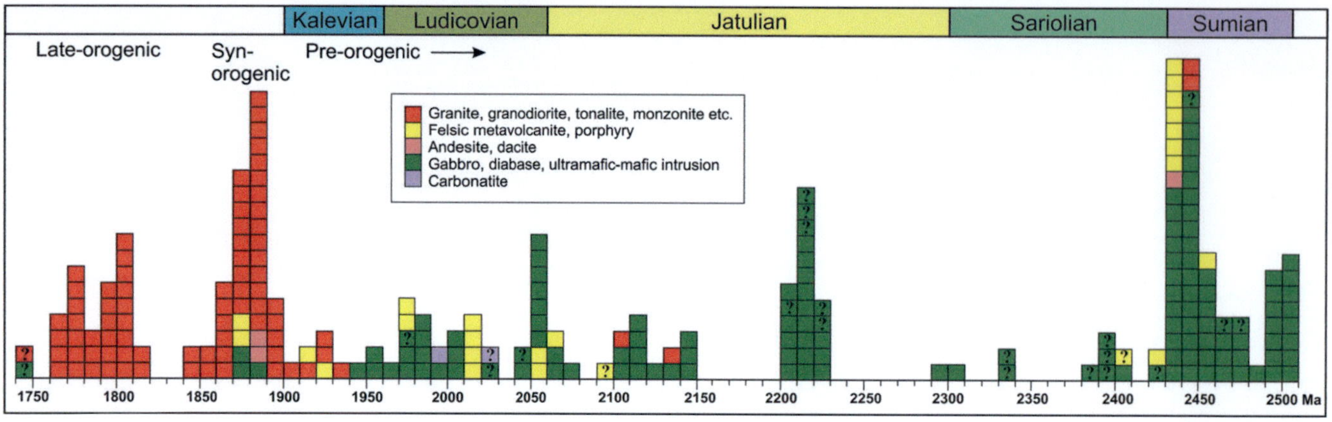

Fig. 3.8 Compilation of U-Pb zircon ± baddelyite dates of igneous rocks from the Karelian and Kola Provinces of the Fennoscandian Shield, including the part of northern Sweden that has been shown by Nd isotopes to be underlain by Archaean crust (cf. Mellqvist et al. 1999). There exist time periods in the evolution of the shield that are characterised by magmatic quiescence, including intervals 2430–2310 Ma, 2290–2230 Ma, 2200–2150 Ma and 1960–1930 Ma (excluding ophiolites). The period between 2400 and 2200 Ma is a globally observed "crustal age gap" when there is little evidence for juvenile continental crust formation or LIP magmatism (Condie et al. 2009); nevertheless, the Sariolian volcanic rocks most probably erupted during this gap. The lack of reliable age determinations, with the exception of few on mafic dykes in the Archaean basement, is at least partly due to the fact that metamorphosed mafic volcanic rocks are commonly not amenable for isotopic dating. Dates labeled with a question mark have a large analytical error or, in a few cases, are otherwise problematic. For classification of granitoids into different orogenic types, see Nironen (2005)

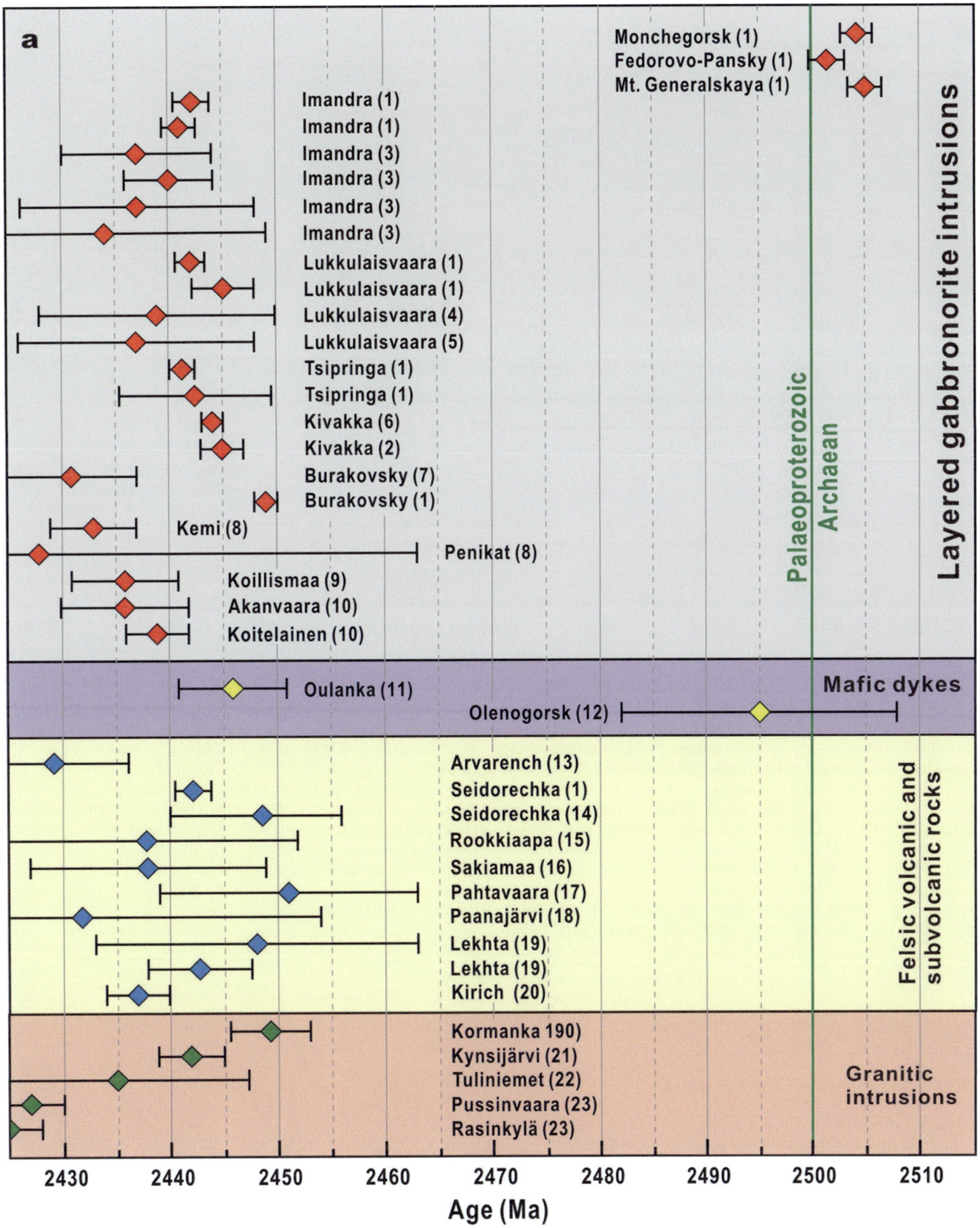

Fig. 3.9 (continued)

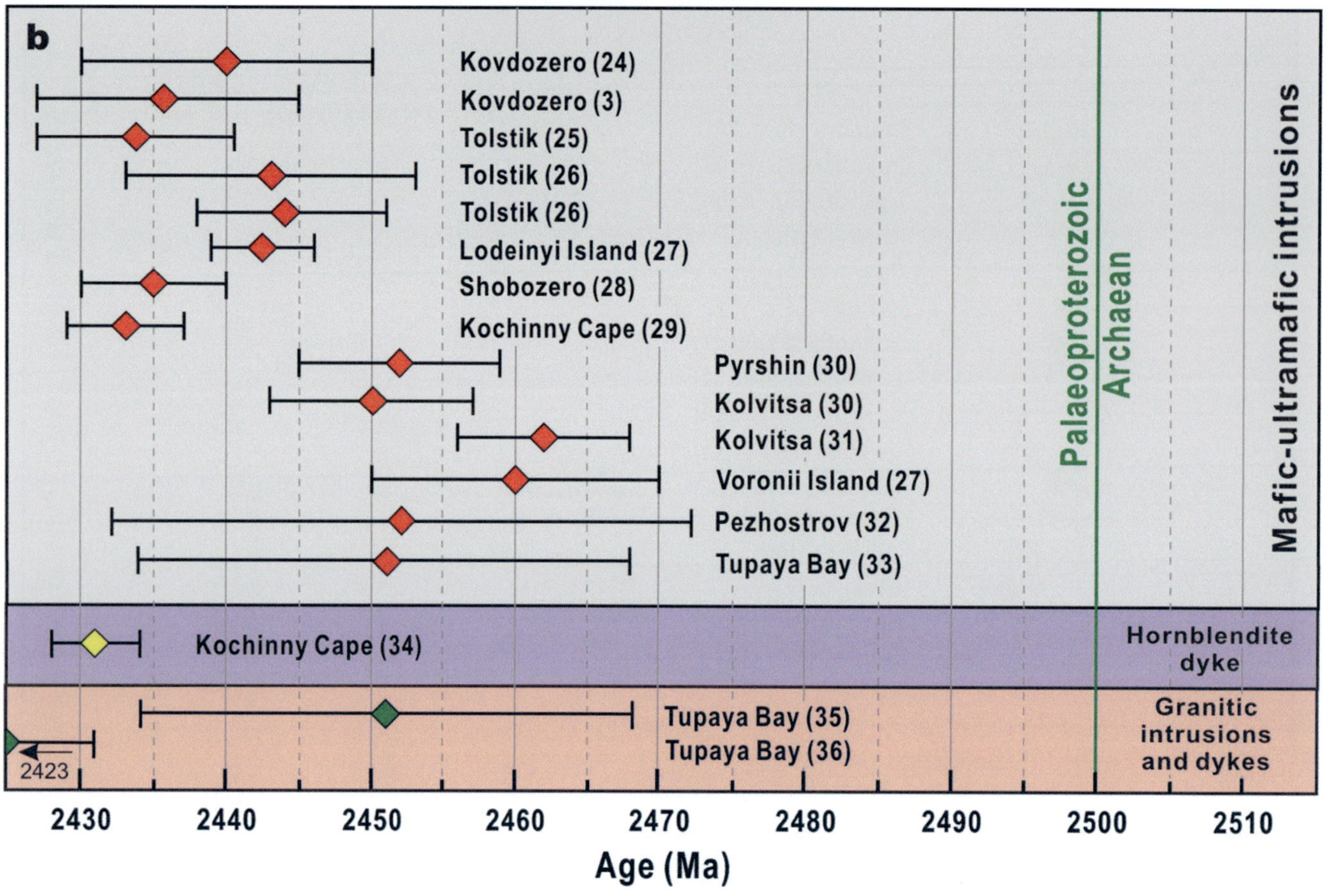

Fig. 3.9 U-Pb zircon ± baddeleyite ages and their error bars measured for Sumian-age intrusive and volcanic rocks. (**a**) Russian and Finnish parts of the Fennoscandian Shield excluding the Belomorian Mobile Belt. (**b**) The Belomorian Mobile Belt (for location see Fig. 3.3 in Chap. 3.1). Data sources: (1) Amelin et al. (1995), (2) Balashov et al. (1993), (3) Bayanova (2004), (4) Bayanova and Mitrofanov (2005), (5) Barkov (1992, cited by Barkov et al. 2001), (6) Barkov (1991, cited by Glebovitsky et al. 2001), (7) Bailly et al. (2009), (8) Perttunen and Vaasjoki (2001), (9) Alapieti (1982), (10) Mutanen and Huhma (2001), (11) Vuollo and Huhma (2005), (12) Smol'kin et al. (2004), (13) Vrevskii et al. (2010), (14) Chashchin et al. (2008), (15) Manninen et al. (2001), (16) Räsänen and Huhma (2001), (17) Nironen and Mänttäri (2003), (18) Buiko et al. (1995), (19) Levchenkov et al. (1994), (20) Puchtel et al. (1997), (21) Lauri and Mänttäri (2002), (22) Luukkonen (1988), (23) Mikkola et al. (2010), (24) Efimov and Kaulina (1997), (25) Bogdanova and Bibikova (1993), (26) Kaulina and Bogdanova (2000), (27) Sharkov et al. (2004), (28) Slabunov et al. (2001), (29) Kaulina (1996), (30) Mitrofanov et al. (1995), (31) Frisch et al. (1995), (32) Alexejev et al. (2000), (33) Lobach-Zhuchenko et al. (1993), (34) Bogdanova et al. (1993), (35) Bibikova et al. (1993), (36) Lobach-Zhuchenko et al. (1998). All analyses made by the conventional TIMS method except those by Bailly et al. (2009), which are LA-ICP-MS data

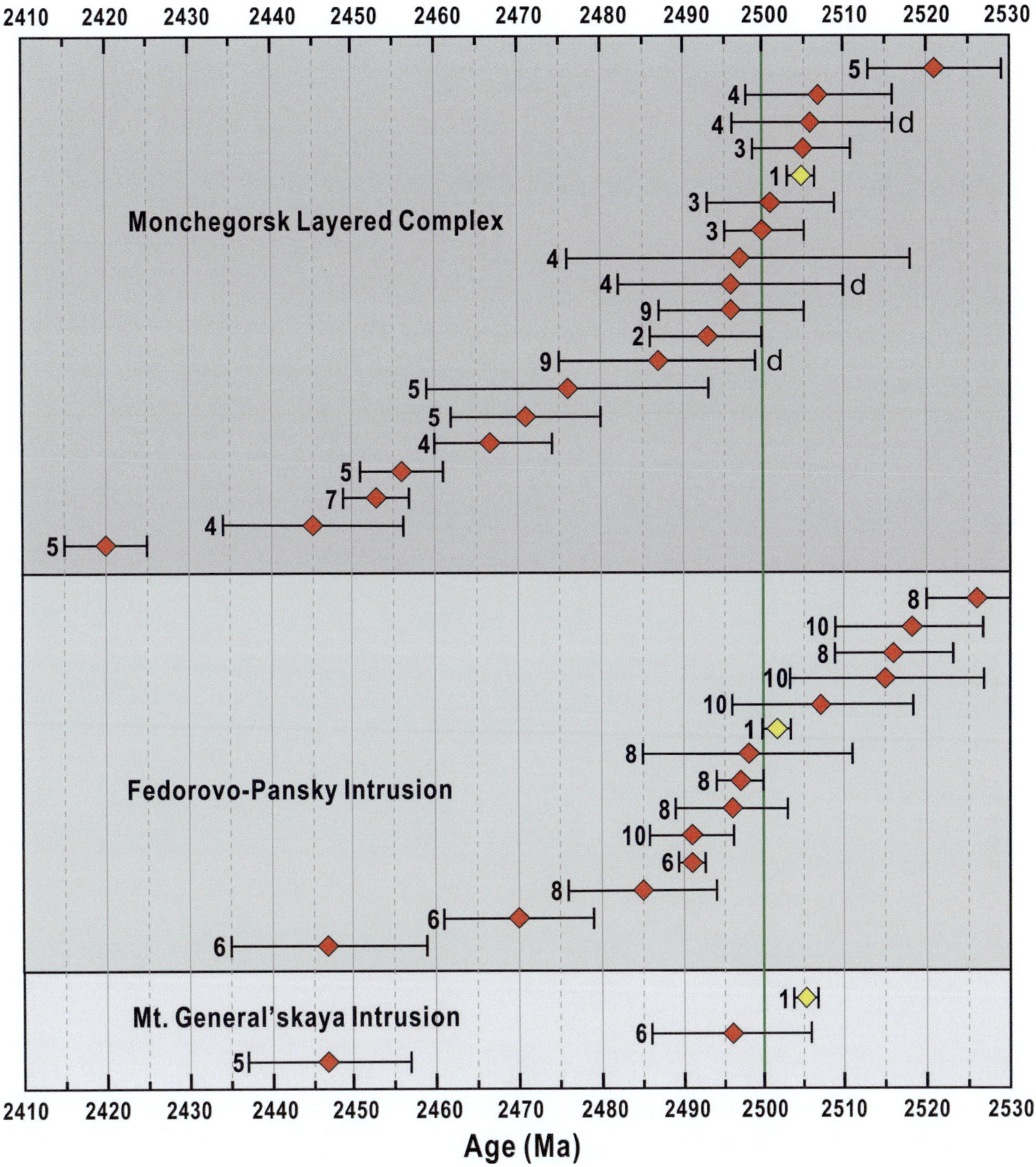

Fig. 3.10 U-Pb zircon ± baddeleyite ages and their error bars for the Monchegorsk Layered Complex, Fedorovo-Pansky Complex and Mt. Generalskaya intrusion. Data sources: (1) Amelin et al. (1995), (2) Balashov et al. (1993), (3) Bayanova and Mitrofanov (2005), (4) Bayanova et al. (2009), (5) Bayanova et al. (2010), (6) Mitrofanov and Bayanova (1999), (7) Mitrofanov et al. (1995), (8) Nitkina (2006), (9) Smol'kin et al. (2004), (10) Grooshan and Nitkina (2008). The dates published by Amelin et al. (1995) are shown in *yellow*; d = within-intrusion noritic dyke

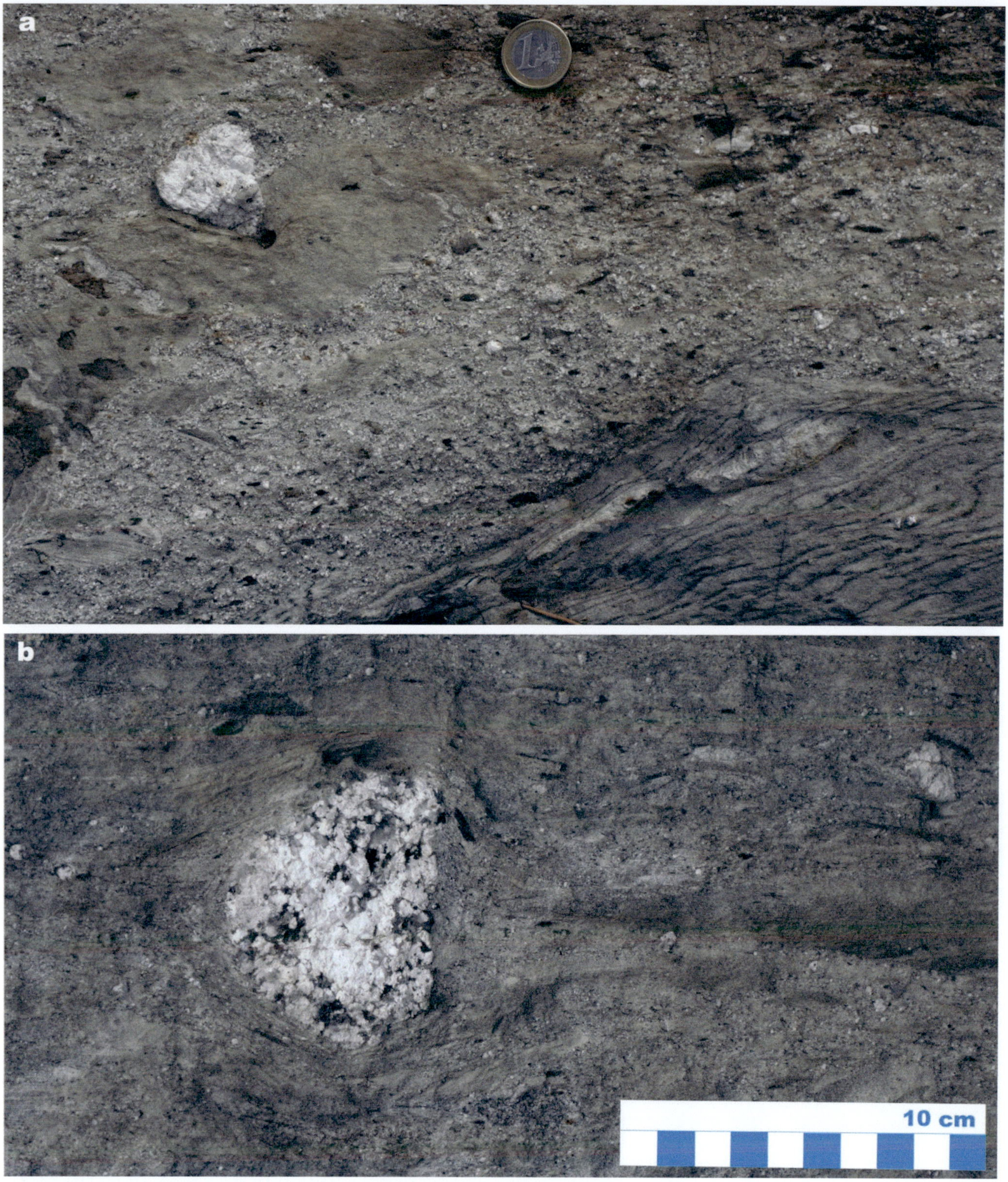

Fig. 3.11 Selected major sedimentary and igneous rock types representing the 2430–2200 Ma Sariolian system. *Urkkavaara Formation, North Karelia Belt*: (**a**) Syndepositionally deformed, bedded, silty sandstone (*lower right corner*) overlain by unsorted sandy gritstone with an oversize granitoid clast (*white*) representing subaerially-deposited glaciogenic sediments. (**b**) Tonalitic clast piercing layers beneath and overlain by layers above in crudely bedded, unsorted, silty, glaciomarine sandstone

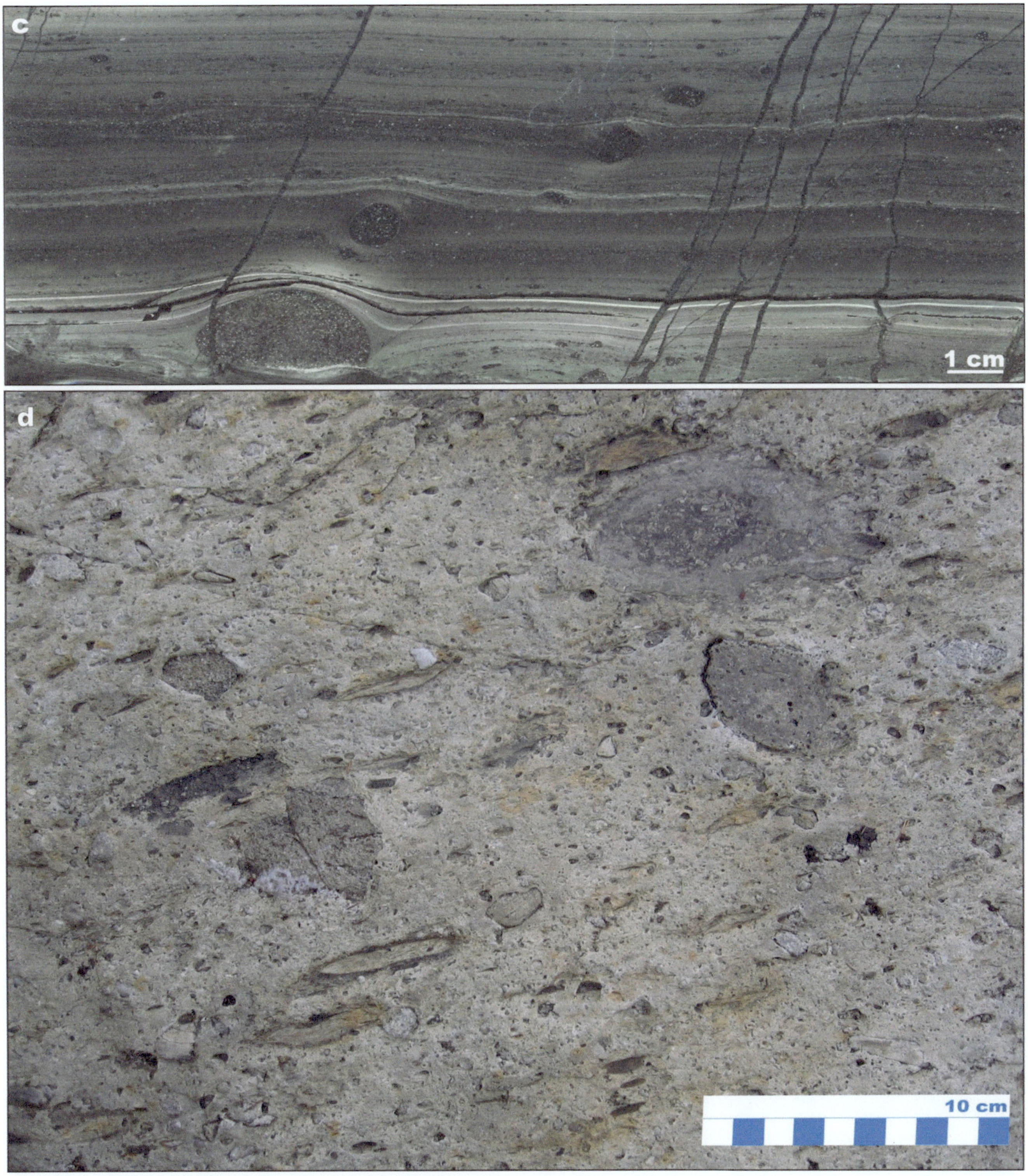

Fig. 3.11 (continued) *Polisarka Sedimentary Formation, Imandra/Varzuga Belt:* (**c**) Dropstones of variable size in finely laminated, varved, clayey glaciomarine siltstone. *Pajozero Formation, Shambozero Belt:* (**d**) Diamictite containing unsorted, polymict clasts in a massive clayey, fine-grained sandstone matrix

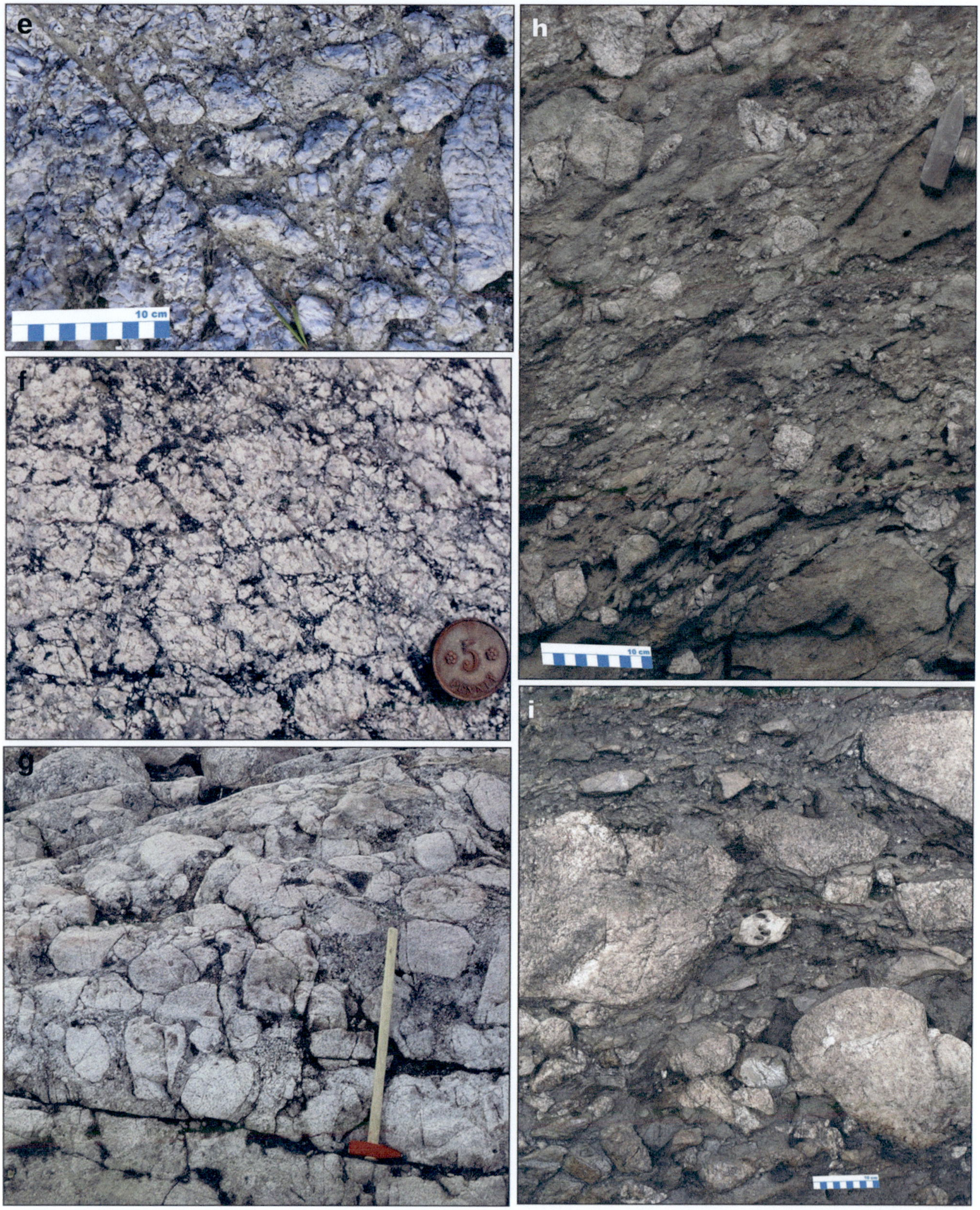

Fig. 3.11 (continued) *Pechenga Belt*: (**e**) Fractured Archaean pegmatite at the base of the Pechenga Belt succession; fracturing is associated either with incipient extensional tectonics or with ice-shattering event. *Peräpohja Belt*: (**f**) Magnetite-filled cracks in jointed Archaean basement granite beneath the basal conglomerate shown in photograph **g**. (**g**) Basal boulder conglomerate (above the hammer head) on Archaean basement granite; boulder composition reflects a local source. Hammer head length is 15 cm. *Ahven-Kivilampi Formation, Kainuu Belt*: (**h**) Conglomerate with polymict, variably rounded and unsorted fragments of granitoid and mafic lava supported by a gritty sandstone matrix. (**i**) Fragment-supported, unsorted, polymict-conglomerate overlain by pebbly gritstone

Fig. 3.11 (continued) *Pjalozero Formation, Central Russian Karelia*: (**j**) Conglomerate with unsorted pebbles and boulders of granites with subordinate clasts of gneisses and crystalline schists supported by a gritty greywacke matrix. (**k**) Fractured Archaean granite overlain by fragment-supported monomict conglomerate with clasts derived from the underlying basement granite. *Pajozero Formation, Shambozero Belt:* (**l**) Bedding surface of volcaniclastic-matrix-supported conglomerate with rounded, unsorted, polymict clasts

Fig. 3.11 (continued) *Neverskrukk Formation, Pechenga-Pasvik Belt*: (**m**) Clast-supported, polymict conglomerate with imbricated boulders. (**n**) Greywacke matrix-supported, plagiogranite-bearing conglomerate interbedded with sandy gritstone. (**o**) Gritstone containing angular plagiogranite clasts derived from the underlying Archaean basement

Fig. 3.11 (continued) *Kaukonen Formation, Central Lapland Belt:* (**p**) Amygdaloidal andesitic lava; leaf for scale is 2.5 cm in diameter. *Polisarka Formation, Imandra/Varzuga Belt*: (**q**) Pyroxene spinifex texture in komatiitic basalt; FAR-DEEP 3A drillcore, core diameter is 5 cm. (**r**) Pillow fragment breccia of ultramafic komatiite. Photographs courtesy of Vesa Perttunen (**f** and **g**) and Pentti Rastas (**p**). Photographs **a, b, h–i** by Eero Hanski and **c–e**, **j–o** and **q–r** by Victor Melezhik

Fig. 3.12 Selected images of palaeosol and chemically mature sedimentary rocks representing the 2200–2060 Ma Jatulian system. *Hokkalampi Formation, North Karelia Belt:* (**a**) Hokkalampi palaeosol with vein quartz fragments as a chemically resistant residue; coin diameter is 2 cm. (**b**) Kyanite-quartz palaeosol with a tourmaline-quartz fragment as a chemically resistant residue; coin diameter is 2 cm. *Reittiö Formation, Nilsiä Belt*: (**c**) Conglomerate containing rounded, unsorted clasts of vein quartz supported by a quartz-gritstone matrix. *Koli Formation, North Karelia Belt*: (**d**) Quartz-sericite schist (in the *middle*) with quartz-pebble conglomerate channels sandwiched between beds of conglomerate composed of poorly sorted pebbles and cobbles of vein quartz

Fig. 3.12 (continued) *Kaisavaara Formation, Peräpohja Belt*: (**e**) Well-rounded and well-sorted clasts of quartzite embedded in a dolarenite matrix. *Koli Formation, North Karelia Belt*: (**f**) Cross- and parallel-bedded quartzite. *Jero Formation, North Karelia Belt*: (**g**) Cross-bedded, arkosic gritstone-matrix conglomerate. *Virttiövaara Formation, Central Lapland Belt*: (**h**) Trough cross-bedded, quartzite with fuchsite-rich layers (*green*). (**i**) Fuchsite-bearing, coarse-grained, quartz sandstone grading into fine-grained sandstone. Photograph (**e**) courtesy of Vesa Perttunen, the rest of the photographs by Eero Hanski

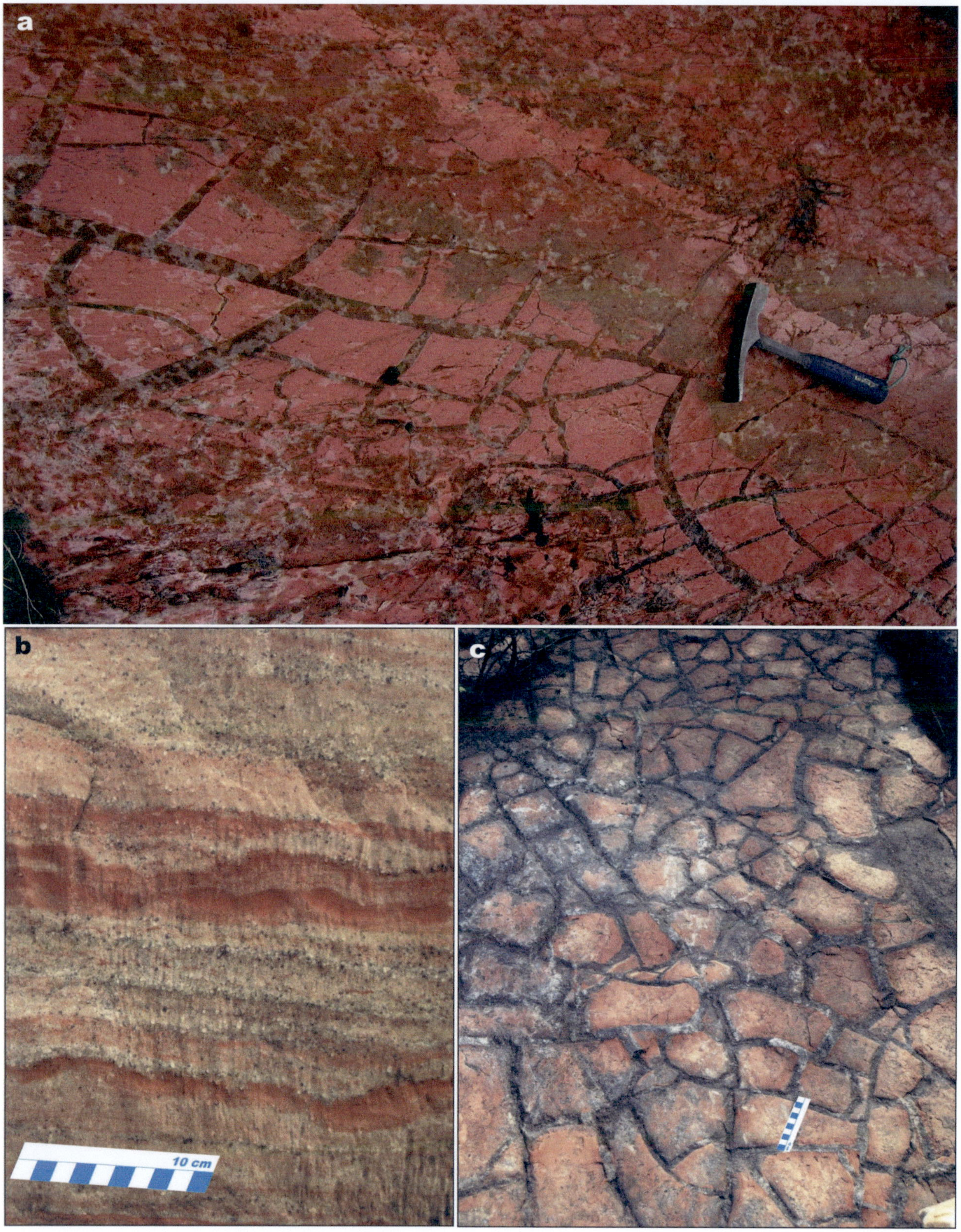

Fig. 3.13 Selected images of Jatulian system "red beds". *Segozero Group, Segozero Belt*: (**a**) *Red* mudstone beds with irregular desiccation cracks; hammer head is 14 cm long. (**b**) Arkosic sandstone with *red*, ripple-bedded mudstone layers. (**c**) Polygonal desiccation cracks developed on a *red* mudstone bed

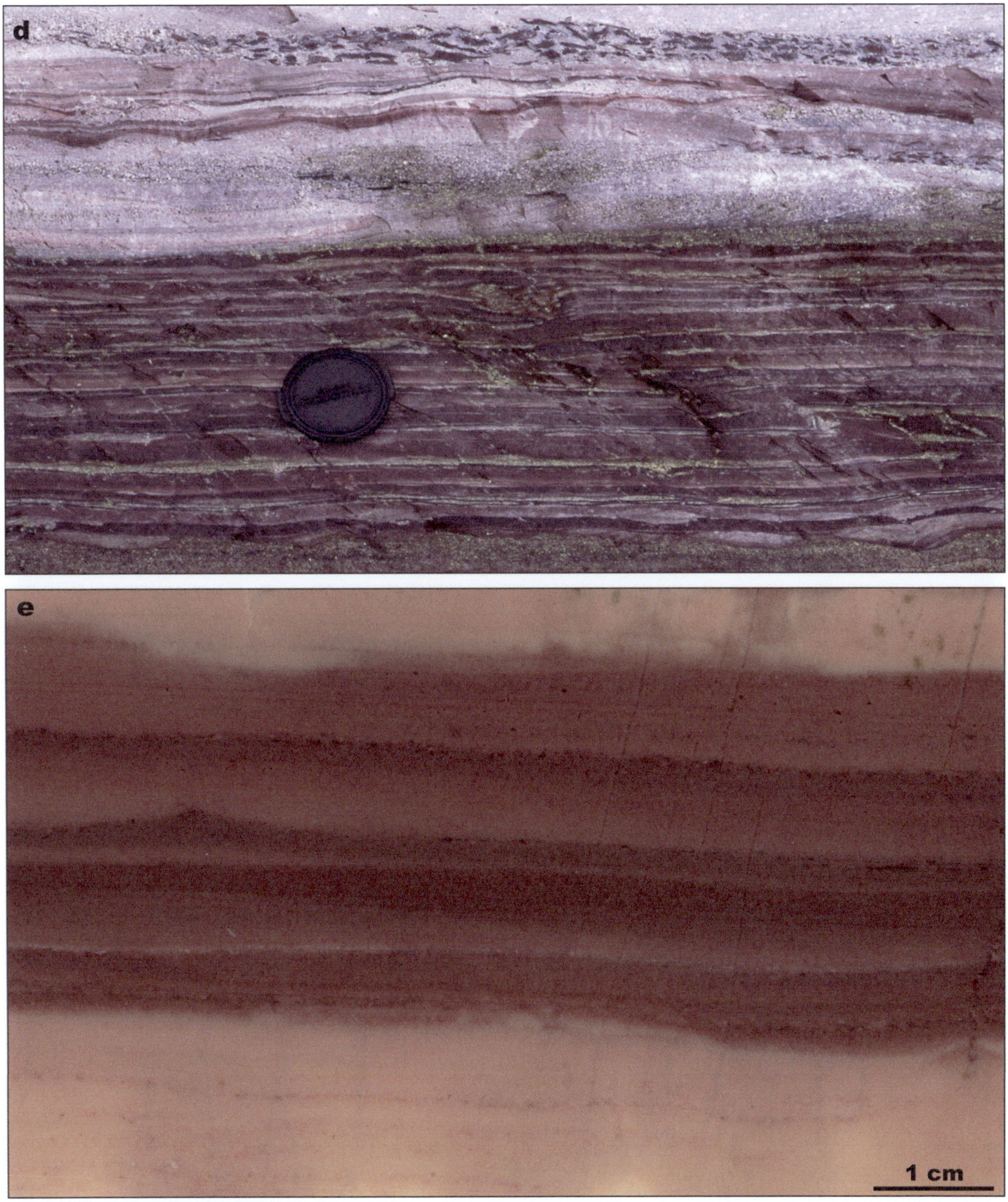

Fig. 3.13 (continued) *Rukatunturi Quartzite Formation, Kuusamo Belt:* (**d**) Lenticular-bedded "red bed" unit (continuous *purple* mudstone layers covering 'lens-like' sandstone ripples) overlain by *pale pink*, low-angle cross-bedded sandstone, followed by flaser-bedded mudstone-siltstone couplets with sandstone layers containing intensively desiccated, dismembered mudstone layers on *top*; lens cap diameter is 6 cm. *Umba Sedimentary Formation, Imandra/Varzuga Belt:* (**e**) Rhythmically bedded, beige and purple dolarenite (Photograph (**a**) courtesy of Kari Strand, the rest of the photographs by Victor Melezhik)

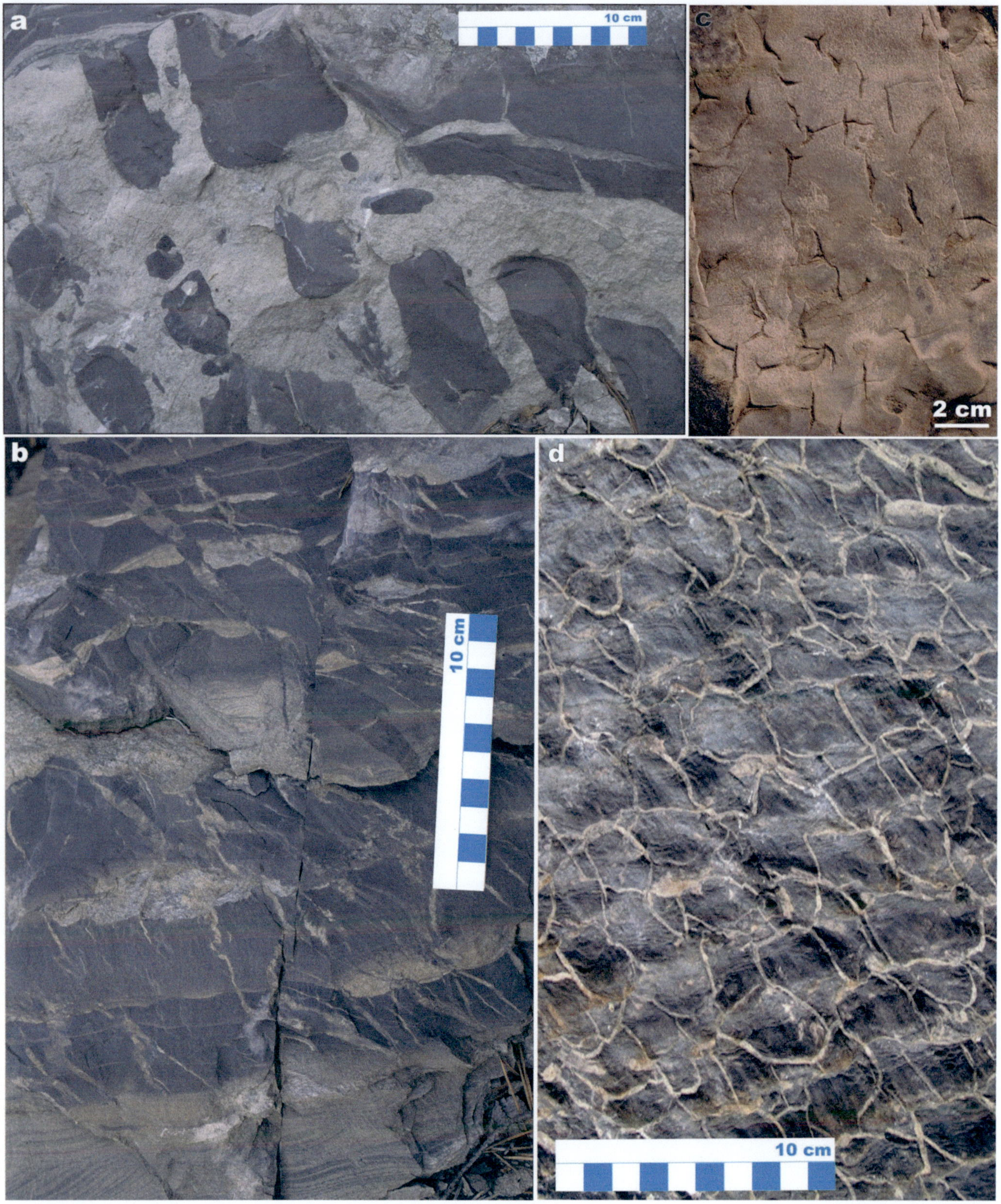

Fig. 3.14 Selected images of desiccated mud beds of the Jatulian system. *Quartzite Schist (Erivaaransuo) Formation, Kuusamo Belt*: (**a**) Intensely desiccated and dismembered, haematite-rich mud layers imbedded in massive arkosic sandstone. (**b**) Cross-bedded sandstone (*bottom*) overlain by mudstone-sandstone couplets occurring as continuous, desiccated, black mudstone layers covering sandstone ripples, thus forming lenticular bedding. *Lower formation, Middle group, Kalix Belt:* (**c**) Incipient desiccation developed on marl bed surface. *Rukatunturi Quartzite Formation, Kuusamo Belt:* (**d**) Siltstone with interfering rhomboidal ripple marks affected by desiccation cracks. Sample (**d**) courtesy of the Geological Museum of the Department of Geosciences, University of Oulu

Fig. 3.14 (continued) *Rukatunturi Quartzite Formation, Kuusamo Belt:* (**e**) Polished slab showing tidal sandstone-mudstone couplets and lenses. Note the influence of waves in the form of rippled tops on some of the sandstone beds and the presence of sand-filled desiccation cracks in many of the mudstone layers as evidence of intermittent subaerial exposure (Sample (**e**) courtesy of the Northern Finland Office of the Geological Survey of Finland. Photographs by Victor Melezhik (**a–b**) and Eero Hanski (**d**); (**c**) reproduced from Wanke and Melezhik (2005) with permission of Elsevier. Photograph (**e**) courtesy of Reijo Lampela)

Fig. 3.15 Selected images of cross-bedded and rippled arenites of the Jatulian system. *Himmerkinlahti Member, southeastern Kuusamo Belt:* (**a**) Parallel- and cross-bedded sandstones interbedded with unsorted conglomerate with polymict clasts. *Segozero Group, Segozero Belt*: (**b**) Tabular, unidirectional cross-bed coset. *Virttiövaara Formation, Central Lapland Belt:* (**c**) Trough cross-bedded, fuchsite-bearing quartzite with clay chips (*yellow* and *dark green*)

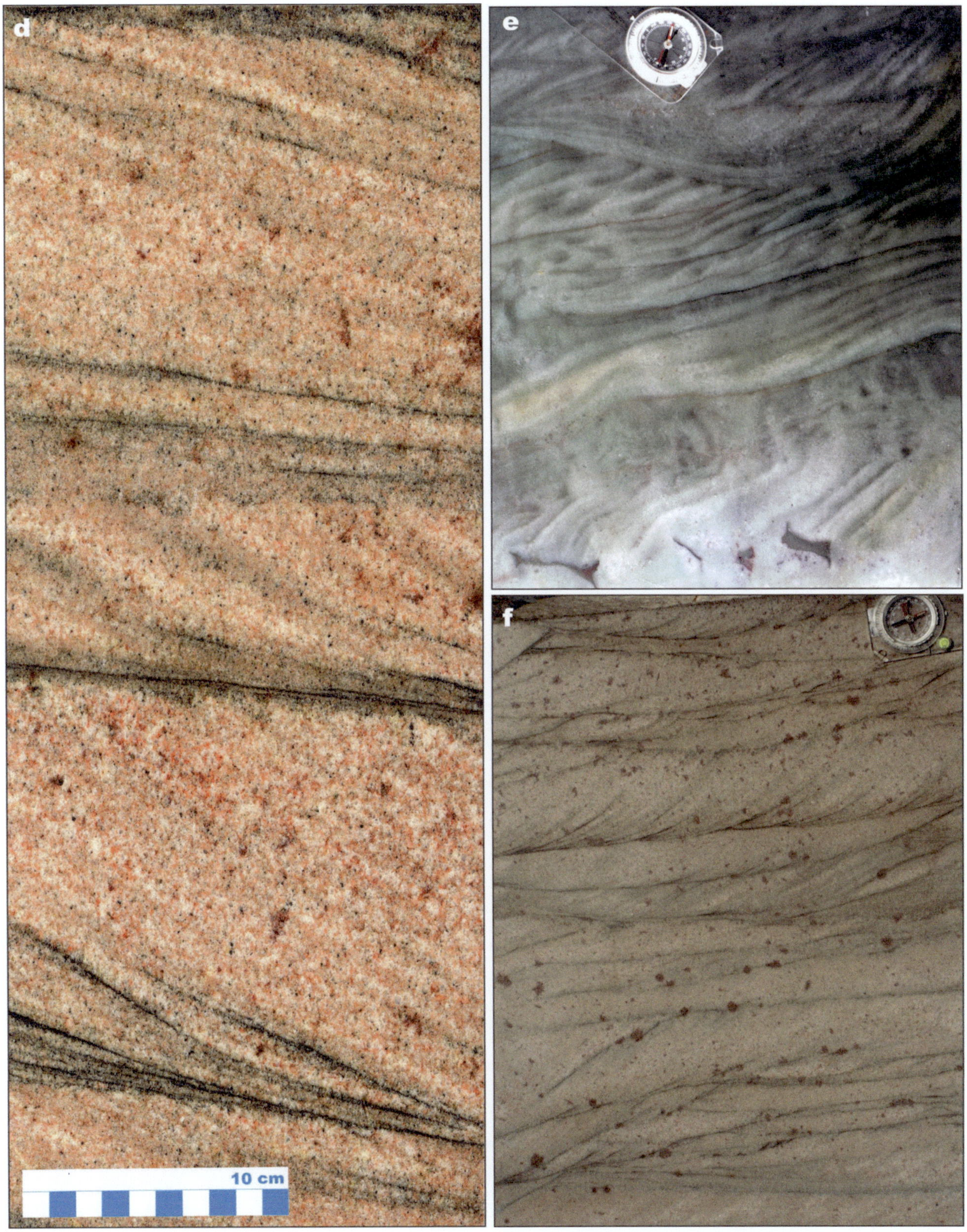

Fig. 3.15 (continued) *Virttiövaara Formation, Central Lapland Belt:* (**d**) Tabular and trough cross-bedded quartz sandstone with some layers showing enrichment in clastic magnetite and haematite (*black*). (**e**) Sets of small-scale trough cross-beds in fuchsite-bearing quartzite; the lower overturned forsets contains *brown* clay chips; Compass width is 5 cm. (**f**) Sets of small-scale tabular and trough cross-beds in quartz sandstone; *red dots* represent post-depositional haematisation; compass width is 5 cm

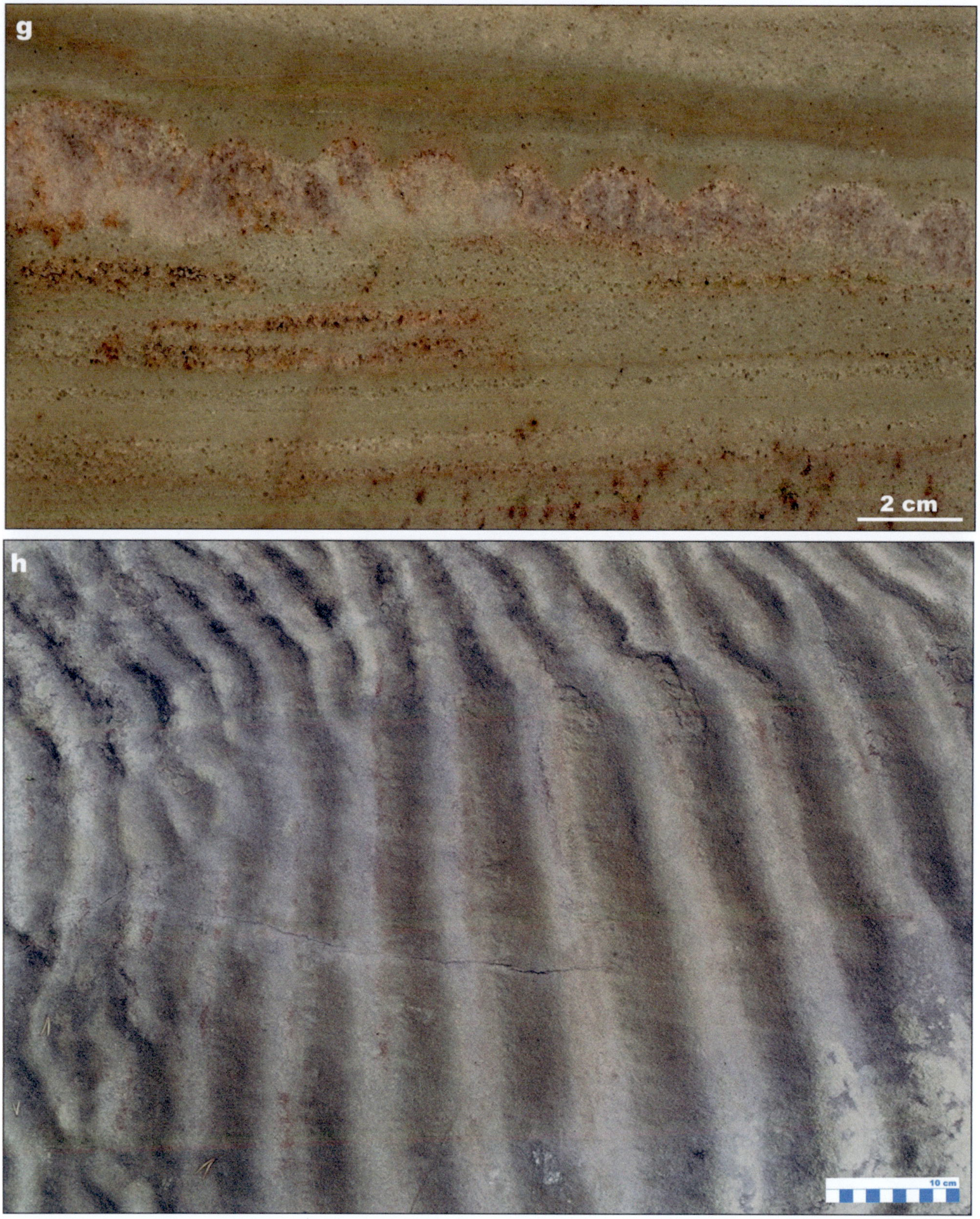

Fig. 3.15 (continued) *Virttiövaara Formation, Central Lapland Belt:* (**g**) Sand ripples in horizontally bedded quartz sandstone. *Palokivalo Formation, Peräpohja Belt:* (**h**) Symmetrical oscillation ripples in quartz sandstone bed (Photographs by Victor Melezhik (**a, b**) and Eero Hanski (**c–g**); photograph (**h**) courtesy of Vesa Perttunen)

Fig. 3.16 Selected images of carbonate lithofacies of the Jatulian system. *Lower formation, Middle group, Kalix Belt:* (**a**) Stromatolitic dolostone (beige) and massive and laminated dolarenite lenses (*pale brown* and *greenish*) encased in mafic tuff (*dark brown*)

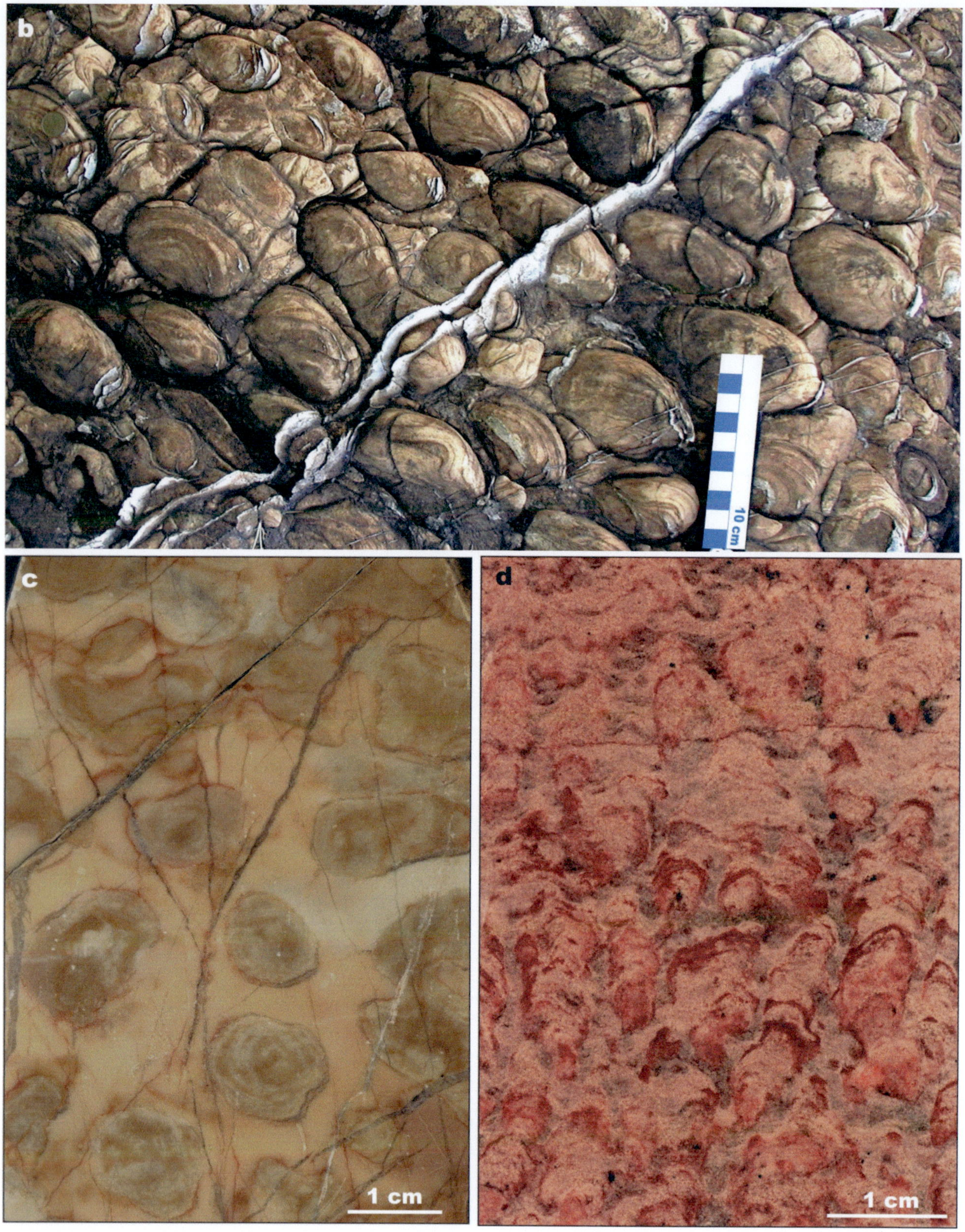

Fig. 3.16 (continued) *Rantamaa Formation, Peräpohja Belt:* (**b**) Bedding surface with exhumed tightly-spaced domal, dolomitic, stromatolites in a dolarenite matrix. (**c**) Columnar dolomitic stromatolites (*pale brown*) in a *yellow* dolarenite matrix. *Tulomozero Formation, Onega Basin:* (**d**) Pink, tightly spaced, branching, dolomitic stromatolites

Fig. 3.16 (continued) *Umba Sedimentary Formation, Imandra/Varzuga Greenstone Belt:* (**e**) *White*, massive to crudely bedded, finely crystalline dolostone; FAR-DEEP drillcore 4A; core diameter is 5 cm. *Kuetsjärvi Sedimentary Formation, Pechenga Belt:* (**f**) Flat-laminated and mini-domal dolomitic stromatolites. (**g**) Clayey dolomitic gritstone with desiccated travertine crust (*pale pink*)

Fig. 3.16 (continued) *Rantamaa Formation, Peräpohja Belt:* (**h**) Quartz sandstone with unsorted, variably rounded dolostone clasts. (**i**) *Grey* greywacke erosively overlain by fine-pebble dolostone conglomerate grading into dolomitic gritstone (Photographs by Victor Melezhik (**a, c–g**) and Eero Hanski (**b, h, i**); photograph (**b**) courtesy of Vesa Perttunen)

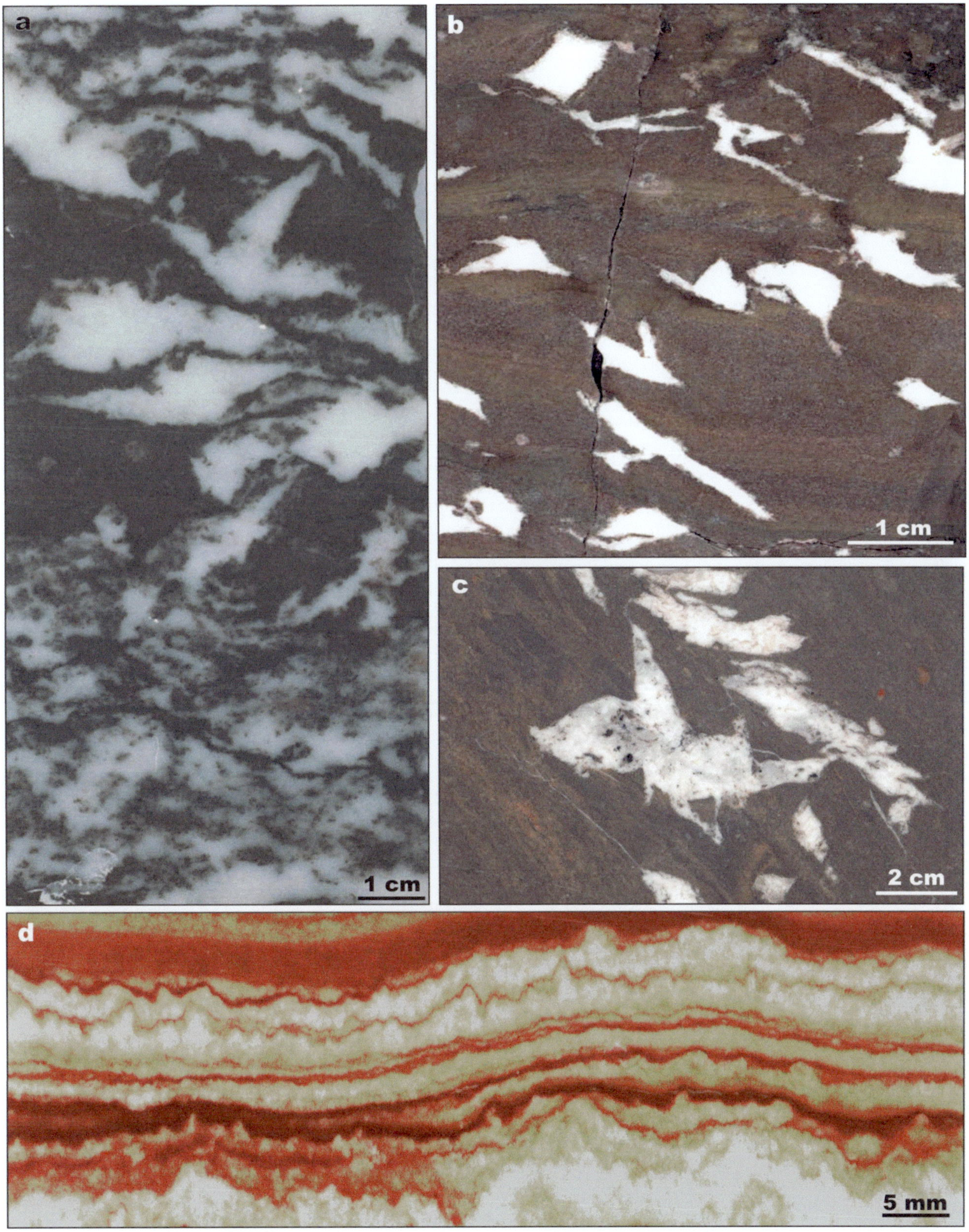

Fig. 3.17 Selected images of Ca-sulphates of the Jatulian system. *Tulomozero Formation, Onega basin*: (**a**) Dolomite-pseudomorphed gypsum rosettes, crystals and impregnation in *dark grey*, haematite-rich marl. (**b**) *Brown*, bedded mudstone with tabular and swallowtail-twined crystals of gypsum partially pseudomorphed by dolomite. (**c**) *Brown*, bedded mudstone with gypsum rosettes partially pseudomorphed by dolomite and quartz. (**d**) Dolomite-pseudomorphed Ca-sulphate (*white*) and *red* mudstone layers plastically deformed into enterolithic structure modified by styolitisation (Photographs (**b–d**) reproduced from Melezhik et al. (2005) with permission from Elsevier)

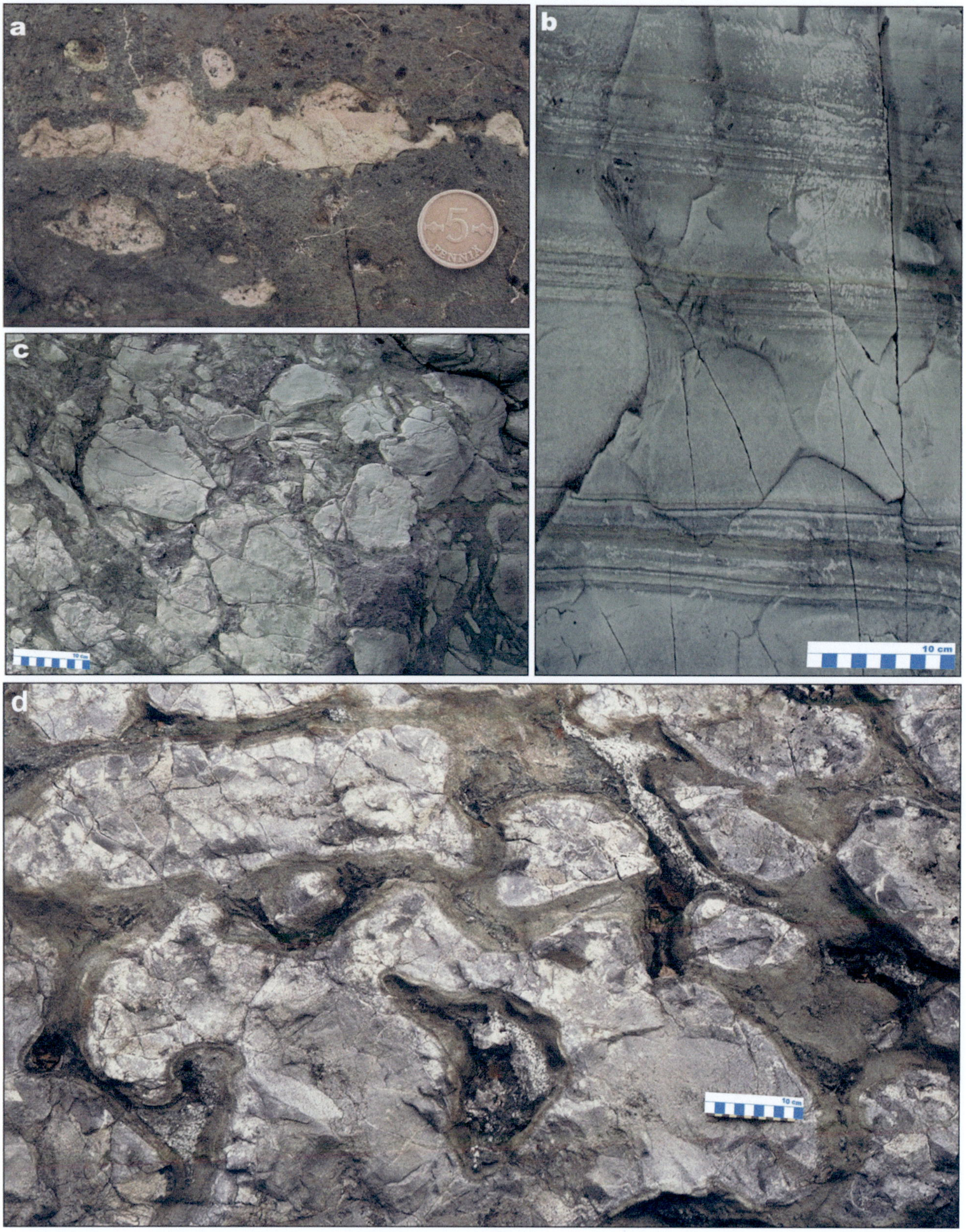

Fig. 3.18 Selected images of volcanic rocks of the Jatulian system. *Jouttiaapa Formation, Peräpohja Belt*: (**a**) Amygdaloidal mafic lava; coin diameter is 2 cm. *Tikanmaa Formation, Peräpohja Belt*: (**b**) Rhythmically bedded mafic tuff. *Onega Basin:* (**c**) Mafic flow breccia. *Santalampi Formation, Peräpohja Belt*: (**d**) Mafic pillow lava.

Fig. 3.18 (continued) *Kuetsjärvi Volcanic Formation, Pechenga Belt*: (**e**) Columnar joints in alkaline basalt with pervasive epidotisation (*bright greenish*) on either side of the joints. (**f**) Ubiquitous quartz-chlorite-epidote-filled amygdales in mafic lava. (**g**) Syndepositionally folded fluidal, rhyolitic, magnetite-haematite-rich (*dark grey* and *black bands*) rhyolitic lava-flow (Photographs courtesy of Vesa Perttunen (**a, b, d**) and Sergey Svetov (**c**); photographs (**e–g**) by Victor Melezhik)

Fig. 3.19 Selected major sedimentary and volcanic rock types representing the 2060–1960 Ma Ludicovian system. *Zaonega Formation, Onega Basin*: (**a**) C_{org}-rich rock, locally termed "shungite", from the type locality near the village of Shunga (total organic carbon content is 62 wt.%, Melezhik et al. 1999). Black, C_{org}-rich rocks comprise several sedimentary successions in the Karelian province, and make the Ludicovian system rocks distinctly different from other early Palaeoproterozoic formations of the Fennoscandian Shield; width of the photograph is 10 cm. *Kolosjoki Volcanic Formation, Pechenga Belt:* (**b**) Tholeiitic pillow lava; width of the view is c. 2 m. Pillow lavas comprise several kilometre-thick formations in the Kola province and in the Central Lapland Belt

Fig. 3.19 (continued) *Zaonega Formation, Onega Basin*: (**c**) Rhythmically bedded, C_{org}-rich siltstone (*black*)-greywacke; FAR-DEEP drillcore 13A. (**d**) Massive organosiliceous rock; FAR-DEEP drillcore 12B. (**e**) Massive organosiliceous rock, locally termed "maksovite" from the Maksovo deposit. (**f**) Black shale-hosted carbonate concretion; FAR-DEEP drillcore 13A. Core diameter in photographs **c–f** is 50 mm

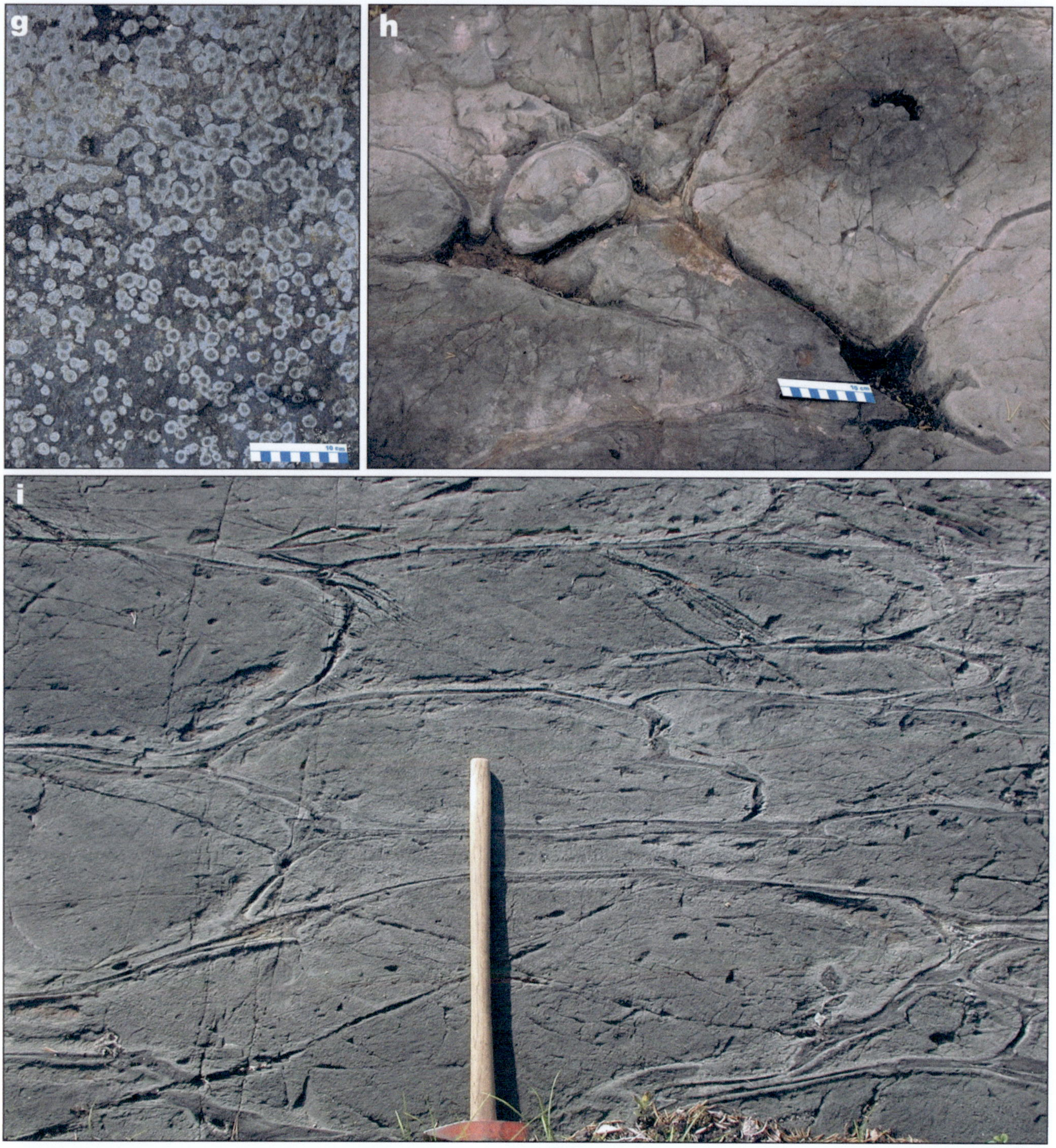

Fig. 3.19 (continued) *Suisari Formation, Onega Basin*: (**g**) Variolitic mafic lava that, together with pillowed and massive lavas and tuffs, forms a c. 500-m-thick unit overlying C_{org}-rich sedimentary rocks. *Väystäjä Formation, Peräpohja Belt*: (**h, i**) Tholeiitic pillow lava

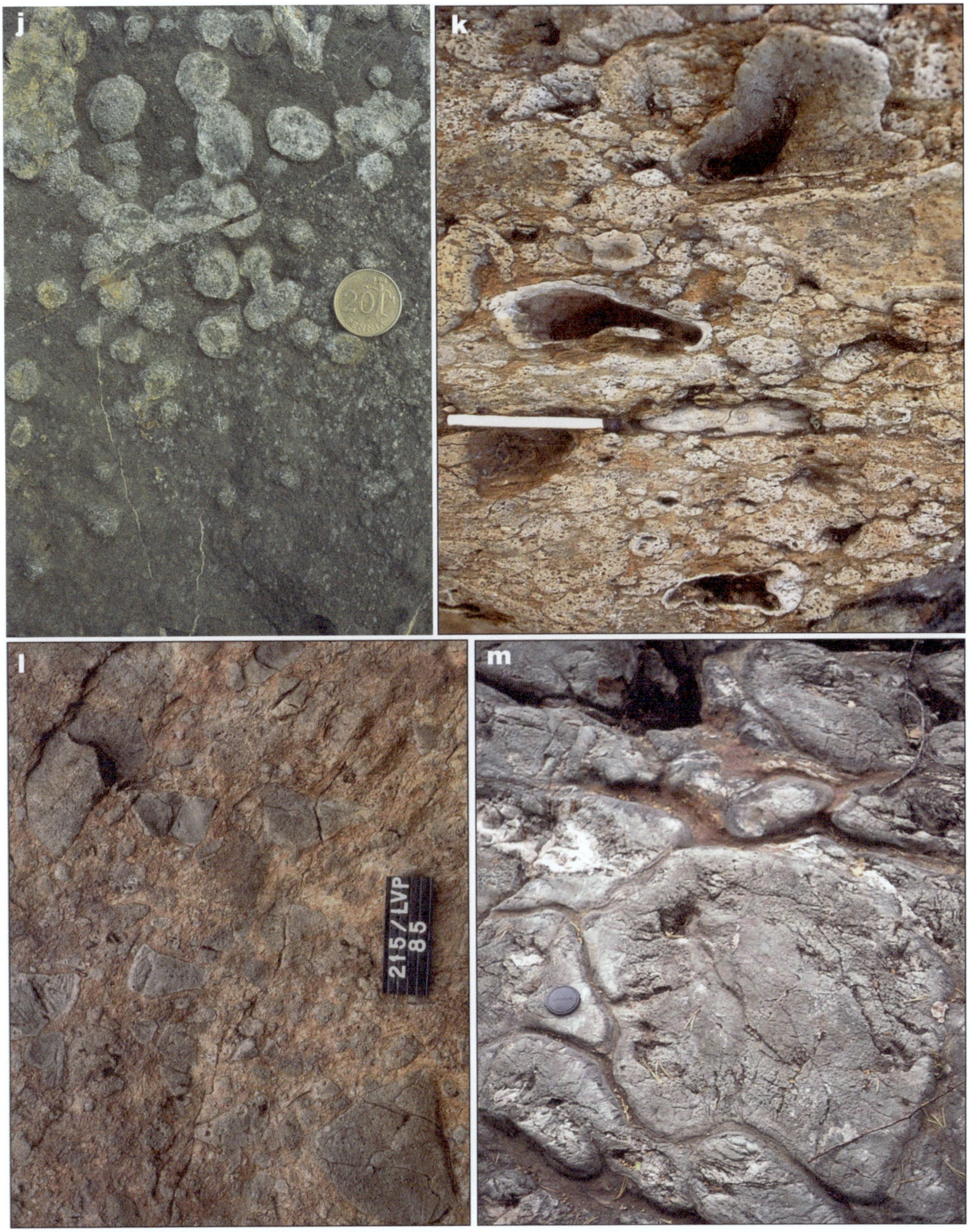

Fig. 3.19 (continued) *Linkupalo Formation, Central Lapland Belt*: (**j**) Variolitic mafic lava; coin diameter is 2 cm. *Sattasvaara Formation, Central Lapland Belt*: (**k**) Zoned fragments of komatiitic lava in a hyaloclastic ultramafic tuff matrix; matchstick length is 4 cm. (**l**) Fragmental komatiitic lava; tag length is 10 cm. *Linkupalo Formation, Central Lapland Belt*: (**m**) Tholeiitic pillow lava; lens cap diameter is 6 cm

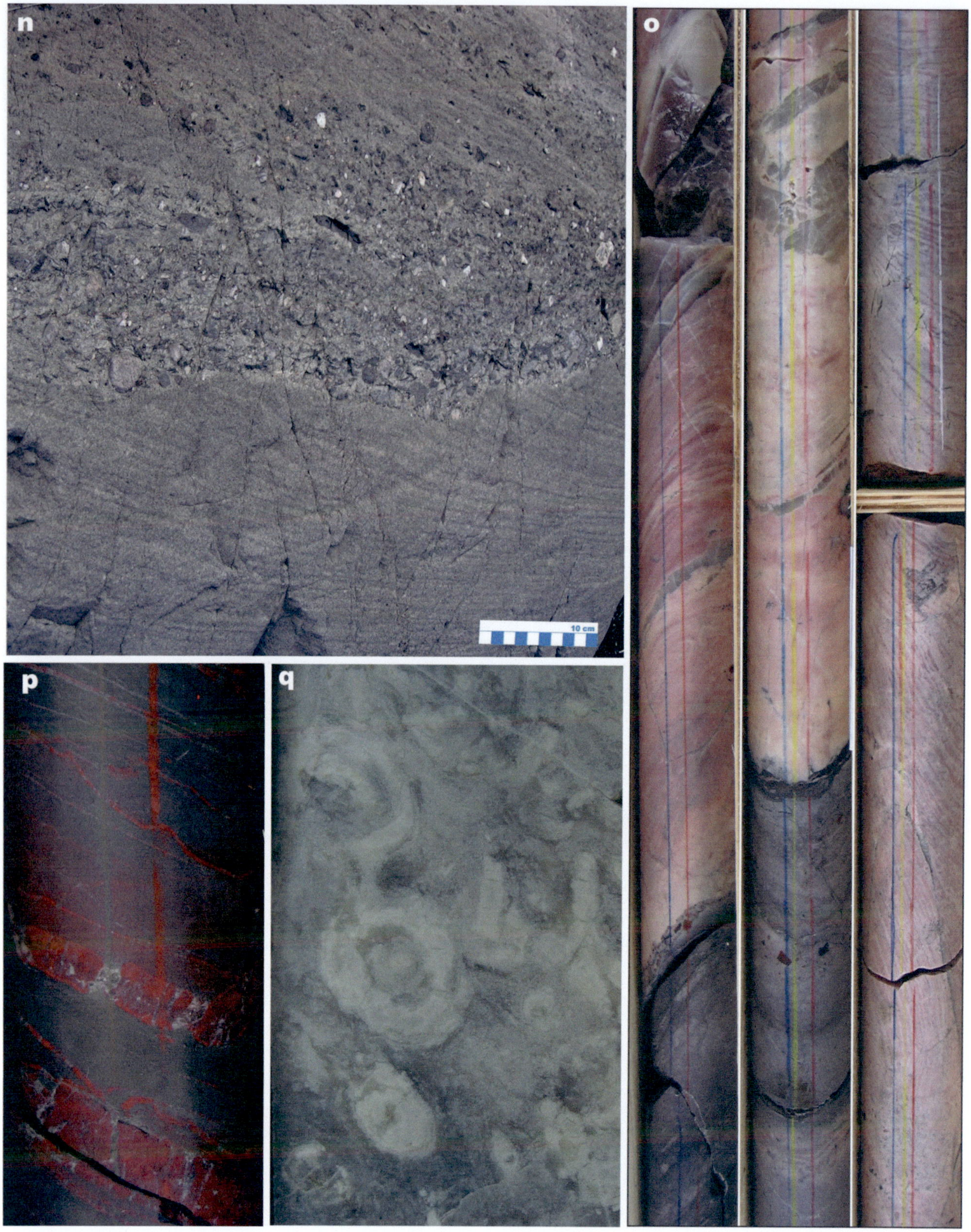

Fig. 3.19 (continued) *Kolosjoki Sedimentary Formation, Pechenga Belt:* (**n**) Cross-bedded volcaniclastic greywacke with a channel filled with polymict volcaniclastic conglomerate grading into gritstone. (**o**) *Pale pink* dolostones interbedded with chert (the *upper* part of the *left* core) and greywacke; FAR-DEEP drillcore 8A. (**p**) Jasper layers and veinlets in haematite-rich greywacke; FAR-DEEP drillcore 8B. (**q**) Silicified (*dark grey*), spheroidal stromatolites (*white*); FAR-DEEP drillcore 8B. Core diameter in photographs **o–q** is 50 mm

Fig. 3.19 (continued) *Kolosjoki Volcanic Formation, Pechenga Belt:* (**r**) Tholeiitic pillow lava. (**s**) Tholeiitic flow breccia. *Pilgujärvi Sedimentary Formation, Pechenga Belt:* (**t**) C_{org}-rich greywacke with a calcite concretionary bed (*top*) and several pyrite layers

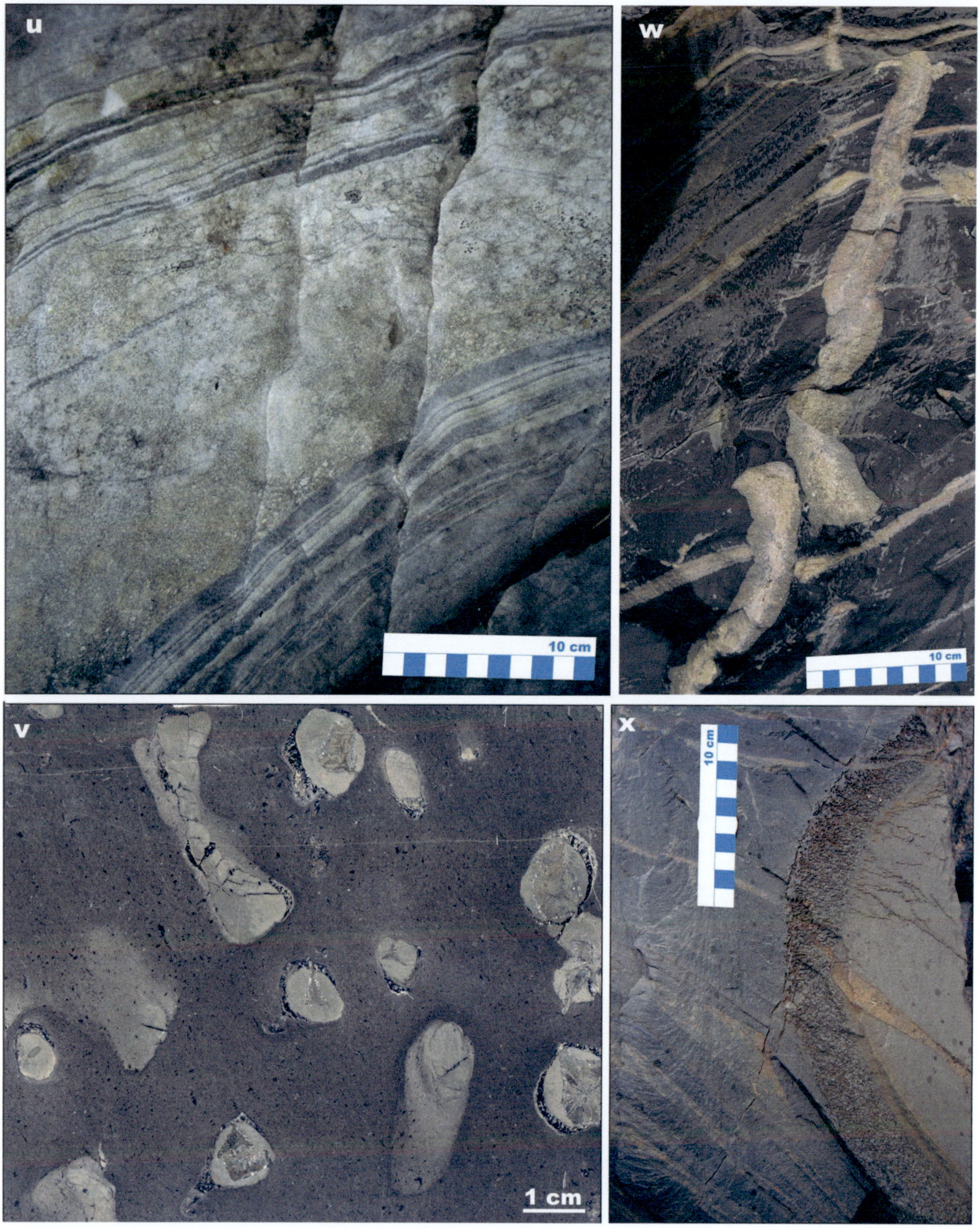

Fig. 3.19 (continued) *Pilgujärvi Sedimentary Formation, Pechenga Belt:* (**u**) Rhythmically bedded greywacke-mudstone (*black*) couplets overlain by gritstone that grades into greywacke, rhythmically bedded greywacke-mudstone and *black* mudstone. (**v**) Zoned, pyrite concretions overgrown by pyrrhotite; two spherical concretions in the centre have septarian cracks. (**w**) *Black* mudstone and rhythmically bedded mudstone-shale injected by pyrite- and pyrrhotite-rich sedimentary dyke, which experienced syndepositional fragmentation. (**x**) Large, zoned, spherical, calcite concretion in sulphide-bearing greywacke

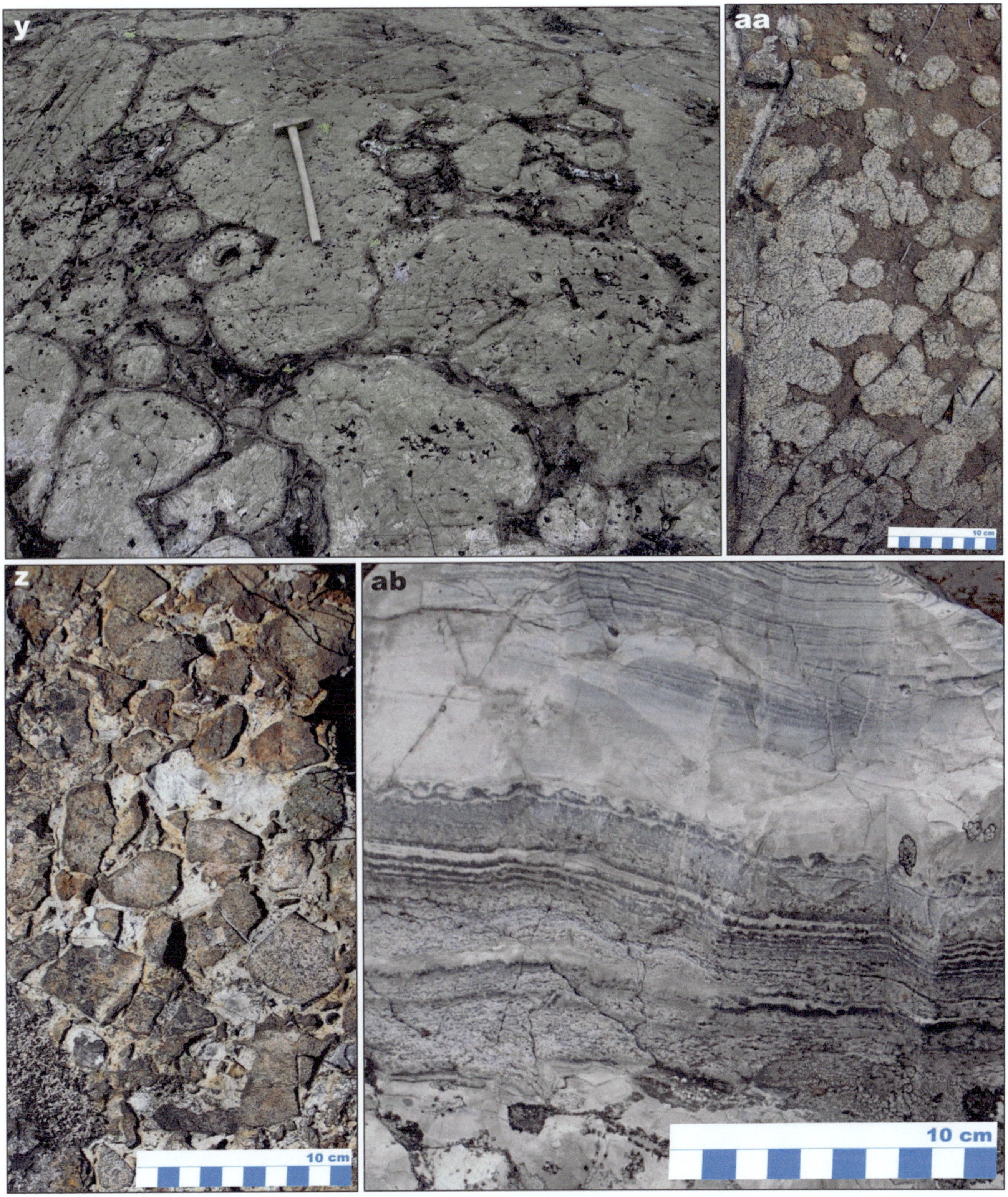

Fig. 3.19 (continued) *Pilgujärvi Volcanic Formation, Pechenga Belt:* (**y**) Pillowed tholeiitic lava with hyaloclastic material (*dark grey*) filling inter-pillow space. (**z**) Mafic lava fragments in a felsic tuff matrix. (**aa**) Felsic globules (ocelli) in ferropicritic lava. (**ab**) Laminated rhyolitic tuff forming a c. 50 cm bed between mafic lavas (Photographs by Victor Melezhik (**a–f, n–x**) and Eero Hanski (**y–ab**). Photograph courtesy of Sergey Svetov (**g**), Vesa Perttunen (**h, i**), Vesa Kortelainen (**j**), Jorma Räsänen (**l, k**) and Reino Kesola (**m**))

Fig. 3.20 Selected major sedimentary and volcanic rock types representing the 1960–1900 Ma Kalevian system. *Kondopoga Formation, Onega Basin*: (**a**) A quarry wall perpendicular to the bedding, exhibiting rhythmically bedded greywacke-mudstone-shale turbidites comprising bulk of the c. 700-m-thick Kondopoga Formation; width of the view is c. 2.5 m

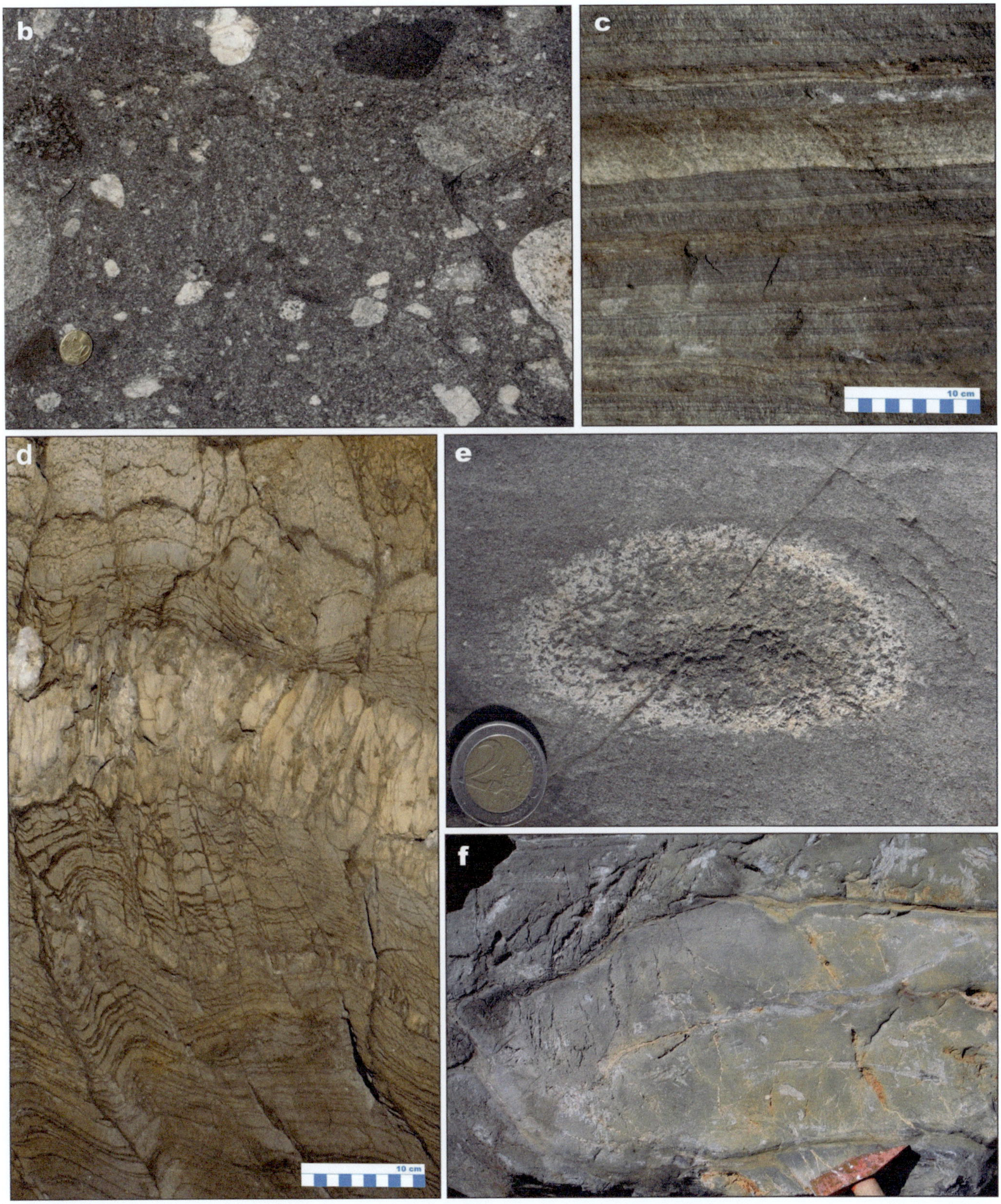

Fig. 3.20 (continued) *Utajärvi Conglomerate Formation, Kiiminki Belt*: (**b**) Basal conglomerate with unsorted, polymict clasts supported by a gritstone matrix. *Yli-Kiiminki Greywacke Formation, Kiiminki Belt*: (**c**) Mica schist retaining primary rhythmic bedding; note sandstone bed (*light grey*) grading into siltstone-shale. *Vepsä Formation, Kiiminki Belt*: (**d**) Rhythmically bedded dolostone-shale units with a thick brecciated dolostone bed in the middle. *Yli-Kiiminki Greywacke Formation, Kiiminki Belt*: (**e**) Former zoned calcite concretion replaced by calc-silicate minerals in mica schist. *Kiiminki Volcanic Formation, Kiiminki Belt*: (**f**) Tholeiitic pillow lava. Hammer head length is 15 cm

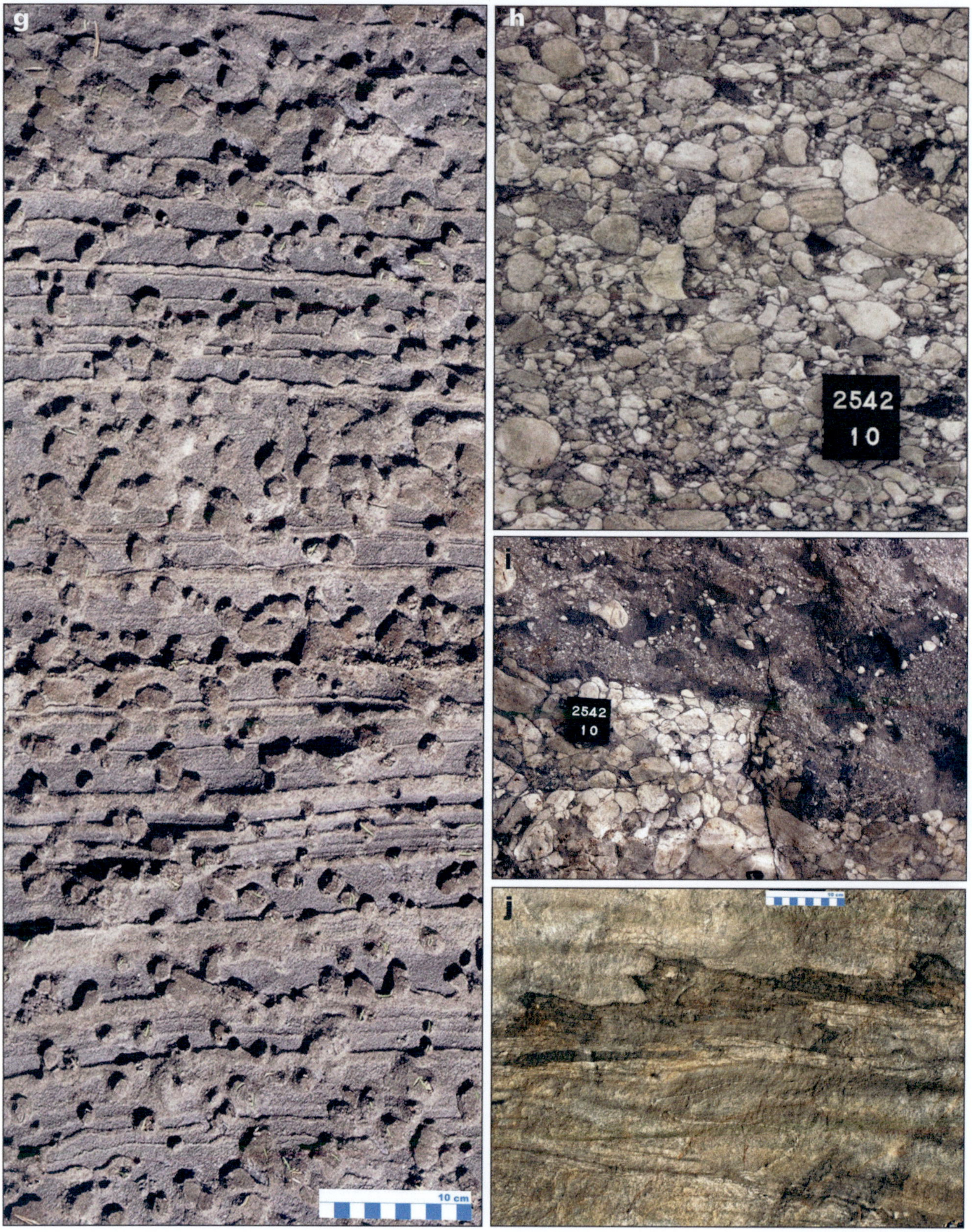

Fig. 3.20 (continued) *Mellajoki Suite, Peräpohja Belt:* (**g**) Bedded mica schist with weathered cordierite porphyroblasts. *Martimo Formation, Peräpohja Belt:* (**h**) Intraformational quartzite-pebble conglomerate occurring within a greywacke succession. (**i**) Quartzite-pebble conglomerate filling a channel in pebbly gritstone. *Nuasjärvi Group, Kainuu Belt:* (**j**) Turbiditic greywacke belonging to allochthonous Kaleva rocks

Fig. 3.20 (continued) *Kainuu Schist belt*: (**k**) Sheeted dyke complex, Jormua ophiolite. (**l**) Pillow lava, Jormua ophiolite

Fig. 3.20 (continued) *Kondopoga Formation, Onega Basin*: (**m**) Rhythmically-bedded, graded sandstone-siltstone-shale turbiditic units. (**n**) Fining-upward turbiditic succession; note partial cementation of a thick sandstone unit with dolomite (*pale brown*). (**o**) Slumping in turbiditic greywacke; coin diameter is 1 cm. (**p**) Diagenetic ankerite concretion (*pale brown*) in rhythmically bedded turbiditic greywacke-mudstone

Fig. 3.20 (continued) *Kondopoga Formation, Onega Basin*: (**q**) Cross-section view of rhythmically and graded-bedded sandstones interbedded with mass-flow deposit containing large, unsorted, angular fragments of laminated sandstone-siltstone

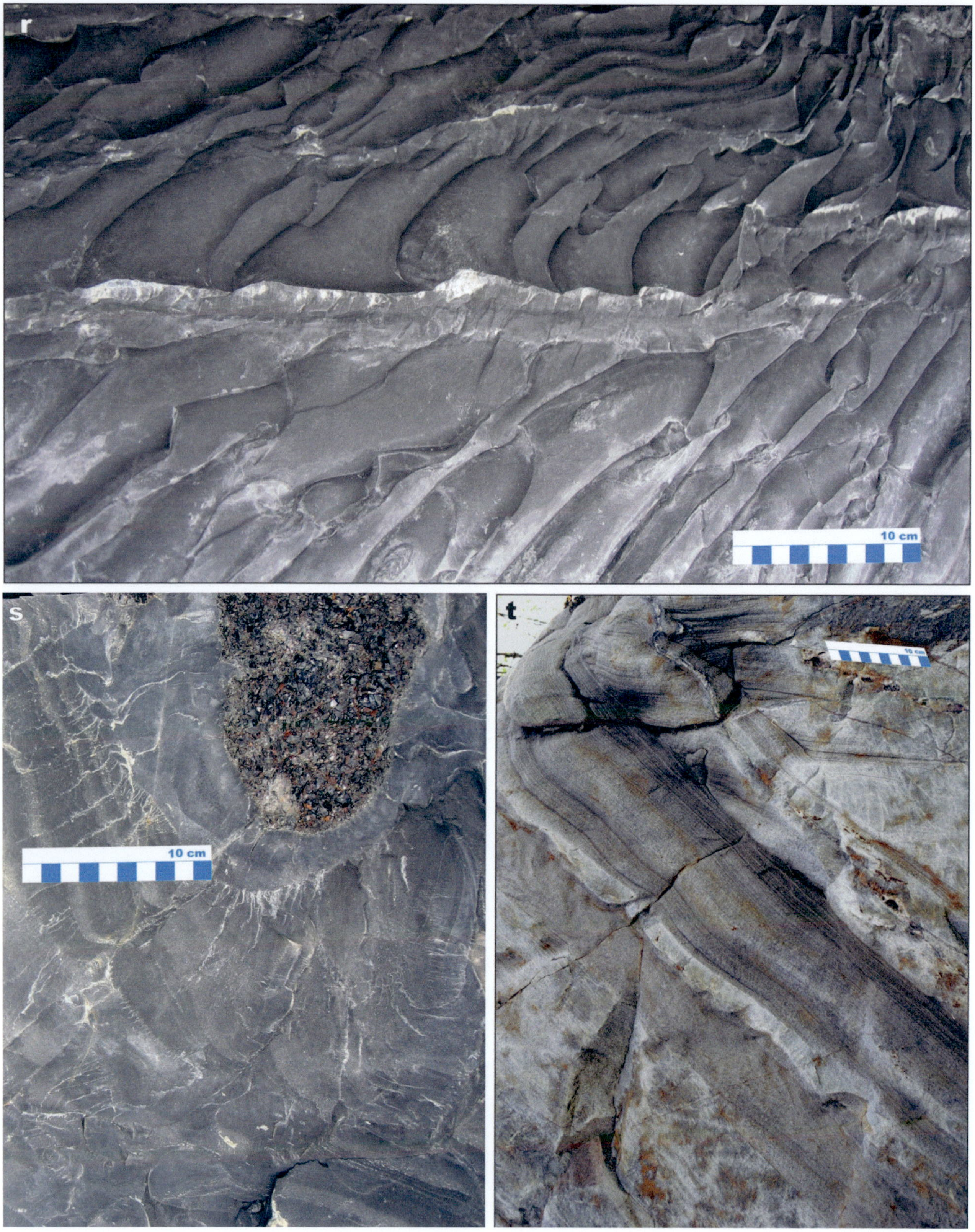

Fig. 3.20 (continued) *Kondopoga Formation, Onega Basin*: (**r**) Small-scale slumping creating a rippled surface; note that some ripples have subvertical or even overturned slopes. (**s**) Clast of pyrobitumen in turbiditic greywacke. *Ladoga Group, Ladoga Lake area:* (**t**) Turbiditic rhythmites preserved in amphibolite-facies crystalline schists

Fig. 3.20 (continued) *Ladoga Group, Ladoga Lake area:* (**u**) Paragneiss-hosted, calcite-diopside-forsterite (originally calcitic) concretion surviving a high-temperature amphibolite-facies metamorphism accompanied by partial melting (Photographs by Victor Melezhik (**a, k, m–u**) and Eero Hanski (**b–f, j, l**). Phographs (**g–i**) courtesy of Vesa Perttunen)

Fig. 3.21 Selected images of post-Karelian sedimentary rocks. *Pyhätunturi Formation, Central Lapland Belt*: (**a**) Cosets of small-scale trough cross-beds in arkosic sandstone with layers rich in hematite and other heavy minerals. *Levi Formation, Central Lapland Belt*: (**b**) Haematite-stained, fragment-supported conglomerate with polymict clasts. Sample (**b**) courtesy of the Geological Museum of the Department of Geosciences, University of Oulu. *Kumputunturi Formation, Central Lapland Belt*: (**c**) Polymict, clast-supported conglomerate. *Shoksha Formation, western Onega Lake area:* (**d**) Subaerial, *pink* quartz sandstone that constitutes most of the Shoksha Formation (Photographs (**a–c**) by Eero Hanski and n(**d**) Victor Melezhik)

References

Alapieti T (1982) The Koillismaa layered igneous complex, Finland: its structure, mineralogy and geochemistry, with emphasis on the distribution of chromium. Bull Geol Surv Finl 319:116

Alapieti T, Filén B, Lahtinen J, Lavrov M, Smol'kin V, Voitsekhovsky S (1990) Early Proterozoic layered intrusions in the northeastern part of the Fennoscandian Shield. Mineral Petrol 42:1–22

Alexejev N, Zinger T, Belyatsky B, Balagansky BB (2000) Age of crystallization and metamorphism of the Pestovoy gabbro-anorthosites, northern Karelia, Russia. In: Proceedings of the 5th SVEKALAPKO workshop, Institute of Geosciences and Astronomy, University of Oulu, Lammi, p 3

Amelin YuV, Heaman LM, Semenov VS (1995) U-Pb geochronology of layered mafic intrusions in the eastern Baltic Shield: implications for the timing and duration of Palaeoproterozoic continental rifting. Precambrian Res 75:31–46

Bailly L, Augé T, Cocherie A, Trofimov NN, Golubev AI, Tkachev AV, Cherkasov SV (2009) New data on the age of the Burakovsky layered intrusion, Karelia. Doklady Earth Sci 426:534–538

Balagansky VV, Timmerman MJ, Kozlova NYe, Kislitsyn RV (2001) A 2.44 Ga syn-tectonic mafic dyke swarm in the Kolvitsa Belt, Kola Peninsula, Russia: implications for early Palaeoproterozoic tectonics in the north-eastern Fennoscandian Shield. Precambrian Res 105:269–287

Balashov Yu, Bayanova T, Mitrofanov F (1993) Isotope data on the age and genesis of layered basic-ultrabasic intrusions in the Kola Peninsula and northern Karelia, northeastern Baltic Shield. Precambrian Res 6:197–205

Barkov AY (1991) A U-Pb dating of zircons from the Kivakka layered intrusion. North Karelia, All-Union Seminar, St. Petersburg, pp 21–23

Barkov AY (1992) Aspects of Petrology and Geochemistry of the Lukkulaisvaara and Kivakka Layered Intrusions, Northern Karelia, Ph.D. thesis, Russian Academy of Sciences, IGEM, Moscow (in Russian)

Barkov AY, Martin RF, Tarkian M, Poirier G, Thibault Y (2001) Pd-Ag tellurides from a Cl-rich environment in the Lukkulaisvaara layered intrusion, northern Russian Karelia. Can Mineralog 39:639–653

Barnes S-J, Often M (1990) Ti-rich komatiites from northern Norway. Contrib Miner Petrol 105:42–54

Bayanova TB (2004) Age of reference geological complexes of the Kola region and the duration of igneous processes. Nauka (Science), St. Petersburg, p 174 (in Russian)

Bayanova TB, Mitrofanov FP (2005) Layered Proterozoic PGE intrusions in Kola region: new isotope data. In: Törmänen TO, Alapieti T (eds) 10th international platinum symposium, 8–11 Aug 2005, Oulu, Extended Abstracts, Geological Survey of Finland, pp 289–291

Bayanova T, Smolkin V, Levkovich N (1999) U-Pb geochronological study of Mount Generalskaya layered intrusion, northwestern Kola Peninsula, Russia. Inst Mining Metall Trans Sect B: Appl Earth Sci 108:B83–B90

Bayanova T, Ludden J, Mitrofanov F (2009) Timing and duration of Palaeoproterozoic events producing ore-bearing layered intrusions of the Baltic Shield: metallogenic, petrological and geodynamic implications. Geol Soc Lond, Spec Pub 323:165–198

Bayanova TB, Nerovich LI, Mitrofanov FP, Zhavkov VA, Serov PA (2010) The Monchetundra basic massif of the Kola Region: new geological and isotope geochronological data. Doklady Earth Sci 431:288–293

Bekasova NB, Dudkin OB (1982) Composition and nature of concretionary phosphorites from the early Precambrian of Pechenga (Kola Peninsula). Lithol Miner Res 16:625–630 (in Russian)

Bibikova EV, Skiöld T, Bogdanova SV, Drugova GM, Lobach-Zhuchenko SB (1993) Geochronology of the Belomorides: an interpretation of the multistage geological history. Geochem Int 10:1393–1412

Bleeker W, Hamilton RE, Ernst RE, Kulikov VS (2008) The search for Archean–Paleoproterozoic supercratons: new constraints on Superior–Karelia–Kola correlations within supercraton Superia, including the first ca. 2504 Ma (Mistassini) ages from Karelia. In: 33rd international geological congress, Oslo, 6–14 Aug 2008, Abstracts www.33igc.org. Accessed May 2010

Bogdanova SV, Bibikova EV (1993) The "Saamian" of the Belomorian Mobile Belt: new geochronological constraints. Precambrian Res 64:131–152

Bogdanova MN, Yefimov MM, Sorokhtin NO, Balashov YuA, Gannibal LF, Ryungenen GI (1993) Evolution of polymetamorphism in the Granulite Belt of the Kola Peninsula (Kolvitsa zone) and U–Pb dating of diaphthoresis of an anorthosite association. Dokl Akad Nauk 331:332–334 (in Russian)

Buiko AK, Levchenkov OA, Turchenko SI, Drubetskoi YeE (1995) Geology and isotopic dating of the early Proterozoic Sumian-Sariolan Complex in northern Karelia (Paanajärvi-Tsipringa structure). Stratigr Geol Correlat 3(4):16–30

Cawthorn RG, Walraven F (1998) Emplacement and crystallization time for the Bushveld Complex. J Petrol 39:1669–1687

Chashchin VV, Bayanova TB, Levkovich NV (2008) Volcanoplutonic association of the early-stage evolution of the Imandra–Varzuga rift zone, Kola Peninsula, Russia: geological, petrogeochemical, and isotope-geochronological data. Petrology 16:279–298

Chistyakov AV, Sharkov EV (2008) Petrology of the Early Proterozoic Burakovka complex, southern Karelia. Petrology 16(1):66–91 (in Russian)

Claesson S, Huhma H, Kinny PD, Williams IS (1993) Svecofennian detrital zircon ages – implications for the Precambrian evolution of the Baltic Shield. Precambrian Res 64:109–130

Cloud P (1987a) Oasis in space: Earth history from the beginning. W. W. Norton and Co., New York, p 515

Cloud P (1987b) Trends, transitions, and events in Cryptozoic history and their calibration: apropos recommendations by the Subcommission on Precambrian Stratigraphy. Precambrian Res 37:257–264

Condie KC, O'Neill C, Aster RC (2009) Evidence and implications for a widespread magmatic shutdown for 250 My on Earth. Earth Planet Sci Lett 282:294–298

Efimov AA, Kaulina TB (1997) Geological features and U–Pb ddating (first data) of the southeastern part of the Kovdozero basic-ultrabasic massif (Puakhta Block). In: Proceedings of the conference on the Belomorian Mobile Belt, (Geology, Geodynamics, Geochronology), Petrozavodsk, KarNTs RAN, Petrozavodsk, p 31 (in Russian)

Eriksson P, Altermann W, Nelson D, Mueller W, Catuneanu O (eds) (2004) The Precambrian Earth: tempos and events. Elsevier, Amsterdam, p 941

Eskola PE (1919) Hufvuddragen av Onega-Karelens geologi. Meddel. geol. fören. i Helsingfors 1917–1918:13–18

Eskola P (1925) On the petrology of Eastern Fennoscandia. 1. the mineral development of basic rocks in the Karelian formations. Fennia 45(19):93

Fallick AE, Melezhik VA, Simonson B (2008) The ancient anoxic biosphere was not as we know it. In: Dobretsov N, Kolchanov N, Rozanov A, Zavarzin G (eds) Biosphere origin and evolution. Springer, New York, pp 169–188

Fedotov ZhA (1985) Evolution of the Proterozoic volcanism in the Western part of the Pechenga-Varzuga Belt (Petrogeochemical Aspects). Kola Science Centre, Apatity, p 120 (in Russian)

Frisch T, Jackson GD, Glebovitsky VA, Yefimov MM, Bogdanova MN, Parrish RR (1995) U-Pb geochronology of zircon from the Kolvitsa gabbro-anorthosite complex, southern Kola Peninsula, Russia. Petrology 3:248–254 (in Russian)

Galdobina LP (1987a) The Ludicovian Superhorizon. In: Sokolov VA (ed) Geology of Karelia. Nauka (Science), Leningrad, pp 59–67 (in Russian)
Galdobina LP (1987b) The Livian Superhorizon. In: Sokolov VA (ed) Geology of Karelia. Nauka (Science), Leningrad, pp 67–71 (in Russian)
Galdobina LP (1987c) The Vepsian Superhorizon. In: Sokolov VA (ed) Geology of Karelia. Nauka (Science), Leningrad, pp 71–74 (in Russian)
Galdobina LP, Melezhik VA (1986) Stratigraphy of the Ludicovian in the eastern part of the Baltic Shield. In: Sokolov VA, Heiskanen KI (eds) Early Proterozoic of the Baltic Shield, Proceedings of the Finnish-Soviet symposium, Petrozavodsk 19–27 Aug 1985, Karelian Science Centre of the USSR Academy of Sciences, Petrozavodsk, pp 226–235
Gaskelberg LA, Robonen WI, Zagorodny VG (1986) Stratigraphy of the Sumian and Sariolian in the Karelia-Kola region. In: Sokolov VA, Heiskanen KI (eds) Early Proterozoic of the Baltic Shield, Proceedings of the Finnish-Soviet symposium, Petrozavodsk 19–27 Aug 1985, Karelian Science Centre of the USSR Academy of Sciences, Petrozavodsk, pp 206–215
Gehör S, Havola M (1988) The depositional environment of the early Proterozoic Tuomivaara iron-formation and associated metasediments, eastern Finland. Geol Surv Finl Spec Pap 5:109–133
Glebovitsky VA, Semenov VS, Belyatsky BV (2001) The structure of the Lukkulaisvaara intrusions, Oulanga Group, northern Karelia: Petrological implications. Can Mineralog 39:607–639
Golubev AI, Ahmedov AM, Galdobina LP (1984) Geochemistry of lower Proterozoic "Black Shale" complexes of Karelia-Kola region. Nauka (Science), Leningrad, p 193 (in Russian)
Golubev AI, Svetov SA, Svetova AI (2002) Sumian (2.55–2.40 Ga) Basaltic Andesite associations of central Karelia, Online journal. Investigated in Russia. pp 903–913 (in Russian) http://zhurnal.ape.relarn.ru/articles/2002/081.pdf
Grooshan NYu, Nitkina NA (2008) Formation of the platinum-bearing mafic-ultramafic Fedorova Tundra massif: new geological and U-Pb isotopic data. In: Mitrofanov FP, Bayanova TB, Voitekhovsky YuL, Pripachkin VA (eds) Geology and ecogeology, investigations of young researchers. Institute of Geology, Kola Science Centre, Russian Academy of Sciences, Apatity, pp 10–17 (in Russian)
Guo Q, Strauss H, Kaufman AJ, Schröder S, Gutzmer J, Wing B, Baker MA, Bekker A, Jin Q, Kim S-T, Farquhar J (2009) Reconstructing Earth's surface oxidation across the Archean-Proterozoic transition. Geology 37:399–402
Hannah JL, Bekker A, Stein HJ, Markey RJ, Holland HD (2004) Primitive Os and 2316 Ma age for marine shale: implications for Paleoproterozoic glacial events and the rise of atmospheric oxygen. Earth Planet Sci Lett 225:43–52
Hanski EJ (1992) Petrology of the Pechenga ferropicrites and cogenetic, Ni-bearing gabbro-wehrlite intrusions, Kola Peninsula, Russia. Bull Geol Surv Finl 367:192
Hanski EJ (1997) The Nuttio serpentinite belt, Central Lapland: an example of Paleoproterozoic ophiolitic mantle rocks in Finland. Ofioliti 22:35–46
Hanski EJ (2001) History of stratigraphic research in northern Finland. Geol Surv Finl Spec Pap 33:15–43
Hanski E, Huhma H (2005) Central Lapland greenstone belt. In: Lehtinen M, Nurmi PA, Rämö OT (eds) Precambrian geology of Finland – key to the evolution of the Fennoscandian Shield. Elsevier, Amsterdam, pp 139–194
Hanski E, Huhma H, Rastas P, Kamenetsky VS (2001a) The Palaeoproterozoic komatiite-picrite association of Finnish Lapland. J Petrol 42:855–876
Hanski E, Walker RJ, Huhma H, Suominen I (2001b) The Os and Nd isotopic systematics of the 2.44 Ga Akanvaara and Koitelainen mafic layered intrusions in northern Finland. Precambrian Res 109:73–102
Hanski E, Huhma H, Vaasjoki M (2001c) Geochronology of northern Finland: a summary and discussion. Geol Surv Finl Spec Pap 33:255–279
Hanski E, Huhma H, Vuollo J (2010) SIMS zircon ages and Nd isotope systematics of the 2.2 Ga mafic intrusions in northern and eastern Finland. Bull Geol Soc Finl 82:31–62
Hedberg HD (ed) (1976) International Stratigraphic guide, international Subcommission on Stratigraphic Classification (ISSG). John Wiley & Sons, New York, p 200
Heiskanen KI (1987a) The Sumi superhorizon. In: Sokolov VA (ed) Geology of Karelia. Nauka (Science), Leningrad, pp 40–44 (in Russian)
Heiskanen KI (1987b) The Sarioli superhorizon. In: Sokolov VA (ed) Geology of Karelia. Nauka (Science), Leningrad, pp 44–51 (in Russian)
Honkamo M (1985) On the Proterozoic metasedimentary rocks of the Northern Pohjanmaa schist area, Finland. Bull Geol Surv Finl 331:117–129
Huhma H (1986) Sm-Nd, U-Pb and Pb-Pb isotopic evidence for the origin of the Early Proterozoic Svecokarelian crust in Finland. Bull Geol Surv Finl 337:48
Huhma H, Cliff RA, Perttunen V, Sakko M (1990) Sm-Nd and Pb isotopic study of mafic rocks associated with early Proterozoic continental rifting: the Peräpohja schist belt in northern Finland. Contrib Mineral Petrol 104:369–379
Karhu JA (1993) Paleoproterozoic evolution of the carbon isotope ratios of sedimentary carbonates in the Fennoscandian Shield. Bull Geol Surv Finl 371:87
Karhu JA (2005) Paleoproterozoic carbon isotope excursion. In: Lehtinen M, Nurmi PA, Rämö OT (eds) Precambrian geology of Finland – key to the evolution of the Fennoscandian Shield. Elsevier, Amsterdam, pp 669–680
Karhu JA, Holland HD (1996) Carbon isotopes and the rise of atmospheric oxygen. Geology 24:867–879
Karhu J, Kortelainen N, Huhma H, Perttunen V, Sergeev S (2007) New time constraints for the end of the Paleoproteroozoic carbon isotope excursion. In: 7th international symposium on applied isotope geochemistry, Stellenbosch, 10–14th Sept 2007, Abstracts, pp 76–77
Kaulina TV (1996) Results of U-Pb dating of rock from the Tolstik Peninsula, northwestern region of the White Sea. In: Golubev A (ed) Problems in the geology of the Karelia and Kola Regions. Academy of Science, Karelian Science Centre, Petrozavodsk, pp 64–71 (in Russian)
Kaulina TV, Bogdanova MN (2000) Major stages of the evolution of the northwestern Belomorian region (from U-Pb isotope data). Lithosphere 12:85–97 (in Russian)
Kohonen J (1995) From continental rifting to collisional crustal shortening – Paleoproterozoic Kaleva metasediments of the Höytiäinen area in North Karelia, Finland. Bull Geol Surv Finl 380:79
Kohonen J, Marmo J (1992) Proterozoic lithostratigraphy and sedimentation of Sariola- and Jatuli-type rocks in the Nunnanlahti-Koli-Kaltimo area, eastern Finland; implications for regional basin models. Bull Geol Surv Finl 364:67
Kontinen A (1986) Early Proterozoic stratigraphy and sedimentation in the Hyrynsalmi area, eastern Finland. In: Sokolov VA, Heiskanen KI (eds) Early Proterozoic of the Baltic Shield, Proceedings of the Finnish – Soviet symposium, Petrozavodsk 19–27 Aug 1985, Karelian Science Centre of the USSR Academy of Sciences, Petrozavodsk, pp 75–103
Kontinen A (1987) An early Proterozoic ophiolite – the Jormua mafic–ultramafic complex, northeastern Finland. Precambrian Res 35:313–341
Kozlova NE, Balagansky VV (1995) Structural evolution of the Southern zone rocks. In: Mitrofanov FP, Smol'kin VF (eds) Magmatism, Sedimentogenes and Geodynamics of the Pechenga Palaeoriftogenic Structure. Kola Science Centre, Apatity, pp 151–163 (in Russian)

Krivolutskaya NA, Belyatskii BV, Smol'kin VF, Mamontov VP, Fanygin AS, Svirskaya NM (2010) Geochemical features of the Drusite Massifs, the central part of the Belomorian Mobile Belt: II. Sm–Nd isotopic system of the rocks and the U–Pb isotopic system of zircons. Geochem Int 48:1064–1083

Kulikov VS (1988) High-magnesian volcanism in the early Proterozoic. In: Bogatikov OA (ed) Komatiitic and high-magnesian volcanics in the early Precambrian of the Baltic Shield. Nauka (Science), Leningrad, pp 20–88 (in Russian)

Kulikov VS, Bychkova YaV, Kulikova VV, Koptev-Dvornikov EV, Zudin AI (2005) Role of deep-seated differentiation in formation of Paleoproterozoic Sinegorie Lava Plateau of Komatiite Basalts, Southeastern Fennoscandia. Petrology 13:469–489

Kulikov VS, Bychkova YaV, Kulikova VA, Ernst R (2010) The Vetreny Poyas (Windy Belt) subprovince of southeastern Fennoscandia; An essential component of the ca. 2.5–2.4 Ga Sumian large igneous province. Precambrian Res 183:589–601

Kullerud K, Skjerlie KP, Corfu F, de la Rosa JD (2006) The 2.40 Ga Ringvassøy mafic dykes, West Troms Basement Complex, Norway: the concluding act of early Palaeoproterozoic continental breakup. Precambrian Res 150:183–200

Laajoki K (1986) The Precambrian supracrustal rocks of Finland and their tectono-ecogenic evolution. Precambrian Res 33:67–85

Laajoki K (1988) Nenäkangas-type conglomerates – evidence of a major break in the sedimentation of the early Proterozoic (Karelia) quartzites in Kainuu, northern Finland. Geol Surv Finl, Spec Pap 5:91–107

Laajoki K (1990) Early Proterozoic tectofacies in eastern and northern Finland. In: Naqvi SM (ed) Precambrian continental crust and its economic resources, vol 8, Developments in Precambrian geology. Elsevier, Amsterdam, pp 437–452

Laajoki K (1991) Stratigraphy of the northern end of the early Proterozoic (Karelian) Kainuu Schist Belt and associated gneiss complexes. Bull Geol Surv Finl 358:105

Laajoki K (2005) Karelian supracrustal rocks. In: Lehtinen M, Nurmi PA, Rämö OT (eds) Precambrian geology of Finland – key to the evolution of the Fennoscandian Shield. Elsevier, Amsterdam, pp 279–342

Laajoki K, Saikkonen R (1977) On the geology and geochemistry of the Precambrian iron formations in Väyrylänkylä, South Puolanka area, Finland. Bull Geol Surv Finl 292:137

Lahtinen R, Huhma H, Kontinen A, Kohonen J, Sorjonen-Ward P (2010) New constraints for the source characteristics, deposition and age of the 2.1–1.9 Ga metasedimentary cover at the western margin of the Karelian Province. Precambrian Res 176:77–93

Lauri LS, Mänttäri I (2002) The Kynsijärvi quartz alkali feldspar syenite, Koillismaa, eastern Finland – silicic magmatism associated with 2.44 Ga continental rifting. Precambrian Res 119:121–140

Lehtonen M, Airo M-L, Eilu P, Hanski E, Kortelainen V, Lanne E, Manninen T, Rastas P, Räsänen J, Virransalo P (1998) The stratigraphy, petrology and geochemistry of the Kittilä greenstone area, northern Finland. A report of the Lapland Volcanite project, Geol Surv Finl, Rep Inv 140: 144 (in Finnish with English summary)

Levchenkov OA, Nikolaev AA, Bogomolov ES, Yakovleva SZ (1994) U-Pb ages of Sumian acid magmatic rocks in northern Karelia. Stratigr Geol Correlat 2(1):3–9 (in Russian)

Lobach-Zhuchenko SB, Bibikova EV, Drugova GM, Belyatskii BV, Gracheva TV, Amelin YuV, Matrenichev VA (1993) Geochronology and petrology of the Tupaya Guba Igneous Complex in Northwestern White Sea Region. Petrology 1(6):657–677

Lobach-Zhuchenko SB, Arestova NA, Chekulaev VP, Levsky LK, Bogomolov ES, Krylov IN (1998) Geochemistry and petrology of 2.40–2.45 Ga magmatic rocks in the north-western Belomorian Belt Fennoscandian Shield, Russia. Precambrian Res 92:223–250

Lukkarinen H (2008) Pre-quaternary rocks of the Siilinjärvi and Kuopio map-sheet areas, Explanation to the maps of pre-quaternary rocks, sheets 3331 and 3242, Geological Survey of Finland, Espoo, p 228 (in Finnish with English summary)

Luukkonen E (1988) Pre-quaternary rocks of the Moisiovaara and Ala-Vuokki map-sheet areas, Explanation to the geological map of Finland 1:100 000, Pre-quaternary rocks, sheets 4421 and 4423 + 4441, Geological Survey of Finland, p 90 (in Finnish with English summary)

Manninen T (1991) Volcanic rocks in the Salla area, northeastern Finland. A Report of the Lapland Volcanite Project, Geol Surv Finl, Rep Investig 104, p 97 (In Finnish with English summary)

Manninen T, Huhma H (2001) A new U-Pb zircon age from the Salla schist belt, northern Finland. In: Vaasjoki M (ed) Radiometric age determinations from Finnish Lapland and their bearing on the timing of Precambrian volcano-sedimentary sequences, Geol Surv Finl, Spec Pap 33: 201–208

Manninen T, Pihlaja P, Huhma H (2001) U-Pb geochronology of the Peurasuvanto area, northern Finland. In: Vaasjoki M (ed) Radiometric age determinations from Finnish Lapland and their bearing on the timing of Precambrian volcano-sedimentary sequences. Geol Surv Finl, Spec Pap 33: 189–200

Marmo J (1986) Sariolan stratigraphy and sedimentation in the Koli-Kaltimo area, North Karelia, eastern Finland. In: Sokolov VA, Heiskanen KI (eds), Early Proterozoic of the Baltic Shield, Proceedings of the Finnish – Soviet symposium, Petrozavodsk 19–27 Aug 1985, Karelian Science Centre of the USSR Academy of Sciences, Petrozavodsk, pp 149–190

Marmo J (1992) The lower Proterozoic Hokkalampi paleosol in North Karelia, eastern Finland area, northern Finland. In: Schildlowski M et al (eds) Early organic evolution: implications for mineral and energy deposits. Springer, Berlin, pp 41–66

Marmo JS, Ojakangas RW (1984) Lower Proterozoic glaciogenic deposits, eastern Finland. Geol Soc Am Bull 95:1055–1062

Martin AP, Condon DJ, Prave AR, Melezhik VA, Fallick A (2010) Constraining the termination of the Lomagundi-Jatuli positive isotope excursion in the Imandra-Varzuga segment (Kola Peninsula) of the North Transfennoscandian Greenstone Belt by high-precision ID-TIMS, AGU meeing, 13–17 Dec 2010, San Francisco, Abstract #U33A-0010

Matrenichev VA, Vrevsky AV, Sergeev SA, Matukov DA (2006) The Ludicovian–Kalevian boundary in the northern Ladoga region: geological relations and isotopic age. Doklady Earth Sci 407A:388–392

Melezhik VA (1992) Early Proterozoic sedimentary and rock-forming basins of the Baltic Shield. Nauka (Science), St. Petersburg, p 258 (in Russian)

Melezhik VA, Fallick AE (1996) A widespread positive $\delta^{13}C_{carb}$ anomaly at around 2.33–2.06 Ga on the Fennoscandian Shield: a paradox. Terra Nova 8:141–157

Melezhik VA, Predovsky AA (1978) Carbonate and silicate concretions in metamorphosed Precambrian formations of the Kola Peninsula. Trans USSR Acad Sci 239:668–871 (in Russian)

Melezhik VA, Predovsky AA (1982) Early Proterozoic sedimentary and rock-forming basins of the Baltic Shield. Nauka (Science), Leningrad, p 256 (in Russian)

Melezhik VA, Sturt B (1994) General geology and evolutionary history of the early Proterozoic Polmak-Pasvik-Pechenga-Imandra/Varzuga-Ust'Ponoy greenstone belt in the northeastern Baltic Shield. Earth-Sci Rev 36:205–241

Melezhik VA, Sturt BA (1998) The Palaeoproterozoic (2.5–1.7 Ga) Midcontinent rift system of the northeastern Fennoscandian shield *versus* the early Proterozoic Pechenga-Varzuga Belt: a case of back-arc spreading, Discussion. Can J Earth Sci 35:720–725

Melezhik VA, Borisov AE, Fedotov ZhA, Prdovsky AA (1982) The Varzuga Series. In: Gorbunov GI (ed) The Imandra/Varzuga Zone of Karelides. Nauka (Science), Leningrad, pp 57–85 (in Russian)

Melezhik VA, Basalaev AA, Predovsky AA, Balabonin NL, Bolotov VI, Pavlova MA, Gavrilenko BV, Abzalov MZ (1988) Carbonaceous rocks of the early stages of Earth evolution. Nauka (Science), Leningrad, p 200 (in Russian)

Melezhik VA, Fallick AE, Makarikhin VV, Lyubtsov VV (1997) Links between Palaeoproterozoic palaeogeography and rise and decline of stromatolites: Fennoscandian Shield. Precambrian Res 82:311–348

Melezhik VA, Grinenko LN, Fallick AE (1998) 2000 Ma sulphide concretions from the 'Productive' Formation of the Pechenga Greenstone Belt, NW Russia: genetic history based on morphological and isotopic evidence. Chem Geol 148:61–94

Melezhik VA, Fallick AE, Hanski EJ, Kump LR, Lepland A, Prave AR, Strauss H (2005) Emergence of the modern Earth system during the Archean-Proterozoic transition: challenges of future research. Geol Soc Am Today 15:4–11

Melezhik VA, Huhma H, Condon DJ, Fallick AE, Whitehouse MJ (2007) Temporal constraints on the Paleoproterozoic Lomagundi-Jatuli carbon isotopic event. Geology 35:655–658

Mellqvist C, Öhlander B, Skiöld T, Wikström A (1999) The Archaean–Proterozoic Palaeoboundary in the Luleå area, northern Sweden: field and isotope geochemical evidence for a sharp terrane boundary. Precambrian Res 96:225–243

Meriläinen K (1980) Stratigraphy of the Precambrian in Finland. Geologiska Föreningens i Stockholm Förhandlingar (GFF) 102:177–180

Mikkola P, Kontinen A, Huhma H, Lahaye Y (2010) Three Paleoproterozoic A-type granite intrusions and associated dykes from Kainuu, East Finland. Bull Geol Soc Finl 82:81–100

Mitrofanov F, Bayanova T (1999) Duration and timing of ore-bearing Paleoproterozoic intrusions of Kola province, In: Stanley CJ et al. (eds) Mineral deposits: processes to processing, Proceedings of the 5th Biennial SGA meeting and the 10th Quadrennial IAGOD symposium, vol 2, London, 22–25 Aug 1999, A. A. Balkema, Rotterdam, pp 1275–1278

Mitrofanov FP, Balagansky VV, Balashov YuA, Gannibal LF, Dokuchaeva VS, Nerovich LI, Radchenko MK, Ryungenen GI (1995) U-Pb age of gabbro-anorthosite massifs in the Lapland Granulite belt. Norges Geologiske Undersøkelse, Special Paper 7:179–183

Mokrousov VA, Smol'kin VF, Skuf'in PK (1995) Stratigraphy of the Karelian (Early Proterozoic) complex. In: Mitrofanov FP, Smol'kin VF (eds) Magmatism, Sedimentogenesis and Geodynamics of the Pechenga Palaeorift. Kola Science Centre, Apatity, pp 8–18 (in Russian)

Mutanen T, Huhma H (2001) U-Pb geochronology of the Koitelainen, Akanvaara and Keivitsa layered intrusions and related rocks. Geol Surv Finl, Spec Pap 33:229–246

Negrutsa TF, Negrutsa VZ (1981) Early Proterozoic Yanisjärvi tilloids, South Karelia, USSR. In: Hambrey MJ, Harland WB (eds) Pre-Pleistocene glacial record. Cambridge University Press, Cambridge, pp 681–682

Negrutsa VZ, Predovsky AA, Sokolov VA (1986) Stratigraphy of the Jatulian in the eastern part of the Baltic Shield. In: Sokolov VA, Heiskanen KI (eds) Early Proterozoic of the Baltic Shield, Proceedings of the Finnish-Soviet symposium, Petrozavodsk 19–27 Aug 1985, Karelian Science Centre of the USSR Academy of Sciences, Petrozavodsk, pp 216–225

Negrutza VZ (1984) Early Proterozoic stages of evolution of the Eastern part of the Baltic Shield. Nedra, Leningrad, p 270

Negrutza VZ, Heiskanen KI, Negrutza TF, Robonen VI (2000) The 3rd interdepartmental regional meeting on the Early Proterozoic of Karelia and the Kola Peninsula. Stratigr, Geol Correlat 8:108–112 (in Russian)

Nironen M (2005) Proterozoic orogenic granitoid rocks. In: Lehtinen M, Nurmi PA, Rämö OT (eds) Precambrian geology of Finland – key to the evolution of the Fennoscandian Shield. Elsevier B.V, Amsterdam, pp 443–480

Nironen M, Mänttäri I (2003) Structural evolution of the Vuotso area, Finnish Lapland. Bull Geol Soc Finl 75:93–101

Nisbet EG (1982) Definition of Archean – comment and a proposal on the recommendations by the Subcommission on Precambrian Stratigraphy. Precambrian Res 19:111–118

Nitkina EA (2006) U–Pb zircon dating of rocks of the platiniferous Fedorova–Pana Layered Massif, Kola Peninsula. Doklady Earth Sci 408:551–554

Nykänen O (1971) On the Karelides in the Tohmajärvi area, eastern Finland. Bull Geol Soc Finl 43:93–108

Ojakangas RW (1985) Evidence for early Proterozoic glaciations: the dropstone unit – diamictite association. Bull Geol Surv Finl 331:51–72

Ojakangas RW (1988) Glaciation: an uncommon "mega-event" as a key to intracontinental and intercontinental correlation of early Proterozoic basin fill, North America and Baltic Cratons. In: Kleinspehn KL, Paola C (eds) New perspectives in basin analysis. Springer, New York, pp 431–444

Ojakangas RW, Marmo JS, Heiskanen KI (2001) Basin evolution of the Paleoproterozoic Karelian Supergroup of the Fennoscandian (Baltic) Shield. Sediment Geol 141–142:255–285

Papineau D, Mojzsis SJ, Coath CD, Karhu JA, McKeegan KD (2005) Multiple sulfur isotopes of sulfides from sediments in the aftermath of Paleoproterozoic glaciations. Geochim Cosmochim Acta 69:5033–5060

Pekkarinen L (1979) The Karelian formations and their depositional basement in the Kiihtelysvaara-Tohmajärvi district, eastern Finland. Bull Geol Surv Finl 301:141

Pekkarinen LJ, Lukkarinen H (1991) Paleoproterozoic volcanism in the Kiihtelysvaara-Tohmajärvi district, eastern Finland. Bull Geol Surv Finl 357:30

Peltonen P (2005) Ophiolites. In: Lehtinen M, Nurmi PA, Rämö OT (eds) Precambrian geology of Finland – key to the evolution of the Fennoscandian Shield. Elsevier, Amsterdam, pp 237–278

Peltonen P, Kontinen A, Huhma H (1998) Petrogenesis of the mantle sequence of the Jormua ophiolite (Finland): melt migration in the upper mantle during Palaeoproterozoic continental break-up. J Petrol 39:297–329

Peltonen P, Kontinen A, Huhma H, Kuronen U (2008) Outokumpu revisited: new mineral deposit model for the mantle peridotite-associated Cu-Co-Zn-Ni-Ag-Au sulphide deposits. Ore Geol Rev 33:559–617

Perttunen V (1985) On the Proterozoic stratigraphy and exogenic evolution of the Peräpohja area, Finland. Bull Geol Surv Finl 380:131–141

Perttunen V, Hanski E (2003) Pre-quaternary rocks of the Koivu and Törmäsjärvi map-sheet areas, Explanation to the geological map of Finland 1:100 000, Pre-quaternary rocks, sheets 3631 and 2633, Geological survey of Finland, p 88 (in Finnish with English summary)

Perttunen V, Vaasjoki M (2001) U-Pb geochronology of the Peräpohja Schist Belt, northwestern Finland. Geol Surv Finl, Spec Pap 33:45–84

Plumb KA (1991) New Precambrian time scale. Episodes 14:139–140

Plumb KA, James HL (1986) Subdivision of Precambrian time: suggestions by the subcommission on Precambrian stratigraphy. Precambrian Res 32:65–92

Predovsky AA, Fedotov ZhA, Ahmedov AM (1974) Geochemistry of the Pechenga complex (Metamorphosed sedimentary and volcanic rocks). Nauka (Science), Leningrad, p 139 (in Russian)

Preiss WV (1982) Lett Austr Geol 42: 8–9

Puchtel IS, Hofmann AW, Mezger K, Schipansky AA, Kulikov VS, Kulikova VV (1996) Petrology of a 2.41 Ga remarkably fresh komatiitic basalt lava lake in Lion Hills, central Vetreny Belt, Baltic Shield. Contrib Mineral Petrol 124:273–290

Puchtel IS, Haase KM, Hofmann AW, Chauvel C, Kulikov VS, Garbe-Schönberg C-D, Nemchin AA (1997) Petrology and geochemistry of crustally contaminated komatiitic basalts from the Vetreny Belt, southeastern Baltic Shield: evidence for an early Proterozoic mantle plume beneath rifted Archean continental lithosphere. Geochim Cosmochim Acta 61:1205–1222

Puchtel IS, Arndt NT, Hofmann AW, Haase KM, Kröner A, Kulikov VS, Kulikova VV, Garbe-Schönberg C-D, Nemchin AA (1998) Petrology of mafic lavas within the Onega plateau, central Karelia: evidence for the 2.0 Ga plume-related continental crustal growth in the Baltic Shield. Contrib Mineral Petrol 130:134–153

Puchtel IS, Brügmann GE, Hofmann AW (1999) Precise Re-Os mineral isochron and Pb-Nd-Os isotope systematics of a mafic-ultramafic sill in the 2.0 Ga Onega plateau (Baltic Shield). Earth Planet Sci Lett 170:447–461

Puchtel IS, Brügmann GE, Hofmann AW, Kulikov VS, Kulikova VV (2001) Os isotope systematics of komatiitic basalts from the Vetreny belt, Baltic Shield: evidence for a chondritic source of the 2.45 Ga plume. Contrib Mineral Petrol 140:588–599

Räsänen J, Huhma H (2001) U-Pb dating in the Sodankylä schist area, Central Finnish Lapland. Geol Surv Finl, Spec Pap 33:153–180

Räsänen J, Hanski E, Lehtonen MI (1989) Komatiites, low-Ti basalts and andesites in the Möykkelmä area, Central Finnish Lapland. Geol Surv Finl Rep Invest 88

Rastas P, Huhma H, Hanski E, Lehtonen MI, Härkönen I, Kortelainen V, Mänttäri I, Paakkola J (2001) U-Pb isotopic studies on the Kittilä greenstone area, central Lapland, Finland. Geol Surv Finl, Spec Pap 33:95

Reuschel M, Strauss H, Lepland A, Melezhik VA, Cartigny P, Kaufmann AJ (2009) Multiple sulfur istotope measurements from the 2.4 Ga old Seidorechka formation. In: 19th annual VM Goldschmidt conference, Davos, 21–26 June 2009, Abstracts, p A1091

Semikhatov M, Shurkin K, Aksenov Y, Bekker Yu, Bibikova Y, Duk V, Yesipchuk K, Karsakov L, Kiselev V, Kozlov V, Lobach-Zhuchenko S, Negrutsa V, Robonen V, Sez'ko A, Filatova L, Khomentovskiy V, Shemyakin V, Shul'diner V (1991) A new stratigraphic scale for the Precambrian of the USSR. Int Geol Rev 33:413–422

Serov PA, Nitkina EA, Bayanova TB, Mitrofanov FP (2007) Comparison of new U–Pb and Sm–Nd isotope data on rocks of the early barren phase and basal ore-bearing rocks in the PGE-bearing Fedorovo–Pana Layered Massif, Kola Peninsula. Doklady Earth Sci 416:1125–1127

Sharkov EV, Krasivskaya IS, Chistyakov AV (2004) Belomorian drusite (coronite) complex, Baltic Shield, Russia: An example of dispersed intrusive magmatism in early Paleoproterozoic mobile zones. Russ J Earth Sci 6(3):185–215

Shul'diner VI, Levchenkov OA, Yakovleva SZ, Makeev AF, Komarov AN, Kononel'ko DL, Baltybaev ShK, Kozyreva IV (2000) The late Karelian in the stratigraphic scale of Russia: determination of its lower boundary and regional units in the stratotype area. Stratigr Geol Correlat 8:544–556

Silvennoinen A (1985) On the Proterozoic stratigraphy of Northern Finland. In: Laajoki K, Paakola J (eds) Proterozoic exogenic processes and related metallogeny. Bull Geol Surv Finl 331: 107–116

Silvennoinen A (1991) Pre-quaternary rocks of the Kuusamo and Rukatunturi map-sheet areas, Explanation to the maps of Pre-quaternary rocks, sheets 4542 + 4542 and 4613, Geological Survey of Finland, Espoo, p 228 (in Finnish with English summary)

Silvennoinen A (1992) General Geological setting and deep fracture structures in the Kuusamo-Kuolajärvi-Panajärvi area. Geol Surv Finl, Spec Pap 13:5–10

Silvennoinen A, Honkamo A, Juopperi H, Lehtonen M, Mielikäinen P, Perttunen V, Rastas P, Räsänen J, Väänänen J (1980) Main features of the stratigraphy of North Finland. In: Silvennoinen A (ed) Jatulian geology in the Eastern part of the Baltic Shield, Proceedings of a Finnish-Soviet symposium, Finland, 21–26 Aug 1979, Rovaniemi, The Committee for Scientific and Technical Co-operation between Finland and Soviet Union, pp 153–162

Simonen A (1986) Stratigraphic studies on the Precambrian in Finland. Bull Geol Surv Finl 336:21–37

Skuf'in PK, Bayanova TB (2006) Early Proterozoic central-type volcano in the Pechenga Structure and its relation to the ore-bearing gabbro-wehrlite complex of the Kola Peninsula. Petrology 14:609–627

Skuf'in PK, Bayanova TB, Mitrofanov FP (2006) Isotope age of subvolcanic graditoid rocks of the Early Proterozoic Panarechka volcanotectonic structure, Kola Peninsula. Proc Russ Acad Sci, Earth Sci 409:774–778

Slabunov AI, Larionov AN, Bibikova EV, Stepanov VS, Kirnozova TI (2001) Geology and geochronology of the Shobozero Massif of the lherzolite-gabbronorite complex in the Belomorian Mobile Belt. In: Geology and useful minerals of Karelia, Issue 3, Karelian Research Centre, Russian Academy of Sciences, Petrozavodsk, pp 3–14 (in Russian)

Smol'kin VF (1992) Early Precambrian Komatiitic and Picritic Magmatism of the Baltic Shield. Nauka (Science), St. Petersburg, p 278 (in Russian)

Smol'kin VF (2005) The Early Proterozoic. In: Glebovitsky VA (ed) The Early Precambrian of the Baltic Shield. Nauka (Science), St. Petersburg, pp 59–123 (in Russian)

Smol'kin VF, Dain AD (1985) Gabbro-wehrlite formation of the Imandra/Varzuga zone of Karelides. Sov Geol 12:94–105 (in Russian)

Smol'kin VF, Skuf'in PK (1995) Volcanic associations of the Northern and Southern zones. In: Mitrofanov FP, Smol'kin VF (eds) Magmatism, sedimentogenesis and geodynamics of the Pechenga Palaeoriftogenic structure. Kola Science Centre, Apatity, pp 37–81 (in Russian)

Smol'kin VF, Fedotov ZhA, Neradovsky YuN, Bayanova TB, Borisova VV, Glaznev VN, Dedyukhin AN, Orsoev DA, Ohnenstetter M, Ohnenstetter D, Raevsky AB, Tolstikhin IN, Chaschin VV, Mokrushin AV, Novikov DD, Ikorcky SV, Kamensky IL, Delenitzin AA (2004) Layered intrusions of the Monchegorsk Ore region: petrology, mineralization, isotopes, and deep structure. Kola Science Centre, Russian Academy of Sciences, Apatity, p 177 (in Russian)

Sokolov VA (1980) Jatulian formations of the Karelian ASSR, In: Silvennoinen A (ed) Jatulian geology in the Eastern part of the Baltic Shield, Proceedings of a Finnish-Soviet symposium, Finland, 21–26 Aug 1979, Rovaniemi, The Committee for Scientific and Technical Co-operation between Finland and Soviet Union, pp 163–194

Sokolov VA (1987) Jatuli Superhorizon. In: Sokolov VA (ed) Geology of Karelia. Nauka (Science), Leningrad, pp 51–59 (in Russian)

Stepanova A, Stepanov V (2010) Paleoproterozoic mafic dyke swarms of the Belomorian Province, eastern Fennoscandian Shield. Precambrian Res 183:602–616

Strand K (1988) Alluvial sedimentation and tectonic setting of the early Proterozoic Kurkikylä and Kainuu Groups in northern Finland. Geol Surv Finl, Spec Pap 5:75–90

Strand K (2005) Sequence stratigraphy of the siliciclastic East Puolanka Group, the Paleoproterozoic Kainuu Belt, Finland. Sediment Geol 176:149–166

Strand K, Laajoki K (1993) Palaeoproterozoic glaciomarine sedimentation in an extensional tectonic setting: the Honkala Formation, Finland. Precambrian Res 64:253–271

Strand K, Köykkä J, Kohonen J (eds) (2010) Guidelines and procedures for naming Precambrian geological units in Finland. Geol Surv Finl, Guide 55: 41

Trendall AF (1984) The Archean/Proterozoic transition as a geologic event – a view from Australian evidence. In: Holland HD, Trendall AF (eds) Dahlem Konferenzen 1984. Springer, Berlin, pp 243–259

Väyrynen H (1933) Über die Stratigraphie der karelischen Formationen. Bull Comm Géologique Finlande 101:54–78

Vetrin VR, Turkina OM, Rodionov NV (2008) U-Pb age and genesis of granitoids of the southern framing of the Pechenga structure, Baltic Shield. Trans Russ Acad Sci 418:806–810

Vrevskii AB, Bogomolov ES, Zinger TF, Sergeev SA (2010) Polychronic sources and isotopic age of the volcanogenic complex (Arvarench unit) of the Imandra-Varzuga structure, Kola Peninsula. Doklady Earth Sci 431(1):386–389

Vuollo J, Huhma H (2005) Paleoproterozoic mafic dikes in NE Finland. In: Lehtinen M, Nurmi PA, Rämö OT (eds) Precambrian geology of Finland – key to the evolution of the Fennoscandian Shield. Elsevier, Amsterdam, pp 195–278

Vuollo J, Piirainen T, Huhma H (1992) Two Early Proterozoic tholeiitic diabase dyke swarms in the Koli-Kaltimo area, Eastern Finland – their geological significance. Bull Geol Surv Finl 363:32

Ward P (1987) Early Proterozoic deposition and deformation at the Karelian craton margin in southeastern Finland. Precambrian Res 35:71–93

Zagorodny VG, Mirskaya DD, Suslova SN (1964) Geology of the Pechenga sedimentary-volcanogenic series. Nauka (Science), Leningrad, p 218 (in Russian)

Zagorodny VG, Predovsky AA, Basalaev AA, Batieva ID, Borisov AE, Vetrin VR, Voloshina ZM, Dokuchaeva VS, Zhangurov ZhA, Kozlova NE, Kravtsov NA, Latyshev LN, Melezhik VA, Petrov VP, Radchenko MK, Radchenko AT, Smol'kin VF, Fedotov ZhA (1982) The Imandra-Varzuga Zone of the Karelides (Geology, Geochemistry, Evolutionary History). Nauka (Science), Leningrad, p 279 (in Russian)

3.3 Palaeotectonic and Palaeogeographic Evolution of Fennoscandia in the Early Palaeoproterozoic

V.A. Melezhik and E.J. Hanski

The Palaeoproterozoic Era spans the time interval between 2500 and 1600 Ma (Plumb 1991) and it is the early part of this time slice, from 2500 to 1900 Ma, that is the focus of our research on the Fennoscandian Shield. Between c. 2505 and 1900 Ma, two main stages of southwest-prograding intraplate rifting led to fragmentation of the Shield and eventually to oceanic crust development, closure of ocean basins, continent-arc collisions and resultant orogenies. Figure 3.22 provides a general guide through the tectonic and associated palaeogeographic stages and Figs. 3.23, 3.24, 3.25, 3.26, 3.27, 3.28, and 3.29 illustrate the main rock types representing the salient features of the early Palaeoproterozoic evolution of the Fennoscandian Shield.

Palaeomagnetic evidence for Palaeoproterozoic plate reconstructions remains ambiguous (Buchan et al. 2000). It is not clear whether a supercontinent formed during the late Archaean or if there were several large, dispersed continental plates (e.g. Bleeker 2003, notes 35 late Archaean cratons that could have been amalgamated in the Palaeoproterozoic). Tentative reconstructions imply the existence of two supercontinents: Kenorland, comprised of the North American, Fennoscandian and Siberian shields, and the less-definite Zimvaalbara, an assembly of the Zimbabwe, Kaapvaal, Pilbara, São Francisco and Indian cratonic blocks (Williams et al. 1991; Roscoe and Card 1993; Heaman 1997; Aspler and Chiarenzelli 1998; Condie 1998). Williams et al. (1991) named the former after the 2720–2680 Ma Kenoran orogeny that formed the Superior craton (Card and Poulsen 1998; Percival et al. 2006). Subsequently, the meaning of "Kenorland" has varied (reviewed by Reddy and Evans 2009), but if it existed, it must have been fragmented during a period of global rifting marked by numerous large igneous provinces at c. 2450 and 2200 Ma (e.g. Heaman 1997; Bleeker and Ernst 2006).

As emphasised by Bleeker and Ernst (2006), short-lived mantle-generated magmatic events, as represented by mafic dyke swarms, provide the most robust tool in reconstructing ancient, pre-Pangaea continental palaeogeographies. Fennoscandia contains at least five main episodes of dyke injections that span c. 2450 and1970 Ma (Vuollo and Huhma 2005). Each set of dyke swarms has different orientations, suggesting several phases of a protracted extensional tectonic regime. It is generally assumed that the initial phase of global rifting began at c. 2450 Ma (Heaman 1997). However, the current geochronological information indicates that in the Fennoscandian Shield rifting started earlier (though there are still problems in the interpretation of the isotopic data, see Chap. 3.1).

3.3.1 Pre-Sumian Time

At c. 2505 Ma, a mantle plume caused incipient uplift, rifting, and emplacement of large gabbro-norite plutons in the Kola craton (e.g. Bayanova et al. 2009), which forms the northeastern part of the Fennoscandian Shield (Fig. 3.22a). There is no robust record of volcanism associated with this early phase of emplacement of the gabbro-norite plutons, and it remains unresolved whether or not any incipient rifts were formed in the response to the mantle plume. The oldest formations in the Vetreny Belt,[1] which is located along the north-eastern margin of the Karelian craton (Fig. 3.1), are the only probable products of 2505 Ma (pre-Sumian) rift basins (Kulikov et al. 2010). These are the Toksha (arkosic arenite and quartzite), Kirich (basalt, andesitic basalt and subordinate komatiitic basalt) and Kozhozero (dolostone

V.A. Melezhik (✉)
Geological Survey of Norway, Postboks 6315 Sluppen, NO-7491 Trondheim, Norway

Centre for Geobiology, University of Bergen, Allegaten 41, N-5007 Bergen, Norway
e-mail: victor.melezhik@ngu.no

[1] In this account the names of the Palaeoproterozoic greenstone belts and schist belts are expressed in a short form, for example, the Vetreny Belt instead of the Vetreny Greenstone Belt.

V.A. Melezhik et al. (eds.), *Reading the Archive of Earth's Oxygenation*, Volume 1: The Palaeoproterozoic of Fennoscandia as Context for the Fennoscandian Arctic Russia - Drilling Early Earth Project,
DOI 10.1007/978-3-642-29682-6_5, © Springer-Verlag Berlin Heidelberg 2013

and tholeiitic basalt) formations. The oldest formation of the Imandra/Varzuga Belt (Fig. 3.1), known as the Purnach Formation (Fig. 4.3), is another probable pre-Sumian sedimentary-volcanic succession. However, this 1,500-m-thick pile of tholeiitic basalts, interbedded with several thick (600 m in total) units of rippled, cross- and parallel-bedded greywacke, has a conformable relationship with the overlying Kuksha Sedimentary Formation which, in places, erosively overlies the 2505 Ma gabbro-norite intrusions. Based on this observation, the Purnach Formation very likely comprises the lowest part of the 2505 and2430 Ma Sumian succession, which is discussed in the following section.

3.3.2 2505–2430 Ma Period

This time interval started with uplift and erosion of the Kola craton. The c. 2500 Ma layered gabbro-norite plutons, emplaced previously along the northern margin of the Imandra/Varzuga Belt were uplifted, faulted, eroded (Figs. 3.22b and 3.23a), and erosionally overlain by basal polymict conglomerates associated with the younger Sumian rift basins (Fig. 3.22c; for details see Chap. 4.1). At least 2–3 km of erosion occurred prior to the deposition of the Sumian supracrustal rocks, as determined by petrological estimates for the depth of crystallisation of the underlying plutonic bodies belonging to the first phase of mafic-ultramafic magmatism at the Archaean-Proterozoic boundary (Latypov et al. 1999). The duration of this phase of uplift and erosion (Fig. 3.22b) is unknown, but currently available data (Amelin et al. 1995) constrain it to between c. 2505 Ma (the age of the eroded gabbro-norite pluton) and 2442 Ma (the age of the Sumian volcanic rocks in the Imandra/Varzuga Belt).

After the c. post-2504 to 2442 Ma tectonic inversion and erosion episode, the Archaean-early Palaeoproterozoic transition was marked on the Fennoscandian Shield by the emplacement of another generation of layered gabbro-norite plutons and numerous smaller intrusive bodies (Fig. 3.23b) at c. 2440–2430 Ma (Fig. 3.22c; Alapieti 1982; Amelin et al. 1995; Vogel et al. 1998; Perttunen and Vaasjoki 2001; Iljina and Hanski 2005; Sharkov 2006). This was accompanied by a new phase of crustal extension and intraplate rifting (Melezhik and Sturt 1994). Two major SE-NW-trending rift systems were generated. One was developed near the southwest boundary of the Kola craton, and the other along the northeast margin of the Karelian craton (Fig. 3.22c). Mafic dykes in the Archaean basement related to this phase of rifting have a variable orientation but most of them are NE-SW-trending (Vuollo and Huhma 2005).

In the Salla, Imandra/Varzuga and Vetreny belts and in the northern Onega Lake area, the Sumian-age rift basins were predominantly filled with volcanic rocks, although several hundreds of metres of sedimentary rocks interbedded with lavas accumulated in several basins (e.g. Heiskanen 1987a). The sedimentary part of the rift infill is mainly composed of polymict conglomerates, greywackes and shales, but mature quartz arenites and carbonate rocks are also present (Fig. 3.23a, c–g). Depositional environments range from alluvial fan and braided fluvial settings in the Finnish part of the Karelian craton (Heiskanen 1987a; Ojakangas et al. 2001) to a clastic marine shelf along the Kola craton. The volcanic rocks are mainly subaerially erupted komatiitic and tholeiitic to calc-alkaline lavas (Melezhik and Sturt 1994; Puchtel et al. 1996, 1997, 2001; Hanski et al. 2001; Hanski and Huhma 2005; Kulikov et al. 2010). In all Sumian rift basins both subaerially erupted amygdaloidal and submarine pillow lavas occur (Fig. 3.23h–m). The volcanic formations reach 5,500 m in thickness in the Imandra/Varzuga Belt (Fedotov et al. 1982) and as much as 8,000 m in the Vetreny Belt (Kulikov et al. 2010). Most of the volcanic successions are considered to be coeval and comagmatic with the c. 2440 Ma layered gabbro-norite complexes (Alapieti 1982; Melezhik and Sturt 1994; Amelin et al. 1995; Vogel et al. 1998; Chashchin et al. 2008; Bogatikov et al. 2010; Kulikov et al. 2010) and are interpreted to be continental flood basalts (Puchtel et al. 1996, 1997, 2001). A common feature of the volcanic rocks is a strong interaction of the magma with the continental crust through which it ascended to the surface (Puchtel et al. 1997; Hanski et al. 2001).

A conservative estimate of the lateral extent of the Sumian flood basalt province, based on the present-day distribution of the layered gabbro intrusions, dykes and coeval volcanic formations, is c. 700,000 km^2 (Melezhik 2006). However, it is likely that it was much greater (Bogatikov et al. 2010), particularly considering that the Kola and the Karelia provinces were once separated by the c. 2100 Ma Kola Ocean, and are now juxtaposed along the Lapland-Kola suture (Daly et al. 2006).

No consistent set of palaeomagnetic data has so far been obtained for the c. 2500 Ma intrusions. The latitudinal position of the Shield has been calculated as within 5–42° from the equator based on measurements from the Shalskiy dyke, Monchegorsk Layered Complex and Mt. Generalskaya intrusion (Mertanen et al. 2006; Pechersky et al. 2004; Arestova et al. 2004). The younger, c. 2440 Ma magmatic event indicates a sub-equatorial positions at c. 30°S as recorded by gabbro-norite dykes in the Lake Pääjärvi area near the Oulanka Group layered intrusions in central Russian Karelia (Mertanen et al. 1999).

The mantle-plume-driven continental uplifts continued on the Fennoscandian Shield until c. 2435 Ma, as indicated by the age of layered gabbro-norite plutons in Finland (Perttunen and Vaasjoki 2001). Daly et al. (2006) suggested that the mantle-plume and related magmatic underplating at

c. 2450 Ma led to crustal thickening and subsequent uplift and erosion. The top of the Sumian continental flood basalts and penecontemporaneous layered intrusions is marked by a regional unconformity that can be traced throughout Fennoscandia (e.g. Sturt et al. 1994; Ojakangas et al. 2001) (Fig. 3.22d). All rocks beneath this unconformity surface have been deeply eroded. In the Peräpohja Belt, the unconformity cuts c. 3,000 m down section into the Penikat layered gabbro-norite intrusion (Laajoki 2005). Assuming that the layered gabbro-norites crystallised at depths of 2–3 km (e.g. Latypov et al. 1999), at least 5–6 km of erosion occurred prior to the deposition of the Sariolian units in the Peräpohja Belt. In the Koillismaa area, the originally large, sheet-like Koillismaa layered intrusion was, after solidification, faulted, tilted and fragmented by block movements into several individual bodies. Karinen (2010) suggested that the tensional stress-field, which was associated with the rifting and emplacement of the intrusion, changed later to a compressional stress field, leading to thrusting of intrusion blocks over each other. However, it is presently not known to which regional tectonic events these compressional movements were connected and when they occurred. Nevetheless, it is clear that the igneous bodies were uplifted and variably eroded before the deposition of Sariolian sedimentary rocks. This implies that a vast volume of the Sumian flood basalt province was subject to deep erosion as well. The duration of the post-Sumian erosional period cannot be precisely constrained, but it is younger than c. 2440–2435 Ma, the ages of the Sumian volcanic rocks in the Imandra/Varzuga and Vetreny belts (Amelin et al. 1995; Puchtel et al. 1996) and Kemi layered gabbro-norite intrusion (Perttunen and Vaasjoki 2001), and older than overlying 2430 Ma Sariolian rocks.

3.3.3 2430–2300 Ma Period

The Sariolian system supracrustal rocks were deposited during this time slice (see Chap. 3.1). Owing to the above-mentioned deep erosional processes, palaeotopography would have been marked by large, rift-separated uplifts and prominent erosional valleys typical of plume-generated continental flood basalt provinces (Melezhik 2006). The physical weathering and deep erosion (as much as 5–6 km) of the Archaean crust and flood basalts occurred over a period of several millions of years. Such processes should have inevitably resulted in an enhanced rate of silicate weathering and a considerable consumption of atmospheric CO_2, and could have been one of the crucial causes responsible for the onset of the Huronian glaciation (Young 1991; Young et al. 2001; Melezhik 2006).

The newly formed rift basins were filled by as much as 500 m (note, though, that thicknesses vary greatly across the shield) of polymict gritstones and conglomerates (Fig. 3.24a–k) with rare greywacke and minor shales (Fig. 3.24l) and, in the Imandra/Varzuga Belt, there are high-Sr limestones. Clasts in the conglomerates are typically derived from local sources including Sumi-age igneous rocks, various older rock types from the basement as well as underlying Sariolian volcanic rocks. In the Pechenga and Peräpohja belts, the conglomerates rest erosionally on, and contain reworked clasts of, the c. 2505 Ma and 2435 Ma layered gabbro-norite intrusions, respectively (Fig. 3.24m). Glaciogenic rocks, such as varved sediments and diamictites, are important and form parts of the sedimentary successions (Fig. 3.24n–q) in the northern North Karelian Belt (Marmo and Ojakangas 1984), and the Shambozero-Lekhta, and Imandra/Varzuga belts (see Chap. 7.2). Associated volcanic rocks commonly overlie the sedimentary successions and reach 1,000 m in thickness. They consist mainly of tholeiitic and komatiitic, amygdaloidal, globular- and spinifex-textured lavas extruded under subaerial or subaqueous conditions (Fig. 3.24r–u).

Several and varied palaeogeographic and palaeotectonic reconstructions have been advanced for the 2430–2300 Ma time frame but all share the common theme of intracontinental rifting (Fig. 3.22e). In the Kola craton, SE-NW-trending rift systems have been predominantly filled by volcanic rocks. A similarly trending rift system formed along the northern margin of the Karelia craton with an apparent ‘rift star’ in its central part (Fig. 3.22e). Such ‘rift stars’ have four branches radiating from a common centre; each is fairly equant and filled with an initial thin basal polymict conglomerate overlain by thick piles of volcanic rocks. In contrast, the central part of the Karelian craton is marked by a basin filled mostly with immature, polymict, alluvial conglomerates and only minor volcanic rocks (Heiskanen 1990, 1987b). In addition, 2403–2200 Ma multiple rifts have been recently recognised on the Norwegian coast (a chain of windows in the Caledonides termed the West Troms Basement Complex, Fig. 3.1) (Bergh et al. 2010). Kullerud et al. (2006) reported on a 2403 ± 3 Ma dyke complex (Ringvassøy dykes) which they correlate broadly with a global Palaeoproterozoic magmatic event that formed extensive bimodal intrusive and extrusive suites in the Fennoscandian Shield. They suggested a distinct trend from the 2505–2490 Ma (Kola Peninsula) through the 2460–2440 Ma (Kola, Russian and Finnish Karelia) to the 2403 Ma dykes in the West Troms Basement Complex, implying that the locus of maximum extension and magmatic activity may have been shifting with time.

Some models suggest that deposition on the Fennoscandian Shield was solely in marine environments (Negrutza 1984; Heiskanen 1990), whereas others infer predominantly subaerial volcanism and sedimentation (Melezhik 1992; Laajoki 2005; Lahtinen et al. 2008).

However, direct evidence for marine deposition has been documented only for the younger part of the Sariolian successions in the northern North Karelian Belt (Marmo and Ojakangas 1984; Laajoki 2005), and possibly in the Imandra/Varzuga Belt, assuming that the glacially associated high-Sr limestones would require precipitation from seawater (see Chap. 6.1.2). Considerable thicknesses of immature polymict conglomerates in both the Kola and Karelian cratons (up to 500 m) imply a palaeogeography of highly dissected relief across the Fennoscandian Shield. Ojakangas et al. (2001) reported diamictite-dropstone associations in 16 widespread localities (Fig. 3.22e shows their spatial distribution), which implies that a significant part of the shield might have been covered with ice during the Huronian-age icehouse (for details see Chap. 7.2). Reliable palaeomagnetic data showing the geographic position of the Karelian craton during the glaciation event are still scarce. The data obtained from late-Sumian igneous rocks (see above) point to a location at low latitudes at c. 2440 Ma (Fig. 3.25). This together with similar palaeomagnetic results from Huronian glacial formations (Williams and Schmidt 1997) can be taken as evidence for a near-equatorial, global glaciation during the ensuing Sariolian time (Evans et al. 1997; Mertanen and Pesonen 2005).

3.3.4 2300–2060 Ma Period

This time slice is marked by deposition of the Jatulian system sedimentary and volcanic rocks. Its base defines an unconformity surface, but one that is less pronounced in comparison with those bounding the two older groups. Most of the unconformity-associated palaeoweathering features beneath the Jatulian rocks (the first documented features were by Metzger in 1924) are thin intervals and horizons (reviewed in Sokolov and Heiskanen 1984; Marmo 1992; Ojakangas et al. 2001), which is particularly the case for the Pechenga and Imandra/Varzuga belts (Melezhik and Predovsky 1982). However, some of the palaeoweathering profiles formed during post-Huronian time are remarkable. Marmo (1992) reported an as much as 80 m thick pre-Jatulian palaeosol developed on granites and Sariolian sandstones in the northern North Karelian Belt (Figs. 3.1 and 3.26a). The most deeply weathered Sariolian arkosic sandstones, now metamorphosed into quartz-kyanite rocks (Fig. 3.26b), contain up to 21 wt.% Al_2O_3, 77 wt.% SiO_2, 0.5 wt.% TiO_2 and 0.5 wt.% K_2O (Marmo 1992). In terms of primary sedimentary mineralogy, such rocks would represent a mixture of quartz and kaolinite.

The strong chemical weathering demonstrated by Jatulian sedimentary rocks has been regarded as evidence for subtropical to tropical, humid climate conditions (Marmo 1992). Palaeomagnetic measurements from the c. 2060 Ma Kuetsjärvi Volcanic Formation, carried out by Torsvik and Meert (1995), yielded a latitudinal position at 23° from the equator but with large errors and therefore this result needs to be tested with further palaeomagnetic studies.

Bleeker and Ernst (2006) compared short-lived, Palaeoproterozoic mafic magmatic events in several cratonic Archaean blocks and produced a palaeogeographic reconstruction in which the Karelia-Kola craton was placed against the southern margin of the Superior craton as part of the supercraton called Superia. This is similar to the reconstruction of Heaman (1997) with the exception that the Karelia-Kola craton is "up-side-down" in Heaman's model. Bleeker and Ernst (2006) admit that these geological reconstructions do not fully comply with the existent palaeomagnetic data, but on the other hand, the complexity of the palaeomagnetic data from the Fennoscandian Shield is not easy to interpret. In their supercraton, Bleeker and Ernst (2006) delineated radiating patterns of dyke swarms of four events (c. 2505, 2450, 2210–2220 and 2110 Ma), with their focal points at the southern and eastern margin of the Superior craton margin. This model suggests that the Superior and Karelia-Kola cratons shared multiple magmatic events until c. 2100 Ma. A subsequent major dyke injection episode that took place at c. 1980–1970 Ma in the Karelia craton is not found in the Superior craton, hence the final break-up of Superia can be estimated to have occurred between 2100 and 1980 Ma.

The regional palaeogeographic and palaeotectonic reconstructions advanced for the Jatulian rocks, like those for the older sequences, vary greatly in detail (Negrutza 1984; Heiskanen 1990; Laajoki 2005; Lahtinen et al. 2008; Akhmedov et al. 2004). However, most agree on a basic template in which a SE-NW-trending intracontinental rift system developed along the Kola craton between c. 2200 and 2100 Ma. The presence of mature sandstones, and alternating sandstone-mudstone assemblages suggest high wave-energy and tidally influenced shorelines and the existence of open sea conditions (e.g. Nikula 1988; Laajoki 2005). A vast, clastic epeiric sea had developed on the Karelian craton that paralleled the Kola rift system, in which three distinct carbonate platforms formed (tentatively named as the Vesterälen, Peräpohja and Onega; Fig. 3.22f).

The sedimentary-volcanic infill of the Kola intracratonic rifts is traceable over a distance of more than 1,000 km (Melezhik and Sturt 1994) and defines a couplet in which the sedimentary rocks comprise the lower and thinner part. The sedimentary units are lithologically diverse, ranging from fluvial-deltaic siliciclastic rocks to shallow-water lacustrine tufas, travertines and stromatolitic carbonates in the Pechenga rift (Fig. 3.26c–m), to fine-grained, clastic marine shelf rocks and deep-marine resedimented carbonate rocks in the Imandra/Varzuga rift (Fig. 3.26n–q). Both the siliciclastic and chemical sedimentary rocks display variegated colour (Fig. 3.26d, h, k, m, o, and p). The volcanic rocks are more than 1,000 m thick and occur as subaerially erupted amygdaloidal, fluidal lavas, lava-breccias and tuffs, minor volcaniclastic rocks, and rare pillow basalts (Fig. 3.26r–u). Peculiar types are highly oxidised alkaline and subalkaline volcanic rocks ranging in composition from picritic to rhyolitic (Fig. 3.26v; see Chap. 6.2.3).

On the central Karelian craton the initial deposits in the epeiric seaway are thick successions (up to 400 m) of alluvial and lacustrine conglomerates, gritstones and sandstones (Fig. 3.26w, x). Prevalent transport directions are south to north and southeast to northwest (Heiskanen 1990). These are followed by similarly thick units (+500 m; Sokolov 1987) of mainly fine-grained, well-sorted, mature arenites deposited under shallow-water conditions (Fig. 3.26y); thin beds of haematite-rich siltstones and mudstones, commonly desiccated (Fig. 3.26x–ac), occur in places. Many rocks are red in colour, hence "red beds" (Fig. 3.26ad). The volcanic rocks are much less significant than those on the Kola craton and are mainly subaerially erupted tholeiitic basalts.

The Finnish part of the Karelian craton shows somewhat similar depositional environments. In the North Karelia Belt, close to the western edge of the Karelian craton, Kohonen and Marmo (1992) described a relatively monotonous supracrustal sequence of clastic sedimentary rocks with a maximum thickness of 2,500 m (Fig. 3.26af, ag). It is noteworthy that the lowermost coarse-grained quartzites and conglomerates are locally rich in haematite that seems to be detrital in origin (Fig. 20c in Kohonen and Marmo 1992). The sedimentary sequence is dominated by mature quartzites with the exception of an arkosic formation in the middle part, which has been interpreted as resulting from a rift phase interrupting a long stable platformal period. Palaeocurrent measurements indicate a general westward transport in the lower part of the sequence. In the middle part, the direction was first variable between southwest to northwest and then toward the south, whilst the highest units in the succession show a transport path toward the northwest. As indicated by the presence of well-sorted massive gravels and sedimentary structures, such as desiccation cracks and large-scale trough cross-bedding (Fig. 3.26ah), the major part of sedimentation took place in a subaerial, fluvial environment. Well-sorted quartz sandstone on top of the sequence suggests a transgression and sedimentation in a nearshore marine environment (Marmo et al. 1988; Kohonen and Marmo 1992).

In the Kainuu Belt deposition of a thick succession of epiclastic sediments was only once interrupted by a volcanic event producing a thin unit of mafic lava. Up to 1,400 m of mature quartz-pebble conglomerates and quartz arenites (Fig. 3.26ai), representing alluvial fans and braided fluvial deposits, were followed by 2,300–2,600 m of proximal shelf sediments related to a passive margin developed to the west and southwest of the Kainuu Belt. The latter sediments were divided by Strand (2005) into four sequences, each recording a long-term transgressive-regressive cycle. The first sequence starts with coarse, clastic, alluvial-plain sediments and shallow-marine siltstones deposited in incised valleys with the palaeocurrent data indicating a general southwesterly or westerly trend of material transport. They are overlain by rippled and horizontally laminated sandstones representing a wave-dominated shoreline setting. The second sequence is separated from the underlying rocks by a 20–50 m thick, subaerially erupted mafic lava, and consists of quartzose arenites with mudstone interbeds in its lower part, representing barrier beach, lagoonal and tidal settings. The upper part is characterised by fluvial and coastal, coarse-grained quartz arenites. The third sequence contains mudstones deposited in a tidal flat environment followed by a highstand sequence of coarse alluvial-plain sandstones. The fourth sequence is dominated by sandstones with large-scale trough cross-bed sets and horizontally laminated, medium-grained sandstones interbedded with fine-grained lithologies, interpreted as sand shoals and sheets of an inner shelf system. The four mentioned sequences form an overall major transgressive succession preceding the development of carbonate platforms and deposition of outer shelf carbonates (Strand 2005).

The general transgressive development from epiclastic sedimentation to carbonate rock deposition occurred widely but differed in detail in different basins. In the Central Lapland and Kuusamo belts, carbonate rocks were preceded by trough-cross-bedded quartzites, and rippled and bipolar cross-stratified sandstones (Fig. 3.26aj–an). In the Kuusamo Belt these arenites are interlaminated with haematite-rich, desiccated mudstones, representing tidal channel and tidal flat settings (Fig. 3.26ao–aq). Well-preserved primary sedimentary structures suggest syndepositional fragmentation of red, haematite-rich, mud layers, thus syndepositional oxidation (for details, see Strand and Laajoki 1999). One peculiarity in the Kuusamo Belt and its continuations in Central Lapland is the presence of quartzitic sandstones enriched in Cr-bearing mica, fuchsite. This mineral occurs now as a matrix mineral between quartz clasts and the enrichment of rocks in fuchsite seems to follow bedding planes (Fig. 3.26aj).

As mentioned previously, major Jatulian carbonate platforms were formed in three areas: Vesterålen, Peräpohja and Onega (Fig. 3.22f). The Vesterålen platform rocks were affected by amphibolite- to granulite-facies metamorphism and the marbles preserve no primary sedimentary structures. They still retain high $\delta^{13}C$ values (Baker and Fallick 1989), which provide a chemostratigraphic correlation to the isotopically similar carbonates on the Karelian craton recording the Lomagundi-Jatuli positive excursion of carbonate carbon (e.g. Karhu 1993; Melezhik et al. 1999a). In contrast, the

Peräpohja and Onega platforms preserve rich original sedimentological characteristics and have attracted the attention of many researchers (Sokolov 1963, 1987; Makarikhin and Kononova 1983; Melezhik et al. 2000; Wanke and Melezhik 2005; Akhmedov et al. 2004; Melezhik and Fallick 2010).

The Onega carbonate platform is marked by large lateral facies variations (Sokolov 1987; Akhmedov et al. 2004), and one of the most detailed studied sections is 800 m thick and comprises several distinct lithofacies (Melezhik et al. 1999a). Figure 3.26ar–ba illustrate some of the main lithofacies and their sedimentological features. The platform carbonate rocks sit on dolomite-cemented sandstones that record a distal fluvial-coastal sand flat system. A significant component of the platform is biostromal and biohermal columnar stromatolites (for details see Chap. 7.8.2) and red, re-sedimented carbonates deposited in shallow-water, low-energy, intertidal zones and barred evaporitic lagoons and peritidal evaporitic environments (Fig. 3.26ar–au). The next abundant lithologies are red, flat-laminated, dolomitic and magnesite stromatolites formed in evaporative ephemeral ponds and coastal sabkhas. Desiccated carbonate beds, red desiccated mudstones with tepees, pseudomorphs after calcium sulphate, and halite casts (Fig. 3.26av–ax) suggest that coastal sabkhas alternated with playa lake environments. Abundant dissolution-collapse breccias in both mudstone and dolostone environments (Fig. 3.26az–ba) are indicative of a considerable volume of once-present sulphates and other soluble salts (Melezhik et al. 2001, 2005a). The overall depositional trend in this evolving platform along the southern Karelian craton is a transition from peritidal and shallow-water tidal to a deeper intertidal environment with intervening sabkhas and playas, then a return to intertidal marine conditions followed by a final phase of emergence and partial erosion (Melezhik et al. 1999a).

The Peräpohja carbonate platform, studied in detail in the Kalix Belt (Wanke and Melezhik 2005; Melezhik and Fallick 2010), shows somewhat different lithofacies (Fig. 3.26bb–by), and a different depositional history. The platform succession is more than 200 m thick and transgressively oversteps a subaerially exposed surface developed on top of a 300-m-thick pile of terrestrial amygdaloidal basalts (Fig. 3.26bb). Sharp juxtaposition of shallow-marine carbonate lithofacies over subaerially extruded lavas suggests rapid subsidence. At an early stage, a shallow-water basin received a significant amount of mafic pyroclastic material (Fig. 3.26bc–bf) accompanied by extensive input of siliciclastic components reworked by waves, storms and tidal currents (Fig. 3.26bg–bj) until the carbonate platform was established and stabilised (Wanke and Melezhik 2005). The platform succession is comprised of laminated dolostones (chemically precipitated and resedimented) and biohermal and biostromal dolostones composed of columnar, spherical and flat-laminated stromatolites (Fig. 3.26bk–bn; for details see Chap. 7.8.2). The platform has experienced frequent episodes of emergence (Fig. 3.26bo–bq) and was affected by synchronous mafic volcanism. Tidal and storm-wave reworking is documented in several beds of ripple- and herringbone-stratified sandstones and abundant storm flakestones (Fig. 3.26br–bx). Resedimented carbonates occur as dolarenite, dolostone breccia and deep-water, ribbon-like dolostone comprised of alternating pale and dark grey couplets (Fig. 3.26by). The dominant carbonate stromatolitic facies have been interpreted as fringing-reef complexes, and the overall depositional setting was categorised as a rimmed carbonate shelf (Wanke and Melezhik 2005). The carbonate platform succession is followed by 2,000 m of deeper-water shales with minor basalts, tuffites (Fig. 3.26bz), limestones and arenites (Lager and Loberg 1990). These siliciclastic sediments overstep the carbonate lithofacies suggesting irreversible drowning of the carbonate shelf and the transition from a rift to drift tectonic setting.

The Jatulian supracrustal successions contain thick accumulations, up to 500 m, of flood basalt-like, subaerial volcanic piles demonstrating continental conditions, though minor units composed of shallow-water pillow lavas are also present. In the south-western part of the shield, as represented by the Peräpohja and Tohmajärvi areas (Fig. 3.1), these mafic magmas are uncontaminated by crustal material, unlike previous magmatic episodes, even though underlain by thinned continental crust (cf. Huhma et al. 1990). In more central parts of the shield, the presence of uncontaminated Jatulian lavas has not been documented. One of the most prominent mafic dyke swarm systems in the Archaean basement with trends varying between E and NW belongs to this stage and indicates severe crustal extension at c. 2100 Ma. Some of these Fe-tholeiitic mafic dykes have been shown to have acted as feeders to Jatulian volcanic rocks (Pekkarinen and Lukkarinen 1991). The previous major magmatic event that took place c. 120 Ma earlier in Finland is not manifested by mafic dykes in the basement but produced c. 2220 Ma semiconformable, sill-like bodies close to the contact zone between the Archaean basement and Palaeoproterozoic bedrock (Hanski et al. 2010), suggesting a less-extensional tectonic regime from that prevailing during rifting at c. 2100 Ma in this part of the shield.

Several protracted continental rifting episodes, as recorded by mafic dyke swarm generations, occurred after the first magmatic event at c. 2505 and, towards the end of Jatulian time, led to further attenuation of the continental crust and deepening of the depositional environment. An initial break-up of the Archaean supercontinent and the formation of the Kola Ocean is roughly dated to 2100–2060 Ma and is marked by arc-related volcanic rocks and back-arc-basin volcaniclastic sediments carrying a significant juvenile signature (e.g. Sharkov and Smol'kin 1997; Daly et al. 2006). The break-up of the Fennoscandian Shield led to the onset of open marine conditions along the fringing continental margin to the northeast of the Kola Ocean and along the Karelian craton. However, whether the extension of the continental crust led to the complete break-up along the south-western margin of the Karelian craton, and development of oceanic crust in Jatulian time, is still uncertain given that ophiolitic rocks are absent in the Jatulian successions (the rocks regarded confidently as slices of oceanic crust are post-Jatulian in age; see Chap. 3.2). The present-day buried SE-NW-trending western boundary of the Karelian Archaean craton (Figs. 3.2 and 3.3 in Chap. 3.1) has been delineated by Nd isotope compositions of felsic plutonic rocks and trends from the Lake Ladoga region, Russian Karelia, through Central Finland to northern Sweden (Huhma 1986; Öhlander et al. 1993; Konopelko et al. 2005). Several models have been presented for the time of generation of this cratonic margin. Nironen (1997) considered two options: (1) the break-up of the continent and development of the pre-Svecofennian ocean took place sometime between 2100 and 2000 Ma (see also Kohonen and Marmo 1992) and later a Red Sea-type marginal basin was formed at the attenuated continental margin, now represented by the c. 1950 Ma crustal section of the Jormua ophiolite; or (2) both the passive continental margin and oceanic crust developed at c. 1950. Recent results of deep reflection seismic sounding across eastern Finland favour the passive continental margin model, in which the Jormua complex and enclosing sedimentary rocks represent remnants of an overthrust sheet transported from the newly formed western margin of the disrupted craton (Kontinen and Paavola 2006). Karhu (1993) suggested that the end of the Lomagundi-Jatuli carbon isotope perturbation at c. 2060 Ma marks the timing of the final continental break-up; this view was also favoured by Wanke and Melezhik (2005) and Lahtinen et al. (2005). So far, the oldest age determination that can be used to constrain the earliest formation of oceanic crust is c. 2015 Ma from the Kittilä Group in Finnish Lapland (see below), which occurs outside the Jatulian time slice (see Chap. 3.2).

In summary, the major environmental components of Jatulian time in the Fennoscandian Shield are: (a) the development of a vast epeiric sea with several extensive carbonate platforms (stable cratonic regime) in a sub-equatorial position; (b) abundant deposition of diverse stromatolites and marine Ca-sulphates (oxic seawater); (c) widespread "red bed" formation (O_2-rich atmosphere); and (d) ^{13}C-rich carbonate formation (Lomagundi-Jatuli perturbation of the global carbon cycle). The geotectonic regime was marked by the break-up of the Fennoscandian Shield and the c. 2100 Ma opening of the Kola Ocean between the Kola and Karelian cratons (Fig. 3.22f).

3.3.5 2060–1950 Ma Period

The current knowledge of the evolution of the Fennoscandian Shield suggests that the complete break-up of the Archaean continental crust and related start of the formation of oceanic crust took place at slightly different times in different parts of the craton. The opening of the pre-Svecofennian ocean led to the generation of a passive margin along the south-western boundary of the Karelian craton, which was later hidden beneath accreted island arc complexes when the ocean was closed during the collision events related to the Svecofennian orogeny. The other half of the split continent drifted away and its current position is unknown.

According to Daly et al. (2006), the opening of the Lapland-Kola Ocean had advanced to a seafloor-spreading stage by c. 2050 Ma, but the exact timing is still poorly constrained (Fig. 3.22f–g). Well-preserved ophiolitic rocks are lacking in the Kola province. Based on geochemistry, potential candidates for metamorphosed oceanic crustal rocks are found in the Tanaelv Mélange bordering the Lapland Granulite Belt. Evidence for the former existence of an oceanic basin comes from arc-derived, juvenile crust recorded by c. 1940–1975 Ma felsic plutons and orthogneisses in the Inari, Tersk and Strel'na Terranes (Daly et al. 2006). In their model, Daly et al. (2006) also proposed another geographically separated, modestly wide ocean, the Pechenga-Imandra-Varzuga Ocean, in which seafloor spreading commenced at 1990–1970 Ma. In our model, the latter is part of the Kola Ocean because there is no robust evidence to infer the existence of a separate ocean. The Pechenga and Imandra/Varzuga belts do not provide information on the exact timing of continental break-up. Although the two uppermost volcanic formations in the North Pechenga Group are mostly composed of submarine pillow lavas (Fig. 3.27a–b), whose magma originated in depleted mantle, a thin acid volcanic unit in the c. 1980 Ma Pilgujärvi Volcanic Formation carries felsic gneiss xenoliths (Hanski 1992) demonstrating the existence of sialic basement under the Pechenga basin at this time. Nevertheless, this does not rule out oceanic spreading elsewhere and further seaward of the continental margin at this time.

Two other seaways did develop. One was in close proximity to the Kola Ocean, termed here the Kittilä Ocean whose opening is dated at c. 2010–2015 Ma (Hanski and

Huhma 2005) (Fig. 3.22g). In the southern part of the shield, the c. 1970 Ma Outokumpu ophiolite complex and the c. 1953 Ma Jormua ophiolites (e.g. Peltonen 2005) are evidence for the second seaway, which we term the Jormua Ocean (Fig. 3.22g). Consequently, in our model the 2060–1950 Ma basinal configuration includes the Kola, Kittilä and Jormua Oceans separating the Fennoscandian Shield into distinct segments. The Pechenga, Imandra/Varzuga and the Central Lapland intraplate rifts remained active until c. 2050 Ma as fringes to the newly formed oceans. In addition, the Onega rift formed at c. 2000 Ma along the southern Karelian continental margin prior to the opening of the Jormua Ocean (Fig. 3.22g).

The rifts fringing the active continental margins of the Kola Ocean accumulated fluvial-deltaic sedimentary and volcaniclastic rocks, jasper beds, redeposited ironstones, shallow-marine stromatolitic dolostones, dolarenites, marls and shales (Fig. 3.27c–i), succeeded by deep-water greywacke turbidites with abundant carbonate and phosphate concretions (Melezhik 1992). These rocks are overlain by more than 1,500 m of tholeiitic pillow lavas and lava breccias (Fig. 3.27a), which are, in turn, overlain by several hundreds of metres of deep-water shelf to turbiditic continental slope organic-matter-rich, sulphide-bearing greywackes and shales with calcite nodules (Fig. 3.27j–n) (e.g. Akhmedov and Krupenik 1990). Repeated submarine volcanism resulted in several kilometres of monotonous tholeiitic massive and pillowed lavas and breccias with minor felsic tuffs and ferropicritic volcanites (Fig. 3.27b, o–s).

In the Central Lapland area, advanced rifting and incipient spreading resulted in deposition of sulphide-bearing "black shales" and tuffites followed by accumulation of primitive volcanic rocks with komatiitic and picritic chemical affinities (Hanski and Huhma 2005). The komatiitic-picritic volcanic association has been traced along strike over a distance of 350 km, and occurs as massive and pillowed (including variolitic varieties) lavas, lava breccia, agglomerates, lapilli tuffs, and reworked tuffs (Fig. 3.27t–w). The reported products of the Kola Ocean per se were volcanic rocks and volcaniclastic sediments associated with volcanic arc and back-arc rift environments (Daly et al. 2006).

The Kittila Ocean features one of the largest accumulation of mafic lavas on the shield, occurring as pillowed and massive lavas and hyaloclastites (Fig. 3.27x–z). They cover an area of more than 2,600 km^2 and form together with associated sedimentary rocks a greenstone complex with a present-day thickness of c. 6 km (Hanski and Huhma 2005). Minor felsic porphyries are present and have been dated at 2015 $\pm$ 15 Ma (Rastas et al. 2001); the age is considered to be the best available to constrain the timing of formation of oceanic crust (Hanski and Huhma 2005). Interbedded sedimentary rocks consist of greywackes, shales, C_{org}- and S-bearing schists, minor jasper breccia (Fig. 3.27aa), Mn-rich carbonate rocks and banded iron formations (Paakkola and Gehör 1988; Lehtonen et al. 1998). Close to the eastern, tectonic contact of the greenstone complex with the cratonic cover sequence, there exists a string of highly refractory, ultramafic lenses, which Hanski (1997) considered dismembered ophiolitic mantle rocks. In the eastern part of the complex, there are also amygdaloidal basalts (Kautoselkä Fm) with a continental geochemical affinity and could represent passive margin volcanic rocks.

The formation of the Jormua Ocean is recorded by EMORB-type pillow basalts, sheeted dyke complexes, gabbros and minor plagiogranites in the Jormua area, Central Finland (Kontinen 1987; Peltonen 2005) (Fig. 3.27ab–ae). The volcanic part of the ophiolite is 100–400 m thick, while the underlying sheeted dyke complex exceeds 1,000 m in thickness. Some of the lavas have erupted directly onto mantle peridotites, and mantle rocks are found as interdyke screens in the sheeted dyke complex. The lherzolitic mantle tectonite bodies are cut by c. 2100 Ma OIB-type dykes carrying a population of Archaean zircon xenocrysts from deeper mantle sources (Peltonen et al. 2003). The ultramafic hosts are also Archaean in age (Tsuru et al. 2000). Thus, in essence, the Jormua ophiolite has two different components, fragments of Archaean subcontinental lithospheric mantle and Palaeoproterozoic oceanic crustal rocks. To explain this kind of lithological combination, Peltonen (2005) maintained that the Jormua Ophiolite represents a transition zone between the continental lithosphere and oceanic regime where subcontinental lithospheric mantle was exhumed by detachment faulting upon development of a passive continental margin.

In the Outokumpu area, dismembered ophiolitic rocks are mainly represented by serpentinised refractory ultramafic lenses with minor gabbroic bodies and gabbroic and basaltic dykes. These are associated with Kalevian sulphide- and C_{org}-bearing schists and turbiditic greywackes (e.g. Koistinen 1981). The Outokumpu-type massive Co-Cu-Zn sulphide ores are hosted by dolomitic carbonate, skarn and quartz rocks that surround ultramafic bodies. The host rocks are not sediments, as was thought previously, but products of extensive metasomatic alteration of ultramafic rock bodies during hydrothermal ore formation on the seafloor (Peltonen et al. 2008). Differing from the Jormua mantle rocks, those of the Outokumpu area are interpreted as asthenospheric abyssal peridotites formed in an oceanic environment (Peltonen 2005). At the time of the formation of the oceanic crust represented by the Outokumpu ophiolitic rocks, rifting in the continental margin continued, as evidenced by the c. 1980–1970 Ma mafic Fe-tholeiitic dyke swarms transecting the Archaean basement in a NW-SE direction (Vuollo and Huhma 2005).

At c. 2000 Ma, along the northeastern Karelian margin towards the future Jormua Ocean, a deep-water, volcanically-active rift formed (Fig. 3.22g). The rift-bound sedimentation and volcanic activity resulted in accumulation of several hundred metres of organic-rich clastic sedimentary rocks, organosiliceous rocks, cherts and carbonates (Fig. 3.27af–ai) interbedded with tholeiitic pillow lavas (e.g. Galdobina 1987a). This succession was sealed beneath several hundred metres of tholeiitic pillow basalts. The C_{org}-rich rocks served as a petroleum source to a supergiant oil-field (Figs. 3.22g and 3.27af, ai). Part of this field has survived later tectono-metamorphic overprinting and preserves many primary features (Melezhik et al. 2009, Fig. 3.27ah, aj). Palaeomagnetic measurements on the Konchozero differentiated sill that is thought to be genetically linked to the volcanism suggest that the position of the shield was near the equator (23°) (Pisarevsky and Sokolov 1999).

To summarise, the major environmental characteristics of the 2060–1950 Ma period include: (*1*) advanced stage of break-up of the Fennoscandian Shield; (*2*) voluminous submarine eruptions of MORB-like pillow basalts; (*3*) the termination of the Lomagundi-Jatuli isotopic excursion; and (*4*) an unprecedented accumulation of C_{org}-rich rocks and formation of a supergiant oil-field linked to the worldwide Shunga Event (Melezhik et al. 1999b, 2004, 2005b).

3.3.6 1950–1900 Ma Period

The post-1950 Ma period lies outside the main focus of this review. Comprehensive summaries of the depositional and tectonic environments of this time are presented in Kähkönen (2005) and Lahtinen et al. (2005). For reasons of completeness, we provide a brief outline of this geological history in the aftermath of the major phase of craton separation.

Development of a continental margin at the western edge of the Karelian craton created slopes that allowed erosional products from the Archaean crust and its Palaeoproterozoic cover to be transported by density currents to deep parts of the ocean. These deposits, known as Kalevian turbidites (Fig. 3.28a, b) (Livian, in Russia), were partly accumulated on attenuated continental crust and partly on newly-formed oceanic crust. Volcanism was subdued at this stage, with the only significant submarine pillow lava formations being found in association with the turbidites in the Kiiminki Belt (Fig. 3.28c) and the Lake Ladoga area. Geochemical and detrital zircon studies have shown that the older flysch-like sediments at the continental margin were derived from the adjacent regions of the Karelian craton but the younger, distal sediments acquired some of their detritus from more distant sources having ages in the range of 1950 to 1920 Ma (Lahtinen et al. 2010). The location of these relatively young sources have been considered to be in a primitive island arc of the Svecofennian composite orogen in the west (Lahtinen 2000) or in the Himalaya-type Lapland-Kola orogen in the northeast (Lahtinen et al. 2010), having its main collisional stage at 1930–1910 Ma (Daly et al. 2006). The latter option suggests that the Karelian and Svecofennian sedimentary rocks are unrelated and separated by a (cryptic) suture zone that is running in a SE-NW direction along the southwestern boundary of the craton.

A northeastern-directed collision of a Palaeoproterozoic microcontinent-arc collage with the Archaean craton at c. 1920–1910 Ma led to obduction of deep-water sediments and their oceanic crust basement and related ore deposits (e.g. Outokumpu) onto the Archaean basement. This was one of the accretionary events that produced the composite Svecofennian orogen in the southwestern part of the shield by amalgamating several microcontinents and island arc complexes in a prolonged period lasting c. 100 Ma (Lahtinen et al. 2009). The island arc complexes, which were formed between c. 1920 and 1890 Ma in unknown locations, comprise voluminous bimodal, arc-type volcanic rocks (Fig. 3.28 d–f) and associated thick successions of shales, volcaniclastic turbiditic greywackes gritstone and polymict conglomerate with some containing limestone beds and abundant carbonate concretions (Fig. 3.28g–l). The shale-greywacke assemblage represents erosional products of volcanic edifices accumulated in fluvial and shallow-water settings as well as in more distal, deep-water turbiditic environments (Kähkönen 2005).

Closure of the Kittilä Ocean in the north, which has been linked by Lahtinen et al. (2005) to accretion of the Norrbotten craton to the Karelian craton (Fig. 3.2), likely took place approximately at the same time (c. 1920 Ma; Hanski and Huhma 2005) as the above-mentioned first collisional event in the south, in the Outokumpu area. Arc-continent and subsequent continent-continent collisions that generated the Lapland-Kola orogen and related high-pressure metamorphic rocks mainly occurred at 1930–1910 Ma (Daly et al. 2006). The U-Pb zircon studies of rocks from the Finnish part of the Lapland Granulite Belt by Tuisku and Huhma (2006) indicate that the deposition of pelitic protoliths of the belt took place later than 1940 Ma, whereas enderbitic plutons representing arc magmatism were emplaced at c. 1920–1905 Ma. Subsequent burial and orogenic metamorphism followed by exhumation up to upper crustal levels was over by c. 1880 Ma. Thus the whole cycle lasted less than c. 60 Ma.

In the southern Karelian craton, the Onega rift succession, hosting a supergiant petroleum field, was subject to uplift and erosion, followed by reactivation of the old rift system, leading to formation of a large lacustrine basin (Fig. 3.22h).

A 500-m-thick succession of lacustrine, turbiditic, volcaniclastic greywackes, siltstones and mudstones, comprising the Kondopoga Formation, sits unconformably (Galdobina 1987b) on the c. 1980 Ma Suisari Formation basalts (Puchtel et al. 1998). The most typical rock type is rhythmically-bedded turbiditic volcaniclastic greywackes with several breccia- and massive sandstone-filled channels (Fig. 3.28m–p). Many greywacke beds contain abundant carbonate concretions and concretionary beds (Fig. 3.28m, n). Channel-filling polymict breccia is commonly matrix-supported and contains large fragments of siltstones, mudstones, sandstones and fragmented carbonate nodules. All clasts show a limited degree of sorting and variable roundness (Fig. 3.28o). Syndepositional faulting is common (Fig. 3.28k). Many beds contain ubiquitous pyrobitumen clasts (Fig. 7.5q–s) derived from eroded surface- and subaqueous oil seeps sourced from the cannibalised petroleum reservoirs (Melezhik et al. 2009).

In the Kola craton, sedimentary-volcanic successions associated with this period of time remain largely undated and poorly studied. In the southern part of the Imandra/Varzuga Belt, thick successions of orogenic rhyolites, rhyodacites and minor basalts, and volcaniclastic greywackes and arkosic sandstones (Saminga and Panarechka formations) are intruded by plagiomicrocline granite and subvolcanic trachydacite whose emplacement has been constrained to 1940 ± 5 Ma and 1970 ± 18 Ma, respectively (Skuf'in et al. 2006). Apparently, several orogenic and post-orogenic basins formed in the southern part of the Pechenga Belt, now represented by tectonically imbricated successions of volcaniclastic andesitic greywackes with interlayers of conglomerate containing andesite pebbles with pre-existing tectono-metamorphic fabrics (Melezhik and Sturt 1994). The rocks of the southern part of Pechenga Belt experienced deformation caused by the emplacement of the Kaskeljavr granodiorite massif dated at 1950.0 ± 9.5 Ma (Kozlova and Balagansky 1995; Vetrin et al. 2008). The crystallisation age of another similar intrusion, the Shuonijavr Massif, has been constrained to 1939 ± 7 Ma (Vetrin et al. 2008). However, the relationship of both massifs with the orogenic andesites remains unresolved.

3.3.7 1900–1850 Ma Period

As a hallmark of the last stages of orogeny, molasse-type, clastic and coarse-clastic sediments were accumulated in foreland basins with the detritus derived from various sources including the obducted slices of oceanic crust, the c. 1880 Ma Svecofennian synorogenic plutonic rocks from uplifted and eroded orogens, and the Archaean basement and its Palaeoproterozoic cover (Fig. 3.29a–g). Erosional remnants of these sedimentary formations accompanied locally with intermediate to felsic volcanic rocks are scattered across a wide area in northern Scandinavia (Hanski et al. 2001; Laajoki and Huhma 2006). In central Finnish Lapland, as much as 2,000 m thick sedimentary accumulations occur (Hackman 1927). Clasts in conglomerates represent a large array of lithologies: metavolcanic and intrusive rocks, schists, arenites, vein quartz, chert, jasper, and rare carbonate rock (Fig. 3.29a–c). Previous work by Kortelainen (1983), Räsänen and Mäkelä (1988) and Nikula (1988) suggested a depositional setting of alluvial fans and braided rivers.

3.3.8 Summary

The early Palaeoproterozoic evolution of the Fennoscandian Shield can be divided into several major stages of rifting, spreading and orogenesis. The initial stages are marked by multiphase, southwest-prograding, intraplate rifting associated with 2505–2430 Ma voluminous continental flood basalts emplaced in low latitudes. These were subject to uplift, deep dissection, weathering and erosion apparently resulting in intense consumption of atmospheric CO_2 by silicate weathering of fresh basalts. This should be considered as one of the crucial causes leading to the onset of the Huronian-age global cooling. In the aftermath of the Huronian icehouse, the great perturbation of the global carbon cycle (Lomagundi-Jatuli isotopic excursion) is recorded in ^{13}C-rich sedimentary carbonates. This is coincident with a significant rise in atmospheric oxygen as evident by widespread accumulation of terrestrial "red beds" and formation of abundant Ca-sulphates.

The earliest phase of continental separation occurred around 2100–2050 Ma leading to the formation of the Kola Ocean and formation of several carbonate platforms along the Karelian craton. Enhanced production and preservation of the organic matter in lacustrine continental margin rift basins resulted in formation of a supergiant oil-field. Between 2015 and 1953 Ma the shield underwent further drifting and separation as indicated by the newly formed Kittilä and Jormua Oceans; these were aborted and closed between 1920 and 1900 Ma. Orogenic processes affected the northeastern part of the shield in response to opening of the large pre-Svecofennian Ocean to the southwest. Subsequent multiple orogenic events continued mainly in the southwest, although they affected the older Palaeoproterozoic rocks across the entire shield area.

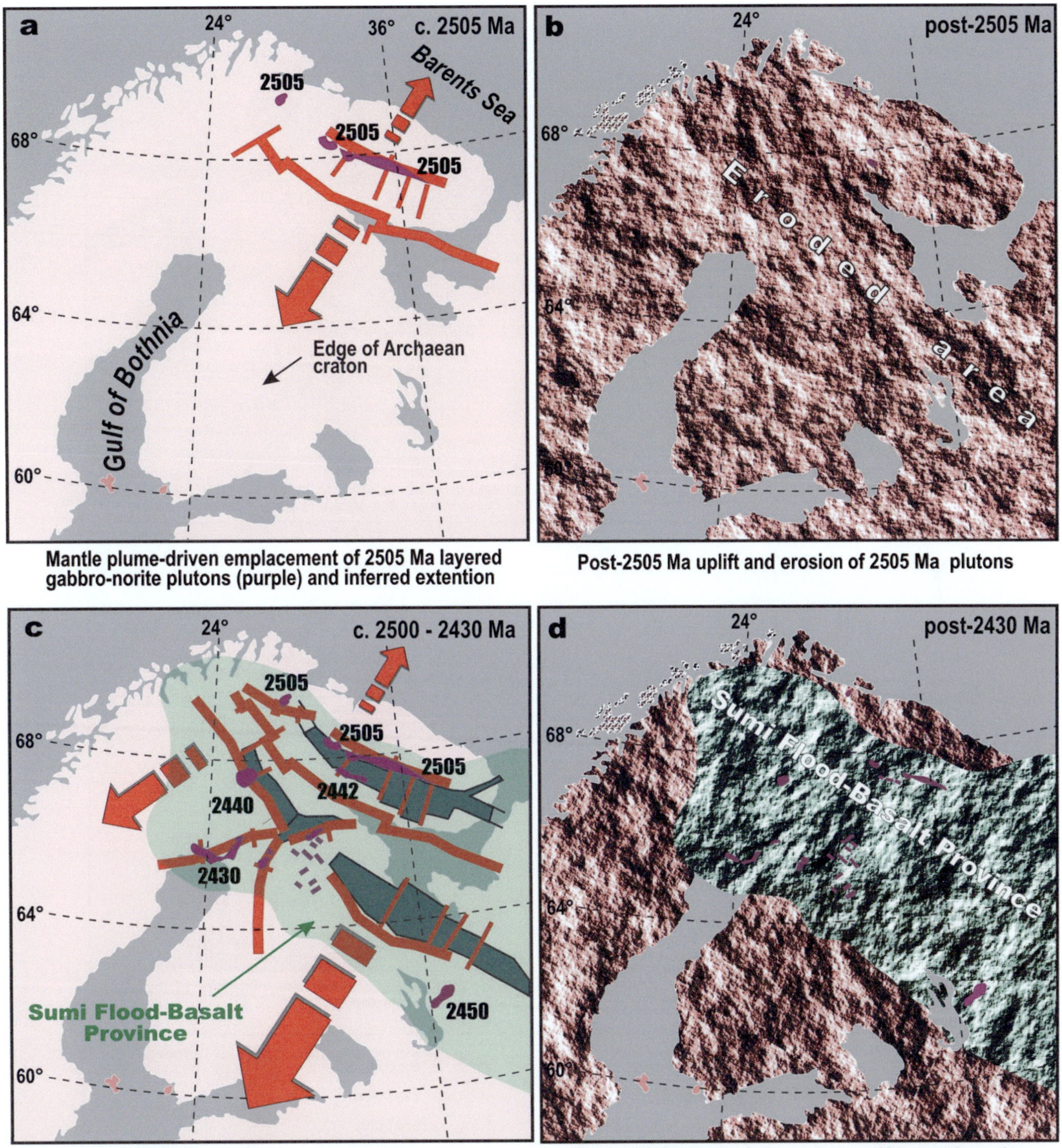

Fig. 3.22 Major stages of palaeotectonic and palaeogeographic evolution of the Fennoscandian Shield in the Early Palaeoproterozoic (**a–h**). Due to the lack of robust palaeomagnetic data, the 2060–1900 Ma time-slice reconstractions (**g** and **h**) were made based on the present-day locations. Position of accreted/obducted, 2100–1950 Ma oceanic crust, volcanic arc complexes and reworked Archaean crust in Filand (**h**) were adapted from Lahtinen et al. (2005). Position of 2000–1900 Ma volcanic arc complexes and reworked Archaean crust in Kola and Karelia (**h**) were adapted from Daly et al. (2006)

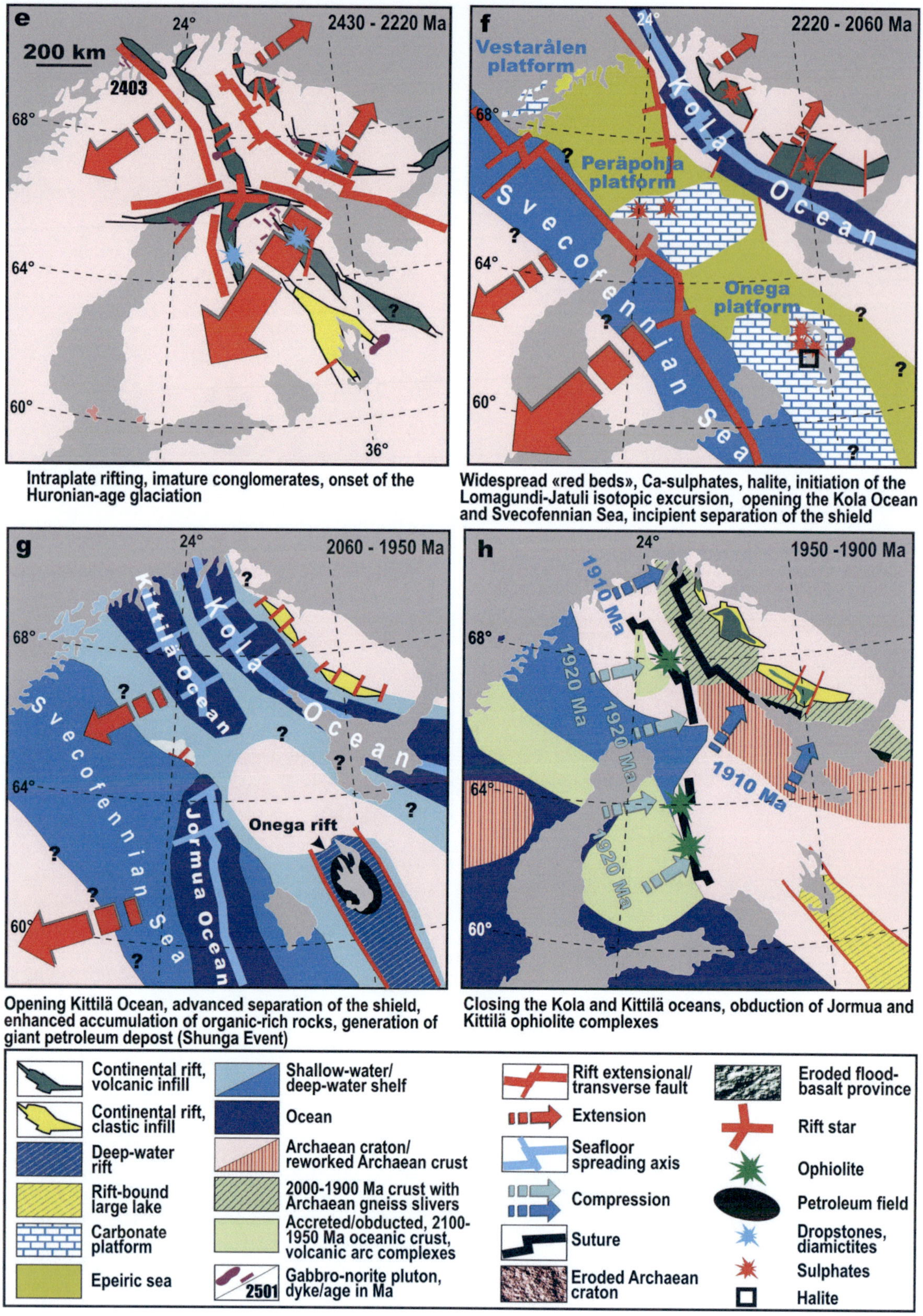

Fig. 3.22 (continued)

Fig. 3.23 Selected images of sedimentary and volcanic rocks representing major evolutionary features of the 2505–2430 Ma period *Kuksha Sedimentary Formation, western Imandra/Varzuga Belt*: (**a**) Polymict conglomerate at base of the formation overlaying 2505 Ma gabbro-norite intrusion in the Monchegorsk area; the conglomerate contains clasts of anorthosite (*1*) and gabbro-diorite (*2*) in dark grey gabbroic matrix; drillhole 2174, core diameter is 6 cm. *Belomorian Belt*: (**b**) Rhythmic layering in c. 2450 Ma minor gabbro-norite intrusion from the (Anisimov island). *Rookkiaapa Formation, Central Lapland Belt:* (**c**) Polymict conglomerate with basement granitoid clasts floating in a greywacke matrix

Fig. 3.23 (continued) *Seidorechka Sedimentary Formation, western Imandra/Varzuga Belt*: (**d**) Sandstone-mudstone with flaser and wavy bedding at the base of the formation. (**e**) Sandstone-mudstone with wavy and lenticular bedding at the base of the formation. (**f**) Cross-bedded quartz sandstone from the middle part of the formation, coin diameter is 2 cm

Fig. 3.23 (continued) *Seidorechka Sedimentary Formation, western Imandra/Varzuga Belt*: (**g**) Siltstone-shale (*dark grey*) overlain by massive dolostone (tan) and dolomite-cemented sandstone (*dark brown*) followed by *pale grey* quartz sandstone. *Kuksha Volcanic Formation, western Imandra/Varzuga Belt*: (**h**) Contacts between three thin, amygdaloidal, basaltic lava flows; lens cap diameter is 6 cm. *Central Karelia:* (**i**) Variolitic komatiitic basalt

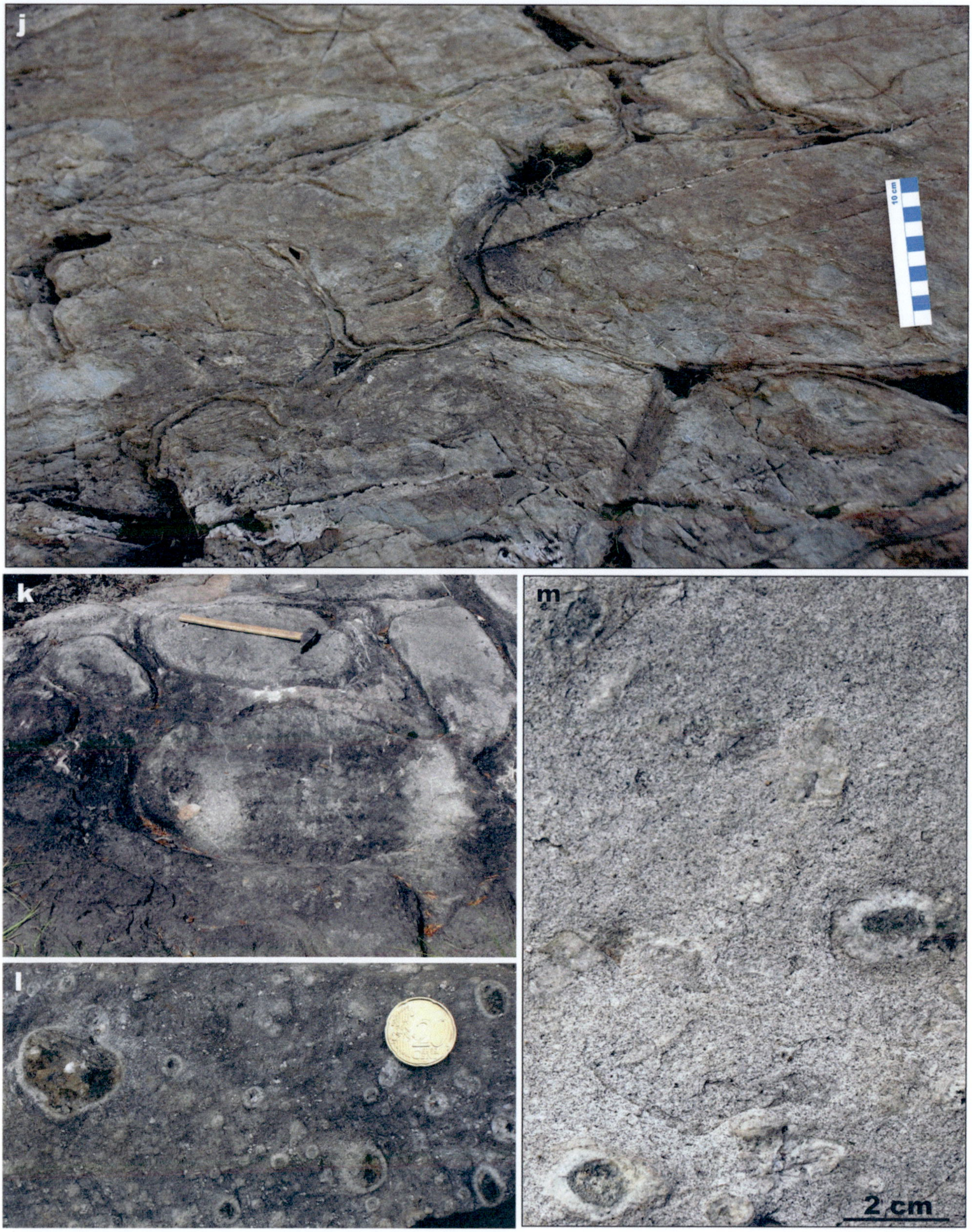

Fig. 3.23 (continued) *Central Karelia:* (**j**) Pillowed komatiitic basalt forming one of the numerous lava flows in a c. 1,500 m thick volcanic formation. *Tunguda Formation, Lekhta Belt:* (**k**) Pillowed basaltic andesite from middle part of c. 2,000 m thick volcanic unit. *Salla Formation, Salla Belt*: (**l**) Lithophysae in dacitic ash-flow tuff; coin diameter is 2 cm. *Seidorechka Volcanic Formation, western Imandra/Varzuga Belt*: (**m**) Amygdaloidal andesitic-dacite from the middle part of the formation (Photographs by Victor Melezhik (**a, d–h, m**) and Eero Hanski (**l**). Photographs courtesy of Petri Peltonen (**c**), Evgeny Sharkov (**b**), Sergey Svetov (**i, j**) and Valery Zlobin (**k**). Sample (**l**) courtesy of Tuomo Maninen)

Fig. 3.24 Selected images of sedimentary and volcanic rocks representing major evolutionary features of 2430–2300 Ma period. *Pechenga/Pasvik Belt, Neverskrukk Formation:* (**a**) Boulder-conglomerate (to the *right* of 10 cm long scale-bar) overlying uneven palaeorelief on top of paragneisses (to *left*)

Fig. 3.24 (continued) *Pechenga/Pasvik Belt, Neverskrukk Formation:* (**b**) Jointed Archaean pegmatite at the base of the formation; joints are filled by fine-grained massive to laminated greywacke. (**c**) Clast-supported polymict conglomerate containing rounded fragments of Archaean granites and diorites

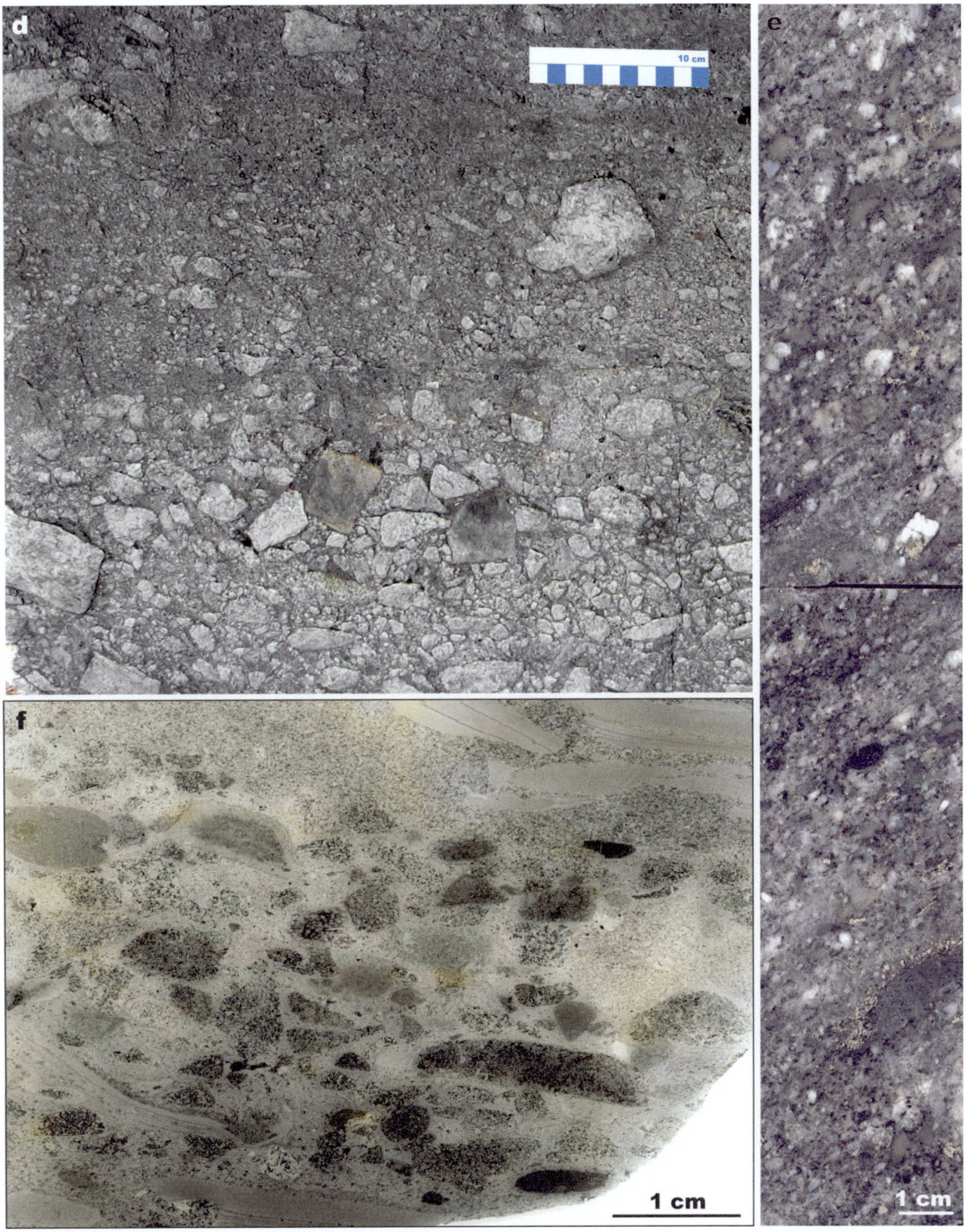

Fig. 3.24 (continued) *Pechenga/Pasvik Belt, Neverskrukk Formation:* (**d**) Conglomerate grading to gritstone and sandstone with scattered clasts of Archaean granites. (**e**) Drillcore of pebbly gritstone comprised of angular and rounded clasts of granites, amphibolites and schists. *Polisarka Sedimentary Formation, Imandra/Varzuga Belt:* (**f**) Polished slab of fine-pebble polymict conglomerate containing rounded fragments of basalts and andesites and platy clasts of laminated siltstone (*upper right* and *lower left* corners)

Fig. 3.24 (continued) *Sompujärvi Formation, Peräpohja Belt:* (**g**) Basal conglomerate deposited on Archaean granitoid basement; large granitoid clasts derived from a very local source; hammer-head is 14 cm long. *Unikumpu Formation, Kuusamo Belt*: (**h**) Conglomerate containing clasts of Archaean basement granitoids and Sumian felsic to intermediate volcanic rocks; hammer-head is 14 cm long. *Ahven-Kivilampi Formation, Kainuu Belt*: (**i**) Matrix-supported, polymict conglomerate containing clasts from underlying mafic volcanic rocks and granitoids from Archaean basement. Hammer-head in (**g**) and (**h**) is 14 cm long

Fig. 3.24 (continued) *Sompujärvi Formation, Peräpohja Belt:* (**j**) Basal polymict conglomerate with magnetite-rich matrix; *Segozero Belt:* (**k**) Basal polymict conglomerate with well-rounded boulders of Archaean granites. *Polisarka Sedimentary Formation, Imandra/Varzuga Belt:* (**l**) Polished drillcore slab of parallel-laminated shale; FAR-DEEP Hole 3A; core diameter is 5 cm. Hammer-head in (**j**) and (**k**) is 14 cm long

Fig. 3.24 (continued) *Peräpohja Belt:* (**m**) C. 2440 Ma layered gabbro-norite intrusion (*1*) erosively overlain by polymict conglomerate of the Sompujärvi Formation (*2*), with large, light-coloured clasts of Archaean granitoids, followed by mafic lava of the Runkaus Formation (*3*). *Polisarka Sedimentary Formation, Imandra/Varzuga Belt:* (**n**) Polished slab of "varved" siltstone and graded sandstone-siltstone beds with numerous different scale dropstones

Fig. 3.24 (continued) *Urkkavaara Formation, Northern Karelia Belt*: (**o**) Granitic dropstone in clayey siltstone. *Pajozero Formation, Shambozero Belt*: (**p**) Rhythmically bedded calcareous sandstone-shale. (**q**) Diamictite composed of unsorted polymict clasts in massive clayey sandstone matrix

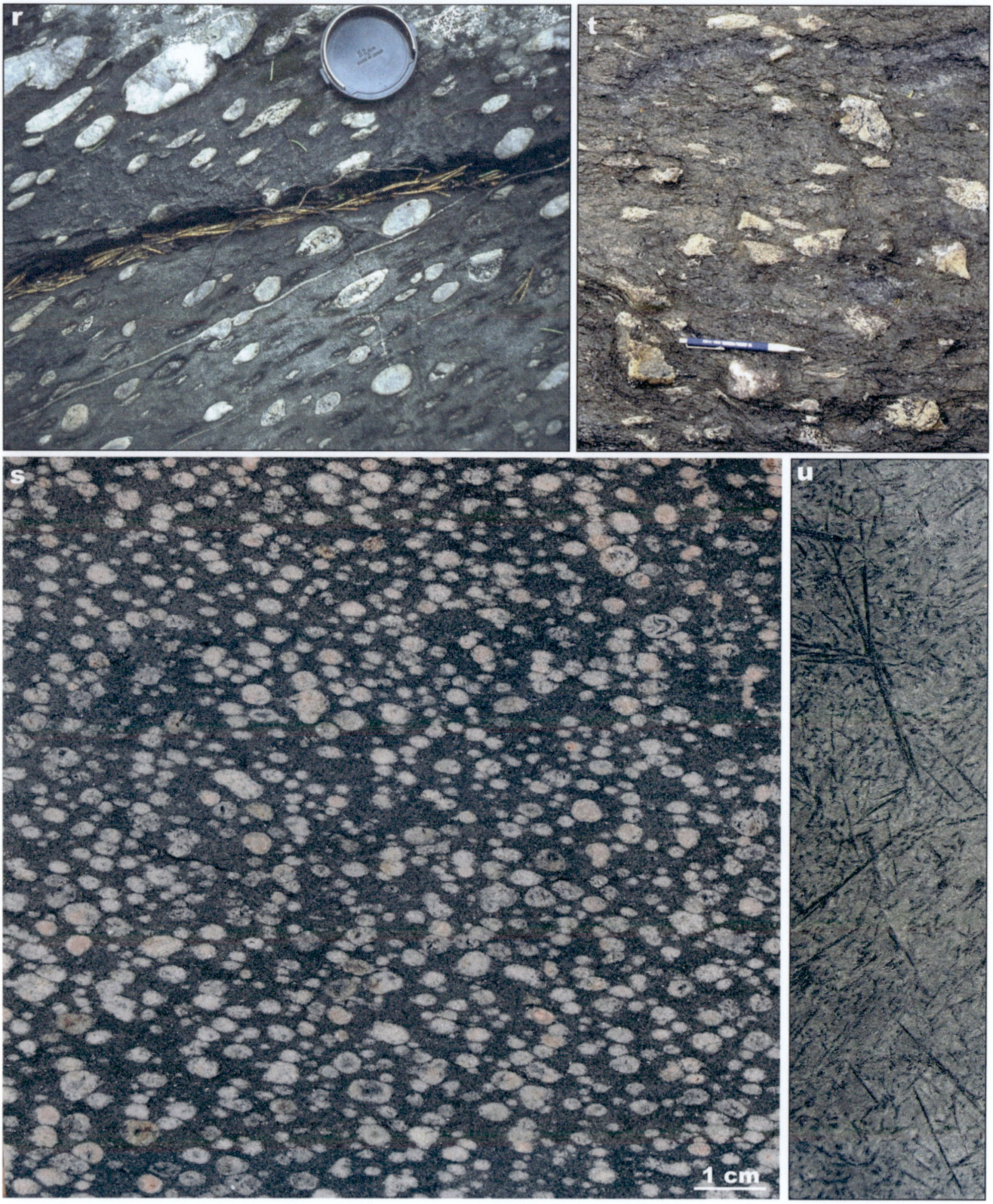

Fig. 3.24 (continued) *Kaukonen Formation, Central Lapland Belt:* (**r**) Amygdaloidal andesitic lava; lens cap diameter is 6 cm. *Möykkelmä Formation, Central Lapland Belt:* (**s**) Highly amygdaloidal, basaltic andesite lava flow. (**t**) Gneiss fragments from Archaean basement in komatiitic tuff; pen length is 12 cm. *Polisarka Volcanic Formation, Imandra/Varzuga Belt:* (**u**) Pyroxene spinifex texture in komatiitic basalt; FAR-DEEP Hole 3A, core diameter is 5 cm (Photographs by Victor Melezhik (**a–f, k–n, p, q, u**) and Eero Hanski (**h, i, o, s**). Photographs courtesy of Vesa Perttunen (**g, j**), Pentti Rastas (**r**) and Jorma Räsänen (**t**))

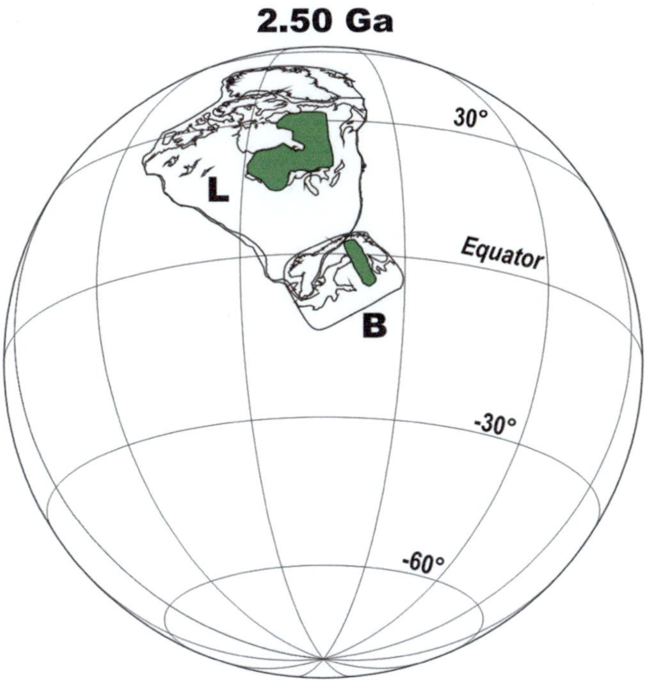

Fig. 3.25 Continental reconstruction at 2500 and 2450 Ma. **B** stands for Baltica (the Fennoscandian Shield) and **L** for Laurentia. The Karelian and Superior cratons are shown in *green* (Figure courtesy of Satu Mertanen (Based on data from Mertanen et al. 2006))

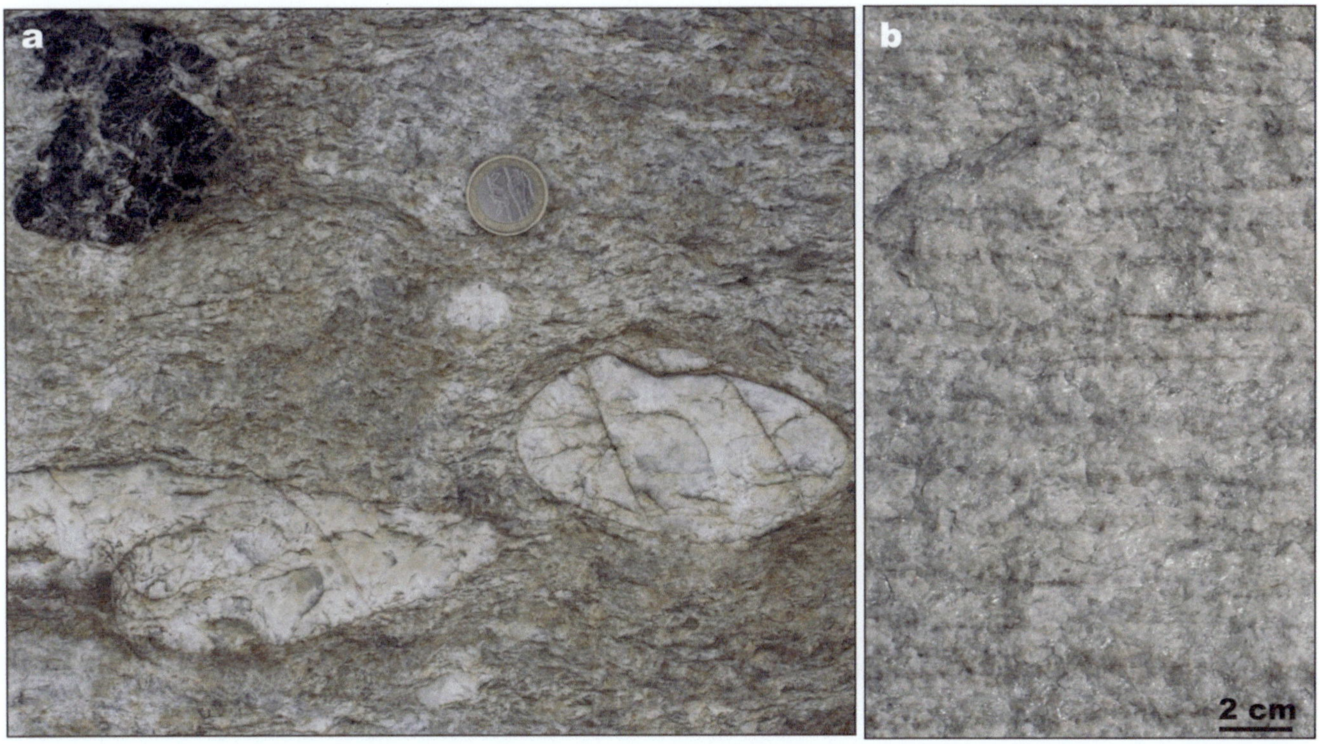

Fig. 3.26 Selected images of sedimentary and volcanic rocks representing major evolutionary features of the 2300–2060 Ma period. *Hokkalampi Formation, North Karelia Belt*: (**a**) Chemically-resistant tourmaline and vein quartz clasts in the Hokkalampi palaeosol; coin diameter 2 cm. (**b**) The Hokkalampi palaeosol comprised of quartz, sericite and kyanite

Fig. 3.26 (continued) *Kuetsjärvi Sedimentary Formation, Pechenga Belt:* (**c**) Stacked cross-bed co-sets in quartzitic sandstone. (**d**) "Ball-and-pillow" structure in arkosic sandstone. (**e**) Sedimentary breccia consisting of dolostone clasts (*pale yellow*) in muddy sandstone matrix. Pen length in (**d**) and (**e**) is 13 cm

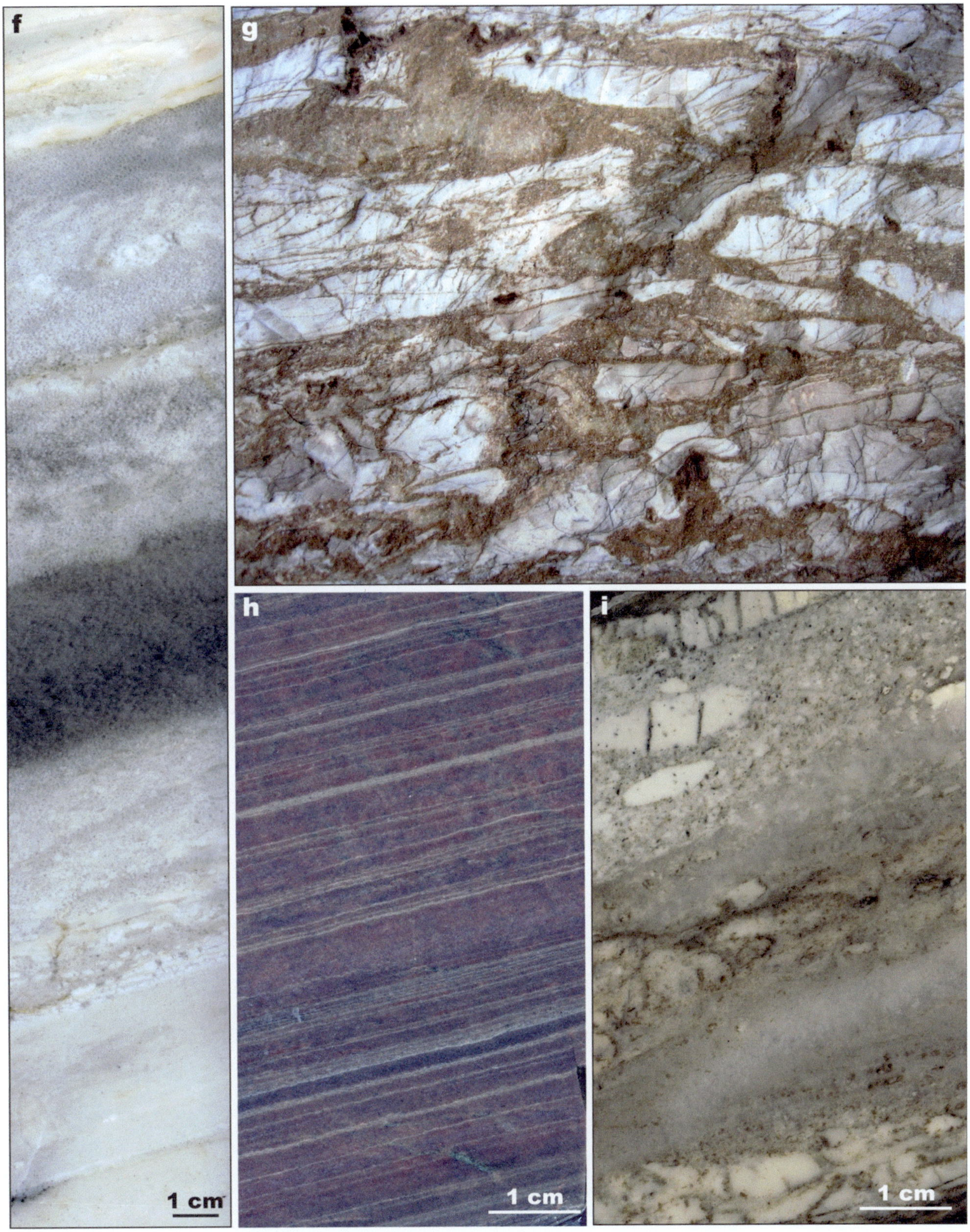

Fig. 3.26 (continued) *Kuetsjärvi Sedimentary Formation, Pechenga Belt:* (**f**) Beds of massive dolostone (*bottom*) overlain by *pale grey* cross-laminated, allochemical dolostone and *dark grey* massive quartzitic sandstone which are followed by dolomite-cemented sandstone and dolarenite caped by light-coloured, faintly laminated dolomicrite. (**g**) Angular, platy clasts of white dolostone encased in a dolomite-cemented sandstone; photograph width is 0.5 m. (**h**) Lacustrine, variegated, haematite-rich siltstone with fine, varve-like lamination. (**i**) Lacustrine, crudely cross-stratified dolomite-cemented sandstone with angular dolostone fragments

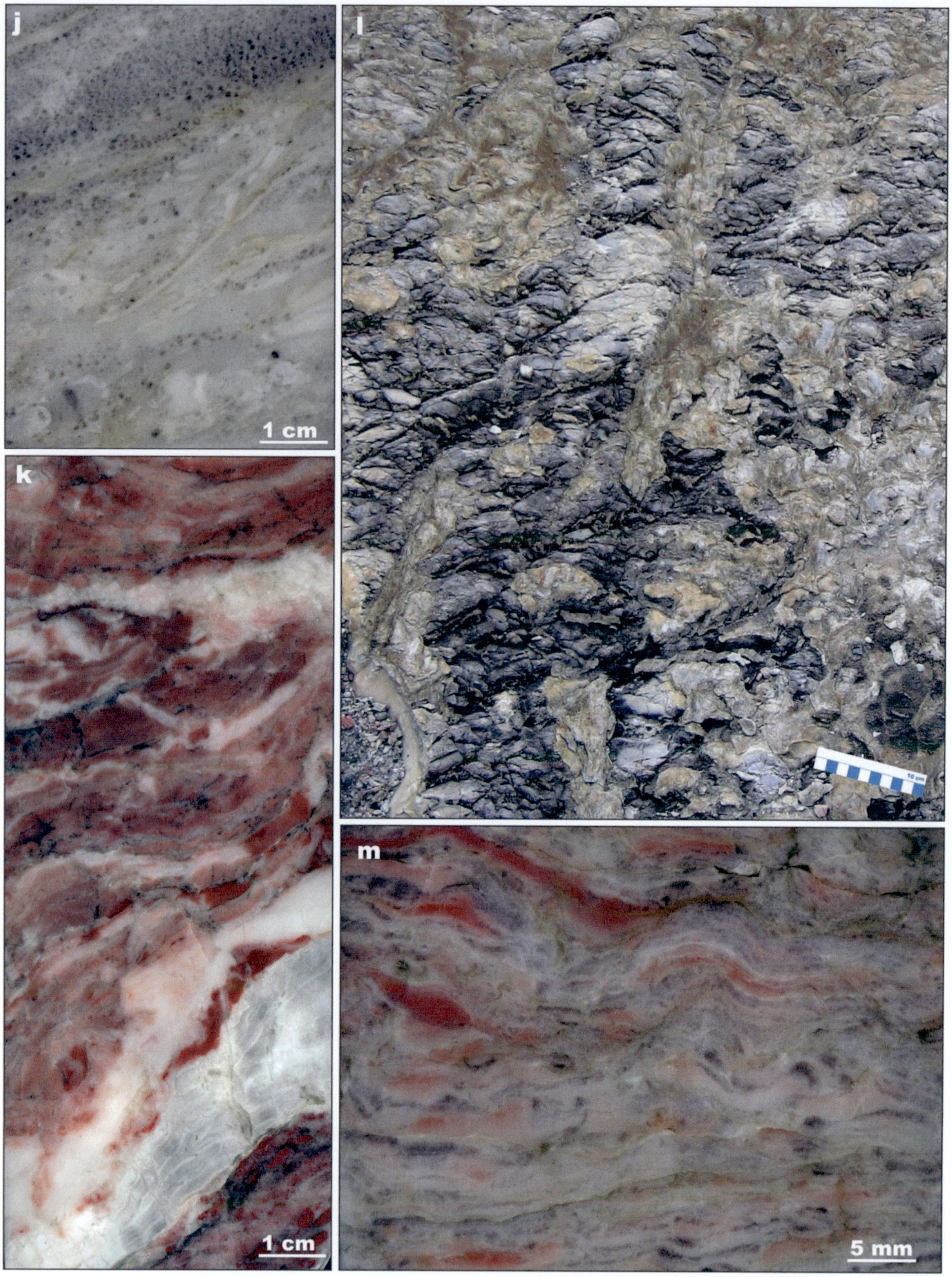

Fig. 3.26 (continued) *Kuetsjärvi Sedimentary Formation, Pechenga Belt:* (**j**) Lacustrine, allochemical sandy dolostone overlain by quartzitic sandstone with dolomite matrix. (**k**) Vertical section through haematite-stained dolocrete with tepee-related deformation and thin laminated travertine crust (light-coloured). (**l**) Bedding surface covered with travertine crust. (**m**) Vertical section through mottled, microbially laminated (stromatolitic) dolostone comprised of flat, curly and blister microbial dolomicrite laminae separated by thin, irregular partings of *pale grey* silica sinters

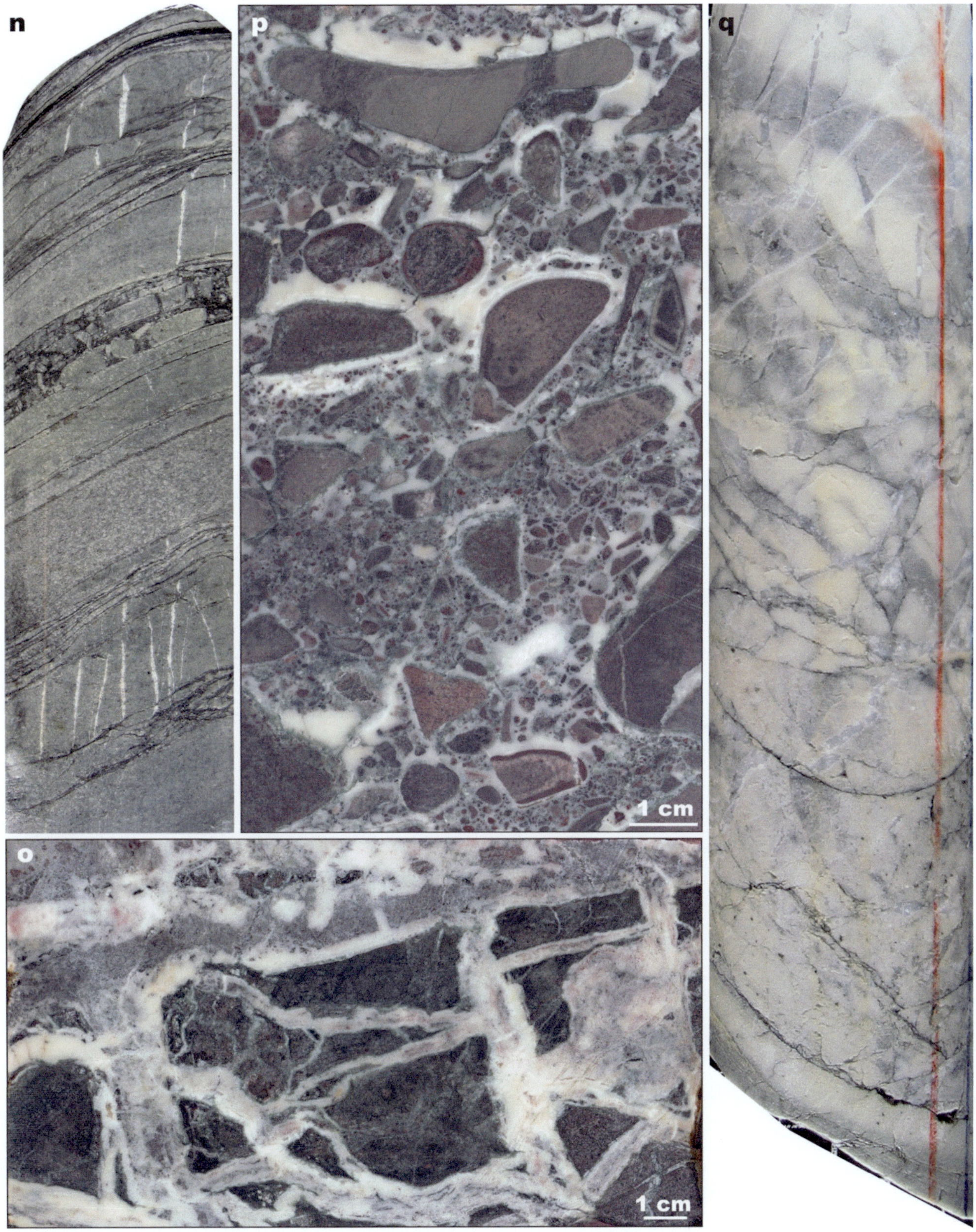

Fig. 3.26 (continued) *Umba Sedimentary Formation, Imandra/ Varzuga Belt,*: (**n**) Low-angle cross-laminated siltstone with thin mudstone drapes and disintegrated bed caused by slumping (**o**) Clasts of ultramafic lava cemented by dolomite. (**p**) Matrix-supported (dolorudite) dolostone conglomerate with pebbles of brown Fe-rich carbonates cemented by white dolomite drusite. (**q**) Massive, microcrystalline dolostone beds with thin mud-drapes overlain by in situ brecciated dolostone beds. (**n**) and (**q**) FAR-DEEP Hole 4A, core diameter is 5 cm

Fig. 3.26 (continued) *Kuetsjärvi Sedimentary Formation, Pechenga Belt:* (**r**) Subaerially-erupted amygdaloidal alkaline basalt – a representative rock type comprising c. 500 m of *upper* part of the formation. (**s**) Mud-flow consisting of large, unsorted fragments of alkaline basalts, andesites and dacites floating in black haematite-rich mudstone matrix; pen length is 13 cm. (**t**) Pillows of alkaline basalts separated by black, haematite-rich, softly-deformed mudstone on which the lava-flow was emplaced

Fig. 3.26 (continued) *Kuetsjärvi Volcanic Formation, Pechenga Belt:* (**u**) Alkaline basaltic lava-flow with scattered pillows; blue scale-bar in the *right-lower* corner is 10 cm. (**v**) Alkaline rhyolite flow-breccia with large, syndepositionally-fragmented, rounded blocks of *pink*, massive rhyolite separated by fluidal lava; joints in rhyolite blocks and interfragmental space is filled with black haematite and magnetite; many bands in fluidal lava are enriched in haematite and magnetite and occur as black layers; hammer head is 16 cm long

Fig. 3.26 (continued) *Tulomozero Formation, Onega Basin*: (**w**) Quartzitic sandstone with stacked, small-scale, trough cross-bed co-sets overlain by a bed with tabular cross-bedding having tangential base; scale-bar in the bottom is 10 cm. (**x**) Matrix-supported conglomerate with sorted and rounded clasts of vein quartz. (**y**) Symmetrical, bifurcating, wave-ripples on sandstone bed surface; hammer head is 16 cm

Fig. 3.26 (continued) *Segozero Formation, Segozero Belt*: (**z**) Cross-section view of a series of *dark brown*, desiccated mudstone beds in parallel-bedded arkosic sandstone. (**aa**) Cross-section through multiple-desiccated and dismembered brown mudstone bed in a coarse-grained arkosic sandstone. (**ab, ac**) Bedding surface with numerous dark-coloured and brown mud-chips in a sandstone-bed surface. (**ad**) *Red*, haematite-stained, intensely desiccated mudstone bed; hammer-head is 20 cm long

Fig. 3.26 (continued) *Koli Formation, North Karelia Belt:* (**af**) Fragment-supported, quartz-pebble conglomerate showing normal grading. *Jero Formation, North Karelia Belt*: (**ag**) Cross-bedded, arkosic gritstone. *Koli Formation, North Karelia Belt*: (**ah**) Large-scale, trough cross-bedding in arkosic sandstone. *Paljakkavaara Formation, Kainuu Belt*: (**ai**) Quartz-pebble conglomerate supported by clayey gritstone matrix

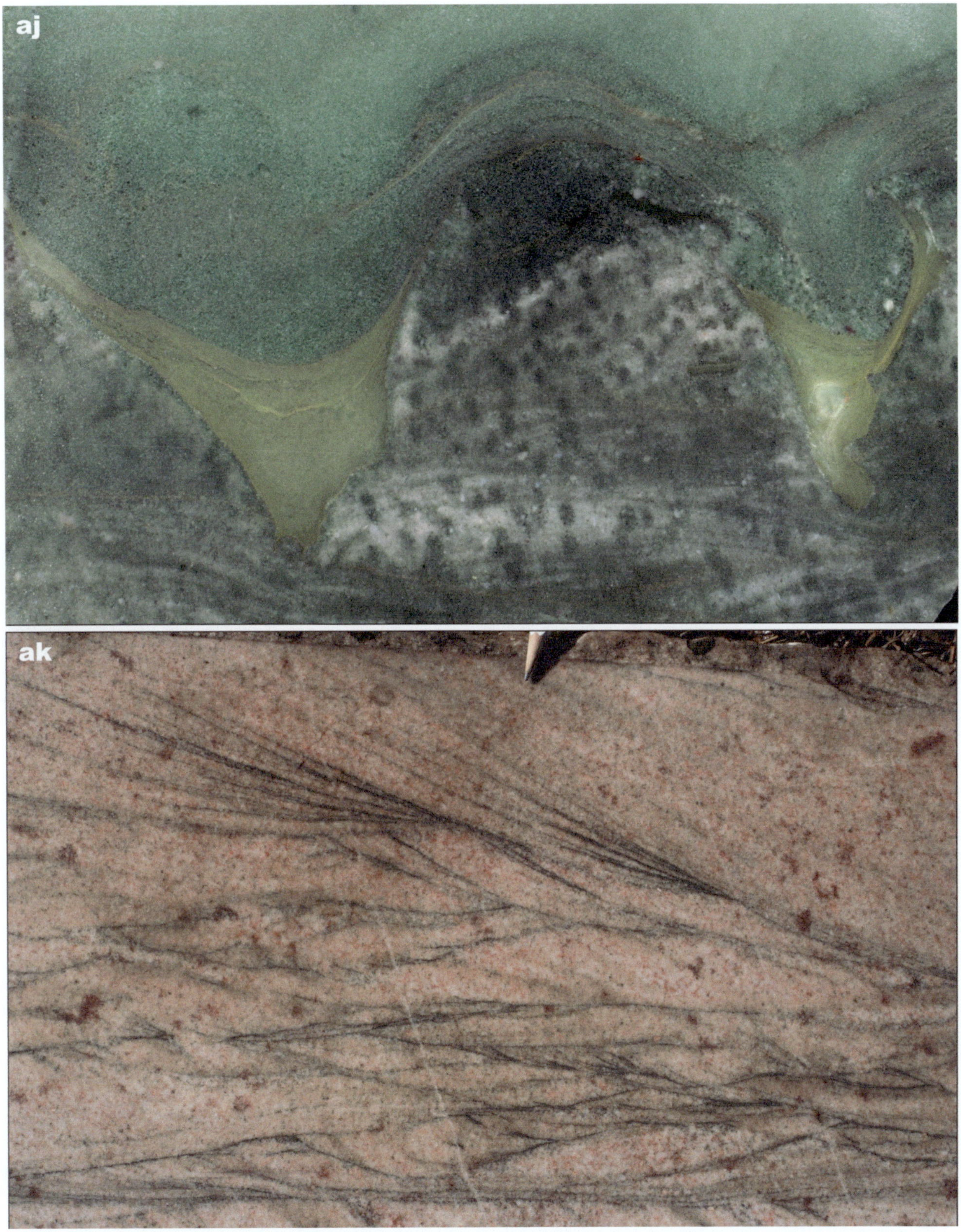

Fig. 3.26 (continued) *Virttiövaara Formation, Central Lapland Belt:* (**aj**) Erosional surface on top of a small-scale, trough cross-bedded, "spotty", fuchsite-rich quartzite; small-scale erosional troughs are filled with light brown claystone followed by laminated and massive fuchsite-rich quartz sandstone; width of view is 25 cm. *Virttiövaara Formation, Central Lapland Belt*: (**ak**) Small-scale, trough cross-beds in arkosic sandstone; several laminae are enriched in clastic haematite (black); pencil tip is 1 cm long

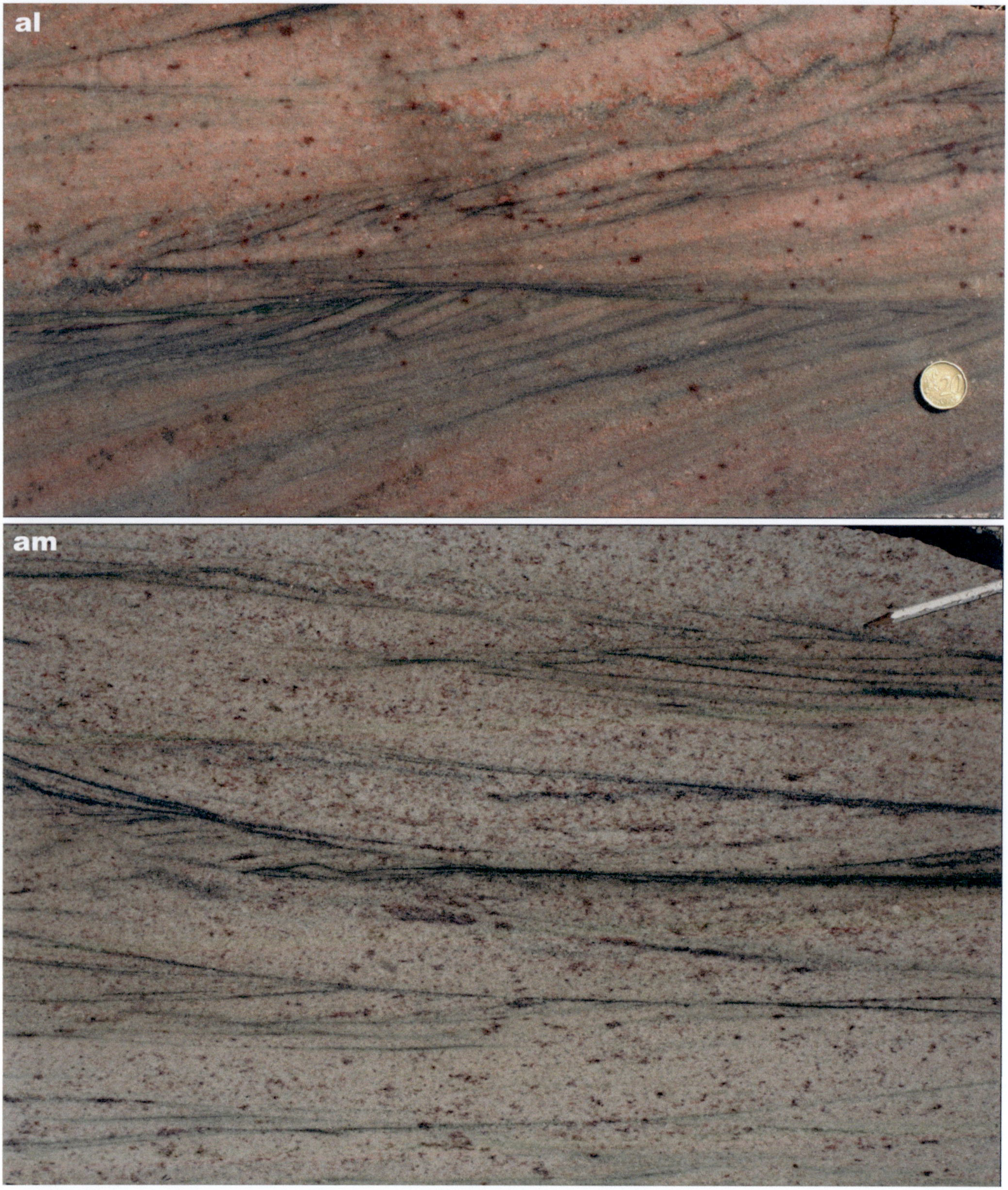

Fig. 3.26 (continued) *Virttiövaara Formation, Central Lapland Belt*: (**al, am**) Small-scale, trough cross-beds in pale pink and light grey arkosic sandstone; several laminae enriched in clay-material and clastic haematite (*black*); coin diameter is 2 cm, pencil length is 14 cm

Fig. 3.26 (continued) *Palokivalo Formation, Peräpohja Belt:* (**an**) Symmetrical, bifurcating wave-ripples in quartz sandstone. (**ao**) Sandstone-filled, polygonal desiccation cracks in a dark brown, haematite-stained, mudstone bed; tag width is 5 cm. *Rukatunturi Quartzite Formation, Kuusamo Belt*: (**ap**) Rhomboidal ripples affected by desiccation cracks in a siltstone bed; pen length is 12 cm. (**aq**) Interbedded wavy sandstone-mudstone succession with numerous brown mudstone rip-ups above erosional surface that progressively cuts down from left to right in to underlying beds; lens cap diameter is 6 cm

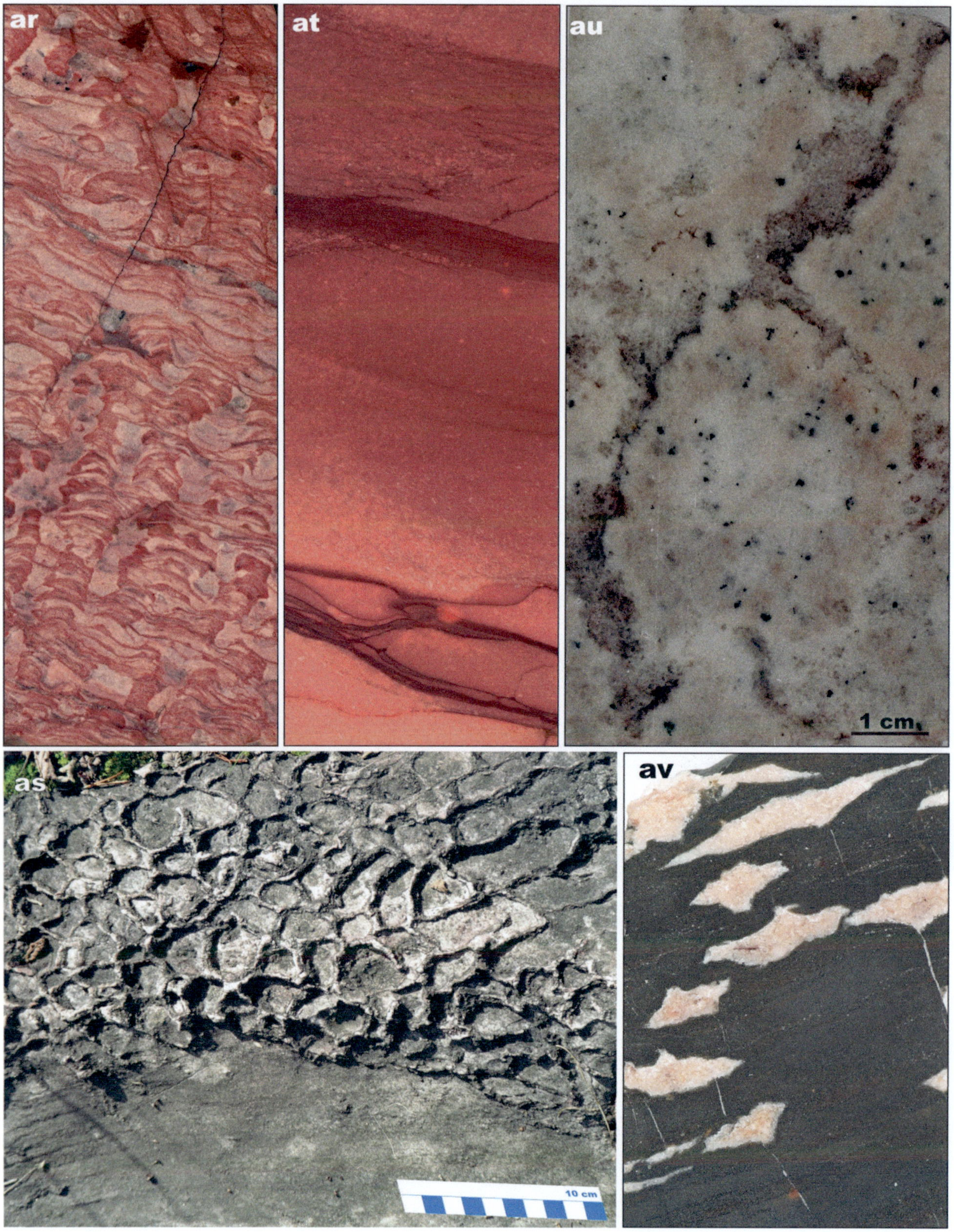

Fig. 3.26 (continued) *Tulomozero Formation, Onega Basin*: (**ar**) *Red* columnar stromatolites showing bifurcate style of beta-branching; intercolumnar space is filled with *pale pink* dolarenite; core diameter is 4 cm. (**as**) Bedding surface showing a subvertical contact between stromatolitic biostrome and crudely bedded dolarenite. (**at**) *Red*, low-angle cross-bedded, fine-grained dolarenite with a few ripples (*bottom*) separated by brown mudstone (wavy bedding); (**au**) Desiccated bedding surface (extending from *upper left* to *lower right* corner) between two *pale pink* dolostone layers. (**av**) Cross-section view of flat, discoidal, dolomite-pseudomorphed, gypsum crystals flattened normal to c-axis in *black*, haematite-rich, mudstone. Drillcore diameter in (**ar**) and (**at–av**) is 4 cm

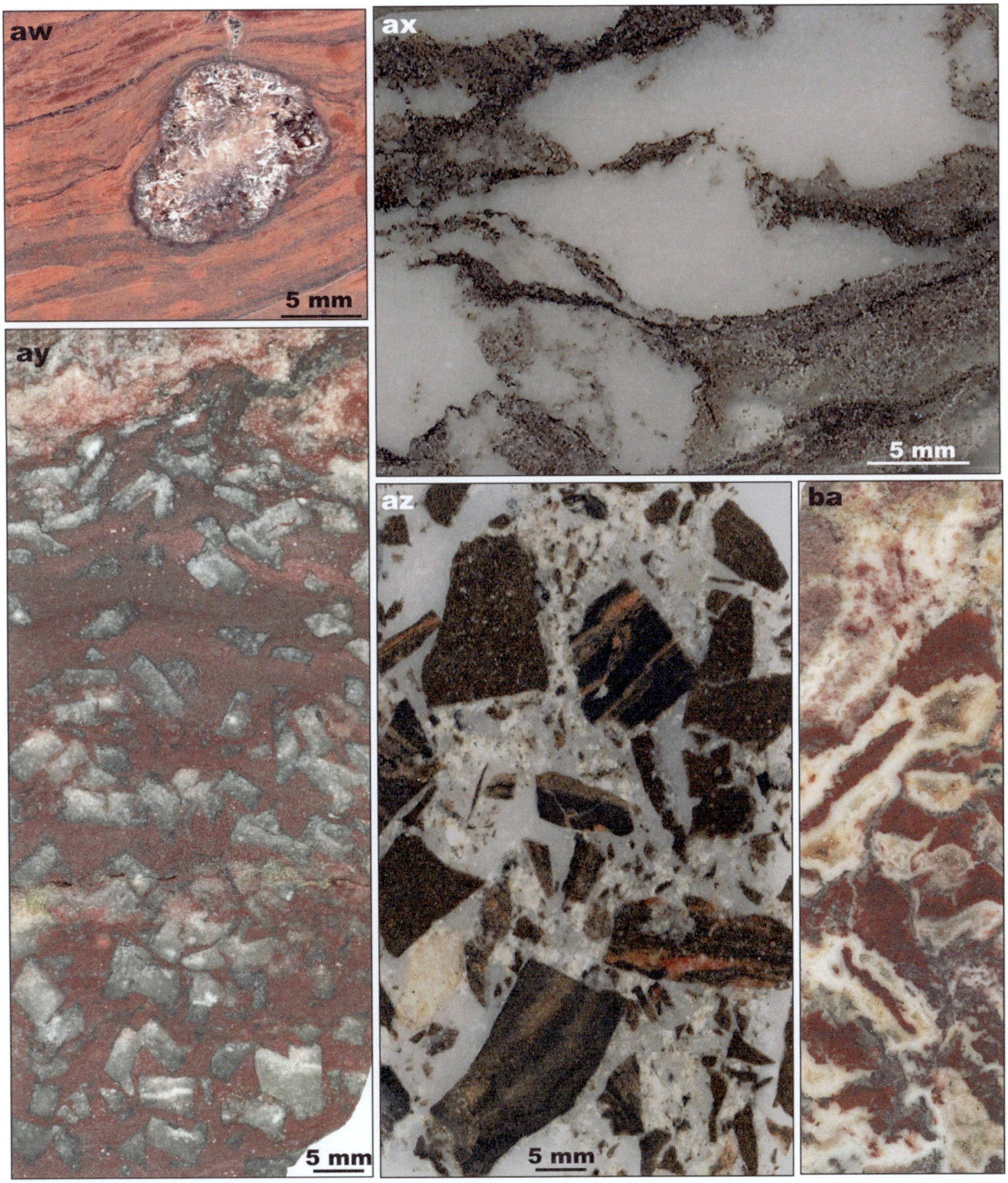

Fig. 3.26 (continued) *Tulomozero Formation, Onega Basin*: (**aw**) Silica-pseudomorphed Ca-sulphate nodule in red mudstone with prominent differential compaction suggesting an early diagenetic origin of the nodule. (**ax**) Dolomarl with white dolomite-pseudomorphed, displacively grown, gypsum nodules and crystals having discoidal or lenticular morphology flattened normal to c-axis; sulphate growth caused plastic distortion of primary lamination. (**ay**) Silica- and dolomite-pseudomorphed halite crystals grown displacively in *red* mudstone. (**az**) Collapse-breccia consisting of brown, haematite-stained mudstone fragments cemented with white sparry calcite. (**ba**) Dissolution breccia consisting of red, haematite-stained mudstone fragments cemented with initial isopachous bladed meteoric phreatic calcite

Fig. 3.26 (continued) *Middle formation, Kalix Belt*: (**bb**) Plan view of mafic, amygdaloidal lava flow with incipient pillow; pen length is 13 cm. *Lower formation, Kalix Belt*: (**bc**) Photomicrograph in transmitted non-polarised light showing cross-section view of small-scale sedimentary dyke developed in volcaniclastic greywacke. (**bd**) Mafic tuff (*dark brown*) interfingering but not mixed with laminated dolomicrite. (**be**) Photomicrograph in transmitted non-polarised light of dolomite-cemented sandstone sharply overlain by arenite with mafic tuff matrix. (**bf**) Photomicrograph in transmitted non-polarised light of dolomite- and tuff-cemented arenite containing layers of pure, fine-grained material

Fig. 3.26 (continued) *Lower formation, Kalix Belt*: (**bg**) Cross-section view through light brown, stuctureless sandstone containing dolostone clasts and a light coloured bed of dolorudite. (**bh**) Cross-section view through light brown, stuctureless sandstone overlain by cross-bedded dolarenite and stuctureless dolarenite with scattered dolostone pebbles. *Rantamaa Formation, Peräpohja Belt:* (**bi**) Cross-bedded quartz sandstone with platy, rounded fragments of tan dolostone clasts; coin diameter is 2 cm. Pen length in (**bg**), (**bh**) and (**bi**) is 15 cm. (**bj**) Cross-bedded quartz sandstone with pale brown dolostone intraclasts

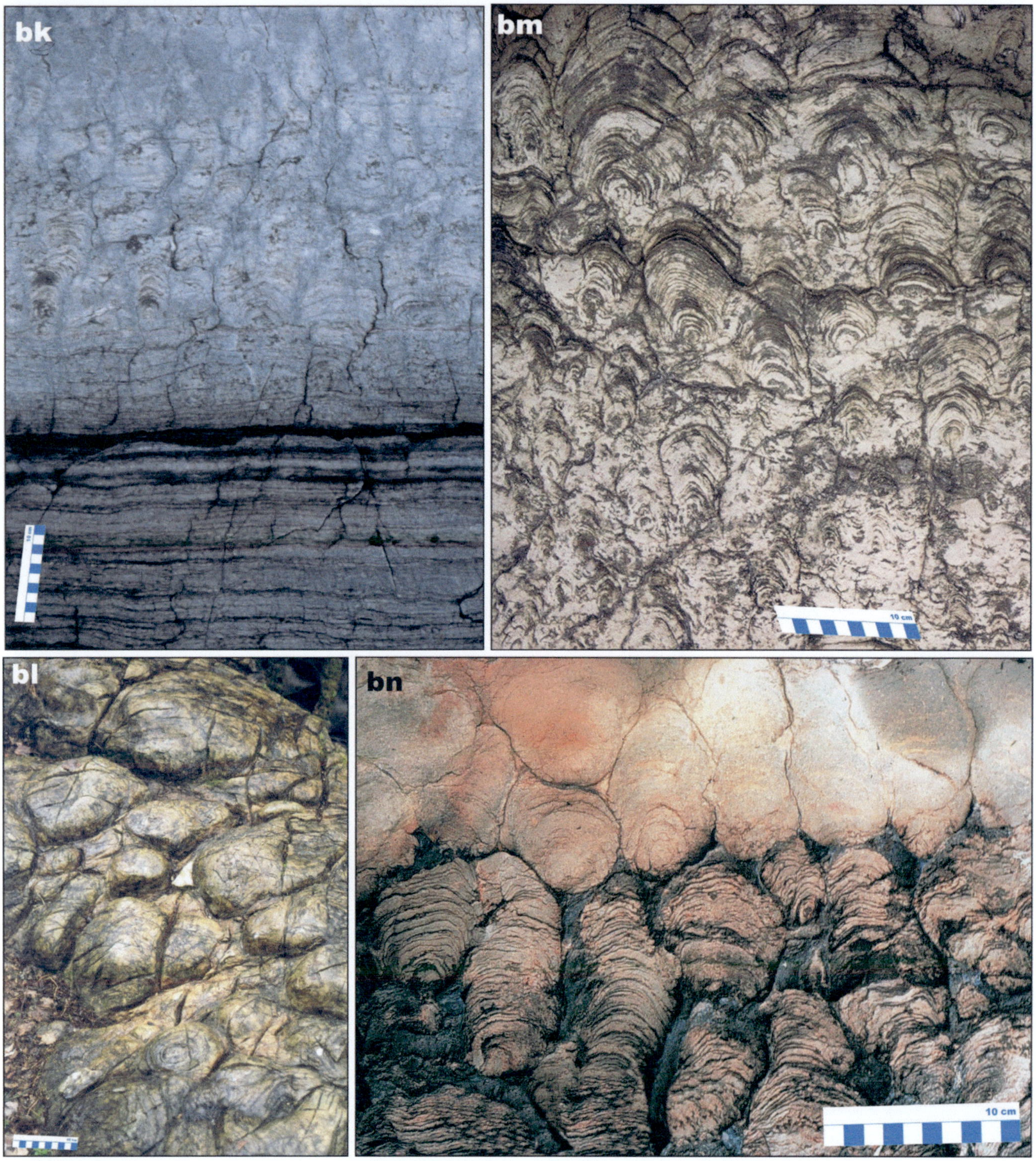

Fig. 3.26 (continued) *Lower formation, Kalix Belt*: (**bk**) Cross-section view of outcrop showing part of a intertidal biostrome comprised of cryptalgal laminated dolostones and beta-branching, columnar stromatolite. *Rantamaa Formation, Peräpohja Belt:* (**bl**) Plan-view of exhumed surface of stromatolite biostrome. (**bm**) Cross-section view of tightly-packed, subspherical stromatolites. (**bn**) Bioherm comprised of loosely-packed stromatolite columns separated by black clayey greywacke and overlain by tightly packed sub-spheroidal stromatolites with dolarenite filling interspherical space

Fig. 3.26 (continued) *Lower formation, Kalix Belt*: (**bo**) Bedding-parallel surface with incipient desiccation cracks in dolomarl. (**bp**) Mafic tuff-filled desiccation cracks in mudstone. (**bq**) Sand-filled desiccation cracks in dolomarl; coin diameter is 2 cm. *Upper formation, Kalix Belt*: (**br**) Interbedded sandstone-dolarenite overlain by a bed of stone-rosettes

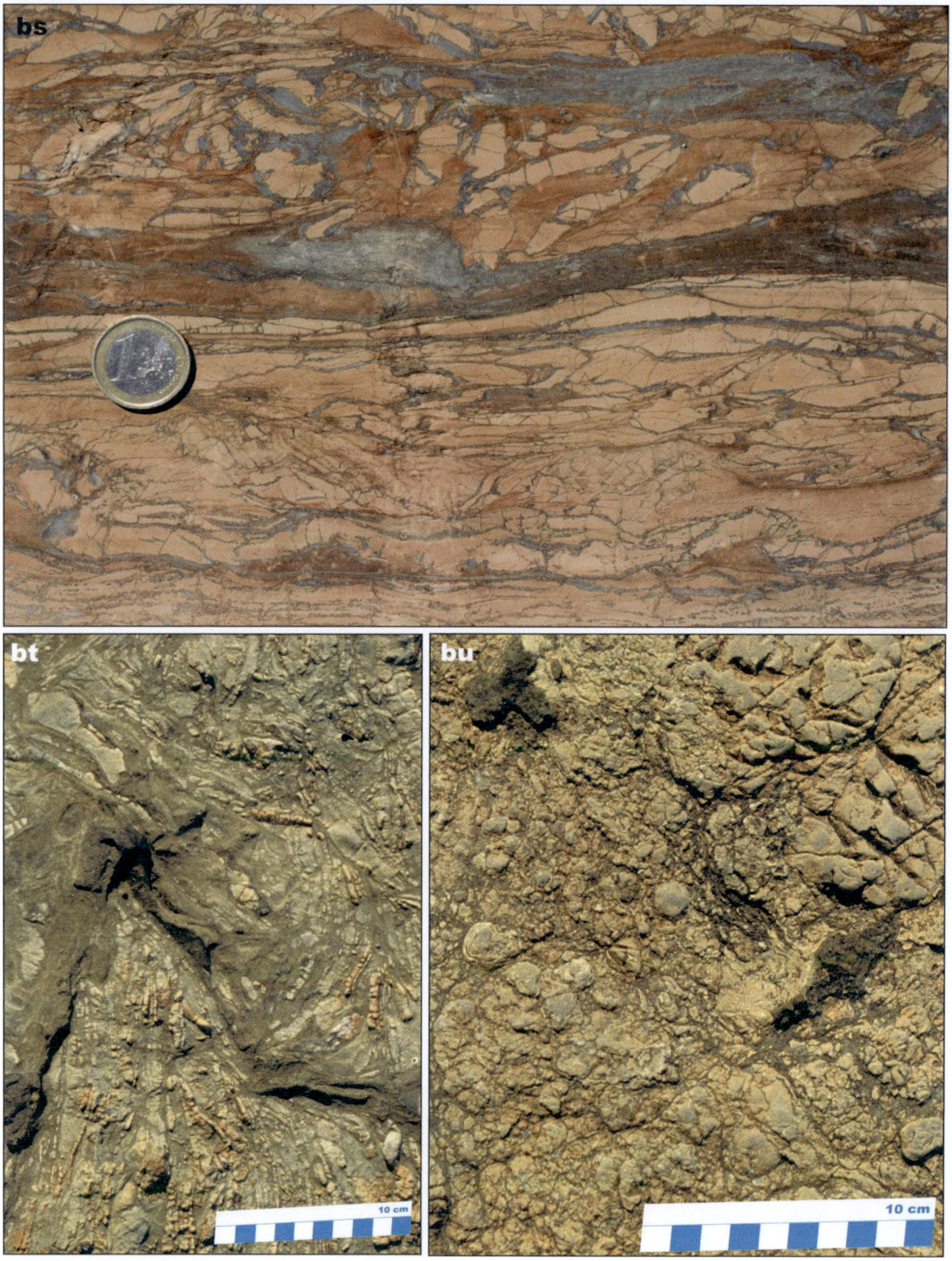

Fig. 3.26 (continued) *Rantamaa Formation, Peräpohja Belt:* (**bs**) In situ-brecciated dolostone with lenses of grey arenite and stone-rosettes; coin diameter is 2 cm. *Upper formation, Kalix Belt*: (**bt**) Bedding surface of dolostone covered by stone-rosettes. (**bu**) Wrinkled surface on dolostone bed

Fig. 3.26 (continued) *Lower formation, Kalix Belt*: (**bv**) Quartz-sandstone with reversals of dip in succeeding cross-stratified units (herring-bone cross-stratification. (**bw**) Quartz-sandstone with imbricated pebbles and herring-bone cross-stratification; pen length is 15 cm. *Rantamaa Formation, Peräpohja Belt:* (**bx**) Dolarenite with quartz sandstone exhibiting herring-bone cross-stratification

Fig. 3.26 (continued) *Rantamaa Formation, Peräpohja Belt:* (**by**) Rhythmically bedded (ribbon) light- and dark-coloured dolostone deposited on a continetal slope; width of the view is c. 5 m. *Tikanmaa Formation, Peräpohja Belt:* (**bz**) Rhythmically bedded mafic tuff (Photographs by Eero Hanski (**a, b, af, ag, ai–am, ap, bj, bs**) and Victor Melezhik (**g, h, l–ad, ar, at, au,y, az, ba, bh, bk, bl, br, bt, bw**). Photographs courtesy of Rein Kesola (**ah**), Vesa Perttunen (**an, ao, bm, bn, by, bz**) and Kari Strand (**aq**). Photographs (**c–f, i–k**) reproduced with permission of Royal Society of Edinburgh from Melezhik and Fallick (2005). Photographs (**au, as**) reproduced from Melezhik et al. (2000), (**av–ax**) from Melezhik et al. (2005a), (**bb–bg, bo–bq**) from Wanke and Melezhik (2005))

Fig. 3.27 Selected images of sedimentary and volcanic rocks representing major evolutionary features of the 2060–1950 Ma period. *Kolosjoki Volcanic Formation, Pechenga Belt*: (**a**) Tholeiitic pillow lava extruded on a seafloor and forming together with flow breccia the main part of the c. 2 km thick formation. *Pilgujärvi Volcanic Formation, Pechenga Belt*: (**b**) Tholeiitic pillow lava extruded on a seafloor and forming the main component of the c. 5 km thick formation

Fig. 3.27 (continued) *Kolosjoki Sedimentary Formation, Pechenga Belt*: (**c**) Tabular cross-bedded greywacke overlying rippled and parallel-bedded sandstone-mudstone. (**d**) Co-sets of trough cross-beds comprised of volcaniclastic greywacke. (**e**) Interbedded gritstone, sandstone and siltstone showing wavy, parallel bedding and a lens of pale green mudstone on *top*; mobile phone length is 10 cm

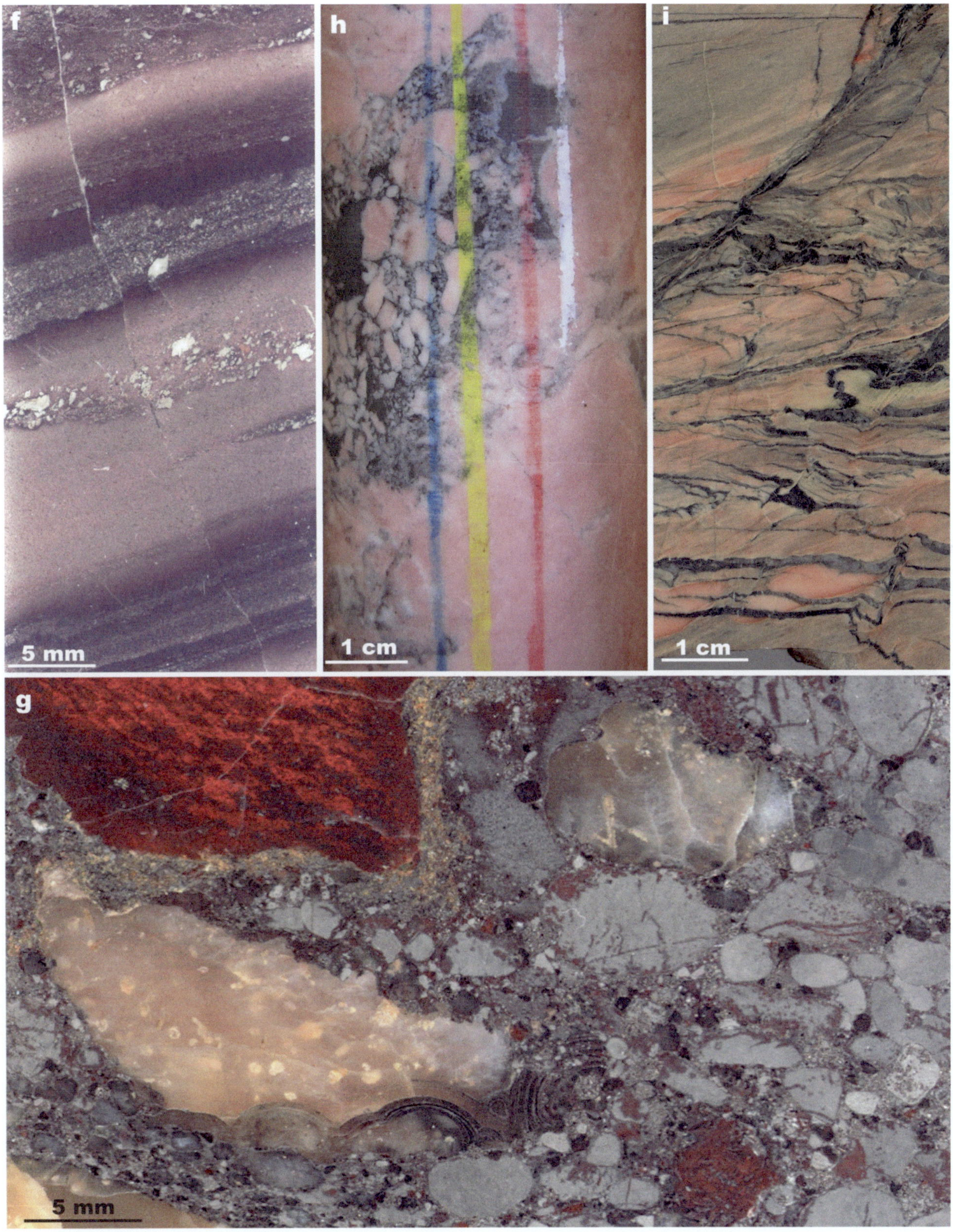

Fig. 3.27 (continued) *Kolosjoki Sedimentary Formation, Pechenga Belt*: (**f**) Scanned image of interbedded, variegated, parallel-bedded, volcaniclastic sandstone-siltstone with scattered grit-size clasts of quartz. (**g**) Scanned image of fine-pebble conglomerate comprised of unsorted clasts of jasper (*red*), weathered-out quartz amygdales (light coloured) and haematite (*grey*). (**h**) Pocket of dolostone conglomerate supported by black mudstone matrix in *pink*, massive dolostone. (**i**) Haematite-rich siltstone, mylonitised and cut by extensional cracks

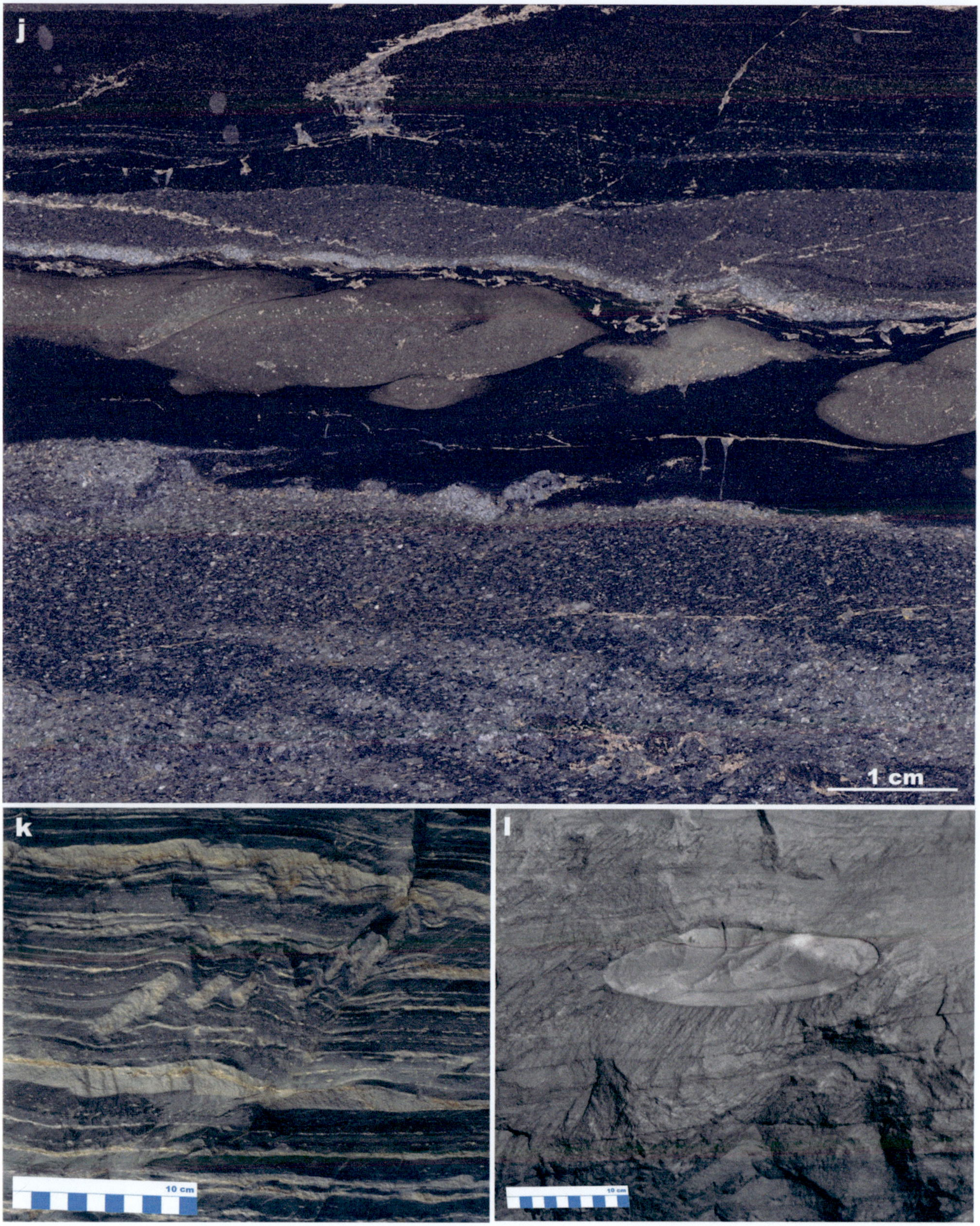

Fig. 3.27 (continued) *Pilgujärvi Sedimentary Formation, Pechenga Belt*: (**j**) Scanned, polished slab exhibiting C_{org}- and sulphide-rich greywacke-siltstone-mudstone with small-scale slump structure within black mudstone bed. (**k**) Cross-section view of rhythmically bedded greywacke-siltstone with dismembered sandstone bed occurring as angular imbricated clasts. (**l**) Calcite concretion in greywacke

Fig. 3.27 (continued) *Pilgujärvi Sedimentary Formation, Pechenga Belt*: (**m**) Cross-section view through rhythmically bedded, C_{org}-rich greywacke (*grey*)-shale (*black*) with layers of massive pyrite (*white arrow*) and pyrrhotite (*red arrow*). (**n**) Cross-section through thick sandstone units interbedded with thin shale (*lower part*) overlain by black, thinly-bedded shale with phosphorite-clast gritstone bed at its base. *Pilgujärvi Volcanic Formation, Pechenga Belt*: (**o**) Exhumed *upper* surface of pillowed lava flow. (p) Pyroclastic material forming thin bed between mafic lava flows

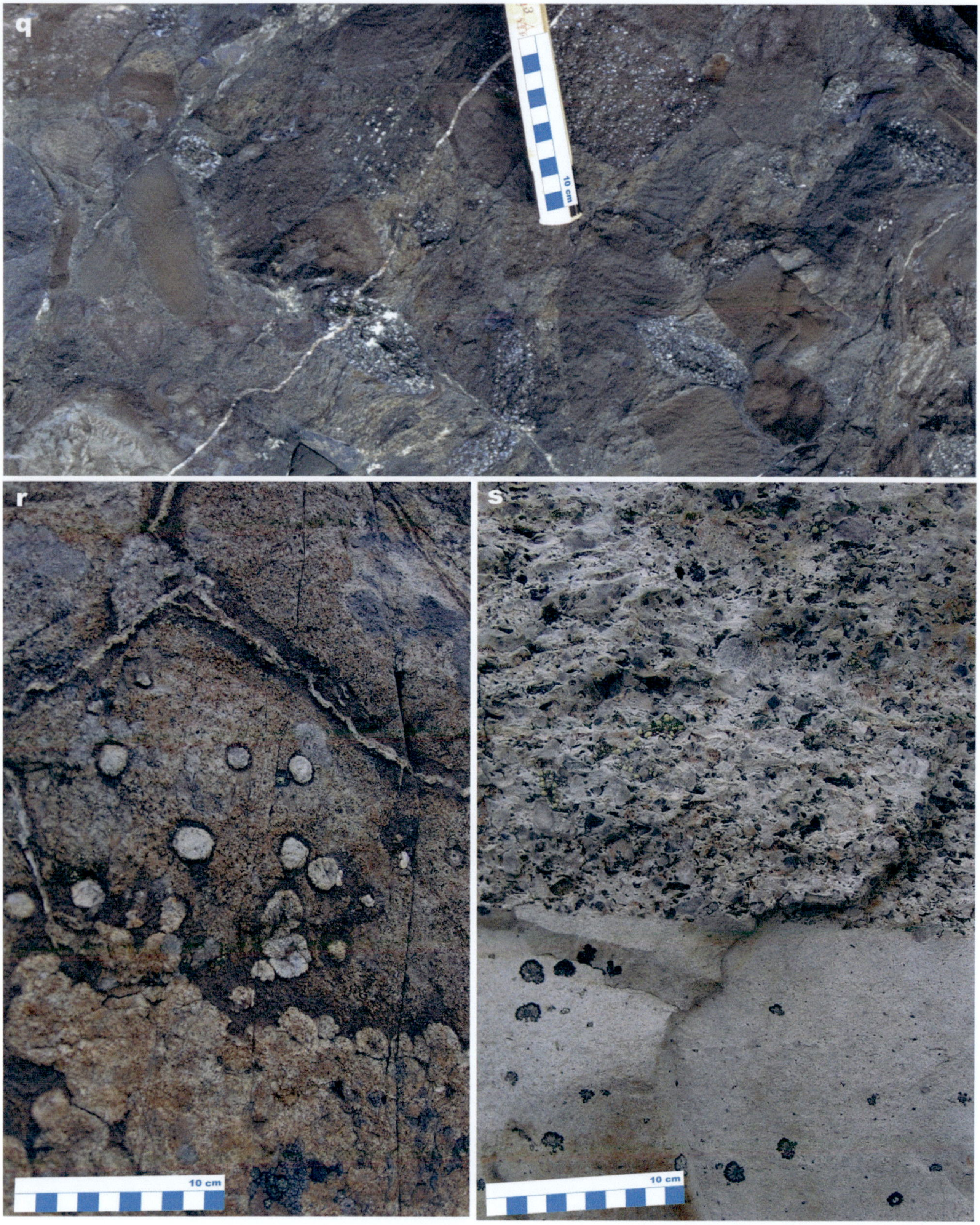

Fig. 3.27 (continued) *Pilgujärvi Sedimentary Formation, Pechenga Belt*: (**q**) Ferropicritic breccia from one of the eruptive centres occurring within the deep-water succession. *Pilgujärvi Volcanic Formation, Pechenga Belt*: (**r**) Ferropicritic lava with felsic globules (ocelli). Partially jointed together to form a layer, and cooling cracks. (**s**) Felsic volcanic rocks showing a contact between massive (glassy) and fragmental *upper* beds; black spots are lichen

Fig. 3.27 (continued) *Peuramaa Formation, Central Lapland Belt:* (**t**) Picritic pillow lava; width of the view is 1.5 m. *Linkupalo Formation, Central Lapland Belt:* (**u**) Variolitic mafic lava. *Sattasvaara Formation, Central Lapland Belt:* (**v, w**) Fragmental komatiitic lava. Lens cap diameter in (**u**) and (**v**) is 6 cm

Fig. 3.27 (continued) *Linkupalo Formation, Central Lapland Belt:* (**x**) Tholeiitic pillow lava forming part of c. 6 km thick sedimentary-volcanic succession deposited on oceanic floor; lens cap diameter is 6 cm. *Köngäs Formation, Central Lapland Belt:* (**y**) Pillowed subalkaline lava extruded on the sea-floor of the Kittilä Ocean

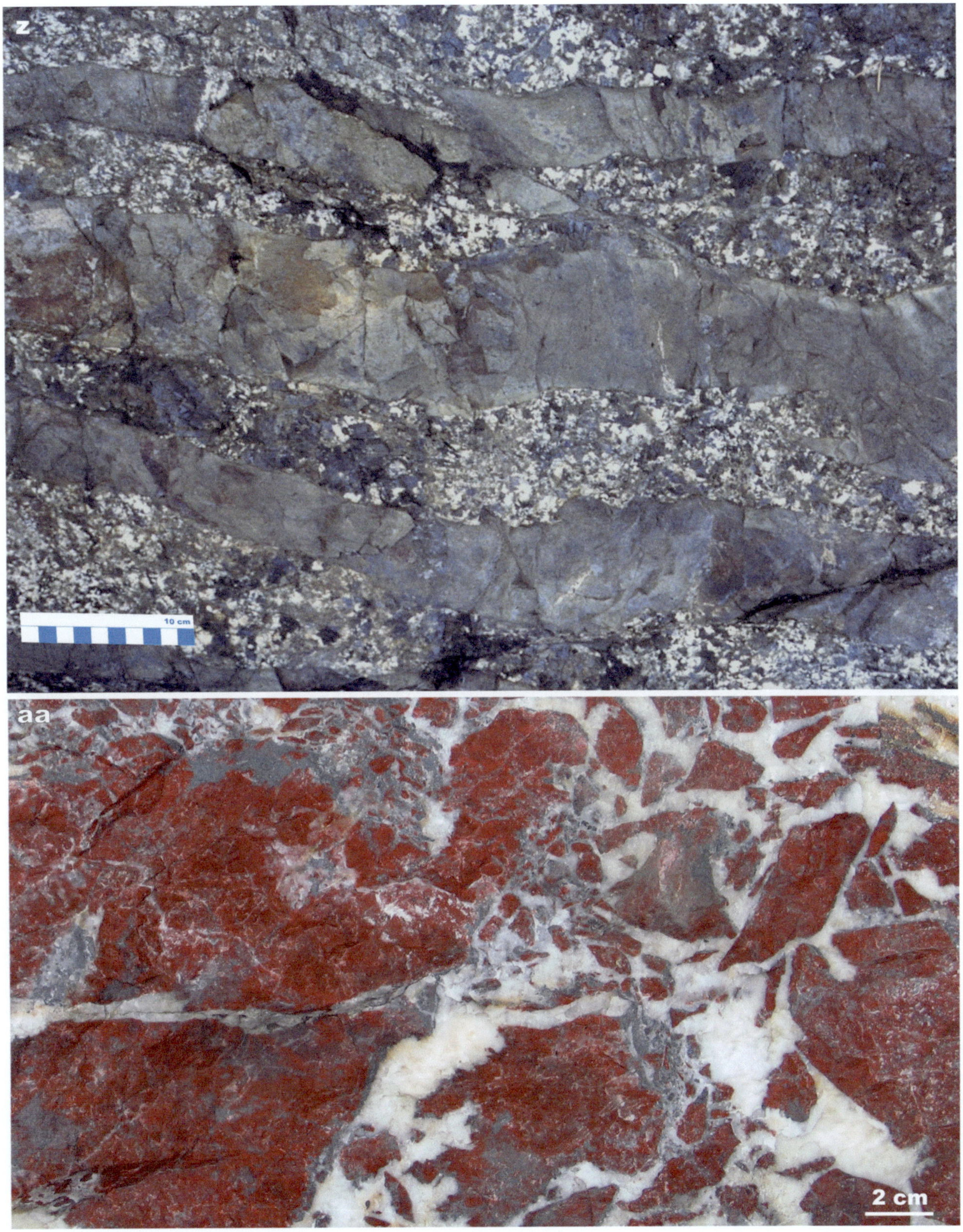

Fig. 3.27 (continued) *Veikasenmaa Formation, Central Lapland Belt:* (**z**) Subvolcanic dyke complex with a set of parallel mafic dykes cutting coarsely crystalline gabbro indicating seafloor spreading. (**z**) *Porkonen Formation, Central Lapland Belt:* (**aa**) Jasper breccia representing seafloor hydrothermal chemical sediments. Sample (**aa**) courtesy of the Geological Museum of the Department of Geosciences, University of Oulu

Fig. 3.27 (continued) *Jormua ophiolite, Kainuu Belt*: (**ab**) Sheeted dyke complex; hammer head is 15 cm. (**ac**) A close-up view of dyke-to-dyke contacts in sheeted dyke complex; lens cap diameter is 6 cm. (**ad**) Tholeiitic pillow lava associated with the sheeted dyke complex; lens cap diameter is 6 cm. (**ae**) Harzburgite mantle tectonite; chrome spinel and former orthopyroxene stand out on weathered surface as black and light-coloured grains, respectively; coin diameter is 2 cm

Fig. 3.27 (continued) *Zaonega Formation, Onega Basin*: (**af**) Rhythmically bedded greywacke-shale units. (**ag**) Massive, organosiliceous rock of enigmatic origin containing c. 40 wt.% total organic carbon. (**ah**) Massive and bedded limestone with pyrobitumen-rich veins (*black*) indicating migration of petroleum. (**ai**) Massive limestone interbedded with black, C_{org}-rich mudstone beds (*left core*), and with thin shale partings (*right core*). (**aj**) Lustrous pyrobitumen (former petroleum) from a c. 5 cm wide interlayer opening (small-scale oil trap); sample was collected from underground mine near the village of Shunga. Core diameter in (**af–ai**) is 5 cm (Photographs by Victor Melezhik (**a, c–s, af–aj**) and Eero Hanski (**b, w, z–ae**). Photographs courtesy of Penti Rastas (**t, u, x**) and Jorma Räsänen (**v**). Photograph (**y**). Reproduced from Hanski and Huhma (2005) with permission from Elsevier)

Fig. 3.28 Selected images of sedimentary and volcanic rocks representing major evolutionary features of the 1950–1900 Ma period. *Vuotto Greywacke Formation, Kiiminki Belt:* (**a**) Thinly-bedded greywacke-shale (mica schist) with carbonate-bearing layers (recessive relief). *Utajärvi Conglomerate Formation, Kiiminki Belt:* (**b**) Massive, poorly-sorted, pebbly greywacke with shale lenses. *Kiiminki Volcanic Formation, Kiiminki Belt:* (**c**) Tholeiitic pillow lava

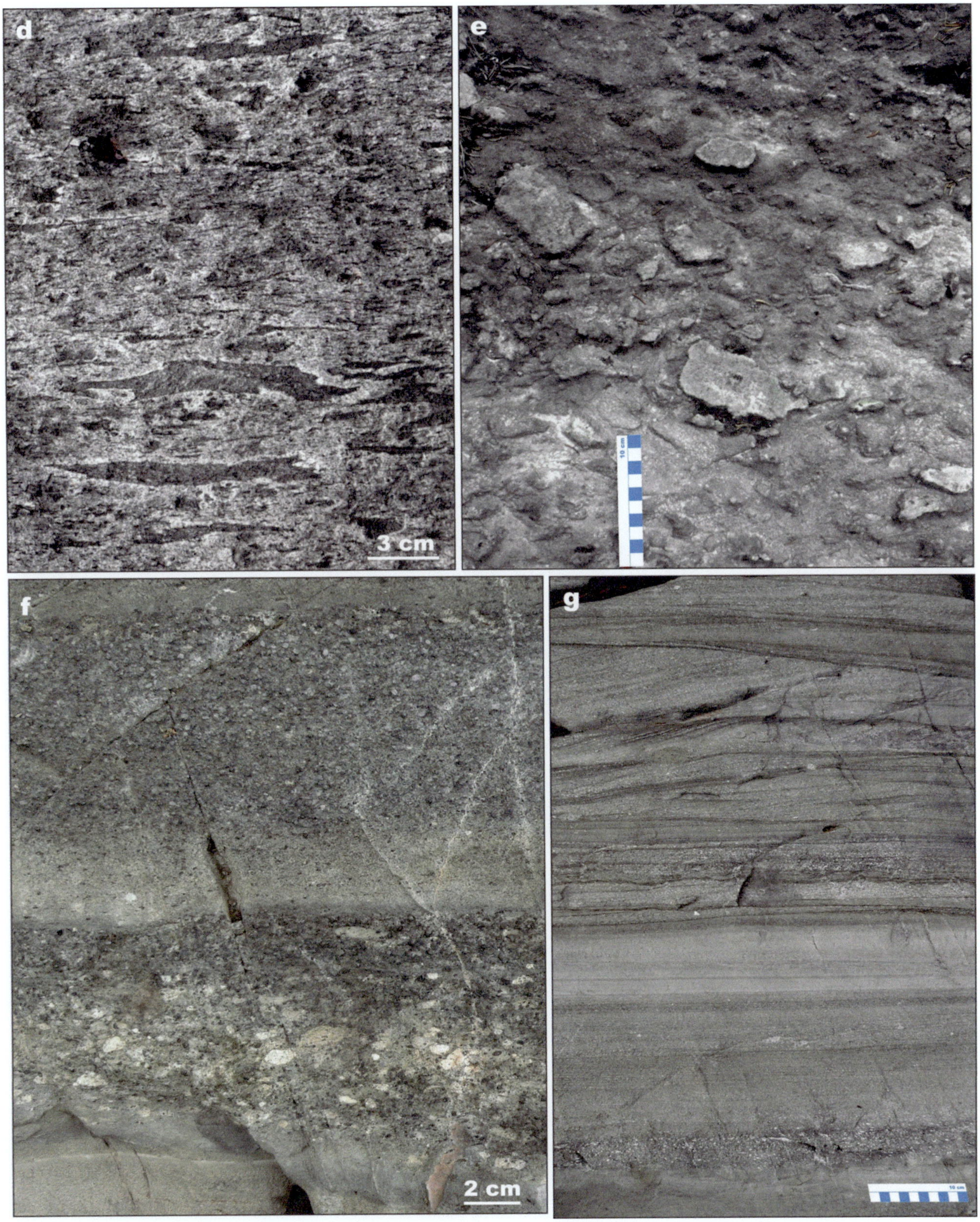

Fig. 3.28 (continued) *Sileäkallio, Tampere area:* (**d**) Ignimbrite with dark coloured fiamme; subaerial, volcanic-arc volcanism. *Takamaa, Ylöjärvi area:* (**e**) Mafic agglomerate/lapilli tuff with non-rounded bombs; subaerial, volcanic-arc volcanism. *Papinvuori north, Tampere area:* (**f**) Lapilli tuff bed (*below*) overlain by pyroxene crystal-rich tuff bed; both units show reverse grading. (**g**) Parallel-bedded greywacke-siltstone overlain by co-sets of trough cross-beds

Fig. 3.28 (continued) *Utajärvi Conglomerate Formation, Kiiminki Belt:* (**h**) Gritstone-matrix-supported basal Kalevian conglomerate with unsorted, polymict clasts. *Vuotto Greywacke Formation, Kiiminki Belt:* (**i**) Polymict gritsone with the lower bed grading into greywacke; coin diameter 2.3 cm. (**j**) Graded beds and Bouma units in greywacke-siltstone-shale; coin diameter is 2 cm. *Martimo Formation, Peräpohja Belt:* (**k**) Metamorphosed carbonate concretions in massive greywacke

Fig. 3.28 (continued) Ladoga Series, northern Ladoga area: (**l**) Metamorphosed (calcite-diopside-forsterite) zoned calcite concretions in paragneiss that underwent high-temperature amphibolite facies metamorphism and partial melting

Fig. 3.28 (continued) *Kondopoga Formation, Onega Basin:* (**m**) Rhythmically-bedded greywacke-siltstone units with ankerite concretionary beds (brown). (**n**) Greywacke with ankerite nodule and concretionary bed (*brown*). (**o**) Mass-flow breccia with unsorted and syndepositionally-deformed clasts of greywacke-siltstone floating in non-bedded greywacke matrix

Fig. 3.28 (continued) *Kondopoga Formation, Onega Basin*: (**p**) Syndepositional faulting and slumping in rhythmically-bedded greywacke-siltstone-shale. (**q**) A fragment of pyrobitumen clast coated by angular particles of fine-grained sandstone. (**r**) Pyrobitumen clasts in massive greywacke. (**s**) Pyrobitumen "pancake" on the bedding surface of greywacke (Photographs by Eero Hanski (**a–c, h–j**) and Victor Melezhik (**l-s**). Photographs courtesy of Yrjö Kähkönen (**d–g**) and Vesa Perttunen (**k**))

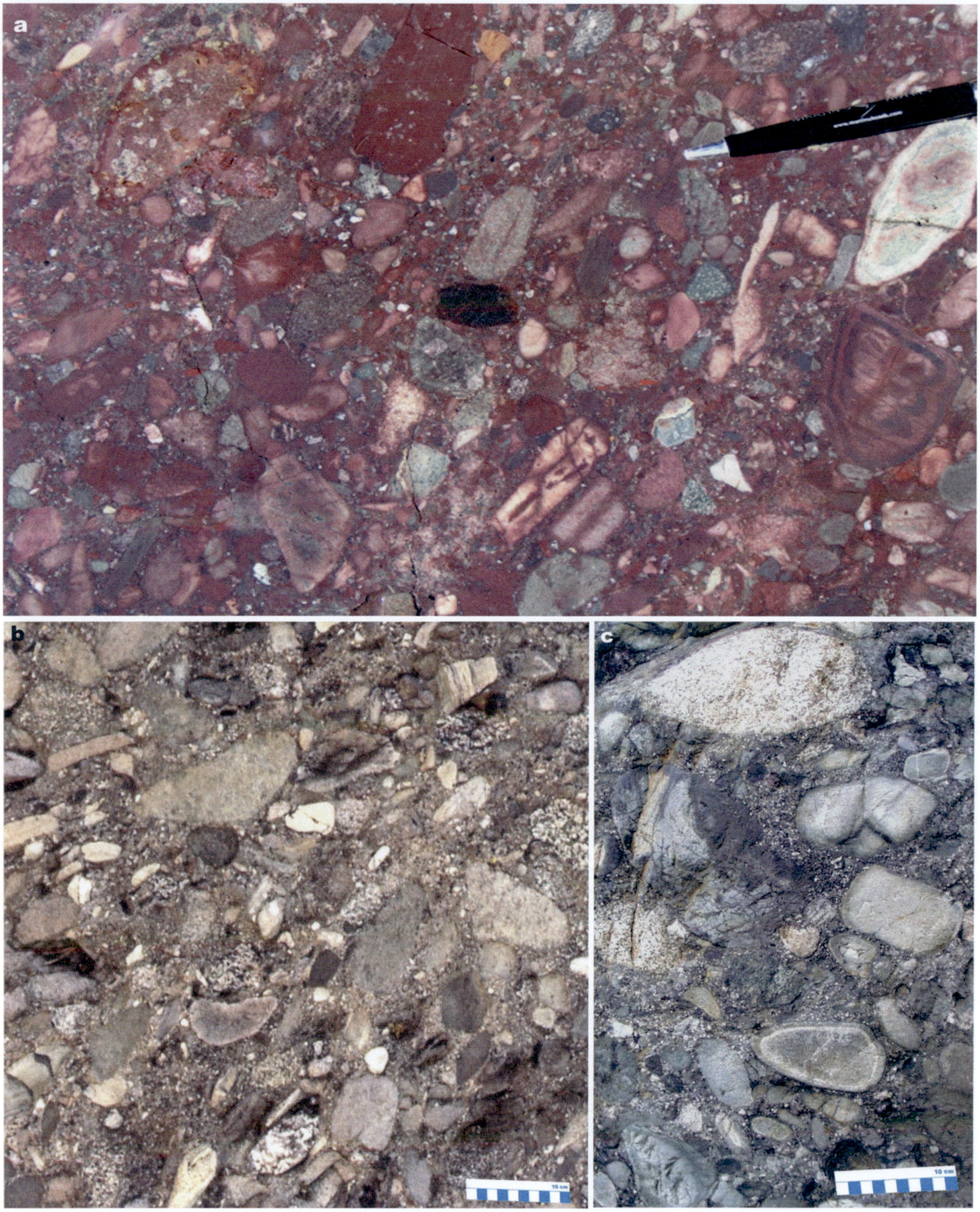

Fig. 3.29 Selected images of sedimentary and volcanic rocks representing the 1900–1850 Ma, post-orogenic period. *Levi Formation, Central Lapland Belt:* (**a**) Subaerially deposited, haematite-stained, fragment-supported conglomerate with polymict clasts. *Kumputunturi Formation, Central Lapland Belt:* (**b, c**) Subaerially deposited, fragment-supported, polymict conglomerate containing unsorted, variably rounded clasts of mafic and felsic lavas and granite

Fig. 3.29 (continued) *Värttiövaara Formation, Central Lapland Belt:* (**d**) Cosets of small-scale trough cross-beds in arkosic sandstone with some layers rich in haematite and other heavy minerals. *Pyhätunturi Formation, Central Lapland Belt:* (**e**) Cosets of small-scale trough cross-beds with layers enriched in haematite (Photographs (**a–c, e**) by Eero Hanski. Photograph (**d**) courtesy of Raimo Ristimäki. Sample (**a**) courtesy of the Geological Museum of the Department of Geosciences, University of Oulu)

References

Akhmedov AM, Krupenik VA (1990) Turbidity conditions of sedimentation and pyrite formation in the Early Proterozoic Pechenga Basin. Sov Geol 11:51–60

Akhmedov AM, Panova EG, Krupenik VA, Sveshnikova KY (2004) Early Proterozoic and Devonian arid palaeobasins from joint zone of the Baltic Shield and Russian Platform. Trans St Petersburg Soc Natural 2(86):1–140 (in Russian)

Alapieti T (1982) The Koillismaa layered igneous complex, Finland: its structure, mineralogy and geochemistry, with emphasis on the distribution of chromium. Bull Geol Surv Finl 319:1–116

Amelin YuV, Heaman LM, Semenov VS (1995) U-Pb geochronology of layered mafic intrusions in the eastern Baltic Shield: implications for the timing and duration of Palaeoproterozoic continental rifting. Precambrian Res 75:31–46

Arestova NA, Khramov AN, Gooskova EG, Iosifidi AG (2004) New paleomagnetic evidence from the A Early Proterozoic (2.5–2.4 Ga) Mount Generalskaya and Imandra layered intrusions, Kola Peninsula. Fizika Zemli 38:65–76

Aspler LB, Chiarenzelli JR (1998) Two Neoarchean supercontinents? Evidence from the Paleoproterozoic. Sediment Geol 120:75–104

Baker AJ, Fallick AE (1989) Heavy carbon in two-billion-year-old marbles from Lofoten-Vesterålen. Norway: implications for the Precambrian carbon cycle. Geochim Cosmochim Acta 53:1111–1115

Bayanova T, Ludden J, Mitrofanov F (2009) Timing and duration of Palaeoproterozoic events producing ore-bearing layered intrusions of the Baltic Shield: metallogenic, petrological and geodynamic implications. In: Reddy SM, Mazumder R, Evans DAD, Collins AS (eds) Palaeoproterozoic supercontinents and global evolution, 323rd edn. Geological Society of London, Special Publications, London, pp 165–198

Bergh SG, Kullerud K, Armitage PEB, Zwan KB, Corfu F, Ravna EJK, Myhre PI (2010) Neoarchaean and Svecofennian tectono-magmatic evolution of the West Troms basement Complex, Northern Norway. Norwegian J Geol 90:1–28

Bleeker W (2003) The late Archaean record: a puzzle in c. 35 pieces. Lithos 71:99–134

Bleeker W, Ernst R (2006) Short-lived mantle generated magmatic events and their dyke swarms: the key unlocking Earth's palaeomagnetic record back to 2.6 Ga. In: Hanski E, Mertanen S, Rämö T, Vuollo J (eds) Dyke Swarms – time markers of crustal evolution. Taylor & Francis, London, pp 3–26

Bogatikov OA, Kovalenko VI, Sharkov EV (2010) Magmatism, tectonics, geodynamics of the Earth. Nauka (Science), Moscow, p 606 (in Russian)

Buchan KL, Mertanen S, Park RG, Pesonen LJ, Elming S-Å, Abrahamsen N, Bylund G (2000) Comparing drift of laurentia and baltica in the Proterozoic: the importance of key palaeomagnetic poles. Tectonophysics 319:167–198

Card KD, Poulsen KH (1998) Geology and mineral deposits of the superior province of the Canadian Shield. In: Lucas SB, St. Onge MR (eds) Geology of the Precambrian Superior and Grenville Provinces and Precambrian Fossils in North America, 7th edn. Geological Survey of Canada, Geology of Canada, Canada, pp 13–194

Chashchin VV, Bayanova TB, Levkovich NV (2008) Volcanoplutonic association of the early-stage evolution of the Imandra–Varzuga rift zone, Kola Peninsula, Russia: geological, petrogeochemical, and isotope-geochronological data. Petrology 16:279–298

Condie KC (1998) Episodic continental growth and supercontinents; a mantle avalanche connection? Earth Planet Sci Lett 163:97–108

Daly JS, Balagansky VV, Timmerman MJ, Whitehouse MJ (2006) The Lapland-Kola orogen: Palaeoproterozoic collision and accretion of the northern Fennoscandian lithosphere. In: Gee DG, Stephenson RA (eds) European lithosphere dynamics, 32nd edn. Geological Society Memoir, London, pp 579–598

Evans DAD, Beukes NJ, Kirschvink JL (1997) Low-latitude glaciations in the Palaeoproterozoic era. Nature 386:262–266

Fedotov ZhA, Basalaev AA, Melezhik VA, Latyshev LN (1982) The Strel'na series. In: Gorbunov GI (ed) The Imandra/Varzuga zone of the Karelides. Nauka (Science), Leningrad, pp 33–56 (in Russian)

Galdobina LP (1987a) The Ludikovi Superhorizon. In: Sokolov VA (ed) Geology of Karelia. Nauka (Science), Leningrad, pp 59–67 (in Russian)

Galdobina LP (1987b) The Livian Superhorizon. In: Sokolov VA (ed) Geology of Karelia. Nauka (Science), Leningrad, pp 67–71 (in Russian)

Hackman V (1927) Studien über den Gesteinsaufbau der Kittilä-Lappmark. Bulletin de la Commission Géologique de Finland 79:1–105

Hanski EJ (1992) Petrology of the Pechenga ferropicrites and cogenetic, Ni-bearing gabbro-wehrlite intrusions, Kola Peninsula, Russia. Geol Surv Finl Bull 367:192

Hanski E (1997) The Nuttio serpentinite belt, central Lapland: an example of Paleoproterozoic ophiolitic mantle rocks in Finland. Ofioliti 22:35–46

Hanski E, Huhma H (2005) Central Lapland greenstone belt. In: Lehtinen M, Nurmi PA, Rämö OT (eds) Precambrian geology of Finland – key to the evolution of the Fennoscandian Shield. Elsevier, Amsterdam, pp 139–194

Hanski E, Huhma H, Vaasjoki M (2001) Geochronology of northern Finland: a summary and discussion. In: Vaasjoki M (ed) Radiometric age determination from Finnish Lapland and their bearing on the timing of Precambrian volcano-sedimentary sequences, 33rd edn. Geological Survey of Finland Special Paper, Espoo, pp 255–279

Hanski E, Huhma H, Vuollo J (2010) SIMS zircon ages and Nd isotope systematics of the 2.2 Ga mafic intrusions in northern and eastern Finland. Bull Geol Soc Finl 82:31–62

Heaman LM (1997) Global mafic volcanism at 2.45 Ga: remnants of an ancient large igneous province? Geology 25:299–302

Heiskanen KI (1987a) The Sumi superhorizon. In: Sokolov VA (ed) Geology of Karelia. Nauka (Science), Leningrad, pp 40–44 (in Russian)

Heiskanen KI (1987b) The Sarioli superhorizon. In: Sokolov VA (ed) Geology of Karelia. Nauka (Science), Leningrad, pp 44–51 (in Russian)

Heiskanen KI (1990) Palaeogeography of the Baltic Shield in Karelian time. Institute of Geology, Karelian Science Centre, Petrozavodsk, p 128

Huhma H (1986) Sm-Nd, U-Pb and Pb-Pb isotopic evidence for the origin of the Early Proterozoic Svecokarelian crust in Finland. Bull Geol Surv Finl 337:48

Huhma H, Cliff RA, Perttunen V, Sakko M (1990) Sm-Nd and Pb isotopic study of mafic rocks associated with early Proterozoic continental rifting: the Peräpohja schist belt in northern Finland. Contrib Mineral Petrol 104:369–379

Iljina M, Hanski E (2005) Layered mafic intrusions of the Tornio-Näränkävaara belt. In: Lehtinen M, Nurmi PA, Rämö OT (eds) Precambrian geology of Finland – key to the evolution of the Fennoscandian Shield. Elsevier, Amsterdam, pp 101–138

Kähkönen Y (2005) Svecofennian supracrustal rocks. In: Lehtinen M, Nurmi PA, Rämö OT (eds) Precambrian geology of Finland – key to the evolution of the Fennoscandian Shield. Elsevier, Amsterdam, pp 343–406

Karhu JA (1993) Palaeoproterozoic evolution of the carbon isotope ratios of sedimentary carbonates in the Fennoscandian Shield. Bull Geol Surv Finl 371:87

Karinen T (2010) The Koillismaa Intrusion, northeastern Finland – evidence for PGE reef forming processes in the layered series. Bull Geol Soc Finl 404:176

Kohonen J, Marmo J (1992) Proterozoic lithostratigraphy and sedimentation of Sariola- and Jatuli-type rocks in the Nunnanlahti-Koli-Kaltimo area, eastern Finland; implications for regional basin models. Bull Geol Surv Finl 364:67

Koistinen TJ (1981) Structural evidence of an early Proterozoic strata-bound Cu-Co-Zn depositis, Outokumpu, Finland. Tran R Soc Edinburgh: Earth Sci 72:115–158

Konopelko D, Savatenkov V, Glebovitsky V, Kotov A, Sergeev S, Matukov D, Kovach V, Zagornaya N (2005) Nd isotope variation across the Archaean-Proterozoic boundary in the North Ladoga area, Russian Karelia. Geologiska Föreningens i Stockholm Förhandlingar: 127:115–122

Kontinen A (1987) An early Proterozoic ophiolite – the Jormua mafic–ultramafic complex, northeastern Finland. Precambrian Res 35:313–341

Kontinen A, Paavola J (2006) A preliminary model of the crustal structure of the Eastern Finland Archaean complex between Vartius and Vieremä, based on constraints from surface geology and FIRE 1 seismic survey. In: Kukkonen IT, Lahtinen R (eds) Finnish reflection experiment FIRE 2001–2005, 43rd edn. Geological Survey of Finland, Special Paper, Espoo, pp 223–240

Kortelainen V (1983) Sedimentology of the Sirkka conglomerate and Levitunturi quartzite in Kittilä, M.Sc. thesis, Department of Geology, University of Helsinki, p 101 (in Finnish)

Kozlova NE, Balagansky VV (1995) Structural evolution of the Southern zone rocks. In: Mitrofanov FP, Smol'kin VF (eds) Magmatism, sedimentogenesis and geodynamics of the Pechenga Paleorift. Kola Science Centre, Apatity, pp 151–163 (in Russian)

Kulikov VS, Bychkova YaV, Kulikova VA, Ernst R (2010) The Vetreny Poyas (Windy Belt) subprovince of southeastern

Fennoscandia; An essential component of the ca. 2.5–2.4 Ga Sumian large igneous province. Precambrian Res 183: pp 589–601
Kullerud K, Skjerlie KP, Corfu F, de la Rosa JD (2006) The 2.40 Ga Ringvassøy mafic dykes, West Troms Basement Complex, Norway: the concluding act of early Palaeoproterozoic continental breakup. Precambrian Res 150:183–200
Laajoki K (2005) Karelian supracrustal rocks. In: Lehtinen M, Nurmi PA, Rämö OT (eds) Precambrian geology of Finland – key to the evolution of the Fennoscandian Shield. Elsevier, Amsterdam, pp 279–342
Laajoki K, Huhma H (2006) Detrital zircon dating of the Palaeoproterozoic Himmerkinlahti Member, Posio, northern Finland; lithostratigraphic implications. Geol Surv Finl Spec Pap 78:177–182
Lager I, Loberg B (1990) Sedimentologisk-bassänganalytisk malmprospekteringsmetodik inom norrbottniska grönstenbälten. STU-Projekt 86-03967P, Slutrapport. Ekonomisk geologi, Högskolan i Luleå, Luleå, p 112 (in Swedish)
Lahtinen R (2000) Archaean–Proterozoic transition: geochemistry, provenance and tectonic setting of metasedimentary rocks in central Fennoscandian Shield, Finland. Precambrian Res 104:147–174
Lahtinen R, Korja A, Nironen M (2005) Paleoproterozoic tectonic evolution. In: Lehtinen M, Nurmi PA, Rämö OT (eds) Precambrian geology of Finland – key to the evolution of the Fennoscandian Shield. Elsevier, Amsterdam, pp 481–532
Lahtinen R, Garde AA, Melezhik VA (2008) Paleoproterozoic evolution of Fennoscandia and Greenland. Episodes 31:20–28
Lahtinen R, Korja A, Nironen M, Heikkinen P (2009) Palaeoproterozoic accretionary processes in Fennoscandia. Geol Soc, Lond, Spec Pub 318:237–256
Lahtinen R, Huhma H, Kontinen A, Kohonen J, Sorjonen-Ward P (2010) New constraints for the source characteristics, deposition and age of the 2.1–1.9 Ga metasedimentary cover at the western margin of the Karelian Province. Precambrian Res 176:77–93
Latypov RM, Mitrofanov FP, Alapieti TT, Kaukonen RJ (1999) Petrology of Upper Layered Horizon (ULH) of West-Pansky intrusion, Kola Peninsula, Russia. Geol Geophys 40:1434–1456
Lehtonen M, Airo M-L, Eilu P, Hanski E, Kortelainen V, Lanne E, Manninen T, Rastas P, Räsänen J, Virransalo P (1998) The stratigraphy, petrology and geochemistry of the Kittilä greenstone area, northern Finland. A Report of the Lapland Volcanite Project, Geol Surv Finl, Rep Invest 140: 144 (in Finnish with English summary)
Makarikhin VV, Kononova GM (1983) Early Proterozoic Phytolites of Karelia. Nauka (Science), Leningrad, p 180 (in Russian)
Marmo JS (1992) The lower Proterozoic Hokkalampi paleosol in North Karelia, eastern Finland. In: Golubic S, Kimberley MM, Trudinger PA, Schidlowski M (eds) Early organic evolution, implications for energy and mineral resources. Springer, Berlin/Heidelberg, pp 41–66
Marmo JS, Ojakangas RW (1984) Lower Proterozoic glaciogenic deposits, eastern Finland. Geol Soc Am Bull 98:1055–1062
Marmo J, Kohonen J, Sarapää O, Äikäs O (1988) Sedimentology and stratigraphy of the Lower Proterozoic Sariola and Jatuli Groups in the Koli-Kaltimo area, eastern Finland. Geol Surv Finl, Spec Pap 5:11–28
Melezhik VA (1992) Early Proterozoic sedimentary and rock-forming Basins of the Baltic Shield. Nauka (Science), St. Petersburg, p 258 (in Russian)
Melezhik VA (2006) Multiple causes of Earth's earliest global glaciation. Terra Nova 18:130–137
Melezhik VA, Fallick AE (2005) The Palaeoproterozoic, rift-related, shallow-water, ^{13}C-rich, lacustrine carbonates, NW Russia – Part I: sedimentology and major element geochemistry. Trans R Soc Edinburgh: Earth Sci 95:393–421
Melezhik VA, Fallick AE (2010) On the Lomagundi-Jatuli carbon isotopic event: the evidence from the Kalix Greenstone Belt, Sweden. Precambrian Res 179:165–190
Melezhik VA, Predovsky AA (1982) Geochemistry of Early Proterozoic lithogenesis. Nauka (Science), St. Petersburg, p 208 (in Russian)
Melezhik VA, Sturt BA (1994) General geology and evolutionary history of the early Proterozoic Polmak-Pasvik-Pechenga-Imandra/Varzuga-Ust'Ponoy Greenstone Belt in the north-eastern Baltic Shield. Earth-Science Reviews 36:205–241
Melezhik VA, Fallick AE, Medvedev PV, Makarikhin VV (1999a) Extreme $^{13}C_{carb}$ enrichment in ca. 2.0 Ga magnesite-stromatolite-dolomite-'red beds' association in a global context: a case for the world-wide signal enhanced by a local environment. Earth-Science Reviews 48:71–120
Melezhik VA, Fallick AE, Filippov MM, Larsen O (1999b) Karelian shungite – an indication of 2000 Ma-year-old metamorphosed oil-shale and generation of petroleum: geology, lithology and geochemistry. Earth-Science Reviews 47:11–40
Melezhik VA, Fallick AE, Medvedev PV, Makarikhin VV (2000) Palaeoproterozoic magnesite–stromatolite–dolostone–'red bed' association, Russian Karelia: palaeoenvironmental constraints on the 2.0 Ga positive carbon isotope shift. Norsk Geologisk Tidsskrift 80:163–186
Melezhik VA, Fallick AE, Medvedev PV, Makarikhin V (2001) Palaeoproterozoic magnesite: lithological and isotopic evidence for playa/sabkha environments. Sedimentology 48:379–397
Melezhik VA, Filippov MM, Romashkin AE (2004) A giant Palaeoproterozoic deposit of shungite in NW Russia: genesis and practical applications. Ore Geol Rev 24:135–154
Melezhik VA, Fallick AE, Rychanchik DV, Kuznetsov AB (2005a) Palaeoproterozoic evaporites in Fennoscandia: implications for seawater sulphate, δ^{13}C excursions and the rise of atmospheric oxygen. Terra Nova 17:141–148
Melezhik VA, Fallick AE, Hanski E, Kump L, Lepland A, Prave A, Strauss H (2005b) Emergence of the modern earth system during the Archean-Proterozoic transition. Geol Soc Am Today 15:4–11
Melezhik VA, Fallick AE, Filippov MM, Lepland A, Rychanchik DV, Deines JE, Medvedev PV, Romashkin AE, Strauss H (2009) Petroleum surface oil seeps from Palaeoproterozoic petrified giant oilfield. Terra Nova 21:119–126
Mertanen S, Pesonen LJ (2005) Drift history of the shield. In: Lehtinen M, Nurmi PA, Rämö OT (eds) Precambrian geology of Finland – key to the evolution of the Fennoscandian Shield. Elsevier, Amsterdam, pp 645–668
Mertanen S, Halls HC, Vuollo JL, Pesonen LJ, Stepanov VS (1999) Paleomagnetism of 2.44 Ga mafic dykes in Russian Karelia, eastern Fennoscandian Shield – implications for continental reconstructions. Precambrian Res 98:197–221
Mertanen S, Vuollo JI, Huhma H, Arestova NA, Kovalenko A (2006) Early Paleoproterozoic-Archean dykes and gneisses in Russian Karelia of the Fennoscandian Shield – New paleomagnetic, isotope age and geochemical investigations. Precambrian Res 144:239–260
Metzger AA (1924) Die jatulishen Bildungen von Suojarvi in Ostfinland. Bulletin de la Commission Géologique de Finland 64:1–86
Negrutza VZ (1984) Early Proterozoic stages of evolution of the Eastern Baltic Shield. Nedra, Leningrad, p 270 (in Russian)
Nikula R (1988) Palaeosedimentology of Precambrian tidal Virttiövaara and fluvial Värttiövaara quartzite formations in Sodankylä, Northern Finland. Geol Surv Finl, Spec Pap 5:177–188
Nironen M (1997) The Svecofennian Orogen: a tectonic model. Precambrian Res 86:21–44
Ohlander B, Skiöld T, Elming S-A, BABEL Working Group, Claesson S, Nisca DH (1993) Delination and character of the Archaean–Proterozoic boundary in northern Sweden. Precambrian Res 64:67–84

Ojakangas RW, Marmo JS, Heiskanen KI (2001) Basin evolution of the Paleoproterozoic Karelian Supergroup of the Fennoscandian (Baltic) Shield. Sediment Geol 141–142:255–285

Paakkola J, Gehör S (1988) The lithofacies associations and sedimentary structures of the iron-formation in the early Proterozoic Kittilä greenstone belt, northern Finland. Geol Surv Finl, Spec Pap 5:213–238

Pechersky DM, Zakharov VS, Lyubushin AA (2004) Continuous record of geomagnetic field variations during cooling of the Monchegorsk, Kivakka and Bushveld Early Proterozoic layered intrusions. Russ J Earth Sci 6:391–456

Pekkarinen LJ, Lukkarinen H (1991) Paleoproterozoic volcanism in the Kiihtelysvaara – Tohmajärvi district, eastern Finland. Bull Geol Surv Finl 357:30

Peltonen P (2005) Ophiolites. In: Lehtinen M, Nurmi PA, Rämö OT (eds) Precambrian geology of Finland – key to the evolution of the Fennoscandian Shield. Elsevier, Amsterdam, pp 237–278

Peltonen P, Mänttäri I, Huhma M, Kontinen A (2003) Archean zircons from the mantle: the Jormua ophiolite revisited. Geology 31:645–648

Peltonen P, Kontinen A, Huhma H, Kuronen U (2008) Outokumpu revisited: new mineral deposit model for the mantle peridotite-associated Cu–Co–Zn–Ni–Ag–Au sulphide deposits. Ore Geol Rev 33:559–617

Percival JA, Sanborn-Barrie M, Skulski T, Scott GM, Helmstaedt H, White DL (2006) Tectonic evolution of the western Superior Province from NATMAP and Lithoprobe Studies. Can J Earth Sci 43:1085–1117

Perttunen V, Vaasjoki M (2001) U-Pb geochronology of the Peräpohja schist belt. Geol Surv Finl, Spec Pap 33:45–84

Pisarevsky SA, Sokolov SJ (1999) Palaeomagnetism of the Palaeoproterozoic ultramafic intrusion near Lake Konchozero, Southerm Karelia, Russia. Precambrian Res 93:201–213

Plumb KA (1991) New precambrian time scale. Episodes 14:139–140

Puchtel IS, Hofmann AW, Mezger K, Schipansky AA, Kulikov VS, Kulikova VV (1996) Petrology of a 2.41 Ga remarkably fresh komatiitic basalt lava lake in Lion Hills, central Vetreny Belt, Baltic Shield. Contrib Mineral Petrol 124:273–290

Puchtel IS, Haase KM, Hofmann AW, Chauvel C, Kulikov VS, Garbe-Schönberg C-D, Nemchin AA (1997) Petrology and geochemistry of crustally contaminated komatiitic basalts from the Vetreny Belt, southeastern Baltic Shield: evidence for an early Proterozoic mantle plume beneath rifted Archean continental lithosphere. Geochim Cosmochim Acta 61:1205–1222

Puchtel IS, Arndt NT, Hofmann AW, Haase KM, Kröner A, Kulikov VS, Kulikova VV, Garbe-Schönberg C-D, Nemchin AA (1998) Petrology of mafic lavas within the Onega plateau, central Karelia: evidence for the 2.0 Ga plume-related continental crustal growth in the Baltic Shield. Contrib Mineral Petrol 130:134–153

Puchtel IS, Brügmann GE, Hofmann AW, Kulikov VS, Kulikova VV (2001) Os isotope systematics of komatiitic basalts from the Vetreny belt, Baltic Shield: evidence for a chondritic source of the 2.45 Ga plume. Contrib Mineral Petrol 140:588–599

Räsänen J, Mäkelä M (1988) Early Proterozoic alluvial deposits in the Pyhätunturi area, northern Finland. Geol Surv Finl, Spec Pap 5:239–254

Rastas P, Huhma H, Hanski E, Lehtonen MI, Paakkola J, Mänttäri I, Härkönen I (2001) U-Pb isotopic studies on the Kittilä greenstone area, Central Lapland, Finland. Geol Surv Finl, Spec Pap 33:95–141

Reddy SM, Evans DAD (2009) Palaeoproterozoic supercontinents and global evolution: correlations from core to atmosphere. In: Reddy SM, Mazumdr R, Evans DAD, Collins AS (eds) Palaeoproterozoic supercontinents and global evolution, 323rd edn. Geological Society of London, Special Publication, London, pp 165–198

Roscoe SM, Card KD (1993) The reappearance of the Huronian in Wyoming: rifting and drifting of ancient continents. Can J Earth Sci 30:2475–2480

Sharkov EV (2006) Formation of layered intrusions and their ore mineralization. Scientific World, Moscow, p 368 (in Russian)

Sharkov EV, Smol'kin VF (1997) The early Proterozoic Pechenga-Varzuga Belt: a case of back-arc spreading. Precambrian Res 82: 135–151

Skuf'in PK, Bayanova TB, Mitrofanov FP (2006) Isotope age of subvolcanic granitoid rocks of the Early Proterozoic Panarechka volcano-tectonic structure, Kola Peninsula. Doklady Earth Sci 409: 774–778

Sokolov VA (1963) Geology and lithology of middle Proterozoic carbonate rocks of Karelia. USSR Academy of Sciences, Moscow-Leningrad, p 184 (in Russian)

Sokolov VA (1987) Jatuli superhorizon. In: Sokolov VA (ed) Geology of Karelia. Nauka (Science), Leningrad, pp 51–59 (in Russian)

Sokolov VA, Heiskanen KI (1984) Stages of formation of chemical weathering crusts in the Precambrian. In: Proceedings of the 27th international geological congress, Invited Papers, vol 5. Book, Geology of the Precambrian, pp 35–43

Strand K (2005) Sequence stratigraphy of the siliciclastic East Puolanla Group, the Palaeoproterozoic Kainuu belt, Finland. Sediment Geol 176:149–166

Strand K, Laajoki K (1999) Application of the parasequence concept to the Paleoproterozoic record of the northern Fennoscandian Shield. Precambrian Res 97:253–267

Sturt BA, Melezhik VA, Ramsay DM (1994) Early Proterozoic regolith at Pasvik, NE Norway: palaeotectonical implications for the Baltic Shield. Terra Nova 6:618–633

Torsvik TH, Meert JG (1995) Early Proterozoic palaeomagnetic data from the Pechenga zone (north-west Russia) and their bearing on Early Proterozoic palaeogeography. Geophys J Int 122:520–536

Tsuru A, Walker RJ, Kontinen A, Peltonen P, Hanski E (2000) Re-Os isotopic systematics of the 1.95 Ga Jormua ophiolite complex, northeastern Finland. Chem Geol 164:123–141

Tuisku P, Huhma H (2006) Evolution of migmatitic granulite complexes: implications from Lapland Granulite Belt, Part II: isotopic dating. Bull Geol Soc Finl 78:143–175

Vetrin VR, Turkina OM, Rodionov NV (2008) U–Pb age and genesis of granitoids in the southern framing of the Pechenga Structure, Baltic Shield. Doklady Earth Sci 419:298–302

Vogel DC, Vuollo JI, Alapieti TT, James RS (1998) Tectonic, stratigraphic, and geochemical comparison between ca. 2500–2440 Ma mafic igneous events in the Canadian and Fennoscandian Shields. Precambrian Res 92:89–116

Vuollo J, Huhma H (2005) Paleoproterozoic mafic dikes in NE Finland. In: Lehtinen M, Nurmi PA, Rämö OT (eds) Precambrian geology of Finland – key to the evolution of the Fennoscandian Shield. Elsevier, Amsterdam, pp 195–278

Wanke A, Melezhik VA (2005) Palaeoproterozoic sedimentation and stromatolite growth in an advanced intracontinental rift associated with the marine realm: a record of the Neoarchaean continent breakup? Precambrian Res 140:1–35

Williams GE, Schmidt PW (1997) Paleomagnetism of the Paleoproterozoic Gowganda and Lorrain formations, Ontario: low paleolatitude for Huronian glaciations. Earth Planet Sci Lett 153: 157–169

Williams H, Hoffman PF, Lewry JF, Monger JWH, Rivers T (1991) Anatomy of North America: thematic geologic portraits of the continent. Tectonophysics 187:117–134

Young GM (1991) The geologic record of glaciations: relevance to the climatic history of Earth. Geosci Can 18:100–108

Young GM, Long DGF, Fedo ChM, Nesbitt HW (2001) Paleoproterozoic Huronian basin: product of a Wilson cycle punctuated by glaciations and a meteorite impact. Sediment Geol 141–142: 233–254

3.4 Evolution of the Palaeoproterozoic (2.50–1.95 Ga) Non-orogenic Magmatism in the Eastern Part of the Fennoscandian Shield

E.J. Hanski

3.4.1 Introduction

In this chapter, a synthesis is presented on the magmatic evolution that occurred in the eastern part of the Fennoscandian Shield between the initial rifting of the Archaean cratons of the shield and the commencement of the Svecofennian orogenic events, covering a prolonged inter-orogenic period from c. 2.5 to 1.95 Ga. The focus is on igneous events that took place on and in the Karelian and Kola cratons. These cratons were stabilised by c. 2.6 Ga, were in close connection or part of the Superia supercraton, started to drift apart after c. 2.45 Ga and were recombined during the Lapland-Kola orogeny at c. 1.93–1.91 Ga (Mertanen and Pesonen 2005; Bleeker and Ernst 2006; Daly et al. 2006; Mertanen and Korhonen 2011). Since the initial rifting of the cratons starting near the Archaean-Proterozoic boundary, the magmatic events over the following time period of c. 500 Ma took place in a continental, within-plate environment. There is no evidence for oceanic crust formation before c. 1.95–2.0 Ga. The 1.95–1.96 Ga age of the Jormua and Outokumpu Ophiolites (Peltonen et al. 1996, 2008) records the definite time limit of the final break-up of the Karelian craton. Before this break-up, both the Karelian and Kola cratons were subjected to extensional stresses and multiphase intraplate rifting, as attested by the presence of many generations of mafic dyke swarms in the Archaean basement (Vuollo and Huhma 2005), but whether any fragmentation and dispersion of the cratons happened between c. 2.45 and 2.0 Ga, is presently unkown.

Kump and Barley (2007) drew attention to the transition of the style of volcanism at the Archaean-Proterozoic boundary, from dominantly submarine eruptions to a mixture of submarine and subaerial eruptions, and proposed that the rise of atmospheric oxygen at c. 2.5 Ga was related to this abrupt change in volcanism. They maintained that the widespread Palaeoproterozoic subaerial volcanism was connected to the formation of tectonically stable, high-standing continental cratons in late Archaean. The evolution of volcanism in the Fennoscandian Shield seems to fit with this conception, as the earliest Palaeoproterozoic volcanic eruptions on the shield were predominantly subaerial in nature with mafic to ultramafic magmas ascending from the mantle through a thick continental crust.

Other globally observed phenomena that coincide within the time interval of 2.5–1.95 Ga include the general paucity of subduction-related igneous rocks and juvenile crustal additions between 2.45 and 2.2 Ga, suggesting a temporal shutdown of the subduction system (Condie et al. 2009). Also large igneous provinces, which are thought to be related to mantle plumes, show a global minimum in their abundance in the same time interval. Condie et al. (2009) discussed potential causes and consequences, including effects on sea level, of the cessation of subduction processes. In this connection, it is only pointed out that despite the global Palaeoproterozoic magmatic age gap between 2.45 and 2.2 Ga, one of the igneous events in the Fennoscandian Shield discussed in this account, i.e. the Sariolian magmatism, took place within this time span.

This review concentrates on Karelian volcanic and intrusive rocks in the eastern and northern parts of the Fennoscandian Shield, that is, in the Kola Peninsula, Russian Karelia and Finland, excluding those occurring in northern Sweden and Norway. For the latter two regions, the reader is referred to Henriksen (1983), Pharaoh and Pearce (1984), Pharaoh (1985), Krill et al. (1985), Pharaoh et al. (1987), Bergh and Torske (1988), Barnes and Often (1990), Lager and Loberg (1990), Kullerud et al. (2006) and Myhre et al. (2011). Volcanic and intrusive events are considered by grouping them into time intervals following the main Karelian chronostratigraphical units (systems): Sumi (2505–2430 Ma), Sariola (2430–2300 Ma), Jatuli (2300–2060 Ma), Ludicovi (2060–1960 Ma) and Kaleva

E.J. Hanski (✉)
Department of Geosciences, University of Oulu, P.O. Box 3000, Oulu 90014, Finland
e-mail: eero.hanski@oulu.fi

V.A. Melezhik et al. (eds.), *Reading the Archive of Earth's Oxygenation*, Volume 1: The Palaeoproterozoic of Fennoscandia as Context for the Fennoscandian Arctic Russia - Drilling Early Earth Project,
DOI 10.1007/978-3-642-29682-6_6,

(1960–1900 Ma) (see Chap. 3.2). In addition, igneous rocks belonging to allochthonous (ophiolitic) complexes are treated separately, and although being regarded as "orogenic", some late volcanic sequences from the Kola Peninsula (South Pechenga, Panarechka), lying presently on Archaean crust, are discussed briefly.

In Fig. 3.30 are shown geographical locations referred to in the following text, particularly those that are not dealt with in other chapters of this and accompanying volumes. Figure 3.31 displays a geochemical summary of volcanic rocks in terms of MgO and TiO_2 contents using the grouping noted above and different symbols for rocks from Russian Karelia, the Kola Peninsula and Finland. In the following treatment of volcanic rocks, these regions are usually handled in the listed order, followed by discussion on intrusive rocks. To eliminate some of the alteration effects and to be able to compare differently documented analytical data from various literature sources, before plotting on diagrams, all major element analyses were recalculated to 100 wt.% on a volatile-free basis with total Fe as FeO_T.

3.4.2 Sumian Magmatism

The Sumian igneous rocks include both volcanic and intrusive rocks with the most significant of the latter being 2.44–2.50 Ga gabbro-noritic layered intrusions. The volcanic rocks are roughly contemporaneous with these intrusions or slightly older, as in places where they are cut by gabbro-noritic intrusive bodies.

Several studies have recently been published on Sumian igneous rocks in the Vetreny Belt, Russian Karelia, occurring in the southeastern corner of the exposed Fennoscandian Shield (Kulikov 1988; Puchtel et al. 1996, 1997, 2001; Sharkov et al. 2005; Kulikov et al. 2010). In that area, volcanic and associated terrigenous sedimentary rocks constitute a 4- to 8-km-thick supracrustal sequence distributed over an area 250 km in length and 15–85 km in width. The sequence is divided into six suites, of which the second one, the Kirich suite, and the sixth one, the Vetreny Poyas suite, are composed solely of volcanic rocks. The Kirich suite contains amygdaloidal and pillowed lava flows, which are predominantly andesitic in chemical composition though compositions vary from komatiitic basalts to dacites. The Vetreny suite with a thickness of more than 3 km is the most voluminous of the six units. It has been estimated that the total volume of the erupted material was c. 30,000 km^3 (Kulikov et al. 2005). The volcanic rocks are composed entirely of komatiitic rocks. They occur as massive or pillowed lava flows, or differentiated, spinifex-textured lava flows (Puchtel et al. 1996), varying in thickness from less than 1 m to as much as 150 m and totaling more than 60 in number (Kulikov et al. 2005). Locally, the rocks are extremely well-preserved by Precambrian standards, containing primary silicate minerals and displaying beautiful quench textures (Fig. 3.32).

The MgO content of the volcanic rocks from the Vetreny Poyas suite has a wide and continuous spread from c. 30 wt. % to as low as 6 wt.%, and in terms of major elements, the rocks appear to form a single petrogenetic trend. The presence of igneous olivine and the low degree of alteration have allowed to confidently estimate the parental magma composition of the rocks. Puchtel et al. (1997) concluded that the komatiitic rocks in different parts of the belt were generated from parental magmas having MgO between 12.8 and 15.5 wt.%, meaning that the most magnesian rocks are enriched in cumulate olivine while the least magnesian basalts are fractionation products of high-MgO komatiitic basalt. The melt compositions determined by Puchtel et al. (1997) are consistent with the weighted average composition of the whole Vetreny Poyas suite having 14.2 wt.% MgO (Kulikov et al. 2010). Analyses of volcanic glass testify that post-eruptional differentiation of interstitial melt proceeded even to dacitic liquid compositions (Sharkov et al. 2005). The most magnesian olivine ($Fo_{89.5}$), resulting in the highest MgO (17.3 wt.%) of the calculated parental magma, was measured from the Vinela dyke, which is a N-S-trending, 25-km-long and 0.5- to 1-km-wide, ultramafic intrusion cutting the Archaean basement on the SW side of the Vetreny Belt and likely represents a feeder chamber for Vetreny Poyas suite volcanic rocks (Puchtel et al. 1997).

The MgO-rich nature of the parental magma and the relatively low TiO_2 and high SiO_2 contents has prompted some researchers to characterise the magma type as boninite-like or belonging to a siliceous high-magnesian series (Sharkov et al. 1997, 2005). In Fig. 3.33, rocks of the Vetreny Belt komatiite-basalt suite are compared with typical Archaean Munro-type komatiitic and Phanerozoic boninitic rock suites. It is evident that, in terms of major elements, the Vetreny Belt rocks are very similar to the Munro komatiites and related fractionation products and, although overlapping with some boninites with respect to silica, are lower in SiO_2 than most boninites and do not show such a high degree of TiO_2 depletion that characterises the boninitic magma series; this difference was also pointed out by Kulikov et al. (2005, 2010). Puchtel and Humayun (2001) provided an in-depth analysis of the komatiite vs. boninite identification problem and argued against the boninitic nature of the Vetreny lavas on the basis of the PGE and chalcophile element behaviour.

In the $[Al_2O_3]$ vs. $[TiO_2]$ diagram of Fig. 3.34, representative compositions of primitive Palaeoproterozoic volcanic rocks from the Fennoscandian Shield are compared to each other and some Archaean and Phanerozoic primitive magma compositions. This diagram was designed by Hanski et al. (2001a) to classify olivine-saturated primitive magmas and

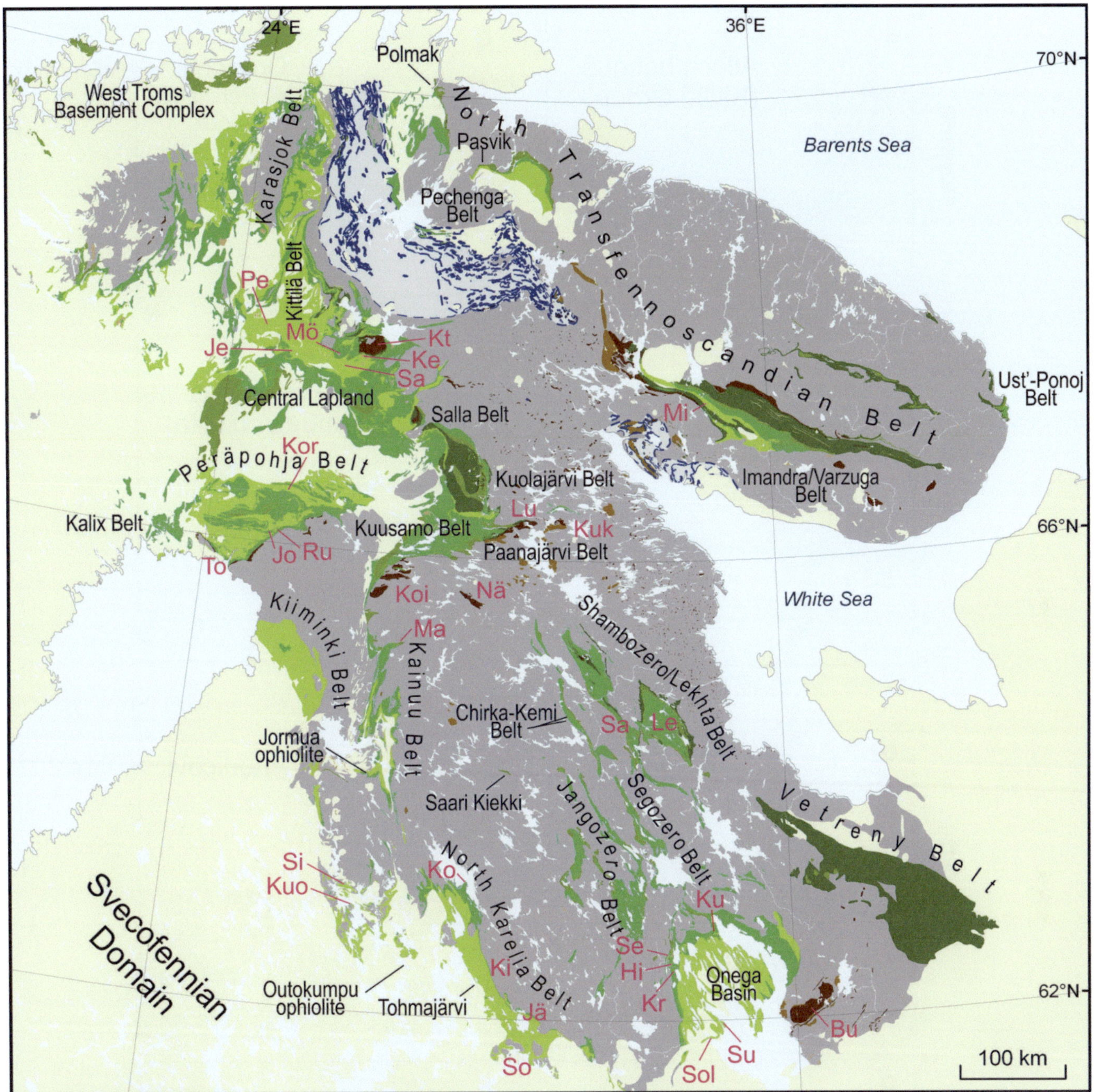

Bu = Burakovka	Koi = Koillismaa	Lu = Lukkulaisvaara	Sat = Sattasvaara
Hi = Hirvas	Kor = Korkiavaara	Ma = Matinvaara	Se = Semch
Jä = Jänisjärvi	Kr = Krasnaya Rechka	Mi = Mittrijärvi	Si = Siilinjärvi
Je = Jeesiörova	Kt = Koitelainen	Mö = Möykkelmä	So = Sortavala
Jo = Jouttiaapa	Ku = Kumsy	Nä = Näränkävaara	Sol = Solomennoe
Ke = Keivitsa	Kuo = Kuopio	Pe = Peuramaa	Su = Suisari
Ki = Kiihtelysvaara	Kuk = Kukasozero	Ru = Runkaus	To = Tornio
Ko = Koli	Le = Lekhta	Sa = Sapovaara	

Fig. 3.30 Map of the eastern part of the Fennoscandian Shield showing 1.96–2.50 Ga supracrustal belts and igneous rocks (*green* and *brown*) and localities mentioned in the text (the geological map modified by Aivo Lepland after Koistinen et al. 2001). *Grey* = Archaean basement, beige = <1.96 Ga sedimentary and igneous rocks. For a more detailed legend of the geological map, see Fig. 3.1 in Chap. 3.1

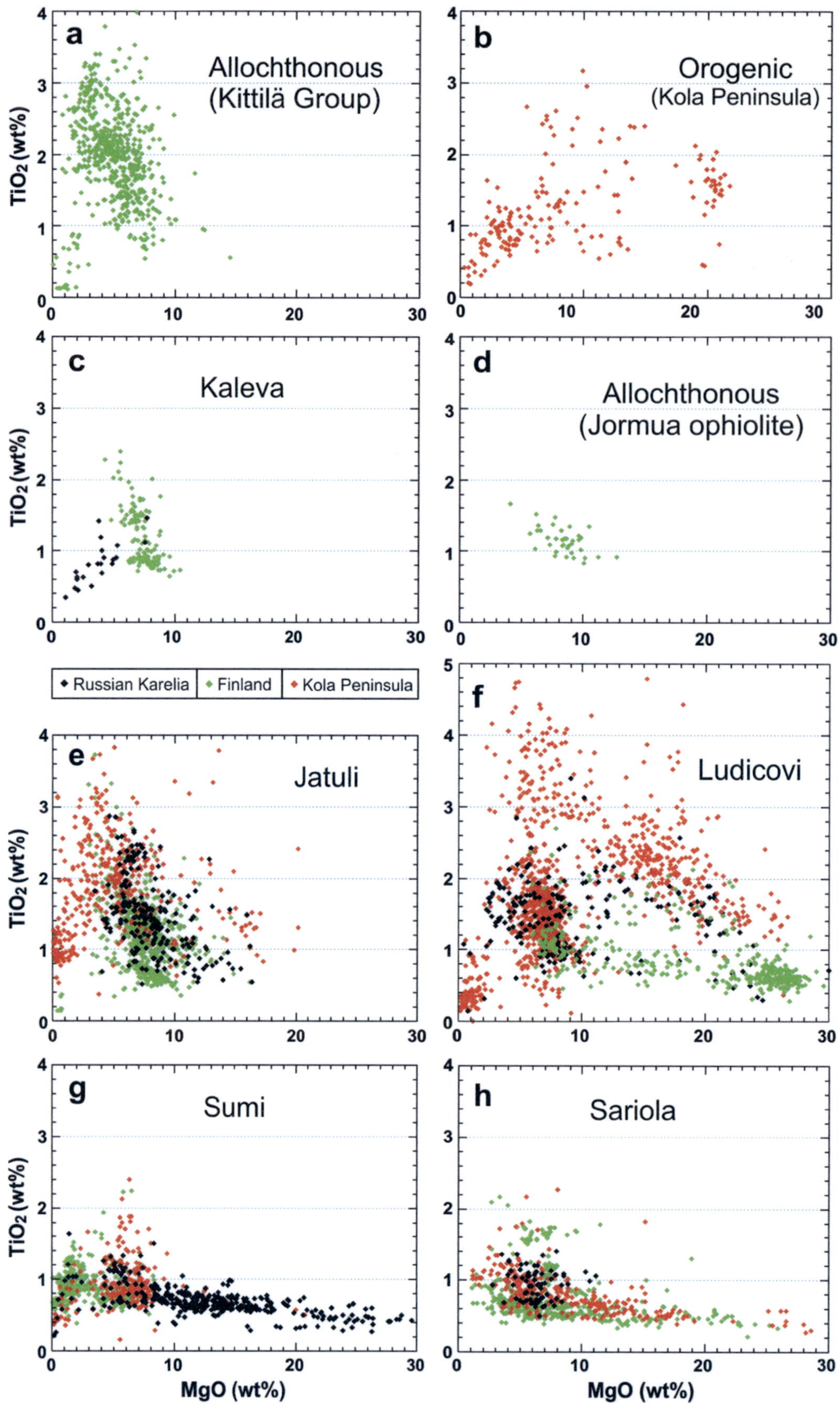

Fig. 3.31 Variation of MgO and TiO_2 in volcanic rocks in the main Palaeoproterozoic stratigraphic units. Analytical data are shown separately with different colours for rocks from the Kola Peninsula (*red*), Russian Karelia (*black*) and Finland (*green*)

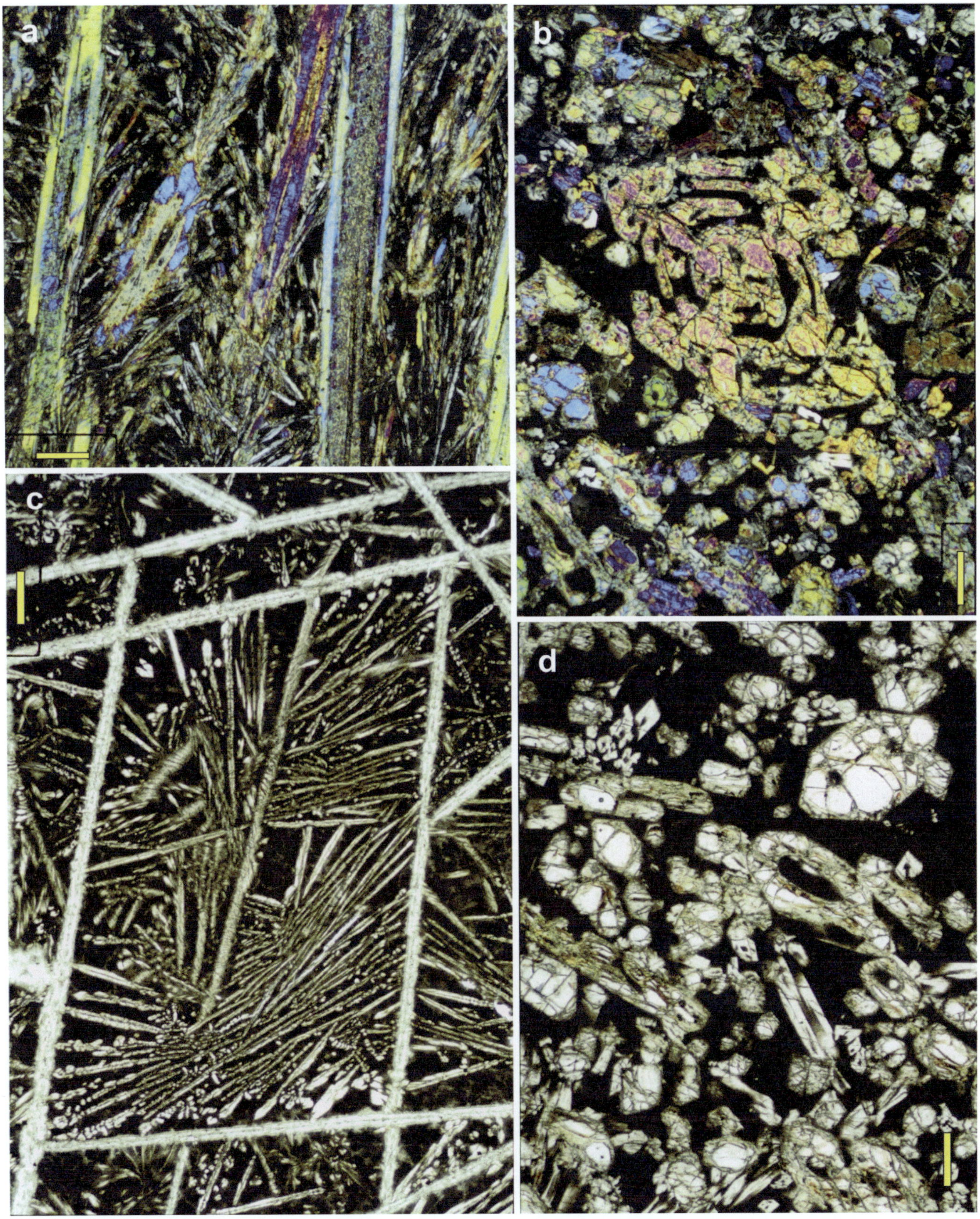

Fig. 3.32 Photomicrographs of Sumian komatiitic lavas from the Vetreny Belt. (**a**) Pyroxene spinifex texture ("string beef" spinifex). Cross-polarised light, scale bar 1.0 mm. (**b**) Skeletal, fresh olivine crystal surrounded by smaller ones. Cross-polarised light, scale bar 0.5 mm. (**c**) Olivine plates replaced by serpentine and fresh clinopyroxene needles in devitrified glass. Parallel-polarised light, scale bar 1.0 mm. (**d**) Olivine crystals in devitrified glass. Note rounded melt inclusions in the largest, euhedral crystal. Parallel-polarised light, scale bar 0.5 mm (Photographs courtesy of Igor Puchtel)

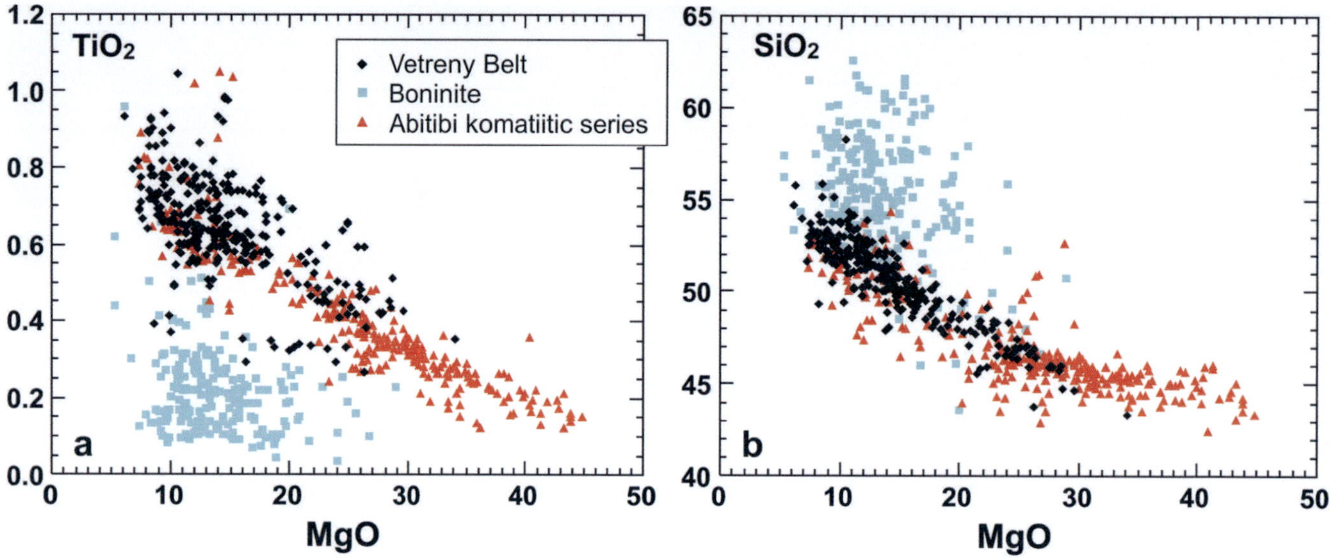

Fig. 3.33 Sumian Vetreny Belt komatiitic to basaltic rocks compared with literature data on Archaean Munro Township-type komatiitic series (from Canada) and Phanerozoic boninitic series (all analyses calculated 100 wt.% volatile-free)

contains molar abundances of TiO_2 and Al_2O_3 projected from olivine composition to eliminate the effects of olivine fractionation or accumulation (for more information, see the caption to Fig. 3.34). On this diagram, the average compositions of the Vetreny Belt komatiitic volcanic rocks and related intrusions plot in the middle of the field of Munro Township komatiites.

Despite the komatiitic major element composition, there are some chemical features that the Vetreny Belt rocks share with boninites, i.e., high contents of large-ion lithophile trace elements (Fig. 3.35a) albeit due to different reasons; in the case of boninites, the high LIL abundances are a result of mantle source enrichment via the activity of subduction-related fluids (e.g. Beccaluva and Serri 1988), whereas the chemical features of the Vetreny Belt volcanic rocks can be best explained by contamination of the magma with crustal material. Puchtel et al. (1997, 2001) have shown convincingly that in the Vetreny Poays suite, there is a good correlation between the variation of the initial Nd, Os and Pb isotopic ratios and incompatible trace element ratios (such as Nb/Th, La/Sm, Zr/Y), which is consistent with the interaction of the komatiitic parental magma with Archaean continental crust. Mineral-whole rock isochron-based initial ε_{Nd} values obtained by Puchtel et al. (1997) are -0.9 and -1.2 for lava flows and -1.4 for the Vinela dyke, while whole-rock analyses have given ε_{Nd} between $+0.4$ and -3.1 for lava flows and between -1.0 and -3.2 for a subvolcanic intrusion (Girnis et al. 1990; Puchtel et al. 1997; Kulikov et al. 2008).

Puchtel et al. (1997) estimated that the uncontaminated primary magma had an initial ε_{Nd} of +2.6. Using a high-MgO komatiitic basalt (MgO 17.6 wt.%; similar to an incompatible element-depleted komatiitic basalt from the Belingwe greenstone belt) as a mixing end-member, they calculated that the compositional variation of the Vetreny Belt rock suite is controlled by fractional crystallisation and 4–15% contamination with older felsic crustal rocks and concluded that the contamination of the magma took place mainly en route to the surface. Kulikov et al. (2005) proposed that the mantle plume-generated primary magma had originally MgO as high as 30 wt.% and was differentiated by olivine fractionation at the base of the crust and later in subvolcanic magma chambers.

Svetov et al. (2004) have studied four key sections (Krasnaya Rechka, Koikary, Semch River and Kumsa) of Sumian volcanic rocks at the northern and northwest rim of the Onega Basin (Fig. 3.30; for the geology of the Onega Basin, see Chap. 4.3). The volcanic rocks occur as massive and amygdaloidal lava flows, pillow lavas and minor agglomerates and tuffs (see Fig. 4.40c–m) with the flow thicknesses varying between 9 and 45 m and the total thickness reaching 1.5 km. In terms of SiO_2, most of the rocks are basaltic andesites and andesites. They include high-Mg varieties with MgO of c. 10 wt.%, which can be classified as komatiitic basalts (or boninites, as done by Svetov et al. 2004). The available chemical data suggest that the rocks contain two subtle types, one having low TiO_2 averaging 0.7 wt.% and the other one being slightly higher in TiO_2 averaging 1.3 wt.%. Only the latter one is found in the Krasnaya Rechka area, while both types occur in the Semch and Kumsy areas. Mafic to intermediate Sumian volcanic rocks display very similar trace element characteristics to those of the Vetreny Belt komatiitic rocks, though having higher abundances. This is illustrated in Fig. 3.35b, which shows chemical data for basaltic

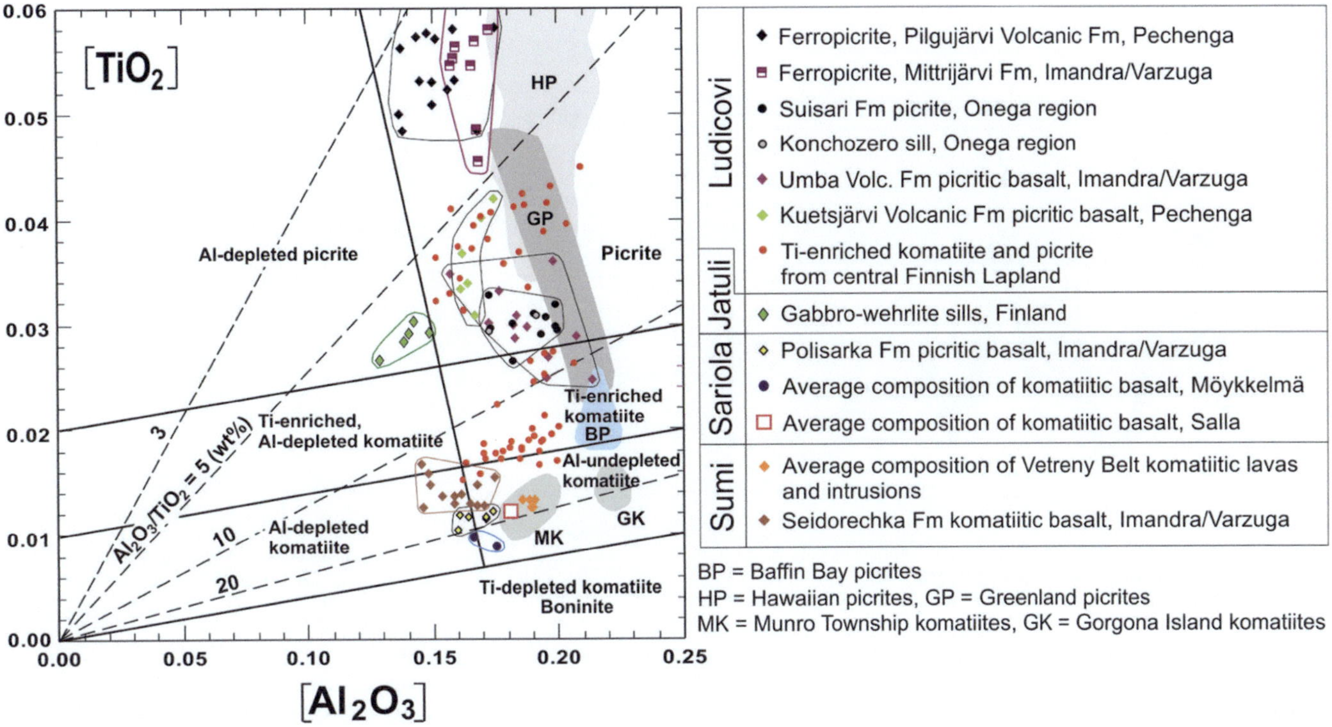

Fig. 3.34 $[Al_2O_3]$ vs $[TiO_2]$ diagram for representative compositions of high-MgO magmas from the eastern part of the Fennoscandian Shield and some comparative Archaean and Phanerozoic primitive magma compositions. $[Al_2O_3]$ and $[TiO_2]$ are Al_2O_3 and TiO_2 projected from olivine composition, calculated in mole proportions (normalised to unity) using the following equations: $[Al_2O_3] = Al_2O_3/(2/3 - MgO - FeO)$ and $[TiO_2] = TiO_2/(2/3 - MgO - FeO)$ (see Hanski 1992; Hanski et al. 2001a) (Literature sources available from the author upon request)

andesites and andesites from the Semch River area (Svetov et al. 2009).

In the Imandra/Varzuga Belt, Kola Peninsula, Sumian volcanic rocks are found in three lithostratigraphic units, the Purnach Formation, Kuksha Volcanic Formation and Seidorechka Volcanic Formation, all belonging to the Strel'na Group. These formations are described in Chap. 4.1 focusing on the geology of the Imandra/Varzuga Belt and dealt with in the reports of FAR-DEEP Holes 1A and 3A (Chaps. 6.1.1 and 6.1.2), and it suffices here to recapitulate the essential features of the volcanic parts of the formations. The Purnach Formation contains a thick sequence of subaerially erupted mafic lavas recrystallised to fine- to medium-grained amphibolites, which locally have amygdaloidal structures as evidence for their supracrustal origin. The Kuksha Volcanic Formation volcanic rocks occur as well-defined, amygdaloidal mafic lava flows (Figs. 3.23h, 4.4c, and 4.5a) and rare pillow lavas (Fig. 4.4d) and, in some areas, there also exist mafic subvolcanic intrusive bodies. The thickest of the units, the Seidorechka Volcanic Formation, has been dated at 2442.2 ± 1.7 Ma (Amelin et al. 1995) and 2448 ± 8 Ma (Chashchin et al. 2008), being coeval with the intruding Imandra layered gabbro-norite lopolith with an age of 2441.0 ± 1.6 Ma (Amelin et al. 1995). The formation is lithologically diverse, containing subaerial, amygdaloidal mafic to intermediate lava flows (Figs. 3.23m and 4.7o), komatiitic basalts and breccias (Fig. 4.7l–m) and numerous gabbroic and pyroxenitic sills and dykes with komatiitic affinities, and rhyolitic and dacitic lava flows (Fig. 4.7p–t) associated with co-magmatic subvolcanic bodies of granophyre.

Figure 3.36a–d compare major element compositions of the three volcanic formations of the Strel'na Group. Mafic and intermediate rocks of the Seidorechka Formation are in general higher in SiO_2 and K_2O and lower in FeO_T than those from the Purnach and Kuksha formations. Some of the Purnach and Kuksha Formation basalts and andesites are exceptional among the Sumian volcanic rocks in being moderately enriched in TiO_2 (>1.5 wt.%) (Fig. 3.36d) as the Sumian mafic to intermediate volcanic rocks have commonly TiO_2 less than 1.4 wt.% and most often less than 1.0 wt.% (Fig. 3.31g). Felsic volcanic rocks from the Seidorechka Formation plot in the tholeiitic field on a Jensen's cation plot (Fig. 3.37a) due to their low Al_2O_3 contents. The Seidorechka komatiitic basalts have commonly MgO between 10 and 15 wt.%, sometimes up to 20 wt.%, and show subchondritic Al_2O_3/TiO_2 as a consequence of a relatively low Al_2O_3 content. The latter feature

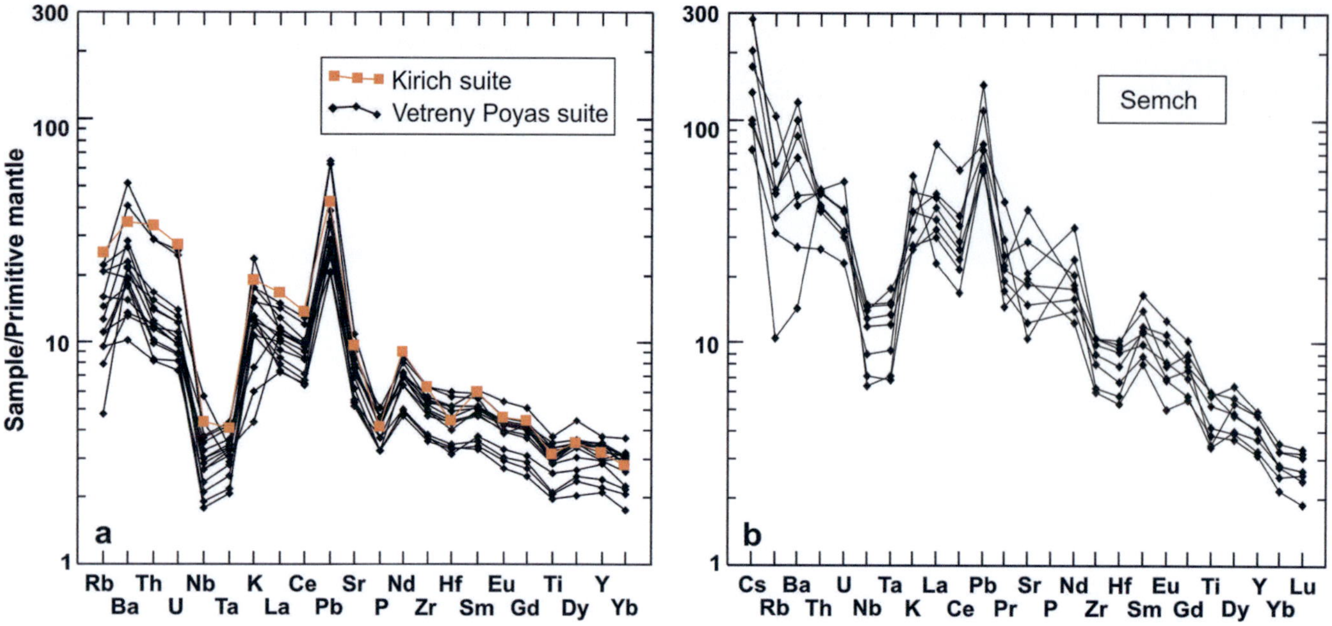

Fig. 3.35 (**a**) Primitive mantle-normalised trace element characteristics of Sumian komatiitic rocks from the Vetreny Poyas and Kirich suites, Vetreny Belt, SE Russian Karelia, showing a strong crustal signature. Analytical data from Puchtel et al. (1997). (**b**) Trace element characteristics of Sumian basaltic andesites and andesites from the Semch area, Central Russian Karelia. Data from Svetov et al. (2009). Normalising values after Sun and McDonough (1989)

makes them straddle the boundary between Al-depleted and Al-undepleted komatiites in the [TiO_2] vs. [Al_2O_3] diagram (Fig. 3.34).

Trace element characteristics of the Seidorechka Formation basalts differ from those of basalts from the Purnach and Kuksha formations as exemplified by chondrite-normalised Ce/Yb ratios in Fig. 3.37b. From this diagram, it can be inferred that the Kuksha and Purnach Formation rocks have only slightly LREE-enriched chondrite-normalised REE patterns, while in the Seidorechka Formation, the patterns are strongly negatively sloping (see Fig. 6.11 in Chap. 6.1.1). In the Seidorechka Formation, one can further distinguish two broad groups, one with $(Ce/Yb)_{CN}$ of c. 4–7 and the other with the ratio close to 10. Interestingly enough, the latter group is formed by samples from FAR-DEEP Hole 1A. According to the Nd isotopic data published by Chashchin et al. (2008), calculated ε_{Nd}(2450 Ma) values for three samples, representing basalts, andesites and rhyolites from the Seidorechka Formation, are clearly negative (−2.3 to −2.8) indicating magma interaction with old continental crust. Additional Nd isotopic data are documented by Vrevskii et al. (2010) who studied the Arvarench Unit, a supracrustal sequence that is correlated with the Seidorechka Formation and consists of a variety of volcanic rocks ranging from komatiites to dacites. They obtained a U-Pb zircon age of 2427 ± 16 Ma for a dacite and used it to calculate the following initial ε_{Nd} values: komatiites −1.6 to −4.1; basalts +0.4 to 1.4 and −1.9; andesites −0.8 to −2.2; dacites −3.3 to −8.4.

In Finnish Lapland, volcanic rocks that can be correlated with Sumian rocks of Russian Karelia were assigned to the Salla Group by Hanski and Huhma (2005). They are found in the Salla Belt close to and across the border with Russia (Manninen 1991) and, as more restricted occurrences, adjacent to the Akanvaara and Koitelainen layered intrusions (Manninen et al. 2001) and on a small Archaean gneissic dome at Tojottamanselkä (Peltonen et al. 1988). Salla Group volcanic rocks served as country rocks for the Akanvaara and Koitelainen intrusions and were erupted at c. 2.44 Ga, being thus indistinguishable in age from the intrusions (Manninen et al. 2001). The thickness of the Sumian sequence in the Salla area is considerable (≤2 km) but difficult to estimate accurately. The volcanic rocks vary from intermediate to felsic. Least evolved magmas formed subaerially erupted, 10- to 30-m-thick, amygdaloidal lava flows (Fig. 3.38a), whereas more felsic magmas produced lapilli tuffs and variously welded ash-flow tuffs (Fig. 3.23l) (Manninen 1991). Figure 3.31g shows a plot of MgO vs. TiO_2 for mostly Finnish Sumian rock analyses from the Salla Belt. The diagram reveals a relatively evolved general nature of the Salla Group rocks and the lack of highly magnesian volcanic rocks, which, in contrast, are typical among Sumian rocks in the Vetreny Belt, Russian SE Karelia. The geochemical and isotopic features [ε_{Nd}(2440 Ma) −3.1, −4.4] of the Salla Group rocks attest that they belong to a cogenetic fractionation series with a parental magma that had experienced considerable crustal

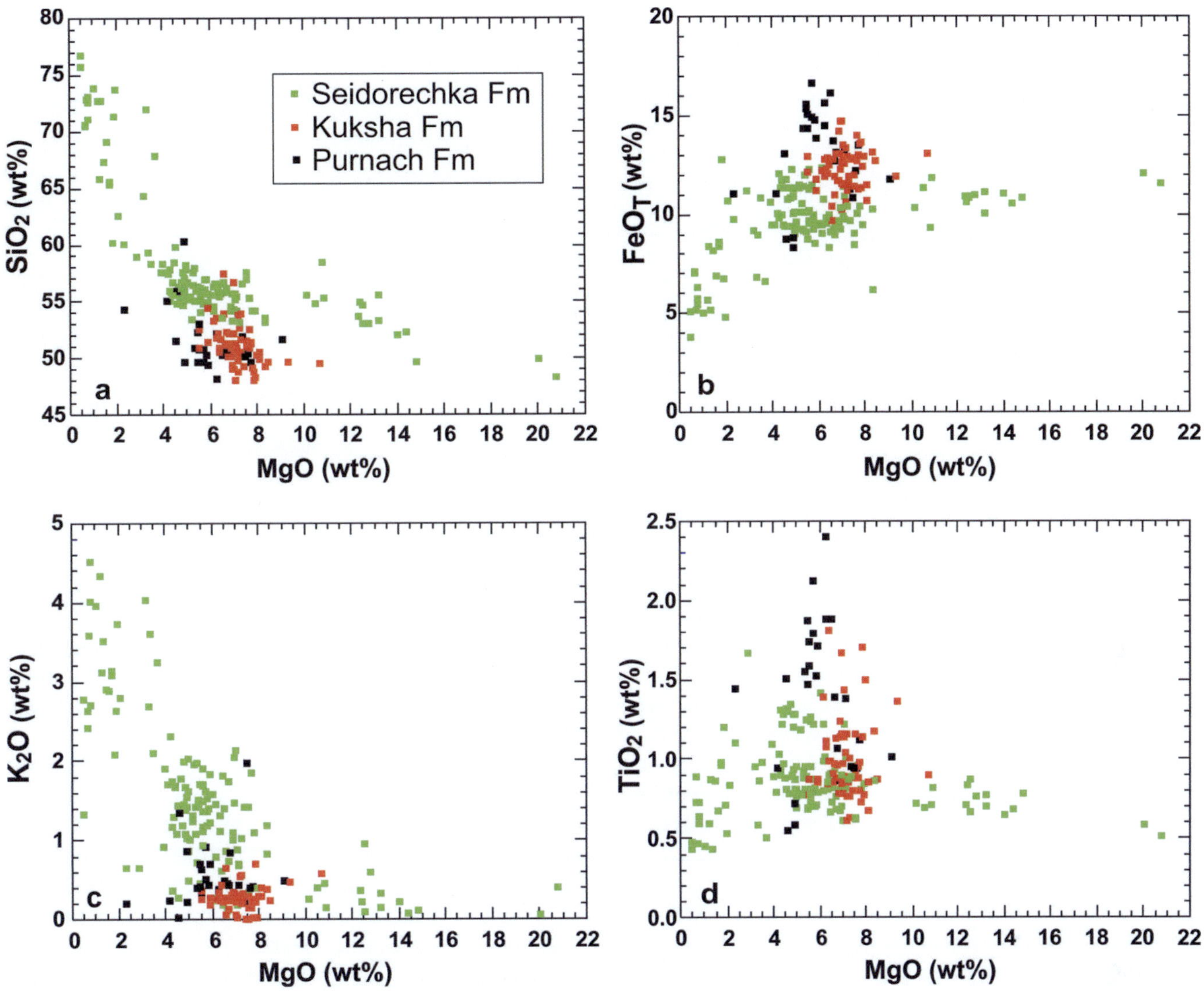

Fig. 3.36 Sumian volcanic rocks from three formations of the Strel'na Group, Imandra/Varzuga Belt, on SiO_2, FeO_T, K_2O and TiO_2 vs. MgO plots. Data sources: Fedotov (1983b, 1985), Zagorodny et al. (1980), Fedotov et al. (1982), Predovsky et al. (1987), Mints (1996), Chashchin et al. (2008), Zh. Fedotov (pers. comm., 2010) and FAR-DEEP Project

contamination before its main fractionation (Hanski and Huhma 2005).

In addition to the areas mentioned above, Sumian rocks in Finland occur farther south at the southern margin of the Kuusamo Belt and in the Koillismaa area. In the former area, quartz porphyry fragments are common in the (Sariolian) basal conglomerate of the belt (Silvennoinen 1972), indicating the former presence of wider occurrences of Sumian felsic volcanic rocks. On the Russian side of the border, in the Paanajärvi Belt forming the eastern continuation of the Kuusamo Belt, felsic volcanic rocks outcrop several hundred metres in thickness as the oldest Palaeoproterozoic formations (Kulikov et al. 1980; Heiskanen 1980). In the Koillismaa area, Sumian volcanic rocks are thought to have acted as hanging wall rocks during the emplacement of the Koillismaa Igneous Complex into the unconformity between the Archaean basement and the overlying Palaeoproterozoic cover (Lauri et al. 2003; Karinen 2010). They are assigned to the Sirviö Group, which is divided into two lithostratigraphic formations (Sirniövaara, Unijoki). The lower one is composed of rhyodacitic lavas and breccia interlayeres. The unit has a granophyric texture generated by recrystallisation due to the thermal effects of the underlying large basaltic magma chamber. The upper formation comprises lavas and crystal tuffs of andesitic to rhyolitic composition and is overlain by a thick conglomerate marking the start of Sariolian-age sedimentation. Figure 3.39 compares the trace element characteristics of Sumian andesitic to rhyolitic rocks from the Salla and Sirniö Groups.

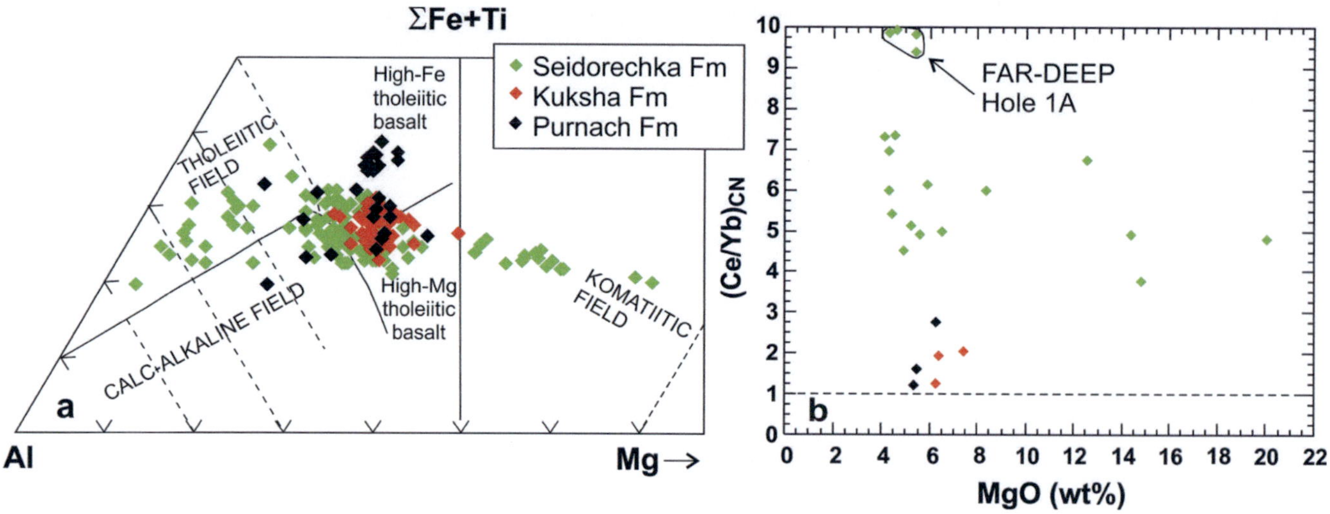

Fig. 3.37 Jensen's cation plot (**a**) and chondrite-normalised Ce/Yb vs. MgO diagram (**b**) for Sumian Strel'na Group volcanic rocks (Data sources as in Fig. 3.36)

Mafic to ultramafic intrusive rocks constitute an essential part of the Sumian magmatism over the whole shield, and together with associated volcanic rocks, have been assigned to the c. 2.5–2.4 Ga Sumian large igneous province by Kulikov et al. (2010). Although there are some problems with the dating of the layered intrusions (see Chap. 3.2), at least a major part of them seem to have been generated coevally (within analytical errors) with Sumian volcanic rocks. The intrusive bodies occur as layered gabbro-noritic intrusions in several parts of the shield, including the Tornio-Näränkävaara belt (Fig. 3.30) and its eastern continuation in northern Russian Karelia (for example, Kemi, Suhanko, Koillismaa, Kivakka, Lukkulaisvaara; Lavrov 1979; Alapieti et al. 1990; Glebovitsky et al. 2001; Iljina and Hanski 2005), central Finnish Lapland (Koitelainen, Akanvaara; Mutanen 1997; Hanski et al. 2001b, c), the Pechenga-Imandra/Varzuga Belt, Kola Peninsula (Mt. General'skaya, Monchegorsk, Imandra, Fedorovo-Pansky; Tel'nov et al. 1996; Smol'kin et al. 2004; Schissel et al. 2002), and southern Russian Karelia (Burakovka; Higgins et al. 1997; Chistyakov and Sharkov 2008). These intrusions were commonly emplaced at or close to the unconformity between the Archaean basement and overlying Palaeoproterozoic cover. Mafic-ultramafic dykes cutting the basement have also been shown to belong to the same magmatic episode (Puchtel et al. 1997; Vuollo and Huhma 2005). Furthermore, Kulikov et al. (2008, 2010) regarded the subvolcanic layered sills and lopoliths that are found in the lowermost Palaeoproterzoic sedimentary rocks in the Vetreny Belt as comagmatic with the volcanic rocks of the Vetreny Poays suite, and hence the Sumian igneous rocks have four different modes of occurrence in southern Russian Karelia. The Belomorian mobile Belt between the Karelian and Kola cratons is known to contain small, c. 2.5–2.4 Ga mafic to ultramafic intrusions that are assigned to the so-called Drusite Complex (Sharkov et al. 2004; Krivolutskaya et al. 2010a, b). High-Mg gabbro-noritic dykes are an essential component of this magmatic phase in the Belomorian Belt (Stepanova and Stepanov 2010). These were preceded by another, undated Mg-tholeiitic dyke generation.

Sumian mafic to ultramafic dykes that can be roughly correlated with the mafic layered intrusions in terms of their time of emplacement are widely distributed in the Archaean basement. They were classified by Vuollo and Huhma (2005) into five lithological types having the following orientations and chemical attributes: (1) NE-trending boninite-norite dykes (high MgO, SiO_2, Cr, Ni, and LREE, low TiO_2 and Zr); (2) NW-trending gabbro-norite dykes (low TiO_2, Cr, and Zr); (3) NW-trending low-Ti tholeiitic dykes; (4) E-trending Fe-tholeiitic dykes; and (5) E-trending orthopyroxene-plagioclase-phyric dykes (high SiO_2 and LREE; calc-alkaline affinity). Precise age data for these dyke generations are still very scarce with one exception being a U-Pb baddeleyite age of 2446 ± 6 Ma measured for a gabbro-noritic dyke from Russian Karelia (Vuollo and Huhma 2005). As revealed by the list above, the chemical composition of the dykes is variable with the type of magma ranging from high-Mg, low-Ti boninite-like compositions to high-Ti, high-Fe tholeiitic basalts. They differ also isotopically; Vuollo and Huhma (2005) reported negative initial ε_{Nd} values (-2.4 and -1.8) for types 1 and 2 and positive values (+0.3 and +1.7) for types 3 and 4.

The Sumian layered mafic-ultramafic intrusions have a great economical importance as they host magmatic ore deposits and potential future resources of various metals. They contain stratiform chromium (e.g. Kemi, Burakovka; Alapieti et al. 1989; Sharkov et al. 1995; Lavrov et al. 2006) and Fe-Ti-V deposits (Mustavaara; Juopperi 1977), and

Fig. 3.38 Field characteristics of Sumian and Sariolian mafic volcanic rocks. (**a**) Sumian amygdaloidal basaltic andesite lava from the Salla Group, Salla Belt. Amygdales filled mainly with plagioclase. Scale bar 10 cm in length. (**b**) Quartz-filled amygdales in andesitic lava from the Sariolian Matinvaara Formation, Kainuu Belt. (**c**) Sariolian basaltic lava from the Runkaus Formation, Peräpohja Belt, with amygdales filled with chlorite. Coin diameter 2.3 cm. (**d**) Andesitic volcanic breccia from the Sariolian Ahmalahti Formation, Pechenga Belt, containing abundant xenolithic fragments from the Archaean basement (Photograph (**a**) reproduced from Hanski and Huhma (2005) with permission from Elsevier (Photograph (**c**) courtesy of Vesa Perttunen; (**b**) and (**d**) taken by Eero Hanski)

PGE-reefs (Pana Tundra; Penikat; Latypov et al. 1999; Alapieti and Lahtinen 2002). Nickel and copper mineralisations with elevated PGE occur as disseminated sulphides and epigenetic veins in ultramafic cumulates (Monchegorsk; Gorbunov et al. 1985). Sulphide disseminations have also been discovered in lower marginal series rocks (Koillismaa; Lahtinen 1985; Iljina et al. 1992). These may be accompanied by massive offset sulphides in footwall rocks (Iljina et al. 1992). In addition, sulphide disseminations have been found in fine-grained, noncumulate gabbro-norites within the layered series (Lukkulaisvaara, Koillismaa; Glebovitsky et al. 2001; Karinen 2010).

Many attempts have been made to estimate the parental magma composition for the c. 2.5–2.4 Ga layered intrusions using different methods (chilled margins; weighted average compositions of intrusions or megacyclic units; fine-grained, noncumulate-textured rocks; associated dykes in the basement). In common with the Sumian volcanic rocks, all estimates have given a low TiO_2 content for the magma, which together with high MgO has allowed to characterise the magma as boninite (Alapieti et al. 1990), siliceous high-magnesian basalt (Saini-Eidukat et al. 1997), or crustally contaminated komatiitic basalt (Puchtel et al. 1997; Hanski et al. 2001; Kulikov et al. 2008). In detail, the analysed or calculated compositions differ considerably as indicated by Fig. 3.40a, b. As reference points, the black symbols in Fig. 3.40a, b represent average bulk compositions of a komatiitic unit and the whole Vetreny Poyas suite from the

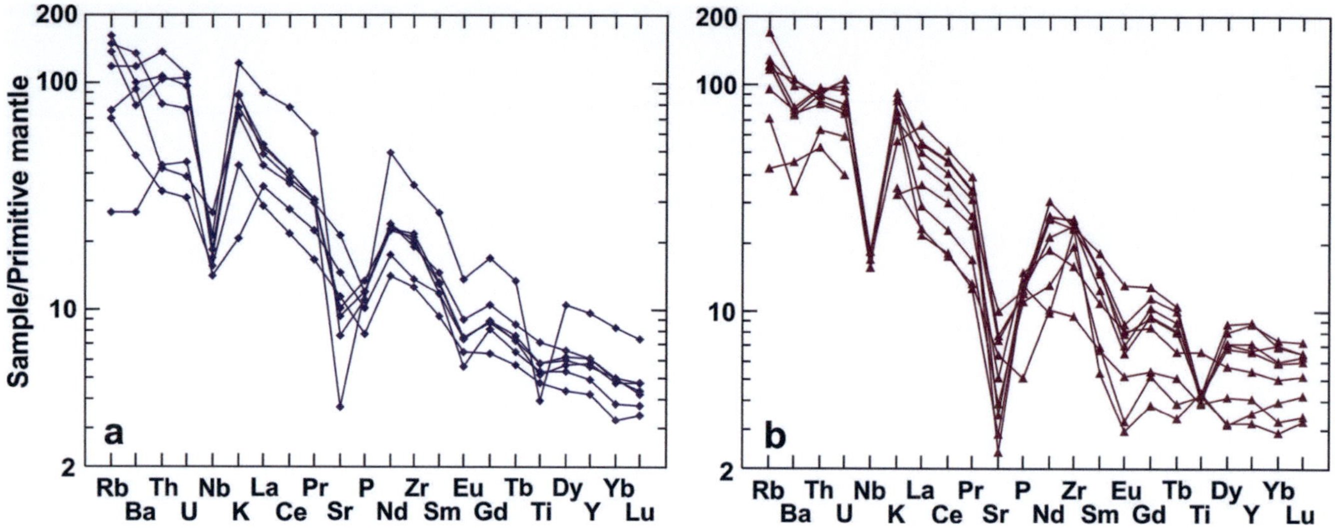

Fig. 3.39 Primitive mantle-normalised trace element patterns for Sumian andesitic to rhyolitic volcanic rocks from the Salla Group, Salla Belt (**a**) and Sirniö Group, Koillismaa (**b**) (Data from T. Manninen (pers. comm. 2011) and Lauri et al. (2003))

Vetreny Belt (Kulikov et al. 2010), and the arrows indicate olivine fractionation and accumulation trends of this kind of komatiitic basalt parental magma. Closest to the komatiitic basalt trend plot: (1) the six dyke samples having TiO_2 between 0.54 and 0.69 wt.% and MgO between 8.3 and 17.3 wt.% (Saini-Eidukat et al. 1997; Vogel et al. 1998); (2) the calculated parental magma compositions of the Burakovsky intrusion (orange symbols, Nikolaev and Ariskin 2005); (3) the weighted average composition of the Ruiga intrusion in the Vetreny Belt (Kulikov et al. 2008); and (4) an upper chilled margin sample from Koillismaa (Karinen 2010). As a whole, the chilled margin analyses or calculated weighted average compositions of intrusions do not represent a single, well-constrained magma type but scatter widely and are mostly more depleted in TiO_2 and other incompatible elements than the Sumian komatiitic basalts.

The green symbols in Fig. 3.40a represent fine-grained, noncumulate-textured gabbro-norites, which occur either as lens-like bodies within layered series (Latypov et al. 2008; Karinen 2010) or as strongly reversely zoned basal marginal zones (Hanski et al. 2001b, c; Latypov et al. 2011; Egorova and Latypov 2012). They can show ultradepletion in incompatible elements: for a rock having c. 10 wt.% MgO, the lowest measured TiO_2 is 0.06 wt.%, Zr beneath the detection limit of XRF (<10 ppm), and LREE below the chondritic levels (Latypov et al. 2011). With decreasing TiO_2, chondrite-normalised REE patterns change from LREE-enriched to flat or even LREE-depleted with the REE concentrations being decoupled from ε_{Nd}(2440 Ma), which is negative (c. −2.0). The microgabbros are too low in incompatible elements to represent magmas that generated the layered series of the Sumian intrusions, as the general trace element characteristics of these intrusions are similar to those of the Vetreny Belt komatiitic basalts (Hanski et al. 2001a, c). Nevertheless, the ultradepleted character and reverse fractionation of marginal zone microgabbros are perplexing petrological problems and probably require multiple igneous processes for their genesis (Latypov et al. 2011). Furthermore, noncumulate-textured bodies within layered series are interesting as they are associated with base metal and PGE mineralisations (Glebovitsky et al. 2001; Karinen 2010).

Besides the presence exotic marginal series rocks, other factors hamper the evaluation of the primary parental magma of the intrusions, including strong crustal contamination during ascent and multiple magma injections and ensuing magma mixing. Also the fractionation stage of the intruding magma, as reflected in their Cr content, for example, seems to have been different in different intrusions or their parts (Fig. 3.40b). It has been demonstrated by studying the mineralogy and chemical composition of cumulates in the Tornio-Näränkävaara intrusion belt that the magma chambers were fed with at least two kinds of parental magmas in terms of their Cr content. Some of the intrusions were generated either from a relatively primitive, higher-Cr magma or a more fractionated, lower-Cr magma, while others experienced influx of both magma types, first the higher-Cr and then the lower-Cr one, resulting in an abrupt drop of the chromium concentration at the boundary of two megacyclic units (Iljina and Hanski 2005).

The parental magmas of the c. 2.5–2.4 Ga layered intrusions were strongly enriched in LREE compared to HREE (e.g., Hanski et al. 2001b; Nikolaev and Ariskin 2005; Karinen 2010), and numerous isotopic analyses of samples

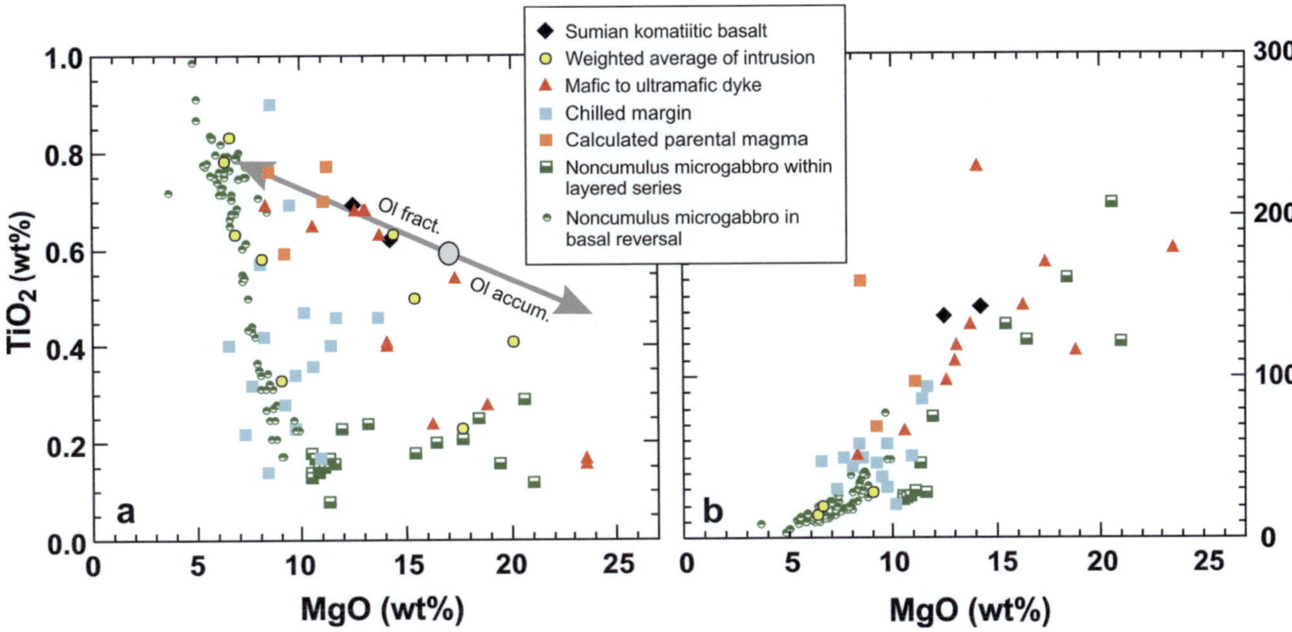

Fig. 3.40 MgO vs. TiO_2 (**a**) and MgO vs. Cr (**b**) plots for Sumian low-Ti lavas and various rocks and average compositions from 2.5 to 2.4 Ga layered intrusions (Data from Alapieti and Piirainen (1984), Sharkov (1984), Laz'ko and Sharkov (1988), Alapieti et al. (1990), Saini-Eidukat et al. (1997), Vogel et al. (1998), Koptev-Dvornikov et al. (2001), Glebovitsky et al. (2001), Alapieti and Lahtinen (2002), Nikolaev and Ariskin (2005), Kulikov et al. (2008, 2010), Karinen (2010), and Latypov et al. (2008, 2011))

from layered series have produced rather consistently an unradiogenic Nd isotopic composition with initial ε_{Nd} commonly falling in a range of −0.5 to −2.5 (Fig. 3.41; e.g. Huhma et al. 1990; Balashov et al. 1993; Amelin and Semenov 1996; Glebovitsky et al. 2001; Hanski et al. 2001b; Chistyakov and Sharkov 2008; Epifanova et al. 2008; Karinen 2010). These features have been interpreted either as evidence for interaction between an originally primitive, mantle-derived, low-Nd magma and continental crust (Amelin and Semenov 1996; Hanski et al. 2001b; Kulikov et al. 2010) or as an enriched lithospheric mantle signature (Vogel et al. 1998; Andersen et al. 1998). Relatively unradiogenic Os coupled with unradiogenic Nd in the magma that formed the Koitelainen and Akanvaara intrusions is consistent with a high-Os, low-Nd primitive parental magma that was contaminated with low-Os, high-Nd lithospheric material (Hanski et al. 2001b). In the most detailed study on the within-intrusion variation of the Nd isotope composition, Amelin and Semenov (1996) noted that in the intrusions of the Oulanka group, the isotope systematics are consistent with a closed-system crystallisation without significant interaction of the magma with country rocks. On the other hand, the Burakovka intrusion shows a more complex behavior of isotopes, which requires multiple magma injections and complex mixing and contamination processes.

The highest ε_{Nd} values (+3 to +4) in Fig. 3.41 are those obtained by Glebovitsky et al. (2001) for a sill-like body of fine-grained, noncumulate microgabbro-norite within the Lukkulaisvaara intrusion. The positive ε_{Nd} values, which were corroborated by isochrons of appropriate ages, differ markedly from the negative values documented for noncumulate rock bodies from Finnish intrusions that are similar or even lower than those measured for associated cumulates (Hanski et al. 2001b; Karinen 2010; Latypov et al. 2011). Epifanova et al. (2005) have also studied the Lukkulaisvaara microgabbro-norites. They performed five Sm-Nd isotope analyses of which four produced initial ε_{Nd} values of −1.7 to −2.6, thus differing from the results of Glebovitsky et al. (2001).

The Sumian magmatism was not only restricted to mantle-derived mafic magmatism but also included some felsic plutonism, as exemplified by several A-type potassic granitic stocks up to 5 km in size and associated felsic dykes in eastern Finland and adjacent parts of Russian Karelia (Levchenkov et al. 1994; Buiko et al. 1995; Lauri et al. 2006; Mikkola et al. 2010). Palaeoproterozoic potassic alkali granites and monzonite-granites also occur in the central Kola Peninsula and small bodies of "red" granites and granodiorites are found in the Belomorian Belt (Sharkov and Bogatikov 2000). The ages cluster at c. 2.44–2.43 Ga, but some are younger as shown by the age of 2389 ± 5 Ma reported by Mikkola et al. (2010), indicating a prolonged magmatic activity. Figure 3.42a shows a trace element composition of the Kynsivaara quartz-alkali feldspar syenite

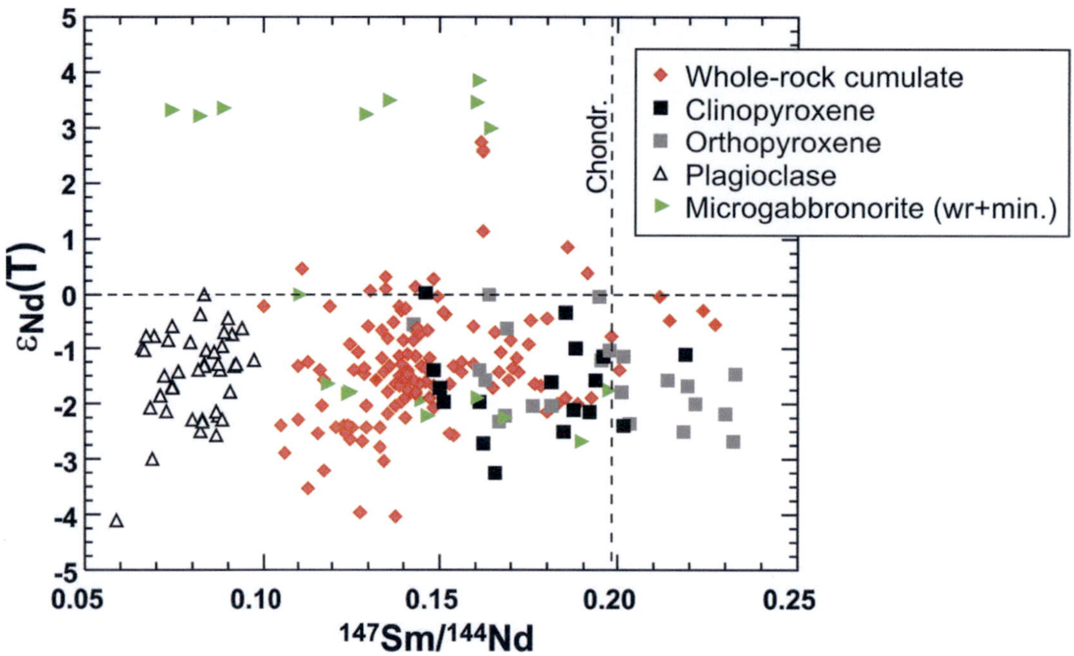

Fig. 3.41 Initial ε_{Nd} values for whole-rock samples and mineral separates from 2.45–2.50 Ga gabbro-noritic layered intrusions. Data from Huhma et al. (1990), Tolstikhin et al. (1992), Balashov et al. (1993), Amelin and Semenov (1996), Saini-Eidukat et al. (1997), Glebovitsky et al. (2001), Hanski et al. (2001b), Smol'kin et al. (2004), Serov (2005), Epifanova et al. (2005, 2008), Karinen (2010) and Latypov et al. (2011). The exceptionally high values for microgabbro-norites are from Glebovistky et al. (2001)

from Koillismaa. Lauri et al. (2006) invoked advanced differentiation of a siliceous high-Mg basalt parental magma via a combined assimilation-fractional crystallisation process to explain the genesis of this syenite body.

3.4.3 Sariolian Magmatism

Sariolian igneous rocks are predominantly subaerial volcanic rocks (e.g. Figs. 3.24r and 3.40p) and are younger than 2.44 Ga as their eruption took place after the gabbro-noritic layered intrusions had uplifted and partly eroded, though field evidence for this kind of relationship is not available in all regions. In general, direct precise dating of Sariolian volcanic rocks has so far been unsuccessful.

In the Imandra/Varzuga Belt, Sariolian volcanic rocks are represented by the 300- to 600-m-thick Polisarka Volcanic Formation (see Chap. 4.1). The rocks vary from basaltic komatiites to basaltic andesites with the primitive volcanic rocks dominating and the evolved ones being located in the upper part of the formation. A submarine eruptional environment is evident by abundant pillow lavas. Komatiitic basalt lavas also display spinifex textures (Fig. 4.9h, j) and variolitic pillow lava (Fig. 4.9i) and pillow breccia structures (Fig. 3.11r). As shown by Fig. 3.43a, the volcanic rocks from the Polisarka Volcanic Formation form a coherent low-Ti magmatic series evolving to andesitic differentiates. Volcanic rocks in the Polisarka Volcanic Formation that have TiO_2 higher than 1.0 wt.% only include intermediate tuffs (Fig. 3.43a) and some minor mafic lavas within the underlying Polisarka Sedimentary Formation, both recovered in FAR-DEEP Hole 3A. This drillhole also intersected highly magnesian, strongly carbonatised rocks, shown as green symbols in Fig. 3.43a, that likely represent a subvolcanic komatiitic intrusion (see Chap. 6.1.2). Immobile trace element characteristics of the komatiite-andesite series show a strong crustal signature, including high LREE/HREE and LREE/HFSE ratios, as the spidergrams in Fig. 3.43b demonstrate. Isotopic data are presently lacking.

Sumian sedimentary and volcanic rocks are lacking in the Pechenga Belt, and the Proterozoic supracrustal sequence starts with Sariolian basal conglomerates and other coarse-clastic sedimentary rocks of the Neverskrukk Formation and subaerial volcanic rocks of the Ahmalahti Formation, deposited unconformably on the Archaean basement and the Sumian Mt. General'skaya gabbro-noritic intrusion (see Chap. 4.2). The uppermost part of the formation was penetrated by FAR-DEEP Hole 5A (see Chap. 6.2.2), and in the Kola Superdeep Drillhole the Ahmalahti Formation was encountered at a depth of interval of 5717–6823 m (Kozlovsky 1987), indicating a total thickness of more than 1 km. The age of the formation is presently not known precisely, but is bracketed between the ages of the Mt. General'skaya intrusion (c. 2.5 Ga) and the Kuetsjärvi Volcanic Formation (c. 2.06 Ga).

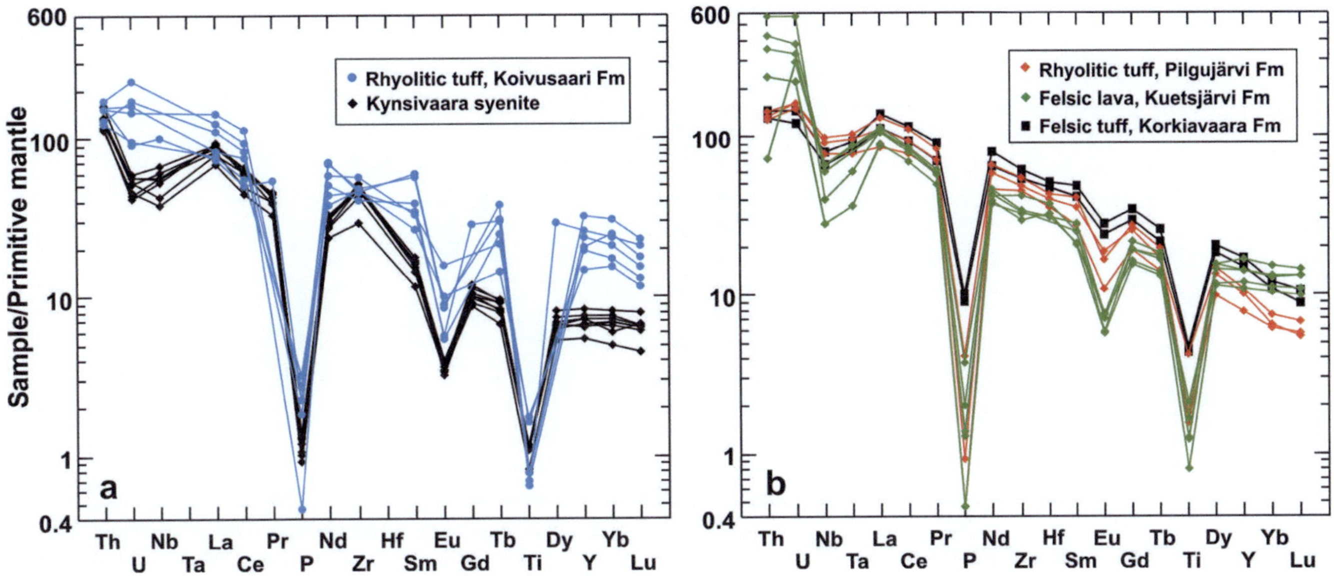

Fig. 3.42 Primitive mantle-normalised trace element characteristics of A-type felsic volcanic and plutonic rocks generated during Sumian, Jatulian and Ludicovian time: (**a**) 2.44 Ga quartz-alkali feldspar syenite, Koillismaa area (Lauri and Mänttäri 2002) and 2.06 Ga felsic tuffs from the Koivusaari Formation, Siilinjärvi Belt (Lukkarinen 2008); (**b**) 2.06 Ga felsic lavas from the Kuetsjärvi Volcanic Formation (FAR-DEEP data), Pechenga Belt, 1.98 Ga felsic tuffs from the Pilgujärvi Volcanic Formation, Pechenga Belt (this study), and 1.98 Ga felsic tuffs from the Korkiavaara Formation, Peräpohja Belt (Hanski et al. 2005)

The most typical rock type of the Ahmalahti Formation is amygdaloidal basaltic andesite (Fig. 4.19a, b) though differentiates to dacitic composition are found. Also more magnesian basalts or even picrites occur locally (Predovsky et al. 1974; Mints 1996; Skuf'in and Theart 2005). The rocks generally form a low-Ti suite with the exception of a few samples from the Kola Superdeep Hole, for example, that have TiO_2 higher than 1.5 wt.% (Skuf'in and Yakovlev 2007). Outcrops within Zapolyarny town bear witness of strong interaction of the magma with sialic basement during its ascent to the surface (Fig. 3.38d). This is also evident in the trace element and isotope geochemistry of the whole volcanic unit. The magma was LREE-enriched and relatively poor in HFSE, as is typical of Sariolian magmas (Fig. 3.43a, b; see also Fig. 6.50). The three Sm-Nd analyses published by Skuf'in and Theart (2005) and Skuf'in and Bayanova (2006) yield low initial ε_{Nd} values of -3.2 to -5.0 calculated for an age of 2350 Ma.

In northern Finland, Sariolian-system volcanic rocks, which are distributed widely from the western to eastern border (Väänänen 1989; Manninen 1991; Lehtonen et al. 1998), were assigned to the Onkamo Group by Hanski and Huhma (2005). This group has recently been renamed the Kuusamo Group (Geological Survey of Finland, 2011, Bedrock of Finland – DigiKP, Digital map database, version 1.0; http://www.geo.fi/en/bedrock.html). An example of well-studied Sariolian volcanic rocks is located at Möykkelmä, where a small Archaean basement dome is surrounded by a 250-m-thick, subaerially erupted succession of komatiites and low-Ti basalts and andesites (Räsänen et al. 1989). The volcanic series, which lies directly on the Archaean basement, is divided into five subunits: the lower and upper ultramafic members and the lower, middle and upper mafic members whose stratigraphical positions are shown in the drillcore section of Fig. 3.44. The ultramafic rocks are volcaniclastic varying from agglomerates to tuffs and commonly carry granitoids xenoliths (Fig. 3.24t) torn from the underlying Archaean basement. The mafic units are composed of variously amygdaloidal lavas (Fig. 3.24s) with intervening tuff layers. They become more primitive upwards in the sequence with the lowermost member being andesitic and the uppermost one basaltic.

The chemical composition of the ultramafic volcanic rocks mostly corresponds to that of komatiitic basalt and, in some cases ($MgO > 18$ wt.%), that of komatiite (Fig. 3.44). A strong crustal signature is evident by suprachondritic Al_2O_3/TiO_2, high $(La/Yb)_{CN}$ of 4.7–6.1 and low $(Th/Ta)_{PM}$ of 2–6 (cf. Fig. 3.45a). A common feature of all five volcanic members is their low TiO_2 content (<0.7 wt. %). A special feature of the uppermost mafic unit, which is basaltic in composition, is its relatively high Cr abundance ranging between 790 and 1,110 ppm (Fig. 3.44). Although all volcanic rocks in the drilled section share trace element compositions indicative of crustal contamination (Fig. 3.45a), the mutual relationships of the members are not explicable by a different degree of contamination or any other simple petrogenetic process. Besides displaying a reversed fractionation through time (height in Fig. 3.44),

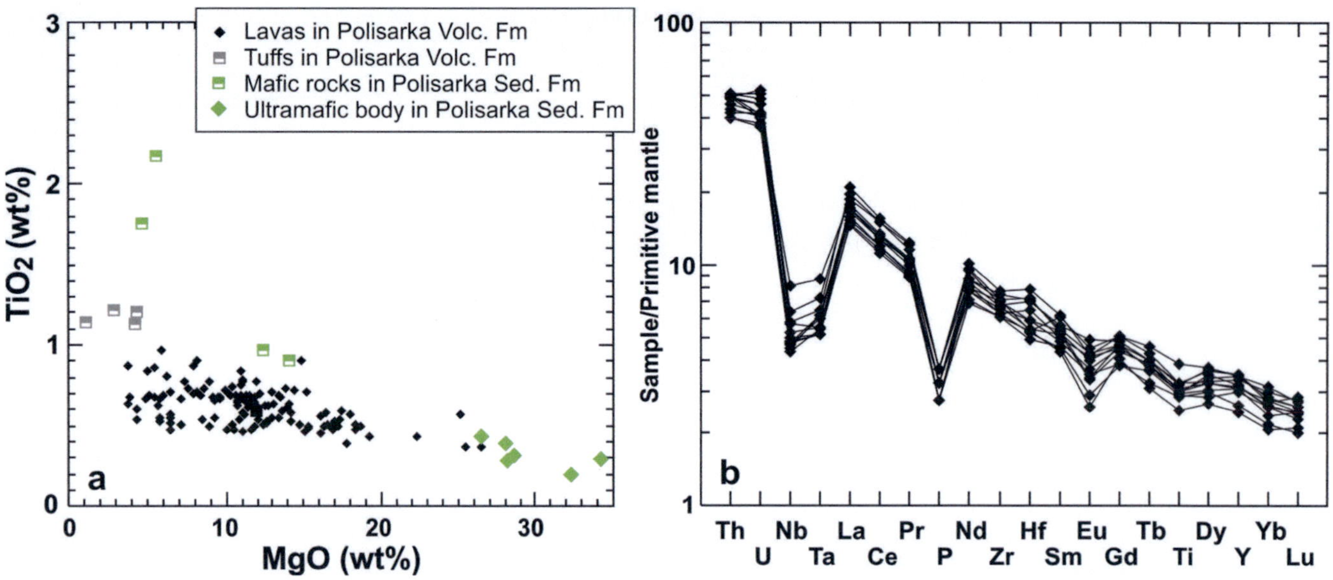

Fig. 3.43 (**a**) TiO_2 vs. MgO plot for volcanic and subvolcanic rocks from the Sariolian Polisarka Sedimentary and Volcanic formations, Imandra/Varzuga Belt. Data sources: Fedotov (1983a, 1985), Zagorodny et al. (1980), Zhangurov and Smol'kin (1982), Melezhik et al. (1982), Mints (1996), Hanski and Smol'kin (unpublished), Zh. Fedotov (pers. comm., 2010) and FAR-DEEP Project. (**b**) Primitive mantle-normalised trace element diagram for komatiitic basalts (MgO 10.2–12.2 wt.%) from the Polisarka Volcanic Formation (E. Hanski and V. Smol'kin, unpubl. data)

the mafic members differ considerably in their REE characteristics (Fig. 3.45a) and do not always show the expected anticorrelation between compatible and incompatible trace elements (for example, Cr and TiO_2 in Fig. 3.44). Furthermore, mixing of komatiitic and andesitic magmas, as represented by the two lowermost members, cannot explain the geochemical features of the basaltic member, because the latter has lower $(La/Yb)_{CN}$ than the potential mixing end-members (Fig. 3.45a).

Another well-documented volcanic sequence of the Sariolian Kuusamo Group is found in the Salla Belt, at the Finnish-Russian border, where komatiitic basalts (MgO 12–18 wt.%) and related high-Mg basalts (MgO 8–12 wt.%) of the Mäntyvaara Formation were deposited on a weathering crust developed on Sumian (Salla Group) volcanic rocks (Manninen 1991). The volcanic rocks occur as amygdaloidal lavas and variolitic pillow lavas, and as pyroclastic deposits such as tuffs, lapilli tuffs and tuff breccias. Chemically, they are similar to the magnesian volcanic rocks of the Sariolian Polisarka Formation in the Imandra/Varzuga Belt in having high LREE/HREE (Fig. 3.45b) and LREE/HFSE ratios. Hanski and Huhma (2005) published Nd isotope data for a suite of magnesian basalts giving an average ε_{Nd}(2350 Ma) of -3.0, which is indicative of a significant contribution from Archaean LREE-enriched lithosphere.

Sariolian mafic volcanic rocks start the magmatic evolution of the Peräpohja and Kuusamo Belts and the northern part of the Kainuu Belt (Puolanka), lying either directly on the Archaean basement or separated from it by a thin basal conglomerate (Silvennoinen 1972; Perttunen 1985, 1989; Laajoki 1991). In the Peräpohja area, partly eroded, c. 2.44 Ga layered intrusions also served locally as the depositional substrate (see Fig. 3.24m). The exact age of these Sariolian volcanic units is still unknown. It is only definite that they are younger than 2.44 Ga but older than 2.22 Ga, the latter figure being the age of cutting mafic sills. Huhma et al. (1990) obtained an imprecise Sm-Nd isochron age of 2330 ± 180 Ma with an initial ε_{Nd} of -0.6 ± 1.1 for the Runkaus Formation in the Peräpohja Belt. Additional constraints for the minimum age come from the U-Pb date of c. 2250 Ma measured for secondary titanite from the same formation (Huhma et al. 1990).

The 300- to 500-m-thick Matinvaara Formation in the Kainuu Belt is composed of amygdaloidal basaltic and andesitic lavas that range in thickness from 5 to 30 m (Fig. 3.38b) (Strand 1985). They are overlain by volcaniclastic conglomerates of the Ahven-Kivilampi Formation (see Fig. 3.40h). The Greenstone Formation I of Silvennoinen (1972) in the Kuusamo Belt, which has recently been renamed the Kuntijärvi Formation (Geological Survey of Finland, 2011, Bedrock of Finland – DigiKP, Digital map database, version 1.0; http://www.geo.fi/en/bedrock.html), has a thickness of several hundred metres and is composed of amygdaloidal and massive lavas, and volcanic breccias with granitoid fragments from the basement and tuffite schists. The Runkaus Formation in the Peräpohja Belt is 40–100 m thick and constitutes 2–5 subaerial, amygdaloidal lava flows and local agglomeratic

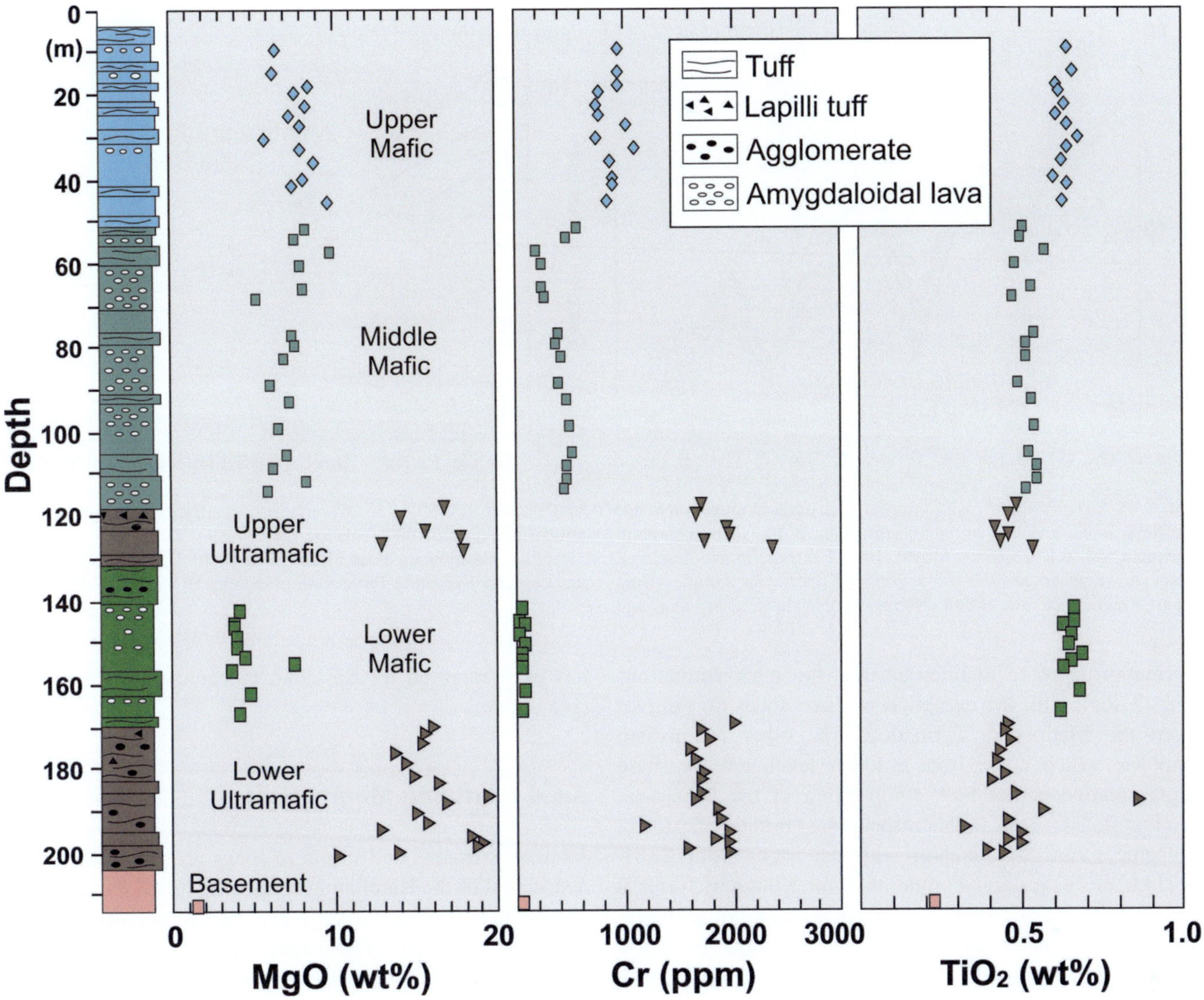

Fig. 3.44 Drillcore section of a Sariolian volcanic sequence (Kuusamo Group) erupted on Archaean basement at Möykkelmä, Central Finnish Lapland (Data from Räsänen et al. 1989)

deposits. The geochemistry of these three volcanic formations is discussed together with that of the lavas of the Saari-Kiekki Belt (see below).

The Kaita-Kiekki Formation in the small Saari-Kiekki Belt crossing the Finnish-Russian border in eastern Finland is the southernmost major Sariolian volcanic unit in Finland (Luukkonen 1989). It contains up to 1,000-m-thick volcanic succession overlying a 500-m-thick sequence of basal conglomerates on rifted Archaean basement. The volcanic pile is composed of magnesian basalts, basaltic andesites and andesites that occur as tuffs, agglomerates and tuffites, and lavas displaying massive, amygdaloidal, variolitic and breccia structures and columnar jointing, indicative of sub-aerial eruptions.

In Fig. 3.46a, the four Sariolian volcanic units from the four mentioned supracrustal belts are compared in a TiO_2 vs. MgO diagram. The most primitive volcanic rocks are found in the Saari-Kiekki Belt where the MgO content of the lavas reaches 11.8 wt.%, and all volcanic rocks are low in TiO_2 (0.4–0.7 wt.%). Analogously with the Sariolian basalts at Möykkelmä discussed above, the Saari-Kiekki basalts and basaltic andesites are rich in Cr (500–1,140 ppm), whereas andesites have a very low Cr level (30–80 ppm). Thus a distinct gap exists in Cr concentrations between the rocks having an MgO content of higher or lower than c. 5.0 wt.% (Fig. 3.46b), suggesting an occurrence of two major pulses of magmas with a different evolutionary history.

The Runkaus Formation volcanic rocks form two distinct groups in terms of the TiO_2 level; drilling has shown that the upper part of the formation is more enriched in TiO_2 and other incompatible elements compared to the lower part (Perttunen and Hanski 2003). In general, the Runkaus

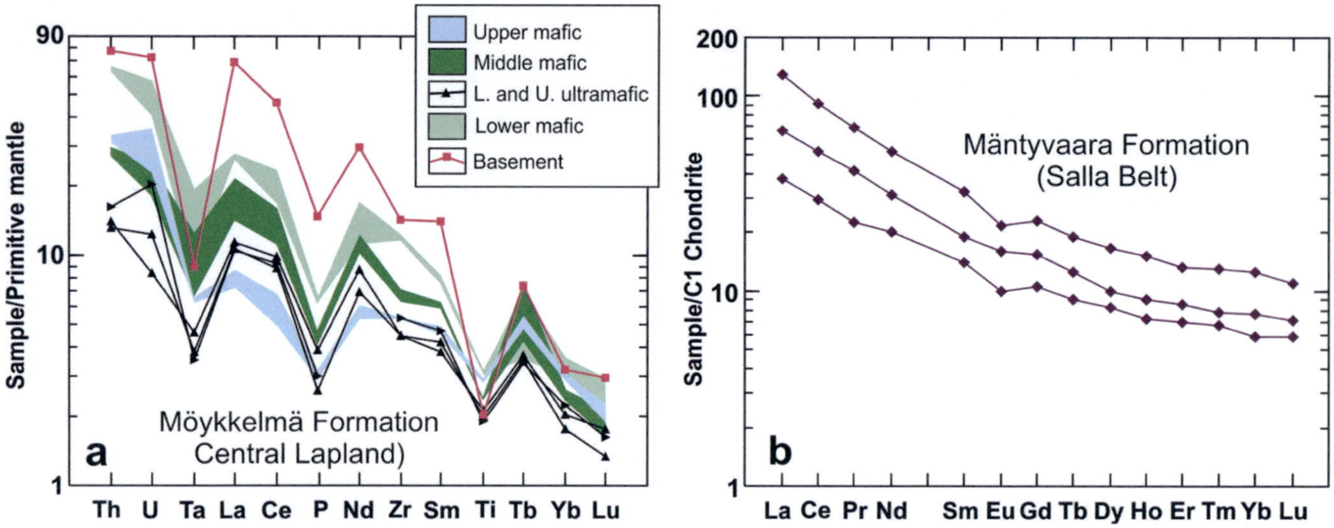

Fig. 3.45 (**a**) Primitive mantle-normalised trace element plots for the three mafic and two ultramafic members in the section through Sariolian volcanic rocks at Möykkelmä, Central Finnish Lapland. Also shown is an analysis of a granitoid basement sample (Data from Räsänen et al. 1989). Normalising values after Sun and McDonough (1989). (**b**) Chondrite-normalised REE patterns of komatiitic basalts and basalts (MgO 4.9–12.9 wt.%) from the Sariolian Mäntyvaara Formation, Salla Belt. Unpublished data of the Lapland Volcanite Project, the Geological Survey of Finland

Formation is more titaniferous than the other formations (Fig. 3.46a), with the exception of three andesitic samples from the Matinvaara Formation. The other Matinvaara samples, which come from a lower level and are more representative of the bulk composition of the formation, are low in TiO_2 and, in this respect, are "normal" Sariolian volcanic rocks. In common with the rocks from Saari-Kiekki, the Kuntijärvi Formation from Kuusamo forms a low-Ti differentiation series, reaching a dacitic composition. Incompatible trace element data are available from the Runkaus and Kuntijärvi formations. As an indication of a strong crustal signature are the sloping REE patterns of both formations (Fig. 3.47b) and their location in the field of calc-alkaline volcanic arc basalts in the Th-Zr/3-Ta diagram of Wood (1980) (Fig. 3.47a).

One of the mafic dyke generations that Vuollo and Huhma (2005) have recognised in the Archaean basement has an age of c. 2.32 Ga, fitting within the timeframe of the Sariolian system. A few EW-trending dykes have been dated from the Iisalmi and Taivalkoski blocks of the basement, but the real abundance of these dykes is difficult to assess because they are chemically indinguishable from younger Fe-tholeiitic dykes. Few available chemical analyses suggest that they differ from most Sariolian mafic rocks in being relatively high in TiO_2 (c. 2.0 wt.%) and, in this respect, match with the anomalous samples from the Matinvaara Formation (see Fig. 3.46a). It is also noteworthy that the 2319 $\pm$ 27 Ma isochron produced for a dyke in the Taivalkoski block gave a positive initial ε_{Nd} of +1.8 (Vuollo and Huhma 2005), which is higher than what has been reported for Sariolian volcanic rocks in general (see above).

3.4.4 Jatulian Magmatism

Jatulian extrusive and intrusive rocks are the most widely distributed of the Karelian igneous rock suites, and Jatulian-age mafic dykes form several generations of swarms in the Archaean basement. These igneous rocks are overwhelmingly mafic in composition with ultramafic and felsic varieties being almost absent (Fig. 3.31e). Most of the evolved compositions in Fig. 3.31e are from a single volcanic formation (Kuetsjärvi) from the Pechenga Belt. Although Jatulian effusive sequences have often been regarded as Precambrian equivalents of Phanerozoic continental flood basalts (CFB) (e.g. Svetov 1979; Malashin et al. 2003), their mode of occurrence does not only imply monotonous subaerial fissure eruptions, but a considerable portion of magma was erupted subaqueously (see Fig. 4.45i), though never in deep water.

In Russian Karelia, the Jatulian volcano-sedimentary successions have traditionally been divided into three main units, Lower, Middle and Upper Jatulian, with volcanic rocks tending to be concentrated in the upper part of these units or forming the ending phase of a unit (e.g. Sokolov et al. 1970; Meriläinen 1980). The stratigraphy is somewhat different in different areas, and five type sections have been defined, for which more detailed information can be found in Sokolov (1980), for example. Palaeovolcanological studies

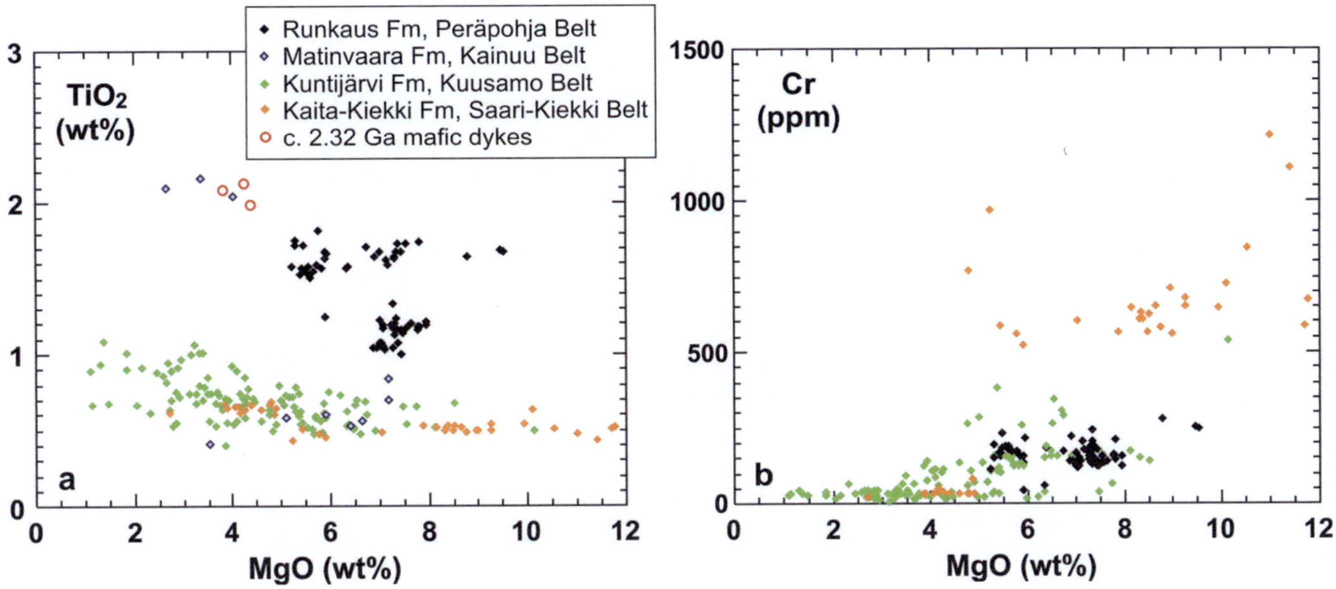

Fig. 3.46 TiO_2 vs. MgO (**a**) and Cr vs. MgO (**b**) diagrams for Sariolian mafic volcanic rocks from the Runkaus, Matinvaara, Kuntijärvi and Kaita-Kiekki formations, representing the lowermost volcanic rocks in four supracrustal belts (Peräpohja, Kainuu, Kuusamo and Saari-Kiekki, respectively) (Data from Strand (1985), Luukkonen (1989), Perttunen (1989), Perttunen and Hanski (2003), and the Lapland Volcanite Project (the Geological Survey of Finland). Also shown in (**a**) are analyses of c. 2.32 Ga mafic dykes from the Archaean basement (J. Vuollo, pers. comm. 2011))

carried out by A.P. Svetov and his coworkers have resulted in maps, which show the geographical distribution of the Lower, Middle and Upper Jatulian volcanic rocks and their eruptive centres (Svetov 1979, 1980; Golubev and Svetov 1983). Such maps are reproduced in a simplified form in Fig. 3.48a, b together with average present-day oxidation states of iron in mafic lavas from different areas. So far, reliable geochronological data of each of the three Jatulian volcanic stages in Russian Karelia are lacking. Philippov et al. (2007) reported mutually consistent U-Pb zircon ages of 1983.4 ± 6.5 Ma and 1984 ± 8 Ma for two iron ore-bearing gabbroic sills (Koikary-Svyatnavolok and Pudozhgora) from the Onega Basin, which have been thought to be comagmatic with Upper Jatulian volcanic rocks (e.g. Svetov 1979; Golubev and Svetov 1983). These ages are younger than expected in the light of the U-Pb zircon ages of c. 2060 Ma and c. 2106 Ma measured for Upper Jatulian volcanic rocks in the Pechenga and Peräpohja belts, respectively (Melezhik et al. 2007; Karhu et al. 2007), but are similar to the age of the Ludicovian magmatism in the Pechenga Belt and Onega Basin (Hanski et al. 1990; Puchtel et al. 1998, 1999).

Lower Jatulian volcanic rocks were deposited on a regressive succession of terrigenous sedimentary rocks or more rarely directly on Archaean basement and were erupted most widely in the Kuolajärvi area and an area north of Lake Onega (Fig. 3.48a). The volcanism produced a small number (1–4) of massive and amygdaloidal lava flows having a total thickness between 20 and 90 m (Svetov 1979, 1980). Svetov (1980) pointed out that, particularly in the southern part of their area of occurrence, the Lower Jatulian mafic lavas are characterised by a high degree of haematisation and thus have high $Fe^{3+}/\Sigma Fe$.

Of all Palaeoproterozoic igneous activity, the Middle Jatulian volcanism was most widely distributed in Russian Karelia (see Fig. 3.48b). The total thickness of the lava successions varies between 250 and 400 m, and the number of individual lava flows often exceeds 15. The lavas display massive, amygdaloidal, breccia and pillow structures (Fig. 4.45g–j) and contain minor tuffogenic interlayers (Sokolov 1978; Svetov 1979). The largest number of oxidised lavas occurs in the southern part of the area where subaerial eruptions are most common whereas in the northern part, submarine volcanism prevailed. Subvolcanic gabbroic bodies are also found in underlying sedimentary rocks. Three to five short breaks in volcanism gave space to fumarolic and solfataric activity resulting in ferriferous and ferriferous-siliceous deposits and minor copper mineralisations (Svetov 1980; Svetov and Sviridenko 2005).

The Upper Jatulian magmatism was most limited in extent (Fig. 3.48b). Upper Jatulian dolomites are overlain in some areas by mafic effusives reaching 300 m in thickness. They occur in various forms including massive or pillowed lavas, pahoehoe basalts (Fig. 3.73a), hyaloclastites, agglomerates and tuffs (Svetov 1980). Gabbro-dolerites

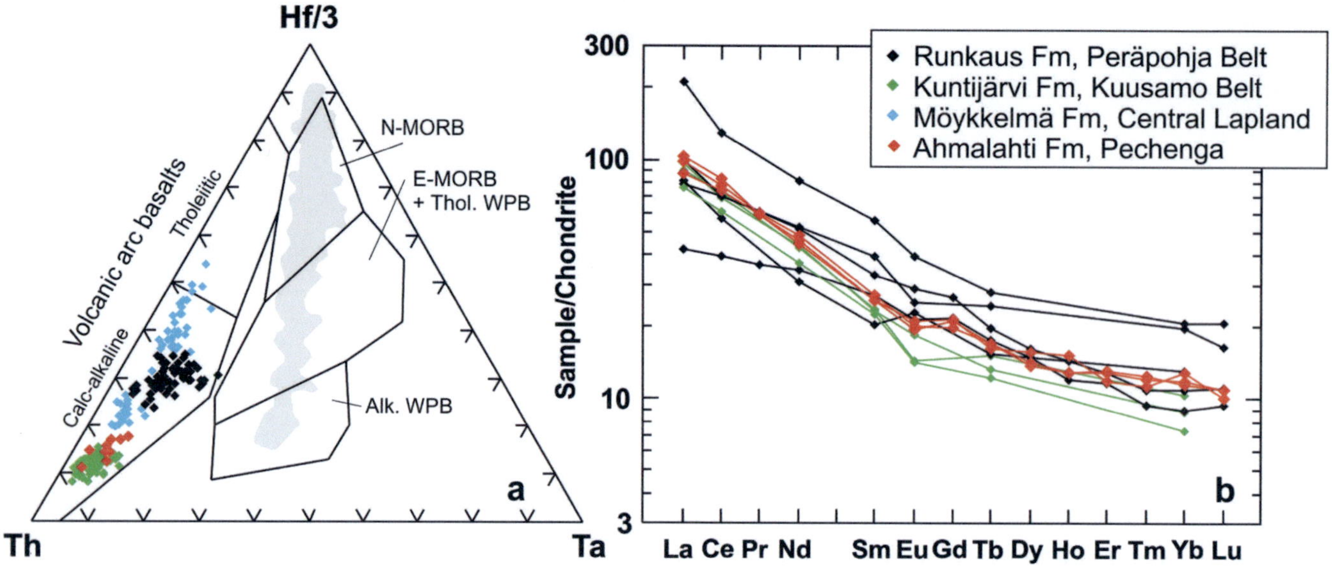

Fig. 3.47 Discrimination diagram of Wood (1980) (**a**) and chondrite-normalised REE diagram (**b**) for the following Sariolian mafic volcanic formations: Runkaus Formation (Peräpohja), Kuntijärvi Formation (Kuusamo), Möykkelmä Formation (central Finnish Lapland) and Ahmalahti Formation (Pechenga). Figure (**a**) was constructed partly based on INAA Th and Ta data, XRF Zr data and calculated Hf abundances assuming primitive mantle-like Zr/Hf. Data sources: Perttunen (1989), Räsänen et al. (1989), unpublished data of the Lapland Volcanite Project (the Geological Survey of Finland), and FAR-DEEP Project. Also shown in (**a**) is a *grey* field of recent MORBs based on the analytical compilation of Arevalo and McDonough (2010)

regarded as Upper Jatulian in age, with some of them hosting titanomagnetite ore deposits (Pudozhgorsk, Koikary-Svyatnavolok; Svetov and Bogachev 1976), cut Middle and Upper Jatulian sedimentary rocks.

In Russian Karelia, erupted magma was restricted in its chemical composition to tholeiitic basalt and basaltic andesite with MgO being higher than 4 wt.% (Fig. 3.50). Highly magnesian volcanic rocks, which are typical of the Sumian, Sariolian and Ludicovian systems, are more limited in abundance and less primitive in the Jatulian successions. Popov et al. (1983) and Makarikhin and Satsuk (1987) presented a set of analyses of Middle to Upper Jatulian basalts having MgO contents of 10–16 wt.%. However, due to the lack of compatible element data (Cr, Ni) for these rocks, the obvious presence of alteration effects, and the general paucity of olivine-rich rocks among Jatulian lavas, it is uncertain whether the erupted magma ever had MgO higher than 10 wt.%. Popov et al. (1983) divided the high-Mg basalts into different classes based on the total alkali content, but as there is no correlation between $Na_2O + K_2O$ and other incompatible components, such as TiO_2, and $Na_2O + K_2O$ correlates negatively with CaO, alkalies and CaO seem to have been mobile and their variation do not represent the original magmatic characteristics. Svetov (1968) reported analyses of chloritic tuffs from Lake Segozero with MgO between 14 and 17 wt.% (15–20 wt.% as volatile-free), but these were also likely affected by secondary processes. In general, Jatulian volcanic rocks in Russian Karelia have undergone various kinds of alteration processes including haematisation, albitisation, biotitisation, epidotisation and propylitisation (Svetov 1980; Malashin et al. 2003).

Middle Jatulian volcanic rocks are basalts while Lower Jatulian rocks are mostly lower in MgO and higher in SiO_2, being classified as basaltic andesites (Fig. 3.50a). Figure 3.49 shows a Jensen cation plot for average compositions of Lower, Middle and Upper Jatulian volcanic rocks from various areas. The Middle Jatulian basalts fall mostly in the high-Mg tholeiitic field, while the Lower and Upper Jatulian volcanics plot both in the high-Mg and high-Fe tholeiitic fields, though generally in the latter. High FeO_T is accompanied by high TiO_2 and V (Fig. 3.50b–d). In terms of TiO_2, volcanic rocks of the Hirvas section can be regarded as representing a typical Jatulian magmatic evolution, starting with Lower Jatulian high-Ti, evolved basalts, changing to more primitive low-Ti basalts of Middle Jatulia and terminating with high-Ti, high-Fe (low-Si) basalts (Fig. 3.50c). However, this kind of evolution is not always the rule as shown by the Kukaozero section, which starts with low-Ti basalt instead of high-Ti basalts (Fig. 3.50c). The most primitive lavas are Middle Jatulian in both mentioned sections.

Malashin et al. (2003) and Ivanikov et al. (2008) published trace element data for Jatulian mafic volcanic rocks from all three stratigraphic levels. Their analytical data are shown in Fig. 3.51a. All samples display a negative Ta-Nb anomaly and, with the exception of two Upper Jatulian samples, all show sloping primitive mantle-normalised REE patterns with the most magnesian Middle

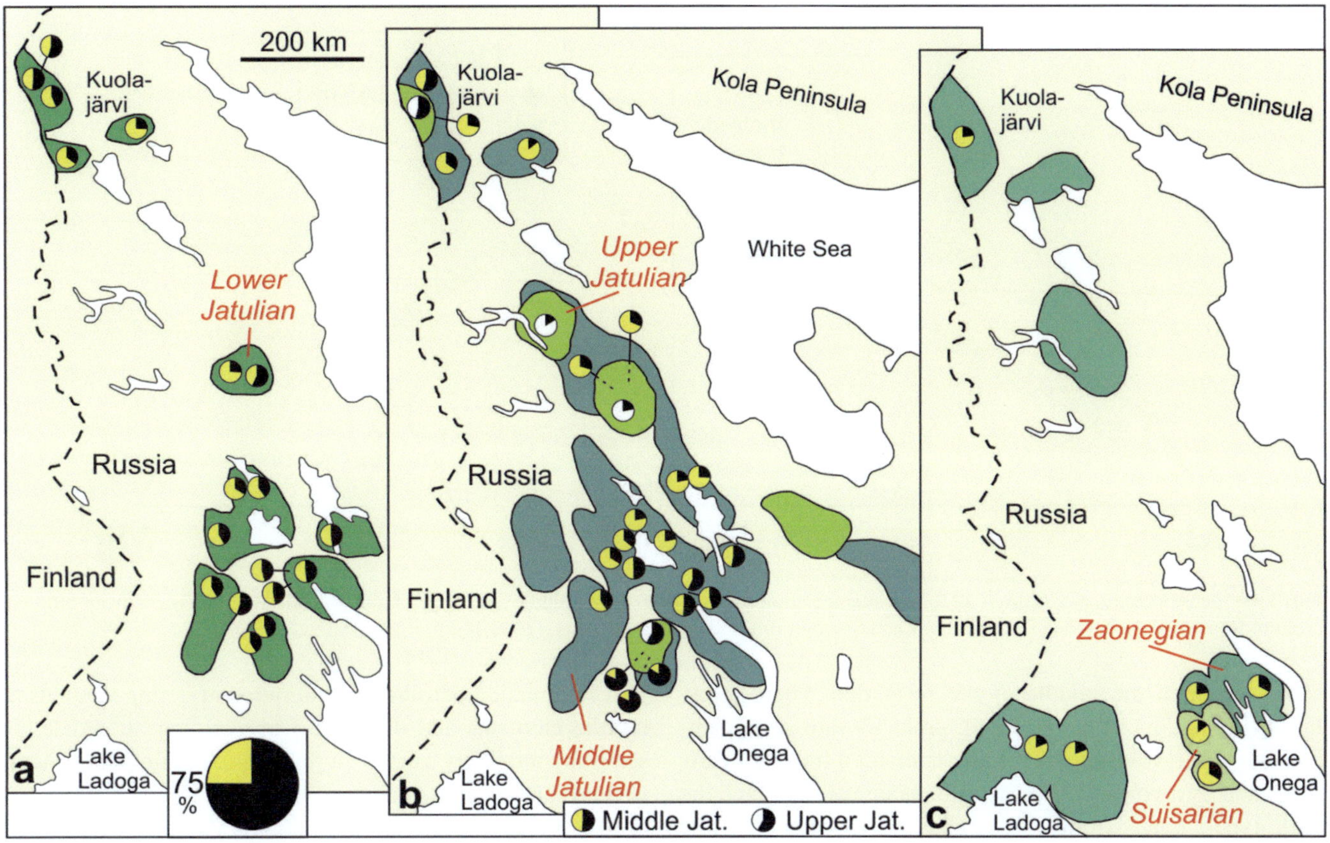

Fig. 3.48 Palaeovolcanological construction of the occurrence of Lower Jatulian (**a**), Middle and Upper Jatulian (**b**), and Ludicovian (Zaonegian and Suisarian) (**c**) volcanic rocks in Russian Karelia. Also shown are average analyses of the oxidation state of iron expressed as percentage of Fe_2O_3 of Fe_2O_3 + FeO (wt.%) (Data from Golubev and Svetov (1983))

Jatulian basalts having the lowest level of LREE. The Upper and Lower Jatulian rocks have similar LREE contents, but HREE are much lower in Lower Jatulian volcanics, resulting in clear differences in La/Yb. When La/Yb and TiO_2 are combined in the same diagram (Fig. 3.51b), rocks from the three stratigraphic levels plot in separate fields. The two exceptional Upper Jatulian samples (from Pyalozero) have gently rising REE patterns (Fig. 3.51a), and according to Ivanikov et al. (2008), resemble in this respect some basalts from the overlying Ludicovian Zaonega Formation.

In the Pechenga and Imandra/Varzuga Belts, Kola Peninsula, Jatulian volcanic rocks are represented by the Kuetsjärvi and Umba Volcanic formations, respectively. Compared to Jatulian volcanic successions in Russian Karelia and Finland, they are more variable, including a wide spectrum of rock types from picrites to rhyolites/trachytes, while in Russian Karelia, the Jatulian volcanic rocks are mainly mafic to intermediate, and a consirable portion of the volcanites show an alkaline affinity in contrast to the tholeiitic affinity of the Jatulian volcanics in Russian Karelia and Finland. In addition, the volume ratio of the volcanic and sedimentary rocks is larger in the Jatulian supracrustal successions of the Kola Peninsula.

Despite mineralogical changes, the Jatulian Kuetsjärvi Volcanic Formation in the Pechenga Belt is one of the best-preserved Precambrian volcanic units in the Fennoscandian Shield. Melezhik et al. (2007) determined a U-Pb zircon age of 2058 $\pm$ 6 Ma for this as much as 2-km-thick unit. The lowermost and uppermost parts of the formation were sampled by FAR-DEEP Holes 5A and 7A, respectively, while Hole 6A represents a section from the middle part of the formation. The Kola Superdeep Drillhole penetrates the entire formation at depth interval of 4,884–5,642 m (Kozlovsky 1987). The formation is described in Chap. 4.2 on the geology of the Pechenga Belt and in the chapters dealing with the above-mentioned FAR-DEEP drillholes (Chaps. 6.2.3, 6.2.4, and 6.2.5). Relevant references of previous studies include Predovsky et al. (1974 and 1987), Skuf'in et al. (1986) and Skuf'in and Yakovlev (2007).

Primary magmatic structures and microscopic textures of the volcanic rocks, including a diversity of amygdale fillings precipitated from hot fluids after eruption, are documented by photographs in several chapters of this and accompanying volumes (Figs. 3.18e–g, 3.52a–c, 4.24a–al, 6.51a, 6.54, 6.55, 6.60a–at, and 6.65j–q) and also in Skuf'in (1980b). These indicate a subaerial environment of eruption excluding a

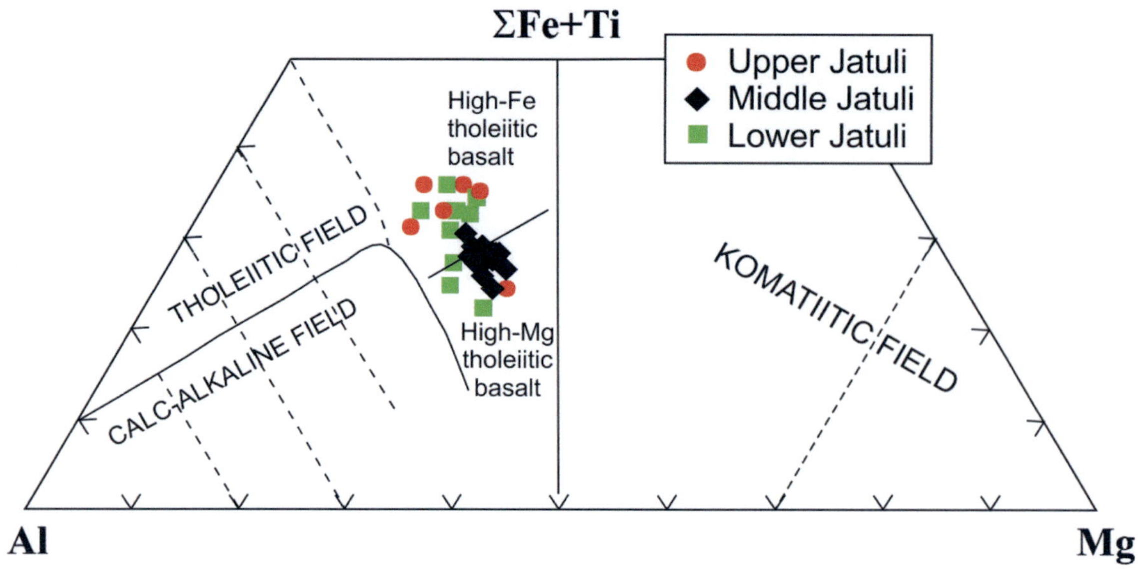

Fig. 3.49 Jensen cation plot (Jensen 1976) for average compositions of Lower, Middle and Upper Jatulian mafic volcanic rocks from different areas of Russian Karelia (Data from Golubev and Svetov (1983) and Malashin et al. (2003))

short period in the middle of the formation when pillow lavas (Fig. 4.24r) were emplaced, probably into an ephemeral lake. The volcanic succession is divided into two parts separated by a volcaniclastic conglomerate member, and the rock types range from picritic basalts to trachytes and rhyolites (for more details, see Chap. 4.2). Figure 3.53a, b summarise the geochemical features of the formation.

The naming of the felsic rocks, in particular, is not straightforward as different classification diagrams give a different nomenclature. For example, Fig. 3.53d, which involves SiO_2, indicates that the most evolved lavas are dacites or rhyodacites, whereas incompatible element ratios shown in Fig. 3.53c classify them mostly as trachyandesites. Moreover, relative low Zr/Ti suggests that almost all rock samples are subalkaline, but if Nb/Y is taken into account, many mafic samples plot in the alkali basalt field and some even in the basanite field. Among the basalts with an obvious alkaline affinity are the lowermost lavas which were recovered by FAR-DEEP Hole 5A (see Chap. 6.2.2) and outcrop in the Luostari area (Fig. 3.53c). All analysed samples have high LREE/HREE (Fig. 3.53a). The high Zr up to c. 700 ppm in felsic lavas is consistent with a peralkaline character of the magma (see Fig. 3.42b).

Using the isotopic measurements of Skuf'in and Theart (2005) and an age of 2060 Ma, initial ε_{Nd} values varying from −2.8 to −4.2 can be calculated for Kuetsjärvi Formation lavas. These results together with unpublished data by H. Huhma and E. Hanski (initial ε_{Nd} from +0.4 to −4.9) demonstrate a significant isotopic heterogeneity and suggest major but variable interaction of the magmas with lithospheric material and ensuing notable changes in the composition of the original magmas. These processes together with late- and post-magmatic hydrothermal alteration affecting contents of mobile elements and difficulties in avoiding amygdales in sampling are likely causes for the hybrid nature of the chemistry of the volcanic rocks.

The volcanic unit in the Imandra/Varzuga Belt that is correlative with the Kuetsjärvi Volcanic Formation is the 1- to 2-km-thick Umba Volcanic Formation. The internal stratigraphy and volcanological features are described in Chap. 4.1. All structures indicate eruptions under subaerial conditions. Most volcanic rocks are amygdaloidal lava flows but pyroclastic deposits including tuffs (Fig. 4.11r) and agglomerates are found as well (Borisov 1990). The rock types vary from picrites via trachybasalts and trachyandesites to even felsic ignimbrites with the most primitive variants occurring in the lower part and most differentiated rocks (Fig. 4.11s) in the upper part of the volcanic succession. The Kuetsjärvi and Umba volcanic rocks resemble chemically each other very well (Fig. 3.53). As in the case of the Kuetsjärvi Formation lavas, the SiO_2 vs. Zr/TiO_2 plot indicate a subalkaline affinity for the Umba rocks (Fig. 3.53d), but the diagram involving Nb/Y clearly argues for an alkaline character for many of the basaltic samples (Fig. 3.53c). The lack of intermediate members in Fig. 3.53c, d is not due to the paucity of this kind of rocks but rather the lack of trace element data. Picritic basalts and picrites have MgO contents between 10 and 20 wt.% and have a somewhat variable TiO_2 contents (0.8–1.5 wt.%) resulting in a large scatter in the $[Al_2O_3]$ vs. $[TiO_2]$ diagram and consequently the rocks are placed partly in the Ti-enriched komatiite but mostly in the picrite field (Fig. 3.34). The chondrite-normalised REE profile of a picritic sample from FAR-DEEP Hole 4A that is shown in Fig. 3.69b is similar to that of ferropicrites but the element concentrations

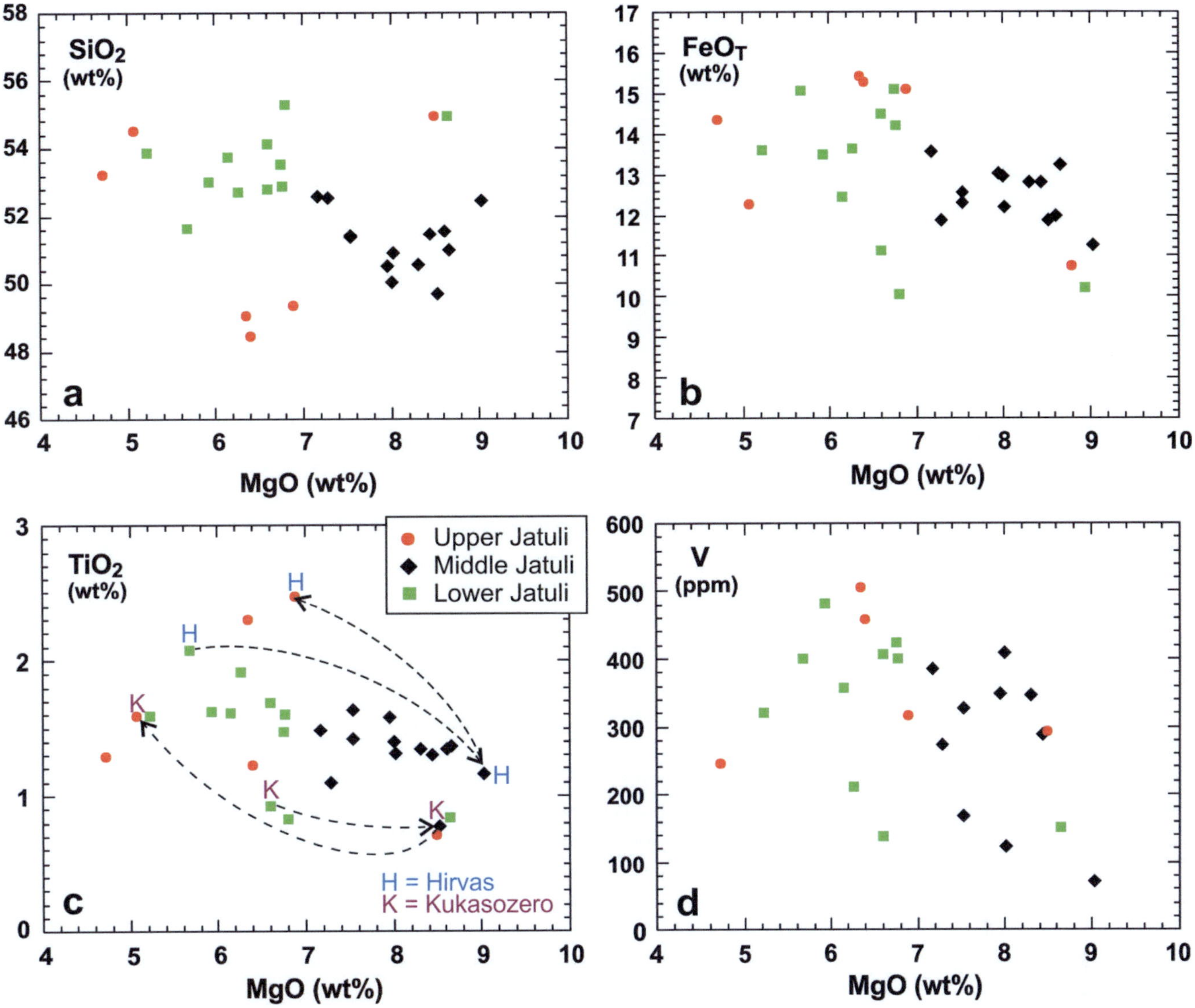

Fig. 3.50 Average compositions of Lower, Middle and Upper Jatulian volcanic rocks from various areas in Russian Karelia. In (**c**) compositional evolution in terms of TiO_2 and MgO from Lower through Middle to Upper Jatulian lavas from the Hirvas and Kukasozero areas are indicated with dashed lines (Data from Golubev and Svetov (1983) and Malashin et al. (2003))

are clearly lower. Part of the Umba picrites and picritic basalts are high in P_2O_5 (0.4–0.9 wt.%) and therefore have high P_2O_5/TiO_2 ratios and differ in this respect from the Kuetsjärvi Formation rocks (Fig. 3.53b). Among basaltic and intermediate rocks there also exist P-rich varieties (P_2O_5 up to 1.1 wt.%). This feature has been observed at Pechenga, too, where high-P, though slightly more evolved, volcanic rocks occur in the upper part of the Kuetsjärvi Volcanic Formation (Flow B, Hole 7A, see Chap. 6.2.4). Published analytical data on the most evolved volcanic rocks (trachydacites) in the Umba Formation are still scarse (Predovsky et al. 1987).

Jatulian volcanic rocks in Finland occur throughout the Karelian formations from the SE border to Lapland. Among them, the Jouttiaapa Formation, which has been imprecisely dated by the Sm-Nd method at 2090 ± 70 Ma by Huhma et al. (1990), represents a thick continental flood basalt-type volcanic sequence in the Peräpohja Belt, reaching 500 m in thickness. The volcanic rocks are mostly amygdaloidal lavas (Figs. 3.54a and 3.47a) with very few thin quartzite, silt or mafic tuff interlayers. Figure 3.55 illustrates a stratigraphic cross-section through the effusive sequence constructed on the basis of two drill holes (R5, R7), one cutting the upper and the other the lower contact of the formation. The obtained section contains at least 26 individual lava flows from less than 1–50 m thick and negligible sedimentary or tuffic interlayers.

Geochemically, the Jouttiaapa Formation is mainly composed of tholeiitic basalts with minor basaltic andesites. Figure 3.55 shows that there is considerable internal

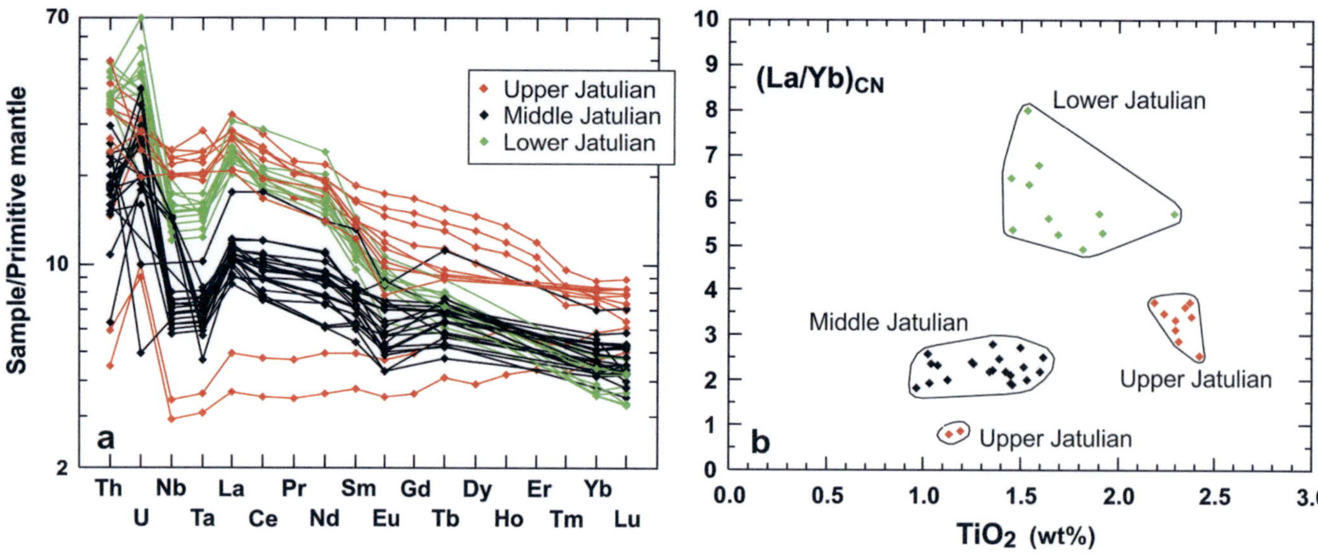

Fig. 3.51 (**a**) Primitive mantle-normalised REE diagram, extended with Th, U, Nb and Ta, and (**b**) chondrite-normalised La/Yb vs. TiO_2 diagram for Lower, Middle and Upper Jatulian volcanic rocks from southern and central Russian Karelia (Data from Malashin et al. (2003) and Ivanikov et al. (2008). Normalising values after Sun and McDonough (1989))

heterogeneity in the volcanic section with MgO varying between 5 and 10 wt.% and TiO_2 between 0.45 and 1.65 wt.%. The sedimentary interlayers seem to mark breaks in the volcanism and more or less significant changes in the chemistry of the erupted lavas. In Fig. 3.54, five chemically different magmatic pulses are distinguished, but the actual number is likely larger as the presented section is probably not quite complete. The variability is also seen in trace elements. Figure 3.55 depicts the most remarkable feature of the Jouttiaapa Formation, an ultradepletion of LREE in chondrite-normalised REE plots. The most depleted lavas are found in the middle, low-Ti part of the volcanic section. As is seen in Fig. 3.56, the least-depleted Jouttiaapa lavas have REE levels similar to those of MORB, while the rest are even more depleted. Low LREE/MREE ratios are consistent with the initial ε_{Nd} value of $+4.2 \pm 0.5$ calculated from the isochron of Huhma et al. (1990). Figure 3.57b displays Jouttiaapa Formation basalts plotted on the TiO_2/Yb vs. Nb/Yb diagram of Pearse (2008) where they fall within the MORB array. The analytical results of the same samples could not be used on the accompanying Th/Yb vs. Nb/Yb diagram (Fig. 3.57a) because Th was under the detection limit (evidently due to the depleted character of the basalts).

The Jouttiaapa Formation lavas were generated from a mantle source having a long-lived depletion in incompatible elements. Exhibiting trace element and isotopic characteristics similar to those of convecting asthenospheric mantle, they passed virtually uncontaminated through the continental crust. Since the beginning of the Palaeoproterozoic era, the Jouttiaapa Formation was the first magmatic unit on the shield whose magma experienced little to no interaction with lithosphere during its ascent to the surface. In the Fennoscandian Shield and worldwide, the formation is exceptional in its combination of MORB-like geochemistry and CFB-type tectonic environment and physical volcanology. The apparent uniqueness is highlighted by the following citation from Thompson et al. (1983): "*We know of no cases of magmas erupted in a continental setting with the precise elemental and isotopic composition of MORB*." However, referring to Baffin Island basalts in Greenland, for example, they continue: "*Nevertheless, there are several suites of CFB which closely resemble MORB . . . at first sight*." Is the situation the same with the Jouttiaapa basalts? To answer this question, it is necessary to scrutinise more closely how well the chemistry of the Jouttiaapa Formation matches with that of recent MORBs. This is done in Fig. 3.58e using FeO_T and TiO_2 abundances. It is evident that the Jouttiaapa basalts differ from reference MORBs, as represented by the large analytical data base of Arevalo and McDonough (2010), in having distinctly higher FeO_T at given TiO_2. The same phenomenon was observed earlier by Hanski (1992) for the Ludicovian "MORB-like" basalts of the Kolosjoki and Pilgujärvi volcanic formations in the Pechenga Belt. Figure 3.59a, b demonstrate that the reason for the observed difference in Fig. 3.58e is not only in the relatively low TiO_2 or high FeO_T abundances of the Jouttiaapa basalts but in both of these factors. Although deviating from MORB in terms of TiO_2-FeO_T relations, the Jouttiaapa basalts were generated from mantle having a long-term REE characteristics similar to those of contemporaneous depleted MORB mantle, as the initial positive ε_{Nd}(2100 Ma) values follow the evolution line of the model depleted mantle of DePaolo (1981) (Huhma et al. 1990). It seems that the residual mantle from which the Jouttiaapa basaltic magma

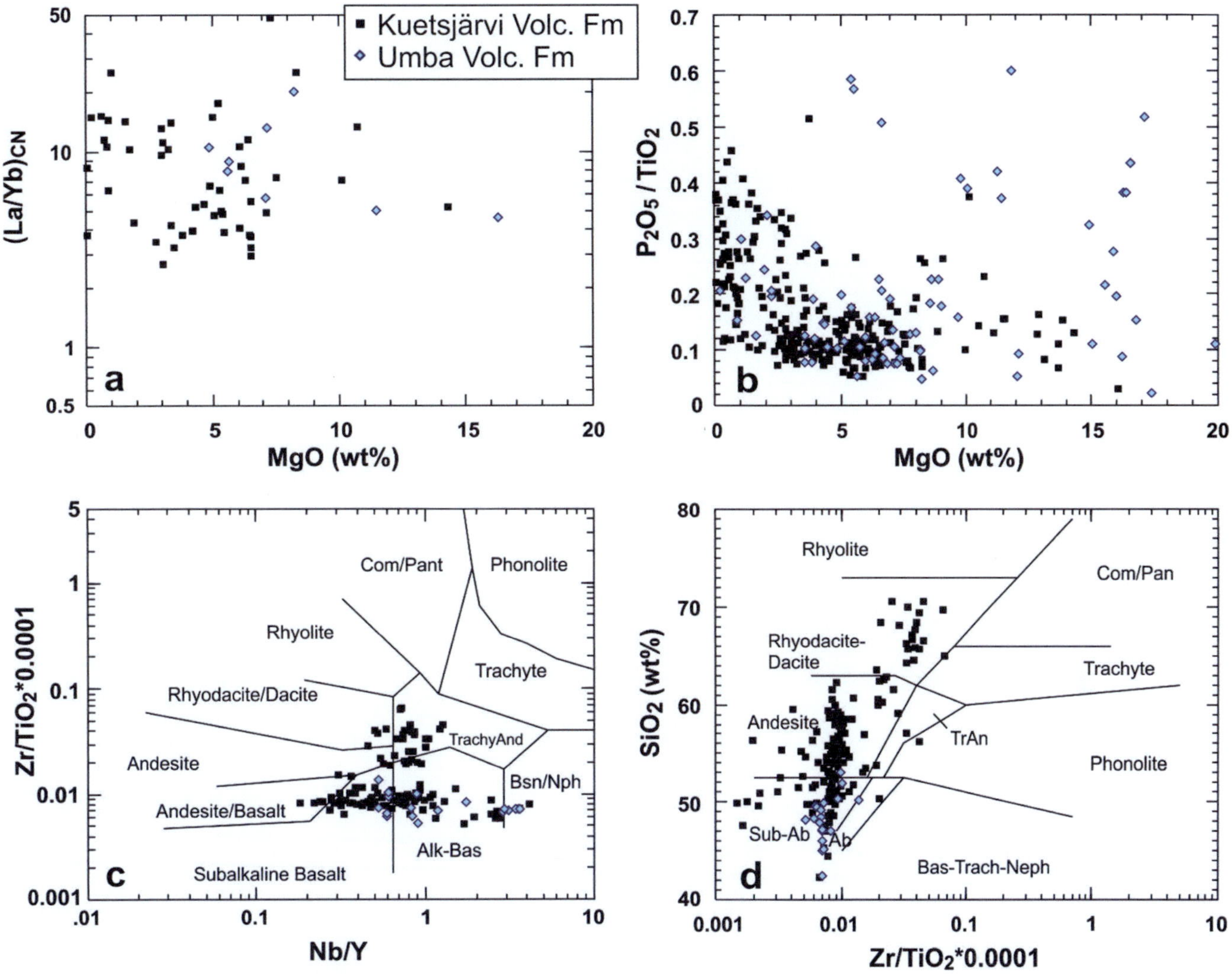

Fig. 3.52 Chemical properties of the Jatulian Kuetsjärvi and Umba Volcanic formations from the Pechenga and Imandra/Varzuga Belts, respectively (Literature sources available from the author upon request)

was generated was itself formed by partial melting of peridotitic material that was more iron-rich than model primitive upper mantle.

Due to its distinctive geochemical characteristics, the Jouttiaapa Formation could be easily followed to the neighbouring supracrustal belts or correlated with other volcanic formations elsewhere in the Jatulian successions or even in other shields. However, it seems that the effusion of this kind of magma was restricted to the Peräpohja Belt and chemically similar rocks are also absent among the dyke rocks in the basement (Vuollo and Huhma 2005). On the Swedish side of the border, the small Kalix Belt (Fig. 3.30) contains basaltic lavas in association with thick deposits of dolomites in a similar fashion as in the Peräpohja Belt. The chemical compositions of the Kalix lavas documented by Lager and Loberg (1990) are higher in TiO_2 than the Jouttiaapa Formation basalts (see Fig. 3.59b). On the other hand, Martinsson and Billström (2006) state that most of the Kalix basaltic lavas have a low to moderate TiO_2 content and resemble MORB in their composition, thus potentially representing analogues of the Jouttiaapa Formation basalts. However, the actual analytical data are so far unpublished, and the correlation hence remains tentative. In the upper part of the Kalix volcanic pile, there also exist extremely TiO_2-rich (TiO_2 up to 6.0 wt.%), likely mantle plume-related basaltic lavas and tuffs, which are overlain by transitional basalt types intervening with stromatolitic dolomites.

The Jouttiaapa Formation is overlain by an orthoquartzite unit, which is in turn followed by the Tikanmaa Formation consisting of mafic tuff and tuffites. This is a widespread formation showing distinct laminar bedding (Figs. 3.26bz, 3.47b, and 3.54b) and, in thickest layers, also graded and current bedding (Perttunen and Hanski 2003). Geochemically, the Tikanmaa tuffs are similar to the least-depleted

Fig. 3.53 Rock types from the Pechenga Belt. (**a**) Synvolcanic flow-banding in a felsic lava from the Jatulian Kuetsjärvi Volcanic Formation. (**b**) Scanned thin section of plagioclase-phyric trachyandesite from the Kuetsjärvi Volcanic Formation. Photo width 2 cm. (**c**) Contact of two alkali basaltic lava flows, Kuetsjärvi Volcanic Formation. (**d**) Amygdaloidal pillow lava from the Ludicovian Kolosjoki Volcanic Formation. (**e**) Large quartz-filled amygdale in a massive basaltic lava flow the Ludicovian Pilgujärvi Volcanic Formation. (**f**) Pillow lava of tholeiitic basalt, Pilgujärvi Volcanic Formation (Photographs taken by Eero Hanski)

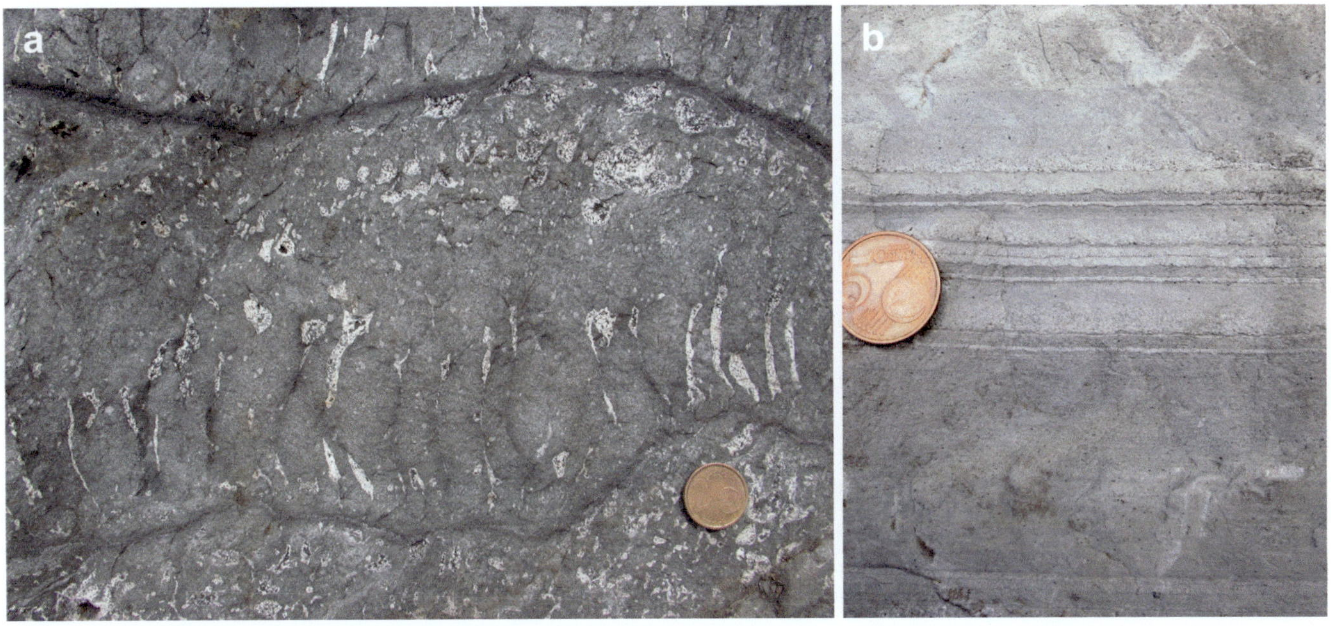

Fig. 3.54 (**a**) Thin basaltic lava flows with subvertical pipe amygdales in the basal part of lavaflows from the Jatulian Jouttiaapa Formation, Peräpohja Belt. Coin diameter 1.8 cm. Photograph courtesy of Vesa Perttunen. (**b**) Parallel-laminated mafic tuff in the Jatulian Tikanmaa Formation, Peräpohja Belt. Coin diameter 2.1 cm (Photograph by Eero Hanski)

Jouttiaapa Formation basalts (Fig. 3.60a). Higher up in the stratigraphy, two other Jatulian formations have been defined, which are composed of mafic tuffs and tuffites (Perttunen and Hanski 2003). The 100- to 400-m-thick Hirsimaa Formation is located beneath the dolomitic Rantamaa Formation exhibiting a positive Lomagundi-Jatuli carbon isotope anomaly (Karhu 1993). Dating of zircon (2106 ± 8 Ma) from this tuff unit has produced valuable information on the timing of the above-mentioned isotope anomaly (Karhu et al. 2007). The other tuff formation, the Lamulehto Formation, in turn occurs on top of the Rantamaa Formation dolomites as a 100-m-thick unit. Geochemically, the two tuff units differ from the older (but roughly coeval) Tikanmaa Formation tuffs in being trachyandesitic and having sloping chondrite-normatised REE profiles Fig. 3.60a. Another volcanic rock unit in the Peräpohja Belt deserves to be mentioned in this context. Despite the fact that Jatulian basaltic lava eruptions took place subaerially, producing the Jouttiaapa Formation flood basalts, mafic pillow lavas are found (see Fig. 3.47d) in the belt, which Perttunen and Hanski (2003) assigned to the Santalampi Formation. However, they are not exposed but have been observed as numerous boulders in close association with carbonate-matrix quartz conglomerates and therefore their place in the stratigraphy is uncertain.

In the Kuusamo Belt, Jatulian volcanic rocks occur at two stratigraphic levels and are known in the geological literature as the Greenstone Formation II and the Greenstone Formation III of Silvennoinen (1972). They have recently been renamed the Petäjävaara and Ruukinvaara formations, respectively (Geological Survey of Finland, 2011, Bedrock of Finland – DigiKP, Digital map database, version 1.0; http://www.geo.fi/en/bedrock.html). These two formations can be distinguished easily from each other as the former comprises pillow lavas 30–50 m in total thickness and the latter is composed of amygdaloidal basaltic lava flows and tuffs forming a pile up to c. 400 m thick. The lavas of the Ruukinvaara Formation were erupted subaerially, and both the underlying and overlying sediments were deposited in shallow water as they exhibit desiccation crack structures (Figs. 3.42d and 3.43d). The Ruukinvaara Formation has been regarded as an ancient continental flood basalt succession due to its uniform thickness and wide distribution from the southern part of the Kuusamo Belt to central Finnish Lapland, indicating eruption on a flat, peniplaned land surface (Silvennoinen 1991).

The Petäjävaara Formation pillow lavas are high-Mg basaltic andesites and andesites and the Ruukinvaara Formation rocks Fe-rich tholeiitic basalts (Fig. 3.61a). The latter are higher in FeO_T and TiO_2, while the former are higher in SiO_2, in spite of being also higher in MgO (Fig. 3.62b). With respect to alkalies, the volcanic rocks in the two formations have undergone different alteration processes: the subaerially erupted lavas have the highest K_2O contents and the

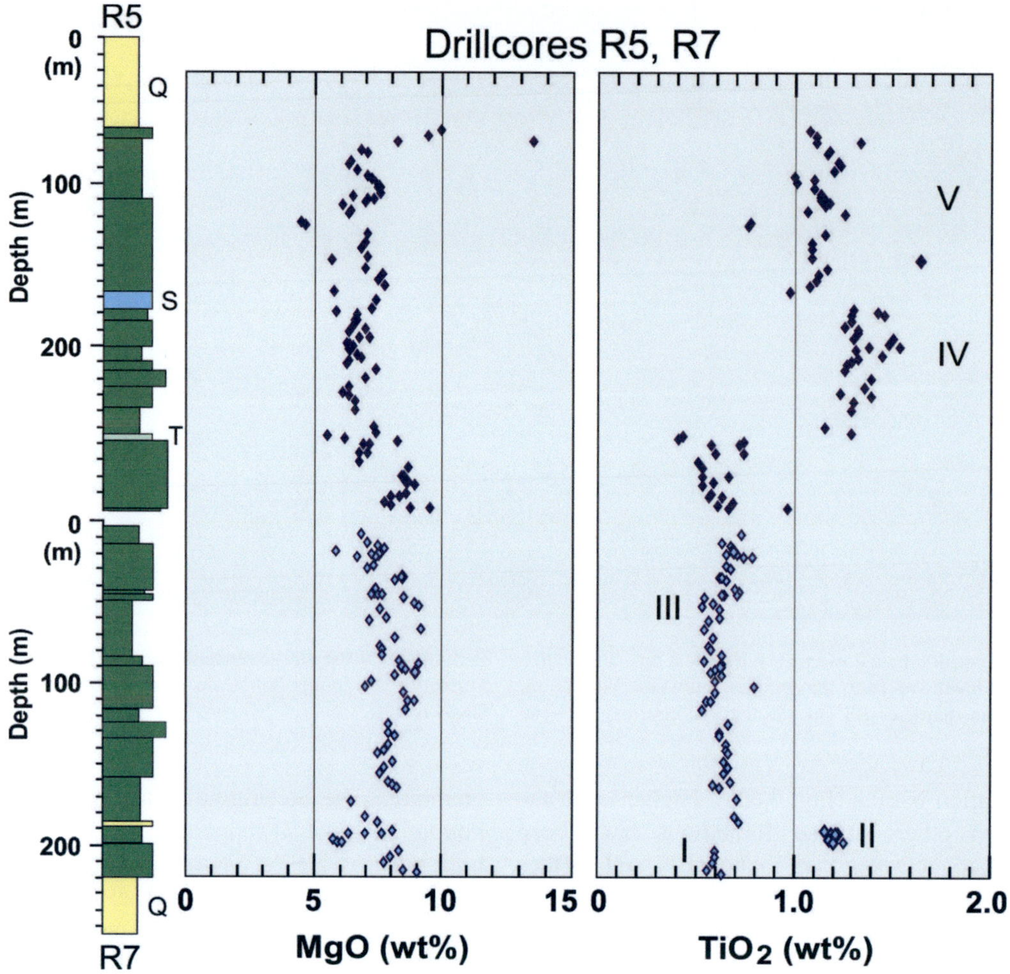

Fig. 3.55 A composite lithological section and chemical variation through the Jatulian Jouttiaapa Formation, constructed on the basis of two drillcores (R5, R7). Q = quartzite, T = mafic tuff, S = siltstone (Modified after Perttunen and Hanski (2003). I–V denote chemically different magma pulses)

subaqueously erupted lavas the highest Na_2O contents (Fig. 3.61c). Immobile trace element analyses reveal higher concentrations of incompatible elements, such as LREE and HSFE, in the high-Fe tholeiites of the Ruukinvaara Formation (Fig. 3.61d). The chondrite-normalised REE patterns of the Petäjävaara Formation basaltic andesites are roughly flat with some erratic variation due to secondary alteration and/or imprecision of the employed INAA techniques.

A place where a connection has been established between Jatulian diabase dykes intruding the Archaean basement and mafic volcanic rocks within the Proterozoic cover sequence is located at Kiihtelysvaara, Finnish North Karelia. Figure 3.62 shows a geological map of this area after Pekkarinen and Lukkarinen (1991). A volcanic unit, the Koljola Formation, is bounded by two Jatulian quartzitic sedimentary formations and varies from 30 to 80 m in thickness. It contains several mafic lava flows and pyroclastic deposits. Many diabase dykes terminate when they reach the level of the Koljola Formation and are interpreted as feeder dykes to the lavas (Fig. 3.62). Diabases can also act as feeders to gabbroic sills further upwards in the stratigraphy. Zircon from two diabase samples was dated giving coherent results (Fig. 3.62) and jointly an age of 2113 ± 4 Ma (Pekkarinen and Lukkarinen 1991). Chemically, the mafic lavas are evolved Fe-rich tholeiitic basalts and andesites with low Cr and Ni. Alteration is usually extensive as shown by low CaO (often 2 wt.% or less) and high total alkalies (>5 wt.%). The Koljola Formation seems to have been formed by two types of magmas in terms of REE contents, the uppermost part by one which was less enriched in LREE than the magma that produced the rest of the formation (Fig. 3.60b). Two samples from a sill (Hyypiä sill) have also low LREE and are more akin to the samples from the upper rather than the lower part of the Koljola Formation. One sample from this sill analysed for Sm-Nd isotopes yielded an $\varepsilon_{Nd}(2100\ \text{Ma})$ value of -1.7 (Huhma 1987), which is similar to the value reported by

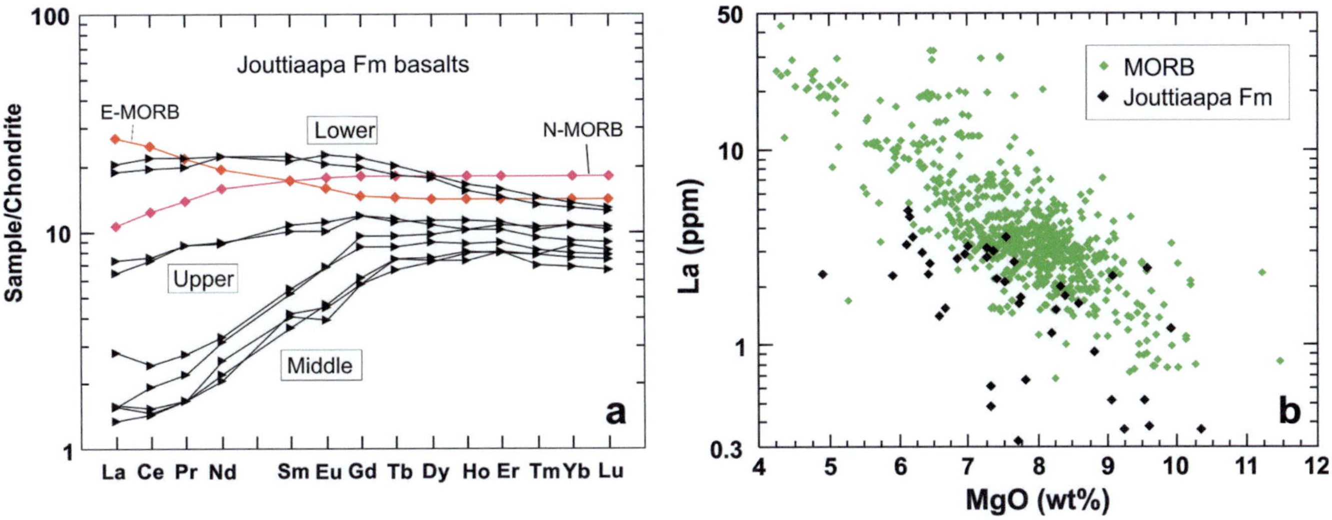

Fig. 3.56 (**a**) Chondrite-normalised REE patterns of basaltic lavas from the lower, middle and upper parts of the Jatulian Jouttiaapa Formation compared to those of typical N-MORB and E-MORB as given by Sun and McDonough (1989). Modified after Perttunen and Hanski (2003). (**b**) La and MgO in Jouttiaapa Formation basalts and different MORBs (N-MORB, T-MORB, E-MORB) from the Atlantic, Indian and Pacific Oceans. MORB data taken from Arevalo and McDonough (2010)

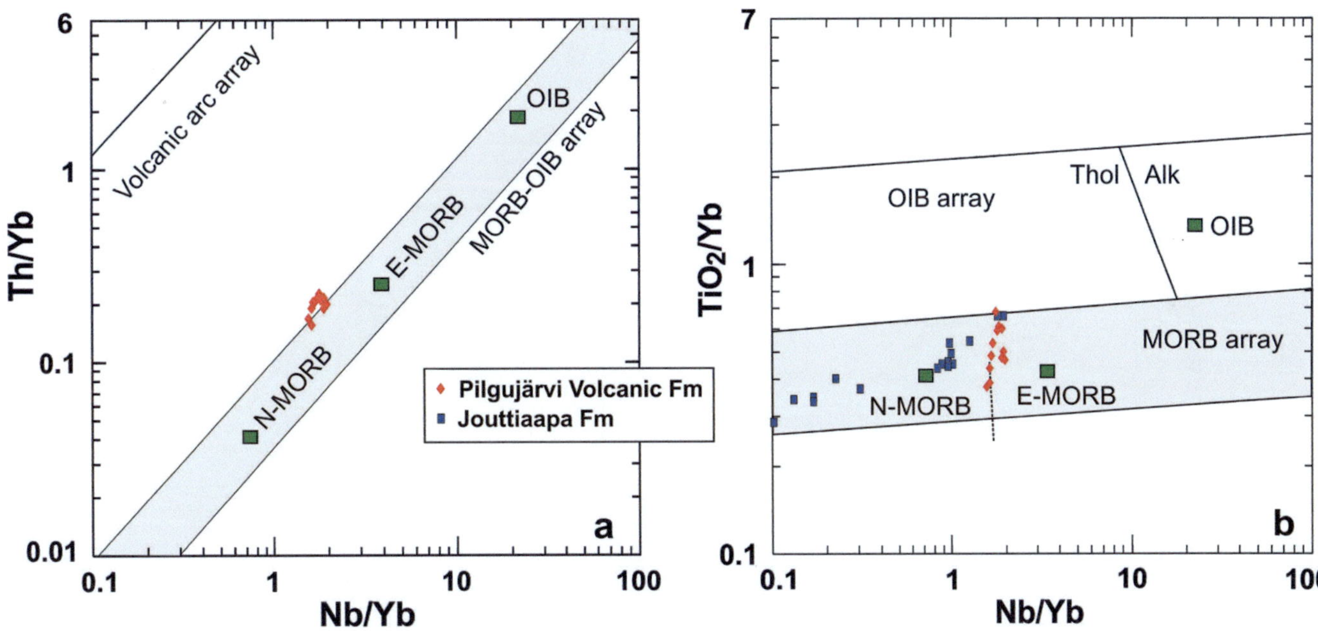

Fig. 3.57 Tholeiitic basalts from the Jouttiaapa Formation and Pilgujärvi Volcanic Formation plotted on the discrimination diagrams of Pearse (2008) (Data from Perttunen and Hanski (2003) and E. Hanski (unpublished data))

Philippov et al. (2007) for a Jatulian sill in the Onega Lake region (see above).

Nykänen et al. (1994) have studied igneous rocks of the Tohmajärvi mafic complex (4 km $\times$12 km in size), which is situated close to the Finnish-Russian border in SE Finland (see index map of Fig. 3.62). It is exposed in a window in the surrounding Kalevian metaturbidites and comprises mostly subvolcanic gabbroic sills but also pillow lavas and mafic tuffs associated with dolomites, black shales and mica schists. This lithological association is typical of the "marine Jatuli", which is currently correlated with Ludicovian supracrustal rocks (see Chap. 3.2). Despite this setting, the Tohmajärvi complex is treated in this context because Huhma (1986) obtained a Jatulian U-Pb zircon age of 2105 $\pm$15 Ma for the complex. They are thus similar in age to the Koljola Formation discussed above but differ in

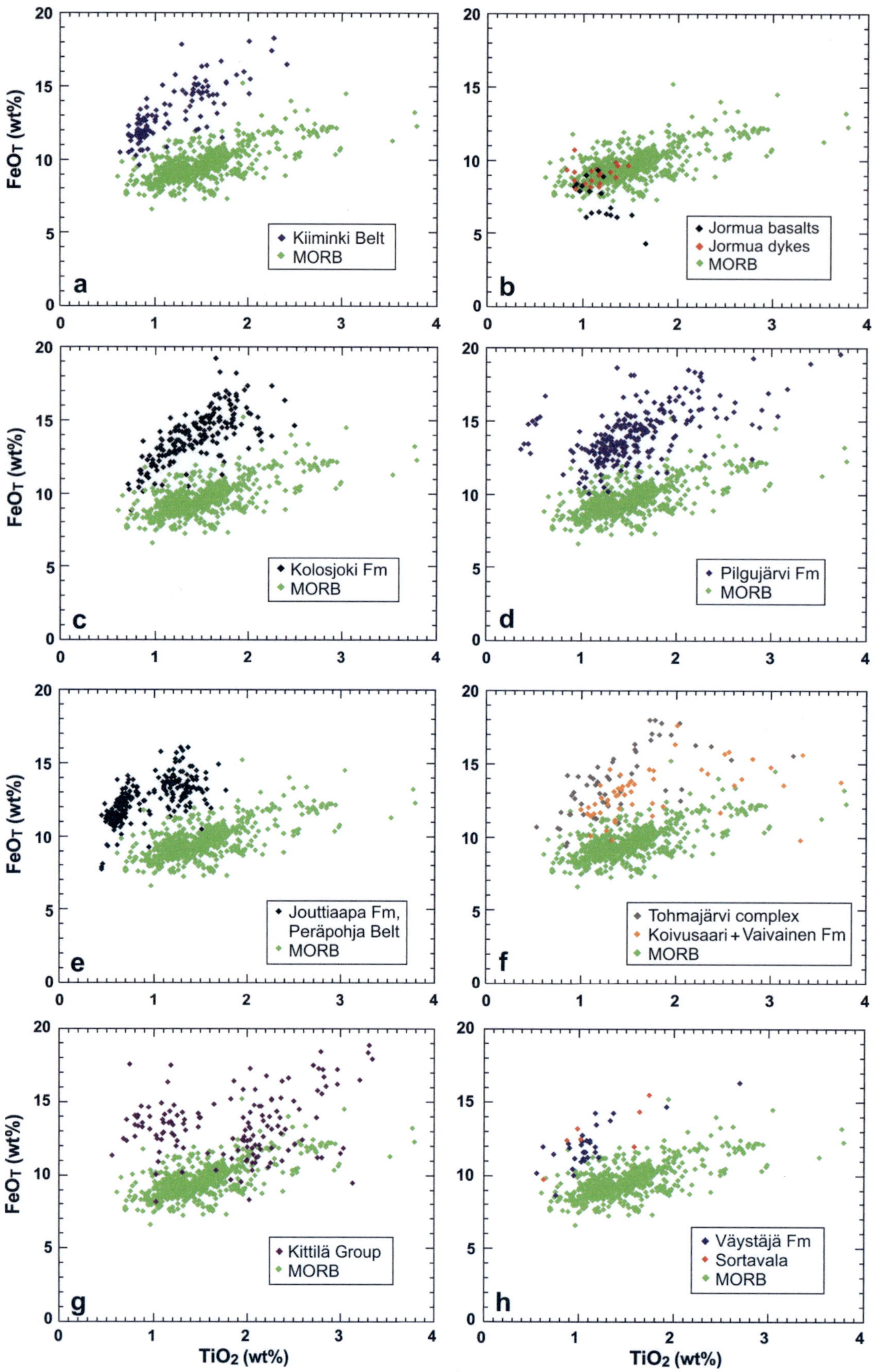

Fig. 3.58 FeO_T vs. TiO_2 plots comparing recent MORBs and those Jatulian and post-Jatulian tholeiitic basalts that have MORB-like REE characteristics and/or Nd isotopic systematics. Reference MORB data constitute c. 700 analyses of N-MORB, T-MORB and E-MORB from the Atlantic, Indian and Pacific Oceans, taken from Arevalo and McDonough (2010). Data sources: Jouttiaapa and Väystäjä formations,

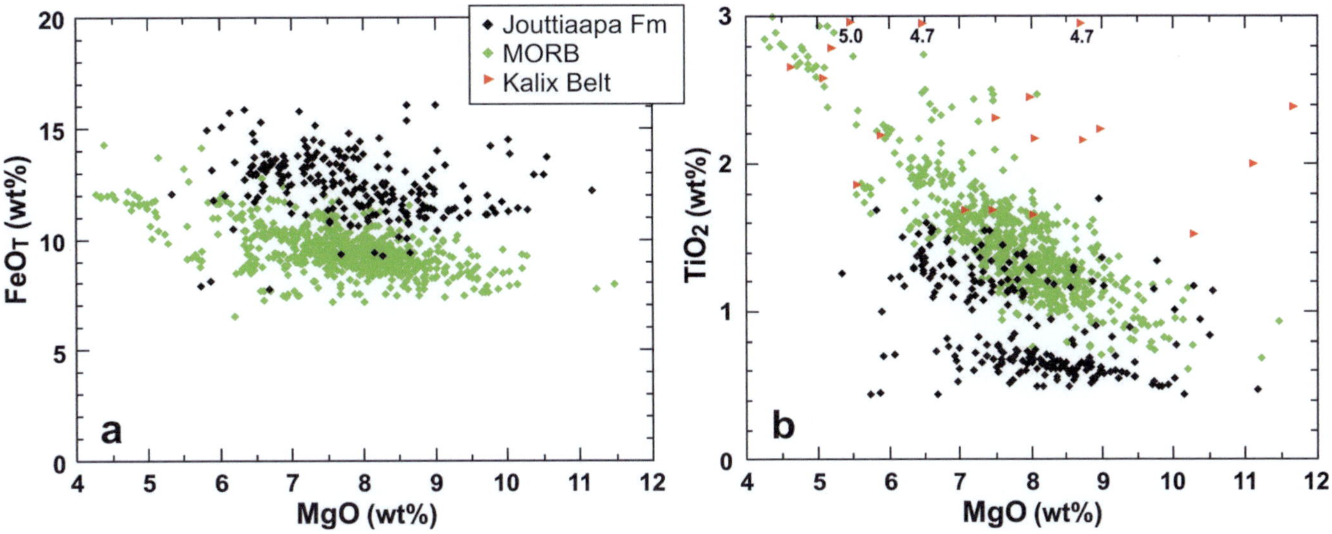

Fig. 3.59 FeO_T vs. MgO (**a**) and TiO_2 vs. MgO (**b**) plots comparing basaltic lavas of the Jatulian Jouttiaapa Formation from the Peräpohja Belt and basaltic lavas from the Kalix Belt with recent MORBs. Reference data constitute c. 700 analyses of different MORBs (N-MORB, T-MORB, E-MORB) from the Atlantic, Indian and Pacific Oceans, taken from Arevalo and McDonough (2010) (Jouttiaapa Formation data mainly from Perttunen and Hanski (2003) and Perttunen (1989) and Kalix analyses from Lager and Loberg (1990))

chemistry. The rocks of the Tohmajärvi complex form a tholeiitic differentiation series from magnesian basalts to Fe-rich andesites and show non-fractionated or slightly LREE-enriched chondrite-normalised REE patterns with $Ce_{CN}/Yb_{CN} = 0.94–1.8$ (Fig. 3.60b) and Nb/Ce similar to or slightly higher than the ratio in primitive mantle (Nykänen et al. 1994). These features together with an initial ε_{Nd} of +2.6 (Huhma 1986) are compatible with negligible or no interaction with continental crust. The volcanic rocks thus resemble MORB but, as in the case of the coeval Jouttiaapa Formation basalts, are too rich in FeO_T at a given TiO_2 to be identical to recent MORBs (Fig. 3.58f).

Rare Jatulian felsic volcanic rocks are found, in addition to the Kuetsjärvi Volcanic Formation in the Pechenga Belt (see below), in the Koivusaari Formation of the small Siilinjärvi Belt, near Kuopio, central Finland, with both formations having similar ages of c. 2.06 Ga (Melezhik et al. 2007; Lukkarinen 2008). A pyroclastic ash-flow tuff member occurs between massive and pillowed mafic volcanic rocks in the bimodal Koivusaari Formation. Geochemically, the felsic rocks are high-siliceous A-type rhyolites with high contents of incompatible trace elements such as Zr (500–630 ppm) and La (50–100 ppm) and low concentrations of Eu, P and Ti. They resemble the felsic tuffs of the Pilgujärvi Volcanic Formation at Pechenga, except having a higher level of HREE (see Fig. 3.42a). Two samples of the Koivusaari Formation felsic tuffs analysed for Sm-Nd isotopes yielded negative but variable initial ε_{Nd} values (−0.7, −4.5) (Lukkarinen 2008).

The mafic volcanic rocks of the Koivusaari Formation reach several hundred metres in thickness and comprise pillowed and massive lavas with minor pyroclastic interbeds (Lukkarinen 2008). Other very similar and likely correlative Jatulian volcanic rocks, assigned to the Vaivainen Formation, occur in the Kuopio area c. 20 km south of the Siilinjärvi Belt. They comprise a sequence up to 600 in thickness and consist of massive, porphyritic and pillowed lavas that form a tholeiitic differentiation series with MgO varying from 8.0 to 4.3 wt.% and TiO_2 from 1.0 to 2.0 wt.% (Fig. 3.63a). The Vaivainen Formation and the bulk of the Koivusaari Formation were generated from a similar parental magma, but the mafic rock compositions of the latter formation show a shift to less magnesian compositions and reach a higher degree of differentiation. The MgO content ranges from 6.9 to 3.0 wt.%, having a gap in the middle part of the range, and TiO_2 rises up to 3.3 wt.% (Fig. 3.63a). This bimodality is seen also in the abundances of incompatible trace elements such as REE (see below). The most evolved and incompatible element-rich lavas occur in the lower part of the formation (Parkkila Member). Furthermore, there is other heterogeneity in the Koivusaari Formation caused by a few pillow lava samples that have high P_2O_5 up to 1.6 wt.%. In fact, based on TiO_2/P_2O_5, three different magma types can be distinguished with most samples plotting around a

Fig. 3.58 (continued) Perttunen and Hanski (2003) and Perttunen (1989); Kiiminki Belt, Kähkönen et al. (1986) and Honkamo (1989), Jormua, Kontinen (1987) and Peltonen et al. (1996); Tohmajärvi, Nykänen (1992); Koivusaari and Vaivainen formations, Lukkarinen (2008); Sortavala, Svetov and Sviridenko (1992); Kolosjoki and Pilgujärvi formations, FAR-DEEP and various literature sources; Kittilä Group, the Lapland Volcanite Project (Geological Survey of Finland, see Lehtonen et al. 1998)

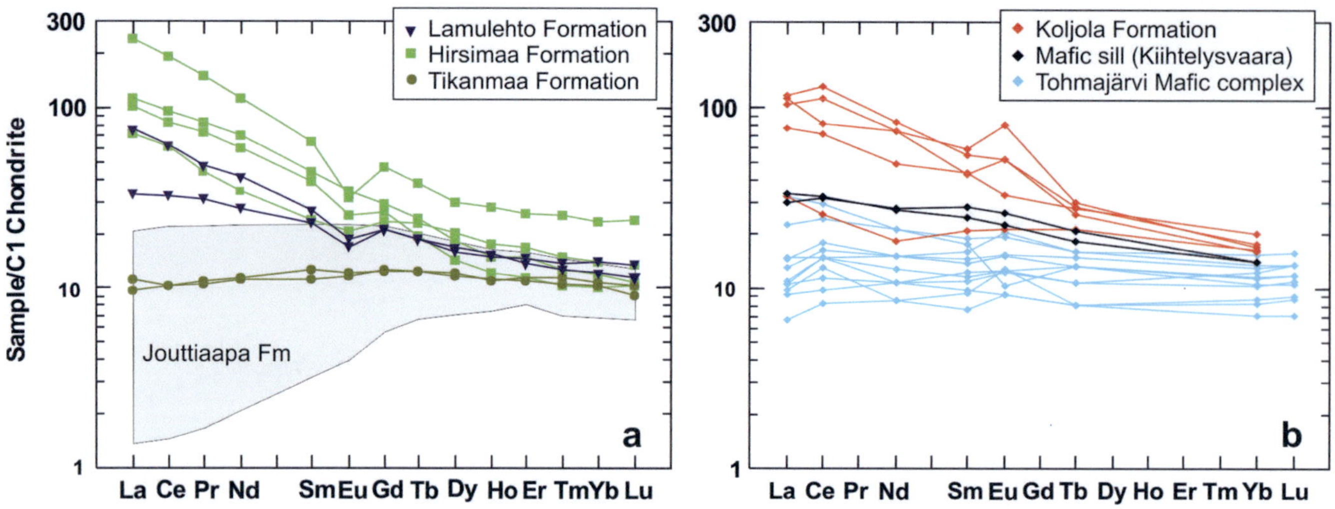

Fig. 3.60 (**a**) Rare earth element characteristics of three volcanic units containing mafic tuffs compared to those of the Jouttiaapa Formation, Peräpohja Belt, all assigned to the Jatuli (modified after Perttunen and Hanski 2003). (**b**) Chondrite-normalised REE characteristics of the Jatulian Koljola Formation and associated mafic sill from the Kiihtelysvaara area and the Tohmajärvi mafic complex. The single sample from the Koljola Formation with a flat REE distribution is from the uppermost part of the formation, while the others are from lower parts (Data from Pekkarinen and Lukkarinen (1991) and Nykänen et al. (1994))

line with TiO_2/P_2O_5 of 8 (Fig. 3.63a). These three groups can be also recognised in the chondrite-normalised REE diagram of Fig. 3.63b. The two types having the highest P_2O_5/TiO_2 ratios also show the highest La/Yb ratios, implying an alkaline affinity. Neodymium isotope data are available only for mafic rocks from the Koivusaari Formation (Lukkarinen 2008). The lower part of the formation containing LREE-enriched lavas has negative low ε_{Nd} suggesting that the differentiation processes leading to their evolution were accompanied by crustal contamination. The other samples analysed from the formation, including least fractionated lavas with the flattest REE patterns (see Fig. 3.63b), one alkaline basalt and three mafic dyke samples, gave positive values, which when regressed together, yielded an ε_{Nd}(2060 Ma) value of $+2.0 \pm 0.5$ (Lukkarinen 2008) indicating less or no contamination. Despite possessing radiogenic Nd, the lavas of the Koivusaari Formation plot above the MORB field in Fig. 3.58f.

At c. 2220 Ma, the Jatulian magmatism produced a set of ultramafic-mafic intrusions that have been assigned to the gabbro-wehrlite association or called karjalitic intrusions (Vuollo and Piirainen 1992; Vuollo and Huhma 2005; Hanski et al. 2010). In Finland, they are encountered in all belts containing Jatulian sedimentary rocks, occurring as differentiated sills that were injected into Jatulian quartzites or the underlying Archaean basement. When found in the granitoid basement, they always occur close to the contact between the basement rocks and the overlying Palaeoproterozoic cover as described, e.g., from the Koli area (Vuollo and Piirainen 1992). Strangely enough, the igneous bodies of this magmatic event also occur within the Archaean Kuhmo greenstone belt, but not in the surrounding gneissic basement area (Hanski et al. 2010). Volcanic counterparts in the Palaeoproterozoic supracrustal sequences have not been identified. In the absence of known feeder dykes in the basement, it seems that the magma travelled a long distance in horizontal magma chambers along the unconformity between Archaean and Proterozoic rocks, leaving open the question of the location of the magma upwelling from the mantle.

The sills are more than 100 km long and a few hundred metres thick and are differentiated into lower olivine cumulates, middle clinopyroxene plus clinopyroxene-magnetite cumulates and upper gabbroic rocks. The ultramafic rocks are characterised by the presence of primary hydrous igneous minerals, edenitic amphibole and titanian phlogopite, as intercumulus phases. The strong modal and chemical differentiation of the layered bodies was a consequence of the parental magma composition, which was hydrous, magnesian low-Al tholeiite with MgO of c. 10 wt.%, resulting in a late appearance of plagioclase as a liquidus phase. In the $[Al_2O_3]$ vs. $[TiO_2]$ diagram of Fig. 3.34, the parental magma composition falls in the field of Al-depleted picrites (and picritic basalts), indicating a rather unusual magma type.

The easily recognisable 2.22 Ga layered sills were generated by a distinct, short-lived magmatic event that spread its products widely in eastern and northern Finland, providing possibilities for intra-shield and even inter-shield correlation of Palaeoproterozoic formations (Hanski et al. 2001c). Yet, in the geological literature, no well-documented examples of analogous differentiated sills have been reported

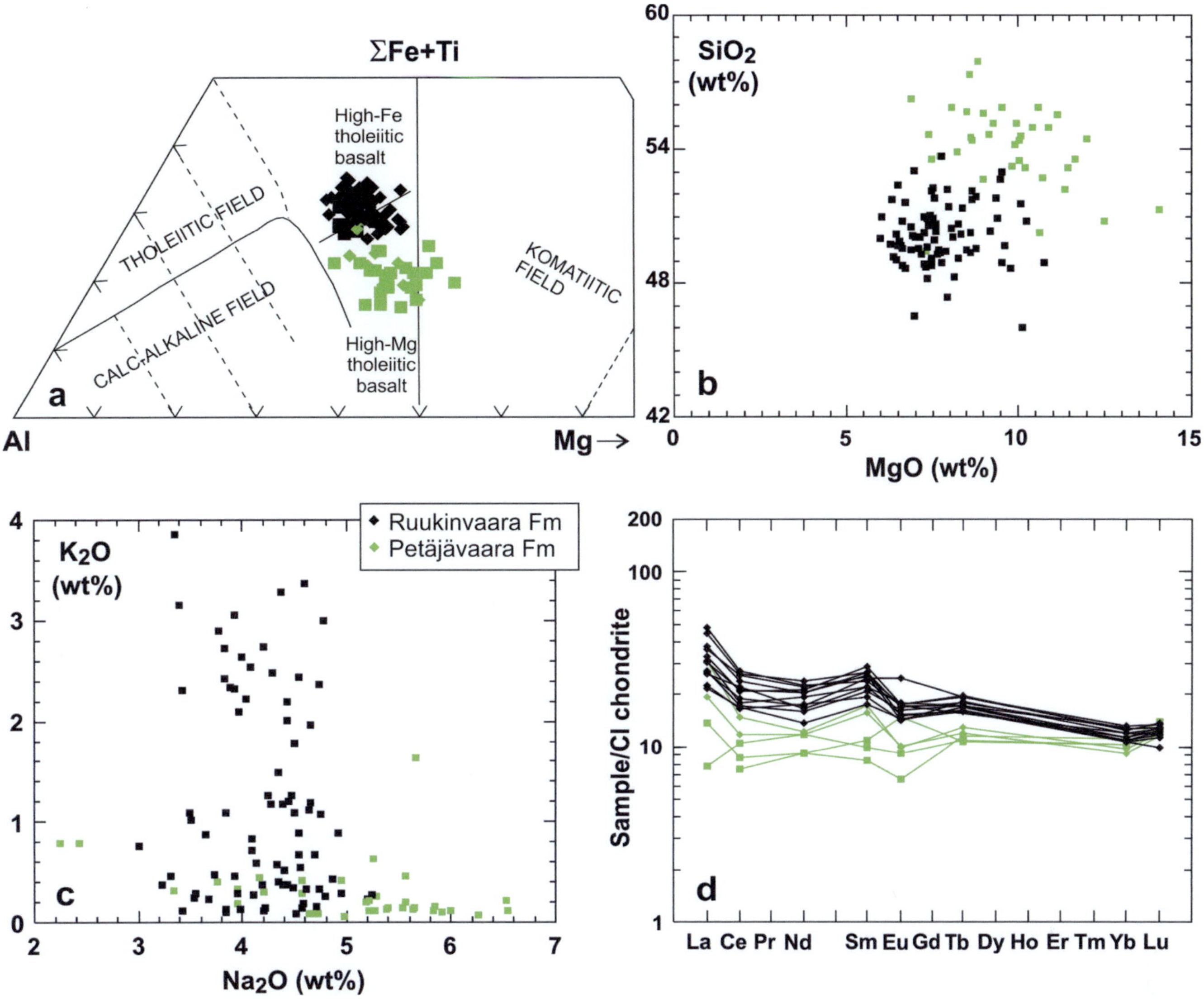

Fig. 3.61 Chemical characteristics of Jatulian volcanic rocks from the Kuusamo Belt. The Petäjävaara and Ruukinvaara formations correspond to the former Greenstone Formation I and II of Silvennoinen (1972), respectively (Data sources: Veki (1991) and unpublished data of the Lapland Research Project (the Geological Survey of Finland)

from the northern or eastern parts of the shield outside Finland (Hanski et al. 2001c), though an identical age (2221 ± 3 Ma) was obtained by Bergh et al. (2007) for a thick diorite sill cutting Palaeoproterozoic Vanna Group sedimentary rocks of the West Troms Basement Complex in northern Norway. Moreover, a potential counterpart, in terms of age, is the c. 2017 Ma Nipissing sills intruding Huronian rocks in the Superior Province of the Canadian Shield (Bleeker and Ernst 2006), but their parental magma was chemically quite different (Lightfoot et al. 1993).

As already noted above, Jatulian dyke swarms cut through the Archaean basement and Jatulian supracrustal rocks. The most important event seems to be that dated at c. 2.1 Ga, which gave rise to a regular set of E-W- or NW-SE-trending mafic dykes with a typical width ranging between 10 and 100 m (Vuollo and Huhma 2005), indicating a significant period of continental extension. Chemically the dykes are rather homogeneous quartz-normative, sub-alkaline, Fe-rich tholeiitic basalts enriched in lithophile trace elements ($La_{CN}/Yb_{CN} = 2–5$) and, based on a single analysis, have initial ε_{Nd} of +0.6 (Vuollo and Huhma 2005). Another group of Fe-tholeiitic dykes was intruded close to the Jatuli-Lucovi boundary at c. 2.05 Ga, especially in Finnish Lapland (Rastas et al. 2001; Räsänen and Huhma 2001).

Palaeoproterozoic felsic plutonic rocks that were generated during the prolonged non-orogenic, 2.5–1.9 Ga period are very rare in the Fennoscandian Shield. In addition to the Sumian A-type granitic intrusions mentioned earlier, they

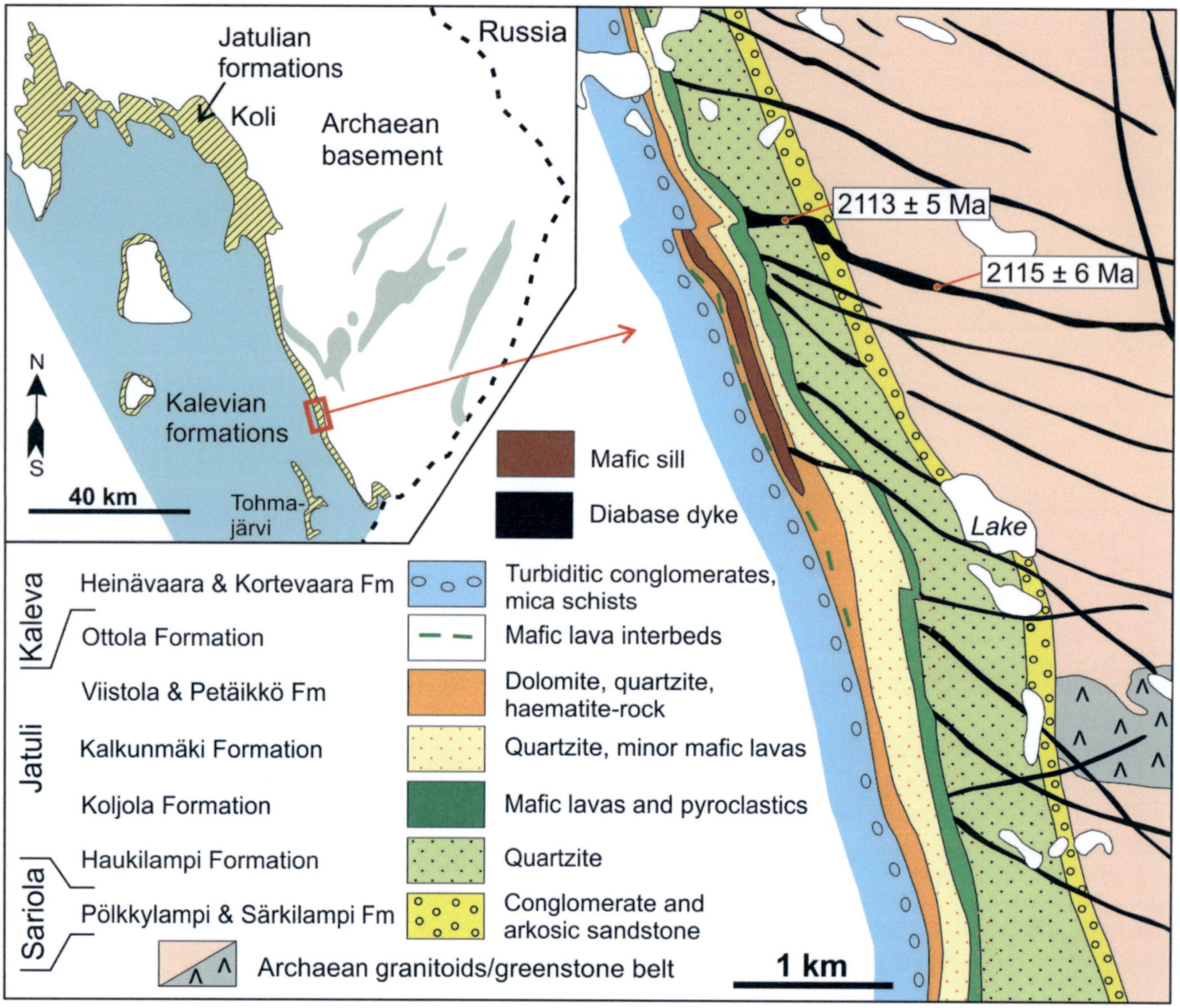

Fig. 3.62 Palaeoproterozoic sedimentary and volcanic rocks in the Kiihtelysvaara area, Finnish North Karelia, after Pekkarinen and Lukkarinen (1991)

include granite and granodiorite intrusions occurring at the northern margin of the Central Lapland Granitoid Complex in Finland (Ahtonen et al. 2007). These bodies were dated by the conventional TIMS method (Rastas et al. 2001), and the ages were reproduced by ion microprobe (Ahtonen et al. 2007). The latter technique yielded ages of 2126 ± 5 Ma and 2105 ± 4 Ma, respectively, indicating that the intrusions were emplaced within the time period of the Jatulian sedimentation and mafic volcanism. The felsic magmatic events have been interpreted to be related to underplating of mafic magmas in the lower crust and ensuing partial melting of Archaean crust, as evidenced by the negative ε_{Nd} values of the intrusions (Ahtonen et al. 2007). This implies that there must have been a considerable amount of coeval mafic magmatism in the shield at c. 2.11–2.13 Ga to be able to generate extensive crustal melting locally. As illustrated by the U-Pb age compilation in Fig. 3.8 (Chap. 3.2), this time interval matches with one of the episodes of the Jatulian mafic magmatism, including, for example, the mafic dykes in the Kiihtelysvaara area in southeastern Finland (see Fig. 3.62).

Prolific magmatic ore deposits of the Jatulian intrusive rocks are limited. As mentioned earlier, titanomagnetite ores (Pudozhgorsk, Koikary-Svyatnavolok) are found in gabbro-dolerite sills in the Lake Onega region (Svetov and Bogachev 1976; Golubev and Trofimov 2011). In addition, Fe-Ti-V oxide deposits are associated with the c. 2.06 Ga gabbro-anorthosite complex of Otanmäki in Central Finland close to the craton margin (Talvitie and Paarma 1980).

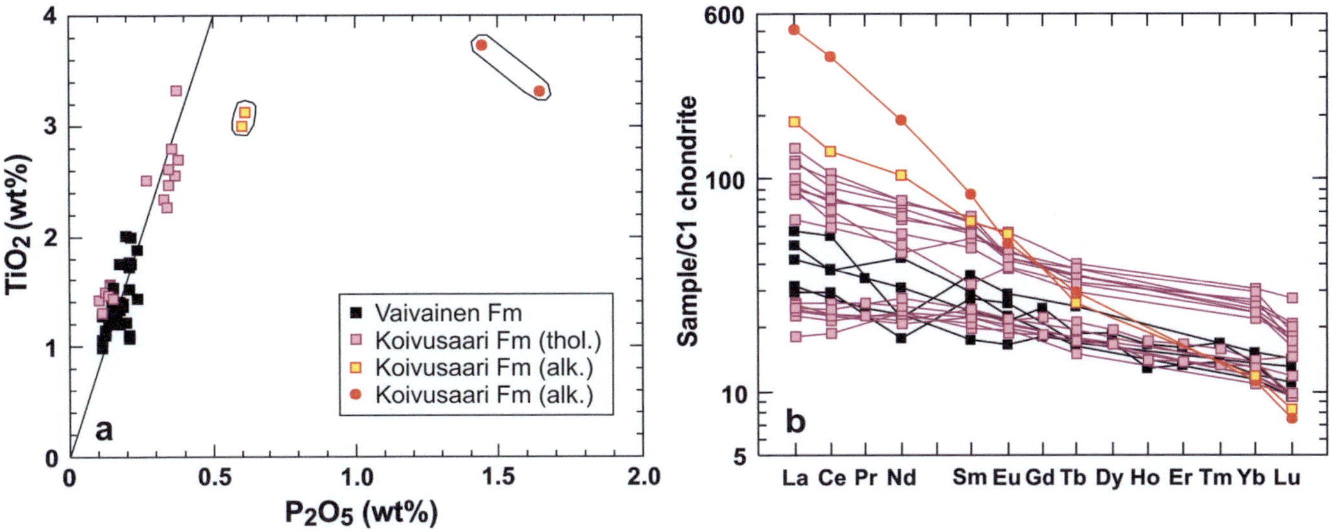

Fig. 3.63 Chemical characteristics of mafic volcanic rocks from the Jatulian Koivusaari and Vaivainen formations, Siilinjärvi Belt and Kuopio area, respectively, central Finland (Data from Lukkarinen (2008))

3.4.5 Ludicovian Magmatism

Figure 3.47c illustrates the geographic distribution of eruptions of Ludicovian volcanic rocks in Russian Karelia. Volcanic rocks of the Zaonega Formation and correlative units occur in the northern and southern parts of the region whereas the Suisarian Formation volcanic rocks are only found in the south. In the Onega area, from where the formation name is derived, mafic volcanic rocks of the Zaonega Formation occur in the upper part of the shungite-bearing black shale and dolomite sequence and reach 300 m in thickness. They display massive, amygdaloidal, pillow and various breccia structures (e.g. Golubev and Kulikov 2011) (Fig. 4.52x, ak–al). The volcanism was accompanied by hypabyssal magmatism in the form of sill-like gabbroic bodies, which were originally injected into soft, carbon-rich sediments as evidenced by the presence of peperite structures (Poleshchuk 2011).

The Zaonega Formation rocks are tholeiitic basalts and rarely andesites with MgO mostly confined in the range of 4–10 wt.%. The lavas form a coherent differentiation series, in which TiO_2 increases with decreasing MgO from slightly less than 1.0 wt.% up to 2.5 wt.%. Published trace element data are so far scarse, being mainly restricted to those reported by Puchtel et al. (1998) from the Solomennoe area and Suisari Island (Fig. 3.64a). The analysed samples are strongly LREE-enriched and, interestingly, display a strong negative phosphorus anomaly. They are also depleted in Ta and Nb compared to La and U. Relatively low HFSE contents are compatible with mostly negative initial ε_{Nd} values, both indicating an influence of crustal contamination (Puchtel et al. 1998). Additional chemical information on the Zaonega Formation basalts is available from some conference abstracts. According to Philippov and Ivanikov (2001), early Ludicovian (Zaonegian) basalts are mainly characterised by almost non-fractionated chondrite-normalised REE patterns with average $(Ce/Yb)_{CN} = 1.57$ and $(La/Sm)_{CN} = 1.35$. Based on the Onega parametric drillcore, Narkisova et al. (2010) also characterized Zaonegian basalts as having slightly fractionated chondrite-normalized REE patterns with $(La/Sm)_{CN} = 1.6–1.7$ and $(La/Yb)_{CN}$ of 2.5–3.0. Ivanikov et al. (2008) showed graphically a chondrite-normalised REE pattern of an average Zaonegian basalt with low $(La/Sm)_{CN}$ (<1), which is reproduced in Fig. 3.64a as normalised to primitive mantle. This average composition and the chemical features reported by Philippov and Ivanikov (2001) and Narkisova et al. (2010) differ markedly from the chemical data of more LREE-enriched basalts from Puchtel et al. (1998) (see Fig. 3.64a). Ivanikov et al. (2008) noted that the Zaonegian basalts they studied are similar to some Upper Jatulian low-Ti basalts with flat chondrite-normalised REE patterns (see Fig. 3.51a). So it seems that the chemistry of the basaltic rocks changes at the Jatulian-Ludicovian transition from LREE-enriched and crustally contaminated to those with little or no LREE enrichment, returning to LREE-enriched varieties higher in the stratigraphy of the Zaonega Formation, as represented by the samples studied by Puchtel et al. (1998).

Palaeoproterozoic volcanic rocks of the Sortavala Group, which occur at the northern shore of Lake Ladoga (see Fig. 3.30), have been regarded as correlative to the Zaonega Formation (Svetov and Sviridenko 1992). Matrenichev et al. (2004) published a study on a section through the stratigraphy of the Sortavala Group in an area between Ladoga and Jänisjärvi Lakes. The section exceeds 1 km in thickness and

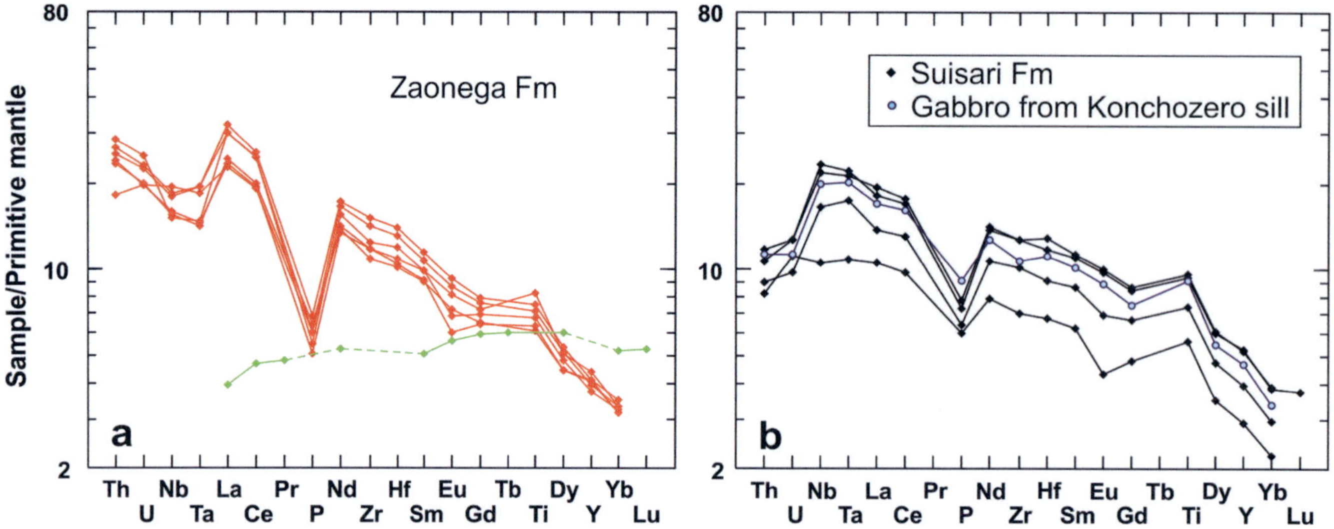

Fig. 3.64 Immobile trace element compositions of Ludicovian volcanic rocks from the Onega region, normalised to model primitive mantle of Sun and McDonough (1989). (**a**) Zaonega Formation. Red symbols: basalts from the Solomennoe area and Suisari Island after Puchtel et al. (1998); *green* symbols: average Zaonegian basalt after Ivanikov et al. (2008). (**b**) Suisarian olivine-clinopyroxene-phyric magnesian basalts and gabbro from the Konchozero sill, all from the Angozero area, after Puchtel et al. (1998, 1999)

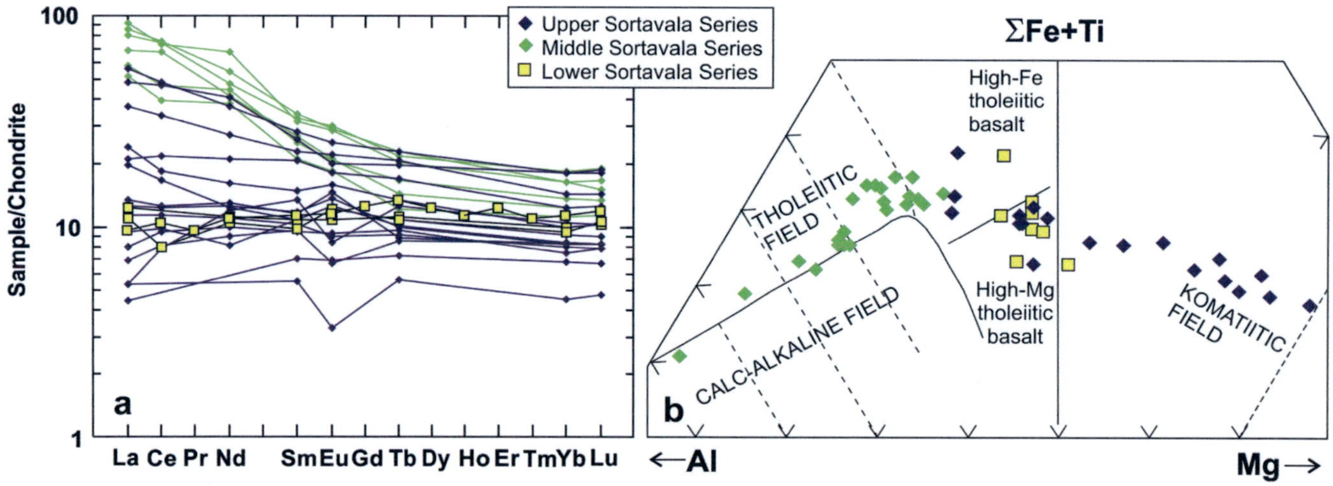

Fig. 3.65 Chemical characteristics of the Ludicovian Sortavala Group volcanic rocks from the northern Ladoga Lake region, which are correlated with volcanic rocks of the Zaonega Formation (Data from Matrenichev et al. (2004) and Ivanikov et al. (1998))

is divided into three volcano-sedimentary parts, the lower, middle and upper units. The lower unit contains mafic, massive and pillowed lavas, and tuffs and tuffites associated with carbonaceous shales and sandstones. The mafic lavas are tholeiitic basalts with low (0.6–1.0 wt.%) or moderate (1.6–1.7 wt.%) TiO_2 contents and nonfractionated or slightly LREE-depleted REE patterns (Fig. 3.65a). Ivanikov et al. (1998) analysed the Sm-Nd isotopic composition for five samples from the lower unit, of which 4 yielded ε_{Nd}(2000 Ma) values between +5.6 and +6.0 and one sample a value of +2.1, indicating a very depleted mantle source.

The middle unit is made up of plagioclase-phyric intermediate and felsic lavas and lava breccias, tuffs, tuffites and polymictic conglomerates. Based on their present alkali-silica relations, the volcanic rocks are mostly classified as trachyandesites and mugearites, rarely rhyolites. They have negatively sloping chondrite-normalised REE patterns (Fig. 3.65a) and plot in the Jensen's cation plot in the tholeiitic field (Fig. 3.65b). In the absence of trace element data other than REE, the mildly alkaline character suggested by high alkalies remains speculative.

The upper unit comprises lavas, lava breccias and agglomerates of komatiites, komatiitic basalts and tholeiitic basalts, mafic to ultramafic tuffs and tuffites, and volcaniclastic conglomerates. Komatiites occur as thin lava flows up to 1.2 m thick with relics of amygdaloidal structures, while the thickness of less magnesian lavas reach 30 m. Komatiitic rocks have high MgO up to 23 wt.%. The character of the komatiitic magma remains somewhat unclear as the Al_2O_3/TiO_2 ratio in the published 8 analyses with MgO >14 wt.% varies between 9.6 and 23.2. The chondrite-normalised REE distributions are flat or slightly LREE-depleted for komatiites, flat or slightly LREE-enriched for komatiitic basalts and more sloping for basalts ($La_{CN}/Yb_{CN} = 1.7–3.1$).

Due to their MORB-like geochemical and isotopic properties, Ivanikov et al. (1998) considered the low-K-Ti tholeiitic basalts of the lower unit analogous to the Jormua Ophiolite basalts. In Fig. 3.78a, REE contents of an average basalt composition from Ivanikov et al. (1998) are compared with those of the Jormua lavas and dykes. Despite the similarity, Matrenichev et al. (2004) still regard the studied rocks as Ludicovian in agreement with Svetov and Sviridenko (1992). Shul'diner et al. (2000) have determined a U-Pb zircon age of 1963 ± 19 Ma for a gabbroic intrusion, which is comparable with the ages obtained for the Finnish ophiolites, but this gabbroic body occurs within Archaean basement gneisses and its connection to the volcanic rocks of the Lake Ladoga region is presently unproven.

In Russian Karelia, volcanic rocks of the Suisari Formation (also known as the Suisaari Formation) are restricted to the Onega Lake area (Fig. 3.48c; Kulikov et al. 2011). The areal extent of the Suisarian rocks delineated in Fig. 3.48c is more limited than what is shown in the maps by Svetov (1979, 1980) and Golubev and Svetov (1983) for the following reasons. Previously the bulk of mafic volcanic rocks occurring on both sides of the border in the Kuolajärvi-Salla area were assigned to Suisarian rocks (Kulikov et al. 1980), but presently they are regarded as pre-Jatulian (Manninen 1991). In a similar way, the highly magnesian volcanic rocks in the Vetreny Belt are now considered Sumian (Puchtel et al. 1997; Kulikov et al. 2010), rather than Suisarian (Kulikov 1988).

An easy access, low degree of deformation and low grade of metamorphism (mostly prehnite-pumpellyite facies) have made the Suisarian volcanic rocks a popular volcanonological research target since the nineteenth century (e.g. Marmo 1949; Svetov 1979). The type locality of Suisarian volcanic rocks is located north of Petrozavodsk in the area of Konchozero and Ukshozero Lakes and Kondopozhskaya Bay of Lake Onega (Fig. 3.66a; Kulikov et al. 1999). Figure 3.66b displays four stratigraphical sections from this area, transecting the contact between the Zaonega and Suisari formations. The contact is marked by the appearance of picritic basalts and olivine-phyric basalts belonging to the Suisari Formation. In addition to these magnesian volcanic rocks, the Suisari Formation contains more evolved pyroxene-phyric and pyroxene-plagioclase-phyric basalts. The total thickness of the volcanic sequence reaches 700 m. The effusives occur as massive and pillow lavas intervening with fragmental rocks including agglomerates, tuffs, tuff conglomerates, and hyaloclastites (see Fig. 4.55a–e). Despite the relatively nonaltered nature of the picritic basalts, olivine is not preserved but is replaced by secondary minerals, including haematite, producing a spotty appearance for the rocks (Svetov 1979). One of the areas containing abundant picritic basalts and studied in detail by Svetov and Sokolov (1978), Svetov (1979) and Golubev and Svetov (1984) is Shidguba on the southwestern shore of the Kondopozhskaya Bay. Figure 3.67 shows a geological map of the area after Svetov (1979), including two volcanic necks close to the Shidguba Bay with oval shapes and diameters of 120 and 290 m on the surface. The southern one acted as the eruption conduit for the lavas of the surrounding area.

In the Suisarian Formation, lava flows and pillow lavas of picritic basalt composition are characterised by the presence of variolitic structures (see Figs. 3.19g and 4.55e), well-described at Yalguba, for example (Svetov 2008; Gudin and Chistyakov 2010). These rocks were the focus of the classical studies by Levinson-Lessing (1935, 1949), in which he proposed silicate liquid immiscibility as the mechanism producing the variolitic structures. Recently, Svetov (2008) and Svetov et al. (2011) performed detailed mineralogical and geochemical studies of these rocks and advocated for the model of liquid immiscibility. It is, however, pertinent to comment here that similar structures in Archaean or Palaeoproterozoic mafic variolitic and ocellar rocks have typically been attributed to spherulitic crystallisation of highly supercooled liquids (e.g. Fowler et al. 2002; Sandstå et al. 2011), while post-eruptional blotchy alteration of glassy mesostasis was suggested by Hanski (1993) to be the cause for the generation of the globular structures observed in Ludicovian ferropicritic rocks at Pechenga.

The Suisarian magmatism also produced subvolcanic, layered ultramafic-mafic sills, which were emplaced into Zaonega Formation pyroclastic volcanic rocks. Among them the most well-known are the Konchozero and Ternavolok sills, which have thicknesses of up to 200 and 60 m, respectively (Kulikov et al. 1976; Puchtel et al. 1995; Trofimov 2010). Using several methods (Sm-Nd, Pb-Pb, Re-Os), the Konchozero sill has been dated at c. 1.98 Ga (Puchtel et al. 1998, 1999) and gives an age for the Suisarian volcanic rocks as well, a genetic link that is unequivocally based on the geochemical and isotopic evidence (Puchtel et al. 1995, 1998, 1999) (see below).

The volcanic rocks of the Suisari Formation constitute an extensive differentiation series with MgO varying from c. 24 to 3 wt.% along with the change of the liquidus silicate minerals from olivine through olivine + clinopyroxene to clinopyroxene + plagioclase. In average compositions from different areas, Golubev and Svetov (1983) and Golubev et al. (1984) report MgO contents (volatile-free) of 15.7–21.4 wt.% for picrites and picritic basalts, 20.8 wt.% for picritic dykes, 8.3–11.3 wt.% for pyroxene-phyric basalts, and 4.1–8.5 wt.% for plagioclase-phyric basalts.

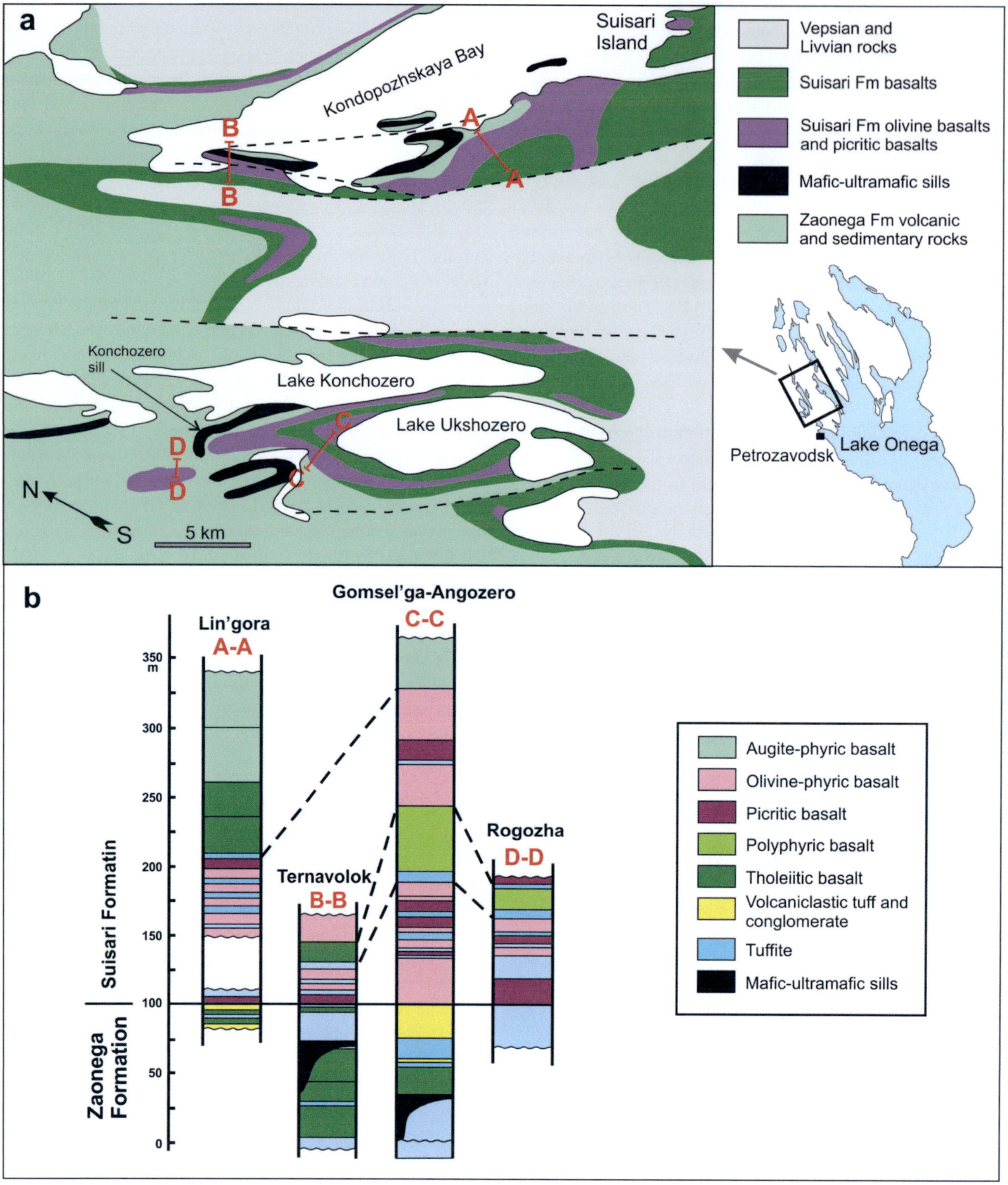

Fig. 3.66 (**a**) Map showing rocks of the Ludicovian Zaonega and Suisari formations in the Kondopozhskaya Bay and Konchozero Lake areas, northwest of Lake Onega. (**b**) Stratigraphical cross-sections of four profiles from the Kondopozhskaya Bay and Konchozero Lake areas, showing the contact of the Zaonega and Suisari formations. See (**a**) for the location of the profiles (Both figures after Kulikov (1988))

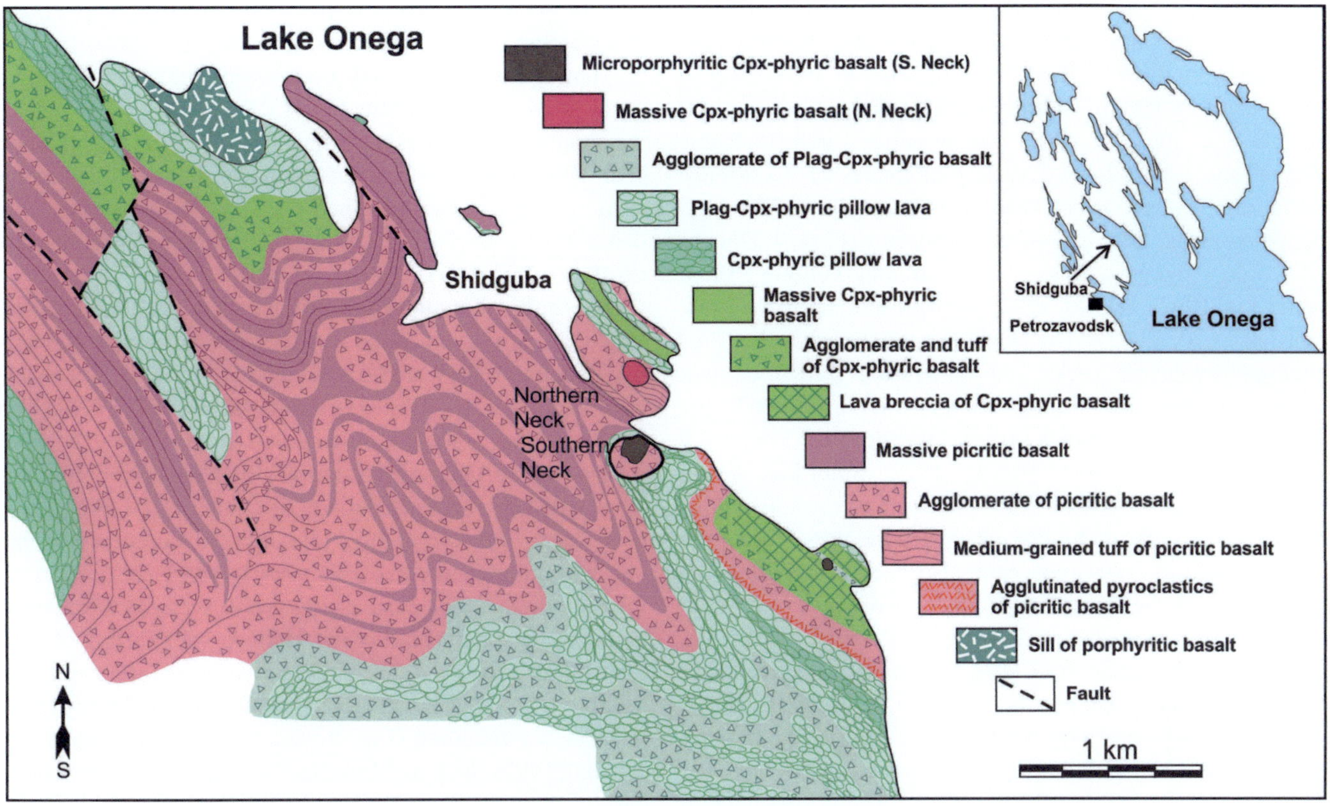

Fig. 3.67 Map showing the occurrence of Ludicovian Suisari Formation volcanic rocks and related volcanic necks in the Shidguba area, Kondopozhskaya Bay, northwestern part of Lake Onega (after Svetov 1979)

The whole-rock compositions are variably affected by the presence of phenocryst phases, and in the absence of preserved primary olivine, the parental melt composition cannot be precisely determined. Chilled margin compositions of the Konchozero sill have MgO contents no higher than 8.1–11.8 wt.% (Golubev et al. 1987a, b), 9.3–11.4 wt.% (Kulikov 1988) and 8.5–9.1 wt.% (Puchtel et al. 1998, 1999). The Suisarian picrites, picritic basalts and ultramafic dykes and the parental magma of the volcanic series as well have low Al_2O_3/TiO_2 of 5–6 and, on the [Al_2O_3] vs. [TiO_2] diagram, plot in the lower part of the picrite division, overlapping the field of the Umba picrites (Fig. 3.34). Figure 3.64b demonstrates the similarity of the trace element characteristics of the olivine-clinopyroxene-phyric basalts and magma that produced the Konchozero sill. The magmas were enriched in LREE and HFSE resembling OIB and had a similar negative phosphorous anomaly as is observed in Zaonegian basalts (see Fig. 3.64a, b). The Zaonegian and Suisarian samples analysed by Puchtel et al. (1998) differ markedly from each other with respect to the mutual behaviour of LIL and HFSE. The lack of negative Nb-Ta anomalies in Suisarian samples is coupled with positive initial ε_{Nd} values for lavas and the Konchozero sill (ε_{Nd}[1975 Ma] = +3.2 ± 0.1 for the sill), suggesting minimal crustal contamination, although interaction with sialic crust is evidenced by the presence of Archaean zircon grains in one of the picritic basalt samples studied by Puchtel et al. (1998). The sample set studied by Philippov and Ivanikov (2001) yielded a Sm-Nd isochron age of 2013 ± 52 Ma and a lower initial ε_{Nd} of −0.5 ± 0.18.

In the Pechenga Belt, Ludicovian volcanic rocks constitute the major portion of the upper part of the North Pechenga Group. They occur as two thick lithostratigraphic units, the Kolosjoki and Pilgujärvi volcanic formations, which are separated from each other by the Pilgujärvi Sedimentary Formation, the "Productive Formation", enclosing Ni-Cu sulphide ore-bearing, mafic-ultramafic sills (Gorbunov et al. 1985; Hanski 1992; Smol'kin and Borisov 1995) (see Chap. 4.2). The age of the Pilgujärvi Volcanic Formation is well-known (c. 1980 Ma; Hanski et al. 1990; Skuf'in and Bayanova 2006) but that of the Kolosjoki Volcanic Formation can only be bracketed between c. 2060 and 1980 Ma (2000 Ma used below in calculating initial ε_{Nd}). Both volcanic formations are mainly composed of alternating pillowed (Figs. 3.19b, r, y, 3.27a, b, o, 3.52d, f, 4.27a, 4.30a, b, and 6.82) and massive mafic lava flows with variable vesicularity (Figs. 3.52c–e and 6.85i–l, n), supplemented by lava breccias and hyaloclastites (Figs. 4.27b and 4.30c), mafic tuffs

(Fig. 3.27p), gabbroic sills and, in the Pilgujärvi Formation, a horizon of rhyolitic tuffs (Figs. 3.19ab and 4.30i–t) (e.g. Predovsky et al. 1974; Rusanov 1981; Hanski 1992; Skuf'in and Theart 2005). In addition, there are ferropicritic rocks, mostly in the Pilgujärvi Sedimentary and volcanic formations, which have a variety of modes of occurrence including layered lava flows, massive (Fig. 4.30d) and pillowed lavas, tuffs and eruptive breccias (Figs. 4.29be–bf and 5.6q), dykes and sills (Skuf'in and Fedotov 1989; Smol'kin 1992; Hanski 1992; Hanski and Smolkin 1995; Spisiak et al. 2007). The cogenetic relationship between ultamafic volcanic rocks and Ni-bearing intrusions was already suggested by Zagorodny and Mirskaya (1967) and has been discussed later by many investigators (Predovsky et al. 1974 and 1987; Hanski 1992; Smol'kin 1992). Ferropicritic pyroclastic rocks occur in two eruption centres in the upper part of the Pilgujärvi Sedimentary Formation (Melezhik et al. 1994) and have also been observed in the lower part of the Kolosjoki Volcanic Formation; a 72-m-thick section was penetrated by FAR-DEEP Hole 8A, containing ultramafic tuffs and tuffites with much ferropicritic material (see Chap. 6.1.5). A set of ultramafic dykes having a thickness of a few tens of metres transects obliquely the Kolosjoki Volcanic Formation. They apparently acted as feeders to magmas that spread in the overlying sedimentary Productive Formation as horizontal sills carrying immiscible sulphide droplets contributing to the formation of the Pechenga Ni-Cu deposits. The extent of the reaction of the ferropicritic magma with the sulphide-bearing host-rock black shales and its role in ore formation is still unclear (Hanski et al. 2011). The sills are commonly 5–250 m thick, but in the case of the Pilgujärvi intrusion, the thickness of the magmatic body reaches 500 m.

Figure 3.68 shows a structural, textural and chemical variation through a 24-m-thick, differentiated ferropicritic lava flow with an upper spinifex-textured zone and a lower olivine cumulate zone, i.e. a vertical structure mimicking that of layered komatiitic lava flows. The spinifex texture was formed by platy olivine crystals in nonfractionated ferropicritic magma (Fig. 3.69c) and needle-like clinopyroxene crystals in evolved ferropicritic (or ferrobasaltic) magma (Figs. 3.69b, 3.70b, and 4.30f, h). A special feature of the flows and narrow dykes is the presence of a zone containing light-coloured globules (ocelli) (Figs. 3.19aa, 3.69a–b, 4.30e–f, and 5.6r), whose genesis was regarded as magmatic by Smol'kin et al. (1987) but as postmagmatic by Hanski (1993). Figure 3.70a displays a microscopic texture of an unfractionated, massive ferropicritic lava.

The geochemical character of ferropicritic magma has been discussed in detail and compared with other primitive magma types by Hanski (1992) and Hanski and Smol'kin (1995). In brief, the ferropicritic magma is Mg- and Fe-rich with low Al_2O_3/TiO_2 due to both high Ti and low Al (Fig. 3.34) and has an OIB-like trace element (Fig. 3.59a) and isotopic signature. Initial ε_{Nd} (+1.4 ± 0.4 Ma) obtained by Hanski (1992) indicates a long-term depletion of LREE in the source in spite of the high LREE/HREE measured for the magma, analogously with Hawaiin picrites (e.g. Sims et al. 1994). Osmium in uncontaminated ferropicritic magma is moderately radiogenic (initial γ_{Os} c. +6) providing evidence for an ancient ^{187}Os-enriched mantle plume source (Walker et al. 1997).

The bulk of mafic lavas from both the Kolosjoki and Pilgujärvi formations have a rather coherent major element geochemistry. They are tholeiitic basalts with MgO varying between c. 9 and 4 wt% (e.g. Predovsky et al. 1987; Skuf'in 1980a). There is some internal variation among the basaltic rocks in the vertical section through the Kolosjoki Formation as revealed by the Kola Superdeep and FAR-DEEP drilling (see Chap. 4.2). Also high-Fe-Ti basaltic differentiates occur but, without trace element analyses, it is not always easy to separate them from ferrobasaltic derivatives of ferropicritic magma. Testing the major element similarity of the Kolosjoki and Pilgujärvi tholeiitic lavas with recents MORBs give clearly negative results when the TiO_2-FeO_T relation is concerned (Fig. 3.58c, d), as already noted by Hanski (1992). Yet, as illustrated in Fig. 3.57a, b, tholeiitic basalts from the Pilgujärvi Volcanic Formation plot at the boundary of the MORB-OIB array or within the MORB array on the discrimination diagrams of Pearse (2008).

A subtle distinction can be made between the Kolosjoki and Pilgujärvi formations using REE abundances, which form flat chondrite-normalised patterns in the case of the Pilgujärvi Formation, while the patterns of the Kolosjoki Formation are slightly LREE-enriched (Fig. 3.71a). Only the uppermost (southernmost) basalts of the Pilgujärvi Volcanic Formation, which Skuf'in and Theart (2005) assigned to a separate unit, the Suppvara Formation, are LREE-enriched and also are isotopically distinct. Tholeiitic basalts of the Pilgujärvi Formation have initial ε_{Nd} values of +3.0 to +3.8, indicative of a long-term depletion in LREE compared to MREE in the mantle source (Hanski et al. 1990; Hanski 1992). Likewise, the Kolosjoki Formation basalts have a depleted Nd isotopic composition with ε_{Nd}(2000 Ma) of +1.5 to +3.0 (Huhma and Hanski unpubl. data). Skuf'in and Theart (2005) have obtained a slightly lower range of ε_{Nd}(2000 Ma) for this formation, falling between +1.3 and +2.3. In any case, the Nd isotopic signature of both Kolosjoki and Pilgujärvi Formation tholeiitic basalts suggests negligible interaction of the magmas with continental crust and isotopically a MORB-like mantle source. The uppermost LREE-enriched basalts mentioned above record a near-chondritic isotopic evolution with ε_{Nd}(1980 Ma) of +0.2 to +0.3 (Skuf'in and Theart 2005).

Among the otherwise mafic rocks of the Pilgujärvi Volcanic Formation, a distinct key horizon of felsic tuff can be

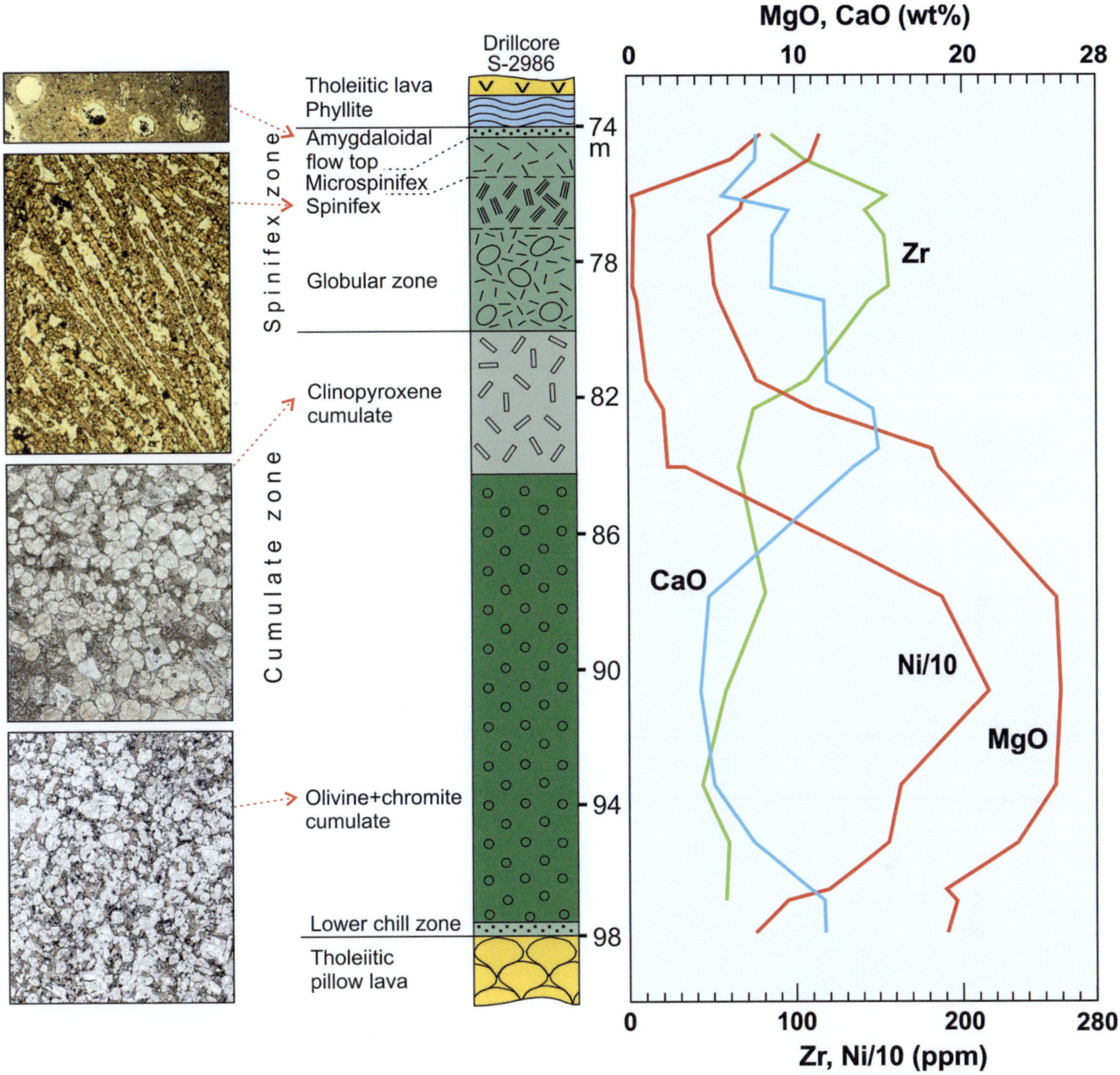

Fig. 3.68 Cross section of a layered ferropicritic lava flow in the Pilgujärvi Volcanic Formation, Pechenga Belt, recovered by a drillhole S-2986

followed for more than 20 km in the middle part of the formation (Borisov and Smol'kin 1992). As described in Chap. 4.2, the felsic rocks occur as several separate layers and show fragmental or very fine-grained, obsidian-like textures (see Fig. 4.30i–r). Chemically, they form a series from A-type, peralkaline dacites to high-silica rhyolites (see Fig. 3.42b). Neodymium isotope compositions with initial ε_{Nd} varying from −1.8 to −1.3 (Skuf'in and Theart 2005) and from −0.7 to 0.6 (Hanski and Huhma unpubl. data) indicate that the silica-rich magma was not generated solely by melting of old Archaean crust, but rather was derived by extensive fractional crystallisation of a hybrid magma.

Several generations of mafic to ultramafic dykes have been recognised in the Archaean basement northeast of the Pechenga Belt (Fedotov 1995). Among them is the NW-trending Nyasyukka dyke complex composed of kaersutite-bearing peridotites with a chemistry similar to that of ferropicrites (Fedotov 1995; Fiorentini et al. 2008). The presence of orthopyroxene, however, indicates that the silica activity was higher than in the magma that produced

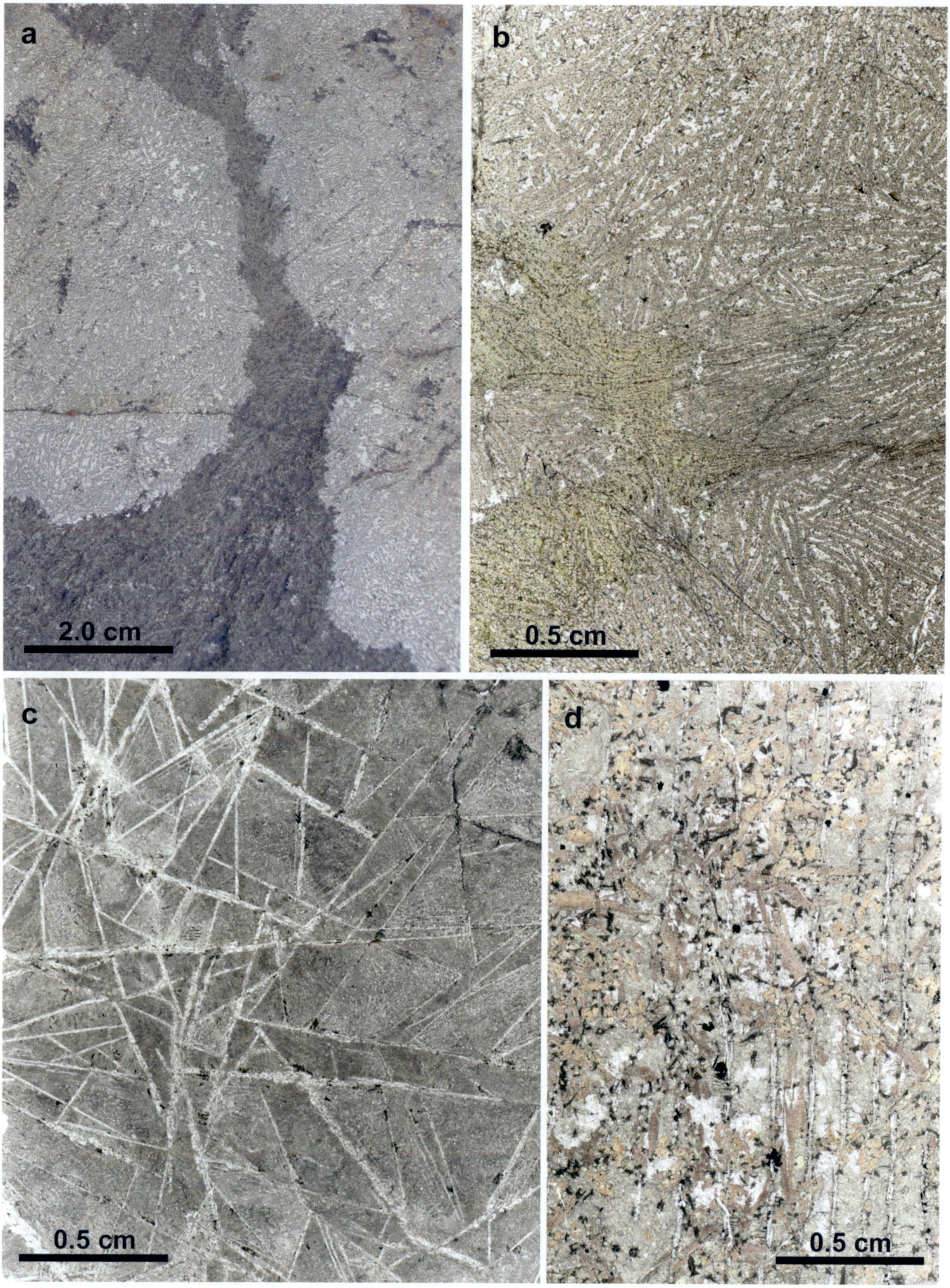

Fig. 3.69 Textures in the upper part of layered ferropicritic lava flows, the Pilgujärvi Volcanic Formation, Pechenga Belt. (**a**) Globular structure. (**b**) Scanned thin section of a globule-bearing rock with needle-like clinopyroxene crystals. (**c**) Scanned thin section showing randomly oriented olivine spinifex texture with olivine replaced by serpentite. (**d**) Scanned thin section exhibiting parallel olivine spinifex texture. Vertical olivine plates are replaced by secondary minerals, but brownish titanaugite is unaltered (Photographs taken by Eero Hanski)

Fig. 3.70 Scanned thin sections of ferropicritic rocks from the Pilgujärvi Volcanic Formation. (**a**) Massive ferropicritic lava containing serpentinised olivine microphenocrysts and microlites in a matrix of *dark brown*, devitrified glass and plumose clinopyroxene. (**b**) Evolved ferropicritic basalt with coarse, acicular or prismatic, zoned and commonly skeletal titanaugite crystals in a matrix of pyroxene-plagioclase intergrowths and scattered amygdales. Note the same scale in both figures (Photographs taken by Eero Hanski)

the ferropicritic rocks in the Pechenga Belt. In addition, the baddeleyite U-Pb age of 1941 ± 3 Ma determined by Smol'kin et al. (2003) is c. 40 Ma younger than the age of the ferropicritic magmatism at Pechenga.

Ludicovian volcanic rocks in the Imandra/Varzuga Belt are found in the Il'mozero and Mittrijärvi formations (see Chap. 4.1). Most volcanic rocks (90 vol%) in the Il'mozero Formation are subaerially erupted low-Ti basaltic andesites with subordinate andesites (Fedotov 1985). Remizova and Barzhitsaya (1984) have published analyses of more Ti-rich basalts (TiO_2 1.5–3.6 wt%), while Mints (1996) reported an analysis of komatiite (MgO 26.3 wt%) from the Il'mozero Formation, but in general, the few primitive volcanic rocks in the lower part of the formation are komatiitic basalts rather than komatiites. Only scarse immobile trace element data are available from the Il'mozero Formation. Those reported by Mints (1996) are plotted in Fig. 3.72b. Skuf'in and Yelizarov (2011) published five Nd isotope analyses on mafic volcanic rocks, of which one yielded an ε_{Nd}(2100 Ma) value of −0.6 and the rest gave values in the range of −2.3 to −4.5. Calculating the initial $^{143}Nd/^{144}Nd$ ratios with a younger age would result in even lower initial ε_{Nd} values. These results are consistent with the high LREE/HREE ratios and relatively low HFSE contents of the low-Ti basalts and andesites (Fig. 3.72a, b).

Ferropicritic volcanic rocks and associated tholeiitic pillow lavas that occur in the Mittrijärvi-Solenoe area of the Imandra/Varzuga Belt has been assigned to the Mittrijärvi Formation (Smol'kin 1992), but so far this formation has been poorly characterised in the geological literature. Geochemically (Figs. 3.34 and 3.71b) and texturally (Fig. 4.14d), the ferropicrites with MgO up to 18 wt% correspond well to those found in the Pilgujärvi Volcanic Formation, providing a good basis for the mutual correlation of these two formations.

Ludicovian ultramafic effusive sequences occur in the Salla area in eastern Finnish Lapland (Manninen 1991) and more abundantly in central Finnish Lapland (Hanski et al.

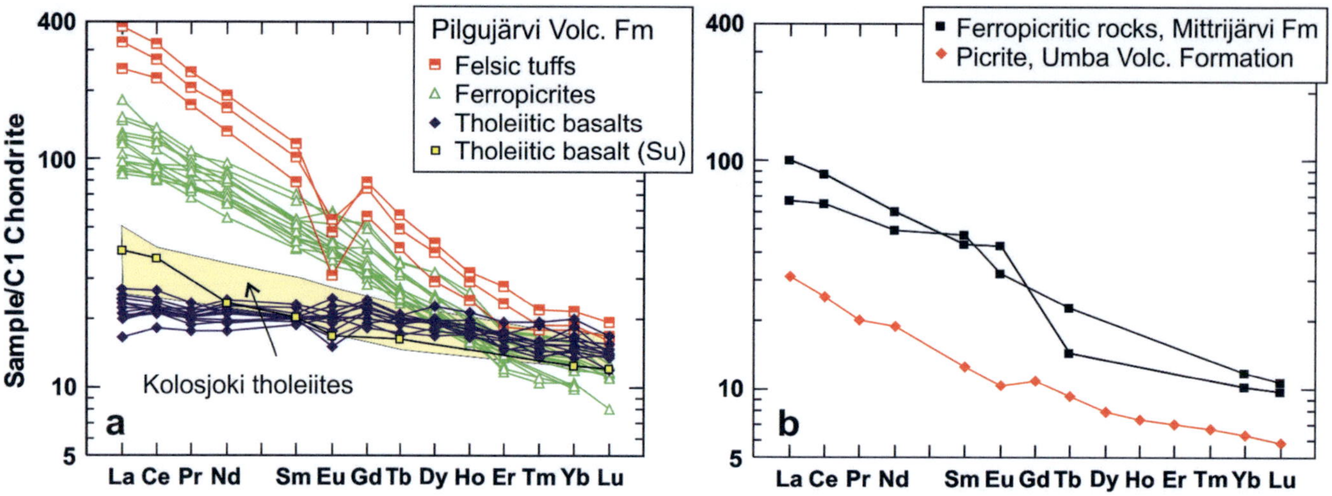

Fig. 3.71 Chondrite-normalised REE patterns for rocks from the Pechenga and Imandra/Varzuga Belts. (**a**) Ludicovian Pilgujärvi and Kolosjoki Volcanic formations, Pechenga. The field of the Kolosjoki basalts is based on data from Hanski (1992) and Mints (1996), and the Pilgujärvi Formation data are from this study, except of an average analysis (*yellow symbol*) for the uppermost part of the formation, which is calculated using the analyses published by Sku'fin and Theart (2005). This part of the formation was assigned to the Suppvara Formation (Su) by Skuf'in and Theart (2005) and differs slightly chemically and isotopically from the rest of the Pilgujärvi Formation. (**b**) Jatulian Umba picrite and Ludicovian Mittrijärvi Formation ferropicritic rocks, Imandra/Varzuga Belt. The Umba data are from FAR-DEEP Hole 4A (see Chap. 6.1.3) and the Mittrijärvi data from this study

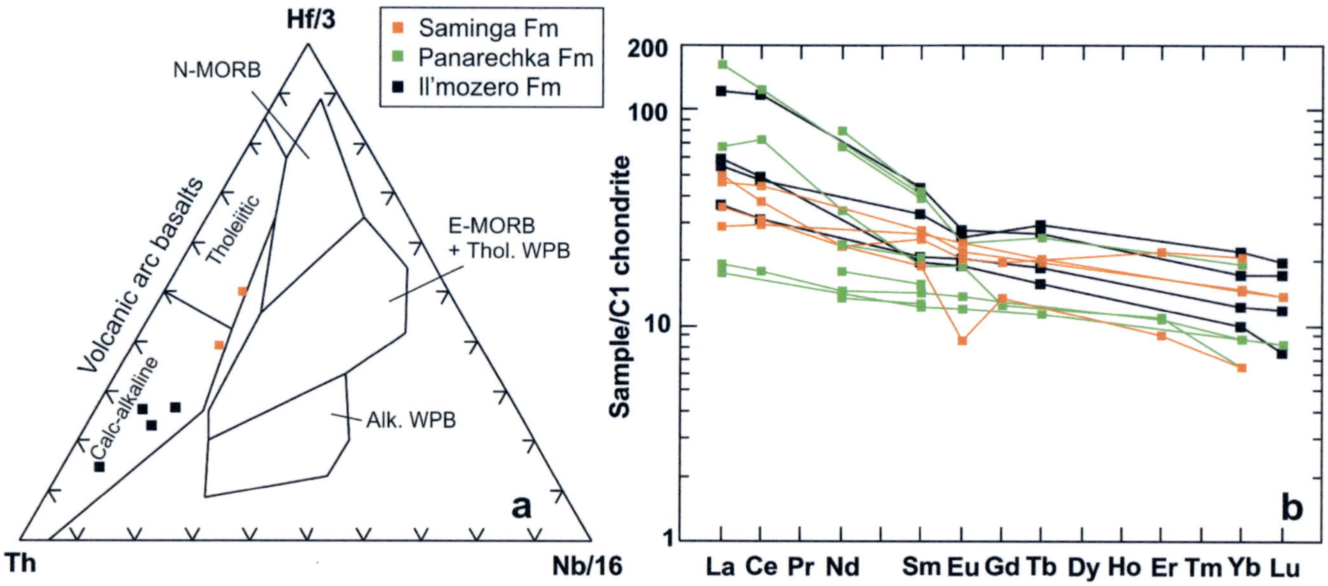

Fig. 3.72 Trace element data from the Ludicovian Il'mozero Formation and "orogenic" Panarechka and Saminga formations, Imandra/Varzuga Belt (Analytical data from Mints (1996) supplemented by Sm-Nd analyses of Skuf'in et al. (2011))

2001a; Gangopadhyay et al. 2006), from where they can be followed further north to the Karasjok belt in northern Norway (Barnes and Often 1990). They were erupted after deposition of a sequence of phyllites and black shales. Chemically, the primitive volcanic rocks are dominated by Ti-enriched komatiites and still more Ti-rich picrites, with no differences in their field characteristics. Also hybrid types between the mentioned end-members are encountered. They are associated with more fractionated, MgO-rich basalts of both main series. The Sm-Nd method has yielded an age of 2056 ± 25 Ma for komatiitic rocks in the Jeesiörova area (Hanski et al. 2001a). Structural and textural preservation is locally very good and clinopyroxene and chromite may be fresh (Fig. 3.73d), but olivine is always replaced by secondary minerals (Fig. 3.73b, c, e). Differing from typical Archaean komatiites, primitive lavas in Lapland rarely display spinifex-textures. Instead, a large part of komatiitic and picritic rocks are volcaniclastic, forming unsorted,

Fig. 3.73 Structures and textures of Ludicovian komatiites from the Sattasvaara (**a**) and Jeesiörova areas (**b–e**), central Finnish Lapland. (**a**) Pyroclastic komatiite. Coin diameter is 2.1 cm. (**b**) Rounded chlorite-filled amygdale rimmed with fine-grained chromite in olivine-phyric komatiite lava, (**c**) Olivine crystals (replaced by serpentine) in dark devitrified glass matrix in komatiite lava. (**d**) Melt inclusions in a chrome spinel grain. (**e**) Pseudomorphs of euhedral olivine set in a matrix of pyroxene needles and brownish devitrified glass, komatiitic lava flow. Photomicrographs (**b**), (**c**) and (**e**): parallel-polarised light, scale bar 1 mm in length; (**d**): reflected light, scale bar 0.1 mm in length (Photograph (**d**) courtesy of Vadim Kamenetsky, other photographs taken by Eero Hanski)

agglomerate-like deposits with rounded or angular fragments ranging from one to tens of centimetres (Figs. 3.19k, l, 3.27v, w, and 3.73a). These deposits can be tens of metres thick, and they are interlayered with fine-grained laminar tuffs. Pyroclastic structures are well observable in the Sattasvaara area where Saverikko (1985) interpreted them as products of explosive, phreatomagmatic volcanism, which is not a common mode of eruption of non-viscous komatiitic magmas. A submarine environment is manifested by the presence of pillow lavas (Fig. 3.27t) and associated pillow breccias, though, according to Saverikko (1985), interlayers of epiclastic volcanic conglomerates also exist indicative of occasional erosional periods.

The komatiitic rocks display a wide range of MgO contents from less than 10 wt% to as much as 31 wt%. In the absence of primary olivine, the MgO content of the parental magma of the komatiitic rocks is not easy to estimate precisely. Pillow lavas and volcaniclastic rocks, which are most representative of the magma composition, have variable MgO between 14 and 27 wt.%, but due to the presence of olivine phenocrysts, the maximum MgO of the liquid was less than the bulk rock MgO content. Also the coexistence of olivine and chrome spinel as liquidus phases constrain the MgO of the liquid to around 20 wt.% with the exact value depending on the Cr content and fO_2 of the liquid. As the name implies, the Ti-enriched komatiitic

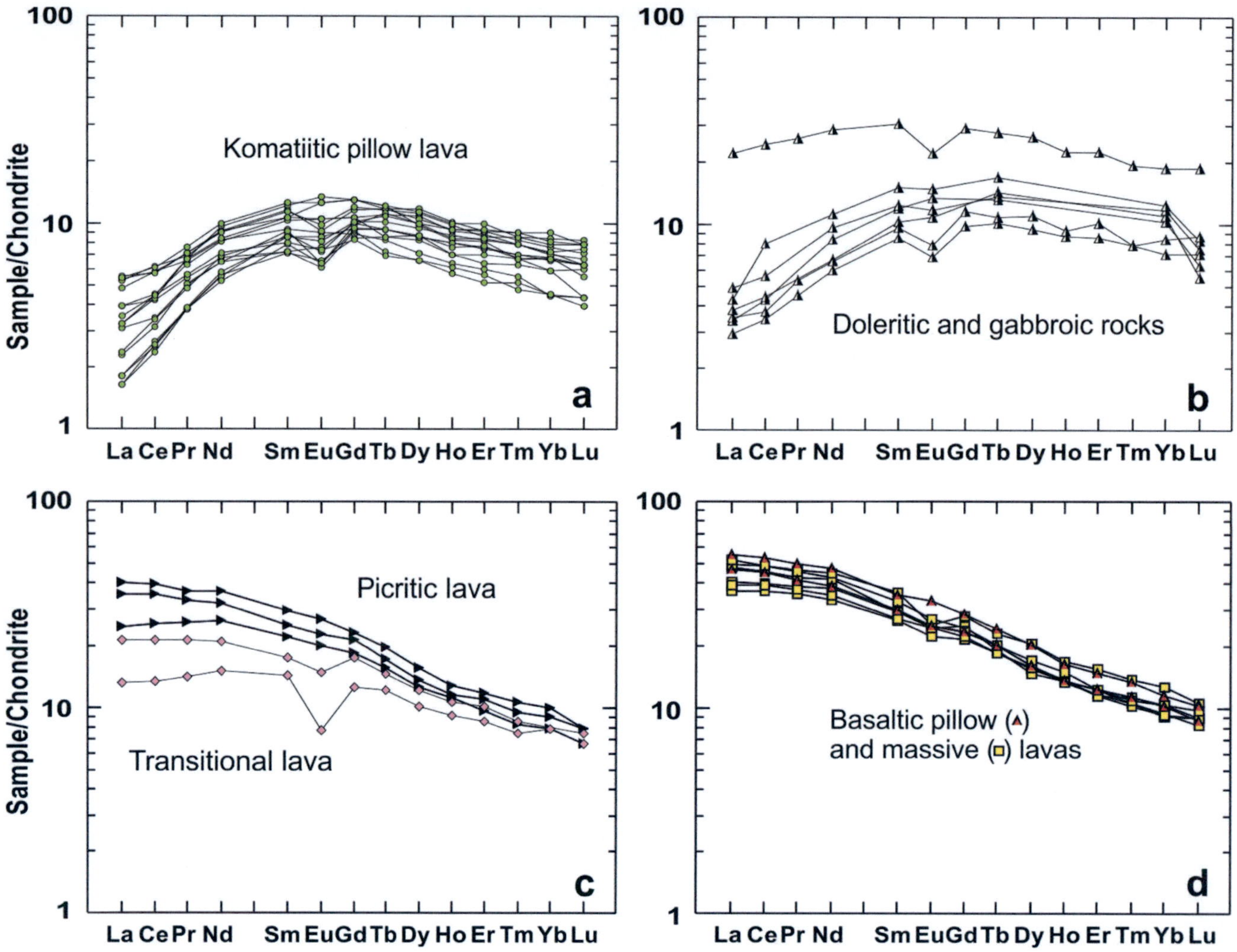

Fig. 3.74 Chondrite-normalised REE profiles of Ludicovian ultramafic lavas and mafic lavas and intrusive rocks (Savukoski Group) from the Jeesiörova and Peuramaa area, central Finnish Lapland (Modified after Hanski et al. 2001a)

rocks are more enriched in moderately incompatible Ti than Munro-type komatiites, resulting in subchondritic Al_2O_3/ TiO_2 (<15) and a shift of the compositions upwards in the [Al_2O_3] vs. [TiO_2] diagram (Fig. 3.34). In contrast, they are depleted in highly incompatible trace elements such as LREE (Hanski et al. 2001a; Gangopadhyay et al. 2006). In the most depleted types, $(La/Sm)_{CN}$ is as low as 0.2. The rocks also have high MREE/HREE ratios and hump-shaped chondrite-normalised REE patterns (Fig. 3.74a). Picritic rocks generally show slightly lower MgO, falling in the range of 11–19 wt.%, are even more enriched in TiO_2 than the komatiitic varieties (Fig. 3.34), have high LREE and display sloping chondrite-normalised REE patterns with $(La/Yb)_{CN}$ up to 4.1 (Fig. 3.74c). Also transitional types can be found having chemical characteristics intermediate between those of Ti-enriched komatiites and picrites (Fig. 3.73c). Basaltic rocks forming lavas and subvolcanic intrusions display REE patterns that are compatible with parental magmas varying from komatiitic to picritic (Fig. 3.74b, d).

Chemical compositions of chromite grains and their melt inclusions (see Fig. 3.73d) demonstrate the coexistence and mixing of komatiitic and picritic magmas at the time of the generation of the komatiite-bearing belts (Hanski and Kamenetsky 2006). Both Ti-enriched komatiites and picrites show a similar depleted Nd isotopic signature (initial ε_{Nd} c. +4.0) apart from a minor variation (up to +2.0) likely due to contamination with pelitic sedimentary material (Hanski et al. 2001a; see also Krill et al. 1985). This indicates a long-term depletion of the mantle source in LREE relative to HREE even though the magmas record a very large range in LREE/HREE. Hanski et al. (2001a) interpreted these features as a consequence of chemical heterogeneities in the mantle source region that were created by complex depletion and enrichment processes shortly prior or related to a dynamic melting process.

Differing from ferropicrites, Ludicovian-age komatiites in the Fennoscandian Shield are not known to host significant Ni-Cu sulphide deposits in spite of their favorable spatial association with sulphide-bearing black shales.

Fig. 3.75 (**a**) Pahoehoe lava from the Hirvas area, Russian Karelia. For the age of these lavas, see Discussion. (**b**) Interbedded 1.98 Ga felsic and mafic tuffs from the Korkiavaara Formation, northern part of the Peräpohja Belt. Coin diameter 2.5 cm (Photograph (**a**) courtesy of Igor Puchtel and photograph (**b**) taken by Eero Hanski)

An occurrence of PGE enrichment in komatiitic rocks is known in central Finnish Lapland (Moilanen 2011). The situation is different when less magnesian magmas are concerned. A large, disseminated Ni-Cu sulphide deposit is related to the c. 2.05 Ga layered intrusion at Keivitsa, central Finnish Lapland (Mutanen 1997; Grinenko et al. 2003). This deposit has a very wide range of metal tenors and peculiar isotopic compositions, including an ore type with primitive-type metal characteristics, such as high Ni and PGE, but unradiogenic, crustal-like Nd isotope systematics (initial ε_{Nd} down to −6.6) (Hanski et al. 1997).

Besides komatiite- and picrite-bearing volcanic successions, Ludicovian rocks in Lapland also consist of independent basaltic units intercalated with phyllites and black shales. They constitute massive and pillowed lavas with frequent variolitic structures, such as those in the Linkupalo Formation (see Figs. 3.19j, m, and 3.27u, x), or are composed of mafic tuffs (the Matarakoski Fm). Chemically, the lavas are Fe-rich tholeiitic basalts or basaltic andesites with LREE-enriched chondrite-normalised REE profiles (Lehtonen et al. 1998).

In the northern and northeastern part of the Peräpohja Belt, a zone of supracrustal rocks (Korkiavaara Formation) exists that is composed of alternating arkosic and amphibolitic components (Fig. 3.75b) and can be followed for more than 100 km. Ion microprobe (SIMS) dating of zircon from several samples revealed only a single zircon population with a Ludicovian age of c. 1975 Ma (Hanski et al. 2005). This together with an A-type granitic composition of the feldspar-rich, felsic rocks (Fig. 3.42b) suggests that they do not represent terrigenous sediments produced by weathering of Archaean basement rocks. Instead, the bimodal rock assemblage is interpreted as a succession of interbedded felsic and mafic tuffs. The correlation of the rock unit with other formations in the shield has not been successful as Ludicovian-age A-type felsic magmatism has not been encountered elsewhere. Nevertheless, it has potential for inter-shield correlation due to its distinctive appearance, chemical composition and probable wide original distribution as a pyroclastic formation.

The youngest Karelian dyke generation occurring throughout the Archaean basement was emplaced in Ludicovian time at c. 1.98 Ga, forming up to 70-m-wide dykes with a rather regular NW-trending strike (Vuollo et al. 1992; Vuollo and Huhma 2005). Their orientation, which is roughly parallel to the hidden Karelian craton margin, suggests that there was a rotation of c. 40–50° in the extensional palaeostress fields in the craton between c. 2.1 and 2.0 Ga. Chemically, the dykes

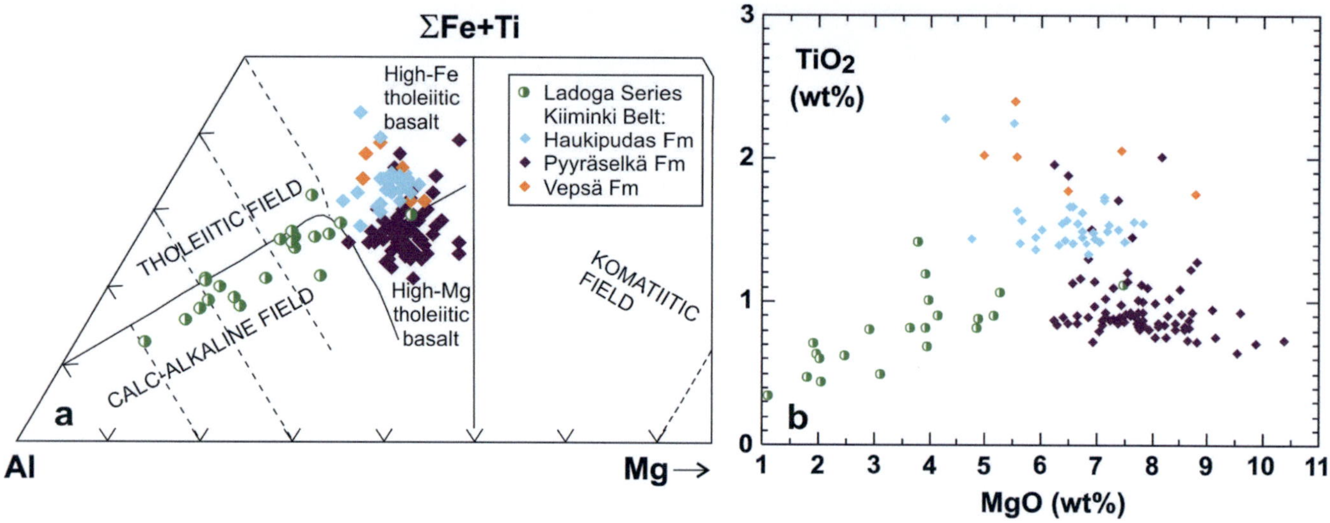

Fig. 3.76 Kalevian volcanic rocks from three formations of the Kiiminki Belt, Finland, and volcanic and subvolcanic rocks from the Ladoga Group, Lake Ladoga area, Russia, plotted on Jensen cation plot (**a**) and TiO_2 vs. MgO (**b**) diagram (Data sources: Kähkönen et al. (1986) and Honkamo (1989) from Kiiminki; Svetov and Sviridenko (1992) from Lake Ladoga)

are not easy to distinguish from older (Jatulian) Fe-tholeiitic dykes in the basement, though Vuollo et al. (1992) reported a weak island arc tholeiitic affinity for dykes in Finnish North Karelia. Few Sm-Nd isotope analyses indicate a near chondritic Nd isotopic signature (Vuollo and Huhma 2005).

Ludicovian time also witnessed restricted carbonatitic and related alkaline magmatism in the northern part of the shield. Recently, Rodionov et al. (2009) reported a SHRIMP baddeleyite age of 1999 ± 2 Ma for the Tiksheozero alkaline complex from the Belomorian Belt, northern Russian Karelia (see also Corfu et al. 2011). It likely belongs to the same magmatic event as the two small carbonatite intrusions (Kortejärvi and Laivajoki; Nykänen et al. 1997) in the SW part of the Kuusamo Belt, c. 175 km SW of Tiksheozero. The age of emplacement of these intrusions is not exactly known, but is thought to have been at c. 2.0 Ga (Vartiainen and Woolley 1974; O'Brien et al. 2005).

3.4.6 Kalevian Magmatism

Volcanic rocks are not abundant within the autochthonous-parautochthonous turbiditic sequences of the Lower Kaleva deposited at the SW margin of the rifted Karelian craton. An exception is in the Kiiminki Belt at the northern coast of the Gulf of Bothnia, where Honkamo (1987, 1989) distinguished several formations bearing mafic volcanic rocks in association with marine pelitic sedimentary rocks. The lowermost of them, the 0.5-km-thick Vepsä Formation, contains mafic lavas, tuffs and volcaniclastic deposits concentrated mainly in its lower part, but they also occur higher in the formation as interlayers with mica schists, black shales, iron formations, cherts and dolomites. The Vepsä Formation is overlain by the volcanic Pyyräselkä Formation, which reaches a thickness of c. 1 km. There are two other volcanic formations (Kiiminki and Martimojoki Fms) in the belt that are correlated with the Pyyräselkä Formation and, in the following account on geochemistry, are combined together with the latter. The Pyyräselkä Formation mostly constitutes massive and pillow lavas (Figs. 3.20f and 3.28c) with minor dolomite intercalations. The Haukipudas Formation, which is located in the western part of the belt, is probably allochthonous, but deposited in the same basin as the other rocks of the belt (Honkamo 1987). It consists of black shales, phyllites, greywackes, and mafic massive lavas and tuffs, with the volcanic rocks forming two, 100- to 300-m-thick units.

The mafic lavas of the Kiiminki Belt are tholeiitic basalts with an MgO content of 5.5–9.0 wt.% (Fig. 3.76b). The lavas of the Haukipudas and Vepsä formations are, on average, slightly more evolved than the Pyyräselkä Formation lavas, and they are also more Fe-rich (Fig. 3.76a) and elevated in TiO_2 ($\geq$1.3 wt.%) (Fig. 3.76b). The Pyyräselkä Formation lavas are Mg-tholeiites, and their TiO_2 in most cases falls below 1.0 wt.%. So far, no precise ICP-MS trace element analyses are available from the Kiiminki Belt, but as is shown in Fig. 3.76a, earlier INAA determinations of La and Sm clearly indicate that the volcanic rocks of the belt have REE characteristics similar to those of MORB (Kähkönen et al. 1986; Honkamo 1987). Most of the Pyyräselkä rocks have chondrite-normalised LREE/MREE lower than unity, whereas the ratio is chondritic or slightly higher in the Haukipudas and Vepsä formations (Fig. 3.77b). Albeit imprecise, the INAA REE spectra of the Haukipudas

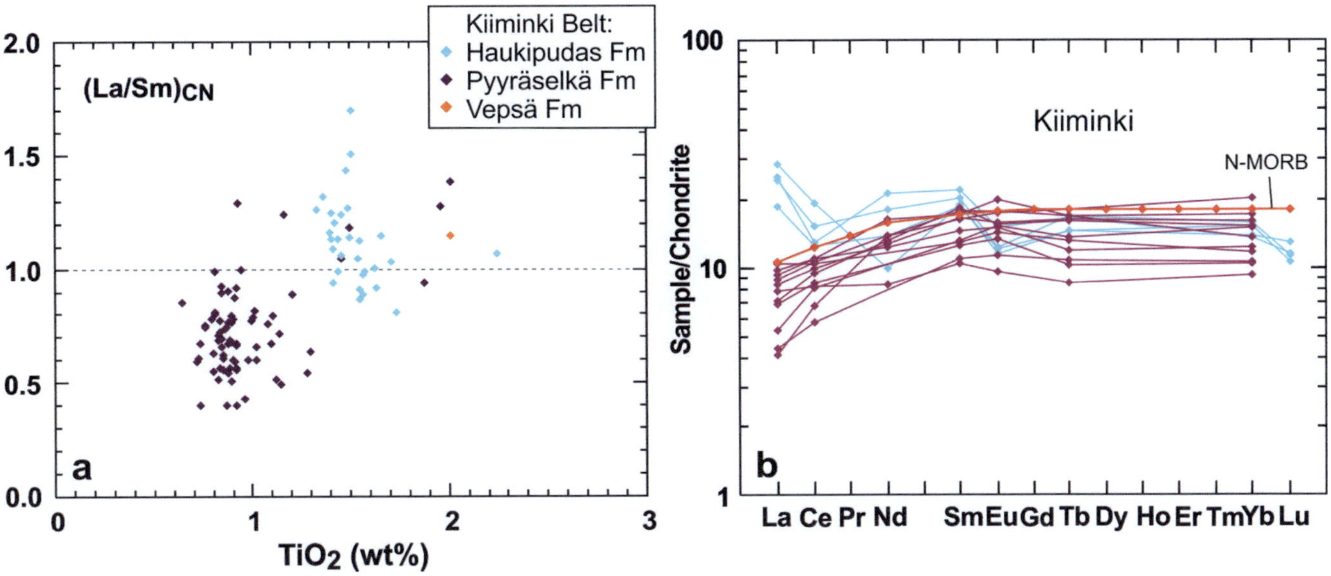

Fig. 3.77 (**a**) Kalevian mafic volcanic rocks from the Vepsä, Pyyräselkä and Haukipudas formations, Kiiminki Belt, Finland, on chondrite-normalised La/Sm vs. TiO_2 diagram. (**b**) Chondrite-normalised REE patterns of mafic volcanic rocks from the Pyyräselkä and Haukipudas formations compared to those of N-MORB of Sun and McDonough (1989). Kiiminki data comprise INAA analyses from Kähkönen et al. (1986) and Honkamo (1989)

and Pyyräselkä formations differ slightly, being flat and LREE-depleted, respectively (Fig. 3.77b). Figure 3.78a illustrates how the volcanic rocks from the Kiiminki Belt plot in the MORB field on the discrimination diagram of Wood (1980), suggesting no or negligible contamination with continental crustal material. Given the association of the Kiiminki Belt lavas with turbidites and black shales and their MORB-like immobile trace element characteristics, the rifting of the basin could have proceeded to the formation of oceanic crust and hence the lavas could represent potential candidates of Palaeoproterozoic analogues of modern MORBs. However, as in the case of the Jouttiaapa basalts, the FeO_T-TiO_2 relations differ considerably from those of modern MORBs (see Fig. 3.58a).

A 30-km-long and 4-km-wide zone of submarine basaltic lavas (see Fig. 3.19h–i) and minor tuffites are found spatially associated with Kalevian greywackes and phyllites in the Peräpohja Belt (Perttunen 1989; Perttunen and Hanski 2003). The lavas are mainly pillowed, low-Ti Mg-tholeiites with MgO between 6 and 12 wt.% and are accompanied by minor dolomites, mica schists and black schists. The currently available trace element data suggest an E-MORB-type chemical character with unfractionated or slightly LREE-enriched REE profiles. Also indications on the presence of more enriched, OIB-type magmas have been observed (Fig. 3.78b). The geological environment, lithology and geochemistry of the Väystäjä Formation lavas are consistent with them being representatives of sea-floor magmatism. However, the correlation of the Väystäjä Formation with Kalevian formations became problematic when Perttunen and Vaasjoki (2001) published a U-Pb zircon age of 2050 ± 8 Ma for a felsic porphyry cropping out in an area of $0.5 \times 2.0\ km^2$ within the Väystäjä volcanic zone. If intruding the associated mafic volcanic rocks, the porphyry would indicate a pre-Kalevian eruption time for the volcanic rocks.

Kalevian volcanic and related plutonic rocks have also been described as part of the Ladoga Group at the northern shore of Lake Ladoga, southern Russian Karelia (Svetov and Sviridenko 1992). In this area, sedimentary rocks of the Ladoga Group have been deposited on folded Ludicovian volcanic rocks of the Sortavala Group (Matrenichev et al. 2006). Volcanic rocks occur with pelitic schists and gneisses as thin (tens of metres) intercalations of mafic to intermediate, pillowed lava flows and rhyodacitic lavas and tuffs. Cross-cutting volcanic necks of gabbroic to dioritic composition are also found. Granulite-facies metamorphism locally hampers recognition of volcanic structures. As shown by Fig. 3.76a, the volcanic and related subvolcanic rocks of the Ladoga Group are chemically rather different from the Kalevian volcanic rocks from the Kiiminki Belt. The former are more fractionated, forming a series from basalts to dacites or even rhyolites, in which TiO_2 decreases with decreasing MgO (Fig. 3.32b). On the Jensen ternary diagram in Fig. 3.76a, most of the analyses published by Svetov and Sviridenko (1992) plot in the calc-alkaline field and in this respect differ markedly from other analyses of Kalevian volcanic rocks.

3.4.7 Post-Ludicovian Allochthonous Units

The Kittilä Group represents one of the largest accumulations of mafic volcanic rocks in the Fennoscandian Shield, forming an essential part of the Central Lapland Greenstone Belt (also knwon as the Kittilä Belt). Besides volcanic rocks, the group

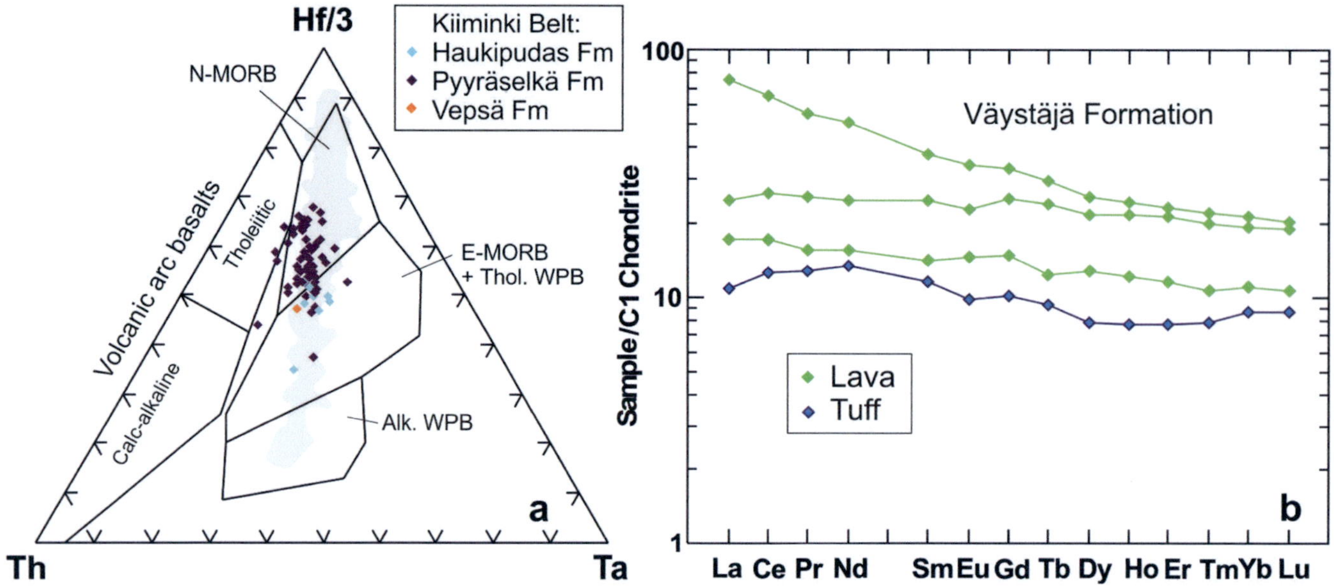

Fig. 3.78 (**a**) Rocks from three Kalevian volcanic formations of the Kiiminki Belt plotted on the discrimination diagram of Wood (1980). The figure was constructed based on INAA Th and Ta data, XRF Zr data and calculated Hf assuming primitive mantle-like Zr/Hf. The grey field represents the area where 99 % of the c. 600 recent MORB analyses of Arevalo and McDonough (2010) are located. Kiiminki data from Honkamo (1989). (**b**) Chondrite-normalised REE patterns of volcanic rocks from the Väystäjä Formation, Peräpohja Belt (Modified after Perttunen and Hanski 2003)

contains marine sedimentary rocks, such as greywackes, phyllites, graphite- and sulfide-bearing schists, banded iron formations and minor carbonate rocks (Lehtonen et al. 1998). The depositional basement of the unit has not been confidently established. This together with the unsuccessful lithostratigraphical correlation with Karelian successions on the shield and the presence of ultramafic lenses close to the eastern margin of the group, interpreted as ophiolitic mantle fragments (Hanski 1997), has provided grounds to consider the volcanic complex an allochthonous unit (Hanski and Huhma 2005). U-Pb zircon ages of mafic and felsic rocks are c. 2010 Ma (Rastas et al. 2001; Hanski and Huhma 2005), and, timewise, the complex thus belongs to the Ludicovian system.

Lehtonen et al. (1998) described several type formations, including the Vesmajärvi, Köngäs and Kautoselkä formations, which differ in their mode of occurrence and geochemistry. The most voluminous Vesmajärvi Formation comprises various types of submarine mafic volcanic rocks, including pillow lavas, pillow breccias and hyaloclastites, and cogenetic dykes (Fig. 3.27z) and gabbroic sills. Also minor felsic dykes and tuffs are encountered amidst the mafic volcanic rocks. The mafic lavas are tholeiitic basalts with a moderate TiO_2 content and their chondrite-normalised REE patterns are usually slightly LREE-enriched, but flat or slightly LREE-depleted patterns also occur. In terms of trace elements, the closest modern analogues are E-MORBs, while N-MORB types are subordinate. No indication of sialic contamination is apparent in the chemistry of these lavas or in their Nd isotope compositions. Whole-rock and pyroxene analyses from the Vesmajärvi Formation yielded a Sm-Nd isochron with an age of 1987 $\pm$ 36 Ma and an initial ε_{Nd} value of +3.8 $\pm$ 0.3 (Hanski and Huhma 2005). The Köngäs Formation, which is found in the western part of the volcanic complex, is also dominated by mafic pillow lavas (Fig. 3.27y) but is chemically more enriched, having high LREE/HREE (La_{CN}/Yb_{CN} 3–6) and other incompatible trace element characteristics typical of enriched within-plate basalts. It shows positive intial ε_{Nd} values (+2.3, +2.7) in spite of relatively low Nd/Sm (Hanski and Huhma 2005). Excluding the Kautoselkä Formation (see below), the total range of ε_{Nd} values in the Kittilä Group lavas is from +2.3 to +4.4, which is consistent with a Nd isotopic evolution similar to that of the contemporaneous model depleted mantle of DePaolo (1981). Felsic dykes and tuff, which can be concluded to be coeval with associated mafic volcanic and subvolcanic rocks based on their field relations (Lehtonen et al. 1998; Rastas et al. 2001), also show similarly positive initial ε_{Nd} (Hanski and Huhma 2005). This suggests that their genesis was not related to melting of old sialic crust but rather can be attributed to extensive fractional crystallisation of mafic magma or remelting of hydrated, newly-formed oceanic crust.

The Kautoselkä Formation mainly occurs in the SE part of the volcanic complex and comprises mafic amygdaloidal lavas, tuffs, and tuffites as the main rock types (Lehtonen et al. 1998). Chemically the lavas are evolved, Fe-rich tholeiitic basalts and andesites. They have high LREE/HREE (La_{CN}/Yb_{CN} ~10), are enriched in incompatible elements

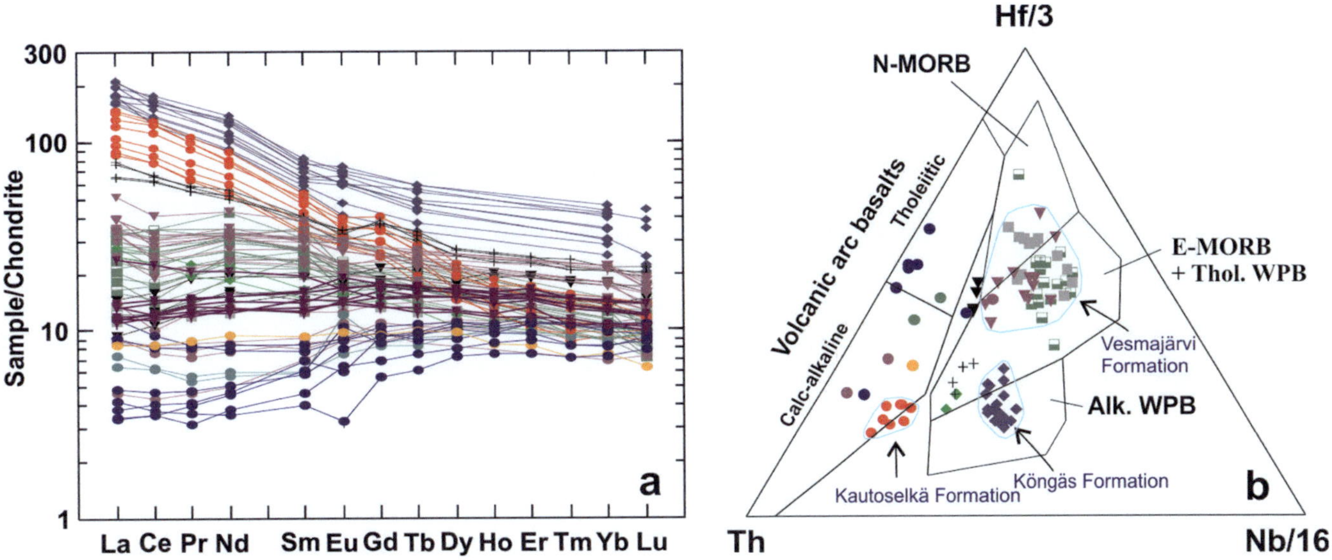

Fig. 3.79 Chondrite-normalised REE patterns (**a**) and Wood's (1980) discrimination diagram (**b**) for mafic volcanic rocks from the Ludicovian-age Kittilä Group, central Finnish Lapland (Modified after Hanski and Huhma 2005)

including TiO_2, and show an alkaline affinity in terms of high Nb/Y. However, Nb is slightly depleted compared to La (primitive mantle-normalised La/Nb = 1.3–2.2), which might indicate some crustal influence. The Nd isotopic composition is less radiogenic than in the other formations discussed above, with initial ε_{Nd} falling in the range of −1.6 to −0.8 (Hanski and Huhma 2005). As a whole, the lavas of the Kautoselkä Formation are transitional between E-MORBs and volcanic arc basalts. Also basaltic lavas, which resemble present-day island arc tholeiites (IAT), have been found in the Kittilä Group, close to the eastern margin of the volcanic complex in association with serpentinite lenses (Hanski 1997). Furthermore, dykes cutting the ophiolitic serpentinites show major and trace element characteristics of magmas occurring in island arc environments, such as IAT tholeiitic and calc-alkaline basalts, and boninites (see Fig. 4.14 in Hanski and Huhma 2005). Figure 3.79 summarises the trace element compositions of the Kittilä Group mafic volcanic rocks, highlighting their geochemical heterogeneity.

The c. 1.95 Ga Jormua Ophiolite in the Kainuu Belt, described by Kontinen (1987) and Peltonen et al. (1996, 1998), is a unique early Precambrian rock assemblage containing mantle tectonites (Fig. 3.27ae), gabbroic bodies, a sheeted dyke complex (Figs. 3.20k and 3.27ab, ac) and other (deeper) mafic dykes, massive and pillowed lavas (Figs. 3.20l and 3.27ad), hyaloclastites and plagiogranites. Tholeiitic pillow lavas and associated sheeted dykes record a MORB-like geochemistry with flat or slightly LREE-enriched or LREE-depleted chondrite-normalised REE patterns (Fig. 3.80a). This feature together with their major element geochemistry (see Fig. 3.58g) makes the Jormua tholeiites the closest analogues of modern MORBs so far found on the Fennoscandian Shield. Compositions of rocks from the Jormua sheeted dyke complex plot in the MORB field in the TiO_2 vs. FeO_T diagram (Fig. 3.58g), but some pillow lava samples are exceptionally low in FeO_T, falling below the MORB field. This is hardly an original feature of the magma. Clinopyroxenetic and hornblenditic dykes deeper in the ophiolitic section, cutting mantle tectonites, are older (c. 2.1. Ga) and show an OIB-like chemistry including a depletion in HREE (Peltonen 2005) (Fig. 3.80a). An interesting feature is that these dykes carry xenocrysts of Archaean zircon (Peltonen et al. 2003), which is consistent with the observed unradiogenic Os in the Jormua mantle rocks (Tsuru et al. 2000) and suggests that the ultramafic rocks in the Jormua Ophiolite are fragments of ancient continental lithospheric mantle.

Igneous rocks of the Cu-Zn sulphide ore-bearing, c. 1.96 Ga Outokumpu Ophiolite, which forms part of the Upper Kalevian allochthonous units in Finnish North Karelia, constitute essentially serpentinised ultramafic rocks subjected to strong sea-floor hydrothermal alteration (Peltonen et al. 2008). In contrast to the Jormua Ophiolite, the crustal igneous section is not well-preserved and volcanic rocks are absent. Neither are there volcanic rocks within the associated Upper Kalevian turbidites, which were likely deposited on the sea-floor (Lahtinen et al. 2009). Mafic rocks are represented by tholeiitic gabbroic bodies and mafic dykes within serpentinites, which in some cases show dyke-in-dyke intrusive structures indicative of an extensional setting (Peltonen 2005). Rare earth element characteristics of these mafic intrusive rocks are illustrated in Fig. 3.80a, b.

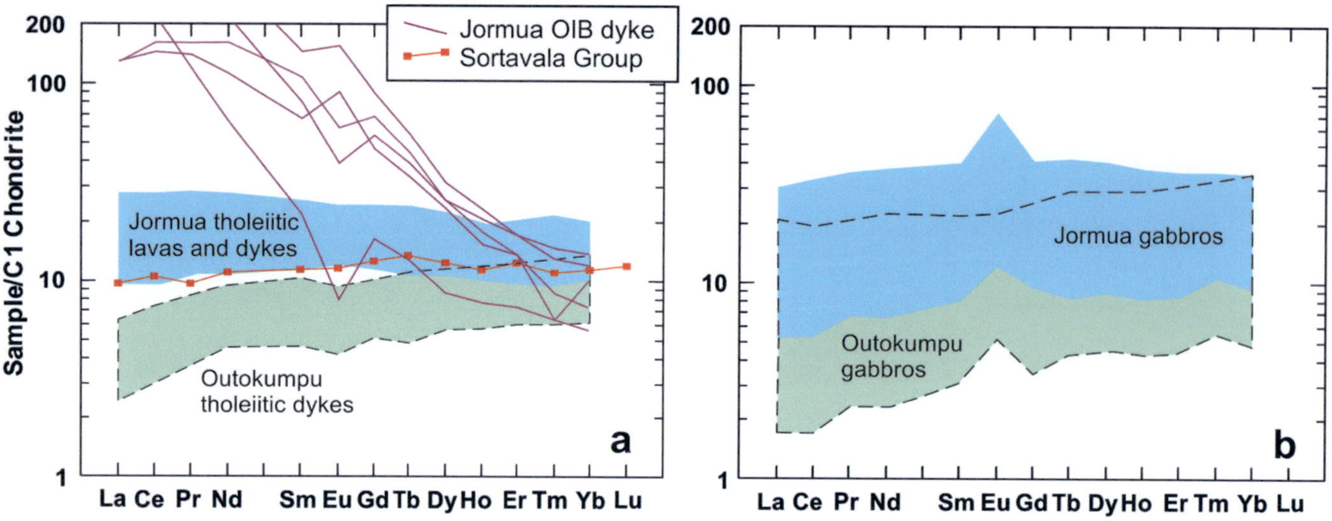

Fig. 3.80 Ranges of chondrite-normalised REE contents of lava flows, dykes and gabbros from the Jormua and Outokumpu Ophiolite and an average composition of basaltic lava from the northern Lake Ladoga area (Data from Peltonen (2005) and Ivanikov et al. (1998))

3.4.8 Orogenic Magmatism in the Kola Peninsula

A thick sequence of Palaeoproterozoic volcanic and sedimentary rocks, forming the South Pechenga Group, are found on the southwestern part of the Pechenga Belt, being separated from the North Pechenga Group by the Poritash fault zone. These rocks include calc-alkaline intermediate to felsic volcanic rocks that are thought to be related to a continent-continent collision event between the Kola and Karelian cratons and ensuing development of a subduction zone (Sharkov and Smol'kin 1997). In Chap. 4.2, a description of the South Pechenga Group is presented emphasising the lack of geochronological and stratigraphical constraints. Several formations have been distinguished with characteristic lithological and geochemical features; Skuf'in and Theart (2005) named the following: the Kallojaur, Ansemjoki (Bragino), Mennel, Kaplya and Kassesjoki formations, and considered them to have been deposited in that order. Effusive and pyroclastic volcanic rocks include a contrasting association of rhyolites, dacites, alkaline and high-Mg andesites, tholeiitic basalts, ferrobasalts and picrites, with mafic and ultramafic types having been erupted mostly subaqueously and intermediate to felsic types subaerially. The Kallojaur Formation is intruded by up to 1.2-km-thick, subvolcanic dacitic bodies of the Poritash Complex (Skuf'in 1995b).

Figures 3.81 and 3.82b present a collection of analyses from different formations of the South Pechenga Group and the Poritash Complex. Together with the Poritash subvolcanic dacites, the Kaplya Formation volcanic rocks form the most evident low-Ti, calc-alkaline trend of differentiation compatible with an island arc environment (Figs. 3.81b and 3.82b). However, the South Pechenga Group also contains volcanic rocks that suggest a non-arc-related setting. This concerns particularly the Menel Formation, which comprises moderately TiO_2-rich mafic lavas and picritic volcanic rocks with MgO reaching 20 wt.% (Fig. 3.81b). They have trace element characteristics typical of within-plate basalts (Fig. 3.81a). On the other hand, the volcanic rocks of the Ansemjoki/Bragino and Kassesjoki formations have a chemical nature that is somewhere between those of the Kaplya and Menel formations (Fig. 3.81a, b). Moreover, the Ansemjoki/Bragino Formation lavas contain some Fe-rich mafic varieties, which are not compatible with an island arc environment. Using the age of 1950 Ma, the Sm-Nd isotope analyses published by Skuf'in and Theart (2005) yield slightly positive initial ε_{Nd} values for the Menel Formation and negative values to -5.1 for the Kaplya Formation and the Poritash Complex. It seems that the South Pechenga Group is a collage of rock units generated in different geotectonic settings.

In the Imandra/Varzuga Belt, the Panarechka and Saminga formations have been regarded as the stratigraphic counterparts of the South Pechenga Group (Smol'kin 1997; Sharkov and Smol'kin 1997). They occur in the southwestern flank of the belt, forming a synform 8 km $\times$ 21 km in size (Smol'kin 1995; Skuf'in et al. 2006). The Panarechka Formation is c. 1 km thick and contains a sedimentary lower part with arkosic and greywacke sandstones and a volcanic upper part with basalts, gabbroic sills and psammitic tuffs of picritic basalt. The overlying Saminga Formation is confined to two caldera-like structures and is composed mainly of dacitic to rhyolitic lavas and rhyolitic ignimbrites. A late intrusive phase is manifested by plagiomicrocline granites and a series of subvolcanic trachydacitic bodies, for which Skuf'in et al. (2006) obtained U-Pb zircon ages of 1940 $\pm$ 5 Ma and 1907 $\pm$ 18 Ma, respectively, and interpreted them as also representing the timing of the volcanism. The former age

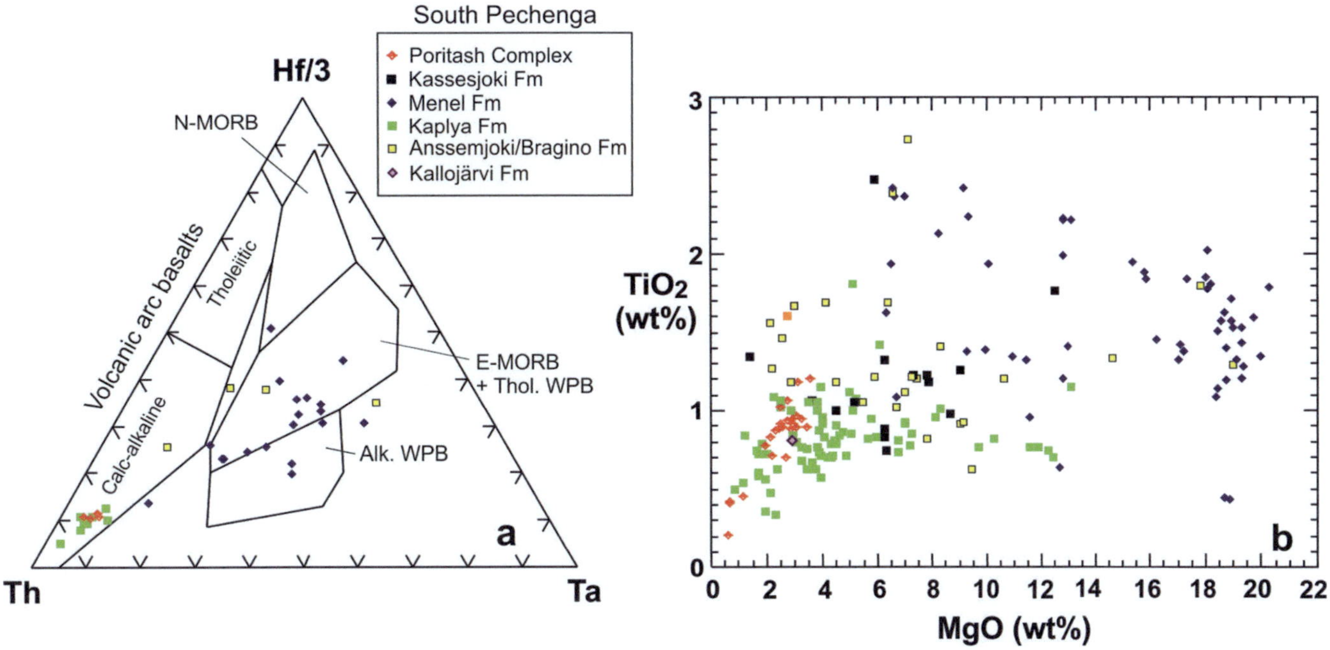

Fig. 3.81 Diagrams depicting chemical characteristics of volcanic rocks from different formations of the South Pechenga Group (Data from Predovsky et al. (1974, 1987), Skuf'in et al. (1988), Skuf'in (1995a, b), Skuf'in and Theart (2005), and P. Skuf'in (pers. comm.))

corresponds to the ages measured for felsic plutons at the southwestern margin of the South Pechenga Zone (Skuf'in et al. 2000; Vetrin et al. 2008).

The available major element analyses of rocks from the Panarechka and Saminga formations are plotted in a Jensen cation plot in Fig. 3.82a, indicating that most of the evolved volcanic rocks have a calc-alkaline affinity (Fig. 3.82a) whereas basalts show a tholeiitic affinity, though the total set of analyses forms a low-Ti, Bowen-type fractionation trend with decreasing FeO_T when MgO decreases below wt.%. Trace element data are presented in Fig. 3.72 based on analyses published by Mints (1996). The Panarechka lavas form two groups with high LREE/HREE (andesitic basalt, rhyolite) and moderately elivated LREE/HREE (andesitic basalt, trachyandesitic basalt). The Saminga Formation rocks also display LREE-enriched chondrite-normalised REE patterns (Fig. 3.72b). Excluding one sample, Skuf'in and Yelizarov (2011) obtained positive ε_{Nd}(1950 Ma) values between +1.7 and +5.0 for the Panarechka Formation basalts, which are compatible with the single analysis of rhyodacite with ε_{Nd} +2.8 at 1950 Ma reported by Daly et al. (2006) from the Panarechka Formation.

3.4.9 Nd Isotope Evolution

Although Nd isotope data are presented in the above discussion together with other geochemical data of each igneous unit, it is pertinent to summarise the available isotope data in a graphical from. This is done in Fig. 3.83 as a ε_{Nd} vs. time plot, excluding the data from Sumian mafic layered intrusions that are presented separately in Fig. 3.41. Sumian and Sariolian ultramafic to felsic volcanic rocks show unradiogenic Nd isotopic compositions with partial overlap with the compositional field of the Archaean basement. No eruptions of volcanic rocks with positive initial ε_{Nd} occurred before c. 2.1 Ga ago. Instead, several pre-2.1 Ga mafic dykes (marked as Md in Fig. 3.83) in the basement have been analysed that show initial ε_{Nd} >0, though never reaching the evolutionary line of the depleted mantle. The 2.22 Ga differentiated sills have a restricted isotopic composition indicating a close to chondritic isotopic evolution.

The first volcanic rocks registering a Nd isotopic composition indicative of depleted mantle source are Jatulian basaltic rocks as represented by the c. 2.1 Ga Jouttiaapa Fomation and the Tohmajärvi Complex (Fig. 3.83). Analogous dyke rocks in the basement have so far not been recognized; the roughly coeval Hyypiä dyke has yielded an initial ε_{Nd} of −1.7 (Huhma 1987). The c. 2.1 Ga granitoids in Central Lapland have also low initial ε_{Nd} but yet higher than shown by the contemporaneous Archaean basement granitoids. As indicated by Fig. 3.83, the 2.05–2.06 Ga magmatism was very diverse in terms of the Nd isotope composition. The most depleted compositions close to that of depleted mantle are recorded by komatiites and picrites (Jeesiörova and Peuramaa formations) in Central Lapland (Hanski et al. 2001a). The Kuetsjärvi Volcanic Formation in the Pechenga Belt displays a large variation in initial ε_{Nd} reflecting the complexity of its composition and evolution and a variable degree of interaction of the magmas with continental crust.

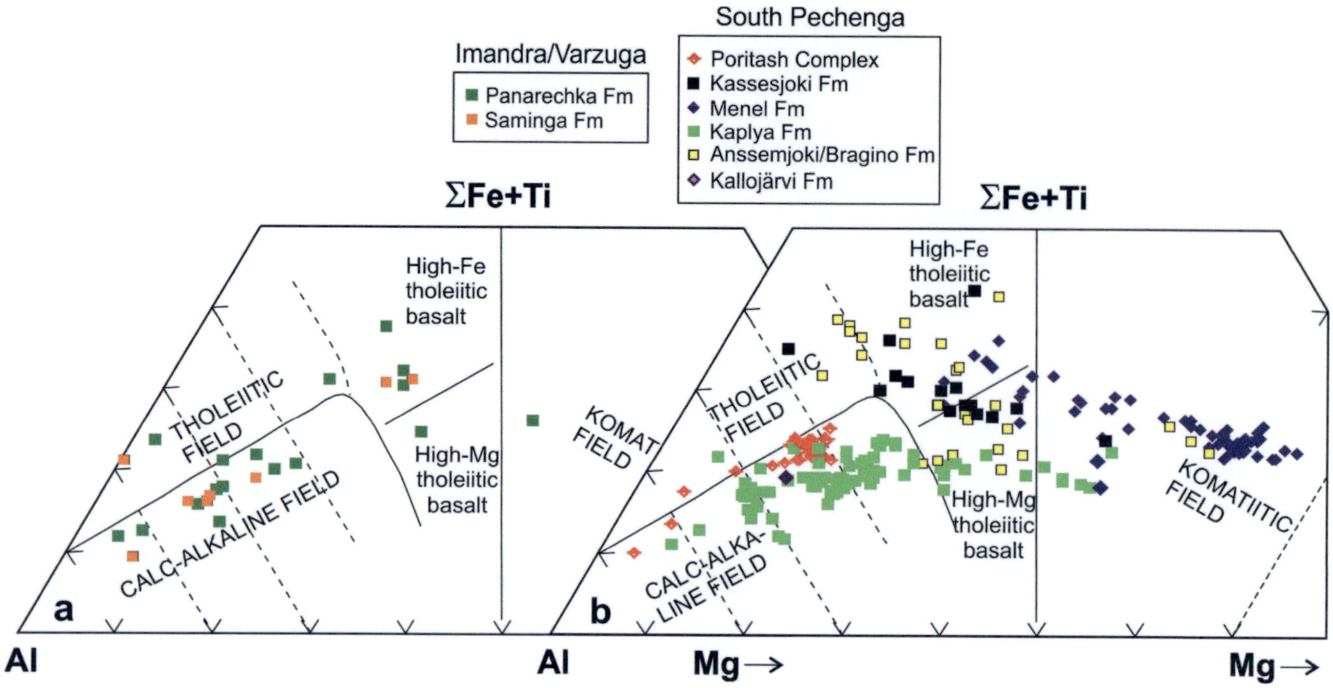

Fig. 3.82 Jensen cation plots for orogenic volcanic rocks from the Panarechka and Saminga formations, Imandra/Varzuga Belt (**a**), and different formations from the Pechenga Group (**b**) (Data sources for (**a**) Predovsky et al. (1987), Mints (1996), Skuf'in et al. (2006) and Skuf'in and Yelizarov (2011), and for (**b**) the same as in Fig. 3.81)

The isotopic composition of the coeval Koivusaari Formation in eastern Finland is also highly variable and the evolutionary lines of the Il'mozero Formation in the Imandra/Varzuga Belt also indicate an unradiogenic Nd isotope signature (Lukkarinen 2008; Skuf'in and Yelizarov 2011). The most extreme case is the Ni-Cu sulphide-bearing Keivitsa mafic-ultramafic intrusion with the initial ε_{Nd} values in some ore types decreasing to −6.6 (Hanski et al. 1997).

Of the Ludicovian volcanic rocks, the Zaonega Formation basalts (Onega Basin) and the Kautoselkä Formation basalts (Kittilä Group, Finnish Lapland) have negative initial ε_{Nd} values (Fig. 3.83). Most other Ludicovian volcanic and subvolcanic rocks have positive initial ε_{Nd} values between +2 and +4 (Fig. 3.83) reflecting a long-term light rare element depletion in their mantle source. These include submarine mafic and felsic volcanic rocks from the Vesmajärvi Formation (Kittilä Group) and basaltic lavas from the Kolosjoki and Pilgujärvi formations (Pechenga) and Suisari Formation (Onega). Only the uppermost part of the Pilgujärvi Volcanic Formation felsic tuffs in the same formation show close to chondritic or slightly negative initial ε_{Nd}, while ferropicritic rocks plot in the middle part between the chondritic and depleted mantle evolutionary lines, which is consistent with their OIB-like geochemical character (Hanski 1992).

Neodymium isotope data are currently lacking from Kalevian volcanic rocks. The Jormua ophiolite basalts are less depleted than the contemporaneous mantle, having ε_{Nd} (1950 Ma) of c. +2 (Peltonen et al. 1996). Figure 3.83 also shows examples of c. 2.1 Ga pyroxenite dyke rocks that occur in the mantle section of the Jormua ophiolite.

The diversity of the rock types in the South Pechenga Group is also reflected in their Nd isotope composition (Skuf'in and Theart 2005). The Kaplya Formation rocks have very low initial ε_{Nd} while, excluding one sample, the Menel Formation rocks have had a close to chondritic or slightly depleted Nd isotope evolution (Fig. 3.83), which is consistent with the inferred different geological settings corresponding to continental arc and within-plate environments, respectively. The Poritash dacitic complex occurring between the Northern and Southern Pechenga Groups is similar to the Kaplya Formation in terms of Nd isotopes (Skuf'in and Theart 2005). The Panarechka Formation of the Imandra/Varzuga Belt, which is regarded as correlative with the South Pechenga Group, shows a depleted Nd isotopic signature, though the initial ε_{Nd} values and the evolutionary lines are variable (Fig. 3.83; Skuf'in and Yelizarov (2011)). The youngest rocks in Fig. 3.83 are syn- and post-orogenic granitoids from northern Finland, exhibiting a strongly sloping trend of decreasing ε_{Nd} with decreasing age (Hanski and Huhma 2005).

3.4.10 Oxidation State of Iron

As already shown in Fig. 3.48, Jatulian volcanic rocks display a high degree of oxidation of iron. This question is addressed in the descriptions of the Pechenga FAR-DEEP

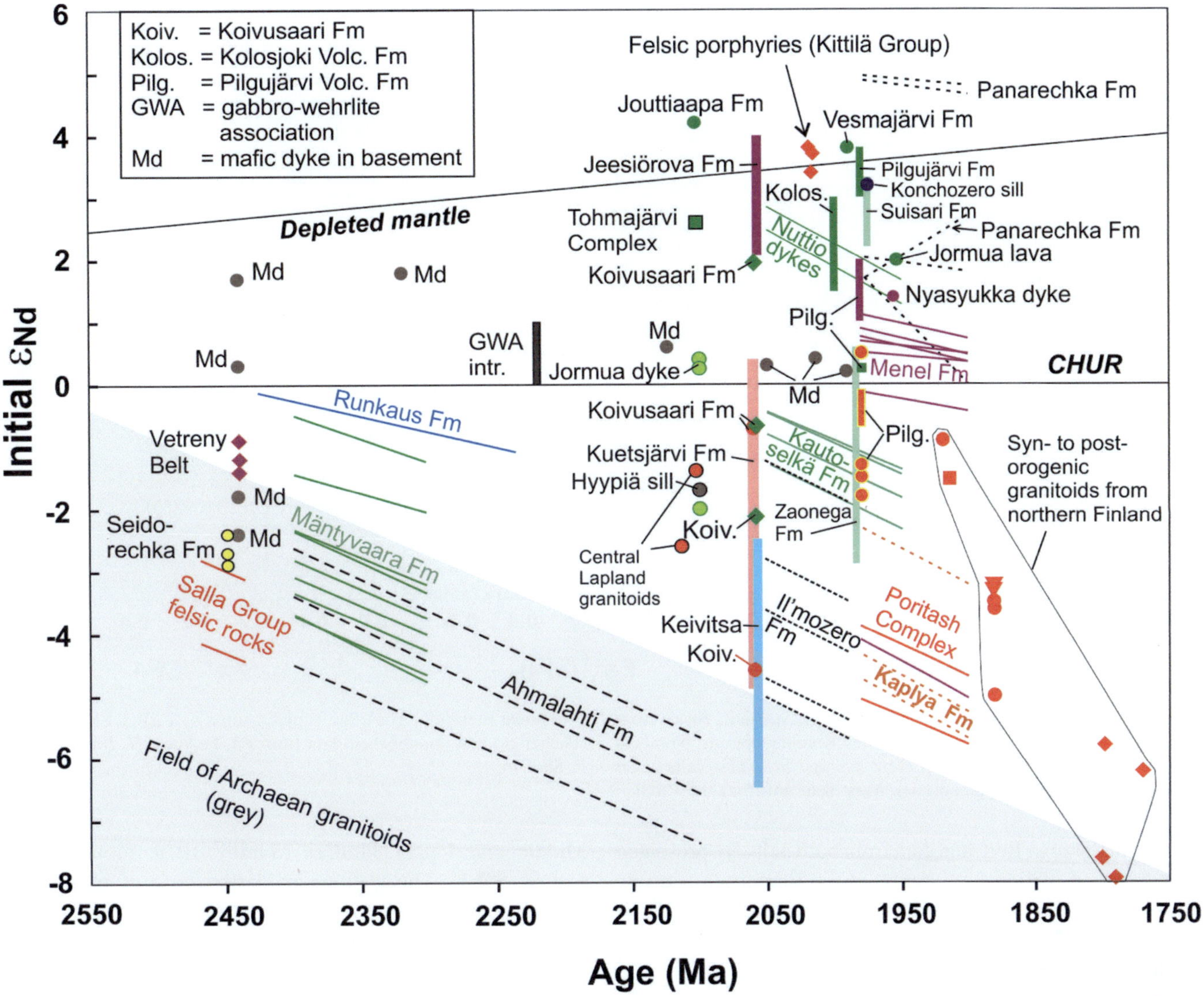

Fig. 3.83 Neodymium isotope evolution of Palaeoproterozoic igneous rocks in the eastern and northen part of the Fennoscandian Shield; note that errors in age and ε_{Nd} values ignored. Data sources: Huhma (1987), Huhma et al. (1990), Hanski (1992), Peltonen et al. (1996, 1998), Hanski et al. (1997), Puchtel et al. (1997, 1998), Hanski et al. (2001a), Hanski and Huhma (2005), Vuollo and Huhma (2005), Skuf'in and Theart (2005), Skuf'in and Bayanova (2006), Chashchin et al. (2008), Hanski et al. (2010), Skuf'in and Yelizarov (2011), and Huhma, Hanski and Smol'kin (unpubl.). Depleted mantle model evolution after DePaolo et al. (1981). For Nd isotopic composition of Sumian plutonic rocks, see Fig. 3.41. Initial values from single measurements or isochrones are shown as equant symbols and ranges of initial values as vertical bars. For samples with an unknown age, evolutionary lines are drawn covering the potential age range

holes (Chaps. 6.2.3 and 6.2.4) and discussed specifically in Chap. 7.4. In this connection, the average current oxidation state of iron in volcanic rocks from the main stratigraphic units is compared. Thousands of samples of Precambrian volcanic rocks have been analysed by wet chemical methods from the Russian part of the Fennoscandian Shield. These analyses include routine determinations of FeO and Fe_2O_3, allowing the present-day redox state of the rocks to be evaluated. Instead, almost all rock analyses from the western part of the shield have been performed by instrumental analytical methods and the redox speciation of iron has mostly remained undetermined. For this reason, the oxidation state of iron in volcanic rocks from different stages of Palaeoproterozoic magmatic evolution will be discussed solely based on analytical data from the Russian part of the shield. These data include individual analyses and calculated average compositions gathered from the literature and determined within the FAR-DEEP project (Pechenga) supplemented by unpublished data provided kindly by Zh. Fedotov, V. Smol'kin and P. Skuf'in.

In Fig. 3.84, the average oxidation state of iron in terms of $Fe^{3+}/\Sigma Fe$ is presented separately for volcanic rocks from the Kola and Karelian provinces. The Kola rocks are further divided into two parts, those from the Pechenga Belt and

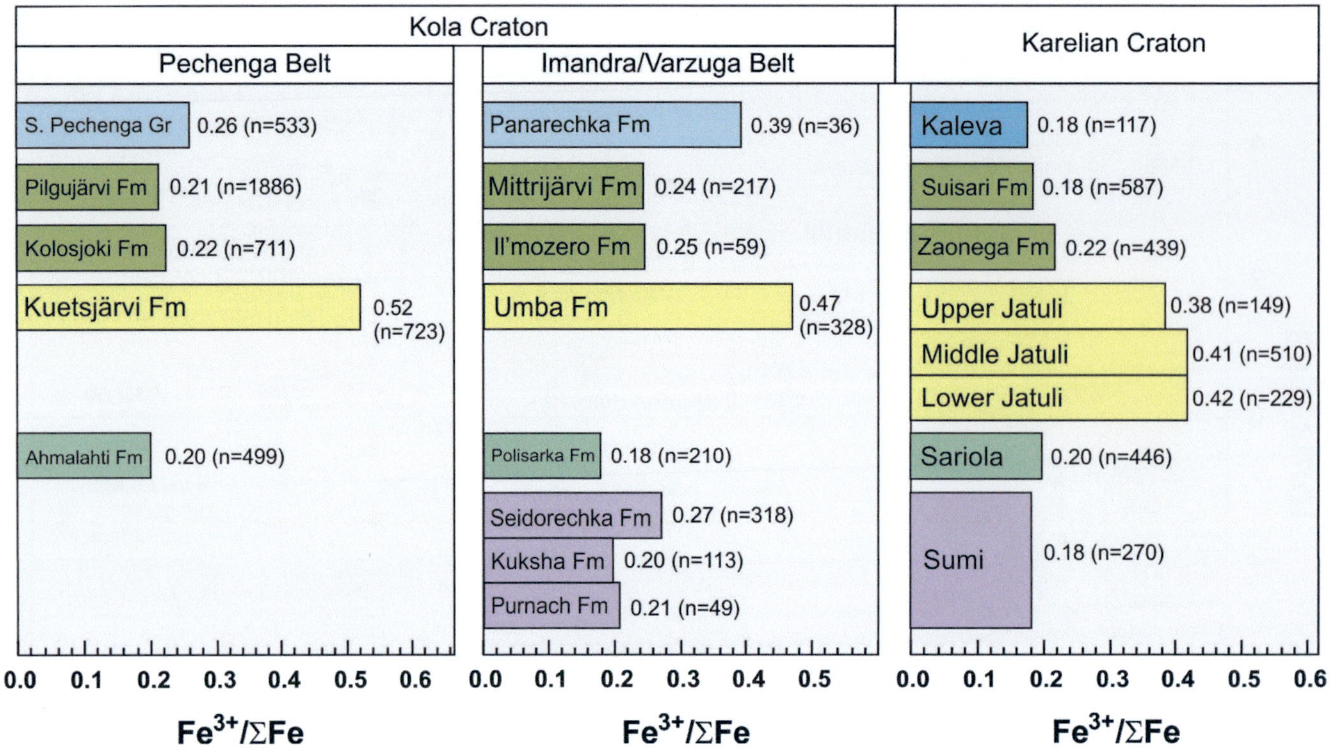

Fig. 3.84 Average oxidation state of iron in volcanic rocks from different stratigraphic units in the Kola and Karelian cratons. Numbers on the right side of the bars represent average $Fe^{3+}/\Sigma Fe$ values and number of analyses (in parenthesis). Most data are from the Russian geological literature. These are supplemented by FAR-DEEP analyses (Pechenga) and unpublished data from Zh. Fedotov, V. Smol'kin and P. Skuf'in

Imandra/Varzuga Belt, and data from both belts are presented as a function of lithostratigraphy at the formation or group level (see Chaps. 4.1 and 4.2). The analytical data from the Karelian province are presented as grouped into general traditional stratigraphic units Sumi, Sariola, Jatuli and Kaleva, with the exception of the Ludicovian rocks, which are divided further into the Zaonega and Suisari formations. The analysed volcanic rocks of these two formations are mainly from the Onega region and the Kalevian rocks from the Lake Ladoga region; the other rocks are more widely spread within the Karelian craton.

The most significant feature in Fig. 3.84 is the high oxidation state of iron in volcanic rocks belonging to the Jatulian system compared to the older or younger rocks. This is true for all three areas shown in Fig. 3.84, but is most strikingly developed in the Pechenga Belt where the 2.06 Ga Kuetsjärvi Volcanic Formation records an average $Fe^{3+}/\Sigma Fe$ ratio of 0.52 (see Chap. 8); the average ratio in other rock units in the Pechenga Belt is less than or equal to 0.25. The Umba Volcanic Formation in the Imandra/Varzuga Belt is also generally highly oxidised (averaging 0.47). Based on their stratigraphic positions, the Jatulian volcanic rocks in the Karelian Block are traditionally divided into Lower, Middle and Upper Jatulian (Svetov 1979; Golubev and Svetov 1983). The average oxidation state of iron is not markedly different at these different levels, and the average $Fe^{3+}/\Sigma Fe$ ratio of all available analyses from those rocks can be calculated to be 0.42 ($n = 1173$), also abnormally high. However, there are significant differences between different areas. For example, the middle Jatulian basaltic lavas in the Hirvas area are highly oxidised with $Fe^{3+}/\Sigma Fe$ averaging 0.78 (Golubev et al. 1984), whereas in the Lekhta or Sapovaara areas, the proportion of ferric iron of total iron is less than 0.30 (Golubev and Svetov 1983). Svetov (1980) commented that pillow lavas are less oxidised than subaerially erupted lavas and hence, the regionally observed differences in the oxidation state of iron is partly explicable by variable types of eruptional setting.

Apart from the Jatulian volcanic units, the other units in Fig. 3.84 mostly display average $Fe^{3+}/\Sigma Fe$ ratios between 0.17 and 0.25. Two units, the Seidorechka Volcanic Formation and the Panarechka Formation from the Imandra/Varzuga Belt, are more oxidised, having average $Fe^{3+}/\Sigma Fe$ values of 0.27 and 0.39, respectively. One reason for this could be the dominance of felsic rocks in these units, as felsic rocks are generally more oxidised than mafic and intermediate ones. The change in the presently observed oxidation state of iron between the Jatulian and pre-Jatulian rocks is generally abrupt. The Sariolian volcanic rocks are typically relatively reduced with the exception of samples from weathering crust that has developed on Sariolian volcanic rocks during the early Jatulian time and may show $Fe^{3+}/\Sigma Fe$ values higher than 0.70 (Heiskanen et al. 1987).

3.4.11 Discussion

The summary that is presented above reveals a great diversity of the Karelian magmatism. It varies in chemical composition from ultramafic to silicic and is manifested in different forms ranging from subaerial or submarine volcanism to deep- to shallow-level plutonism. It seems that each belt has its own lithological and geochemical characteristics and it is not always easy to find counterparts for a stratigraphic unit elsewhere within the same geochronological system. There are many features that prevent a straightforward correlation of the magmatic events between different belts. It was already mentioned earlier that the Jatulian successions differ in many ways between the Karelian and Kola cratons. Of the inconsistences, some examples are listed below.

- When the whole geochemical data sets are considered from different regions, as is done in Fig. 3.31, a great similarity is observed between the Sumian and Sariolian lavas when high-Mg lavas are concerned. They form a differentiation series leading to low-Ti basalts. However, the rocks in these high-MgO series belong mostly to the Sumi in the Russian part of the Karelian craton and the Sariola in Finland and the Kola Peninsula. Given the geochemical similarity, it is tempting to suggest that they are all roughly coeval, but this can hardly be proved without reliable age determinations.
- Among the Sariolian volcanic rocks, the Polisarka Formation in the Imandra/Varzuga Belt contains abundant basaltic komatiites that were erupted subaqueously, whereas the Ahmalahti Formation lavas in the Pechenga Belt are evolved basalts and andesites that were erupted subaerially. This raises a question about the age of the Ahmalahti Formation; the only thing we know now is that it is younger than the c. 2.5 Ga Mt. General'skaya intrusion and younger than the c. 2.06 Ga Kuetsjärvi Volcanic Formation.
- The same question can be asked when discussing the age of the lowermost mafic lavas (the Sariolian Runkaus Formation) in the Peräpohja Belt. As is evident from Fig. 3.46a, they are chemically different from the lowermost volcanic rocks of the Kainuu and Kuusamo Belts. It is known that the Runkaus Formation is older than 2.22 Ga, hence belonging to the lower Jatuli cannot be ruled out. Apart from the imprecise Sm-Nd isotope data, there is no geochronological evidence that would be in conflict with this interpretation, which was also favoured by Perttunen (1980, 1985).
- Using in-situ SHRIMP U-Pb analysis on zircon, Puchtel et al. (1998) made an effort to date a mafic volcanic unit (Fig. 3.75a) at Hirvas, Russian Karelia, which has been regarded as a typical representative of subaerially erupted Upper Jatulian lavas (Svetov 1979). The age determination was performed from a coarse-grained, leucocratic lens in a sheet-like body with an amygdaloidal upper part and massive lower part. The obtained age, 1976 ± 9 Ma, was surprisingly young, Ludicovian, and thus not in concert with what are thought to be the age of the Upper Jatulian volcanic rocks. Puchtel et al. (1998) discussed possible solutions to this dilemma, including the possibility that the dated body is a hypabyssal intrusion belonging to the Ludicovian magmatism, even though it has a geochemical correspondence with the Jatulian lavas of the area, and emphasised the importance of further geochronological studies of Jatulian igneous rocks. An opposite problem has appeared in studies of the Tohmajärvi Mafic Complex, which is associated with a Ludicovian-type of environment with black shales and dolomites. Dating of this complex has resulted in a Jatulian U-Pb zircon age of 2105 ±15 Ma (Huhma 1986).
- The Ludicovian primitive magmas differ chemically and partly also in age in their three main areas of occurrence. Those found in the Onega region are picrites and picritic basalts, while the Central Lapland rocks form a bimodal series of Ti-enriched komatiites and picrites and the Pechenga ferropicrites are unique in having higher FeO_T than in the other types. A common feature of all these magmas is that they were generated from a depleted mantle source. The most obvious correlation can be made between the ferropicrites of the Pechenga and Imandra/Varzuga Belt. It is worth mentioning that the Ludicovian (Zaonegian) komatiitic rocks discovered in the Sortavala area seem to have no analogues among other Ludicovian primitive magmas.
- The existing geological literature does not give a clear picture of the development of the environment of eruptions of mafic lavas, though submarine environments are more prevailing during the formation of the latest, Ludicovian successions. Highly amygdaloidal mafic lavas erupted on dry land form thick units, particularly within Sariolian and Jatulian sections, but pillow lava-bearing formations are found in every geochronological system from the Sumi to Ludocovi. They include the Sumian Kuksha Volcanic Formation (Fig. 4.4d), Arvarench Unit (Vrevskii et al. 2010), Vetreny Suite (Puchtel et al. 1997) and Sumian lavas in the northern part of the Onega basin (Svetov et al. 2004). Sariolian pillow lavas are found in the Polisarka Volcanic Formation (Predovsky et al. 1987) and the Mäntyvaara Formation (Manninen 1991). In Russian

Karelia, pillow lavas are present among the Lower, Middle and Upper Jatulian mafic volcanic rocks (Svetov 1979; Sokolov 1980). In Finland, like in the Kuusamo Belt (Silvennoinen 1972), the start of volcanism with subaerial (Sariolian) lavas can be followed by a Jatulian pillow lava-bearing unit before the appearance of a flood basalt-like unit higher in the stratigraphy. The overall impression is that there was considerable relative sea level fluctuation before Ludicovian time when a more permanent highstand of the sea level developed.

- Initial ε_{Nd} values of magmas show a general tendency to change from negative to positive at the Jatulian-Ludicovian transition (Fig. 3.83), which is, at least partly, related to the attenuation of the lithosphere due to prolonged tensional stresses and resultant decreasing degree of crustal contamination. However, there are some exceptions such as the Jatulian Jouttiaapa Formation with depleted mantle-like Nd isotopic characteristics with no indications of interaction with continental crust. On the other hand, the Ludicovian Il'mozero Volcanic Formation shows very low initial ε_{Nd} values, which are not observed elsewhere in volcanic rocks associated with Ludicovian black shales, though there is some overlap with the Zaonega Formation (Fig. 3.83).

To solve the existing correlation problems, the significance of reliable age determinations cannot be overestimated. Dating of Sumian intermediate to felsic volcanic rocks has been partly successful, but Sariolian volcanic rocks have so far not been dated by the U-Pb method. Also U-Pd-dated Jatulian volcanic rocks are very few, being limited to two c. 2.06 Ga felsic volcanic rock units (Pechenga Belt, Siilinjärvi Belt) and one c. 2.1 Ga mafic tuff (Peräpohja Belt). The age of the Ludicovian volcanic rocks is known reasonably well in some areas (Pechenga) but, in other areas (e.g. Onega), is based on less precise methods (e.g. Sm-Nd) and still needs to be confirmed by the U-Pb technique. Much of our current understanding on the timing of the volcanic events relies on geochronological data obtained from cross-cutting intrusive rocks (e.g. Vaasjoki 2001). Although they are usually more amenable for dating than volcanic rocks, problems may appear when more than one generation of zircon (or baddelyite) is encountered within an intrusive body. This concerns at least the Sumian layered mafic-ultramafic complexes (for more details, see Chap. 3.2).

The relationship between the Sumian and Sariolian volcanic rocks has long been a controversial issue among geologists studying them in Russian Karelia (Sokolov et al. 1987). For example, Heiskanen et al. (1977) assigned pre-Jatulian basaltic and andesitic volcanic rocks to the Sariolian andesite-basalt complex and felsic volcanic rocks to the Sumian dacite-liparite complex. Apparently, the same mafic to intermediate volcanic rocks from the Kumsy area, for example, which Heiskanen et al. (1977) regarded as Sariolian, Svetov et al. (2004) considered Sumian (see above). One way to avoid this kind of discrepancy is to group the volcanic rocks together and call them Sumi-Sariolian (Kulikov et al. 1978; Sokolov et al. 1987). However, this is not a desirable solution because, as advocated in Chap. 3.2, there is an extensive erosional period between the deposition of the Sumi and Sariola system rocks that allowed the c. 2.4–2.5 Ga layered intrusions to become uplifted and partly eroded. In the absence of reliable age determinations or field evidence on the relationship between the volcanic rocks and Sumian intrusive rocks, the confusion is understandable. This is also why potential Sariolian volcanic rocks in Russian Karelia are not considered in detail here. It should also be mentioned that, due to the different opinions in the literature, there are two alternative ways to plot some rock compositions from Russian Karelia in the summary diagrams of Fig. 3.31 (red dots in Fig. 3.31g, h).

Due to the preponderance of andesitic compositions, Sumian-Sariolian volcanic rocks were previously regarded as resulting from orogenic volcanism (Heiskanen et al. 1977). Later the general view has been that the Sumian magmatism is related to initial rifting of continental crust after late Archaean cratonisation, potentially associated with a mantle plume acitivity (Sharkov 1984; Amelin et al. 1995; Puchtel et al. 1997; see also Chap. 3.1). The orogenic hypothesis has later been re-adopted by Mints (1996) and Svetov et al. (2004) based on the chemical similarity of the Sumian rocks and Andean basaltic andesites. Svetov et al. (2009) found a match between Sumian high-Mg andesites and bajaite series rocks from Mexico and favoured a convergent geotectonic setting. On the other hand, Vrevskii et al. (2010) compared Sumian volcanic rocks from central and eastern Russian Karelia with island arc volcanic rocks and observed that the former are different in being lower in Al_2O_3 at the same SiO_2. As noted above, the boninite-like characteristics of the komatiitic rocks can be explained by crustal contamination, and similar incompatible element characteristics of more evolved basalts and andesites testify that they were derived via fractional crystallisation of a contaminated komatiitic magma in a deep magma chamber in continental crust.

A glance at Fig. 3.31 reveals that, excluding the "orogenic" South Pechenga Group, primitive magmas were erupted broadly in two stages, during the Sumi-Sariolian and Ludicovian time, representing different geotectonic stages of the evolution of the shield. In Fig. 3.34, these magmas are compared with Phanerozoic primitive magmas ranging from Gorgona Island komatiites to picrites from Iceland, Greenland and Hawaii. The latter form a more or less linear trend of increasing [TiO_2] with decreasing [Al_2O_3]. It seems that, in general, Palaeoproterozoic primitive magmas from the

Fennoscandian Shield plot to the left of this trend, being lower in Al_2O_3 than Phanerozoic primitive magmas, though the scatter in the composition of individual magma types can be large, likely due to alteration and analytical artifacts. The same phenomenon, different levels of Al_2O_3 between Mesoarchaean and Neoarchaean komatiites, has been known for decades and has been interpreted as resulting from different pressures of melt segregation, primarily due to the differences in the thermal state of the mantle in different times (e.g. Herzberg and O'Hara 1998). Following this reasoning, one could argue from the distribution of the analyses in Fig. 3.34 that, until c. 2.0 Ga, the cooling of the mantle temperatures had not yet proceeded to the level that the mantle had during Phanerozoic time.

Puchtel et al. (1998) proposed that the generation of the Ludicovian primitive magmas, including picrites in the Onega Plateau, ferropicrites in the Pechenga Belt and komatiites in central Finnish Lapland, was related to a single, 1.98 Ga mantle plume having a head diameter of c. 2,000 km. However, as noted previously, the Central Lapland komatiites seem to be older than the c. 1.98 Ga Onega picrites, having an age of c. 2.05 Ga, and the Pechenga ferropicrites are located on a separate craton (Kola), which collided with the Karelian craton after the formation of the Ludicovian rocks (e.g., Lahtinen et al. 2005; Daly et al. 2006). Hence, without reliable palaeomagnetic information, the palaeogeographical relationships between the Onega Plateau and Pechenga Belt at 1.98 Ga are presently unknown. Thus the connection of the three primitive magmatic events to the same plume event may not be valid. Another problem with the plume hypothesis, which is shared by all Ludicovian-age primitive magmas and was discussed by Hanski (1992), Puchtel et al. (1998) and Hanski et al. (2001a), is that these magmas were erupted subaqueously and were temporally and spatially associated with deposition of black shales and other pelitic sediments indicative of deepening of the respective depositional basins. In all these cases, there is no definitive evidence for oceanic crust formation; instead, the geological record indicates that the magmas ascended through continental lithosphere, though likely an attenuated one. This phenomenon, eruption of mafic-ultramafic magmas onto submerged continental crust, is in conflict with the common assumption of a large-scale crustal uplift less than 1–20 Ma prior to the onset of mantle plume head-generated magmatism (e.g. Campbell 2005). Field evidence for such uplifting has been presented from Phanerozoic flood basalt provinces (e.g. Saunders et al. 2007). The same problem also applies to the pillow lava-bearing Vetreny Suite as well as older Archaean greenstone belts worldwide where the generation of submarine komatiitic lavas has been related to a mantle plume. Arndt (1999) addressed this problem and discussed potential reasons why sea levels were high during periods of Precambrian flood basalt volcanism. One of these could be that large parts of ocean basins were occupied by oceanic plateaux or reactivated ocean ridges flooding displaced ocean water to low-lying continents. Recently, Leng and Zhong (2010) proposed another explanation. They carried out numerical modelling showing that if a mantle plume head is temporarily ponding below the 660-km phase change boundary and undergoes a phase transformation from ringwoodite to perovskite, this will cause significant subsidence at the Earth's surface tens of millions of years before the volcanism and further subsidence later due to volcanic loading.

References

Ahtonen N, Hölttä P, Huhma H (2007) Intracratonic Palaeoproterozoic granitoids in northern Finland: prolonged and episodic crustal melting events revealed by Nd isotopes and U-Pb ages on zircon. Bull Geol Soc Finl 79:143–174

Alapieti TT, Lahtinen JJ (2002) Platinum-group element mineralization in layered intrusions of northern Finland and the Kola Peninsula, Russia, In: Cabri LJ (ed) The geology, geochemistry, mineralogy and mineral beneficiation of Platinum-Group Elements, Can Inst Min, Metall Petrol, Spec Vol 54: 507–546

Alapieti T, Piirainen T (1984) Cu-Ni-PGE mineralization in the marginal series of the early Proterozoic Koillismaa layered igneous complex. In: Buchanan DL, Jones MJ (eds) Sulphide deposits in Mafic and ultramafic rocks. The Institution of Mining and Metallurgy, London, pp 123–131

Alapieti TT, Kujanpää J, Lahtinen J, Papunen H (1989) The Kemi stratiform chromitite deposit, northern Finland. Econ Geol 84:1057–1077

Alapieti T, Filén B, Lahtinen J, Lavrov M, Smol'kin V, Voitekhovsky S (1990) Early Proterozoic layered intrusions in the northeastern part of the Fennoscandian Shield. Mineral Petrol 42:1–22

Amelin YuV, Semenov VS (1996) Nd and Sr isotopic geochemistry of mafic layered intrusions in the eastern Baltic shield: implications for the evolution of Paleoproterozoic continental mafic magmas. Contrib Mineral Petrol 124:255–272

Amelin YuV, Heaman LM, Semenov VS (1995) U-Pb geochronology of layered mafic intrusions in the eastern Baltic Shield: implications for the timing and duration of Palaeoproterozoic continental rifting. Precambrian Res 75:31–46

Andersen JCØ, Thalhammer OAR, Prevec S (1998) The Mount General'skaya layered intrusion: evidence for an enriched mantle reservoir beneath the Pechenga area, Kola Peninsula, Russia, In: 8th international platinum symposium, extended abstracts, Geological Society of South Africa and South African Institute of Mining and Metallurgy Symposium Series S18: 13–15

Arevalo R Jr, McDonough WF (2010) Chemical variations and regional diversity observed in MORB. Chem Geol 271:70–85

Arndt N (1999) Why was flood volcanism on submerged continental platforms so common in the Precambrian? Precambrian Res 97:155–164

Balashov YuA, Bayanova TB, Mitrofanov FP (1993) Isotope data on the age and genesis of layered basic-ultrabasic intrusions in the Kola Peninsula and northern Karelia, northeastern Baltic Shield. Precambrian Res 64:197–205

Barnes S-J, Often M (1990) Ti-rich komatiites from northern Norway. Contrib Mineral Petrol 105:42–54

Beccaluva L, Serri G (1988) Boninitic and low-Ti subduction-related lavas from intraoceanic arc-back arc systems and low-Ti ophiolites: a reappraisal of their petrogenesis and original tectonic setting. Tectonophysics 146:291–315

Bergh SG, Torske T (1988) Palaeovolcanology and tectonic setting of a Proterozoic metatholeiitic sequence near the Baltic Shield margin Northern Norway. Precambrian Res 39:227–246

Bergh SG, Kullerud K, Corfu F, Armitage PEB, Davidsen B, Johansen HW, Pettersen T, Knudsen S (2007) Low-grade sedimentary rocks on Vanna, North Norway: a new occurrence of a Palaeoproterozoic (2.4–2.2 Ga) cover succession in northern Fennoscandia. Norwegian J Geol 87:301–318

Bleeker W, Ernst R (2006) Short-lived mantle generated magmatic events and their dyke swarms: the key unlocking Earth's Palaeoproterozoic record back to 2.6 Ga. In: Hanski E, Mertanen S, Rämö T, Vuollo J (eds) Dyke Swarms – time markers of crustal evolution. Taylor & Francis, London, pp 3–26

Borisov AE (1990) Volcanism and native copper mineralisation in the Palaeoproterozoic of the Kola Peninsula. Kola Science Centre, Apatity, p 70 (in Russian)

Borisov AE, Smol'kin VF (1992) On the genesis of high-Si rocks of the forth volcanogenic unit of the Pechenga structure. Proc USSR Acad Sci, Geol Ser 7:66–78 (in Russian)

Buiko AK, Levchenkov OA, Turchenko SI, Drubetskoi YeE (1995) Geology and isotopic dating of the early Proterozoic Sumian-Sariolan Complex in northern Karelia (Paanajärvi-Tsipringa structure). Stratigr Geol Corr 3(4):16–30

Campbell IH (2005) Large igneous provinces and the mantle plume hypothesis. Elements 1:265–269

Chashchin VV, Bayanova TB, Levkovich NV (2008) Volcanoplutonic association of the early-stage evolution of the Imandra–Varzuga rift zone, Kola Peninsula, Russia: geological, petrogeochemical, and isotope-geochronological data. Petrology 16(3):279–298

Chistyakov AV, Sharkov EV (2008) Petrology of the Early Proterozoic Burakovka Complex, southern Karelia. Petrology 16(1):66–91 (in Russian)

Condie KC, O'Neill C, Aster RC (2009) Evidence and implications for a widespread magmatic shutdown for 250 My on Earth. Earth Planet Sci Lett 282:294–298

Corfu F, Bayanova TB, Shchiptsov VV, Frantz N (2011) U-Pb ID-TIMS age of the Tiksheozero carbonatite: expression of 2.0 Ga alkaline magmatism in Karelia, Russia. Cent Eur J Geosci 3:302–308

Daly JS, Balagansky VV, Timmerman MJ, Whitehouse MJ (2006) The Lapland-Kola orogen: Palaeoproterozoic collision and accretion of the northern Fennoscandian lithosphere. In: Gee DG, Stephenson RA (eds) Eur Lithos Dynam, Geol Soc Memoir 32: 579–598

DePaolo DJ (1981) Nd in the Colorado Front Range and implications for crust formation and mantle evolution in the Proterozoic. Nature 291:193–196

Egorova V, Latypov RM (2012) Processes operating during the initial stage of magma chamber evolution: insights from a marginal reversal of the Imandra Layered Intrusion, Russia. J Petrol 53:3–26

Epifanova TA, Kazanov OV, Savatenkov VM (2005) The source material of the microgabbronorites in the layered Lukkulaisvaara massif. In: Mitrofanov FP, Pripachkin VA, Voitekhovsky YuL, Neradovsky YuN (eds) Geology and ecogeology, investigations of young scientists, Proceedings of the 16th conference of young scientists. Institute of Geology, Kola Science Centre, Russian Academy of Sciences, Apatity, pp 32–35 (in Russian)

Epifanova TA, Kazanov OV, Savatenkov VM (2008) Neodymium isotopic composition of the ore-bearing interval in the section of the Northern Pana layered intrusion. In: Mitrofanov FP, Bayanova TB, Voitekhovsky YuL, Pripachkin VA (eds) Geology and ecogeology, investigations of young scientists, Proceedings of the 19th conference of young scientists. Institute of Geology, Kola Science Centre, Russian Academy of Sciences, Apatity, pp 21–22 (in Russian)

Fedotov ZhA (1983a) Volcanic rocks of the komatiite series in the Proterozoic Imandra-Varzuga sedimentary-volcanic complex. In: Predovsky AA, Bolotov VI, Melezhik VA, Gavrilenko BV, Fedotov ZhA (eds) Sedimentary Basins and volcanic zones in the Precambrian of the Kola Peninsula. Nauka (Science), Apatity, pp 81–99 (in Russian)

Fedotov ZhA (1983b) On the acidic volcanism ending the first Strel'na stage of the evolution of the Imandra-Varzuga synclinorium zone (in Russian). In: Predovsky AA, Bolotov VI, Melezhik VA, Gavrilenko BV, Fedotov ZhA (eds) Sedimentary Basins and volcanic zones in the Precambrian of the Kola Peninsula. Nauka (Science), Apatity, pp 99–107 (in Russian)

Fedotov ZhA (1985) Evolution of Precambrian volcanism of the Eastern part of the Pechenga/Varzuga Belt. Kola Science Centre, Apatity, p 105 (in Russian)

Fedotov ZhA (1995) Correlation of dike complexes outside the margin of the Pechenga area and its volcanic sequences. In: Mitrofanov FP, Smolkin VF (eds) Magmatism, sedimentogenesis and geodynamics of the Pechenga Paleorift. Russian Academy of Sciences, Kola Science Centre, Apatity, pp 82–94 (in Russian)

Fedotov ZhA, Basalaev AA, Melezhik VA, Latyshev LN (1982) The Srel'na group. In: Gorbunov GI (ed) The Imandra/Varzuga Zone of Karelids. Nauka (Science), Leningrad, pp 33–56 (in Russian)

Fiorentini ML, Beresford SW, Deloule E, Hanski E, Stone WE, Pearson NJ (2008) The role of mantle-derived volatiles in the petrogenesis of Palaeoproterozoic ferropicrites in the Pechenga Greenstone Belt, northwestern Russia: insights from in-situ microbeam and nanobeam analysis of hydromagmatic amphibole. Earth Planet Sci Lett 268:2–14

Fowler AD, Berger B, Shore M, Jones MI, Ropchan J (2002) Supercooled rocks: development and significance of varioles, spherulites, dendrites and spinifex in Archean volcanic rocks, Abitibi Greenstone belt, Canada. Precambrian Res 115:311–328

Gangopadhyay A, Walker RJ, Hanski E, Solheid PA (2006) Origin of Paleoproterozoic Ti-enriched komatiitic rocks from Jeesiörova, Kittilä Greenstone Complex, Finnish Lapland. J Petrol 47:773–789

Girnis AV, Ryabchikov IV, Suddaby P, Kulikov VS, Kulikova VV (1990) Precambrian mantle evolution: Sm-Nd isotope data on East Karelian komatiites. Geochem Int 28(5):11–19

Glebovitsky VA, Semenov VS, Beyatsky BV, Koptev-Dvornikov EV, Pchelintseva NF, Kiteev BS (2001) The structure of the Lukkulaisvaara intrusion, Oulanka group, northern Karelia: petrological implications. Can Mineral 39:607–637

Golubev AI, Kulikov VS (2011) The Zaonega dolerite-basalt complex. In: Glushanin LV, Sharov NV, Shchiptsov VV (eds) The Onega Palaeoproterozoic structure (geology, tectonics, deep structure and minerageny). Institute of Geology, Karelian Research Centre of RAS, Petrozavodsk, pp 87–91 (in Russian)

Golubev AI, Svetov AP (1983) Geochemistry of platformal Basaltic volcanism in Karelia. Karelian Branch of the Academy of Sciences of the USSR, Petrozavodsk, p 192 (in Russian)

Golubev AI, Svetov AP (1984) Itenerary: Petrozavodsk – Solomennoye – Chertov Stul – Island of Suisari – Cape Pippuri – Petrozavodsk. In: Sokolov VA (ed) Guide-book, geological field trips in Karelia. Institute of Geology, Karelian Branch of the Academy of Sciences of the USSR, Petrozavodsk, pp 79–94

Golubev AI, Trofimov NN (2011) The Podozhgorsk dolerite-gabbro-dioritic complex. In: Glushanin LV, Sharov NV, Shchiptsov VV (eds) The Onega Palaeoproterozoic structure, Geology, tectonics, deep structure and minerageny. Institute of Geology, Karelian Research Centre of RAS, Petrozavodsk, pp 63–67 (in Russian)

Golubev AI, Svetov AP, Heiskanen KI (1984) Itinerary: Petrozavodsk – Krasnaya Rechka – Girvas – Petrozavodsk. In: Sokolov VA (ed) Guide-book, geological field trips in Karelia. Institute of Geology, Karelian Branch of the Academy of Sciences of the USSR, Petrozavodsk, pp 46–61

Golubev AI, Svetov AP, Kulikov VS (1987a) Itinerary: Petrozavodsk-Hirvas-Konchozero-Petrozavodsk. In: Kulikov VS, Rybakov SI, Golubev AI, Svetov AP (eds) Guidebook to geological excursion to Karelia. Institute of Geology, Karelian Science Centre, Russian Academy of Sciences, Petrozavodsk, pp 66–80 (in Russian)

Golubev AI, Svetov AP, Kulikov VS (1987b) Itinerary: Petrozavodsk-Solomennoe-Chertov Stul-Petrozavodsk. In: Kulikov VS, Rybakov SI, Golubev AI, Svetov AP (eds) Guidebook to geological excursion to Karelia. Institute of Geology, Karelian Science Centre, Russian Academy of Sciences, Petrozavodsk, pp 81–89 (in Russian)

Gorbunov GI, Yakovlev YuN, Goncharov YuV, Gorelov VA, Tel'nov VA (1985) The nickel areas of the Kola Peninsula. Bull Geol Surv Finl 333:42–109

Grinenko LN, Hanski E, Grinenko VA (2003) Formation conditions of the Keivitsa Cu-Ni deposit, northern Finland: evidence from S and C isotopes. Geochem Int 41(2):154–167

Gudin AN, Chistyakov AV (2010) Geochemistry of Palaeoproterozoic variolitic basalts of the Suisari suite in central Karelia. In: All-Russian petrographic meeting, magmatism and metamorphism in the history of the Earth, Ekaterinburg, Abstracts, 24–28 Aug 2010, pp 193–194 (in Russian)

Hanski EJ (1992) Petrology of the Pechenga ferropicrites and cogenetic, Ni-bearing gabbro-wehrlite intrusions, Kola Peninsula, Russia. Bull Geol Surv Finl 367:192

Hanski E (1993) Globular ferropicritic rocks at Pechenga, Kola Peninsula: immiscibility versus alteration. Lithos 29:197–216

Hanski E (1997) The Nuttio serpentinite belt, Central Lapland: an example of Paleoproterozoic ophiolitic mantle rocks in Finland. Ofioliti 22:35–46

Hanski E, Huhma H (2005) Central Lapland greenstone belt. In: Lehtinen M, Nurmi PA, Rämö OT (eds) Precambrian geology of Finland – key to the evolution of the Fennoscandian Shield. Elsevier, Amsterdam, pp 139–194

Hanski E, Kamenetsky VS (2006) Spinel-hosted melt inclusions in the Paleoproterozoic primitive volcanic rocks from northern Finland: evidence for coexistence and mixing of komatiitic and picritic magmas. In: International conference on continental volcanism – IAVCEI 2006, Guangzhou, Abstracts & Program, 14–18 May 2006, p 116

Hanski EJ, Smolkin VF (1995) Iron- and LREE-enriched mantle source for early Proterozoic intraplate magmatism as exemplified by the Pechenga ferropicrites, Kola Peninsula, Russia. Lithos 34:107–125

Hanski E, Huhma H, Smolkin VF, Vaasjoki M (1990) The age of the ferropicritic volcanics and comagmatic Ni-bearing intrusions at Pechenga, Kola Peninsula. Bull Geol Soc Finl 62:123–133

Hanski E, Huhma H, Suominen IM, Walker RJ (1997) Geochemical and isotopic (Os, Nd) study of the Keivitsa intrusion and its Cu-Ni deposit, northern Finland. In: Papunen H (ed) Mineral deposits: research and exploration – where do they meet? Proceedings of the fourth Biennial SGA meeting, Turku, 11–13 Aug 1997, A. A. Balkema, Rotterdam, pp 435–438

Hanski E, Huhma H, Rastas P, Kamenetsky VS (2001a) The Palaeoproterozoic komatiite-picrite association of Finnish Lapland. J Petrol 42:855–876

Hanski E, Walker RJ, Huhma H, Suominen I (2001b) The Os and Nd isotopic systematics of the 2.44 Ga Akanvaara and Koitelainen mafic layered intrusions in northern Finland. Precambrian Res 109:73–102

Hanski E, Huhma H, Vaasjoki M (2001c) Geochronology of northern Finland: a summary and discussion. In: Vaasjoki M (ed) Radiometric age determinations from Finnish Lapland and their bearing on the timing of Precambrian volcano-sedimentary sequences. Geol Surv Finl, Spec Pap 33: 255–279

Hanski E, Huhma H, Perttunen V (2005) SIMS, Sm-Nd isotopic and geochemical study of an arkosite-amphibolite suite, Peräpohja Schist Belt: evidence for 1.98 Ga A-type felsic magmatism in northern Finland. Bull Geol Soc Finl 77:5–29

Hanski E, Huhma H, Vuollo J (2010) SIMS zircon ages and Nd isotope systematics of the 2.2 Ga mafic intrusions in northern and eastern Finland. Bull Geol Soc Finl 82:31–62

Hanski E, Luo Zh-Y, Oduro H, Walker RJ (2011) The Pechenga Ni-Cu sulfide deposits, NW Russia: a review with new constraints from the feeder dikes. Rev Econ Geol 17:145–162

Heiskanen KI (1980) Karelian geosyncline (Lithological analysis). Nauka (Science), Leningrad, p 166 (in Russian)

Heiskanen K, Golubev AI, Bondar LF (1977) Orogenic volcanism in Karelia. Nauka (Science), Leningrad, p 216 (in Russian)

Heiskanen KI, Bondar LF, Golubev AI (1987) Itinerary: Petrozavodsk-Krasnaya Rechka-Hirvas-Petrozavodsk. In: Kulikov VS, Rybakov SI, Golubev AI, Svetov AP (eds) Guidebook to a geological excursion to Karelia. Institute of Geology, Karelian Science Centre, Russian Academy of Sciences, Petrozavodsk, pp 53–57 (in Russian)

Henriksen H (1983) Komatiitic chlorite-amphibole rocks and mafic metavolcanics from the Karasjok Greenstone Belt, Finnmark, Northern Norway: a preliminary report. Bulletin of Norges Geologiske Undersøkelse 382:17–43

Herzberg C, O'Hara MJ (1998) Phase equilibrium constraints on the origin of basalts, picrites, and komatiites. Earth-Sci Rev 44:39–79

Higgins SJ, Snyder GA, Mitchell JN, Taylor LA, Sharkov EV, Bogatikov OA, Grokhovskaya TL, Chistyakov AV, Ganin VA, Grinevich NG (1997) Petrology of the Early Proterozoic Burakovsky layered intrusion, southern Karelia, Russia: mineral and whole-rock major-element chemistry. Can J Earth Sci 34:390–406

Honkamo M (1987) Geochemistry and tectonic setting of early Proterozoic volcanic rocks in northern Ostrobothnia, Finland. In: Pharaoh T, Beckinsale RD, Rickard D (eds) Geochemistry and mineralization of Proterozoic volcanic suites. Geol Soc, Spec Pub 33: 59–68

Honkamo M (1989) Volcanic rocks in northern ostrobothnia, A report of the Lapland Volcanite Project. Geol Surv Finl, Rep Inves 89: 46 (in Finnish with English summary)

Huhma H (1986) Sm-Nd, U-Pb and Pb-Pb isotopic evidence for the origin of the Early Proterozoic Svecokarelian crust in Finland. Bull Geol Surv Finl 337:48

Huhma H (1987) Provenance of early Proterozoic and Archean metasediments in Finland: a Sm-Nd isotopic study. Precambrian Res 35:127–143

Huhma H, Cliff RA, Perttunen V, Sakko M (1990) Sm-Nd and Pb isotopic study of mafic rocks associated with early Proterozoic continental rifting: the Peräpohja schist belt in northern Finland. Contrib Mineral Petrol 104:369–379

Iljina M, Hanski E (2005) Layered mafic intrusions of the Tornio-Näränkävaara belt. In: Lehtinen M, Nurmi P, Rämö T (eds) Precambrian Bedrock of Finland – key to the evolution of the Fennoscandian Shield. Elsevier, Amsterdam, pp 103–138

Iljina M, Alapieti T, McElduff BM (1992) Platinum-group element mineralization in the Suhanko-Konttijärvi intrusion, Finland. Aust J Earth Sci 39:303–313

Ivanikov V, Philippov N, Beliatsky B (1998) Geochemistry of the metavolcanic rocks from the Ladoga region: evidence for an early Proterozoic oceanic crust, In: Hanski E, Vuollo J (eds) International Ophiolite symposium and field excursion: generation and emplacement of ophiolites through time, 10–15 Aug 1998, Oulu, Abstracts and Excursion Guide, Geol Surv Finl, Spec Pap 26: 30

Ivanikov VV, Malashin MV, Golubev AI, Philippov NB (2008), New geochemical data on Jatulian basalts from central Karelia. Vestnik St. Petersburg University, Series 7: Geology and Geography 4: 31–45 (in Russian)

Jensen LS (1976) A new cation plot for classifying subalkalic volcanic rocks. Ont Dept Mines, Misc Pap 66:22

Juopperi A (1977) The magnetite gabbro and related Mustavaara vanadium ore deposit in the Porttivaara layered intrusion, north-eastern Finland. Bull Geol Surv Finl 288:68

Kähkönen Y, Mattila E, Nuutilainen J (1986) A revision of the Preterozoic-Archean boundary of the Northern Bothnian schist belt with a discussion of the geochemistry of related basic metavolcanics. Bull Geol Soc Finl 58:109–129

Karhu JA (1993) Paleoproterozoic evolution of the carbon isotope ratios of sedimentary carbonates in the Fennoscandian Shield. Bull Geol Surv Finl 371:87

Karhu J, Kortelainen N, Huhma H, Perttunen V, Sergeev S (2007) New time constraints for the end of the Paleoproteroozoic carbon isotope excursion. In: 7th symposium on applied isotope geochemistry, Stellenbosch, 10–14 Sept 2007, Abstracts, pp 76–77

Karinen T (2010) The Koillismaa intrusion, northeastern Finland – evidence for PGE reef forming processes in the layered series. Bull Geol Surv Finl 404:176

Koistinen T, Stephens MB, Bogatchev V, Nordgulen Ø, Wenneström M, Korhonen J (comps.) (2001) Geological map of the Fennoscandian Shield, scale 1:2 000 000. Espoo/Trondheim/Upsala/Moscow

Kontinen A (1987) An early Proterozoic ophiolite – the Jormua mafic–ultramafic complex, northeastern Finland. Precambrian Res 35:313–341

Koptev-Dvornikov EV, Kireev BS, Pchelintseva NF, Khvorov DM (2001) Distribution of cumulative mineral assemblages, major and trace elements over the vertical section of the Kivakka intrusion, Olanga Group of intrusions, northern Karelia. Petrology 9(1):3–27

Kozlovsky YeA (ed) (1987) The superdeep well of the Kola Peninsula. Springer, Berlin, p 558

Krill AG, Bergh S, Lindahl I, Mearns EW, Often M, Olerud S, Olesen O, Sandstad JS, Siedlecka A, Solli A (1985) Rb-Sr, U-Pb and Sm-Nd isotopic dates from Precambrian rocks of Finnmark. Bulletin of Norges Geologiske Undersøkelse 403:37–54

Krivolutskaya NA, Smol'kin VF, Svirskaya NM, Mamontov VP, Fanygin AS, Belyatskii BV, Roshchina IA (2010a) Geochemical features of the Drusite Massifs, the central part of the Belomorian Mobile Belt: I. Distribution of major and trace elements in the rocks. Geochem Int 48:465–491

Krivolutskaya NA, Belyatskii BV, Smol'kin VF, Mamontov VP, Fanygin AS, Svirskaya NM (2010b) Geochemical features of the Drusite Massifs, the central part of the Belomorian Mobile Belt: II. Sm–Nd isotopic system of the rocks and the U–Pb isotopic system of zircons. Geochem Int 48:1064–1083

Kulikov VS (1988) High magnesian volcanism in the early Proterozoic. In: Bogatikov OA (ed) Komatiites and high-magnesian volcanic rocks in the early Precambrian of the Baltic Shield. Nauka (Science), Leningrad, pp 20–88 (in Russian)

Kulikov VS, Slyusarev VD, Kochnev-Pervukhov VI, Kravchenko AN (1976) Suisarian mafic-ultramafic complex in the Onega synclinorium. In: Kratts KO (ed) Precambrian intrusive Mafic-Ultramafic complexes in Karelia. Nauka (Science), Leningrad, pp 98–109 (in Russian)

Kulikov VS, Svetov AP, Golubev AI (1978) Main features of petrochemical evolution of the volcanism. In: Sokolov VA (ed) Volcanic structures of the Proterozoic in Karelia. Nauka (Science), Leningrad, pp 33–42 (in Russian)

Kulikov V, Galdobina L, Voinov A, Golubev A, Kashpirov S, Polehovsky Yu, Svetov, A (1980) Jatulian geology of the Paanajärvi-Kuolajärvi synclinorium, In: Silvennoinen A (ed) Jatulian geology in the Eastern part of the Baltic Shield, Proceedings of a Finnish-Soviet symposium, Finland, 21–26 Aug 1979, Rovaniemi, The Committee for Scientific and Technical Co-operation between Finland and Soviet Union, pp 73–96

Kulikov VS, Lavrov BS, Pisarevsky SA, Puchtel IS, Sokolov SYa (1999) Palaeoproterozoic Suisarian Picrite-Basalt Complex in Karelia (reference section and petrology). Institute of Geology, Karelian Science Centre, Russian Academy of Sciences, Petrozavodsk, p 96

Kulikov VS, Bychkova YaV, Kulikova VV, Koptev-Dvornikov EV, Zudin AI (2005) Role of deep-seated differentiation in formation of Paleoproterozoic Sinegorie lava plateau of komatiite basalts, southeastern Fennoscandia. Petrology 13:469–489

Kulikov VS, Bychkova YaV, Kulikova VV, Kostitsyn YuA, Pokrovsky OS, Vasil'ev MV (2008) The Ruiga intrusion: a typical example of a shallow-facies Paleoproterozoic peridotite–gabbro-komatiite–basaltic association of the Vetreny Belt, Southeastern Fennoscandia. Petrology 16:531–551

Kulikov VS, Bychkova YaV, Kulikova VV, Ernst R (2010) The Vetreny Poyas (Wind Belt) subprovince of southeastern Fennoscandia: an essential component of the c. 2.5–2.4 Ga Sumian large igneous province. Precambrian Res 183:589–601

Kulikov VS, Svetov AP, Golubev AI (2011) The Suisari Horizon (Suisari picrite-basalt complex). In: Glushanin LV, Sharov NV, Shchiptsov VV (eds) The Onega Palaeoproterozoic structure (geology, tectonics, deep structure and minerageny). Institute of Geology, Karelian Research Centre of RAS, Petrozavodsk, pp 91–101 (in Russian)

Kullerud K, Skjerlie KP, Corfu F, de la Rosa JD (2006) The 2.40 Ga Ringvassøy mafic dykes, West Troms Basement Complex, Norway: the concluding act of early Palaeoproterozoic continental breakup. Precambrian Res 150:183–200

Kump LR, Barley ME (2007) Increased subaerial volcanism and the rise of atmospheric oxygen 2.5 billion years ago. Nature 448: 1033–1036

Laajoki K (1991) Stratigraphy of the northern part of the early Proterozoic (Karelian) Kainuu Schist Belt and associated complexes, Finland. Bull Geol Surv Finl 358:105

Lager I, Loberg B (1990) Sedimentological and basin analysis ore prospecting methods in the North Bothian greenstone belt, STU-Project 86-03967P, Final report, Economic geology, Luleå Technical University, Luleå, p 112 (in Swedish)

Lahtinen J (1985) PGE-bearing copper-nickel occurrences in the marginal series of the Early Proterozoic Koillismaa layered intrusion. Bull Geol Surv Finl 333:165–178

Lahtinen R, Korja A, Nironen M (2005) Paleoproterozoic tectonic evolution. In: Lehtinen M, Nurmi PA, Rämö OT (eds) Precambrian geology of Finland – key to the evolution of the Fennoscandian Shield. Elsevier, Amsterdam, pp 481–532

Lahtinen R, Korja A, Nironen M, Heikkinen P (2009) Palaeoproterozoic accretionary processes in Fennoscandia. Geol Soc, Lond, Spec Pub 318:237–256

Latypov RM, Mitrofanov FF, Alapieti TT (1999) Petrology of the lower layered horizon of the Western Pansky Tundra intrusion, Kola Peninsula. Petrology 7:509–538

Latypov RM, Chistyakova SYu, Alapieti TT (2008) Fine-grained mafic bodies as preserved portions of magma replenishing layered intrusions: the Nadezhda gabbronorite body, Lukkulaisvaara intrusion, Fennoscandian Shield, Russia. Mineral Petrol 92:165–209

Latypov RM, Hanski E, Lavrenchuk A, Huhma H, Havela T (2011) A 'three-increase model' for the origin of the marginal reversal of the Koitelainen layered intrusion, Finland. J Petrol 52:733–764

Lauri LS, Mänttäri I (2002) The Kynsijärvi quartz alkali feldspar syenite, Koillismaa, eastern Finland – silicic magmatism associated with 2.44 Ga continental rifting. Precambrian Res 119:121–140

Lauri LS, Karinen T, Räsänen J (2003) The earliest Paleoproterozoic supracrustal rocks in Koillismaa, northern Finland – their petrographic and geochemical characteristics and lithostratigraphy. Bull Geol Soc Finl 75:29–50

Lauri LS, Rämö OT, Huhma H, Mänttäri I, Räsänen J (2006) Petrogenesis of silicic magmatism related to the 2.44 Ga rifting of Archean crust in Koillismaa, eastern Finland. Lithos 86:137–166

Lavrov MM (1979) Ultramafic bodies and layered Peridotite-Gabbro-Norite intrusions in the Precambrian of Northern Karelia. Nauka (Science), Leningrad, p 136 (in Russian)

Lavrov MM, Trofimov NN, Golubev AJ (2006) Geochemistry and mineralogy of the Burakovsky layered intrusion. Natl Geol No. 1: 3–12 (in Russian)

Laz'ko EE, Sharkov EV (1988) Magmatic rocks, 5th edn, Ultramafic rocks. Nauka (Science), Moscow, p 502 (in Russian)

Lehtonen M, Airo M-L, Eilu P, Hanski E, Kortelainen V, Lanne E, Manninen T, Rastas P, Räsänen J, Virransalo P (1998) The stratigraphy, petrology and geochemistry of the Kittilä greenstone area, northern Finland. A Report of the Lapland Volcanite Project, Geological Survey of Finland, Report of Investigation 140: 144 (in Finnish with English summary)

Leng W, Zhong S (2010) Surface subsidence caused by mantle plumes and volcanic loading in large igneous provinces. Earth Planet Sci Lett 291:207–214

Levchenkov OA, Nikolaev AA, Bogomolov ES, Yakovleva SZ (1994) U-Pb ages of Sumian acid magmatic rocks in northern Karelia. Stratigr Geol Correlat 2(1):3–9 (in Russian)

Levinson-Lessing FYu (1935) On a peculiar type of differentiation represented by the variolites of Yalguba in Karelia. Proc Petrogr Inst 5:21–27 (in Russian)

Levinson-Lessing FYu (1949) Selected works, 1st edn. Academy of Sciences of the USSR, Moscow, p 346 (in Russian)

Lightfoot PC, de Souza H, Doherty W (1993) Differentiation and source of the Nipissing Diabase intrusions, Ontario, Canada. Can J Earth Sci 30:1123–1140

Lukkarinen H (2008) Pre-quaternary rocks of the Siilinjärvi and Kuopio map-sheet areas, Explanation to the maps of pre-quaternary rocks, sheets 3331 and 3242, Geological survey of Finland, Espoo, p 228 (in Finnish with English summary)

Luukkonen E (1989) The early Proterozoic Saari-Kiekki greenstone belt: a representative of the Sariola group at Kuhmo, eastern Finland. Bull Geol Soc Finl 61:161–187

Makarikhin VV, Satsuk YuI (1987) Itinerary: Petrozavodsk-Raiguba-Petrozavodsk. In: Kulikov VS, Rybakov SI, Golubev AI, Svetov AP (eds) Guidebook to a geological excursion to Karelia. Institute of Geology, Karelian Science Centre, Russian Academy of Sciences, Petrozavodsk, pp 58–65 (in Russian)

Malashin MV, Ivanikov VV, Golubev AI, Filippov NB (2003) Geochemistry and petrology of mafic volcanic complexes of lower Proterozoic in Karelia. I. Jatulian volcanic complexes, Vestnik St. Petersburg University, Series 7: Geol Geogr 1(7): 3–32 (in Russian)

Manninen T (1991) Volcanic rocks in the Salla area, northeastern Finland. Geol Surv Finl, Rep Invest 104:97 (in Finnish with English summary)

Manninen T, Pihlaja P, Huhma H (2001) U-Pb geochronology of the Peurasuvanto area, northern Finland. Geol Surv Finl, Spec Pap 33:189–200

Marmo V (1949) Early Jotnian Suoju volcanic complex in eastern Karelia, Annales Academiae Scientarum Fennicae, Series A III, No. 19: 77 (in German)

Martinsson O, Billström K (2006) Plume related MORB and high-Ti basalts in the Kalix greenstones, northern Sweden. 27th Nordic geological winter meeting, 9–12 Jan, Oulu, Abstracts, Bull Geol Soc Finld, Spec Issue 1: 46

Matrenichev VA, Stepanov KI, Pupkov OM (2004) Stratigraphy and chemical characteristics of Early Proterozoic volcanic rocks of the Sortavala uplift (northern Ladoga), Vestnik St. Petersburg University, Series 7: Geol Geogr 2(2): 1–44 (in Russian)

Matrenichev VA, Vrevsky AV, Sergeev SA, Matukov DA (2006) The Ludicovian–Kalevian boundary in the northern Ladoga region: geological relations and isotopic age. Doklady Earth Sci 407A:388–392

Melezhik VA, Borisov AE, Fedotov ZhA, Predovsky AA (1982) The Varzuga series. In: Gorbunov GI (ed) The Imandra/Varzuga Zone of Karelides. Nauka (Science), Leningrad, pp 57–85 (in Russian)

Melezhik VA, Hudson-Edwards KA, Green AH, Grinenko LN (1994) Pechenga area, Russia – Part 2: nickel-copper deposits and related rocks. Trans Inst Min Metall: Sect B: Appl Earth Sci 103: B146–B161

Melezhik VA, Huhma H, Condon DJ, Fallick AE, Whitehouse MJ (2007) Temporal constraints on the Paleoproterozoic Lomagundi-Jatuli carbon isotopic event. Geology 35:655–658

Meriläinen K (1980) Stratigraphy of the Precambrian in Finland. Geologiska Föreningens i Stockholm Förhandlingar (GFF) 102:177–180

Mertanen S, Korhonen F (2011) Paleomagnetic constraints on an Archean-Paleoproterozoic Superior-Karelia connection: new evidence from Archean Karelia. Precambrian Res 186:193–204

Mertanen S, Pesonen LJ (2005) Drift history of the shield. In: Lehtinen M, Nurmi PA, Rämö OT (eds) Precambrian geology of Finland – key to the evolution of the Fennoscandian Shield. Elsevier, Amsterdam, pp 645–668

Mikkola P, Kontinen A, Huhma H, Lahaye Y (2010) Three Paleoproterozoic A-type granite intrusions and associated dykes from Kainuu, East Finland. Bull Geol Soc Finl 82:81–100

Mints MV (1996) The sedimentary-volcanic Pechenga-Imandra-Varzuga belt. In: Leonov YuG (ed) The early Precambrian of the Northeastern Baltic Shield: paleogeodynamics, Crustal structure and evolution. Scientific World, Moscow, pp 77–112 (in Russian)

Moilanen M (2011) The Komatiite-Related Lomalampi Platinum Prospect in Sodankylä. M.Sc. thesis, Department of Geosciences, University of Oulu, p 109 (in Finnish)

Mutanen T (1997) Geology and ore petrology of the Akanvaara and Koitelainen mafic layered intrusions and the Keivitsa-Satovaara layered complex, northern Finland. Bull Geol Surv Finl 395:233

Myhre PI, Corfu F, Bergh S (2011) Palaeoproterozoic (2.0–1.95 Ga) pre-orogenic supracrustal sequences in the West Troms Basement Complex, North Norway. Precambrian Res 186:89–100

Narkisova VV, Krupenik VA, Sveshikova KYu (2010) Petrogenesis of Ludicovian basalts in the Onega structure (based on the results of the Onega parametric drillhole), In: All-Russian petrographic meeting, Magmatism and metamorphism in the history of the Earth, 24–28 Aug 2010, Ekaterinburg, Abstracts, pp 87–88 (in Russian)

Nikolaev GS, Ariskin AA (2005) Burakovo–Aganozero layered massif in the Trans-Onega area: II. Structure of the marginal series and the estimation of the parental magma composition by geochemical thermometry techniques. Geochem Int 43:646–665

Nykänen V (1992) Tectonic setting and differentiation of the Tohmajärvi igneous complex, Res Terrae, Series B, No. 17: 74 (in Finnish)

Nykänen VM, Vuollo JI, Liipo J, Piirainen T (1994) Transitional (2.1 Ga) Fe-tholeiitic-tholeiitic magmatism in the Fennoscandian Shield signifying lithospheric thinning during Palaeoproterozoic extensional tectonics. Precambrian Res 70:45–65

Nykänen J, Laajoki K, Karhu J (1997) Geology and geochemistry of the early Proterozoic Kortejärvi and Laivajoki carbonatites, central Fennoscandian Shield, Finland. Bull Geol Soc Finl 69:5–30

O'Brien HE, Peltonen P, Vartiainen H (2005) Kimberlites, carbonatites, and alkaline rocks. In: Lehtinen M, Nurmi PA, Rämö OT (eds) Precambrian geology of Finland – key to the evolution of the Fennoscandian Shield. Elsevier B.V, Amsterdam, pp 605–644

Pearse JA (2008) Geochemical fingerprinting of oceanic basalts with application to ophiolite classification and the search for Archean oceanic crust. Lithos 100:14–48

Pekkarinen LJ, Lukkarinen H (1991) Paleoproterozoic volcanism in the Kiihtelysvaara-Tohmajärvi district, eastern Finland. Bull Geol Surv Finl 357:30

Peltonen P (2005) Ophiolites. In: Lehtinen M, Nurmi PA, Rämö OT (eds) Precambrian geology of Finland – key to the evolution of the Fennoscandian Shield. Elsevier, Amsterdam, pp 237–278

Peltonen PT, Manninen T, Pihlaja P (1988) The late Archaean volcaniclastic Rookkiaapa Formation in Peurasuvanto, northern

Finland. In: Laajoki K, Paakkola J (eds) Sedimentology of the Precambrian formations in Eastern and Northern Finland, Geol Surv Finl, Spec Pap 5: 165–176

Peltonen P, Kontinen A, Huhma H (1996) Petrology and geochemistry of metabasalts from the 1.95 Ga Jormua Ophiolite, northeastern Finland. J Petrol 37:1359–1383

Peltonen P, Kontinen A, Huhma H (1998) Petrogenesis of the mantle sequence of the Jormua Ophiolite (Finland): melt migration in the upper mantle during Palaeoproterozoic continental break-up. J Petrol 39:297–329

Peltonen P, Mänttäri I, Huhma H, Kontinen A (2003) Archean zircons from the mantle: the Jormua ophiolite revisited. Geology 31:645–648

Peltonen P, Kontinen A, Huhma H, Kuronen U (2008) Outokumpu revisited: new mineral deposit model for the mantle peridodite-associated Cu-Co-Zn-Ni-Ag-Au sulphide deposits. Ore Geol Rev 33:559–617

Perttunen V (1980) Stratigraphy of the Peräpohja schist area, In: Silvennoinen A (ed) Jatulian geology in the Eastern part of the Baltic Shield, Proceedings of a Finnish-Soviet symposium, Finland, 21–26 Aug 1979, Rovaniemi, The Committee for Scientific and Technical Co-operation between Finland and Soviet Union, pp 139–144

Perttunen V (1985) On the Proterozoic stratigraphy and exogenic evolution of the Peräpohja area, Finland. In: Laajoki K, Paakkola J (eds) Proterozoic exogenic processes and related metallogeny. Bull Geol Surv Finl 331: 131–141

Perttunen V (1989) Volcanic rocks in the Peräpohja area, northern Finland, A report of the Lapland Volcanite Project, Geological Survey of Finland, Report of Investigations 94: 40 (in Finnish with English summary)

Perttunen V, Hanski E (2003) Pre-quaternary rocks of the Koivu and Törmäsjärvi Map-Sheet Areas, Explanation to the Geological Map of Finland 1:100 000, Pre-quaternary rocks, sheets 3631 and 2633, Geological survey of Finland, p 88 (in Finnish with English summary)

Perttunen V, Vaasjoki M (2001) U-Pb geochronology of the Peräpohja Schist Belt, northwestern Finland. In: Vaasjoki M (ed) Radiometric age determinations from Finnish Lapland and their bearing on the timing of Precambrian volcano-sedimentary sequences. Geol Finl Spec Pap 33: 45–84

Pharaoh T (1985) Volcanic and geochemical stratigraphy of the Nussir group of Arctic Norway – an early Proterozoic greenstone suite. J Geol Soc Lond 142:259–278

Pharaoh TC, Pearce JA (1984) Geochemical evidence for the geotectonic setting of Early Proterozoic metavolcanic sequences in Lapland. Precambrian Res 25:283–308

Pharaoh TC, Warren A, Walsh NJ (1987) Early Proterozoic metavolcanic suites of the northernmost part of the Baltic Shield. In: Pharaoh T, Beckinsale RD, Rickard D (eds) Geochemistry and mineralization of Proterozoic volcanic suites. Geol Soc, Spec Pub 33: 41–58

Philippov N, Ivanikov V (2001) Evolution of the Karelian Palaeo-Proterozoic (2.25–1.95 Ga) volcanic associations (eastern part of the Fennoscandian Shield): geodynamic significance, EUG General Assembly XI, Strasbourg, 8–12 Apr 2001, Abstracts, p 191

Philippov NB, Trofimov NN, Golubev AI, Sergeev SA, Huhma H (2007) New geochronological data on the Koikary-Svjatnavolok and Pudozhgora gabbro-dolerite intrusives, In: Golubev VI, Shchiptsov VV (eds) Geology and Mineral Resources of Karelia 10: 49–68 (in Russian with English abstract)

Poleshchuk AV (2011) Sill genesis in the Paleoproterozoic tectonic evolution of the Onega trough, Baltic Shield. Doklady Earth Sci 439:939–943

Popov MG, Kishchenko HT, Satsuk YuI, Gorbik HA (1983) Jatulian high-magnesian basalts with different alkalinity (Central Karelia). In: Bogachev AI, Svetov AP, Stepanov VS (eds) Magmatism and metallogenesis of the Precambrian formations in Karelia. Institute of Geology, Karelian Branch of the Academy of Sciences, Petrozavodsk, pp 26–31 (in Russian)

Predovsky AA, Fedotov ZhA, Akhmedov AM (1974) Geochemistry of the Pechenga Complex. Nauka (Science), Leningrad, p 139 (in Russian)

Predovsky AA, Melezhik VA, Bolotov VA, Fedotov ZhA, Basalaev AA, Kozlov NE, Ivanov AA, Zhangurov AA, Skuf'in PK, Lyubtsov VV (1987) Volcanism and sedimentology of the Precambrian in the Northeastern Part of the Baltic Shield. Nauka (Science), Leningrad, p 185 (in Russian)

Puchtel IS, Humayun M (2001) Platinum group element fractionation in a komatiitic basalt lava lake. Geochim Cosmochim Acta 65:2979–2993

Puchtel IS, Bogatikov OA, Kulikov VS, Kulikova VV, Zhuravlev DZ (1995) The role of crustal and mantle sources in the petrogenesis of continental magmatism: isotope and geochemical evidence from early Proterozoic picrites of the Onega Plateau, Baltic Shield. Petrology 3:357–378

Puchtel IS, Hofmann AW, Mezger AW, Shchipansky AA, Kulikov VS, Kulikova VV (1996) Petrology of a 2.41 Ga remarkably fresh komatiitic basalt lava lake in Lion Hills, Central Vetreny Belt, Baltic Shield. Contrib Mineral Petrol 124:273–290

Puchtel IS, Haase KM, Hofmann AW, Chauvel C, Garbe-Schönberg C-N (1997) Petrology and geochemistry of crustally contaminated komatiitic basalts from the Vetreny belt, southeastern Baltic Shield: evidence for an early Proterozoic mantle plume beneath rifted Archean continental lithosphere. Geochim Cosmochim Acta 61:1205–1222

Puchtel IS, Arndt NT, Hofmann AW, Haase KM, Kröner A, Kulikov VS, Kulikova VV, Garbe-Schönberg C-D, Nemchin AA (1998) Petrology on mafic lavas within the Onega plateau, central Karelia: evidence for 2.0 Ga plume-related continental crustal growth in the Baltic Shield. Contrib Mineral Petrol 130:134–153

Puchtel IS, Brügmann GE, Hofmann AW (1999) Precise Re-Os mineral isochron and Pb-Nd-Os isotope systematic of a mafic-ultramafic sill in the 2.0 Ga Onega plateau (Baltic Shield). Earth Planet Sci Lett 170:447–461

Puchtel IS, Brügmann GE, Hofmann AW, Kulikov VS, Kulikova VV (2001) Os-isotope systematics of komatiitic basalts from the Vetreny belt, Baltic Shield: evidence for a chondritic source of the 2.45 Ga plume. Contrib Mineral Petrol 140:588–599

Räsänen J, Huhma H (2001) U-Pb datings in the Sodankylä schist area, the central Finnish Lapland. Geol Surv Finl, Spec Pap 33:153–188

Räsänen J, Hanski E, Lehtonen M (1989) Komatiites, low-Ti basalts and andesites in the Möykkelmä area, Central Finnish Lapland. Geol Surv Finl Rep Invest 88:41

Rastas P, Huhma H, Hanski E, Lehtonen MI, Paakkola J, Mänttäri I, Härkönen I (2001) U-Pb isotopic studies on the Kittilä greenstone area, Central Lapland, Finland. In: Vaasjoki M (ed) Radiometric age determinations from Finnish Lapland and their bearing on the timing of Precambrian volcano-sedimentary sequences. Geol Surv Finl, Spec Pap 33: 95–142

Remizova AM, Barzhitsaya SM (1984) New data on the geological structure of the southern part of the Imandra-Varzuga Belt. In: Kozlov MT, Zagorodny VG, Makievisky SI, Petrov VP, Radchenko AT (eds) Precambrian geology of the Kola Peninsula. Institute of Geology, Kola Branch of the Academy of the USSR, Apatity, pp 87–95 (in Russian)

Rodionov NV, Belyatsky BV, Antonov AV, Presnyakov SL, Sergeev SA (2009) Baddeleyite U–Pb SHRIMP II age determination as a tool for carbonatite massifs dating. Doklady Earth Sci 428:1166–1170

Rusanov MS (1981) Tholeiite-komatiite formation of the Pechenga complex. Sov Geol 2:98–112 (in Russian)

Saini-Eidukat B, Alapieti TT, Thalhammer OAR, Iljina MJ (1997) Siliceous, high-magnesian parental magma compositions of the

PGE-rich Early Paleoproterozoic layered intrusion belt of northern Finland, In: Rongfu P (ed) Proceedings of the 30th international geological Congress, Beijing, 4–14 Aug 1996, vol 9, Energy and mineral resources for the 21st century, geology of mineral deposits, mineral economics, VSP, Utrecht, pp 177–197

Sandstå NR, Robins B, Furnes H, de Wit M (2011) The origin of large varioles in flow-banded pillow lava from the Hooggenoeg Complex, Barberton Greenstone belt, South Africa. Contrib Mineral Petrol 162:365–377

Saunders AD, Jones SM, Morgan LA, Pierce KL, Widdowson M, Xu YG (2007) Regional uplift associated with continental large igneous provinces: the roles of mantle plumes and the lithosphere. Chem Geol 241:282–318

Saverikko M (1985) The pyroclastic komatiite complex at Sattasvaara in northern Finland. Bull Geol Soc Finl 57:55–87

Schissel D, Tsvetkov AA, Mitrofanov FP, Korchagin AU (2002) Basal platinum-group element mineralization in the Federov Pansky layered mafic intrusion, Kola Peninsula, Russia. Econ Geol 97:1657–1677

Serov PA (2005) New Sm-Nd isotope geochemical data on rocks and minerals from the Fedorovo-Pansky intrusion (Kola Peninsula), In: Mitrofanov FP, Pripachkin VA, Voitekhovsky Yu L, Neradovsky YuN (eds) Geology and Ecogeology, Investigations of Young Researchers, Proceedings of the 16th conference of young scientists, Institute of Geology, Kola Science Centre, Russian Academy of Sciences, Apatity, pp 149–152 (in Russian)

Sharkov YeV (1984) Lower Proterozoic continental rift magmatism in the Karelia-Kola region. Geotectonics 18:123–132

Sharkov EV, Bogatikov OA (2000) Early Proterozoic magmatism and geodynamics – evidence of a fundamental change in the earth's evolution. In: Bogatikov OA (ed) Magmatism and geodynamics, terrestrial magmatism throughout the Earth's history. Gordon and Breach Science, Amsterdam, pp 219–294

Sharkov EV, Smol'kin VF (1997) The early Proterozoic Pechenga-Varzuga Belt: a case of back-arc spreading. Precambrian Res 82:135–151

Sharkov EV, Bogatikov OA, Grokhovskaya TL, Chistyakov AV, Ganin VA, Grinevich NG, Snyder GA, Taylor LA (1995) Petrology and Ni-Cu-Cr-PGE mineralization of the largest mafic pluton in Europe: the Early Proterozoic Burakovsky layered intrusion, Karelia, Russia. Int Geol Rev 37:509–525

Sharkov EV, Smol'kin VF, Krassivskaya IS (1997) Early Proterozoic igneous province of siliceous high-Mg boninite-like rocks in the eastern Baltic Shield. Petrology 5:503–522

Sharkov EV, Krasivskaya IS, Chistyakov AV (2004) Belomorian drusite (coronite) complex, Baltic Shield, Russia: an example of dispersed intrusive magmatism in early Paleoproterozoic mobile zones. Russ J Earth Sci 6(3):185–215

Sharkov EV, Evseeva KA, Krasivskaya IS, Chistyakov AV (2005) Magmatic systems of the Early Proterozoic Baltic large igneous province of siliceous high-magnesian (boninite-like) series. Russ Geol Geophys 46:968–980

Shul'diner VI, Levchenkov OA, Yakovleva SZ, Makeev AF, Komarov AN, Konopel'ko DL, Baltybaev ShK, Kozyreva IV (2000) The upper Karelian in the stratigraphic scale of Russia: determination of its lower boundary and regional units in the stratotype area. Stratigr Geol Corr 8(6):20–33 (in Russian)

Silvennoinen A (1972) On the stratigraphy and structural geology of the Rukatunturi area, northeastern Finland. Bull Geol Surv Finl 257:48

Silvennoinen A (1991) Pre-quaternary rocks of the Kuusamo and Rukatunturi map-sheet areas, Explanation to the Geological Map of Finland 1:100 000, Pre-quaternary rocks, sheets 4524 + 4542 and 4613, Geological Survey of Finland, p 62 (in Finnish with English summary)

Sims KWW, DePaolo DJ, Murrell MT, Baldridge WS, Goldstein SJ, Clague DA (1994) Mechanisms of magma generation beneath Hawaii and mid-ocean ridges: Uranium/thorium and samarium/neodymium isotopic evidence. Science 267:508–512

Skuf'in PK (1980a) Characteristics of the volcanism in the Proterozoic Pechenga structure. Bull Mosc Soc Invest Nat 55:120–130 (in Russian)

Skuf'in PK (1980b) The volcanogenic complex of the Pechenga Zone (Lower Proterozoic). In: Zagorodny VG, Predovsky AA, Skuf'in PK (eds) Early Precambrian volcanic rocks in the Kola Peninsula, Atlas of textures and structures. Nauka (Science), Leningrad, pp 49–55 (in Russian)

Skuf'in PK (1995a) On the picritic volcanism in the South Pechenga structural zone (Kola Peninsula). Bull Mosc Soc Nat Geol Sect 70 (6):28–38

Skuf'in PK (1995b) Geology of the Poritash volcanic centre: new results. Norges Geologiske Undersøkelse, Special Publication 7:111–116

Skuf'in PK, Bayanova TB (2006) Early Proterozoic central-type volcano in the Pechenga structure and its relation to the ore-bearing gabbro-wehrlite complex of the Kola Peninsula. Petrology 14:609–627

Skuf'in PK, Fedotov ZhA (1989) Picritic pillow lavas in the section of Early Proterozoic volcanic rocks in the Pechenga structure. Doklady Acad Sci USSR 306:956–962

Skuf'in PK, Theart HFJ (2005) Geochemical and tectono-magmatic evolution of the volcano- sedimentary rocks of Pechenga and other greenstone fragments of the Kola Greenstone Belt. Precambrian Res 141:1–48

Skuf'in PK, Yakovlev YuN (2007) Geological position and geochemistry of volcanic rocks of the Majärvi, Pirttijärvi and Orshoaivi suites from the Kola Superdeep Hole and near-surface zones. Proc Murmansk State Tech Univ 10(2):173–197 (in Russian)

Skuf'in PK, Yelizarov DV (2011) Geochemistry of basaltoids of the Il'mozero and Panarechka suites of the Imandra-Varzuga Complex, In: Voytekhovsky YuL (ed) Mineralogy, petrology and ores of the Kola Region, Proceedings of the VIII All-Russian Fersman scientific session dedicated to the 135th anniversary of birth of academician D.S. Belyankin, Publishing House K & M, Apatity, pp 150–154 (in Russian)

Skuf'in PK, Pushkarev YuD, Kravchenko MP (1986) Volcanic rocks of the mugearite-trachyte formation in the Pechenga volcano-tectonic palaeodepression. Invest Acad Sci USSR, Ser Geol 1:18–29 (in Russian)

Skuf'in PK, Gavrilenko BV, Fedotov ZhA (1988) Porojärvi volcanic cupola-like structure in the southern Pechenga zone (Kola Peninsula). Bull Mosc Soc Nat Geol Sect 63(5):98–107

Skuf'in PK, Bayanova TB, Mitrofanov FP, Apanasevich EA, Levkovich NV (2000) The absolute age of granitoids from the Shuoniyavri pluton in the southern framework of the Pechenga structure, the Kola Peninsula. Doklady Earth Sci 370:114–117

Skuf'in PK, Bayanova TB, Mitrofanov FP (2006) Isotope age of subvolcanic granitoid rocks of the Early Proterozoic Panarechka volcanotectonic structure, Kola Peninsula. Doklady Earth Sci 409:774–778

Smol'kin VF (1992) Early Precambrian Komatiitic and Picritic magmatism of the Baltic Shield. Nauka (Science), St. Petersburg, p 278 (in Russian)

Smol'kin VF (1995) Early Proterozoic. In: Mitrofanov FP (ed) Geology of the Kola Peninsula. Geological Institute, Kola Science Centre, Russian Academy of Sciences, Apatity, pp 38–75

Smol'kin VF (1997) The Paleoproterozoic (2.5–1.7 Ga) Midcontinent rift system of the northeastern Fennoscandian Shield. Can J Earth Sci 34:427–449

Smol'kin VF, Borisov VV (1995) Nickel-bearing intrusive magmatism (gabbro-wehrlite association). In: Mitrofanov FP, Smol'kin VF (eds) Magmatism, sedimentogenes and geodynamics of the Pechenga Palaeoriftogenic structure. Kola Science Centre, Apatity, pp 183–219 (in Russian)

Smol'kin VF, Borisov AE, Marakushev AA (1987) Evidence for differentiation and liquid immiscibility in picritic basalts in Pechenga. Doklady Acad Sci USSR 294:669–673 (in Russian)

Smol'kin VF, Bayanova TB, Fedotov ZhA (2003) Ore-bearing mafic-ultramafic rocks of the Pechenga–Allarechka area, Kola region: isotopic dating. In: Kosakov IK, Kotov AB (eds) Proceedings of the II Russian conference on isotope geochemistry. Institute of Precambrian Geology and Geochronology, Russian Academy of Sciences, St. Petersburg, pp 467–470 (in Russian)

Smol'kin VF, Fedotov ZhA, Neradovsky YuN, Bayanova TB, Borisova VV, Glaznev VN, Dedyukhin AN, Orsoev DA, Ohnenstetter M, Ohnenstetter D, Raevsky AB, Tolstikhin IN, Chaschin VV, Mokrushin AV, Novikov DD, Ikorcky SV, Kamensky IL, Delenitzin AA (2004) Layered intrusions of the Monchegorsk ore region: petrology, mineralization, isotopes, deep structure. Kola Science Centre, Russian Academy of Sciences, Apatity, p 177 (in Russian)

Sokolov VA (ed) (1978) Volcanogenic structures in the Proterozoic of Karelia. Nauka (Science), Leningrad, p 168 (in Russian)

Sokolov VA (1980) Jatulian formations of the Karelian ASSR, In: Silvennoinen A (ed) Jatulian geology in the Eastern part of the Baltic Shield, Proceedings of a Finnish-Soviet symposium, Finland, 21–26 Aug 1979, Rovaniemi, The Committee for Scientific and Technical Co-operation between Finland and Soviet Union, pp 163–194

Sokolov VA, Galdobina LP, Ryleev AV, Satsuk YuI, Svetov AP, Heiskanen KI (1970) Geology, lithology and Palaeogeography of the Jatulian formations in Central Karelia. Institute of Geology, Karelian Science Centre, Academy of Sciences of the USSR, Petrozavodsk, p 366 (in Russian)

Sokolov VA, Stenar' MM, Kulikov VS (1987) Description of the geological structure of Karelia. In: Kulikov VS, Rybakov SI, Golubev AI, Svetov AP (eds) Guidebook to geological excursion to Karelia. Institute of Geology, Karelian Science Centre, Russian Academy of Sciences, Petrozavodsk, pp 3–20 (in Russian)

Spisiak J, Dostal J, Hovorka D, Meres S, Smirnov YuP (2007) Geochemical and mineralogical patterns of Proterozoic metapicrites from the Kola Superdeep Borehole (SG-3). Proc Murmansk State Tech Univ 10(2):198–210 (in Russian)

Stepanova A, Stepanov V (2010) Paleoproterozoic mafic dyke swarms of the Belomorian Province, eastern Fennoscandian Shield. Precambrian Res 183:602–616

Strand K (1985) Lithostratigraphy, palaeosedimentology and structure of the Lower Proterozoic Kurkikylä Group and overlying Karelian formations. Res Terrae, Ser B, No. 7, p 13 (in Finnish)

Sun S-S, McDonough WF (1989) Chemical and isotopic systematics of oceanic basalts; implications for mantle composition and processes. In: Saunders AD, Norry MJ (eds) Magmatism in the Ocean Basins. Geol Soc Spec Pub 42: 313–345

Svetov AP (1968) Some petrochemical features of Jatulian volcanic complexes in central Karelia. In: Sokolov VA, Alekseev BYa, Biskye GS, Bogachev AI, Pekki AS, Robonen VI, Eliseev MA (eds) Volcanogenic and ultrabasic complexes in the Proterozoic of Karelia. Institute of Geology, Karelian Science Centre, Academy of Sciences of the USSR, Petrozavodsk, pp 137–150 (in Russian)

Svetov AP (1979) Platformal Basaltic volcanism in the Karelides of Karelia. Nauka (Science), Leningrad, p 208 (in Russian)

Svetov AP (1980) The Jatulian volcanism in Karelia. In: Silvennoinen A (ed) Jatulian Geology in the Eastern Part of the Baltic Shield, Proceedings of a Finnish-Soviet symposium, Finland, 21–26 Aug 1979, Rovaniemi, The Committee for Scientific and Technical Co-operation between Finland and Soviet Union, pp 195–220

Svetov SA (2008) Liquation differentiation in basaltic systems (examples from Suisari variolites, Jalguba ridge). Geol Use Mineral Karelia 11:120–134 (in Russian)

Svetov AP, Bogachev AI (1976) Sill-like intrusive gabbro-diabases in central and southern Karelia. In: Kratts KO (ed) Precambrian intrusive Mafic-Ultramafic complexes in Karelia. Nauka (Science), Leningrad, pp 89–98 (in Russian)

Svetov AP, Sokolov VA (1978) The Shidguba volcanic structure. In: Sokolov VA (ed) Volcanic structures of the Proterozoic in Karelia. Nauka (Science), Leningrad, pp 111–119 (in Russian)

Svetov AP, Sviridenko LP (1992) Precambrian stratigraphy of Karelia, Sortavala series of Lake Ladoga Svecokarelides. Institute of Geology, Karelian Science Centre, Academy of Sciences of the USSR, Petrozavodsk, p 152 (in Russian)

Svetov AP, Sviridenko LP (2005) Centres of endogenous magmatic activity and ore-formation in the Fennoscandian Shield (Karelian Region). Institute of Geology, Karelian Science Centre, Russian Academy of Sciences, Petrozavodsk, p 357 (in Russian)

Svetov SA, Golubev AI, Svetova AI (2004) Geochemistry of Sumian basaltic andesites of Central Karelia. Geochem Int 42:630–640

Svetov SA, Svetova AI, Nazarova TN (2009) Do Sumian high-Mg andesite-basalts belong to the bajaite series? Geol Use Miner Karelia 12:112–124 (in Russian)

Svetov SA, Smol'kin VF, Paramonov AS (2011) Liquation differentiation in silicate systems: distribution of trace and rare earth elements, In: Voytekhovsky YuL (ed) Mineralogy, Petrology and Ores of the Kola Region, Proceedings of the VIII All-Russian Fersman Scientific Session Dedicated to the 135th anniversary of birth of academician D.S. Belyankin, Publishing House K & M, Apatity, pp 202–206 (in Russian)

Talvitie J, Paarma H (1980) Precambrian basic magmatism and the Ti-Fe ore formation in central and northern Finland. In: Siivola J (ed) Metallogeny of the Baltic Shield. Bull Geol Surv Finl 307: 98–107

Tel'nov VA, Bolshakov AN, Distler VV, Tourovtsev DM, Melezhik VA (1996) Early Proterozoic layered mafic-ultramafic massif of Mt. Generalskaya, In: Melezhik VA (ed) The geology and ore deposits of the Pechenga Greenstone Belt, IGCP Project 336, Field conference and symposium, field trip guidebook, Part III, Norges Geologiske Undersökelse, Trondheim, pp 51–69

Thompson RN, Morrison MA, Dickin AP, Hendry GL (1983) Continental flood basalts ... Arachnids rule OK? In: Hawkesworth CJ, Norry MJ (eds) Continental Basalts and Mantle Xenoliths. Shiva, Nantwich, pp 158–185

Tolstikhin SI, Semenov VS, Amelin YuV, Levchenkov LA, Neymark LA, Buiko AK, Koptev-Dvornikov YeV (1992) Juvenile helium in ancient rocks: II. U-He, K-Ar, Sm-Nd and Rb-Sr systematics in the Monche Pluton. $^{3}He/^{4}He$ ratios frozen in uranium-free ultramafic rocks. Geochimical et Cosmochimica Acta 56:987–999

Trofimov NN (2010) Ternavolok sill – conditions of formation and petrochemical characteristics. Geol Use Miner Karelia 14:107–115 (in Russian)

Tsuru A, Walker RJ, Kontinen A, Peltonen P, Hanski E (2000) Re-Os isotopic systematics of the 1.95 Ga Jormua ophiolite complex, northeastern Finland. Chem Geol 164:123–141

Väänänen J (1989) Volcanic rocks of the Kolari area, western Finnish Lapland, A report of the Lapland volcanite project, Geological survey of Finland, Report of investigations 86: 79 (in Finnish with English summary)

Vaasjoki M (2001) Radiometric age determinations from Finnish Lapland and their bearing on the timing of Precambrian volcano-sedimentary sequences. Geol Surv Finl Spec Pap 33:279

Vartiainen H, Woolley AR (1974) The age of the Sokli carbonatite, Finland, and some relationships of the North Atlantic alkaline igneous province. Bull Geol Soc Finl 46:81–91

Veki A (1991) On the Palaeoproterozoic volcanic and sedimentary rocks of the Kuusamo Schist Belt, Licenciate thesis, Department of Geology, University of Helsinki, p 81 (in Finnish)

Vetrin VR, Turkina OM, Rodionov NV (2008) U-Pb age and genesis of granitoids in the southern framing of the Pechenga structure, Baltic Shield. Doklady Earth Sci 419:298–302

Vogel DC, Vuollo JI, Alapieti TT, James RS (1998) Tectonic, stratigraphic, and geochemical comparisons between ca. 2500–2440 Ma mafic igneous events in the Canadian and Fennoscandian Shields. Precambrian Res 92:89–116

Vrevskii AB, Bogomolov ES, Zinger TF, Sergeev AA (2010) Polychronic sources and isotopic age of the volcanogenic complex (Arvarench Unit) of the Imandra-Varzuga structure, Kola Peninsula. Doklady Earth Sci 431:386–398

Vuollo J, Huhma H (2005) Paleoproterozoic mafic dikes in NE Finland. In: Lehtinen M, Nurmi PA, Rämö OT (eds) Precambrian geology of Finland – key to the evolution of the Fennoscandian Shield. Elsevier, Amsterdam, pp 195–278

Vuollo J, Piirainen T (1992) The 2.2 Ga old Koli layered sill: the low-Al tholeiitic (karjalitic) magma type and its differentiation in northern Karelia, eastern Finland. Geologiska Föreningens i Stockholm Förhandlingar (GFF) 114:131–142

Vuollo J, Piirainen T, Huhma H (1992) Two Early Proterozoic tholeiitic diabase dyke swarms in the Koli-Kaltimo area, eastern Finland – their geological significance. Bull Geol Surv Finl 363:32

Walker RJ, Morgan JW, Hanski EJ, Smolkin VF (1997) Re-Os systematics of early Proterozoic ferropicrites, Pechenga Complex, Russia: evidence for ancient ^{187}Os enriched plumes. Geochim Cosmochim Acta 61:3145–3160

Wood DA (1980) The application of a Th-Hf-Ta diagram to problems of tectonomagmatic classification and to establishing the nature of crustal contamination of basaltic lavas of the British Tertiary volcanic province. Earth Planet Sci Lett 50:11–30

Zagorodny VG, Mirskaya DD (1967) On the ultramafic effusives of the fourth volcanic unit of the Pechenga Series. In: Ivanov AM (ed) Mafic and Ultramafic rocks of the Kola Peninsula. Nauka (Science), Leningrad, pp 3–10 (in Russian)

Zagorodny VG, Kozlov MT, Predovsky AA, Skuf'in PK (1980) The mode of occurrence of early Proterozoic volcanism and some features of its evolution. In: Zagorodny VG, Predovsky AA, Skuf'in PK (eds) Early Precambrian Volcanic Rocks in the Kola Peninsula, Atlas of textures and structures. Nauka (Science), Leningrad, pp 63–85 (in Russian)

Zhangurov AA, Smol'kin VF (1982) Mafic and ultramafic rocks of the middle and late Karelian and Caledonian stages. In: Gorbunov GI (ed) The Imandra/Varzuga Zone of Karelides. Nauka (Science), Leningrad, pp 145–191 (in Russian)

Part IV

Geology of the Drilling Sites

4.1 The Imandra/Varzuga Greenstone Belt

V.A. Melezhik

4.1.1 Introduction

The FAR-DEEP drilling operations were performed in three spatially separated areas in the eastern part of the Fennoscandian Shield, stretching over a distance of c. 800 km from the Barents Sea in the north to Lake Onega in the south (Fig. 4.1). In geological terms, these three areas are the Pechenga Greenstone Belt in the north, the Imandra/Varzuga Greenstone Belt in the northeast, and the Onega Basin in the south. These well-explored and well-mapped areas of the Fennoscandian Shield have rather similar geology, but the preservation of the different formations in terms of how they can be used to assess global geological events. Thus each region provides complementary geological information that, when combined, produces a composite but comprehensive record of the global palaeoenvironmental perturbations across the Archaean-Palaeoproterozoic transition known from Fennoscandia. The Imandra/Varzuga Greenstone Belt (Figs. 4.1 and 4.2) has been the target for continuous exploration for Pt, Ni, Cu, and Au, which has allowed to establish a regional geological framework including information on the stratigraphy, regional geochemistry, metamorphic zoning, and igneous petrology. Main regional geological features were established during the 1:200,000 scale mapping in the 1960s (Potrubovich and Simon 1966) and subsequently complemented by academic research (Kharitonov 1966; Kozlov 1971; Gilyarova 1972; Radchenko 1972a; Bekasova and Pushkin 1972). Later results of the 1:50,000 scale mapping were combined with considerable contributions from various research programmes, as summarised in four monographs devoted to the general geology and stratigraphy (Zagorodny et al. 1982), sedimentology (Melezhik and Predovsky 1982), and geochemistry of sedimentary and volcanic rocks of the Imandra-Varzuga Belt (Fedotov 1985; Predovsky et al. 1987; Borisov 1990). Geotectonic and geodynamical aspects have been addressed in several papers by Zagorodny and Radchenko (1988), Mints (1993, 1966) and Mints et al. 1996, Melezhik and Sturt (1994), and Mintz et al. (1996). There is only a limited number of geochronological measurements constraining the depositional time of sedimentary and volcanic rocks (Amelin et al. 1995; Skuf'in et al. 2006; Vrevsky et al. 2010), though a considerable effort has been focussed on dating the emplacement of economically important layered gabbronorite complexes (Bayanova et al. 2009).

V.A. Melezhik (✉)
Geological Survey of Norway, Postboks 6315 Sluppen, NO-7491 Trondheim, Norway

Centre for Geobiology, University of Bergen, Allegaten 41, N-5007 Bergen, Norway
e-mail: victor.melezhik@ngu.no

4.1.2 General Geological Features

The Imandra/Varzuga Greenstone Belt is a southeastern part of the c. 800-km-long, discontinuously developed system of supracrustal belts extending from northern Norway through northern Finland to NW Russia where it runs across the Kola Peninsula. The composite belt was termed the North Transfennoscandian Greenstone Belt by Melezhik (1996). It has been interpreted as representing a long, several times aborted and reactivated rifting, which was initiated around 2500 Ma in the easternmost part of the Kola Peninsula. The earliest, though undated, rifting episode is recorded in the lowermost unit of the Imandra/Varzuga Greenstone Belt, which is known as the Purnach Formation (Fig. 4.3).

Structurally, the Imandra/Varzuga Greenstone Belt has been termed "a synclinorium" (Polkanov 1936), "an asymmetrical syclinorium" (Kharitonov 1966), "a compound trough" (Simon 1967), "an asymmetrical fault-bound trough" (Gilyarova 1972), or "an asymmetrical synclinorium fold-belt with a reduced southern limb" (Zagorodny and Radchenko 1982, 1988). These different designations reflect a complex structural organisation of the belt and the lack of consensus on its geotectonic constrains. The Imandra/

V.A. Melezhik et al. (eds.), *Reading the Archive of Earth's Oxygenation*, Volume 1: The Palaeoproterozoic of Fennoscandia as Context for the Fennoscandian Arctic Russia - Drilling Early Earth Project,
DOI 10.1007/978-3-642-29682-6_7, © Springer-Verlag Berlin Heidelberg 2013

Varzuga Greenstone Belt belt has been divided into the Northern and Southern Zones separated by a WE-trending and south-dipping fault. The Northern Zone is further divided by NS-trending transversal faults into a few blocks (Zagorodny et al. 1982) with some of them showing a syndepositional nature (e.g. the Tsaga and Strel'na faults, Fig. 4.2b) (Zagorodny and Radchenko 1982; Melezhik and Predovsky 1982; Melezhik 1992). Each fault-bound, transversal block has its characteristic structural organisation. However, the structural features of the belt have so far been poorly studied. In the Central transversal block, the Northern and Southern zones are separated by a longitudinal fault dipping steeply southwards (Fig. 4.2c). Here, the Northern Zone represents a NW-SE-trending, simple half-graben filled predominantly with volcanic rocks dipping 20–60° to SW. In contrast, the Southern Zone appears to be a strongly folded and apparently tectonically imbricated nappe complex overthrust by Archaean rocks from the southwest (Daly et al. 2006); rocks dip 40–90° both south- and northwards (Fig. 4.2c).

In the Eastern block, only rocks belonging to the Northern Zones occur, forming a west-plunging synform closed in the east (Fig. 4.2b). The rocks of the northern limb of this block have a tectonic contact with the Archaean basement rocks, whereas a depositional contact has been reported in places in the southern limb (Zagorodny and Radchenko 1982). However, Daly et al. (2006) maintained that the contact is tectonic throughout of the southern margin and that the Imandra/Varzuga rocks are overthrust by Archaean and younger Palaeoproterozoic rocks from the southwest; just in a similar way as in the central block. Rocks of the Eastern block dip steeply both to the south and north (50–80°).

In the Western block, the belt has a tectonically reduced width (Fig. 4.2c), and the rocks of the Northern and Southern Zones form an east-plunging synform, which closes in the west. The rocks of the northern limb have a depositional contact with Archaean basement rocks, whereas the nature of the contact of the southern limb with the basement rocks remains unclear (Zagorodny and Radchenko 1982).

4.1.3 Stratigraphic Subdivision and Age

The stratigraphic subdivision of the Imandra/Varzuga Greenstone Belt has been established in the 1960s during the 1:200,000 mapping and prospecting programmes (Potrubovich and Simon 1966). It was refined in the 1970s (Bekasova and Pushkin 1972) and has remained essentially unchanged since Zagorodny et al. (1982) published a summary monograph with detailed descriptions of the regional geology, tectonics, stratigraphy, geochemistry, and history of development.

The original stratigraphic names presented by Zagorodny et al. (1982) has been utilised in this account, but the Russian unit terms have been replaced with those commonly used in the western literature; i.e., "series" is substituted by "group", "suite" by "formation", etc. Lithostratigraphically, the Imandra/Varzuga Greenstone Belt comprises the Imandra/Varzuga Supergroup (Melezhik and Sturt 1994), which was originally subdivided into the Strel'na, Varzuga and Tominga series (Zagorodny et al. 1982). The series have later been renamed as the corresponding groups (Melezhik and Sturt 1994). The Strel'na and Varzuga groups are located within the Northern (structural) Zone, whereas the Tominga Group occurs only within the Southern Zone. Although the Tominga Group is presently separated from two older groups by the thrust fault, their chronostratigraphic relationship may have been more complex. This cannot be resolved unless precise radiometric ages, constraining their depositional time, are available.

The lower age limit of the Strel'na Group is constrained by the presence of gabbronorite pebbles in the basal conglomerates of the Kuksha Sedimentary Formation (Fig. 4.3). The pebbles are derived from the gabbro-notite intrusions, which have been dated at 2504.4 ± 1.5 Ma (U-Pb-zircon, Amelin et al. 1995). The upper age limit of the Tominga Group, the youngest in the belt, is older than 1907 ± 18 Ma (U-Pb-zircon, Skuf'in et al. 2006), the age obtained for subvolcanic trachydacite emplaced into rhyolites and dacites of the group.

The Strel'na Group

The cumulative stratigraphic thicknesses of the sedimentary and volcanic rocks of the Strel'na Group (Fig. 4.3) reach 1,400 m and 5,300 m, respectively. The group was originally subdivided into three "suites" (Fedotov et al. 1982), which are from base to top: Purnach, Kuksha and Seidorechka. The Purnach suite comprises alternating sedimentary and volcanic units, whereas the Kuksha and Seidorechka suites comprise a sedimentary-volcanic pair. Each pair begins with sedimentary and ends with volcanic rocks. Further, each sedimentary-volcanic cycle is separated from the overlying one by a nondepositional disconformity, generally marked by palaeoweathering surfaces (Melezhik and Sturt 1994). Boundaries between the cycles are sharp. The sedimentary units of the cycles form formally defined sedimentary "subsuites", whereas the volcanic units comprise formally defined volcanic "subsuites" (Fedotov et al. 1982). The subsuites have been redefined later as formations (Melezhik and Sturt 1994). Thus, the Strel'na Group embraces five formal lithostratigraphic units in the rank of formation (Fig. 4.3).

The Purnach Formation

The Purnach Formation (Fig. 4.3) was originally defined as a suite of polymict sandstones intruded by amphibolite sills (Potrubovich and Simon 1966). The amphibolites were later reinterpreted as volcanic rocks and combined together with the polymict sandstones to form the Purnach suite (Zagorodny et al. 1972) that corresponds to the Purnach Formation of of Melezhik and Sturt (1994).

The western extension of the Purnach Formation is confined by the Strel'na syndepositional fault (Fig. 4.2b), from which it extends eastwards over a distance of 135 km, though the formation thins completely out before reaching the eastern margin of the belt. The formational contact with the various underlying Archaean rocks is tectonic. The formation is more than 2,000 m thick and comprises mainly amphibolites accompanied by four sedimentary units, each 100–200 m in thickness; three of them are situated in the lower part of the formation and one in the uppermost part. The sedimentary rocks have been documented only in the easternmost part of the belt, and their occurrence is limited by a distance of c. 25 km along strike. The rocks have undergone amphibolite facies metamorphism (Petrov and Voloshina 1982). The sedimentary rocks are biotite-quartz-feldspar sandstones with horizontal bedding in two lower units, whereas tabular multidirectional and trough cross-bedding have been documented in two upper units. The amphibolites occur as several interbedded varieties: fine-grained and medium-grained, massive, and amygdaloidal. Geochemically they represent a weakly differentiated series of tholeiitic basalts (Fedotov et al. 1982).

There are no radiometric age data available that could be used to constrain the depositional age of the Purnach Formation. The formation is wedging out along strike from the eastern part of the belt in both directions and is conformably overstepped by the overlying Kuksha Sedimentary Formation, which also rests on various Archaean rocks. This relationship may suggest that either the Purnach Formation was originally deposited in a localised site and succeeded by the Kuksha Sedimentary Formation, or it was considerably eroded prior to the deposition of the Kuksha Sedimentary Formation. Field observations do not allow to discriminate between these two alternatives.

The Kuksha Sedimentary Formation

The Kuksha suite, divided into the Lower (sedimentary) and Upper (volcanic) subsuites, was defined to substitute the Rizhguba (sedimentary) and Strel'na (volcanic) suites in the eastern part of the belt, whereas in its western part, it equates to the so-called "first" sedimentary and "first" volcanic units (Melezhik 1976; Fedotov et al. 1982; Fedotov 1985). Later, the Lower and Upper subsuites were renamed as formations, namely the Kuksha Sedimentary and Volcanic formations (Melezhik and Sturt 1994) (Fig. 4.3).

At the western end of the Western block, the Kuksha Sedimentary Formation occurs as a thin (less then 40 m), discontinuous unit along the northern margin of the belt in an area west of Lake Imandra (Fig. 4.2c; the formation is not shown in the map due to its small thickness). The formation reappears in the Eastern block as a thick, continuous unit stretching from the Strel'na syndepositional fault eastwards, until it is cut by a fault (Fig. 4.2b). Thus, the lateral extent of the formation within the Eastern block is similar to that of the Purnach Formation and was apparently controlled by the Strel'na syndepositional fault (Zagorodny and Radchenko 1982). The thickness of the formation is variable and reaches 400 m. The rocks rest either with a depositional contact on the underlying Purnach amphibolites in the east or erosively on diverse Archaean rocks in the west. The formation has been metamorphosed under epidote-actinolite and epidote-amphibolite facies conditions (Petrov and Voloshina 1982).

In the Eastern block, the lower part of the formation comprises greywackes, whereas the upper part is composed of arkosic sandstones. Both the greywackes and arkosic sandstones exhibit either tabular, unidirectional cross-bedding or horizontal, parallel bedding (Fedotov et al. 1982). In the Western block the formation comprises volcaniclastic sandstones with tabular cross-bedding and horizontal, parallel bedding, and polymict conglomerates, whose clasts commonly reflect the composition of underlying rocks (Melezhik 1976; Fedotov et al. 1982) (Fig. 4.4a, b). In the vicinity of the drilling site of FAR-DEEP Hole 1A (Fig. 4.2c), the formation has a considerably reduced thickness and rests either on an Archaean diorite or on the 2504.4 ± 1.5 Ma (Amelin et al. 1995) layered gabbronorite intrusion. In the former case, the formation is composed of diorite-clast breccia supported by a mafic tuff matrix, whereas in the latter case, it is represented by polymict (mainly gabbro boulders) conglomerate overlain by shales and amygdaloidal basalt (Fig. 4.5). The nature of this gabbroic conglomerate, known as the "Vurechuajvench" conglomerate, has been a matter of continued debate with opinions favouring a tectonic or sedimentary origin (Kozlov and Latyshev 1974; Sidorenko and Luneva 1961; Fedotov et al. 1975; Melezhik 1976). However, recent intensive drilling has shown that the gabbroic boulder-conglomerate overlies depositionally different layers of the layered gabbronorite intrusion (Fig. 4.6), and the base of the conglomerate cuts down the section, from east to west, by more than 300 m over a distance of approximately 2 km. Here, the conglomerate consisting of angular and rounded clasts of gabbro, norite, anorthosite, granophyre and diorite (Fig. 4.5c–f) also appears to seal faulted surface of the gabbronorite body. This provides an important time constraint implying that the Kuksha Sedimentary Formation was deposited later than 2504 Ma ago.

The Kuksha Volcanic Formation

The Kuksha Volcanic Formation (Fig. 4.3) occurs as a 200- to 1,000-m-thick unit continuously developed along the northern margin of the belt from the western side of Lake Imandra to the Eastern block, where it forms the centrocline closure of a larger synform and then re-appears in its southern limb. The formation rests conformably on the Kuksha Sedimentary Formation. In the Western block, where the underlying sedimentary unit is absent, the formation lies directly on Archaean diorite or gabbronorite.

The formation comprises mafic volcanic rocks metamorphosed under amphibolite facies conditions. The volcanic rocks occur as numerous 10- to 40-m-thick, well-defined lava flows with brecciated and amygdaloidal upper parts (Fig. 4.4c). Pillows are rare (Fig. 4.4d) implying essentially subaerial eruptions. In the Western block, on top of Mt. Arvarench (Fig. 4.2b), the volcanic suite is intruded by coarse-grained, mafic, subvolcanic bodies. Some lava flows contain thin screens of bedded carbonate rocks, and xenoliths and dismembered beds of cross-bedded arkosic sandstones (Fig. 4.4e). Based on a significant database, Fedotov (1985) demonstrated that the chemical composition of the mafic lavas is monotonous and similar throughout the belt. He assigned the Kuksha Volcanic Formation to an undifferentiated series of rocks resembling plateau basalts (Fedotov et al. 1982).

The Seidorechka Sedimentary Formation

The Seidorechka Sedimentary Formation (Fig. 4.3), originally known as the Lower Seidorechka subsuite (e.g. Zagorodny et al. 1972), occurs through the Eastern and Central blocks and is cut by a fault in the Western block, though reappears in the northwestern part of the belt (Fig. 4.2b). Along a strike length of 350 km, the formational thickness remains constantly around 300–350 m in the Central block, but is reduced to 150 m towards both the Western and the Eastern blocks. At both ends of the belt, the formation forms the periclinal closure of a large synform, which shows reduced extensions of the southern limbs; the latter in most cases is overthrust by Archaean and younger Palaeoproterozoic rocks. The formation underwent regional metamorphism grading from biotite-chlorite-actinolite facies in the Western block, through biotite-epidote-actinolite facies in the Central block to epidote-amphibolite facies in the Eastern block (Petrov and Voloshina 1982). The actual contact with the underlying Kuksha volcanic rocks has been described only from the Western block, in an area west of Lake Imandra (Fig. 4.2b), where the Seidorechka Formation shales rest with a weathering crust on Kuksha basalts (Fedotov et al. 1975).

Originally the formation was subdivided into four lithostratigraphic members (Melezhik 1976), being from base to top: A (sericite and quartz-sericite shist), B (quartzite), C (sericite-quartz chlorite schist, subordinate dolostone) and D (metaporphyrite, chlorite-albite-quartze schist and picritic tuff). However, a different subdivision is used in this contribution based on the FAR-DEEP drilling results (see Chap. 6.1.1). Member D is considered to represent the lowermost part of the overlying Seidorechka Volcanic Formations. The remaining part of the succession has been subdivided into the Sandstone-Siltstone member (equivalent to Member A of Melezhik (1976)), Dolostone member (new), Quartzite member (equivalent to B), and Limestone-Shale and Shale members (collectively equating to Member C).

The first detailed sedimentological description of the Seidorechka sedimentary rocks was published by Sidorenko and Luneva (1961), which has later been supplemented by results of litho-geochemical research (Melezhik 1976; Fedotov et al. 1982). In the most-detailed studied western part of the belt, the lower part of the formation is dominated by sandstone-siltstone with lenticular, flaser and wavy bedding (Sandstone-Siltstone member, 10–50 m; Fig. 4.7a–c), followed by a thin dolostone unit (Dolostone member, 0–2 m; Fig. 4.7d) and cross-bedded quartzite (Quartzite member, 15 m; Fig. 4.7e–i), whereas calcareous schists and shales of the upper part (Limestone-Shale, 30 m; and Shale, 50 members) exhibit flaser and thin parallel bedding (Fig. 4.7j–k), respectively.

The formation retains all major lithological subdivisions throughout the belt, and sedimentological features of the rocks reflect a depositional setting within a shallow-water clastic shelf through an incipient carbonate platform to a progressively deepening continental shelf. Clasts provenance and sedimentological studies suggest a NS-transport direction for clastic material (Bekasova and Pushkin 1977). Carbonate rocks exhibit $\delta^{13}C$ ranging from -5.4 ‰ to -1.8 ‰ (Melezhik and Fallick 1996). Other details on the depositional framework based on preliminary investigation of FAR-DEEP core and associated new material are presented in Chap. 6.1.1.

The formation is intruded by numerous dykes and sills of pyroxenite and melanocratic gabbro (termed the Kuksha pyroxenite-gabbro complex), which are also known to crosscut the older Purnach and Kuksha formations as well as the Archaean basement (Dokuchaeva et al. 1982). Based on their stratigraphic distribution and geochemical composition, Fedotov (1978) regarded the igneous bodies as feeders for the overlying Seidorechka Volcanic Formation. Further, Dokuchaeva et al. (1982) reported geochemical similarity of the gabbro with the Imandra lopolith and suggested their comagmatic nature, thus linking together the Kuksha pyroxenite-gabbro complex, the Imandra lopolith and the

Seidorechka Volcanic Formation. This inference has later been corroborated on the basis of REE geochemistry (Melezhik and Sturt 1994) and a precise age determination (2441 Ma; Amelin et al. 1995), which is discussed in some detail below.

The Seidorechka Volcanic Formation

The Seidorechka Volcanic Formation (Fig. 4.3), originally known as the Upper Seidorechka subsuite (e.g. Zagorodny et al. 1972), rests conformably on the Seidorechka Sedimentary Formation. It is the thickest formation in the Imandra/Varzuga Greenstone Belt and, together with its sedimentary counterpart, is the only formation occurring throughout the entire belt length of 350 km. The formational thickness is close to 3,000 m in the Central block, though being significantly reduced in the western and eastern parts of the belt. Similar to the sedimentary formation, the volcanic unit forms the periclinal closure of the large synform at both ends of the belt. In the southern limb of the Western block, the thickness of the formation is tectonically reduced. In the Central block, it is overthrust by Archaean and younger Palaeoproterozoic rocks (e.g. Daly et al. (2006). Also similarly to its sedimentary counterpart, the formation underwent regional metamorphism grading from biotite-chlorite-actinolite zone in the west to epidote-amphibolite facies in the east (Petrov and Voloshina 1982).

The Seidorechka Sedimentary Formation starts with a 100-m-thick volcano-sedimentary unit (Member D of Melezhik 1976), which is composed in the Central and Eastern blocks of several mafic lava flows overlain by tuff-matrix sandstone, conglomerate, and komatiitic tuff. The upper part of this unit is also known as "volcanic breccia" (Kharitonov 1966), "xenoclastolava" (Radchenko 1972b) or a "Liht-Luht conglomerate" (Figs. 4.4b and 4.7l–n). This horizon is traceable over a distance of 300 km and serves as a marker unit within the Seidorechka Volcanic Formation (Fedotov 1985).

The remaining part of the formation has been subdivided into three members. The lower member has a thickness of 800 m and comprises a series of lava flows of melanocratic komatiitic basalts and amygdaloidal basaltic andesites (Fedotov et al. 1982; Fedotov 1985). The member contains numerous sills and dykes of melanocratic gabbro and pyroxenite with komatiitic affinities (the Kuksha pyroxenite-gabbro complex; Dokuchaeva et al. 1982), which also intrude the 2505 Ma gabbronorite and the Purnach, Kuksha and Seidorechka Sedimentary formations, but are unknown in the overlying units. They are similar in major, trace and rare-earth element geochemistry with the Seidorechka melanocratic komatiitic basalts, and thus were interpreted as feeder dikes/channels for the lower member of the Seidorechka Volcanic Formation (Fedotov 1985; Melezhik and Sturt 1994).

The middle member has a maximum thickness of 1,500 m and is made up of homogeneous basaltic andesites (Fig. 4.7o), which can be traced throughout the belt. In contrast, the upper member has a variable thickness ranging from zero to 300 m and mostly occurs in the Western block, where it forms the periclinal closure, and in both limbs of the synform (Fedotov. 1985). The upper member is composed of amygdaloidal, rhyolitic and dacitic lava flows (Fig. 4.7p, q). In the Western block, they contain large synvolcanic and apparently co-magmatic bodies of granophyre, locally termed "imandrites" (Ramsay and Hackman 1894). These have been dated at 2442.2 ± 1.7 Ma (Amelin et al. 1995) and provide an age constraint for the upper part of the Seidorechka Volcanic Formation (for more information see Chap. 3.2).

In the Western block, in the area of the Mountain Arvarench (Fig. 4.2b), a probable stratigraphic equivalent of the Seidorechka Volcanic Formation is represented by a komatiitic basalt-andesite-dacite-rhyolite association (known as the Arvarench Formation, e.g. Radchenko et al. 1994). Previously these rocks have been regarded as late Archaean (Pozhilenko et al. (1999), but the recently obtained U-P age of 2429 ± 6.6 Ma (Vrevsky et al. 2010) is consistent with the age of 2442.2 ± 1.7 Ma reported previously by Amelin et al. (1995) for the Seidorechka Volcanic Formation from a different locality. In the Mountain Arvarench area, the exposed part of the formation starts with massive, pillowed and amygdaloidal basalt, mafic agglomerate and flow breccia with several 5- to 12-m-thick bodies of ultramafic and mafic komatiite and bedded tuff. These are succeeded by a c. 1,000-m-thick package of amygdaloidal andesitic, dacitic and rhyolitic lavas (Fig. 4.7r–s) with thin interbeds of agglomerate and accretional lapilli tuffs (Fig. 4.7t).

In the Western block, on the either side of Lake Imandra, the Seidorechka Volcanic Formation has been intruded by the chromite-bearing Imandra layered gabbronorite lopolith tectonically dismembered into a series of fragments (Dokuchaeva et al. 1980). The lower contact of the lopolith with the Seidorechka Volcanic Formation is exposed on either side of Lake Imandra and has been intersected by drillholes in an area south of the FAR-DEEP Hole 1A site (Fig. 4.2c). The contact has an intrusive nature and dips 30–80º to the south and southeast, while the upper contact is not exposed. The lopolith is intersected by peridotite-pyroxenite dykes, which have been reported to be comagmatic with the Polisarka Volcanic Formation komatiites (Dokuchaeva et al. 1982); the latter are stratigraphically above the Seidorechka formations (Fig. 4.3). The emplacement time of the Imandra lopolith has been constrained to 2441.0 ± 1.6 Ma (Amelin et al. 1995). Major, trace and REE geochemistry of the layered gabbronorite complex closely corresponds to that of the volcanic rocks of the Seidorechka Volcanic Formation. This together with the similarity in the ages of emplacement and deposition has been taken as evidence that the intrusive and effusive rocks form a genetically related volcano-plutonic association (Melezhik and Sturt 1994; Chashchin et al. 2008).

Very little has been published on the palaeovolcanology of the Seidorechka Volcanic Formation. Widespread, laterally continuous lava flows with a limited development of pyroclastic facies over the entire beltlength indicate predominantly fissure-type eruptions and have been assigned to a flood-basalt setting (Fedotov 1985). An exception is rhyolitic and dacitic lavas occurring only in the western part of the belt where they show highly variable thicknesses. They might have been erupted from a central type of volcano (Fedotov 1985). The mafic komatiitic "volcanic breccia" or "xenoclastolava" may indicate a phase of uplift followed by erosion and revoking of pyroclastic material. Based on geochemical data, Mints (1996) assigned the Seidorechka volcanism to an active continental margin and interpreted it as a volcanic-arc association.

The Varzuga Group

The Varzuga Series (Zagorodny et al. 1972; Melezhik et al. 1982), later redefined as the Varzuga Group (Melezhik and Sturt 1994) (Fig. 4.3), has cumulative stratigraphic thicknesses of 1,200 m and 3,800 m for its sedimentary and volcanic rocks, respectively. The group was originally subdivided into three "suites" (Zagorodny et al. 1972; Melezhik et al. 1982), named from base to top as Polisarka, Umba and Il'mozero. All of them comprise sedimentary-volcanic pairs with each pair beginning with sedimentary and ending with a volcanic unit. The sedimentary units of the cycles form formally-defined sedimentary "subsuites", whereas the volcanic units of the cycles comprise formally defined volcanic "subsuites" (Zagorodny et al. 1972; Melezhik et al. 1982), both later redefined as formations (Melezhik and Sturt 1994). Thus, the Varzuga Group consists of six formal lithostratigrphic units being, from oldest to youngest, the Polisarka Sedimentary and Volcanic, Umba Sedimentary and Volcanic, and Il'mozero Sedimentary and Volcanic formations (Fig. 4.3).

The Polisarka Sedimentary Formation

Originally in the 1960s, the Polisarka Sedimentary Formation (Fig. 4.3) was defined as the Lower Polisarka subsuite by Sokolova (unpublished report cited in Zagorodny et al. 1972) as the result of 1:200,000 mapping of the central part of the Imandra/Varzuga Belt. Later, the subsuite was redefined as the Polisarka Sedimentary Formation (Melezhik and Sturt 1994). The formation is known to occur in the Central block and immediately east of the Strel'na fault in the Eastern block (Figs. 4.2b and 4.8a). Its thickness varies from 20 to 115 m. The thickness variation has been interpreted to have been controlled by several syndepositional blocks and faults (Melezhik and Predovsky 1982), which is schematically illustrated by a longitudinal lithological profile in Fig. 4.8b. The formation comprises diverse sedimentary rocks including polymict conglomerates, varved siltstones with dropstones, shales, marls and limestones (Fig. 4.9a–g).

Zagorodny et al. (1972) reported that the Polisarka basal conglomerate rests erosionally and with an angular unconformity on palaeoweathered rocks of the Seidorechka Volcanic Formation. Up to 20-m-thick sericite-chlorite-carbonate shale with relics of a porphyritic structure has been reported from the Lower Pana Lake section (Sect. 2 in Fig. 4.8b) and interpreted as representing intensively weathered andesites of the underlying Seidorechka Volcanic Formation (Potrubovich and Simon 1966). Further research (Melezhik and Predovky 1982; Melezhik et al. 1982) has shown that the overlying Polisarka basal conglomerate contains variably weathered clasts derived from the underlying dacites, andesites, and basaltic andesites with the most weathered clasts located in Sect. 3 (Fig. 4.8b), thus confirming the existence of weathered rocks prior to the erosion and deposition of the conglomerate.

The fluctuation of formational thicknesses is accompanied by a considerable lithofacies variation (Fig. 4.8b). To the east of the Strel'na fault (Sect. 6; Fig. 4.8b), the formation is only 15 m thick and consists of komatiitic and basaltic tuffs (Melezhik and Predovsky 1982). On the other hand, to the west of the Strel'na fault (Sects. 4 and 5; Fig. 4.8b), the tuffs are rapidly overstepped by an up to 29-m-thick unit of polymict, poorly-sorted, boulder- to pebble-conglomerate overlain by massive, bedded and varved greywacke and volcaniclastic siltstone with slump structures, abundant lonestones and dropstones, and small pockets of polymict, fine-pebble conglomerate (Fig. 4.9b–g). Although, in Sects. 4 and 5 (Fig. 4.8b), clasts in the basal conglomerate differ in composition, the majority is represented by dacitic rhyolite and quartz-plagioclase rocks with granophyric and bostonitic textures (Melezhik and Predovsky 1982). The lower part of the greywacke-varved siltstone unit contains a c. 10-m-thick body of spinifex-textured ultramafic komatiite (Fig. 4.9h). It remains unclear whether this is a sill or a lava flow, though the overlying massive greywacke is injected by 1- to 5-cm-thick "veinlets" of fine-grained ultramafic komatiite. However, their connection to the main komatiitic body has not been observed.

Further to the west, in Sect. 3, the lithofacies are somewhat similar to those in Sects. 4 and 5. The formational base is composed of polymict, poorly-sorted boulder- to pebble-conglomerate containing rounded clasts of variably weathered andesites and basaltic andesites reflecting the lithology of the immediate local source (Fig. 4.9a). However, the lower part of the greywacke-varved siltstone unit is overstepped here by finely-laminated, varved, clayey siltstones. Similarly to the varved volcaniclastic siltstones, they contain abundant dropstones. The clayey siltstones were previously called "pudding conglomerate" (Melezhik and Predovsky 1982) but were later reinterpreted as glaciogenic mudstones with "dropstones" (Melezhik et al. 2005). These rocks are overlain by massive, bedded and laminated (varved) greywacke and volcaniclastic siltstone with dropstones.

In Sect. 2, the basal conglomerate, with the thickness reduced to 10 m, lies on the weathered andesites as discussed above. The mudstone unit has a considerably reduced thickness though it is overlain by the greywacke-varved siltstone having a thickness similar to that in Sect. 3.

The westernmost part of the Polisarka Sedimentary Formation, located west of the Tsaga syndepositional fault, has been intersected by FAR-DEEP Hole 3A (Sect. 1; for details, see Chap. 6.1.2). The section exhibits an outstanding difference in its lithological composition and thickness compared to those known from all other sections (Fig. 4.8b). The true thickness is twice as large as elsewhere and approaches 120 m, and the main lithofacies are limestones, varved marls with rare dropstones, carbonate breccias and gritstones, diamictite, and several mafic lava flows. These lithofacies rest directly on rhyodacites, thus the basal conglomerate, as observed in others section, is missing here. The c. 100-m-thick, limestone-dominated unit contains a c. 40-m-thick sill or flow of ultamafic komatiite and is overlain by a c. 25-m-thick succession of massive diamictites, parallel- and cross-laminated sandstones, and siltstones with abundant lonestones and dropstones (for details, see Chap. 6.1.2).

The overall thicknesses and lithofacies variations observed in Sects. 2, 3, 4, 5, and 6 were previously attributed to a differential subsidence/uplift of transversal blocks bound by syndepositional Tsaga and Strel'na faults (Melezhik and Predovsky 1982; Melezhik et al. 1982). New data provided by FAR-DEEP Hole 3A shows even more drastic thickness and lithofacies variations, which we tentatively interpret as the result of out-of-phase, rapid subsidence of the transversal block located east of the Tsaga fault (Fig. 4.8). If the boundary between the Polisarka Volcanic and Sedimentary Formations is taken as a reference line, then it appears that a considerable sea-floor palaeotopography has been evened by the deposition of the uppermost sandstone-siltstone, which have similar lithological composition and thickness in all sections except in Sect. 6. The lithofacies documented in Sects. 2, 3, 4, 5, and 6 were previously assigned to a lacustrine, rift-bound basin and no glacial rocks were recognised, though the presence of the unusual "pudding conglomerates" has been acknowledged (Melezhik and Predovsky 1982). The Polisarka glacial rocks were later identified based on re-interpretation of old material (Melezhik et al. 2005). The new FAR-DEEP Core 3A data have not only confirmed the presence of glacial depositis (diamictite, varves, shales with dropstones) but also have broader palaeoenvironmental implications. The thick succession of various limestone lithologies, including those with a high Sr abundance (Appendices 5 and 6), suggest that at least the western part of the Polisarka basin was very likely marine. Other details on the depositional framework based on preliminary investigation of the FAR-DEEP drillcore are presented in Chap. 6.1.2.

The Polisarka Volcanic Formation

The Polisarka Volcanic Formation (Fig. 4.3) was originally in the 1960s defined as the Upper Polisarka subsuite (Zagorodny et al. 1972). The formation rests with a sharp contact on the Polisarka Sedimentary Formation and occurs as a 300- to 600-m-thick unit continuously developed from the eastern side of Lake Imandra towards the east where it is terminated by the Strel'na fault (Fig. 4.2b, c). In the west, the Polisarka volcanic rocks also occur in the southern limb of the synform, south of the Khibiny alkaline massif (Fig. 4.2c).

The formation consists of komatiitic series rocks ranging in composition from melanocratic basalts (80 vol.%) to basalts and basaltic andesites (Fedotov 1985). The former occupy the lower part of the formation and occur either as thin massive lava flows with flow-top breccias or thick piles of pillowed lavas. The upper part of the formation is represented by melanocratic basalts, basalts, and basaltic andesites in the form of thick flow breccias (Fig. 4.9i), pillowed lavas, massive lava flows, and tuffs. The basaltic units show considerable thickness variations. In the Central block (Fig. 4.2b) they form a thin pile of flows or are even totally absent, whereas in the Eastern and Western blocks they occur predominantly as thick piles of flow breccia and show the most diverse facies variations (Melezhik et al. 1982; Fedotov 1985).

The melanocratic basalts are often spinifex-textured, variolitic or globular (Fig. 4.9i–j). They are composed of actinolite and preserve relics of magmatic textures such as chlorite pseudomorphs after olivine phenocrysts and spinifex pyroxene (for details, see Chap. 6.1.2). Basaltic andesites are commonly ophitic or micro-porphyritic rocks with plagioclase phenocrysts reaching up to 5 vol.% in flow-breccia clasts (Fedotov 1985). The Polisarka Volcanic Formation rocks form a differentiated series with MgO contents ranging between 4 and 19 wt.%. Although the MgO content shows a continuous range, the overall data show a bimodal distribution with maxima at 12–15 wt.% and 4–6 wt.%. Similarly to many Palaeoproterozoic komatiitic series, the Polisarka volcanic series is relatively rich in SiO_2 and depleted in TiO_2 (Fedotov 1985).

The Polisarka volcanism was assigned to submarine settings (abundant pillowed lavas) with a dominance of fissure eruptions during early stages (thick continuous mafic and ultramafic lava flows). A central type of eruption was a feature of the late stage when basaltic andesites were dominant volcanic products (Fedotov 1985).

The Umba Sedimentary Formation

The Umba Sedimentary Formation is part of the former Umba suite, which was first defined in 1962 by V.N. Sokolova (cited in Zagorodny et al. 1972). The follow-up work by Bekasova and Pushkin (1972) led to an adjustment of its upper boundary and subdivision of the suite into the Lower Umba subsuite (sedimentary rocks) and Upper Umba subsuite (volcanic rocks); later they were renamed as formations (Melezhik and Sturt 1994).

The Umba Sedimentary Formation extends from the eastern margin of and throughout the Central block, and then further to the Western block, where it is eventually cut by the Khibiny alkaline massif (Fig. 4.2b, c). The formational thickness is only 50–60 m in the Western block (Sects. 1 and 2, Fig. 4.10a, b) though exceeds 120 m throughout the Central block (Sects. 3, 4, 5, 6, and 7). The formation rests with a depositional contact on various volcanic rocks of the Polisarka Volcanic Formation. In Sect. 1 (Fig. 4.10b), a c. 10-m-thick basal bed of marl with clayey siltstone and dolostone lenses rests with a sharp contact on massive, melanocratic komatiitic basalts. In Sect. 2, komatiitic basalts are overlain by crudely bedded dolostone (Fig. 4.10b). In Sect. 5, variegated dolostones lie on a komatiitic flow breccia, whereas in Sect. 6, they lie on a komatiitic pillow lava (Melezhik and Predovsky 1982).

Siliciclastic lithofacies dominate west of the Tsaga faults (Sects. 1, 2, 3, 4, and 5), whereas carbonate rocks are the main lithofacies of the formation in the eastern Central block (Sects. 6 and 7). In all measured sections, the carbonate lithofacies comprise the lower part of the formation except in Sect. 3 (Fig. 4.11a), where they mainly occur in the middle part interbedding with siltstones, shales and mafic lavas (for details see Chap. 6.1.3). It has been proposed that both the thickness and lithofacies variations are controlled by transversal and probably axial syndepositional faults (Melezhik and Predovsky 1982).

The carbonate lithofacies vary greatly in thickness (20–100 m) and their physical appearance. In Sects. 1, 2 and 4, 5, 6, and 7, the carbonate rocks are commonly variegated, fine-grained (micritic) dolostones (Fig. 4.11b–d). They contain lenses of dolorudite (Fig. 4.11e), jasper and variegated cherts (Fig. 4.11f–h), and dolomite-cemented ultramafic breccia (Fig. 4.11i–j). In Sects. 3, 4, 5, and 7, the dolostones are characterised by indistinct, horizontal, thick bedding (Fig. 4.11c). In contrast, the dolostones in Sects. 2 and six exhibit distinct rhythmic bedding (Fig. 4.11b) disrupted by numerous erosional features. Variable-scale erosional channels are filled with polymict conglomerate (Fig. 4.11k–l), dolorudite (Fig. 4.11m–n), and matrix-supported dolostone conglomerate (Fig. 4.11o). The upper part of Sect. 6 shows a crude rhythmic interbedding of 0.2-to 1-m-thick beds of dolomite-cemented, polymict conglomerate, and breccia with thicker units of dolarenite and micritic dolostones containing lenses of ultramafic breccia (Fig. 4.11i–o). Dolorudites contain abundant clasts of jasper and weathered ultramafic material (for details, see Chap. 6.1.4). Marls form 5- to 20-m-thick lenses occurring in different stratigraphic positions but are always bound to and interbedded with other dolostone lithologies with which they have transitional contacts. Marls have a variegated colour and thin parallel bedding (Fig. 4.11p), and they contain lenses of red dolostones.

The Umba Formation dolostones are enriched in ^{13}C and thus record the Lomagundi-Jatuli positive excursion of $\delta^{13}C$ in sedimentary carbonates. Interestingly, the dolostones exhibit a great variation of $\delta^{13}C$ ranging from −2‰ to +7‰ within individual sections as well as between the sections (e.g. Melezhik and Fallick 1996). The reason for such fluctuations remains unclear. The jasper and dolostone contain barite and pyrite (Melezhik and Fetisova 1989) with $\delta^{34}S$ ranging between +27.8‰ and +34.2‰ and between +16.6‰ and +17.3‰, respectively (Grinenko et al. 1989).

Siliciclastic lithofacies include greywacke, interbedded sandstone-siltstone-shale, and quartz sandstone. The greywacke comprises 15- to 20-m-thick units in a lower stratigraphic position in Sects. 1 and 3, and thin lenses in the middle of Sect. 3 where they intercalate with quartz sandstone and shale. In both sections, the greywacke is characterised by thin rhythmic and fine lenticular bedding. Normal grading, reworked beds and soft-sediment deformation features are common. The sandstone-siltstone-shale assemblage in most cases overlies the carbonate lithofacies except in Sect. 3, where it is interbedded with dolostone lithofacies rocks (Fig. 4.10b) and contains thin beds of massive quartz sandstone and greywacke. Thin siltstone-shale couplets mostly display a fine parallel or low angle cross-lamination (Fig. 4.11q).

Quartz sandstones occur mainly in the western part of the belt (Sects. 1, 2, and 3), where they comprise up to 25-m-thick unit in the upper part of the formation; several small lenses interbedded with other carbonate and siliciclastic lithologies have also been documented throughout Sect. 3 (Fig. 4.10b). The quartz sandstones are massive, parallel-bedded and low-angle cross-bedded rocks having sharp contacts with underlying and overlying shales. Some beds are enriched in magnetite. In Sect. 4, the uppermost portion of quartz sandstones contains a lens of jasper and a thin dolostone bed in it, whereas a thin bed of polymict conglomerate occurs in a similar stratigraphic position in Sect. 1.

The overall depositional framework of the Umba Sedimentary Formation includes a rift-bound basin, whose geometry and depth have been controlled by several transversal and axial syndepositional faults (Melezhik and Predovsky 1982; Melezhik 1992). The observed lithofacies variations and sedimentological features of the rocks were assigned to a marine setting with the basin deepening from west to east. Bekasova and Pushkin (1977) suggested a N-S transport direction of clastic material and the Seidorechka volcanic rocks as the main provenance. Syndepositional subaerial and submarine hydrothermal processes have been suggested to be responsible for the formation of jaspers associated with carbonate and siliciclastic lithofacies (Melezhik and Predovsky 1982). Other details on the depositional framework based on preliminary investigation of the FAR-DEEP drillcore and associated new material are presented in Chaps. 6.1.3 and 6.1.4.

The Umba Volcanic Formation

The Umba Volcanic Formation rests depositionally on the siliciclastic rocks of the Umba Sedimentary Formation. The formation is a 1,000- to 2,000-m-thick unit continuously developed from the central Eastern block to the Western block where it is cut by the Khibiny alkaline intrusion (Fig. 4.2c). The formation has been subdivided into four groups of "volcanic flows" (Borisov 1990).

The lowermost group is 150–500 m thick and comprises a series of 15- to 35-m-thick lava flows. The main rock types are trachybasalts, picrites, and picrobasalts. The presence of picrites and picrobasalts is a specific feature of the first group. Their volume may reach 30–50 % and they occur as massive and amygdaloidal porphyritic rocks with olivine and diopside-augite phenocrysts, forming subvolcanic bodies and 15- to 20- m-thick lava flows that are traceable over long distances. The trachybasalts have a porphyritic (with albite phenocrysts) and less commonly an aphyric texture and occur as a series of 10- to 35-m-thick lava flows with up to 1.5-m-thick, brown, oxidised and brecciated flow tops. Trachyandesites are rare and only known in the Eastern block. Tuffs are also rare and occur as thin (<0.5 m) lenses between lava flows (Fig. 4.11r).

The second group of "volcanic flows" forms a 400- to 600-m-thick sequence. Its base is composed of psephitic tuffs, mafic agglomerate, tuffites, volcaniclastic conglomerates, and shales. Massive, homogeneous trachybasaltic porphyrites are dominant rocks, though trachyandesitic porphyrites, tuffs, and flow breccias are present in the Eastern block (Borisov 1990).

The third group has a thickness of 400–500 m and a gradational contact with the underlying volcanic suite, though, in places, a thin bed of psephitic-psammitic tuffs defines the base. Intensely epidotised diabase-porphyrites, diabases, trachyandesitic lavas, and tuffs are characteristic rock types of the group (Borisov 1990). In the Eastern block, volcaniclastic conglomerates, trachyandesitic flow breccias, and tuffs comprise a considerable part of the group.

The fourth group of "volcanic flows", forming the top of the Umba Volcanic Formation, only occurs in the Central block, where it has a thickness of 400–500 m. A thick, continuously-developed tuff unit defines the base of the group. The tuff unit contains rhythmically interbedded agglomerates, psephitic and psammitic tuffs, and ash-beds. The rest of the group is composed of trachybasaltic porphyrites, fluidal trachyandesitic lavas, brown and variegated flow breccias (Fig. 4.11s–t), and black ignimbrites; all are interbedded with lenses and thick beds of tuffs. The overall thickness of the pyroclastic beds and the size of pyroclastic particles decrease westwards (Borisov 1990).

The rocks of the Umba Volcanic Formation form a continuous differentiation series, which has a geochemical nature that distinguishes them from all other volcanic formations of the Imandra/Varzuga Supergroup. The characteristic features include high soda content ($Na_2O > K_2O$) and high contents of Ti, Fe and P. Similarly to the Kuetsjärvi Volcanic Formation at Pechenga, the volcanic rocks are highly oxidised (Melezhik et al. 1982; Fedotov 1985). For example, Borisov (1990) reported an average $Fe^{3+}/\Sigma Fe$ ratio of 0.45 for 62 analyses of basalts and trachybasalts. Also typical are widespread occurrences of native copper (Zilber 1972; Bekasova and Pushkin 1975; Borisov 1990). Nuggets of native copper, up to 20 × 15 × 5 cm in size (Fig. 4.11u), are bound to the first group of lava flows (Borisov 1990). The accumulation of the Umba volcanic rocks took place in an intracratonic rift setting as subaerial (lack of pillow lavas), fissure- and central-type eruptions (Borisov 1990; Predovsky et al. 1987).

The Il'mozero Sedimentary Formation

The Il'mosero suite, divided into the Lower (sedimentary) and Upper (volcanic) subsuites, was defined by Bekasova and Pushkin (1972). Later, the Lower subsuite was renamed as the Il'mozero Sedimentary Formation and the Upper subsuite as the Il'mozero Volcanic Formation (Melezhik and Sturt (1994). The sedimentary formation rests with an erosional contact on the Umba volcanic rocks and occurs only in the Central block (Fig. 4.12a). The western extension is cut by NW-SE trending thrust, whereas in the east, the formation forms the tectonically-fragmented, periclinal closure of a large synform whose southern limb is truncated by another WE-trending thrust (Fig. 4.12a). Thus, the formation extends over a distance of 60 km, while its thickness varies from 400 to 1,200 m. The actual depositional age of the formation remains unknown, though its maximum age has been constrained to 2052 Ma by dating clastic zircons from greywacke beds (Martin et al. 2010).

The formation has been subdivided into four lithostratigraphic units (members), which are informally termed, from bottom to top, as the Tuffite, Greywacke, Dolostone-Chert, and Black Shale members (Melezhik and Predovsky 1982). Melezhik and Predovsky (1982) recognised three types of lithological successions. The first occurs in the west (Sect. 1, Fig. 4.12b), has a thickness of c. 1,000 m and is mainly composed of turbiditic greywackes and "black shales" with subordinate carbonate rocks and mafic tuffs. The second type of succession is a characteristic feature of the central part of the block (Sects. 2 and 3, Fig. 4.12b). The succession has a reduced thickness of 400–600 m and consists of volcaniclastic greywackes, tuffites, and "black shales" with subordinate carbonate rocks. The third type occurs in the eastern part of the Central block (Sects. 4, 5, and 6, Fig. 4.12b). It has a thickness of c. 1,200 m and is characterised by a thick development of reefal stromatolitic dolostones, "black shales", and subordonate turbiditic greywackes.

The Tuffite member is only 9–30 m thick, occurs locally (Sects. 2 and 4) and rests erosionally on an alkali basaltic flow breccia with a small-scale, uneven palaeotopography (Bekasova and Pushkin 1972; Melezhik 1981; Melezhik and Predovsky 1982). The base of the member is defined by a 0.1-to 1.5-m-thick, discontinuously developed bed (Sect. 2) or a series of thin beds (Sect. 4) of matrix-supported, volcaniclastic conglomerate (Fig. 4.13a). Platy clasts of volcanic rocks are commonly uniformly imbricated, though a few beds show a reverse imbrication of small clasts. This together with alternating cross-beds indicates a NS transport direction of clastic material with subsequent limited reworking by low-energy tides (Melezhik and Predovsky 1982). The overlying tuffites form trough cross-beds passing upward into to horizontally-bedded, volcaniclastic greywackes enriched in haematite and magnetite. Some sandstone beds are rich in authigenic tourmaline (Fig. 4.13b). Both the tuffites and the greywackes contain scattered clasts of volcanic rocks. The upward structural change is accompanied by overall fining as well as by an increase in roundness of pebble-size clasts.

The Greywacke member is characterised by highly variable thicknesses ranging from 20 to 300 m (Fig. 4.12b). It rests either on the Tuffite member rocks or directly on the underlying volcanic rocks of the Umba Volcanic Formation. In Sects. 2 and 3, the base of the member is defined by a 1- to 2-m-thick bed of horizontally-bedded greywacke with calcite concretions (Fig. 4.13c); the bed has been traced over a distance of 20 km (Melezhik and Predovsky 1978). Concretions show a considerable depletion in ^{13}C with $\delta^{13}C_{carb}$ ranging between −3.9‰ and −20.5‰ (Melezhik and Fallick 1996). In Sect. 6, the concretionary bed reappears as a c. 30-m-thick unit, which also contains abundant (10–15 vol.%) phosphate-bearing concretions (Fig. 4.13d–e). The greywackes comprise the bulk of the member. The lower part of the member consists of haematite-rich (up to 20 vol.%), rhythmically bedded turbiditic greywacke-siltstone couplets with graded sandstone units. These rocks pass gradually upward into indistinctly laminated or massive, fine-grained sandstone with numerous small-scale erosional and slump features, flame structures, and small-scale troughs filled with medium-grained sandstone rapidly grading into fine-grained, massive sandstone (Fig. 4.13f, g). Thin beds of shales are present only in the upper part of the member.

The Dolostone-Chert member has a greatly variable thickness (Fig. 4.12b). Its base is defined by a c. 20-m-thick bed of horizontally-bedded limestone and sandy limestone, which has been observed in all exposed sections. The limestone bed is followed by crystalline, indistinctly-bedded, grey dolostones. In Sects. 4 and 5, they are laterally overstepped by pale grey and white dolostones (Fig. 4.13h) and intensely silicified stromatolitic dolostones (Fig. 4.13i, j), comprising a series of 0.5–1-m-thick bioherms and biostromes, which are interbedded with oncolitic dolostones, syndepositional stromatolitic breccias (Fig. 4.13k) and chert beds (Melezhik and Predovsky 1982). Biogenic dolostones form a c. 250-m-thick and 6 × 8-km-wide, isometric build-up resembling a reef. In the central part, it rests directly on the Umba Volcanic Formation. Marginal part of this reefal body is composed of syndepositional stromatolite breccias(Fig. 4.13k). Both the limestones and dolostones exhibit near normal $\delta^{13}C_{carb}$ values ranging between -0.8‰ and +2.4‰ (Melezhik and Fallick 1996).

Black Shale member is a c. 1,000-m-thick, poorly exposed unit. In the east (Sect. 6), its base is defined by a thin bed of dolostone-pebble conglomerate and in the west (Sect. 1), by a bed of mafic tuff. The member is mostly composed of C_{org}-bearing siltstone-shale with rare lenses of black chert. The rocks are very homogeneous and mainly exhibit small-scale flaser and wavy bedding. The upper contact with the overlying Il'mozero Volcanic Formation is tectonic.

The overall depositional environment of the Il'mozero Sedimentary Formation can be defined as a change from an alluvial (trough cross-beds) through a tide-influenced delta (reverse clasts imbrication in wavy and cross-stratified greywackes) to a deep-water shelf (fine-grained rhythmically-bedded turbidites) followed by the establishment of a carbonate shelf (limestones and marls) with reefal build-up (stromatolitic bioherms and biostromes) in the eastern part of the basin. The dolostone-pebble conglomerate at the base of the overlying, thick unit of homogeneous, C_{org}-bearing, flaser- and wavy-bedded siltstone-shale suggests a flooding event followed by the development of a shallow-water clastic shelf.

The Il'mozero Volcanic Formation

The Il'mozero Volcanic Formation has a limited development and is restricted to the Central block. Both the lower and upper contacts of the formation are defined by faults. Similarly to its sedimentary counterpart, the western extension of the formation is cut by a NW-SE-trending thrust, whereas in the east, the formation forms the periclinal closure of a large synform, whose southern limb is truncated by another WE-trending thrust. The formation extends over a distance of 50 km, while its thickness is less than 600 m.

The lower part of the formation is intensely sheared and calcitised. Otherwise, the formation has a homogeneous composition with fine-grained, porphyritic basaltic andesites being the main rock type (Melezhik et al. 1982; Fedotov 1985). They comprise 90 vol.% of the formation and occur as a series of 20- to 40-m-thick lava flows with indistinct interflow boundaries unless amygdaloidal tops are developed. A few thin (<10 m) lava flows of dacitic andesites and basaltic komatiites have been observed in the middle part of the formation (Fedotov 1985). A single horizon of volcanic breccia, 40 m thick and 300 m long, occurs in the upper part of the formation. It contains fragments of andesitic basalt and gabbro-diabase in schistose, psammitic, lithic tuff matrix. The size of the fragments ranges between 3 cm and 1.5 m. Geochemically, the Il'mozero volcanic rocks belong to a komatiitic series and resemble similar rocks in the Seidorechka and Polisarka Volcanic formations (Fedotov 1985).

The Tominga Group

The Tominga Group was originally defined as a "Series" by Zagorodny et al. (1978). The group only occurs within the Western and Central blocks of the Southern Zone (Fig. 4.2c), and has thrust contacts along both its northern (with the 'Northern Zone') and southern (supposedly Archaean rocks) boundaries. It has a maximum thickness exceeding 5 km and a poorly-understood internal stratigraphic subdivision (Radchenko et al. 1982) and may represent a series of tectonically imbricated units rather than a coherent succession.

The Tominga Group contains a large array of lithologies metamorphosed under biotite-actinolite-chlorite zone conditions. Sedimentary rocks include thick units of volcaniclastic greywacke, arkosic sandstones, various schists (Fig. 4.14a–c), "black shales", cherts, marls and carbonate rocks. Volcanic rocks are represented by tholeiitic pillow basalts, picritic lavas, andesites, and thick units of dacites and rhyolites. A few attempts have been made to subdivide the group into formations and even to correlate them with stratigraphic units of the Northern Zone (e.g. Melezhik 1978; Radchenko et al. 1982; Melezhik and Predovsky 1982). This resulted in establishment of several formations, such as the Solenozero, Mittrijärvi (e.g. Smolkin and Dain 1985), Rouksa, Panarechka, Saminga and others (Radchenko et al. 1982), but boundaries and chronostratigraphic positions remain poorly understood. However, picritic lavas (Fig. 4.14d) and associated tholeiitic pillowed basalts of the Mittrijärvi formation have been correlated with the Pilgujärvi Volcanic Formation in the Pechenga Greenstone Belt (Smolkin 1995) because of the geochemical similarities of the ferropicrites in these two areas (Fedotov 1983; Predovsky et al. 1987; Skuf'in and Theart 2005). The Soleonozero Formation, which occurs beneath rocks of the Mittrijärvi formation, and consists of sulphide-carbonaceous schists, phosphorite pebble conglomerates, tuffs and ferropicritic lava breccias, has also been correlated with the Pilgujärvi Sedimentary Formation (Smolkin 1995; Sharkov and Smolkin 1997) based on geochemical and lithological similarities. The occurrence of ferropicrites in the Imandra/Varzuga Belt is important not only for correlative purposes but also for ore exploration as those in the Pechenga Greenstone Belt are coeval and comagmatic with gabbro-wehrlite intrusions hosting commercial Ni-Cu ores (e.g. Hanski 1992).

Although the Tominga Group rocks have been considered to represent the youngest rocks in the Imandra/Varzuga Belt (Zagorodny et al. 1978), their age remains unknown. This has tentatively confirmed by geochronological data on the so-called Panarechka volcano-tectonic structure (Skuf'in et al. 2006), a small (21 × 8 km), NW-SE-trending, tectonically isolated fault-bound synform within the southern limb of the outcrop belt (Fig. 4.2b). The synform comprises a >1,000-m-thick unit of greywacke, arkosic sandstones schists (Fig. 4.14a–c) and basalts (Panarechka Formation) overlain by a 500-m-thick pile of dacites, rhyolites and minor andesites (Saminga formation). The arkosic sandstones are intruded by a plagiomicrocline granite dated at 1940 ± 5 Ma and the rhyodacites are cut by subvolcanic bodies of trachydacite with an age of 1907 ± 18 Ma (Skuf'in et al. 2006). The latter date provides a maximum depositional age for the rhyodacitic succession. Interestingly, Melezhik and Predovsky (1982) reported an occurrence of high-Al (22 wt.% Al_2O_3) andesites (Fig. 4.14e) in the upper part of the Panarechka Formation. This provides a link to similar andesites known in the South Pechenga Group; these andesites have been interpreted to represent an orogenic stage in the development of the belt (Melezhik and Sturt 1994).

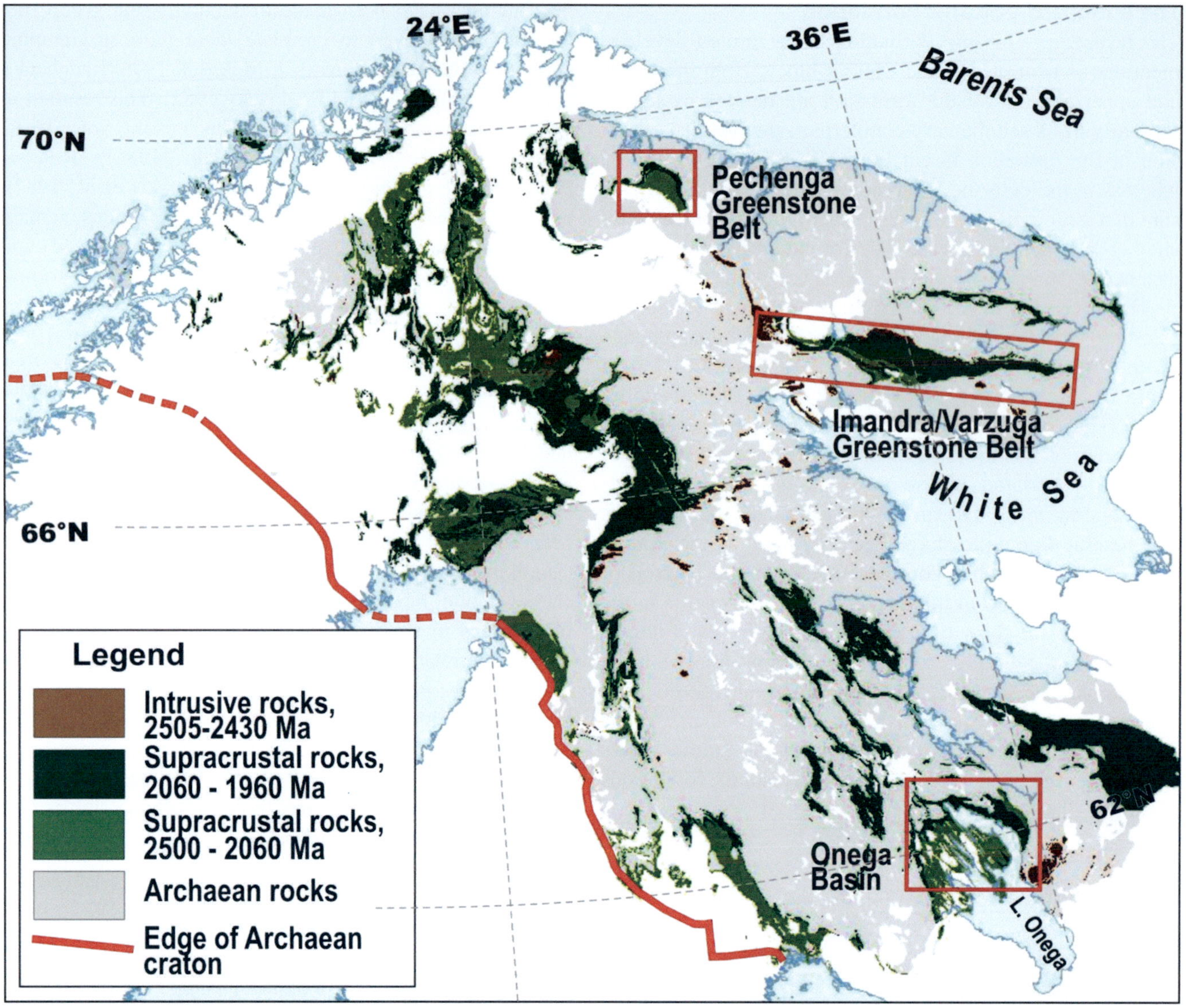

Fig. 4.1 Geological map showing the occurrence of early Palaeoproterozoic supracrustal complexes and 2505–2430 Ma layered gabbronorite intrusions in the eastern part of the Fennoscandian Shield. Framed are the three supracrustal belts where the FAR-DEEP drilling operations were carried out. Geological map modified after Koistinen et al. (2001)

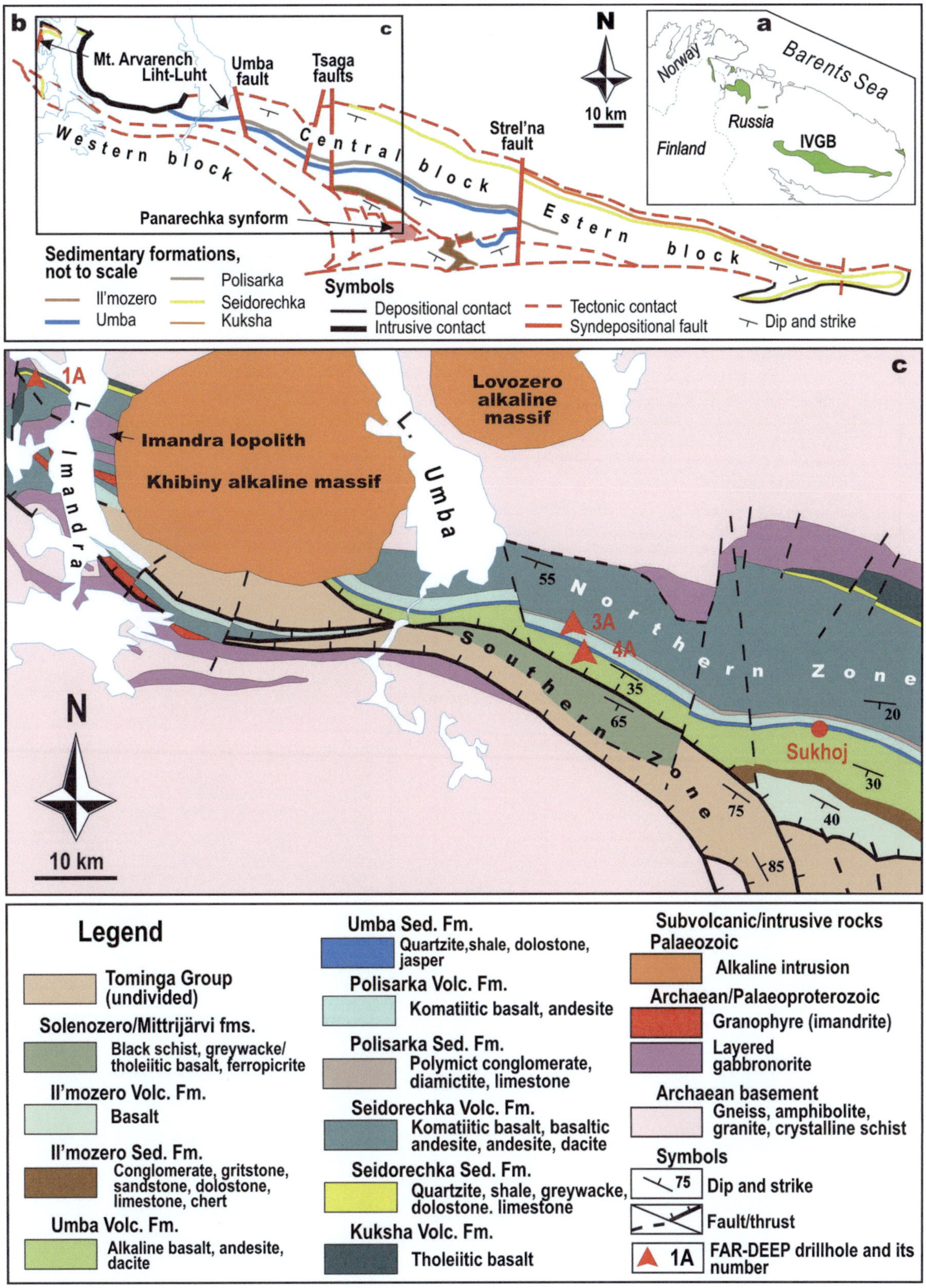

Fig. 4.2 Geographic location (**a**) and major structural features (**b**) of the Imandra-Varzuga Greenstone Belt (IVGB). (**c**) Simplified geological map of the western part of the IVGB showing the locations of the FAR-DEEP Holes 1A, 3A, 4A and the Sukhoj section. Geological map modified after Zagorodny et al. (1982)

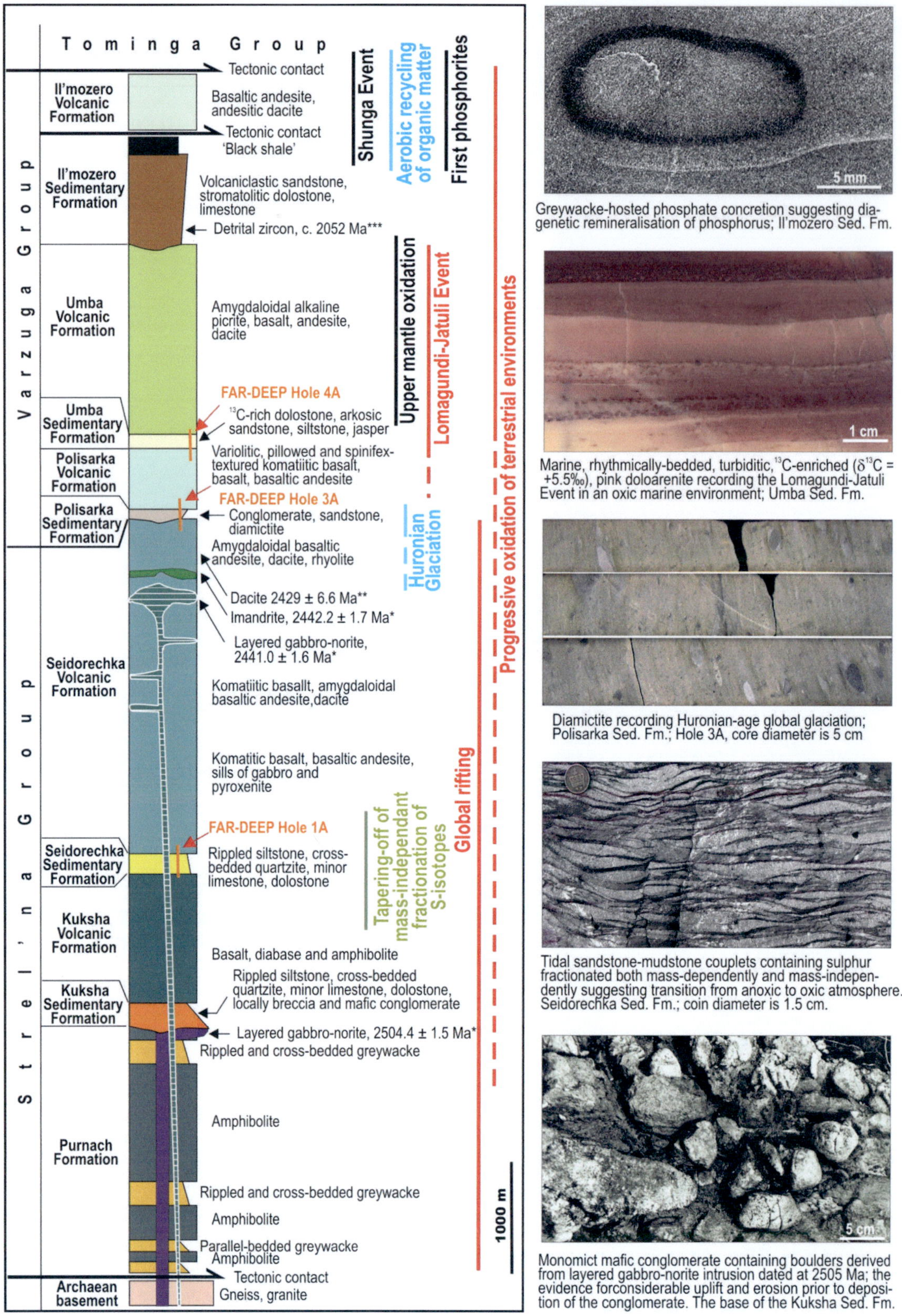

Fig. 4.3 Simplified lithological column through the Stre'lna and Varzuga groups of the Imandra/Varzuga Belt. Also shown is how the evolution of the Pechenga Greenstone Belt is related to global palaeoenvironmental events. Asterisks denote radiometric ages from (*) Amelin et al. (1995), (**) Vrevsky et al. (2010) and (***) Martin et al. (2010)

Fig. 4.4 Selected images of sedimentary and volcanic rocks of the Kuksha Sedimentary and Volcanic formations. (**a**) Parallel-bedded, arkosic sandstone overlain by fragment-supported polymict conglomerate; scale-bar length is 10 cm. (**b**) Unsorted, gritty, polymict conglomerate. (**c**) A series of thin, amygdaloidal, mafic lava flows. (**d**) Mafic pillow lava. (**e**) A thin bed of bedded arkosic sandstone sandwiched between two amygdaloidal, mafic lava flows. Lens cap diameter in (**c**) to (**e**) is 6 cm

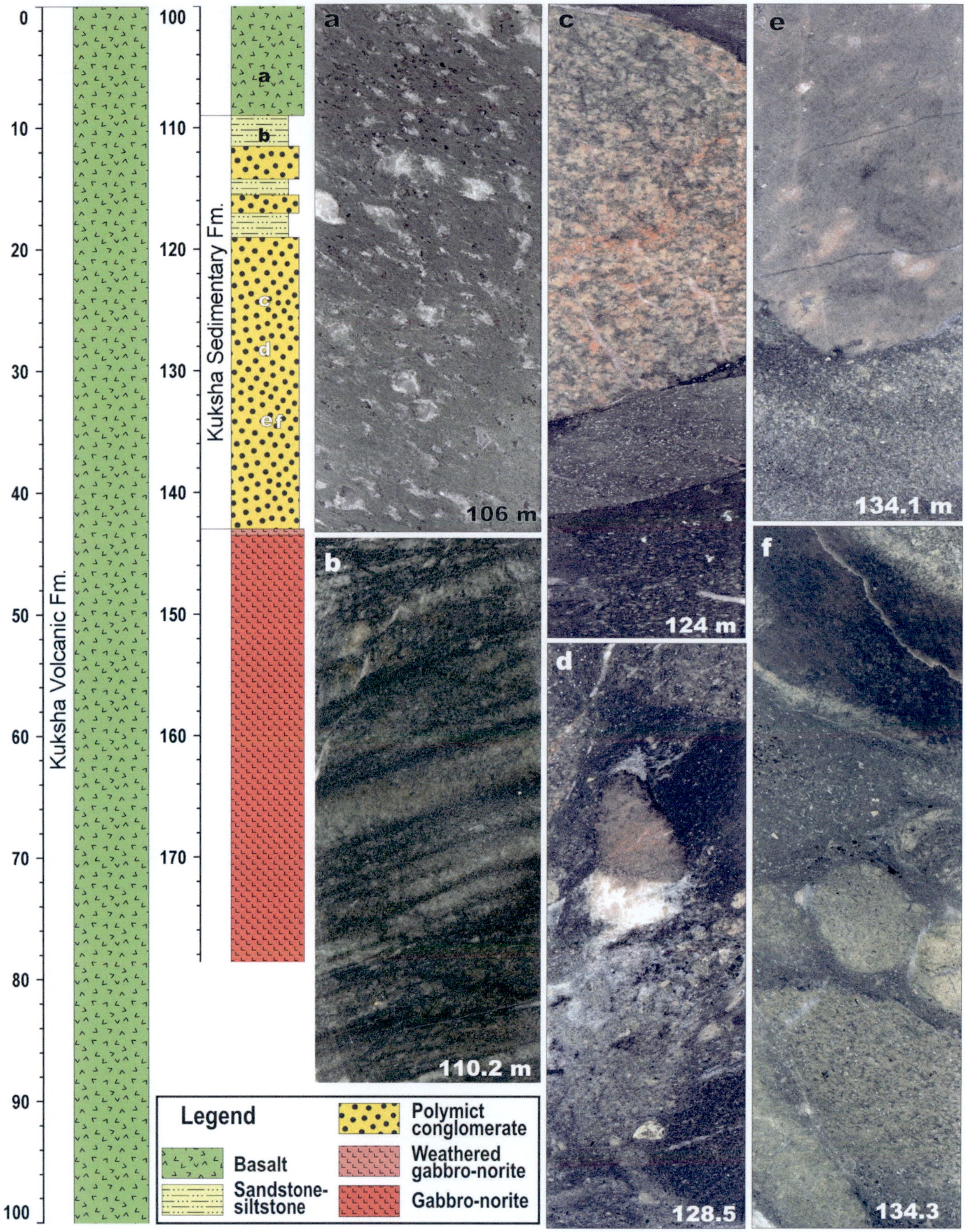

Fig. 4.5 A section of the Kuksha Volcanic and Sedimentary formations and the underlying 2504 Ma gabbronorite intrusion, as revealed by drillcore 2174, and selected images of sedimentary and volcanic rocks deposited on the eroded surface of the intrusion. Drillhole location is shown in Fig. 4.6. *Kuksha Volcanic Formation:* (**a**) Amygdaloidal mafic lava. *Kuksha Sedimentary Formation:* (**b**) Interbedded siltstone-shale couplets. (**c**) Unsorted, rounded clasts of granite (pale pink) and biotite schist (*pale grey*) in a gabbroic sand matrix (*dark grey*). (**d**) Gritty, unsorted, polymict conglomerate. (**e**) Anorthosite fragment (*bright*) in a gabbroic sand matrix (*pale grey*). (**f**) Clast-supported polymict conglomerate. The log is based on unpublished data of Melezhik and Prave. Computer-generated lithological column made by Aivo Lepland. Drillcore diameter 6 cm

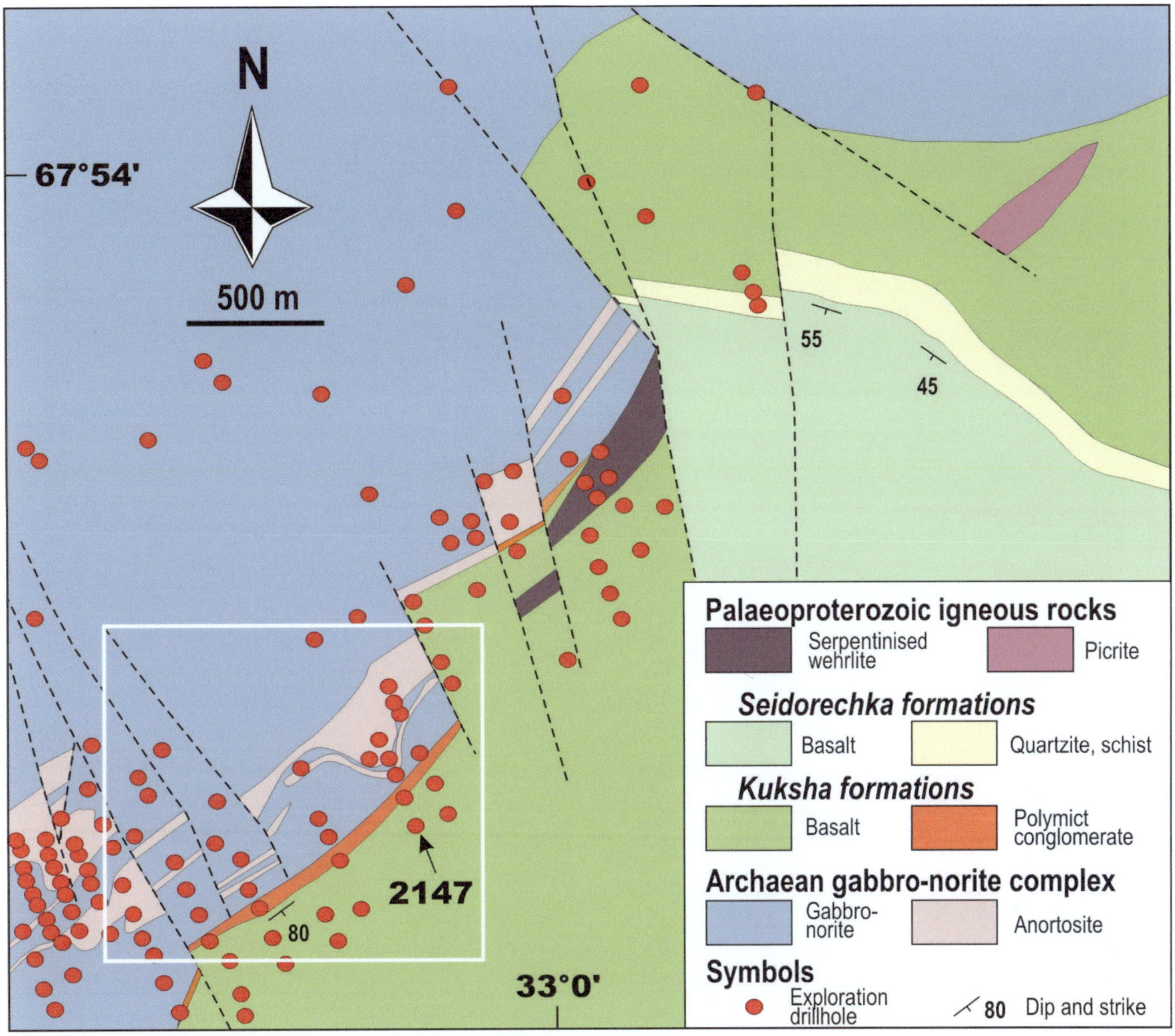

Fig. 4.6 Simplified geological map of the Monchegorsk Peninsula, east of Lake Imandra, showing the contact between the 2504 Ma layered gabbronorite intrusion and the Imandra/Varzuga Belt supracrustal rocks. Note that within the white rectangle, the Kuksha Sedimentary Formation conglomerate seals the faulted surface of the gabbronorite intrusion. A lithological log of Drillhole 2147 is presented in Fig. 4.5. Map courtesy of the Kola Geological Laboratory-Informational Centre

Fig. 4.7 Selected rock images illustrating lithology and structural features of the Seidorechka Sedimentary and Volcanic formations. *Seidorechka Sedimentary Formation, the westernmost part of the belt*: (**a**) Sandstone-siltstone with lenticular and wavy bedding from the base of the Sandstone-Siltstone member. (**b**) Flaser-bedded sandstone-shale from the lower part of Sandstone-Siltstone member. (**c**) Sandstone dyke cutting through dark grey shale and gradually vanishing in a wavy-bedded sandstone-shale unit of the Sandstone-Siltstone member

Fig. 4.7 (continued) (**d**) Laminated dolarenite (base) overlain by semi-massive to thickly bedded dolostone from the Dolostone member. (**e**) Platy, angular, dolostone clasts in a quartz sandstone matrix from the base of the Quartzite member. (**f**) Large clasts of dolostone (*bright brown*) in a cross-bedded, dolomite-cemented sandstone bed with a rippled surface overlain by parallel-bedded quartz sandstone from the base of the Quartzite member. (**g**) Wavy-bedded, dolostone-cemented quartz sandstone overlain by tabular cross-bedded quartz sandstone from the lower part of the Quartzite member

Fig. 4.7 (continued) (**g**) Wavy-bedded, dolostone-cemented quartz sandstone overlain by tabular cross-bedded quartz sandstone from the lower part of the Quartzite member. (**h**) Trough cross-bedded quartz sandstone from the lower part of the Quartzite member. (**i**) Parallel-bedded quartz sandstone from the middle part of the Quartzite member. (**j**) Bedded marl overlain by shale from the Limestone-Shale member. (**k**) Marl-shale interbeds from the Limestone-Shale member

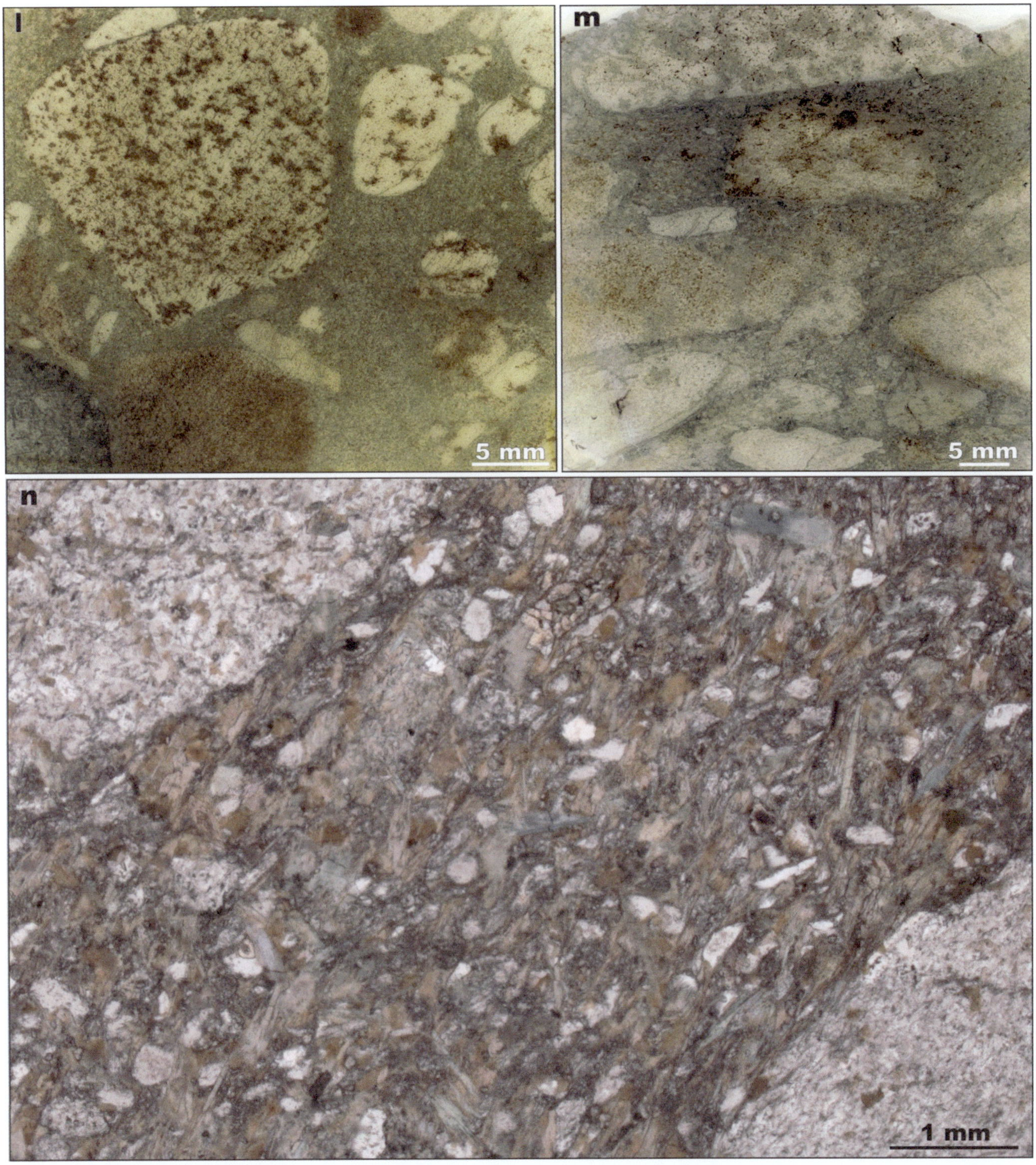

Fig. 4.7 (continued) *Seidorechka Volcanic Formation, westernmost part of the belt:* (**l**, **m**) "Volcanic breccia" containing andesitic clasts supported by a komatiitic tuffite matrix; this breccia serves as a marker horizon in the lower part of the formation. (**n**) Photomicrograph in non-polarised light of komatiitic tuffite matrix from the "volcanic breccia" shown in (**l** and **m**); the matrix consists of amphibole framework containing rounded particles of plagioclase

Fig. 4.7 (continued) *Seidorechka Volcanic Formation, central part of the belt:* (**o**) Photomicrograph in polarised light of ophitic basaltic andesite from the middle part of the formation. (**p**) Photomicrograph in polarised light of fine-grained dacite with a zoned calcite (bright rim)-quartz (grey-white mantle) amygdale; upper part of the formation. (**q**) Photomicrograph in polarised light of a calcite amygdale in fine-grained dacite from the upper part of the formation. *Seidorechka Volcanic Formation, westernmost part of the belt:* (**r**) Dacitic lava with numerous xenoliths of felsic rock (*bright*); lens cap diameter is 6 cm. (**s**) Amygdaloidal, dacitic lava. (**t**) Accretionary lapilli in rhyolitic tuff forming a thin bed in a dacitic lava pile; sample courtesy of Vladimir Pozhilenko

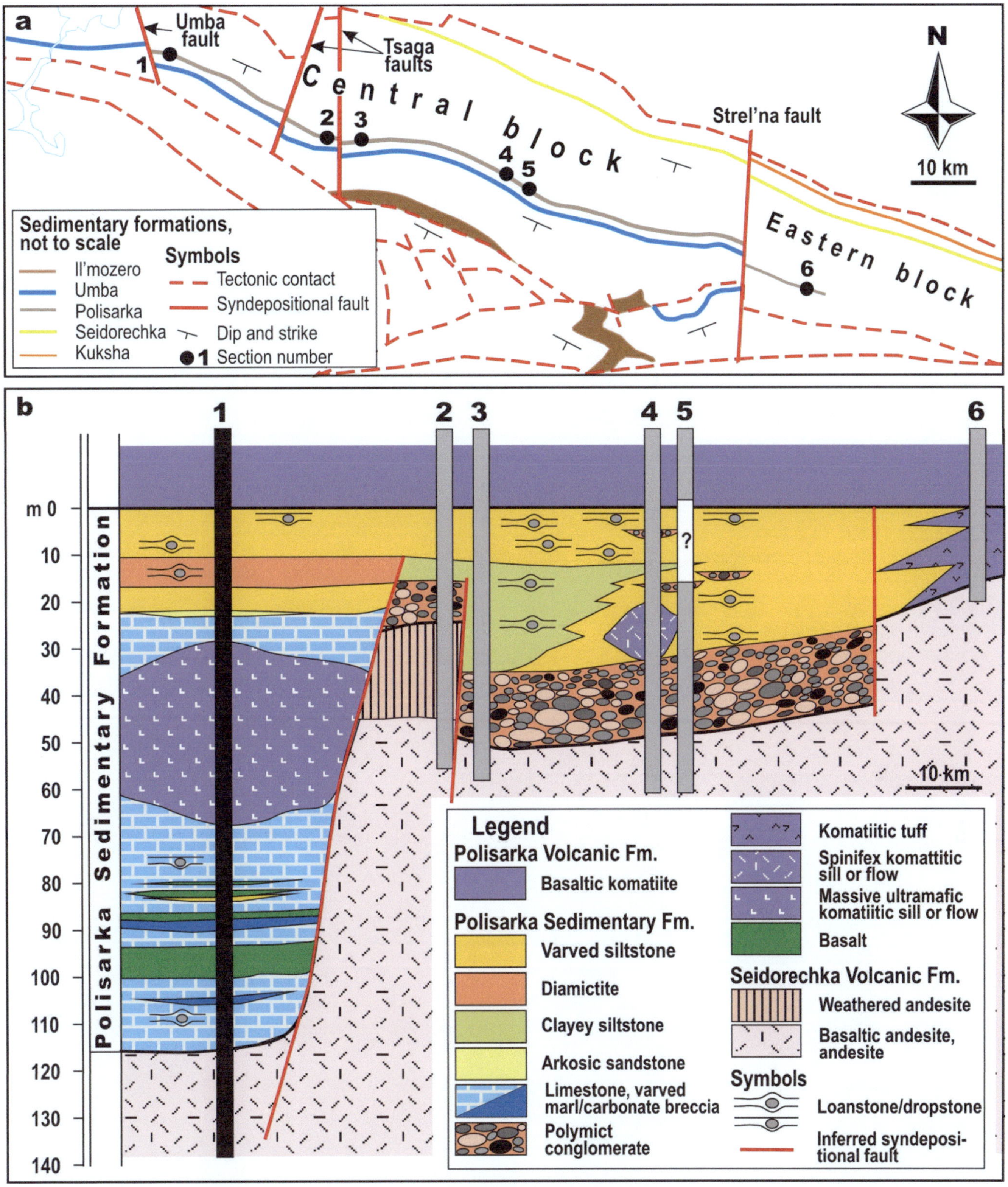

Fig. 4.8 A simplified longitudinal lithological profile through the Polisarka Sedimentary Formation based on data from Melezhik and Predovsky (1982) and FAR-DEEP Hole 3A. Sections 2, 3, 4, 5, and 6 are based onoutcrop observations, while Sect. 1 represents a simplified log of FAR-DEEP Hole 3A (for details, see Chap. 6.1.2 Polisarka Sedimentary Formation: FAR-DEEP Hole 3A)

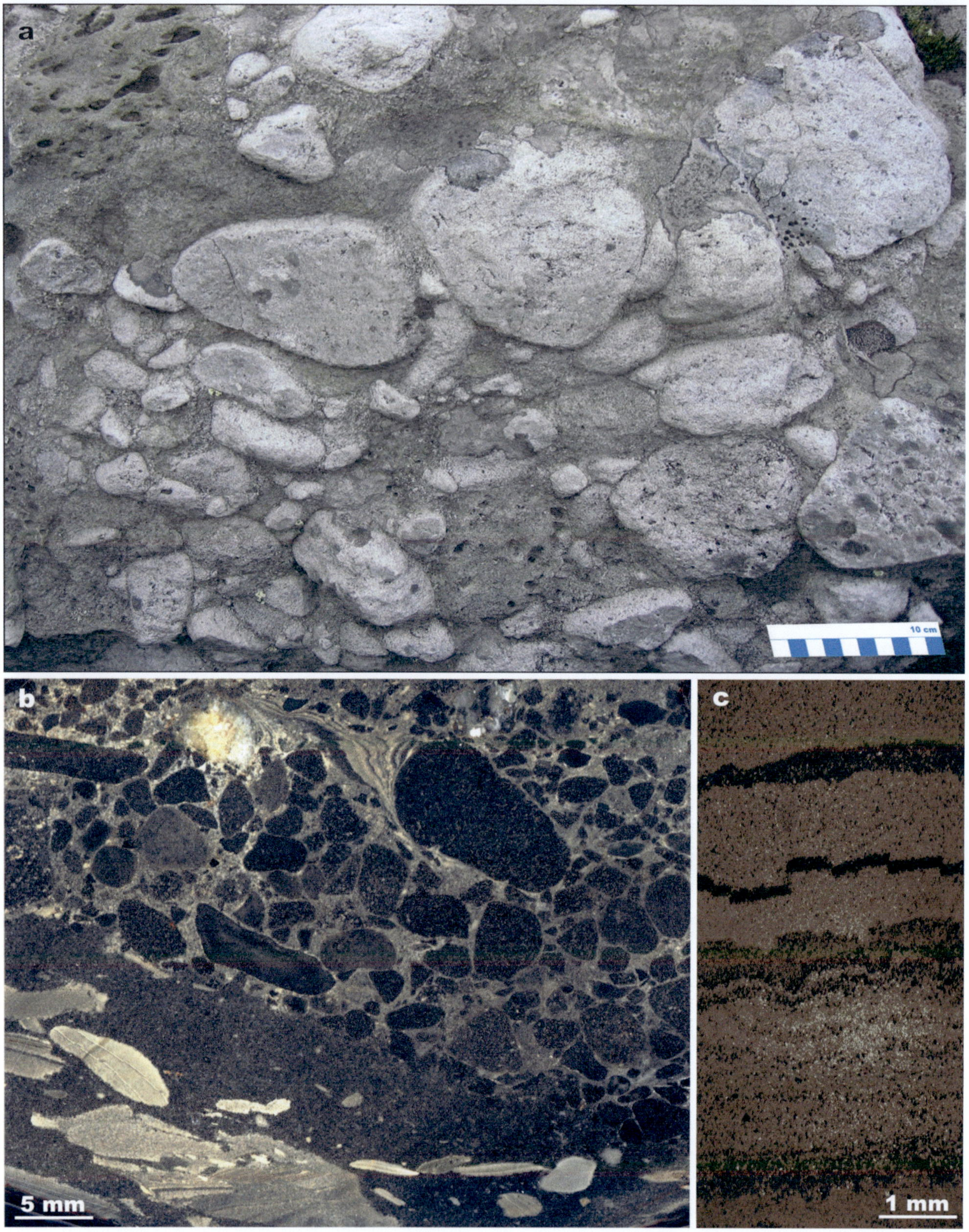

Fig. 4.9 Selected rock images illustrating lithology and structural features of the Polisarka Sedimentary and Volcanic formations. *Polisarka Sedimentary Formation*: (**a**) Polymict, volcaniclastic conglomerate composed of rounded and unsorted clasts of dacites, andesites and basalts from the base of Sect. 3 shown in Fig. 4.8. (**b**) A *dark grey* bed with bright siltstone fragments supported by a matrix overlain by fine-pebble conglomerate containing dark grey andesite clasts in a bright clayey greywacke matrix; note soft deformation in a laminated greywacke lens against large, rounded, andesite clasts (*upper right corner*). (**c**) Scanned image of bedded siltstone (*grey*)-mudstone (*black* to *dark grey*) with several syndepositionally faulted layers in the middle.

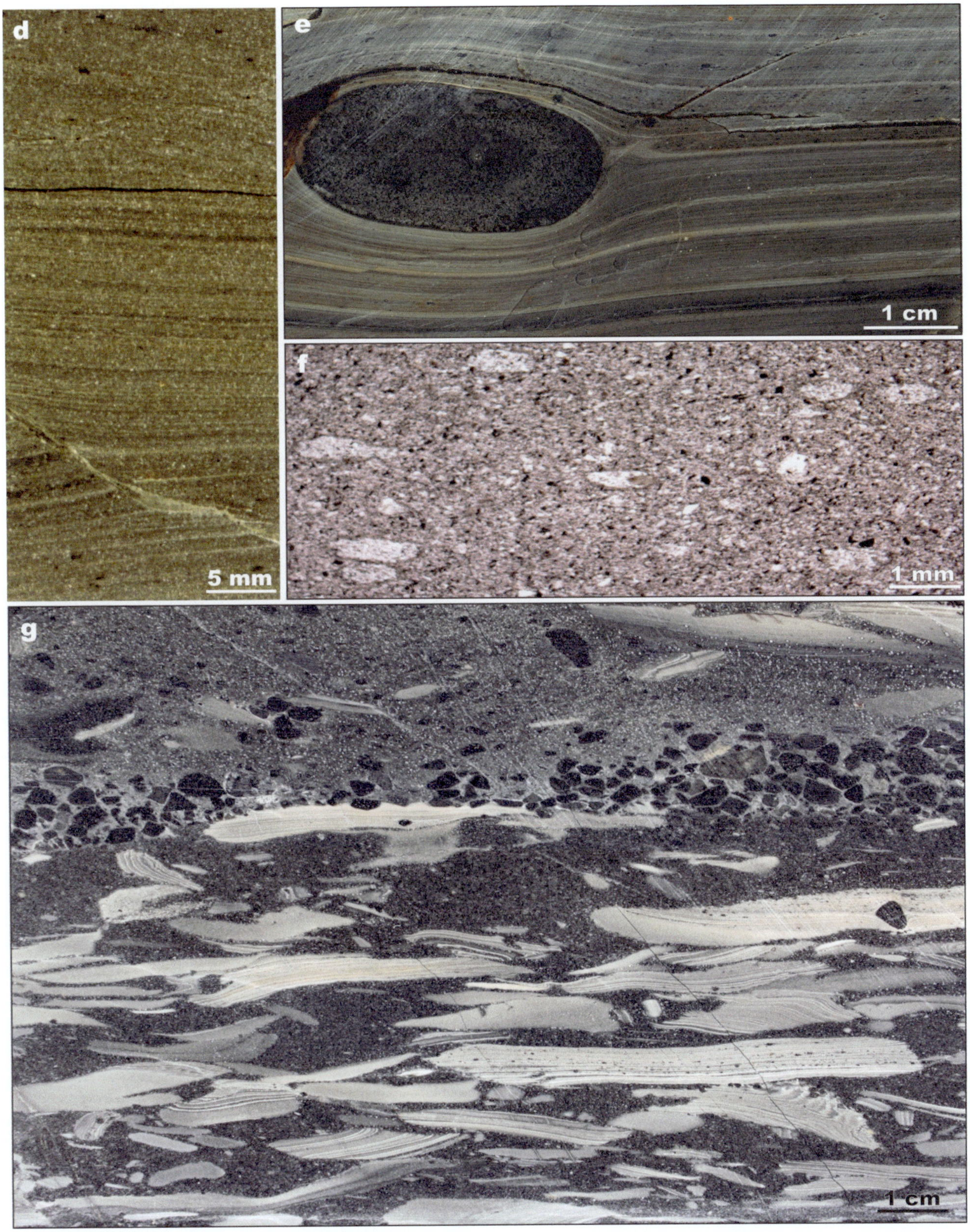

Fig. 4.9 (continued) (**d**) Laminated, fine-grained siltstone, resembling a varved sediment, from the middle part of Sect. 4. (**e**) Dropstone in finely laminated varved, glaciomarine clayey siltstone from the middle part of Sect. 4. (**f**) Massive, fine-grained siltstone with scattered, small dropstones apparently representing a rain-out sediment; middle part of Sect. 4. (**g**) A dark grey, massive siltstone bed containing rip-ups of bright, varved shale with small dropstones in them, followed by a layer of matrix-supported fine-pebble conglomerate grading into a bed of massive greywacke with scattered polymict clasts; upper part of Sect. 4

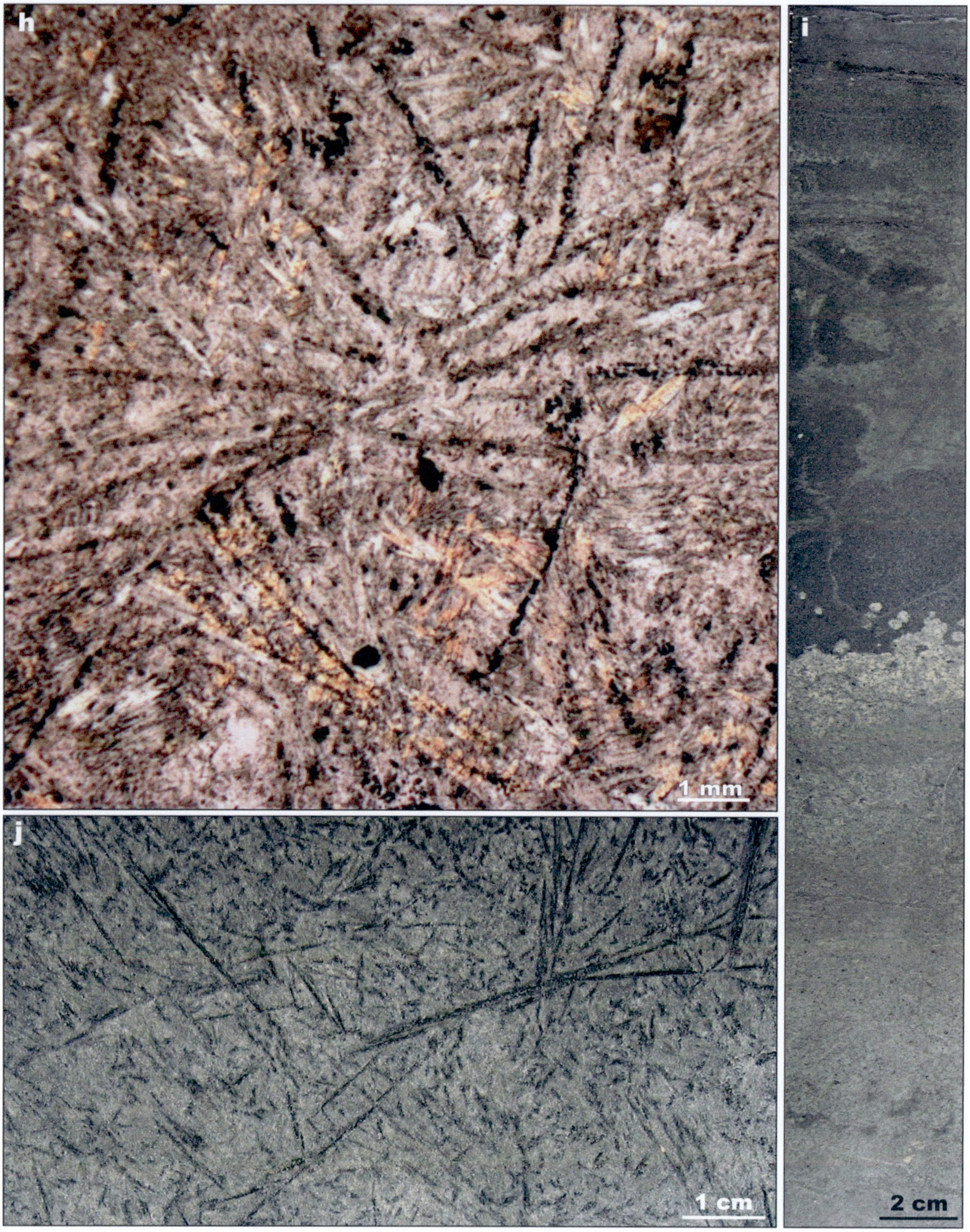

Fig. 4.9 (continued) (**h**) Ultramafic komattite with an olivine spinifex texture; a sill or a lava flow from middle part of Sect. 4. *Polisarka Volcanic Formation*: (**i**) Pale green, globular komatiitic basalt overlain by a flow breccia from the lower part of the formation; Sect. 1 in Fig. 4.8. (**j**) Komatiitic basalt a with pyroxene spinifex texture from the lower part of the formation; Sect. 1 in Fig. 4.8 (Photographs (**i**) and (**j**) are from FAR-DEEP Core 3A)

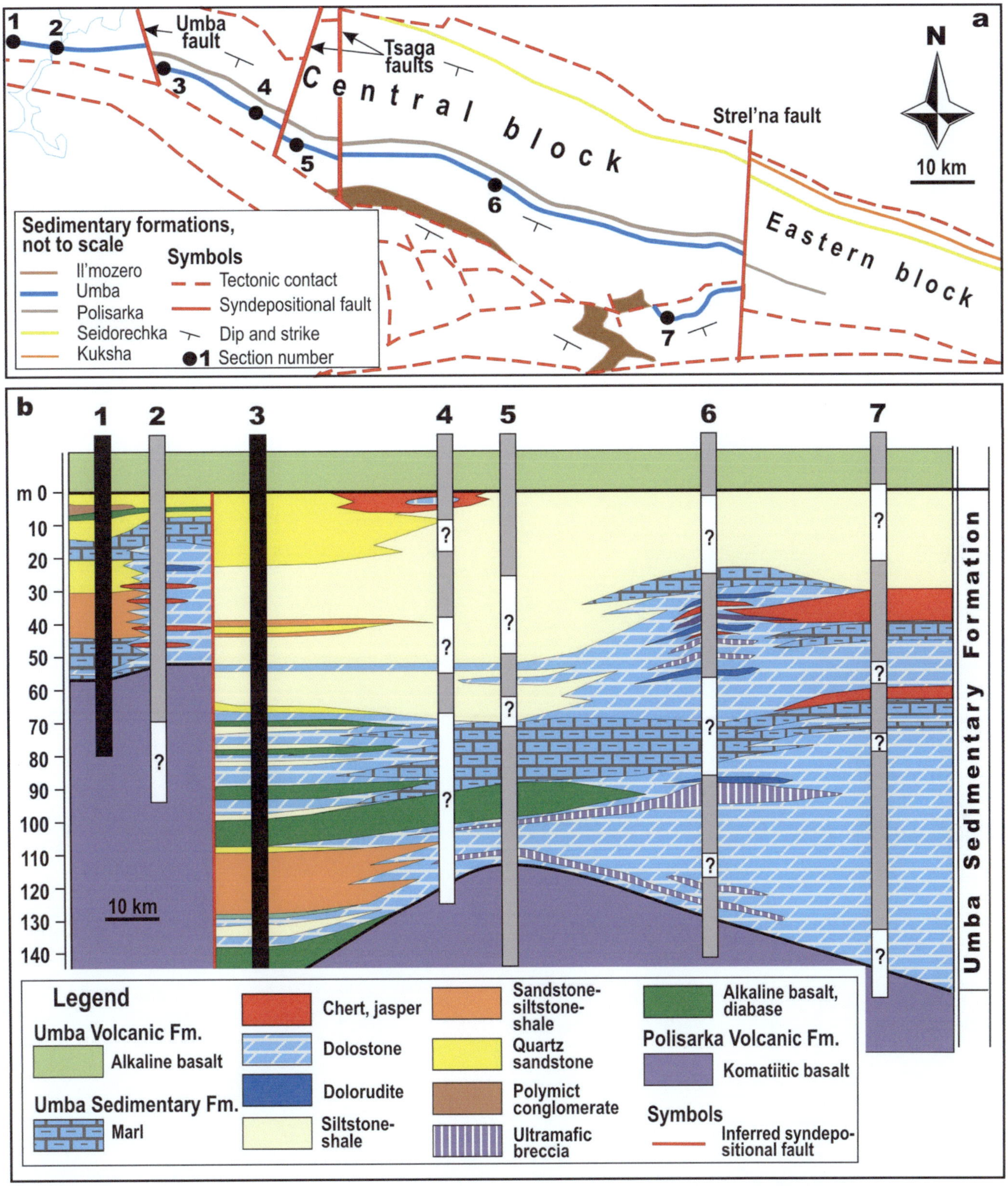

Fig. 4.10 A simplified longitudinal lithological profile through the Umba Sedimentary Formation based on data from Melezhik and Predovsky (1982) and FAR-DEEP Hole 4A. Sections 2 and 4, 5, 6, and 7 are based on outcrop observations, Sect. 3 represents a simplified log of FAR-DEEP Hole 4A (for details, see Chap. 6.1.3 Umba Sedimentary Formation: FAR-DEEP Hole 4A) and Sect. 1 is based on logging of exploration drillholes (Zilber 1972)

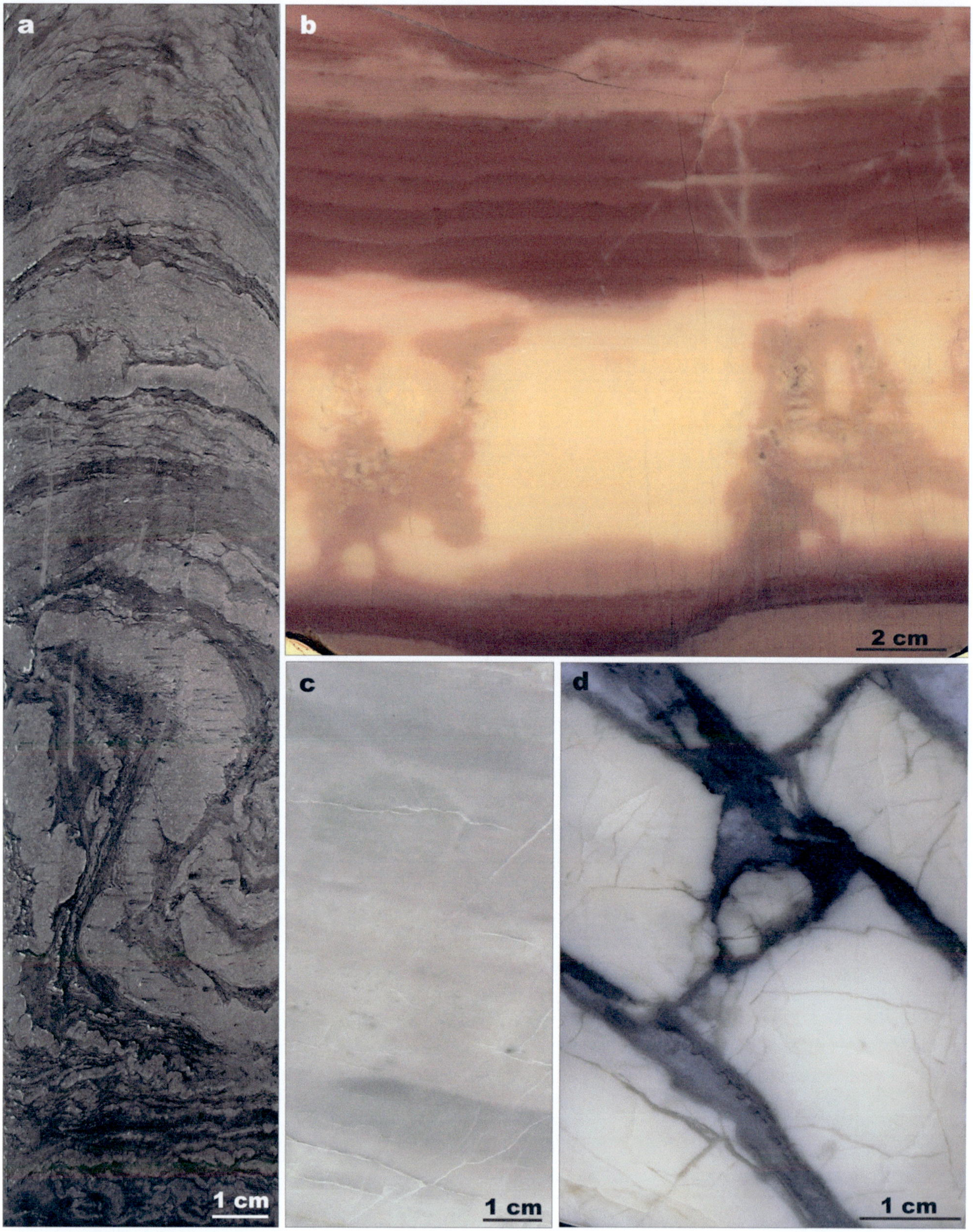

Fig. 4.11 Selected rock images illustrating lithology and structural features of the Umba Sedimentary and Volcanic formations. *Umba Sedimentary Formation*: (**a**) Syndepositional slumping in a sandstone-siltstone-shale unit from the lower part of Sect. 3. (**b**) Selectively bleached, red dolarenite with relics of rhythmic bedding in the upper part of the slab; upper part of Sect. 6. (**c**) Vaguely bedded *white*, fine-grained dolostone from the lower part of Sect. 3. (**d**) White, massive dolostone with extensional cracks filled with *dark grey* quartz; middle part of Sect. 4

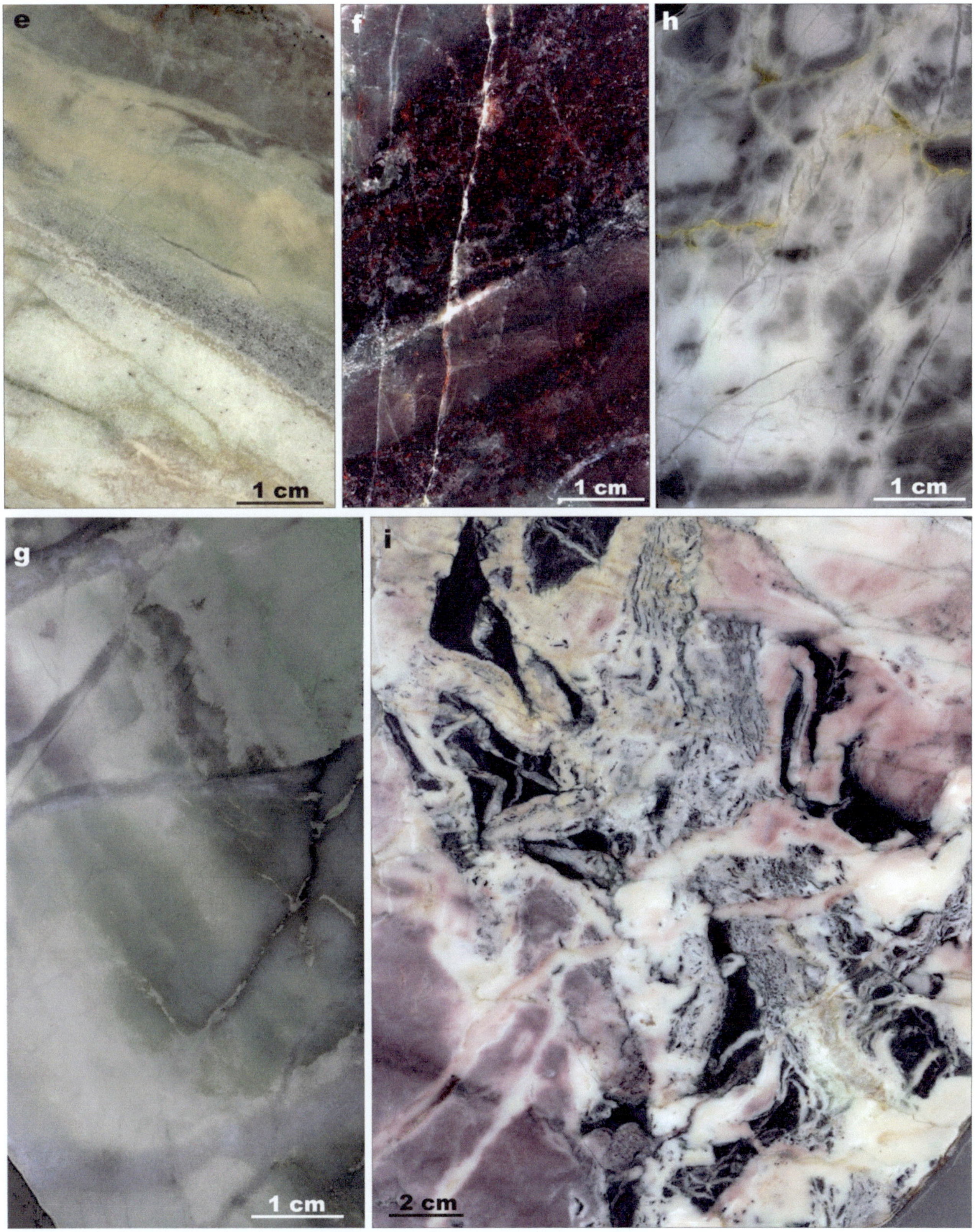

Fig. 4.11 (continued) (**e**) Bedded, siliceous dolorudite from the middle part of Sect. 4. (**f**) Jasper and (**g, h**) variegated chert forming lenses in the upper part of Sect. 4. (**i**) Variegated dolostone with disintegrated clasts of ultramafic lava (*black*); middle part of Sect. 6

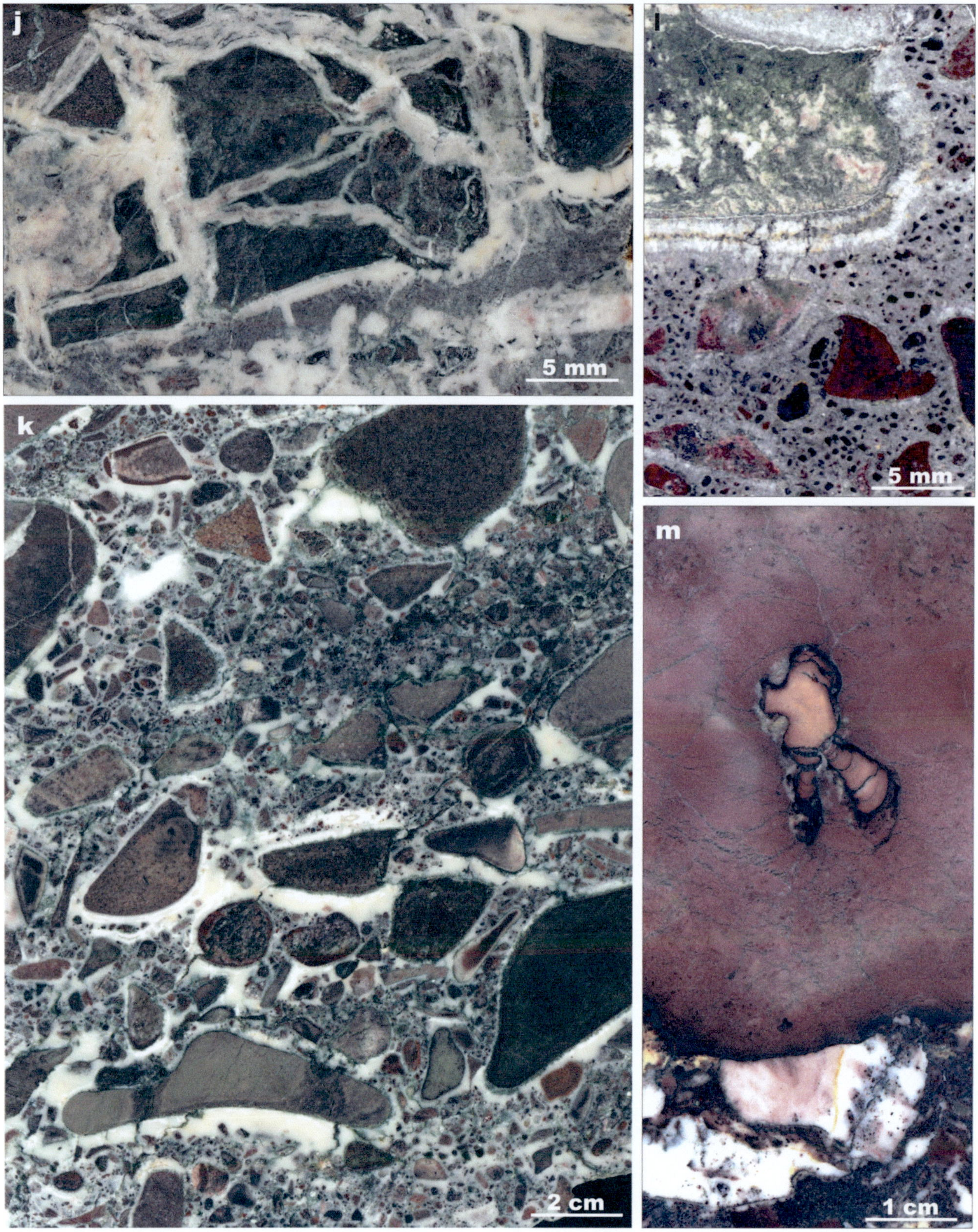

Fig. 4.11 (continued) (**j**) Breccia consisting of altered, angular clasts of ultramafic lava (*dark green-brown*) cemented by *white* dolomite; middle part of Sect. 6. (**k**) Polymict conglomerate consisting of unsorted, rounded clasts of altered ultramafic lava (*dark brown*), Fe-rich dolostones (*pale brown* and *pink*) supported by a polymict gritty matrix and cemented by *white*, isopachous dolomite; middle part of Sect. 6. (**l**) Scanned polished section of polymict conglomerate showing intensively haematised (*red*) and chloritised (*green*) ultramafic clasts cemented by *white* isopachous dolomite; middle part of Sect. 6. (**m**) *Pink*, massive dolarenite with a clast of beige dolostone deposited on an eroded surface of dolostone conglomerate; upper part of Sect. 6

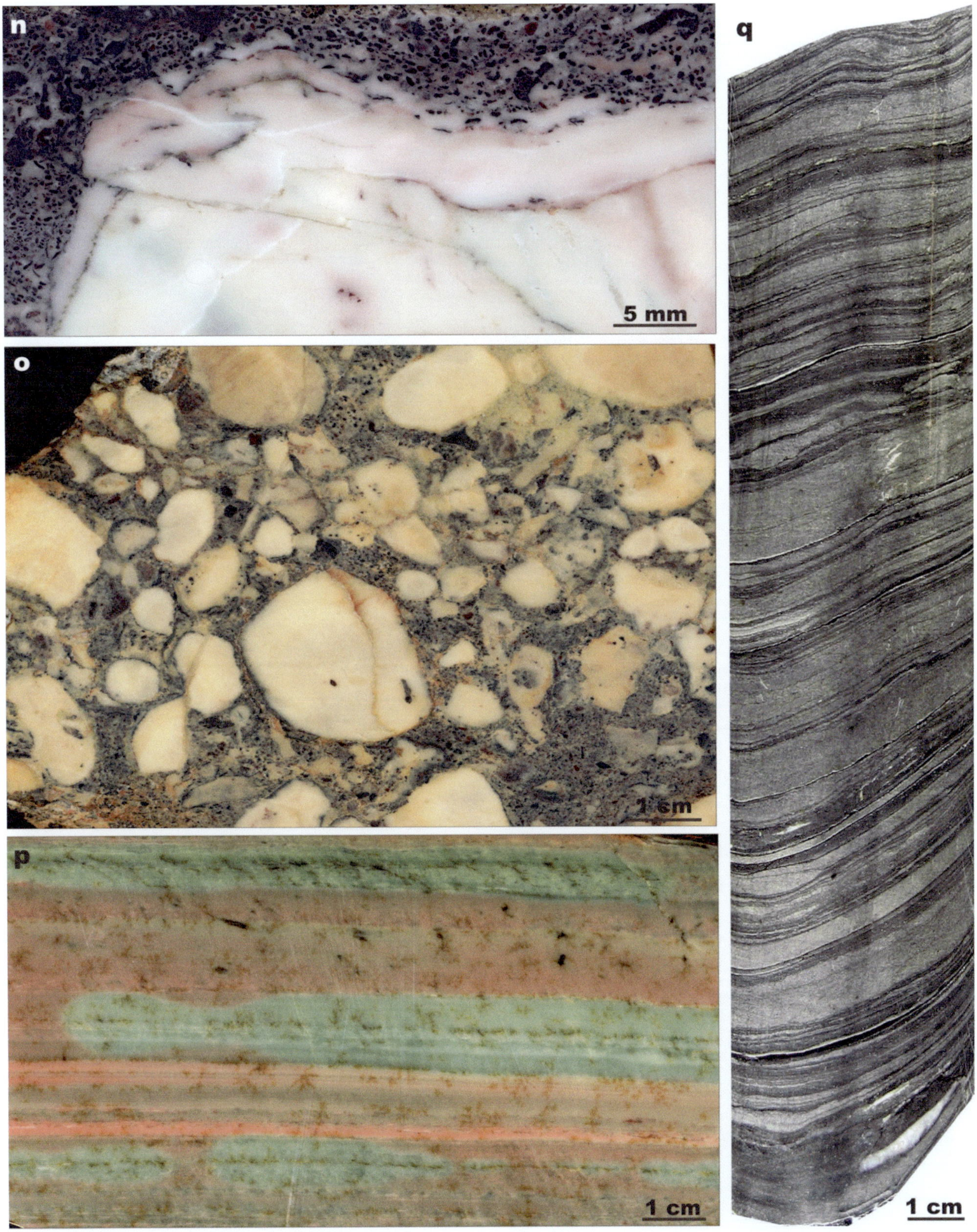

Fig. 4.11 (continued) (**n**) Polymict, volcaniclastic gritstone deposited on an eroded surface of pale beige, massive dolostone; upper part of Sect. 6. (**o**) Conglomerate consisting of rounded clasts of pale beige, massive dolostone supported by a polymict gritstone matrix containing abundant particles of altered ultramafic material; upper part of Sect. 6. (**p**) Variegated dolomarl with a bedding-parallel, pale green "role structure" caused by postdepositional infiltration of reduced fluids; upper part of Sect. 5. (**g**) Thinly laminated sandstone-siltstone-shale with parallel- and low-angle cross-lamination and small-scale slump structures; middle part of Sect. 3

Fig. 4.11 (continued) *Umba Volcanic Formation, central part of the belt*: (**r**) Cross-bedded, carbonate-cemented mafic tuff from the lower part of the formation. (**s**) Flow breccia composed of *dark brown-green* trachyandesite fragments embedded in pale grey dacite with a fluidal structure; upper part of the formation. (**t**) Trachyadesitic flow breccia from the uppermost part of the formation. (**u**) Native copper nugget from the lower part of the formation; sample courtesy of the Geological Museum of the Geological Institute, Kola Science Centre, Apatity; photograph courtesy of Vladimir Voloshin

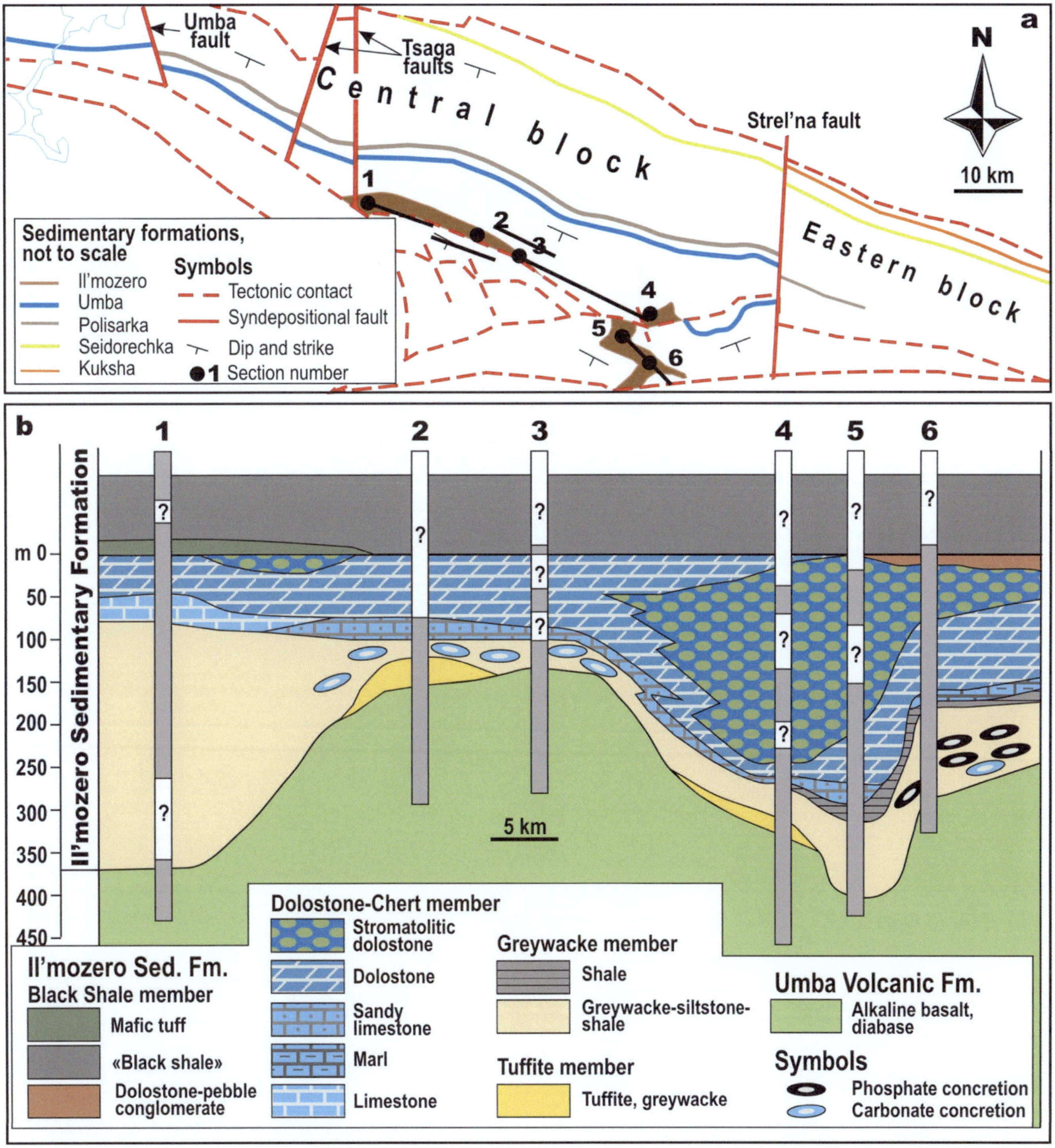

Fig. 4.12 A simplified longitudinal lithological profile through the Il'mozero Sedimentary Formation based on data from Melezhik and Predovsky (1982)

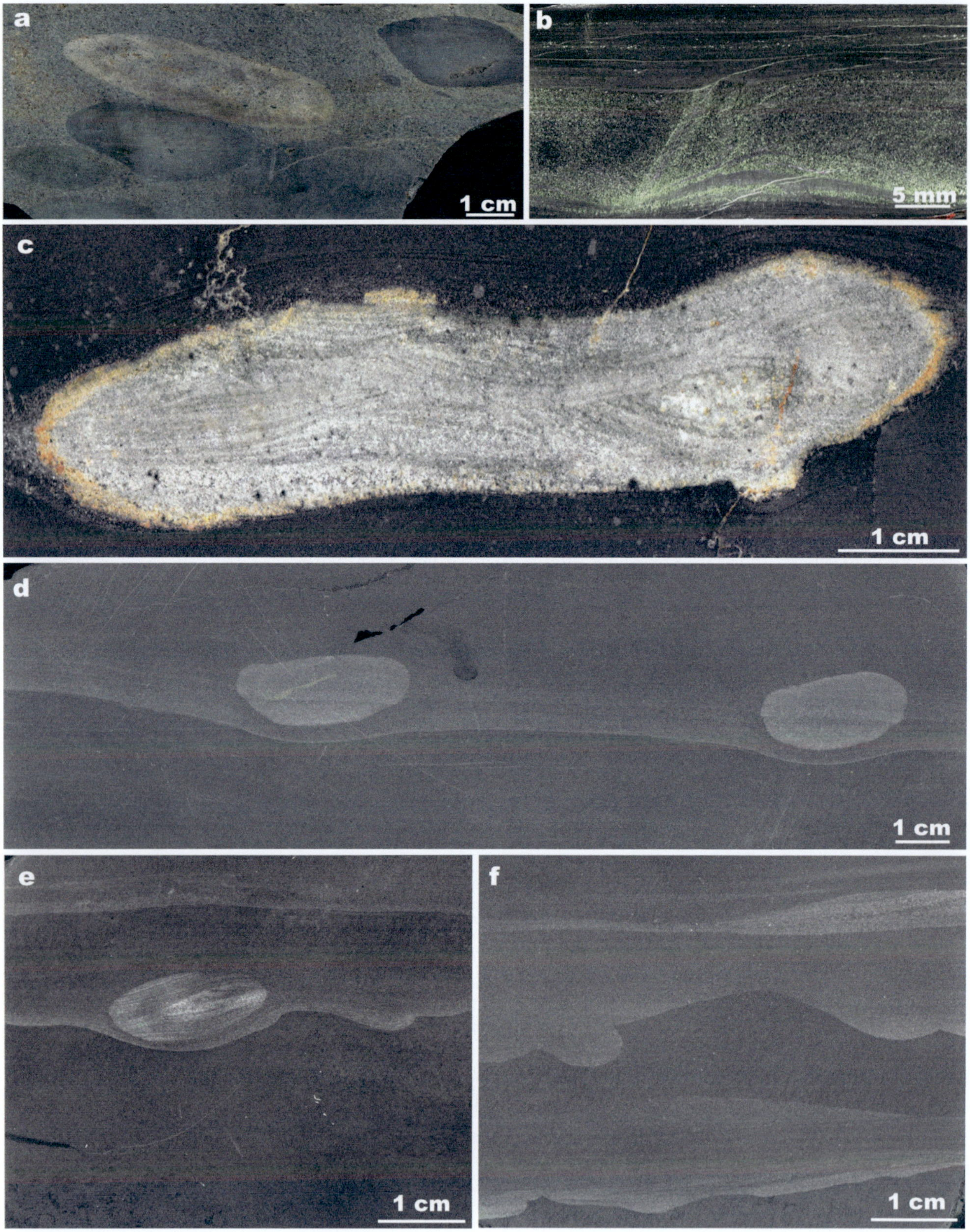

Fig. 4.13 Selected rock images illustrating lithology and structural features of the Il'mozero Sedimentary Formation. (**a**) Basal, polymict, volcaniclastic conglomerate consisting of rounded clasts of trachyandesites and trachybasalts supported by a tuffite matrix; base of Sect. 2. (**b**) Parallel-bedded shale and rippled greywacke rich in authigenic tourmaline (*green*) from the lower part of Sect. 3. (**c**) Early diagenetic calcite concretion preserving original through cross-bedding of a sand bed in which it was formed; lower part of Sect. 3. (**d**) Two phosphate-bearing, diagenetic concretions in a graded sandstone-siltstone bed. (**e**) Phosphate-bearing, diagenetic concretions at the base of sandstone grading in to siltstone deposited on an erosional surface; note that the layering preserved in the concretion is foreign to the hosting graded sandstone bed, thus suggesting redeposition by a turbidity current; lower part of Sect. 6. (**f**) Small-scale erosional surfaces and a flame structure in graded sandstone-siltstone; lower part of Sect. 6

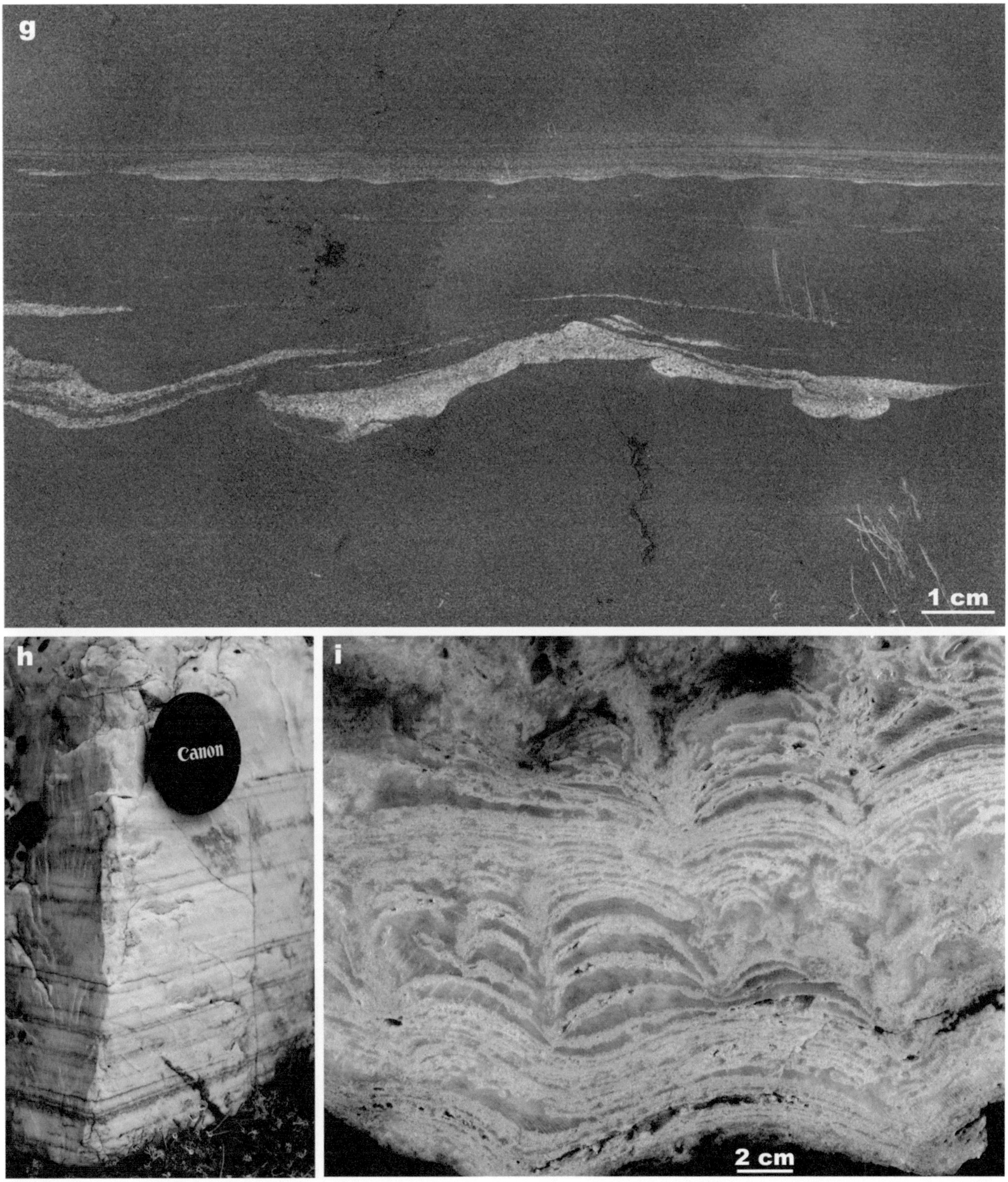

Fig. 4.13 (continued) (**g**) Small-scale erosional surfaces, sand-filled channel and a flame structure in otherwise massive homogeneous siltstone from the lower part of Sect. 6 (**h**) Parallel-bedded, silisified dolostone from middle part of Sect. 4. (**i**) Cross-section view of tightly-packed, columnar to sub-spherical, intensely silicified stromatolitic dolostone from the lower part of Sect. 4

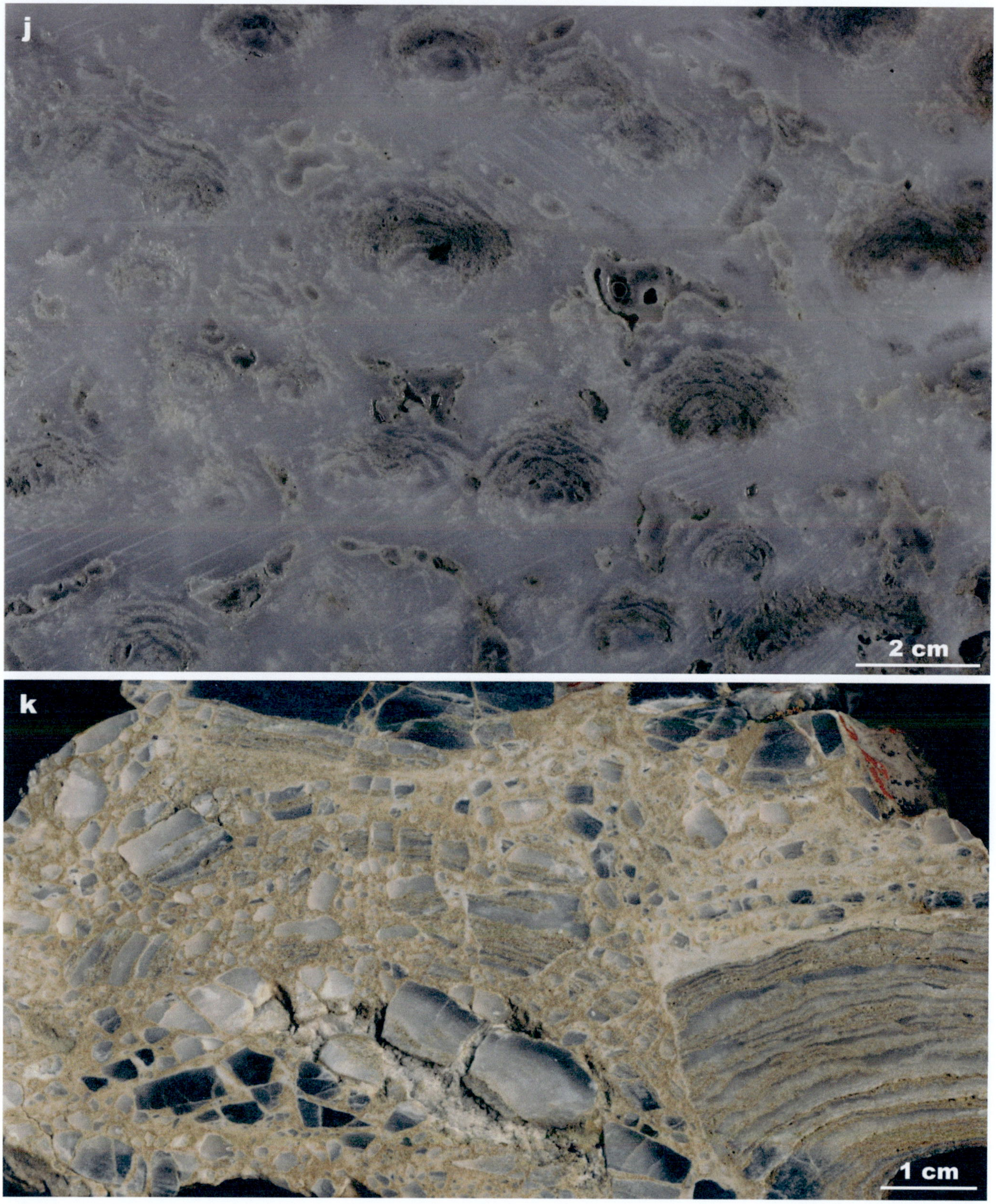

Fig. 4.13 (continued) (**j**) Bedding-parallel surface showing relics of intensely silicified, columnar stromatolitic dolostone from the lower part of Sect. 4. (**k**) Syndepositional stromatolitic breccia from the upper part of Sect 5

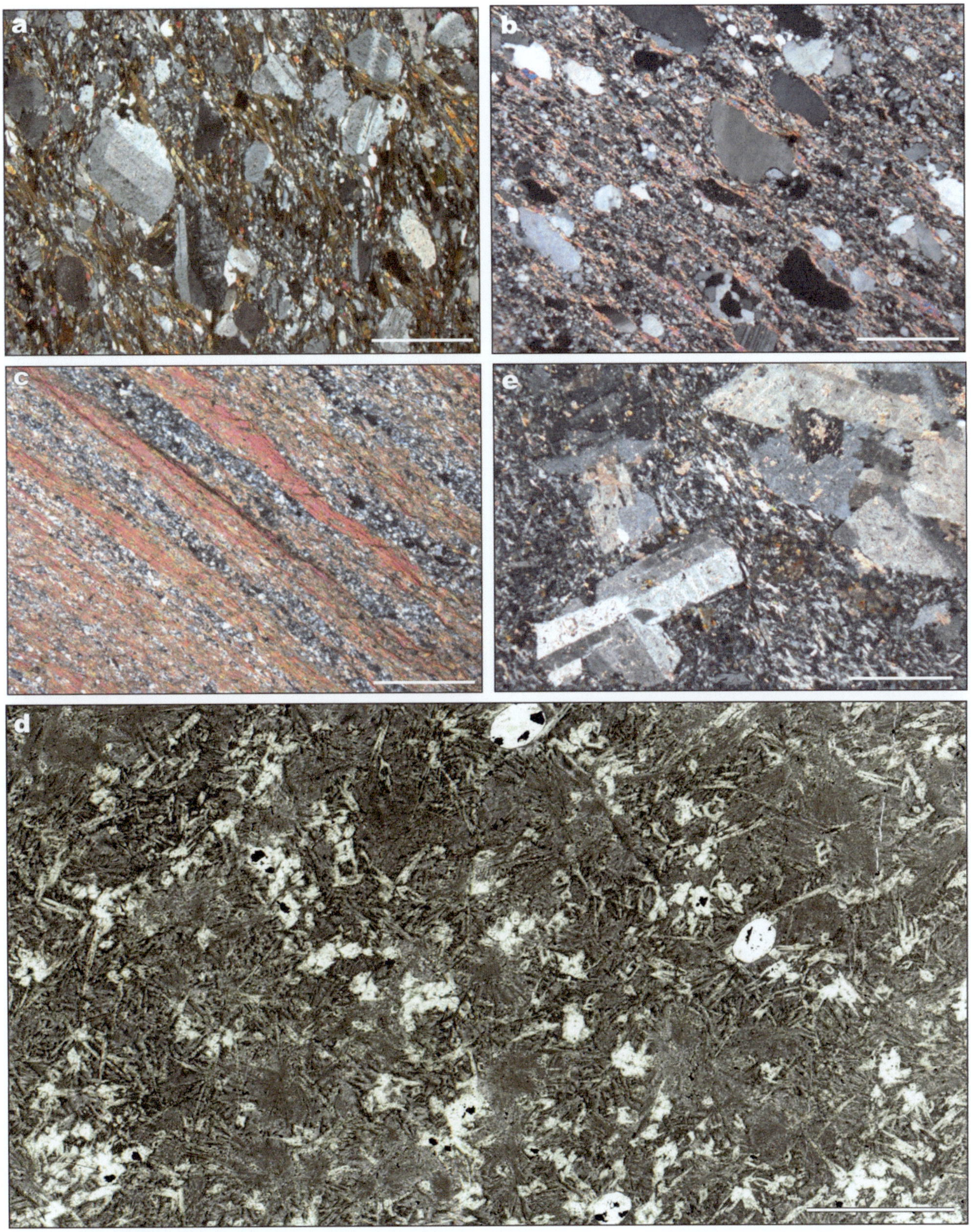

Fig. 4.14 Photomicrographs in cross-polarised light (**a–c, e**) and scanned image (**d**) of key-rocks of the Tominga Group. (**a**) Andesitic volcaniclastic sandstone from the Rouksa unit. (**b**) Sericitic, arkosic sandstone with bimodal grain-size distribution; the Panarechka formation (**c**) Rhythmically bedded sericite schist-arkosic siltstone from the Panarechka formation. (**d**) Sparsely amygdaloidal ferropicritic lava from the Mittrijärvi formation. (**e**) High-Al, porphyritic andesite from the Saminga formation. Scale-bar in (**a–c, e**) is 5 mm; in (**d**) – 5 mm. Photograph (**d**) courtesy of Eero Hanski

References

Amelin YuV, Heaman LM, Semenov VS (1995) U-Pb geochronology of layered mafic intrusions in the eastern Baltic Shield: implications for the timing and duration of Palaeoproterozoic continental rifting. Precambrian Res 75:31–46

Bayanova T, Ludden J, Mitrofanov F (2009) Timing and duration of Palaeoproterozoic events producing ore-bearing layered intrusions of the Baltic Shield: metallogenic, petrological and geodynamic implications. Geol Soc Lond, Spec Pub 323:165–198

Bekasova NB, Pushkin GY (1972) Stratigraphy of the central part of the Imandra/Varzuga structure. In: Zagorodny VG (ed) Geology and metallogeny of the Kola Peninsula. Kola Science Centre, Apatity, pp 28–34 (in Russian)

Bekasova NB, Pushkin G (1975) On cupriferous rocks of the Imandra/Varzuga Zone. In: Geology and geochemistry of metasedimentary complexes of the Kola Peninsula. Kola Science Centre, Apatity, pp 141–149 (in Russian)

Bekasova NB, Pushkin GY (1977) Allothigenic minerals of heavy fraction from metaterrigeneous rocks of the Imandra/Varzuga Zone. In: Sidorenko AV (ed) Lithology and geochemistry of the early precambrian. Kola Science Centre, Apatity, pp 33–44 (in Russian)

Borisov AE (1990) Volcanism and native copper mineralisation in the Palaeoproterozoic of the Kola Peninsula. Kola Science Centre, Apatity, p 70 (in Russian)

Chashchin VV, Bayanova TB, Levkovich NV (2008) Volcanoplutonic association of the early-stage evolution of the Imandra–Varzuga rift zone, Kola Peninsula, Russia: geological, petrogeochemical, and isotope-geochronological data. Petrology 16:279–298

Daly JS, Balagansky VV, Timmerman MJ, Whitehouse MJ (2006) The Lapland-Kola orogen: Palaeoproterozoic collision and accretion of the northern Fennoscandian lithosphere. In: Gee DG, Stephenson RA (eds) European lithosphere dynamics, Geological Society Memoir vol 32. pp 579–598

Dokuchaeva VS, Zhangurov ZhA, Fedotov ZhA (1980) Chromite-bearing norite-gabbronorite complex of the Imandra/Varzuga structural zone. In: Bel'kov IV (ed) Magmatic formations and metallogeny of the mafic-ultramafic complexes of the Kola Peninsula. Kola Science Centre, Apatity, pp 36–51

Dokuchaeva VS, Zhangurov ZhA, Radchenko MK, Fedotov ZhA (1982) Early Karelian mafic-ultramafic complexes. In: Gorbunov GI (ed) The Imandra/Varzuga Zone of the Karelides. Nauka (Science), Leningrad, pp 114–145 (in Russian)

Fedotov ZhA (1978) Chemical composition of basaltic series rocks in light of basalt experimental melting data. Geochemistry 5:680–690 (in Russian)

Fedotov ZhA (1983) Volcanic rocks of the komatiite series in the in the proterozoic Imandra-Varzuga sedimentary-volcanic complex. In: Predovsky AA (ed) Sedimentary basins and volcanic zones in the precambrian of the Kola Peninsula. Kola Science Center, Apatity, pp 81–99 (in Russian)

Fedotov ZhA (1985) Evolution of precambrian volcanism of the Eastern Part of the Pechenga/Varzuga Belt. Kola Science Centre, Apatity, p 105 (in Russian)

Fedotov ZhA, Melezhik VA, Predovsky AA (1975) On two levels of metamorphosed proterozoic weathered crust in the Monchegorsk area. In: Sidorenko AV (ed) Precambrian weathered crusts. Nauka (Science), Moscow, pp 36–39 (in Russian)

Fedotov ZhA, Basalaev AA, Melezhik VA, Latyshev LN (1982) The Strel'na Series, In: Gorbunov, GI (ed.) The Imandra/Varzuga Zone of Karelides, Nauka (Science), Leningrad, p 33-56 (in Russian)

Gilyarova MA (1972) Stratigraphy and structure of Precambrian rocks in Karelia and the Kola Region. Nauka (Science), Leningrad, p 216 (in Russian)

Grinenko LN, Melezhik VA, Fetisova OA (1989) The first finding of barite in Precambrian sedimentary rocks of the Baltic Shield. Doklady (Trans) USSR Acad Sci Miner 304(6):1453–1455

Hanski EJ (1992) Petrology of the Pechenga ferropicrites and cogenetic, Ni-bearing gabbro-wehrlite intrusions, Kola Peninsula, Russia. Geol Surv Finl Bull 367:192

Kharitonov LY (1966) Structure and stratigraphy of the karelides of the Baltic Shield. GOSGEOLTEKHISDAT, Moscow, p 360 (in Russian)

Koistinen T, Stephens MB, Bogatchev V, Nordgulen Ø, Wennesström M, Korhonen J (Comps.) (2001) Geological Map of the Fennoscandian Shield, Scale 1:2 000 000, Espoo/Trondheim/Uppsala/Moscow

Kozlov MT (1971) Stratigraphy of the Imandra/Varzuga mobile zone. In: Zagorodny VG (ed) Stratigraphical subdivision and correlation of the Precambrian in the Northeastern Part of the Baltic Shield. Kola Science Centre, Apatity, pp 99–111 (in Russian)

Kozlov MT, Latyshev LN (1974) New data on the geology of the Imandra part of the Monchegorsk area. In: Bel'kov IV (ed) Regional geology, metallogeny and geophysics, 5th edn. Kola Science Centre, Apatity, pp 67–74 (in Russian)

Martin AP, Condon DJ, Prave AR, Melezhik VA, Fallick A (2010) Constraining the termination of the Lomagundi-Jatuli positive isotope excursion in the Imandra-Varzuga segment (Kola Peninsula, Russia) of the North Transfennoscandian Greenstone Belt by high-precision ID-TIMS, AGU, 13–17 Dec 2010

Melezhik VA (1976) Composition and evolution of mid-proterozoic sedimentary rocks of the Monche Peninsula. In: Bel'kov IV (ed) Geochemical evolution of Precambrian metamorphic complexes of the Kola Peninsula. Kola Science Centre, Apatity, pp 38–50 (in Russian)

Melezhik VA (1978) Stratigraphic and structural status of the supracrustal rocks from the Rouksa River area. In: Zagorodny VG (ed) Stratigraphic subdivisions and correlation of Precambrian rocks of the Kola Peninsula. Kola Science Centre, Apatity, pp 100–108 (in Russian)

Melezhik VA (1981) Palaeogeography and geochemical conditions of sedimentation during Il'mozero time (Early Proterozoic of the Kola Peninsula). Lithol Miner Dep 2:56–73 (in Russian)

Melezhik VA (1992) Early Proterozoic sedimentary and rock-forming basins of the Baltic Shield. Nauka (Science), St. Petersburg, p 258 (in Russian)

Melezhik VA (1996) The North Transfennoscandian Greenstone Belt: and overview. ICDP Project 336 symposium, Rovaniemi, 21–23 Aug 1996, University of Turku, Finland, Program and Abstracts, pp 23–24

Melezhik VA, Fallick AE (1996) A widespread positive $\delta^{13}C_{carb}$ anomaly at around 2.33–2.06 Ga on the Fennoscandian Shield: a paradox? Terra Nova 8:141–157

Melezhik VA, Fetisova OA (1989) Discovery of syngenetic barium sulphates in Precambrian rocks of the Baltic Shield. Doklady (Trans) USSR Acad Sci Geol 307(2):422–425

Melezhik VA, Predovsky AA (1982) Geochemistry of early Proterozoic lithogenesis. Nauka (Science), St. Petersburg, p 208 (in Russian)

Melezhik VA, Predovsky AA (1987) Carbonate and silicate concretions in Precambrian metasedimentary rocks of the Kola Peninsula. Doklady (Trans) USSR Acad Sci Geol 239(3):668–671 (in Russian)

Melezhik VA, Sturt BA (1994) General geology and evolutionary history of the early Proterozoic Polmak-Pasvik-Pechenga-Imandra/Varzuga-Ust'Ponoy Greenstone Belt in the north-eastern Baltic Shield. Earth-Sci Rev 36:205–241

Melezhik VA, Borisov AE, Fedotov ZhA, Predovsky AA (1982) The Varzuga series. In: Gorbunov GI (ed) The Imandra/Varzuga Zone of the Karelides. Nauka (Science), Leningrad, pp 57–85 (in Russian)

Melezhik VA, Fallick AE, Hanski E, Kump L, Lepland A, Prave A, Strauss H (2005) Emergence of the aerobic biosphere during the Archean-Proterozoic transition: challenges of future research. Geol Soc Am Today 15:4–11

Mints MV (1993) Palaeotectonic reconstructions of the early Precambrian of the eastern part of the Baltic Shield. Geotectonics 1:39–56 (in Russian)

Mints MV (1996) The Pechenga-Imandra/Varzuga Sedimentary-Volcanic Belt. In: Leonov YuG (ed) The early Precambrian of the northeastern Baltic Shield: paleogeodynamics, crustal structure and evolution, Scientific World, Moscow, pp 77–112 (in Russian)

Mints MV, Glaznev VN, Konilov AN, Kunina NM, Nikitichev AP, Raevsky AB, Sedikh YuN, Stupak VM, Fonarev VI (1996) The early Precambrian of the Northeastern Baltic Shield. Scientific World, Moscow, p 287 (in Russian)

Petrov VP, Voloshina ZM (1982) Metamorphism. In: Gorbunov GI (ed) The Imandra/Varzuga Zone of the Karelides. Nauka (Science), Leningrad, pp 192–212 (in Russian)

Polkanov AA (1936) Geological essay of the Kola Peninsula. Proc Arctic Inst 53:171

Potrubovich LN, Simon AK (1966) Stratigraphy of the Imandra-Varzuga-Sosnovka flexure. Herald Mosc State Univ, Geol Ser 3:37–45 (in Russian)

Pozhilenko VI, Bayanova TB, Bogachev VA, Philippov NB, Gogol' OV (1999) Relationship and age the Arvarench and Kuksha formations (Mt. Arvarench, Kola region, Baltic Shield), In: Abstract, III All-Russian conference. General problems of subdivision of the Precambrian, Kola Science Centre, Apatity, pp 215–218

Predovsky AA, Melezhik VA, Bolotov VA, Fedotov ZhA, Basalaev AA, Kozlov NE, Ivanov AA, Zhangurov AA, Skuf'in PK, Lyubtsov VV (1987) Volcanism and sedimentology of the precambrian in the Northeastern Part of the Baltic Shield. Nauka (Science), Leningrad, p 185 (in Russian)

Radchenko AT (1972a) Tectonics of the Eastern Part of the Imandra/Varzuga Graben-Syncline. Ph.D. thesis, Leningrad University, Leningrad, p 120 (in Russian)

Radchenko AT (1972b) Fragments of the southern limb of the eastern margin of the Imandra/Varzuga graben-synform. In: Bel'kov IV (ed) Materials on the geology and metallogeny of the Kola Peninsula, 3rd edn. Kola Science Centre, Apatity, pp 39–43 (in Russian)

Radchenko AT, Kravtsov NA, Borisov AE, Latyshev LN, Melezhik VA, Fedotov ZhA (1982) The Tominga series. In: Gorbunov GI (ed) The Imandra/Varzuga Zone of the Karelides. Nauka (Science), Leningrad, pp 85–111 (in Russian)

Radchenko AT, Balagansky VV, Basalaev AA, Beljaev OA, Pozhilenko VI, Radchenko MK (1994) Explanatory Note to 1:500,000 Geological Map of the Northeastern Baltic Shield. Kola Science Centre, Apatity, p 95 (in Russian)

Ramsay W, Hackman V (1894) Das Nefelinsyenitgebiet auf of der Halbinsel Kola. Fennia 14:1–225

Sharkov EV, Smolkin VF (1997) The early Proterozoic Pechenga-Varzuga Belt: a case of back-ark spreading. Precambrian Res 82:135–151

Sidorenko AV, Luneva OI (1961) On the lithological investigation of metamorphic complexes. Nauka (Science), Moscow-Leningrad, p 200 (in Russian)

Simon AK (1967) Geology and Evolution of the Precambrian Imandra/Varzuga/Sosnovka Trough (Kola Peninsula). Ph.D. thesis, Moscow State University, p 22 (in Russian)

Skuf'in PK, Theart HFJ (2005) Geochemical and tectono-magmatic evolution of the volcano-sedimentary rocks of Pechenga and other greenstone fragments of the Kola Greenstone Belt. Precambrian Res 141:1–48

Skuf'in PK, Bayanova TB, Mitrofanov FP (2006) Isotope age of subvolcanic graditoid rocks of the early Proterozoic Panarechka volcanotectonic structure, Kola Peninsula. Proc Russian Acad Sci, Earth Sci 409:774–778

Smolkin VF (1995) Early Protrerozoic. In: Mitrofanov FP (ed) Geology of the Kola Peninsula (Baltic Shield). Kola Science Centre, Apatity, pp 38–75

Smolkin VF, Dain AD (1985) Gabbro-wehrlite formation of the Imandra/Varzuga Zone of the Karelides. Soviet Geol 12:94–105 (in Russian)

Vrevsky AB, Bogomolov ES, Zinger TF, Sergeev SA (2010) Polychronic sources and isotopic age of the volcanogenic complex (Arvarench unit) of the Imandra-Varzuga structure, Kola Peninsula. Dokl (Trans) Russ Acad Sci 43(1):386–389

Zagorodny VG, Radchenko AT (1982) Main geological features of the Imandra/Varzuga Zone and its frame. In: Gorbunov GI (ed) The Imandra/Varzuga Zone of the Karelides. Nauka (Science), Leningrad, pp 8–31 (in Russian)

Zagorodny VG, Radchenko AT (1988) Tectonics of the Karelides, NE Baltic Shield. Nauka (Science), Leningrad, p 111 (in Russian)

Zagorodny VG, Bekasova NB, Pushkin GYu, Radchenko AT (1972) On the geology of middle Proterozoic rocks of the Imandra/Varzuga Zone. In: Bel'kov IV (ed) Materials on the geology and metallogeny of the Kola Peninsula, 3rd edn. Kola Science Centre, Apatity, pp 11–17 (in Russian)

Zagorodny VG, Kravtsov NA, Latyshev LN, Radchenko AT (1978) On the stratigraphy of the upper part of the Imandra/Varzuga complex. In: Zagorodny VG (ed) Stratigraphic subdivisions and correlation of precambrian rocks of the Kola Peninsula. Kola Science Centre, Apatity, pp 79–88 (in Russian)

Zagorodny VG, Predovsky AA, Basalaev AA, Batieva ID, Borisov AE, Vetrin VR, Voloshina ZM, Dokuchaeva VS, Zhangurov AA, Kozlova NE, Kravtsov NA, Latyshev LN, Melezhik VA, Petrov VP, Radchenko MK, Radchenko AT, Smolkin VF, Fedotov ZhA (1982) Imandra/Varzuga Zone of the Karelides (Geology, geochemistry, history of development). Nauka (Science), Leningrad, p 280 (in Russian)

Zilber ME (1972) On the stratigraphy and copper shows in Umba Formation volcanic rocks. In: Bel'kov IV (ed) Materials on the geology and metallogeny of the Kola Peninsula, 3rd edn. Kola Science Centre, Apatity, pp 44–50 (in Russian)

4.2 The Pechenga Greenstone Belt

V.A. Melezhik and E.J. Hanski

4.2.1 Introduction

The Pechenga Greenstone Belt (Fig. 4.15) forms the northwestern part of a NW-SE-trending Palaeoproterozoic belt that extends for c. 800 km from Russia through Norway to Finland and back to Norway. About 600 km of that belt occurs in the Russian part of the Fennoscandian Shield and is commonly called the Pechenga-Imandra/Varzuga Greenstone Belt. In order to embrace the entirety of the belt, Melezhik and Sturt (1994a) introduced the term Polmak-Pasvik-Pechenga-Imandra/Varzuga-Ust'Ponoj Greenstone Belt. This name is cumbersome for practical use, and still misses the Finnish part, the Opukasjärvi Greenstone Belt. In an attempt to resolve the terminology, Melezhik (1996) suggested the name Transfennoscandian Greenstone Belt (Fig. 3.1); however, it remains to be seen whether or not this will be used in practice.

The Pechenga Greenstone Belt is the most intensely studied geological structure in the Kola Peninsula, in part owing to its economic importance as a host of world-class Ni-Cu-sulphide ore deposits but also because the Kola Superdeep Drillhole (SG-3) is located in the region. The first detailed description of the geology of Pechenga dates back to the ore camps of the 1930–1940s (Väyrynen 1938). Later work by Zagorodny et al. (1964), Gilyarova (1967) and Gorbunov (1968) defined key elements of the regional geology and stratigraphy. Predovsky et al. (1974) elaborated on the geochemistry of sedimentary and volcanic rocks, and Zak et al. (1982) refined or reconfirmed previous knowledge on the regional geology. Sedimentary environments and palaeogeographic aspects of the Pechenga Greenstone Belt have been addressed by Melezhik and Predovsky (1982), Negrutsa (1984), Bekasova (1985), Akhmedov and Krupenik (1990) and Melezhik (1992). A comprehensive geochemical and petrological research on the ferropicrites and associated Ni-Cu-bearing gabbro-wehrlite intrusions was made by Hanski (1992) and Smol'kin (1992). A 1:50,000 scale mapping and associated research project was summarised by Mitrofanov and Smol'kin (1995).

During the last decade, a certain effort has been made to obtain radiometric dates on the emplacement of igneous rocks (Amelin et al. 1995; Smol'kin et al. 2003; Vetrin et al. 2008; Bayanova et al. 2009) and the timing of sedimentation and volcanism (Hanski et al. 1990; Melezhik et al. 2007) in the Pechenga area, which has led to a full appreciation of the prolonged duration of the sedimentary-volcanic processes, covering a time span from c. 2500 to 1940 Ma.

Specifically targeted research was also carried out on sedimentology and isotope geochemistry of carbonate rocks (Karhu 1993; Melezhik and Fallick 1996, 2001, 2005; Melezhik et al. 2004, 2005) and sulphidic 'black shales' (Grinenko et al. 1967; Akhmedov and Krupenik 1990; Grinenko and Smol'kin 1991; Melezhik et al. 1998; Barnes et al. 2001) to define a general chemostratigraphic correlation during the 2330–2060 Ma global perturbation of the carbon cycle. Other detailed accounts on the general geology and geochemistry, and palaeotectonic acpects of the Pechenga Greenstone Belt can be found in Mints et al. (1996), and Orlov and Laverov (1998).

Despite the intensive research, no consensus has been reached concerning the geotectonic model (Predovsky et al. 1974; Negrutsa 1984; Mints 1993; Mints et al. 1996; Melezhik and Sturt 1994a; Sharkov and Smol'kin 1997; Smol'kin 1997), though most researchers tend to interpret the evolution of the Pechenga Greenstone Belt as the result of several aborted and reactivated phases of intracontinental rifting, which eventually resulted in a short-term drifting, and finally an arc-continent collision (Mints 1993, 1996; Melezhik and Sturt 1994a; Sharkov and Smol'kin 1997).

V.A. Melezhik (✉)
Geological Survey of Norway, Postboks 6315 Sluppen, NO-7491 Trondheim, Norway

Centre for Geobiology, University of Bergen, Allegaten 41, N-5007 Bergen, Norway
e-mail: victor.melezhik@ngu.no

V.A. Melezhik et al. (eds.), *Reading the Archive of Earth's Oxygenation*, Volume 1: The Palaeoproterozoic of Fennoscandia as Context for the Fennoscandian Arctic Russia - Drilling Early Earth Project,
DOI 10.1007/978-3-642-29682-6_8, © Springer-Verlag Berlin Heidelberg 2013

4.2.2 General Geological Features

The Pechenga Belt was originally subdivided into the Northern and Southern Sub-Zones (Zagorodny et al. 1964). These were later renamed the Southern and Northern Zones, of which the former was considered largely allochthonous and the latter partially autochthonous and partially allochthonous (Melezhik et al. 1994a, b). The Northern and Southern Zones are separated by the major NW-SE-trending Poritash fault zone, dipping 50–60° to the southwest, which is intruded by the Poritash subvolcanic andesites to rhyolites (Zagorodny et al. 1964) (Fig. 4.15). The Northern Zone represents a simple, NW-SE-trending half-graben filled predominantly with volcanic rocks dipping to the SE and SW (20–60°). The basal contact of the Pechenga succession with the Archaean basement rocks, as well as all contacts (except one, see below) between the formations within the Northern Pechenga Zone are largely stratigraphic, though, in places, somewhat tectonically overprinted. The exception is the contact separating the Kolosjoki Sedimentary and Volcanic formations (Fig. 4.16), which is marked by the Luchlompolo fault (Kozlovsky 1984; Gorbatsevich et al. 2010). This fault was originally defined as a thrust but, in contrast to the geometry of thrust faults, it emplaces younger rocks on top of older rocks. Hence, within the Pechenga Greenstone Belt it is recognised as an example of a bedding-parallel fault (for details see *the Kolosjoki Volcanic Formation*).

Several transverse fault systems separate the Northern Zone half-graben into NW-SE- and NE-SW-trending blocks or smaller grabens, which controlled deposition and thicknesses of sedimentary and volcanic formations (Melezhik et al. 1994b). However, a considerable thickness reduction and wedging-out of some sedimentary and volcanic formations towards the eastern and western flanks of the belt (Fig. 4.15) were caused by shearing associated with the northeast-directed faulting of Archaean rocks upon the Pechenga Belt.

The Southern Zone is a strongly folded and tectonically imbricated nappe overthrust by Archaean rocks from southwest. The sedimentary and volcanic rocks of the zone dip 60–85° to the southeast.

The 12,262-m-deep Kola Superdeep Drillhole (SG-3), the deepest by far in the world, intersects the Palaeoproterozoic succession of the Northern Zone from the surface down to a depth of 6,842 m (for the location of the drilling site, see Fig. 4.15). The obtained information from that core, together with surface geology and deep-seismic data, has enabled developing a three-dimensional geological model of the belt (Kozlovsky 1984; Kazansky 1992; Mints et al. 1995; Orlov and Laverov 1998). A NE-SW-trending seismic profile across the central part of the Pechenga Belt revealed that the base of the Palaeoproterozoic succession of the North Pechenga Zone descends gradually to the SW (Mints et al. 1996; Sedykh et al. 1998) such that, by 30 km south of the Superdeep drilling site (when approaching the Poritash fault zone), it is at a depth of 10–11 km. SG-3 revealed that the metamorphic grade increased significantly with depth (amphibolite facies appearing at depth of c. 6,000 m), accompanied by the associated modification of primary mineralogy and geochemistry (Melezhik et al. 2003). On the surface, the rocks of the Northern Zone record zonal metamorphism from prehnite-pumpellyite to greenschist facies in the central part of the Pechenga Belt to amphibolite facies towards the eastern and western margins (Petrov and Voloshina 1995). The rocks of the Southern Zone (South Pechenga Group) have undergone a more complicated tectonic evolution, with the metamorphic grade varying from chlorite-actinolite to amphibolite facies (Petrov and Voloshina 1995).

4.2.3 Stratigraphic Subdivision and Age

The stratigraphic subdivision of the Pechenga Greenstone Belt (Fig. 4.16) has remained essentially unchanged since Zagorodny et al. (1964) published the first general description of the regional geology, tectonics and stratigraphy of what they called the Pechenga Series. Several attempts have been made to revise the stratigraphical nomenclature (e.g. Kozlovsky 1984; Mitrofanov and Smol'kin 1995). These, though, failed to adhere to the International Stratigraphic Code (ISSC 1976; see discussion in Melezhik and Sturt 1994b), hence the international scientific community has continued to employ Zagorodny's original stratigraphy (e.g. Hanski 1992; Melezhik and Sturt 1994b; Green and Melezhik 1999; Barnes et al. 2001), as we also do herein. Although we generally utilise the original geographical unit names, the Russian terms "subseries" and "subsuite" are substituted with "group" and "formation", rspectively (cf. Melezhik and Sturt 1994a; Melezhik et al. 1994a).

Lithostratigraphically, the Pechenga Greenstone Belt comprises the Petsamo Supergroup (Melezhik and Sturt 1994a), previously known as the Pechenga Series (Zagorodny et al. 1964). The series was originally subdivided into the Northern Pechenga and Southern Pechenga subseries, which were later renamed the North Pechenga and South Pechenga groups, respectively (Melezhik and Sturt 1994a; Melezhik et al. 1994). The former is located within the Northern Zone and the latter within the Southern Zone. Being presently separated by the Poritash fault zone (Fig. 4.15), the primary chronostratigraphic relationship between these two groups remains unresolved. Nevertheless, the South Pechenga Group has commonly been regarded as younger than the

North Pechenga Group (e.g. Predovsky et al. 1987; Skuf'in and Theart 2005). Precise radiometric ages are still needed from the former to constrain its depositional timing.

The North Pechenga Group

The North Pechenga Group consists of an impressive sedimentary-volcanic succession that reaches a composite total thickness of more than 12 km (Fig. 4.16). Although the rocks have experienced polyphase metamorphism ranging from prehnite-pumpellyite to upper amphibolite facies (primarily related to the Svecofennian orogeny at 1900–1800 Ma), superb preservation of many original sedimentary and volcanic features is common, especially in the central part of the Northern Zone.

A remarkable feature of the North Pechenga Group stratigraphy (similar to that of the Imandra/Varzuga Greenstone Belt) is the cyclic build-up of sedimentary and volcanic formations. The group is composed of four, paired sedimentary-volcanic cycles, with some separated by non-depositional unconformities (Fig. 4.16) that, in places, are marked by palaeosols (Predovsky et al. 1974; Sturt et al. 1994). However, one of the sedimentary/volcanic cycles, namely the Kolosjoki sedimentary/volcanic pair (Fig. 4.16), is, as noted above, separated by the Luchlompolo fault (Kozlovsky 1984; Mints 1996; Gorbatsevich et al. 2010), a bedding-parallel fault whose magnitude of displacement/transport and overall significance remain to be studied. Another questionable pair is the Pilgujärvi couplet (Fig. 4.16) whose lower, upper and internal lithostratigraphic boundaries remain contested with opinions ranging from largely in situ, syn-depositional contacts (Zagorodny et al. 1964; Predovsky et al. 1974) to entirely tectonic ones (e.g. Mints 1996 and references therein).

The lowermost couplet rests depositionally on Archaean basement rocks, and each successive couplet begins with a sequence of sedimentary rocks and ends with a thick succession of volcanic rocks. Both the sedimentary and volcanic componets of each cycle/couplet have been formally defined as formations. Thus, the North Pechenga Group includes four sedimentary and four volcanic formations. These eight formal lithostratigrphic units (previously known as "subsuites"; Zagorodny et al. 1964) are, from oldest to youngest, the Ahmalahti, Kuetsjärvi, Kolosjoki and Pilgujärvi Sedimentary and Volcanic formations. One modification to this scheme has been the replacement of the term, Ahmalahti Sedimentary Formation, with its Norwegian equivalent in the northern Pasvik area, the Neverskrukk Formation, which is better exposed and better developed than its Russian counterpart (Melezhik et al. 1994a).

Existing geochronological constraints on the North Pechenga Group indicate that its evolution spans some 600 Ma of Earth history. A maximum age of deposition is constrained by the presence of gabbro-norite pebbles in the basal Neverskrukk conglomerates, which have been derived from the Mt. General'skaya layered gabbro-norite intrusion (Figs. 4.15 and 4.16) dated at 2505 ± 1.6 Ma (U-Pb-zircon, Amelin et al. 1995). The most robust minimum age for the group is 1970 ± 5 Ma (U-Pb zircon, Hanski et al. 1990), obtained from felsic tuffs of the Pilgujärvi Volcanic Formation in the uppermost part of the group (Fig. 4.16).

The Neverskrukk Formation

The Neverskrukk Formation (the Lower Ahmalahti subsuite of Zagorodny et al. 1964) forms the basal part of the North Pechenga Group (Figs. 4.15 and 4.16). The contact between the Neverskrukk Formation and the underlying Archaean rocks represents a first-order unconformity, marked by a patchily preserved, palaeo-weathered surface containing an irregular fracture network filled with locally derived sedimentary breccia, sandstone and calcareous-sericitic schist (Sturt et al. 1994). The formation rests discordantly on various Archaean basement rocks (Sturt et al. 1994) including the above-mentioned 2505 Ma layered gabbro-norite intrusion. The formation has a thickness that varies greatly over short distances because the rocks fill a series of palaeovalleys bound to small grabens orientated transverse to the major rift zone. Some of the valleys accommodate more than 300 m of conglomerates and sandstones (for details, see Chap. 6.2.1). The maximum thickness of c. 300 m documented by drilling in the Mt. General'skaya area decreases to c. 80 m in a depositional-dip direction over a distance of c. 800 m. This declines to <20 m over a distance of 19 km as documented in the Kola Superdeep Drillhole (Fig. 4.17).

Buried palaeotopography is preserved in many places (Sturt et al. 1994). At Brattli (Fig. 4.15), basement rocks exhibit fractures that extend 2 m in depth and are filled by angular fragments of fractured rocks and pale brown, fine-grained, massive or weakly laminated rockflour (Fig. 4.18a). These are commonly capped by disintegrated basement rocks appearing as angular, unsorted clasts of different sizes (Fig. 4.18b). The textures and compositions of this 'regolith' (Fig. 4.18a, b) indicate that it experienced little to no chemical weathering.

The basal fill of a palaeovalley at Brattli is composed of immature, closely-packed, unsorted, framework-supported conglomerates. Clasts vary greatly in composition and include gneisses, amphibolites, crystalline schists, banded ironstones, diorites and granites, which can be shown to have local sources from the underlying Archaean provenance. These conglomerates grade upward into ones showing imbricated and more rounded, reworked clasts

(Fig. 4.18c). Although the matrix is composed of sand- and grit-size polymict clasts, the lowermost 50 m contain several 0.5-m-thick beds of calcite-cemented conglomerates ($\delta^{13}C = -17$ ‰, Melezhik and Fallick 1996). The same interval contains oversized granite block reaching 2 m in diameter.

The remaining part of the formation is marked by both fining- and coarsening-upward cycles varying from 0.5 to 10 m in thickness. The former grade from pebble conglomerate to sandstone (Fig. 4.18d, e), whereas the latter grade from siltstone to pebble conglomerate. Some siltstone and fine-grained sandstone beds contain lonestones. The immature and unsorted nature of the clasts, however, prevails throughout the section (Fig. 4.18f). Conglomerates, gritstones and sandstones typically display variably thick trough cross-beds and normal grading (Fig. 4.18d, e).

In the Mt. General'skaya area (Fig. 4.15), the Neverskrukk Formation fills a palaeovalley with an estimated depth of more than 300 m. Here, the formation has been informally divided into 6 units based on drillhole logs (see Chap. 6.2.1). From bottom to top, these are: the Clayey Siltstone unit (0 to >10 m), Conglomerate unit (55 m), Gritstone unit (30 m), Gritstone-Conglomerate-Sandstone unit (125 m), Sandstone unit (13 m) and Conglomerate-Sandstone unit (c. 80 m).

In proximal sections, the Conglomerate unit defines the formational base, whereas the Clayey Siltstone unit represents basal strata in distal sections. The former unit is composed of massive to crudely stratified, clast-supported, polymict conglomerates (Fig. 4.18g, h) resting sharply on the 2505 Ma gabbro-norite intrusion. The latter unit also rests directly on the gabbro-norite substrate and consists of three, 1- to 5-m-thick, coarsening- and thickening-upward cycles from siltstone-mudstone through sandstone-gritstone to fine-pebble conglomerate. The siltstone contains lonestones and numerous oversized clasts resembling dropstones, whereas the sandstone contains mud chips.

The basal conglomerates and coarsening-upward siltstone-sandstone-gritstone-conglomerate cycles are overlain by the Gritstone unit, a package dominated by polymict gritstones (Fig. 4.18i) with abundant mud chips. Above this, the Gritstone-Conglomerate-Sandstone unit starts with a polymict conglomerate followed by massive or crudely stratified gritstones interbedded with polymict conglomerates and massive or normally-graded sandstones with large, partially or entirely pyritised mud chips and small nodular pyrite (for details, see Chap. 6.2.1). Up-section, the sandstones dominate over other lithologies to form the Sandstone unit. The sandstones, both arkosic and greywacke, exhibit cross- or plane-parallel bedding and contain pockets of pebbly material.

In both proximal and distal parts, the Neverskrukk Formation ends with the Conglomerate-Sandstone unit consisting of alternating beds of thick, crudely stratified conglomerates (1–13 m) and numerous thin (0.1–0.9 m), massive or cross- and parallel-bedded, poorly sorted sandstone layers (Fig. 4.18j) with scattered clasts and lonestones. Pebble imbrications (Fig. 4.18c) and cross bedding suggest a south-directed stream flow. Some beds resemble diamictites.

Depositional palaeoenvironments of the Neverskrukk Formation include alluvial channels and fans, and debris flows deposited in fault-bounded extensional basins (Melezhik 1992; Sturt et al. 1994). Regional correlations imply that the basal 'regolith' and overlying Neverskrukk Formation rocks are potentially time equivalent to the global Palaeoproterozoic (Huronian) glacial deposits. However, robust evidence for proving the glacial origin of the Neverskrukk Formation is yet to be documented, though several sedimentological features, such as lonestones and oversized blocks, give hints to possible deposition by glacigenic processes.

The Ahmalahti Formation

The Ahmalahti Formation (Figs. 4.15 and 4.16; originally defined as the Upper Ahmalahti subsuite in the 1960s by Zagorodny et al. 1964) rests depositionally with a sharp contact either on sedimentary rocks of the Neverskrukk Formation or, in places where the Neverskrukk Formation is absent, directly on Archaean basement rocks. The total thickness of the Ahmalahti Formation is 1,100–1,300 m, and it occurs as a continuous unit along the northern and north-eastern margin of the belt (Fig. 4.15). The formation consists of massive and amygdaloidal, subaerially erupted basalts, basaltic andesites and andesitic dacites (Fig. 4.19). Minor magnesian basalts, mafic and ultramafic tuffs have also been reported from the Norwegian part of the belt (Melezhik et al. 1994b). Mafic rocks occur predominantly in the lower part of the formation whereas basaltic andesites and andesitic dacites form the upper part (Smol'kin and Skuf'in 1995). The thicknesses of lava flows vary between 0.5 and 25 m. The formation does not show any volcanological features suggesting deposition in a marine environment (Predovsky et al. 1974). The eruption age of the Ahmalahti Formation is not precisely constrained. The unit is younger than the 2505 Ma Mt. Generalskaya intrusion and older than the c. 2.06 Ga Kuetsjärvi Volcanic Formations (see below). A whole-rock Rb-Sr age of 2324 ± 28 Ma determined by Balashov (1996) is consistent with this age interval but its accuracy is questionable due to the mobility of the elements involved.

The volcanic rocks range from basalts to dacites in their silica content and plot on various discrimination diagrams at the boundary between the tholeiitic and calc-alkaline magma series. They are enriched in LREE and low in high-field strength elements (HFSE), resulting in low Ti/Zr ratios of about 50; these features also typify the Archaean basement complex (Hanski 1992). The Ahmalahti rocks are also characterised by negative initial ε_{Nd} values (Skuf'in and Yakovlev 2007). All geochemical properties suggest a significant crustal contamination. Melezhik and Sturt (1994a) and Mints (1996) recognised komatiitic affinities of some of the Ahmalahti Formation volcanic rocks. The former authors suggested subaerial extrusion within intraplate environments, whereas the latter proposed deposition in continental margin or volcanic-arc settings. However, the volcanic-arc environment is at odds with deposition of the Ahmalahti volcanic rocks on the Archaean basement.

The Kuetsjärvi Sedimentary Formation

The Kuetsjärvi Sedimentary Formation (Figs. 4.15 and 4.16; defined originally as the lower Kuetsjärvi subsuite by Zagorodny et al. 1964) occurs continuously throughout the belt. It has a variable thickness (15–150 m) controlled in the central part of the belt by syndepositional faults (Predovsky et al. 1974; Melezhik and Predovsky 1982) and is reported (Predovsky et al. 1974) to rest depositionally on weathered basaltic andesites of the underlying Ahmalahti Formation. However, the thinning and eventual wedging out of the formation in the western and eastern flanks of the belt (Fig. 4.15) is the result of postdepositional tectonics associated with northeast-directed faulting of Archaean rocks upon the Petsamo Group rocks.

A basal unit comprises calcitised biotite-sericite schist, which has been interpreted as a redeposited and "calichified" weathering crust (Melezhik and Fallick 2005). The formation as a whole consists of siliciclastic-dominated, mixed siliciclastic-dolostone and dolostone-dominated parts, with an upward increase in dolostone at the expense of siliciclastic strata (Fig. 4.20). The succession has been informally subdivided into the lower Quartzite and upper Dolostone members (Melezhik et al. 2004). Sedimentological features of the main rock types from both members are illustrated in Fig. 4.21.

The Quartzite Member is c. 45 m thick and consists of three units (Melezhik and Fallick 2005). The Lower Mixed Siliciclastic-Dolostone unit is 12.6 m thick. Its lower part is interbedded arkosic sandstone, siltstone and magnetite- and haematite-rich mudstone arranged in variably thick Bouma sequences (Fig. 4.21a–f). Lamination is commonly disrupted by soft-sediment deformation in the form of flame, small-scale slump and loading structures (Fig. 4.21a, b). Ripple cross-laminations display unidirectional migration. The upper part of the unit consists of interbedded arkosic siltstones-mudstone with beds of dolomite-cemented sandstones, intraformational dolostone breccia and conglomerate, and thin layers of dolostone (Fig. 4.21g–o). The most common sedimentary structures are parallel sandstone and siltstone laminations showing graded bedding with small-scale, unidirectional, ripple cross-lamination (Fig. 4.21l, n); slump horizons occur in places (Fig. 4.21o). The bases of sandstone beds commonly exhibit numerous small scour-and-fill structures (Fig. 4.21l) and, locally, larger channels (Fig. 4.21k). The dolostone layers are thin and are laterally discontinuous (Fig. 4.21i). Many dolostone beds exhibit vertical and planar sheet cracks (Fig. 4.21g, h) and some contain spherical or polygonal inclusions of dolomite resembling pseudomorphed sulphate nodules (Fig. 4.21j). Mudstone layers, where present, are haematite-rich and generally thin with a low lateral continuity and show desiccation cracks (Fig. 4.21m).

The Middle Siliciclastic-Dominated unit of the Quartzite Member is 17.5 m thick. Its lower part is composed of pale pink, fine-grained, arkosic sandstones with faint parallel bedding and subordinate, thin layers of sericite-rich shales (Fig. 4.21p). These are overlain by purple and brown, fine-grained, haematite-rich, arkosic sandstones with subordinate thin layers of haematite-rich, sericitic shales. These rocks typically appear structureless but rare decimetre-scale trough cross-lamination, horizontally laminated sandstone-mudstone interbeds and small-scale, soft-sediment deformation, including ball-and-pillow structures, can be observed (Fig. 4.21q, r). Dismembered, randomly oriented, black, haematite-rich mudstone layers with flame structures and fluid escape structures are also common (Fig. 4.21s, t). Predovsky et al. (1970) reported small quartzite lenses containing up to 45 wt.% Fe_2O_3 bound to haematite. The upper part of the unit is arkosic and quartzitic sandstone with trough and unidirectional tabular cross-bedding (Fig. 4.21u, v).

The Upper Mixed Siliciclastic-Dolostone unit of the Quartzite Member is 15.7 m thick. Its lower part is composed of pink, brown, violet and purple, haematite-rich, and pale green arkosic sandstones and siltstones (Fig. 4.21w, z) with subordinate haematite-rich mudstones and sandy limestone and dolostone. Overall, the rocks are distinguished by thin horizontal, varve-like lamination. The red colour of the rocks is due to the presence of haematite particles, whereas lighter colours are a result of intensive 'bleaching' and discolouration yielding a 'mottled' (Fig. 4.21y) and variegated (Fig. 4.21w, z) appearance. The upper part of the unit consists of quartzitic sandstones with unidirectional planar cross-bedding (Fig. 4.21aa) interbedded with minor dolostone-cemented arkosic sandstones and thin beds of sandy dolostone showing indistinct, crude, horizontal bedding.

The Dolostone Member is also formed by three units: the Lower Dolostone-Dominated, the Siliciclastic-Dominated and the Upper Dolostone-Dominated units. The Member has a total thickness of c. 80 m. The Lower Dolostone-

Dominated unit is 47.8 m thick and consists in its lower part of interbedded sparry, micritic and allochemical dolostones with rare beds of quartzitic sandstone (Fig. 4.21ab). The sparry and micritic dolostones are characterised by mm-scale lamination, whereas the allochemical dolostones show cm-scale, horizontal lamination and planar cross-bedding. Micritic dolostones contain relics of algal laminae, including millimetre-scale encrusting films. The middle and upper parts of the unit consists of interbedded white, algal, micritic dolostones and silty to sandy, allochemical dolostones; both contain abundant travertine crusts (Melezhik and Fallick 2001; Fig. 4.21ac). In some cases, the allochemical dolostones grade into dolomite-cemented quartzitic sandstones (Fig. 4.21ad). Although flat-laminated stromatolites dominate among algal dolostones, rare columnar stromatolites are also present (Lybtsov et al. 1971; Lybtsov 1979; for details see Chap. 7.8.2). Fenestrae may be very abundant. Brecciation and deformation related to tepees have been observed in both drillcores and outcrops.

The Siliciclastic-Dominated unit occurs in the middle part of the Dolostone Member and is 2.2 m thick. It consists of carbonate breccia interbedded with sericite-chlorite mudstones followed by quartz sandstone. The breccia is red and clast-supported and consists of dismembered dolostone beds overlain by angular, unsorted and closely-packed fragments of red and pink dolostone in a chlorite matrix (Fig. 4.21ae). The mudstone is marked by planar lamination and contains a series of small-scale (3×0.5 mm), siltstone-filled, erosional scours. Angular clasts of quartz and microcline appear for the first time in the formation (Melezhik and Fallick 2005).

The Upper Dolostone-dominated unit is c. 30 m thick and consists of four types of interbedded rocks: (*1*) microbial dolostone, (*2*) sandy, sparry allochemical dolostone, (*3*) micritic dolostone, and (*4*) abundant travertine crusts and small-scale travertine mounds. All lithologies have either pink or variegated colour with a mottled appearance (Fig. 4.21af–ah, ak, al–at).

The microbial dolostones display flat-laminated, undulatory or weakly-domed structures (Fig. 4.21af–ah). In many cases, the microbial lamination is highly disrupted by desiccation features and formation of vugs and micronodules of probable evaporites (Fig. 4.21aj–am). Consequently, microbial dolostones display microbrecciation and are the source of clasts in the interbedded, allochemical dolostones. The stromatolitic dolostones also contain tepee structures (Melezhik and Fallick 2005). Lybtsov et al. (1971) and Lybtsov (1979) reported subspherical stromatolites (for details, see Chap. 7.8.2).

The sandy, sparry allochemical dolostones are pink or mottled and contain both rounded and platy, unsorted intraclasts of micritic and algal dolostones. A few beds contain spherical dolomite particles resembling recrystallised oolites (Fig. 4.21an). Particles of similar morphology and size, though retaining radial and concentric structures, have been described from the Dolostone Member by Lybtsov et al. (1971) who interpreted them as oncolites.

The micritic dolomites form 2- to 10-cm-thick beds interlaminated with allochemical and algal dolostones. They contain dolomite-pseudomorphed micronodules of probable evaporites (Fig. 4.21ai). All carbonate lithologies, excluding travertines, show high $\delta^{13}C$ values with a limited variation between +5.8 and +8.9 ‰ and several short-term stratigraphic excursions (Fig. 4.22).

The travertines occur at various depths throughout the formation as crusts and small-scale mounds (Fig. 4.21ao–ap) composed of 1- to 5-mm-thick multicoloured bands (Fig. 4.21aq–as) of radiating fibrous and blade dolomite crystals. Some mounds exhibit several phases of growth separated by dissolution processes, desiccation, and precipitation of silica sinters (for details, see Chap. 7.9.4). Multiple-phase crusts/mounds show several $\delta^{13}C$ cycles ranging from −6.1 ‰ to +7.7 ‰ and suggesting mixing between mantle and ambient lake-water sources of carbon (Melezhik and Fallick 2001).

In the Dolostone Member, bedding planes affected by formation of vugs and microcavities have been observed developing on different carbonate substrates. The travertine crusts have been most affected by dissolution processes; vugs and microcavities are veneered by finely crystalline silica sinters (Melezhik and Fallick 2001). The microcavities have also been observed in sandy micritic dolostones where they are commonly veneered by thin, irregularly developed, red dolomicrite, silcrete or travertine crust (Fig. 4.21at, au).

In the upper part of the Dolostone Member, the size of voids and cavities reaches a scale of several tens of centimetres and they have been categorised as epikarst (Melezhik et al. 2004). In most cases, the floor and the roof of cavities are coated with beige and yellowish, finely laminated travertine crusts and small ‘stalagmites’ and ‘stalactites’. The remaining space is filled with a chlorite-rich, structureless, dolarenite mass containing fragments of yellowish, laminated travertine (Fig. 4.21av). The karstification is associated with development of sandstone-filled cracks occurring as a network penetrating beds over a depth of 0.5–2 m (Fig. 4.21aw, ax).

Another prominent feature of the upper part of the Dolostone Member is the occurrence of numerous, 0.5- to 5-cm-thick caliche crusts (Melezhik et al. 2004) consisting of (*1*) red to brown, non-laminated dolomicrite; (*2*) ‘vuggy’, dolospar-cemented dolomicrite breccia; and (*3*) a zone of orbicular silicified rocks (Fig. 4.21ay).

Depositional environments of the Kuetsjärvi Sedimentary Formation were previously interpreted as shallow marine (e.g. Predovsky et al. 1974). However, multiple lines of evidence point to its largely non-marine origin (Melezhik and Fallick 2001, 2005; Melezhik et al. 2004). The depositional framework is schematically illustrated in Fig. 4.23 and is inferred to an intraplate rift setting on a deeply, subaerially weathered surface mantling the underlying volcanic rocks. The lower part, consisting of ripple cross-laminated and trough cross-bedded quartzitic and arkosic sandstones-siltstone, was deposited in fluvial-deltaic and lacustrine shoreline environments. The thin, parallel-bedded, variegated to mottled, fine-grained siliciclastic rocks that overlie the lower part of the formation are interpreted as lacustrine deposits. The origin of the middle and upper parts of the formation, containing various dolostone lithologies, abundant desiccation and evaporative features and hydrothermal travertine, is most consistent with a shallow lacustrine setting. The Sr isotope study by Melezhik et al. (2005) suggests a short-term invasion of seawater in the uppermost part of the dolostone succession. Lacustrine dolostones having a high $\delta^{13}C$ isotope signature figure prominently in debates on the global C-cycling in Deep Time (e.g. Lomagundi-Jatulian Event; Melezhik et al. 1999; Melezhik and Fallick 2005).

The Kuetsjärvi Volcanic Formation

The Kuetsjärvi Volcanic Formation (Figs. 4.15 and 4.16; originally defined as the Upper Kuetsjärvi subsuite by Zagorodny et al. 1964) has a transitional contact with the underlying Kuetsjärvi Sedimentary Formation. It occurs continuously throughout the belt and has a maximum thickness of c. 2 km. The volcanic rocks are varied and include amygdaloidal basalts, intermediate and felsic lava flows, picrites, ignimbrites, brecciated flows, bedded tuffs and volcaniclastic sedimentary rocks (Predovsky et al. 1974; Skuf'in et al. 1986). Melezhik et al. (2007) determined a U-Pb zircon age of 2058 $\pm$ 6 Ma for the Kuetsjärvi Volcanic Formation.

A distinctive, 1- to 30-m-thick, volcaniclastic conglomerate unit, the Conglomerate member, capped with pillowed and columnar mafic lavas (Fig. 4.16) separates the formation into two parts of approximately equal thickness. Skuf'in et al. (1986) termed the lower part the mugearite-trachyte unit and the upper part the basalt unit. Melezhik et al. (1994a) subdivided the lower part informally and arbitrarily into the Lower Basalt member and the Lower Trachyandesite member and the upper part was split into the Upper Basalt member and the Upper Trachyandesite member.

The Lower Basalt member is basalts and alkaline basalts with minor trachydacites, trachyandesites and picritic basalts and tuffs. The tuffs occur mainly at the base of the member and are interbedded with thin lava flows (Predovsky et al. 1987). The basalts and alkaline basalts comprise a series of lava flows, whose thickness ranges from a few centimetres up to 25 m. The lava flows show diffuse boundaries unless an amygdaloidal top is well developed (Fig. 4.24a, b).

The Lower Trachyandesite member represents a diverse set of rocks ranging in composition from picritic basalts to rhyodacite, though trachyandesite, trachydacite and rhyodacite are dominant lithologies. The volcanic rocks occur as banded (fluidal) lavas, breccia flows and pyroclastic beds (Fig. 4.24c–g). Amygdaloidal lavas are widespread but massive lavas and columnar-jointed lavas are also present. Vesicles are filled with jasper, quartz, chlorite, calcite and feldspar and range in size from 1 mm to 4 cm (Fig. 4.24h–k). Thicknesses of the lava flows vary between 1 and 16 m. Some of the lava flows have brecciated flow tops, whereas others are separated by beds of pyroclastic material up to 4 m thick. Although there is no regular trend in the stratigraphic distribution of different volcanic rocks, the upper part of the member comprises several flows of distinctive olivine-phyric basalts and picritic basalts. The Lower Trachyandesite member contains rare sills of tholeiitic basaltic andesites. It has been intersected by FAR-DEEP Hole 6A, and more details on the section are presented in Chap. 6.2.3.

The Conglomerate member rests erosively on various rocks of the Lower Trachyandesite member. The thickness of the unit reaches 30 m though locally reducing down to a few centimetres. In the most complete sections, the member comprises a series of fining-upward cycles. The cycles differ in thickness, ranging from 0.2 to 7 m. They are also variable in composition, varying from volcaniclastic sandstone-siltstone to a combination of volcaniclastic conglomerate and volcaniclastic gritstone/sandstone/siltstone. Gritstones are indistinctly bedded and sandstones are massive or indistinctly bedded with scattered, rounded clasts of lava. In contrast, siltstones show distinct parallel-bedding emphasised by the irregular appearance of dark brown and dark violet, haematite-rich layers (Fig. 4.24l). However, in most of the cycles, conglomerate is the dominant lithology (Fig. 4.24m). All conglomerate units are clast-supported with a volcaniclastic sandstone matrix (Fig. 4.24m–o). Clasts are well-rounded, moderately-sorted fragments of lavas ranging in size from 15 to 40 cm, in composition from mafic to felsic, and in structure from massive to amygdaloidal or fluidal. All these features suggest that the underlying volcanic pile was eroded to a considerable depth. Many volcanic clasts display haematite rims or liesegang rings formed by haematite banding (Fig. 4.24n, o). The rings are similar to those found in the underlying volcanic rocks, thus implying their formation prior to deposition of the conglomerate.

In places, the conglomerate-sandstone-gritstone-sandstone-siltstone association is overstepped by black, haematite-rich, mudstone-matrix, non-bedded, volcaniclastic conglomerate

(Fig. 4.24p), itself overlain by a c. 1-m-thick unit of dark brown, haematite-rich, thinly-laminated siltstone-mudstone (Fig. 4.24g). Boulders are well rounded and composed of rhyodacitic, trachydacitic, trachyandesitic and basaltic lavas.

The Upper Basalt member lies either on the Conglomerate member or, when the latter is absent, on various volcanic rocks of the Lower Trachyandesite member. However, in both cases, the basal unit is represented by a distinctive, c. 25-m-thick, Fe-rich, homogeneous, mafic (tholeiitic) lava flow with large amygdales in its top part. The flow appears to be pillowed when it overlies dark brown, haematite-rich, thinly laminated siltstone-mudstone (Fig. 4.24q, r); the siltstone-mudstone also occurs in interpillow space and shows soft-deformation and thermal alteration implying extrusion of the flow onto non-consolidated sediments (Fig. 4.24q). In the case when the flow extruded onto volcaniclastic conglomerate or volcanic rocks, it appears as a homogeneous, green basalt with columnar jointing (Fig. 4.24s) and abundant large vesicles filled with various minerals such as white calcite and quartz, bright red adularia, bright green epidote, and purple axinite (Fig. 4.24t–u). The rest of the Upper Basalt member section mainly comprises thick (20–30 m) lava flows of amygdaloidal alkaline basalt (Fig. 4.24v–x) with minor tuff beds and a single horizon containing volcanic bombs embedded in a chlorite matrix (Predovsky et al. 1974). Many flow tops are characterised by black and brown, highly oxidised, haematite-rich breccia, with a thickness ranging between a few tens centimetres to a few metres (Fig. 4.24y–z). Haematite cement is also common in post-volcanic, extensional breccias (Fig. 4.24aa).

The Upper Trachyandesite member has a transitional contact with the underlying unit, and its lower boundary has been conventionally defined by the first appearance of trachyandesitic lavas (Fig. 4.24ab). The volcanic rocks comprising the member range from amygdaloidal basalts to intermediate lavas (Fig. 4.24ab–af), with a few basaltic flows exhibiting incipient pillows (Fig. 4.24ag). Amygdales are mainly composed of quartz, chlorite and calcite (Fig. 4.24ac–af) and may be plentiful, comprising 50–80 % of the rock volume (Fig. 4.24ac–ad). Close to the contact with the overlying Kolosjoki Sedimentary Formation rocks, the vesicles are filled with red adularia, calcite and chlorite (Fig. 4.24aj); some jasper- and quartz-filled vesicles reach 25 cm in length (Fig. 4.24ai). Jasper veins and veinlets are abundant (Fig. 4.24ah). The uppermost part of the member, when preserved from erosion, comprises andesitic lava breccia with solitary clasts of carbonate rock, and potassium-rich (13 wt.% K_2O) dacitic lava breccia with lava fragments separated by black, haematite-magnetite-rich (25 wt.% Fe_{tot}) bands (Fig. 4.24al). Unusually high enrichment in K_2O and Fe_{tot} is indicative of either postvolcanic surface hydrothermal alteration or palaeoweathering processes.

The volcanic rocks of the Kuetsjärvi Volcanic Formation form a differentiation series ranging from basalts to rhyolites. They have high LREE/HREE ratios and the basalts, relative to other basalts in the Pechenga region, are slightly richer in TiO_2 and P_2O_5 (Kremenetsky and Ovchinnikov 1986; Roman'ko et al. 1991; Melezhik and Sturt 1994; Mints 1996; Smol'kin and Skuf'in 1995; Sharkov and Smol'kin 1997; Skuf'in and Theart 2005).

Available Nd isotope data show negative initial ε_{Nd} values indicating a contribution from crustal sources (Skuf'in and Theart 2005). Depending on the elements employed, the rocks plot into the field of intraplate volcanites or island arc rocks on various discrimination diagrams (e.g. Hanski 1992), reflecting a hybrid nature of their chemical composition due to crustal contamination of the parental magma with an initial within-plate basalt character. The overall geochemical features of the Kuetsjärvi volcanic rocks and their subaerial emplacement suggest an intraplate rift environment (Hanski 1992; Melezhik and Sturt 1994a; Smol'kin and Skuf'in 1995; Mints 1996; Sharkov and Smol'kin 1997). Ephemeral, shallow-water, lacustrine conditions only occurred during deposition of the Conglomerate member and basal pillow lava of the Upper Basalt member.

The Kuetsjärvi volcanic rocks are highly oxidised relative to other volcanic rocks in the Pechenga Belt, having $Fe^{3+}/(Fe^{2+} + Fe^{3+})$ ratios averaging 0.52 (see Chap. 3.4). This is in sharp contrast to the other volcanic units: 89 % of the Ahmalahti, 100 % of the Kolosjoki and 96 % of the Pilgujärvi volcanic rocks have ratios <0.25 (as based on data presented in Predovsky et al. 1974). Importantly, many amygdales are jasper-rich in their mineralogical composition implying high O_2 fugacities in the circulating, post-magmatic fluids from which the vesicle-filling minerals precipitated. Another post-magmatic phenomenon reported from the Kuetsjärvi Volcanic Formation is the presence of native copper (Akhmedov and Chekushin 1977; Predovsky et al. 1987), a feature shared by the stratigraphically correlative volcanic unit, the Umba Volcanic Formation, in the Imandra/Varzuga Greenstone Belt (Borisov 1990).

The Kolosjoki Sedimentary Formation

The Kolosjoki Sedimentary Formation (Figs. 4.15 and 4.16; originally termed the Lower Kolosjoki subsuite by Zagorodny et al. 1964) occurs continuously throughout the belt and has a variable thickness (50–300 m, Fig. 4.25), which is influenced in the central part of the Northern Zone by syndepositional faults (Predovsky et al. 1974; Melezhik and Predovsky 1982). The formation is tectonically attenuated and wedges out completely towards both the western and eastern flanks of the belt.

The formation comprises a heterolithic succession of volcaniclastic greywackes, arkosic sandstone and gritstone, haematitic sandstones and shales, carbonates, jasper and sulphide- and C_{org}-bearing tuffaceous shales. Melezhik et al. (1994a) informally subdivided the succession into the Red Bed, Dolostone and Black Shale members. However, a somewhat different subdivision has been dictated by the FAR-DEEP 8A and 8B core logs (for details see Chap. 6.2.5). Here, the Kolosjoki sedimentary succession has been subdivided from bottom to top into the Sandstone, Lower Greywacke, Gritstone, Haematite, Ferropicritic, Dolostone and Upper Greywacke members.

The formation rests with a well-preserved erosional contact on the Kuetsjärvi Volcanic Formation (Fig. 4.26a), whereas its upper contact is tectonic (Fig. 4.26b). Detailed logs of drillcores from several deep holes, including the Kola Superdeep Drillhole, reveal significant excision of the Dolostone, Upper Greywacke and Black Shale members along the tectonic contact over the length of the Northern Pechenga Zone.

The Sandstone member rests erosionally on the Kuetsjärvi Volcanic Formation rocks (Fig. 4.26a). Its composition and thickness vary greatly along strike due to its infilling of variably developed palaeovalleys. Pale pink or grey, fine-grained, massive volcaniclastic sandstones are the main lithology. In places, they are parallel laminated and normally graded with rare, small-scale trough cross-beds. Clastic material was derived solely from the underlying volcanic rocks.

The Lower Greywacke member has a transitional contact with the underlying massive sandstones and contains at its base a series of channels filled with polymict, volcaniclastic conglomerates (Fig. 4.26c). Its thickness reaches c. 80 m in Sect. 6 shown in Fig. 4.25 though reducing to a few metres elsewhere and the member is absent in Sects. 2 and 7. Volcaniclastic greywackes with a few beds of arkosic sandstone and volcaniclastic conglomerate are the main lithology. The greywackes display planar and trough cross-beds occurring as variably thick cosets (Fig. 4.26d–h) commonly separated by well-developed erosion surfaces (Fig. 4.26i). In places, both the trough and the tabular cross-beds contain overturned, soft-sediment deformed and slumped cross-bedding, small-scale syndepositional folds and faults, and ball-and-pillow structures (Fig. 4.26g, h). Trough and tabular cross-beds may pass laterally into massive or normally graded gritstone-sandstone, or sandstone-black mudstone beds, all containing scattered fragments of volcanic rocks (Fig. 4.26j, k). Many beds of the member contain abundant rip-up intraclasts of black, haematite-rich mudstone (Fig. 4.26l). In the upper part of the member, numerous laminae in cross-beds are enriched in leucoxene, resulting in pale yellow colouration, and cross-bedded greywackes appear to be "spotted" (Fig. 4.26m) due to patchy cementation of iron oxides (Fig. 4.26n). The greywackes are composed of angular clasts of quartz, microcline, rare albite, fragments of intensively altered mafic to intermediate lavas, particles of haematite, magnetite and leucoxene, all in a fine-grained matrix of a similar composition but enriched in sericite. The degree of sorting and rounding does not show any clear stratigraphic trend. In places, the uppermost part of the Lower Greywacke member is marked by the occurrence of a distinctive unit of dark brown, haematite-rich, parallel-laminated siltstone-shale (Fig. 4.26o), though the latter has been only observed in Sects. 1, 3, 4, 5 and 7 (Fig. 4.25).

The Gritstone member comprises distinctive arkosic sandstone-gritstone (Fig. 4.26p, q) and has a variable thickness reaching 70 m in Sect. 5; it is absent in Sect. 1 (Fig. 4.25). It rests with a sharp contact on either dark brown siltstone-shales or cross-bedded greywacke, or directly on the Kuetsjärvi volcanic rocks. Syndepositional movement on horsts and grabens generated variable and local depositional/erosional processes (Melezhik 1992). The base of the member is commonly defined by a thin, polymict, volcaniclastic conglomerate observed in several logged sections. The conglomerate contains unsorted fragments of lavas and large quartz and jasper amygdales reworked from the Kuetsjärvi volcanic rocks (Fig. 4.26o). Similar conglomerates, though containing mainly fine pebbles and grading upward into gritstones (Fig. 4.26r–s), fill numerous small-scale channels throughout the member. They also contain clastic haematite (Fig. 4.26t).

The arkosic sandstones and gritstones are commonly red in colour and massive (Fig. 4.26p–q) or crudely stratified. They also contain lag deposits of clastic haematite, abundant large fragments of haematite "ores" (Fig. 4.26u–w), first described by Akhmedov (1972a, b). Numerous thin, parallel-laminated mudstone and claystone layers are present, and their amount gradually increases upward, defining an overall fining-upward succession and transition to the overlying member. The arkosic gritstones are notable by high Sr (1,150 ppm) and Ba (57,700 ppm) contents (Appendices 20 and 21).

The Haematite member per se has only been documented in Sect. 6, where it is c. 45 m thick (Fig. 4.25). The base of the member is defined by the first appearance of parallel-laminated, haematite-rich sandstone-siltstone (Fig. 4.26x–y). The lowermost part of the member comprises thinly-laminated siltstone units interbedded with graded, greywacke/parallel-laminated siltstone couplets containing numerous small-scale gritstone channels. In some graded beds, the lowermost coarser units may contain platy fragments of massive haematite "ore" (Fig. 4.26z). On the bedding plane, some haematite beds exhibit a wrinkled surface resembling elephant skin texture (Fig. 4.26aa). Some conglomerate channels grade upward into gritstone-sandstone capped by haematitic siltstone (Fig. 4.26ab–ac).

Fine haematite particles are an important part of the sandstone and gritstone matrix. Abundant platy fragments of haematite "ore" are common in the lower part of the member and some contain claystone (sericite) laminae (Fig. 4.26d).

Red, jasperised dolostones, jaspers with colloform structures (Fig. 4.26e, af) and polymict breccia form the base of the overlying succession (c. 30 m thick in Sect. 6) of spectacular haematite-rich, laminated sandstone-siltstones with jasper (Fig. 4.26ag). The rocks are predominantly red, violet and brick-brown with thin, parallel lamination. Fine-grained sandstone-siltstone couplets, 3–4 cm thick and disturbed by syn-depositional folding and slumping, are the dominant features. Several intervals contain jasper that occurs either as 1-mm- to 2-cm-thick, well-defined layers with sharp contacts, or as thicker beds with diffuse boundaries; all are cross-cut by thin jasper veinlets. Many graded beds contain resedimented clasts of jasper (Fig. 4.26ah).

The Ferropicrite member (Fig. 4.25) is preserved only in Sect. 6 (for details see Chap. 6.2.5). Its lower boundary is nondescript and is been defined geochemically, based on high abundances of MgO (14.2–16.6 wt.%), Cr (722–1,100 ppm) and Ni (690–1,050 ppm) (see Appendices 20 and 21). The member comprises massive and indistinctly bedded, coarse-grained ferropicritic tuffs with a clotted microfabric (Fig. 4.26ai), and parallel-laminated siltstones and rhythmically bedded sandstone-siltstone couplets, which are apparently tuffites. All lithological varieties contain beds, lenses, layers and intraclasts of red Fe-rich claystones and thin jasper veinlets. Some of the thicker ferropicritic tuffs show reverse grading. The ferropicritic tuffs do not contain primary minerals and are composed of rock fragments replaced by chlorite, serpentine, talc, phlogopite, apatite and magnetite. The oxidation state of iron is very high; comparable to that of associated haematite-bearing sedimentary rocks. Geochemically, the ferropicritic tuffs are similar to ferropicrites of the Pilgujärvi Volcanic Formation (cf. Hanski 1992).

The Dolostone member documented in all exposed and drilled sections (Fig. 4.25) has a constant thickness of c. 30 and is separated from the underlying ferropicritic tuff by a transitional bed composed of cross-bedded greywacke interbedded with greywacke-matrix-supported breccia with fragments of dolostones and cherts (Figs. 4.26aj, ak). The member comprises interbedded dolarenites and stromatolitic dolostones, with rare limestones in the lower part, and several beds of greywacke, arkosic sandstone, conglomerate and breccia in the upper part. Primary depositional features in all lithological varieties have been partially modified by post-sedimentary recrystallisation and silicification.

The dolarenites are massive or show cross-bedding, crude horizontal and wavy lamination (Fig. 4.26al, am) with small erosional surfaces and pockets of stone rosettes. Some beds exhibit normal grading, while others have mud drapes.

Dolarenite- or mudstone-supported breccia beds, less than 1 m in thickness, contain chert and silicified dolostone fragments. Their contacts with the underlying and overlying carbonate beds appear to be irregular with numerous embayments; a sign of dissolution processes of evaporite-bearing beds or due to surface or subsurface karstification.

"Classic" columnar stromatolites are rare (Sect. 7). They were first identified and described by Ljubtsov (1979). Other forms occur mainly as globular stromatolites in Sect. 6 (for details see Chap. 6.2.5). Biogenicity of many laminated dolostones is questionable and these are likely to be in part tufa or travertine deposits. The bulk of the carbonate rocks are dolostone with the Mg/Ca weight ratio ranging between 0.50 and 0.62 (Appendices 19 and 22) and $\delta^{13}C$ ranging from +1.2 ‰ to +2.4 ‰ (n = 28, Melezhik et al. 2007).

The Upper Greywacke member has a sharp contact with the underlying dolostones. Its actual thickness cannot be determined because the upper contact of the member is tectonic and defined by the Luchlompolo thrust-fault (Sharov et al. 1998; Gorbatsevich et al. 2010; Fig. 4.25). The member is composed of interbedded shale, siltstone, greywacke, clayey greywacke and arkosic sandstone showing parallel and wavy lamination, and many of the sandstone layers are mud-draped. A 3-m-thick unit of "black shale" is present in the upper part of Sect. 4 and contains up to 0.6 wt.% total organic carbon and up to 3.5 wt.% total sulphur (V. Melezhik, unpubl. data). $\delta^{34}S$ obtained from sulphides of the black shales ranges between +20.6 ‰ and +23.6 ‰ (V. Melezhik and A. Fallick, unpubl. data).

In summary, the Kolosjoki Sedimentary Formation varies greatly in thickness and lithofacies and comprises a heterolithic succession of volcaniclastic greywacke, arkosic arenites and conglomerates, haematitic sandstones and siltstone, ferropicritic tuffs, clastic and microbial carbonates, jasper, and black, sulphide-bearing shales. This complexity is compounded by the syn- to post-depositional, intermittent activation, and reactivation, of local horst and graben structures. The formation defines an overall fining-upward succession and its base blankets an irregular palaeotopography displaying many metres of relief as a result of deep erosional incision into the underlying Kuetsjärvi volcanic rocks. Deposition started within an intraplate rift setting, in alluvial and deltaic environments progressing into marine systems (e.g. Predovsky et al. 1974; Melezhik 1992). The deposition was accompanied by an intensive hydrothermal deposition of iron oxides and jasper. Of particular interest are the Kolosjoki carbonate rocks that, in contrast to the isotopically heavy Kuetsjärvi dolostones, exhibit near zero $\delta^{13}C$ values, thus signifying the end of the Lomagundi-Jatuli isotopic excursion (Melezhik et al. 2007). The overlying shales indicate the onset of the Shunga Event, that is an episode of enhanced accumulation of organic matter (e.g. Melezhik et al. 2009).

The Kolosjoki Volcanic Formation

The Kolosjoki Volcanic Formation (Figs. 4.15 and 4.16; the Lower Kolosjoki subsuite of Zagorodny et al. 1964) is in tectonic contact with the underlying sedimentary rocks and the lowermost part of the formation is marked by thrust-fault generated blastomylonites (e.g. Mints 1998). Mapping over the length of the Northern Pechenga Zone and investigations of cores from several deep holes, including the Kola Superdeep Drillhole, have revealed significant excision of the upper part of the Kolosjoki Sedimentary Formation along this contact. The fault has also been recorded at depth by seismic measurements; in fact, the apparent seismic boundary that was regarded before the Kola Superdeep drilling operations as the probable boundary between the Palaeoproterozoic and Archaean complexes turned out to represent the lower contact of the Kolosjoki Volcanic Formation (Kozlovsky 1984; Gorbatsevich et al. 2010). Comparison of the North Pechenga Group succession exposed on the surface with the section intersected 10 km to the south-east at a depth c. 5 km by the SG-3 drillhole (e.g. Melezhik et al. 2003) suggests that both the Kolosjoki Sedimentary and the Kolosjoki Volcanic formations retain their thicknesses. Consequently, the Luchlompolo fault exposed at the surface in the Northern Zone remains largely bedding-parallel, separating older rocks in its footwall from younger rocks in its hanging wall.

The exact age of the Kolosjoki Volcanic Formation is poorly constrained. Given that the underlying Kuetsjärvi Volcanic Formation has been dated at c. 2060 Ma, the previously determined Rb-Sr date of c. 2100 Ma (Balashov 1996) seems to be a poor estimate of the eruption time for the Kolosjoki volcanic rocks. The Kola Superdeep Drillhole intersected a 110-m-thick body of partially mylonitised rhyodacitic porphyry at the tectonic contact (thrust) between Kolosjoki Volcanic Formation ferropicritic tuffs and Kolosjoki Sedimentary Formation dolostones (Kozlovsky 1984), implying that the upper part of the sedimentary formation (Greywacke and Black Shale members) was truncated by the thrust. The rhyodacitic porphyry was dated at 2043 $\pm$ 18 Ma (U-Pb on zircon) by Mitrofanov et al. (2001) having thus, within errors, the same age as the Kuetsjärvi Volcanic Formation, though differing from the latter in terms of trace element geochemistry. Since the igneous body was affected (mylonitised) by the thrust that separates the Kolosjoki Sedimentary and Volcanic formations, and the primary mutual contact relationship of the rocks remains uncertain, the implications of that age are not clear.

The Kolosjoki Volcanic Formation reaches a thickness of c. 2,000 m in the central part of the belt and pinches out in the eastern and western flanks of the belt (Fig. 4.15). The formation has been informally subdived into three subunits: the Lower Basalt, Black Shale and Upper Basalt members (Melezhik et al. 1994a). The volcanic pile is a monotonous sequence of tholeiitic basalts with minor ferropicritic rocks (altered to chlorite-actinolite schist) in the basal part breaking the monotony. Both the Lower and Upper Basalt members mainly comprise massive (72 vol.%) and pillowed (23 vol.%) mafic lavas with subordinate amygdaloidal lavas, lava breccias (Fig. 4.27a–c), hyaloclastites and tuffs (Smol'kin and Skuf'in 1995). The lava flows range from 0.5 to 20 m in thickness, while the maximum thickness of tuff beds is restricted to a few metres. The pillows are amygdale-bearing with most quartz-filled vesicles being concentrated close to the rims and larger ones (up to 2 cm) disseminated in the core (Fig. 4.27a), excluding a very deep water depth during eruption (cf. Jones 1969). The basalt succession is intruded by numerous comagmatic gabbroic sills. In addition, cross-cutting, NW-trending ultramafic dykes have been observed that carry magmatic Ni-Cu sulphides (Gorbunov et al. 1985). They apparently served as feeders for the magma chambers that produced ore-bearing gabbro-wehrlite intrusions within the overlying Pilgujärvi Sedimentary Formation (Hanski et al. 2011).

The Black Shale member, part of which was intersected by FAR-DEEP Hole 9A, is from 70 to 300 m in thickness and comprises thinly interbedded greywacke, mafic tuff, tuffite and S-bearing black shale (Fig. 4.27d) with subordinate chert and limestone beds. The chert beds contain probable coccoidal microfossils (Fig. 4.27e; Ivanova et al. 1988), and the limestone beds show a depletion in ^{13}C ($\delta^{13}C_{carb}$ = −7.8 ‰ to −7.6 ‰; Melezhik and Fallick 1996), though it remains unclear whether such depletion is related to organic diagenesis or reflects the carbon isotopic composition of ambient seawater. $\delta^{34}S$ values obtained from sulphides of the black shale layers range from +3 ‰ to +5.2 ‰ (V. Melezhik and A. Fallick, unpubl. data).

The mafic rocks of the Kolosjoki Volcanic Formation have generally been considered relatively homogeneous iron-rich tholeiitic basalts. However, the SG-3 and FAR-DEEP Hole 9A (see Chap. 6.2.6) have shown that, although the MgO content remains approximately constant (7.0 wt.% on average) as a function of stratigraphic height, some chemical variations can be seen, particularly in the HFSE contents. Three stratigraphically different intervals can be distinguished on the basis of the TiO_2 content. The Lower Basalt member is relatively high in TiO_2 (2.0 wt.% on average) in its lower part and low in TiO_2 (1.1 wt.% on average) in its upper part, while the Upper Basalt member has an intermediate level of TiO_2 (1.6 wt.% on average). The basaltic rocks exhibit flat or slightly LREE-enriched chondrite-normalised REE patterns (Kremenetsky and Ovchinnikov 1986; Roman'ko et al. 1991; Hanski 1992; Smol'kin 1992; Melezhik and Sturt 1994a; Smol'kin and Skuf'in 1995; Mints 1996; Sharkov and Smol'kin 1997) and have positive initial ε_{Nd} values indicating a mantle derivation without significant crustal interaction (Skuf'in and Yakovlev 2007). Several researchers have emphasised the similarity between the Kolosjoki tholeiites and MORBs (Kremenetsky and Ovchinnikov 1986; Sharkov and Smol'kin 1997; Melezhik and Sturt 1994a; Melezhik et al. 1994b; Mints 1996). Hanski (1992), however, discussed several geochemical features (e.g. high FeO_T/TiO_2) that make the Kolosjoki volcanic rocks different from modern MORBs and also from approximately coeval tholeiitic pillow lavas of the Jormua ophiolite complex.

Different depositional models have been advanced for the Kolosjoki Volcanic Formation including an intraplate rift (Smol'kin 1997), back-arc basin (Sharkov and Smol'kin 1997), active continental margin (an interplate rift, Melezhik and Sturt 1994; Melezhik et al. 1994b), and a Red Sea-type basin (Mints 1996). The Black Shale member rocks exhibit LREE-enriched patterns (Melezhik et al. 1994b), thus suggesting a significant influx of continental crustal material.

Pilgujärvi Sedimentary Formation

The Pilgujärvi Sedimentary Formation (Figs. 4.15 and 4.16; originally defined as the Lower Pilgujärvi subsuite by Zagorodny et al. 1964) is the thickest sedimentary unit in, and occurs continuously throughout, the belt and has a variable thickness (500–1,000 m) due to the influence of syndepositional faults (Melezhik et al. 1994c), partially as the result of the tectonic imbrications as suggested by Mints (1996). The formation shows a significant tectonic attenuation towards the eastern and western parts of the belt. The nature of the lower contact of the formation remains controversial and understudied. Zagorodny et al. (1964) and Predovsky et al. (1974) reported that it rests with a depositional contact on weathered basalts of the Kolosjoki Volcanic Formation. However, later, the lower contact was interpreted as tectonic based on the SG-3 data (Kozlovsky 1984; Orlov and Laverov 1998; Mints 1996). The upper boundary of the Pilgujärvi Sedimentary Formation is tectonically modified by thrust faulting in places (Fig. 4.15) but elsewhere appears to be largely stratigraphic into the overlying thick pile of tholeiitic basalts of the Pilgujärvi Volcanic Formation (Zagorodny et al. 1964; Predovsky et al. 1974). However, Mints (1996) concluded that the upper contact is everywhere tectonic and referred to observations from the SG-3 core and investigations of the nearby areas ("near-SG-3 space"). He also pointed out that the Pilgujärvi Sedimentary Formation is intensely folded and faulted and may not represent a coherent sedimentary succession but instead can be interpreted together with underlying and overlying tholeiitic basalts as a tectonically imbricated assemblage. The controversies over the interpretation of contact relationships arise from the general lack of good outcrops and their robust investigation. However, as demonstrated in the following sections, whilst tectonically modified, the formation retains grossly its integrity as a major lithostratigraphic unit across the belt, and the rocks preserve a rich resource of syndepositional information.

The formation hosts sill-like gabbro-wehrlite intrusions containing many economic Ni-Cu sulphide deposits (e.g. Gorbunov 1968; Brügmann et al. 2000; Barnes et al. 2001); for this reason, the sedimentary unit has also been named the "Productive Formation" (Gorbunov et al. 1985). The intrusions have been dated by various methods at c. 1990–1980 Ma (Hanski 1992) with the most precise constraints being ages of 1987 $\pm$ 5 Ma and 1980 $\pm$ 10 Ma provided by the U-Pb technique on zircon and baddeleyite, respectively (Skuf'in and Bayanova 2006). The gabbro-wehrlite intrusions are thought to be coeval and cogenetic with ferropicritic volcanic rocks (Zagorodny and Mirskaya 1967; Hanski 1992; Smol'kin 1992). As the latter are also coeval with the deposition of the "Productive Formation" (Hanski 1992; Gorbunov et al. 1989), the obtained ages also provide time constraints for the sedimentary rocks of the formation.

The Pilgujärvi Sedimentary Formation consists mostly of rhythmically bedded greywackes and black, carbonaceous and sulphidic shales, as well as ferropicritic tuffs and minor ferropicritic lavas (Fig. 4.29). The succession is intruded by numerous gabbroic sills, and more than 200 gabbro-wehrlite intrusive bodies have been documented within the formation (Zak et al. 1982). The gabbroic sills are considered to represent subvolcanic bodies and feeders to the overlying tholeiitic basalts of the Pilgujärvi Volcanic Formation (Zagorodny et al. 1964; Predovsky et al. 1974, 1987; Fedotov 1985; Smol'kin and Skuf'in 1995). The gabbro-wehrlite intrusions generated from ferropicritic magma typically form largely conformable, differentiated bodies a few tens of metres in thickness. The largest of them, the ore-bearing Pilgujärvi intrusion, reaches c. 500 m in thickness and is currently the main source of nickel and copper in the Pechenga ore camp. In the upper part of the formation, there are Ni-Cu ore-bearing thin magmatic units, which show structures indicating rapid cooling and hence emplacement as very shallow-level sills or even lava flows (Gorbunov et al. 1989; Hanski 1992). Some contacts of the intrusive bodies with the host sedimentary rock were affected by shearing whereas others remain intact. In the latter case, the contact-zone alteration is expressed as bleaching and solid-state recrystallisation of the organic-rich sedimentary rocks resulting in formation of hornfels.

The sedimentary formation represents an overall fining-upward sequence and has been informally subdivided by Bekasova (1985) into three horizons, from base to top, the lower, middle and upper horizon (corresponding to members A, B and C of Melezhik et al. 1994a, b; Fig. 4.28). The lower two members (horizons) are distinguished on the basis of overall bed thicknesses, with Member A characterised by thicker and coarser clastic components with respect to

Member B (Bekasova 1985). Although this approach has proven useful for the purpose of sequence analysis (Bekasova 1985), it is difficult to apply objectively in the field. Member C (the Lammas member) was recognised based on its high Cr and Ni content . It is most widely developed in the eastern part of the belt and is reduced in thickness in central and western parts. Overall, the formation shows significant lateral and vertical lithofacies variations (e.g. Bekasova 1985; Melezhik et al. 1994b, 1998), and its subdivision into finer lithostratigraphic subdivisions is not always straightforward. Although the formation was affected by folding, faulting and shearing, its rocks grossly retain primary sedimentary features, as shown in the following sections.

Member A is up to 300 m thick. Its basal shales occur as discontinuous lens-shaped units ranging in thickness from 0 to 140 m and resting with a depositional contact on the Kolosjoki lavas (Predovsky et al. 1974; Bekasova 1985). The basal bed has been interpreted as a redeposited palaeoweathering crust (Predovsky et al. 1974). The rest of the member consists mainly of thick-bedded, organic carbon- and sulphur-bearing gritstone, arkosic sandstone and greywacke, siltstones and subordinate shales and fine-pebble polymict conglomerate (Fig. 4.29a–g). All lithologies occur as Bouma sequences (Fig. 4.29a–f) dominated by T_a and T_b divisions, although complete $T_{a\text{-}d}$ beds can be observed. The thickness of the Bouma sequences reaches 50 cm in the lowermost part of the member and fines upwards. Both fining/thinning and coarsening/thickening upward cycles are present (Fig. 4.29b, c), as are the typical features of the Bouma sequences such as normally graded, rippled, and cross- and parallel-laminated units. Convoluted beds and other sediment-gravity flow deposits are common (Fig. 4.29g). The clastic material is polymict, poorly sorted and angular. Calcite concretionary beds, 1–10 cm in thickness and depleted in ^{13}C ($\delta^{13}C_{carb} = -11$ ‰ to -8 ‰; Melezhik and Fallick 1996), are abundant in the upper part of the member.

Member B varies from 50 to 300 m in thickness and has a transitional contact with the underlying sedimentary succession. The member is composed of thick beds of gritstone-sandstone and C_{org}-bearing (up to 5.3 wt.% TOC; Melezhik et al. 1988) laminated sandstone-siltstone-mudstone rhythmites with abundant diagenetic sulphides and carbonates (Fig. 4.29h–av). Thick beds of basaltic tuffs are common. The rhythmites represent Bouma sequences, commonly 1–10 cm in thickness and of T_a and T_b divisions. T_e has been observed in the upper part of the member section. Thin conglomerates and cross-bedded sandstone units filling scours at the base of Bouma sequences are rare. Figure 4.29h–t illustrate the most common lithofacies.

Regularly bedded rhythmites commonly show syn-sedimentary slumping and brecciation (Fig. 4.29y–ac), which can be attributed to rapid depositional rate, slope instabilities and/or seismicity. Beds with slumped structures range in thickness from a few centimetres to 0.5 m. Breccia beds are from 1 to 2 m thick and contain polymict clasts represented by greywacke and siltstone but obviously are dominated by black mudstones. All clast lithologies are compositionally similar to Pilgujärvi rhythmites, and many are soft-sediment deformed, reflecting their derivation from local, unlithified substrates. In some beds, clasts are cemented with calcite sparite. Mass-flow deposits, occurring as thick (up to 2 m) intervals, are present throughout the member (Fig. 4.29ad–ag). They have an erosional base, sharp upper contact and commonly are dismembered and soft-sediment deformed containing fragmented black mudstones floating in a massive, coarse-grained greywacke matrix.

Variably sized sedimentary dykes are common. Many are composed of coarse-grained sandstone but also contain large clasts of sedimentary rocks and fragments of diagenetically formed pyrite (Fig. 4.29u, v) and some are composed largely of clastic pyrite (Fig. 4.29w, x). All observed dykes have been intruded upwards and then along bedding-parallel space, thus having been formed by overpressure release, indicative of a high depositional rate.

The gritstones-sandstones of Member B (Fig. 4.28) are particularly noteworthy in that they contain clastic phosphorites (up to 8 wt.% P_2O_5; Bekasova and Dudkin 1981) and sulphides. The gritstones-sandstones occur as massive or normally-graded, channelised beds and lenses, varying from 5 cm to 2 m in thickness, within pyrite- and pyrrhotite-rich, sandstone-siltstone turbiditic rhythmites (Figs. 4.28 and 4.29ai, aj). The P_2O_5-bearing strata, with a large proportion of gritstone and coarse-grained sandstone, are confined to the central part of the belt where they have a thickness of 50–200 m and extend over a strike-length of c. 10 km. They pinch-out over a depsoitional-dip distance of c. 400 m and are overstepped by fine-grained siltstone-shale rhythmites.

All the gritstone-sandstone beds have a sharp lower boundary (Fig. 4.29ai, aj). Although the upper contact is commonly sharp, several gritstone beds exhibit grading into fine-grained sandstone with outsized, angular or rounded and soft-sediment deformed fragments of siltstone and mudstone (Fig. 4.29an, ao). Some gritstone beds and lenses are capped with massive black mudstone beds (1–20 cm in thickness) occurring as repeated gritstone-mudstone couples, in which gritstone beds have an erosive base (Fig. 4.29am). The gritstone is composed of rounded and sorted particles of phosphorite, pyrite and pyrrhotite and clasts of bedded sandstones, laminated shales, mafic lavas, vein quartz, quartzites and carbonate rocks (Fig. 4.29ah, ak, al). The gritstone beds with a high concentration of sulphide clasts were locally named "golden gritstones" due to their colour.

Various diagenetic features provide additional peculiarities to primary structures of Member B rocks. Diagenetic calcite in the form of cement and concretionary layers is abundant throughout the member, whereas spherical zoned and non-zoned calcite concretions occur mainly in its upper part (Fig. 4.29ap). All diagenetic carbonates are depleted in ^{13}C ($\delta^{13}C_{carb}$ = −13 ‰ to −7 ‰; Melezhik and Fallick 1996) with no differences between different occurrences of diagenetic calcite. Iron sulphide is found throughout the stratigraphy as layers of massive pyrite and pyrrhotite, and zoned and non-zoned concretions (Fig. 4.29aq–av). $\delta^{34}S$ in all forms of diagenetic sulphides ranges between −8 ‰ and +23 ‰ (Melezhik et al. 1998).

Member C (Lammas member) can be confidently identified only on a geochemical basis by its enhanced concentrations of Cr and Ni (e.g. Predovsky et al. 1974). The original total thickness of the member remains unclear. It varies from a few metres near the Russian-Norwegian border, to 250 m in the central part of the Belt and and from there to c. 600 m in eastern areas. The member comprises ferropicritic tuffs and tuffites (Fig. 4.29aw) with subordinate rhythmically-bedded ferropicritic black shales (8–11 wt.% MgO, up to 1,600 ppm Cr, and up to 850 ppm Ni), mafic tuffs, ferropicritic lava flows and eruptive breccias (Melezhik et al. 1994c). Mass-flow deposits are present (Fig. 4.29ax, ay) but less common than in Member B. In the western part of the belt, where the member has the greatest thickness, it consists mainly (75 vol.%) of ferropicritic tuffs containing stacked lenses of coarser clastic material. The section also includes blocks of granite, quartzite and diabase up 1.5 × 3.0 m in size (e.g. Zagorodny et al. 1964; Gilyarova 1967; Negrutsa 1984; Predovsky et al. 1987). The origin of such blocks was interpreted by some workers as glacial (Gilyarova 1967).

In the central part of the belt, more distal lithofacies (intersected by several deep drillholes including 2400, 2700 and 2900; Fig. 4.28) consists of massive or indistinctly laminated fine-grained greywacke beds (Fig. 4.29az) and basaltic tuffs. Boundaries between thicker (up to 2 m) beds are cryptic. Sand-sized clasts are poorly sorted and angular. Some massive sandstone beds show parallel- or low-angle cross-lamination on top.

Sulphides are abundant as disseminated particles, layers and pyrite and pyrrhotite concretions (Fig. 4.29ba, bb) with $\delta^{34}S$ varying between +1 ‰ and +24 ‰ in all forms of synsedimentary and diagenetic iron sulphides (Melezhik et al. 1998). Diagenetic carbonates are widespread and, in contrast to underlying units, occur mainly as lenses or spherical, and commonly zoned, calcite concretions (Fig. 4.29bc, bd). Their $\delta^{13}C_{carb}$ values range from −13 ‰ to −7 ‰ (Melezhik and Fallick 1996).

In the central-western area, the formation is intruded by a ferropicritic vent (Fig. 4.28), named the Kaula Eruptive Centre (Melezhik et al. 1994c). In the Kotselvaara open pit, the centre crops out within the Member C rocks as a c. 100-m-thick and c. 50-m-wide body comprising a basal, fragment-supported breccia with a matrix of ferropicritic lapilli tuff (Fig. 4.29be, bf), which passes gradually upwards through graded lapilli tuff into ferropicritic black shale. The breccia clasts are angular fragments of C_{org}-rich, laminated shale, basalt, partially melted/recrystallised chert, limestone, and phosphorus-bearing limestone (4.3. wt% P_2O_5; Melezhik et al. 1994c). All limestone fragments are characterised by low $\delta^{13}C_{carb}$ (from −8.7 ‰ to −7.7 ‰; Melezhik and Fallick 1996). The phosphatic, low $\delta^{13}C_{carb}$ limestones are unknown in the underlying formations of the Northern Pechenga Group. They are also unknown within the Pilgujärvi Sedimentary Formation hosting the eruptive breccia, implying that the current position of the formation is very likely allochthonous.

The depositional framework of the Pilgujärvi Sedimentary Formation mostly consists of rhythmically bedded greywackes and black, carbonaceous and sulphidic shales representing sedimentary gravity flow deposits in a marine basin. Members A and B represent a fining-upward sequence with a gradual stratigraphic decrease of the thicknesses of Bouma sequences, and the preservation of T_e divisions towards the top, suggesting basinal deepening. The dominance of arkosic sandstones in the Bouma sequences in Member A is indicative of relative proximity to the continental margin. This together with the thick and coarse-grained nature of the Bouma divisions suggests a continental slope as the most probable depositional setting. Thin- to medium-size Bouma sequences, with some preserving uppermost T_e mudstone units (background sediments), in Member B suggest deposition along the base of the continental slope. The enriched LREE concentrations in the greywacke-mudstone-shale turbiditic rhythmites indicate that continental crust remained as the main source of detritus (Melezhik et al. 1994b).

A series of beds, lenses and small-scale channels filled with P_2O_5-bearing gritstone-sandstone have been variably interpreted as deltaic (Bekasova 1985), submarine slope-slide facies (Akhmedov and Krupenik 1990) and as part of a submarine fan system (Melezhik et al. 1998). The latter interpretation envisages the phosporite and carbonate clasts as being eroded from a nearby shelf and transported to the continental slope through long-lived submarine channels. Laterally continuous breccias and slumped lithologies represent submarine slides and debris flows.

Member C, consisting in part of thick, fine-grained, structureless divisions of homogenous sandstones, resembles sediments which are common in deep-marine turbiditic settings (Shanmugam 1996; Hiscott, et al. 1997) that were deposited by either high-density turbidity currents (Lowe 1982) or a sandy debris-flow mechanism (Shanmugam 1996). Both mechanisms could be applicable for the deposition of the Member C beds, and are compatible with inferring deposition in a base-of-slope setting at the mouth of submarine canyons. Such a depositional scenario can also be applied to the stacked lenses of coarse-grained material and large blocks of various rocks hosted by ferropicritic tuffites (Negrutsa 1984; Akhmedov and Krupenik 1990).

The presence of conspicuous diagenetic sulphide and carbonate concretions (Fig. 4.29at–av, ba–bd), together with their isotopic compositions, provide evidence for the operation of bacterial sulphate reduction (Melezhik et al. 1998) and the onset of aerobic recycling of organic matter in the sedimentary column (Fallick et al. 2008).

Pilgujärvi Volcanic Formation

The Pilgujärvi Volcanic Formation (also known as the Upper Pilgujärvi subsuite or Matert formation of Zagorodny et al. 1964) is the most voluminous in the Belt. It reaches its maximum thickness in the central part of the belt, and from there becomes thinner and more sheared to the west and southeast (Fig. 4.15). Although the contact with the underlying sedimentary unit is typically interpreted as gradational (Zagorodny et al. 1964; Predovsky et al. 1974), in places it has been apparently faulted onto the Pilgujärvi Sedimentary Formation. In such places, the thickness of the Pilgujärvi Sedimentary Formation has been considerably reduced due to apparent tectonic excision (Fig. 4.15). The bulk of the Pilgujärvi Volcanic Formation is made up of submarine tholeiitic basalts, occurring as massive and pillowed lavas, lava breccias and hyaloclastites (Fig. 4.30a–c). Thin phyllite and chert intercalations form a minor sedimentary component. As in the case of the Kolosjoki pillow lavas, the vesicularity of the Pilgujärvi pillow implies emplacement in shallow water depths during their eruption. Pillow lava rims and interpillow hyaloclastites (Fig. 4.30b) in the lower part of the formation contain rare septate-tubular textures 15–20 μm in diameter and up to several hundred μm long. These have been interpreted as results of Palaeoproterozoic sub-seafloor bioalteration (Fliegel et al. 2010).

Some attempts have been made to subdivide the volcanic formation into subunits, each resulting in different lithostratigraphic schemes (Rusanov 1981; Melezhik et al. 1994a; Smol'kin and Skuf'in 1995). Figure 4.31 demonstrates a map and a vertical cross-section of the volcanic succession (divided into six members) from the central part of the belt as constructed by Rusanov (1981). The map shows three major faults (Luotna, Lammas and Poritash) that separate the volcanic rocks into four main tectonic blocks. The volcanic members vary in thickness from 300 to 800 m, with a total composite thickness of 3,500–4,000 m. Borisov (1995) suggested that such great thicknesses could be partially due to thrusting and tectonic imbrication. Rusanov (1981) calculated the following volume percentages of different volcanic rock types in the whole formation: tholeiitic basalt 87 % (including pillowed units totalling 40 %), mafic tuffs 5 %, ferropicrites 3 % (which he called komatiites), ultramafic tuffs 0.5 %, and felsic lavas and tuffs 3 %. The first member (500–600 m thick) is the best documented subunit because of its easy accessibility and penetration by many exploration drillholes. It contains the highest proportion of ferropicritic rocks (11 %), which are also found in members 3 and 5 (Fig. 4.31). The felsic rocks are solely confined to the third member.

All geochemical, isotopic and volcanological features indicate a similar origin for the Pilgujärvi and underlying Kolosjoki formation mafic metavolcanites. The major element composition of the Pilgujärvi rocks is very close to that of the Kolosjoki tholeiites and both possess T-MORB-like trace element chemistry with flat chondrite-normalized REE patterns and a depleted Nd isotopic signature with initial ε_{Nd} of 2–3.5 (Hanski 1992; Walker et al. 1997). The only exception is formed by the southernmost (uppermost) lavas of the Pilgujärvi Volcanic Formation, occurring close to the Porirash fault zone that were assigned to the Suppvaara formation by Skuf'in and Theart (2005). They are more enriched in LREE and have a near-chondritic initial Nd isotopic composition (Skuf'in and Theart 2005).

Ferropicritic rocks were generated from a similar parental magma that produced the Pechenga ore-bearing intrusions, being rich in magnesium and iron (MgO c. 15 wt.% and FeO_t >14 wt.%; Hanski 1992). They were erupted as massive and pillowed lavas, layered lava flows with spinifex-textured upper parts and ultramafic tuffs, and intruded as shallow-level sills and dykes (Fig. 4.30d–g). Both olivine spinifex and pyroxene spinifex textures (Fig. 4.30h) have been observed. The ferropicritic lava flows commonly range in thickness from 3 to 25 m, in a few cases up to 50 m. A characteristic mineralogical feature of the ferropicritic rocks is the presence of the igneous amphibole, kaersutite, giving evidence for a hydrous nature of the magma (Hanski 1992; Fiorentini et al. 2008). In layered lava flows, amphibole occurs as in situ crystallised rims around spinifex pyroxene needles (Fig. 4.30h), which indicates a high water pressure during crystallisation, raising a so-far-unresolved problem of the palaeo-water-depth interpretation, as discussed by Hanski and Smol'kin (1995). Also typical of evolved ferropicritic lavas and dykes is the occurrence of more felsic material in the form of rounded globules (ocelli) (Fig. 4.30e–g) or irregular layers when the globules amalgamate. Their genesis was examined by Hanski (1993), who interpreted the structures as developed by post-eruptional alteration of glassy mesostasis, as opposed to a magmatic origin related to silicate liquid immiscibility (cf. Smol'kin et al. 1987).

The ferropicritic magma has trace element compositions and isotopic (Nd, Os) characteristics analogous to those of modern ocean island basalts, suggesting a mantle plume-related origin (Walker et al. 1997). The U-Pb, Sm-Nd and Pb-Pb systems have yielded ages of c. 1980 Ma for ferropicritic and related intrusive rocks (Hanski et al. 1990; Hanski 1992; Skuf'in and Bayanova 2006).

The felsic rocks form several beds/lenses composed of primarily high-SiO_2 tuffs interbedded with tholeiitic basalts (Mirskaya and Zagorodny 1966; Borisov and Smol'kin 1992). These felsic beds, which vary in thickness from a few cm to 13 m, occupy a rather narrow stratigraphic interval spanning a few hundred meters and provide a conspicuous marker horizon that is easily distinguished in the field (see Fig. 4.30i), having been traced along strike for more than 20 km. The horizon is exposed, for example, in the vicinity of the Kola Superdeep drilling site, but was not intersected by the hole and hence the stratigraphic position of the felsic unit is some 900 m above the base of the Pilgujärvi Volcanic Formation (cf. Kozlovsky 1984). Several tens of meters of the tholeiitic basalts occurring below the felsic tuff beds do not show pillows, instead they are either massive or amygdaloidal (Fig. 4.32). The basaltic flow that overlies the thickest felsic tuff bed exhibits an erosive base and columnar joints (Fig. 4.30f).

The felsic rocks vary from extremely fine-grained, massive or thinly laminated varieties resembling obsidian to a fragmental rock with lapilli-size fragments (Fig. 4.30j–r). There are both coarsening- and fining-upward cycles. Some beds show sedimentary dykes and soft-sediment deformation (Fig. 4.30s–t). The fragments comprise felsic volcanic rocks, resorbed quartz and feldspar crystals, basaltic lava, and some are highly vesicular shards and, importantly, granite and tonalitic (fine-grained) gneiss. The latter has undergone amphibolite facies metamorphism before incorporation into the felsic magma.

Near the Luotna fault (Fig. 4.31), the thickest felsic unit reaches a thickness of c. 13 m. Here, the unit contains 1-cm- to 1-m-diameter fragments of limestone, calcareous sandstone, basaltic ferropicrite, ferrobasalt, ferroandsite and tholeiitic basalt, as well as 1- to 10-m-sized blocks of flow-folded rhyolite and thinly bedded, soft-sediment deformed rhyolitic tuff. All these irregularly distributed fragments are emplaced in a matrix composed of fragmental rocks with ash- and lapilli-size fragments. Such an occurrence was interpreted by Melezhik et al. (1994c) as a cold- and hot-lahar representing a remnant of the dacite-rhyolite eruptive centre, whereas Mints (1996) suggested oceanic island volcanoes. Negrutsa (1995) appealed to a long-distance transport by submarine mass-flow and deposition of clastic material on a continental slope.

Geochemically, the felsic tuffs correspond to relatively iron-rich, A-type, peralkaline dacites to high-silica rhyolites (Hanski 1992). Their REE-pattern is similar to that of the ferropicrites (Melezhik et al. 1994c). Zircon from a felsic tuff has yielded a U-Pb age of c. 1970 Ma (Hanski et al. 1990), providing an important age constraint for the upper part of the North Pechenga Group.

The reason for the appearance of products of a short-lived, acidic magmatic episode in the middle of a thick, basalt-dominated volcanic sequence remains enigmatic. Among the proposed, nonconventional processes are silicate liquid immiscibility (Skuf'in 1990) and a meteorite impact (Jones et al. 2003). Negrutsa (1995) suggested that all exotic igneous material, including felsic tuffs and lavas, were transported by mass-flow and/or turbidity currents over a long distance. In spite of the fact that gneissic fragments exist among the exotic fragments, the close-to-zero or slightly negative initial ε_{Nd} values measured for the felsic tuffs (Skuf'in and Theart 2005; E. Hanski and H. Huhma, unpubl. data) indicate that a simple origin involving partial melting of Archaean sialic basement can be ruled out (values of less than -7 are typical of Archaean crust at c. 2000 Ma).

The Pilgujärvi Volcanic Formation volcanic rocks exhibit features (abundant pillow lavas, monotonous composition and great thickness) suggesting eruption on a seafloor. Reconstructions of the palaeotectonic setting have relied primarily on geochemical characteristics, resulting in a variety of models. Negrutsa (1995) suggested an oceanic setting, and a similar model was advanced by Mints (1993, 1996) who linked the Pilgujärvi volcanic rocks to a back-arc spreading. Melezhik and Sturt (1994a) concluded that the environment was transitional from an intra- to interplate rifting with possible short-term spreading (a Red Sea-type basin). Smolkin (1997) favoured an intraplate rift environment. This currently remains unresolved.

The South Pechenga Group

The South Pechenga Group is separated from the North Pechenga Group by the Poritash fault zone that dips 50–60° to the southwest. The group has an apparent cumulative thickness of more than 10 km (e.g. Smol'kin and Skuf'in 1995). It does not have a well-established stratigraphy and apparently comprises several tectonically imbricated lithological units (e.g. Melezhik and Sturt 1994a). However, acknowledging a limited degree of exposure, Kozlova and Balagansky (1995) reported that some units still retain their structural integrity, though exhibiting swell-and-pinch structures along strike. The South Pechenga Group has been informally subdivided into a set of units composed of diverse lithologies and named by geographic localities. Sedimentary rocks include "black shales", andesitic volcaniclastic turbidites, various schists and subordinate cherts and dolostones, and volcanic rocks varying from rhyolites, dacites, alkaline and high-Mg andesites, tholeiitic basalts, ferrobasalts, picrites to alkaline picrites (Kravtsov

1984; Melezhik and Sturt 1994a; Melezhik et al. 1994b; Smol'kin and Skuf'in 1995; Skuf'in 1995a). Genetic, spatial and chronostratigraphic relationships between the various volcanic rocks of the South Pechenga Group remain unclear, although several attempts have been made to interpret their tectonic settings (e.g. Melezhik and Sturt 1994a) and correlations to the North Pechenga Group rocks (Kravtsov 1984; Smol'kin and Skuf'in 1995). Apart from Rb-Sr methods (Balashov 1996), no direct age determinations are available for the South Pechenga sedimentary and volcanic rocks. In correlations between the Pechenga and Imandra/Varzuga greenstone belts, the formations of the South Pechenga Group have been thought to correspond to the Panarechka and Saminga formations (Smol'kin 1995; Sharkov and Smol'kin 1997; see Chap. 4.1).

Smol'kin (1995), Smol'kin and Skuf'in (1995) and Skuf'in and Theart (2005) subdivided the group and distinguished the following formations: Kallojaur, Ansemjoki (Bragino), Mennel, Kaplya and Kassesjoki (Fig. 4.33). These units are briefly described below, proceeding from the Poritash fault zone across the belt to the southwest.

The north-easternmost unit (the oldest according to Smol'kin and Skuf'in 1995) is the Kallojaur formation (c. 1,200-m-thick) and is composed of rhythmically-bedded black schists, phyllites and calcareous phyllites with subordinate andesitic tuffs, cherts, tholeiitic basalts and picritic basalts. These freatures imply a deep-water depositional setting (e.g. Predovsky et al. 1974, 1987). Based on unpublished exploration report(s), Smol'kin and Skuf'in (1995) suggested that Kallojaur formation schists rest depositionally on the uppermost basalts (their Suppvaara Formation) of the Pilgujärvi Volcanic Formation. The Kallojaur formation is intruded by up to 1.2-km-thick, subvolcanic andesitic to rhyolitic bodies of the Poritash Complex (Skuf'in 1995b) (Fig. 4.15).

The next unit to the south, the Ansemjoki formation (renamed by Skuf'in and Theart 2005, after the former Bragino formation), is up to c. 2,500-m-thick. It is dominated by various volcanic rocks including massive and pillowed basalts, ferrobasalts, and high-Mg basalts, hyaloclastites, and andesitic, basaltic and picrobasaltic tuffs. Sedimentary rocks are represented by black schists and calcareous schists.

The next two units, the Mennel and Kaplya formations, occur as tectonic klippen in the central part of the belt (Fig. 4.33) and have been considered coeval, although composed of different types of volcanic rocks (Melezhik and Sturt 1994; Skuf'in and Theart 2005). The Mennel formation is up to 2,000 m thick and contains picrites and basalts, which occur as massive, pillowed, amygdaloidal and variolitic lava flows, lava breccias and hyaloclastites interbedded with tuff beds of a similar composition and volcaniclastic black schists; rhyodacites and rhyolites are minor. The picritic rocks, which occupy 60 % of the volume of the formation, are moderately rich in TiO_2 and their chondrite-normalised REE patterns are enriched in LREE (Melezhik and Sturt 1994a; Skuf'in and Theart 2005). They resemble the ferropicrites of the North Pechenga Group, though are slightly lower in FeO_T and TiO_2.

The Kaplya formation (the Porojärvi formation of Melezhik et al. 1994b) has a thickness of up to 2,000 m and is dominated by andesites, but also includes other volcanic rocks such as dacites, andesitic basalts and high-Mg basalts. The volcanic rocks occur as massive and breccia flows, extrusive bodies, eruptive breccias and tuffs. The Kaplya formation is best studied in the Porojärvi area in the northwestern part of the South Pechenga zone, where andesitic volcanic rocks form a structure (palaeovolcano?) known as the Porojärvi volcanic centre (Suslova 1966; Skuf'in et al. 1988; Skuf'in and Theart 2005). Melezhik and Sturt (1994) interpreted the assemblage of massive andesitic flows, flow breccias, agglomerates and bedded tuffs as cold and hot lahars. The most primitive volcanic rocks are low-Ti magnesian basalts and basaltic andesites (Skuf'in and Theart 2005), which according to Skuf'in et al. (1988) have major element geochemistry reminiscent of rocks of the modern boninitic volcanic series (e.g. Hickey and Frey 1982). However, though showing an island arc affinity, they have REE patterns that are significantly different from those of boninites per se (Melezhik and Sturt 1994a). In addition, they are not as depleted in HFSE (e.g. TiO_2) as modern boninites. Some andesitic rocks can be classified as trachyandesites based on their silica-alkalies relations, but, as a whole, the Kaplya formation has a calc-alkaline nature due to low TiO_2 and FeO_T.

Smol'kin and Skuf'in (1995) and Skuf'in and Theart (2005) suggested that the Portitash subvolcanic andesitic to rhyolitic bodies, along the contact between the North and South Pechenga groups (Fig. 4.15), could be coeval with the intermediate and felsic volcanic rocks of the Kaplya formation. However, the Poritash subvolcanic andesites and dacites are nonfoliated and have undergone only greenschist facies metamorphism, in contrast to the epidote-amphibolite grade observed in the Kaplya andesites. In addition, according to Smol'kin (1995) and Smol'kin et al. (1995), the Poritash Complex porphyries intrude rocks of both the Kassesjoki formation and the Pilgujärvi Volcanic Formation, implying that their emplacement took place after the tectonic juxtaposition of the North and South Pechenga groups. Different Rb-Sr dates, 1718 ± 29 Ma and 1855 ± 54 Ma, have been obtained for the Poritash and Kaplya rocks (Balashov 1996), but the geochronological significance of these ages is uncertain.

The southernmost unit, the Kassesjoki formation (c. 400 m), has been considered the youngest rocks of the South Pechenga Group (Smol'kin 1995; Skuf'in and Theart 2005). It is represented by rhythmically-bedded, andesitic volcaniclastic sandstone-siltstone with intermittent, 10–50-m-thick conglomerate layers at the base and subordinate basalts in the middle and upper parts. The conglomerates contain fragments of banded-iron formation and numerous clasts of andesite, some of which contain an older tectono-metamorphic fabric implying that such andesitic rocks were subjected to deformation and metamorphism, probably related to a compressional orogenic event, prior to their uplift and erosion (Melezhik et al. 1994b). Sharkov and Smol'kin (1997) interpreted the conglomerates as molasse deposits.

On the southwestern side of the Southern Pechenga Zone, there are several domal structures formed by large granitoid plutons (Figs. 4.15 and 4.33); these have been assigned to the Kaskeljavr Complex (Vetrin et al. 2008). The complex was formed by two intrusive phases, the first being quartz dioritic, dioritic and tonalitic, and the second granodioritic and trondhjemitic. Both phases have yielded similar U-Pb zircon ages of c. 1940–1950 Ma (Skuf'in et al. 2000; Vetrin et al. 2008). The Sm-Nd isotopic characteristics of the first magmatic pulse show a major juvenile component, while the influence of the Archaean basement is larger in the second phase (Vetrin et al. 2008). The youngest Palaeoproterozoic magmatic rocks in the South Pechenga area are 1.71 Ga lamprophyre dykes (Skuf'in et al. 1999).

The relationship between the Kaskeljavr Complex plutons and the sedimentary and volcanic rocks of the South Pechenga Group has not yet been reliably established. Skuf'in and Theart (2005) considered it possible that some of the felsic volcanic rocks (Kalpya formation) and the Kaskeljavr intrusions have a genetic relationship and share the same magmatic source. Kozlova and Balagansky (1995) reported D_3 deformation in South Pechenga Group rocks, which they interpreted as having been generated by the emplacement of the Kaskeljavr granodiorite massif, suggesting that the sedimentary and volcanic rocks are older than c. 1950 Ma. According to Vetrin et al. (2008), the plutons were overthrust from the southwest onto feldspathic and garnet amphibolites, which they regard as Archaean in age. As the South Pechenga Group sedimentary and volcanic rocks are separated from the granitoids by a narrow zone of amphibolites, their depositional age cannot be directly constrained by the geochronological data from the plutons. However, granitic and dioritic dykes up to 20 in thickness are known to cut rocks of the South Pechenga Group, and if correlated with the plutons of the Kaskeljavr Complex (Skuf'in and Theart 2005), would provide a minimum age for the South Pechenga Group. The situation is further complicated by different opinions on the age of the amphibolites framing the Kaskeljavr Granitoid Complex. Vetrin et al. (2008) and Tel'nov (1986), considered the amphibolites Archaean, whereas Skuf'in (1995b) and Skuf'in and Theart (2005) regarded them as highly metamorphosed equivalents of volcanic rocks belonging partly to the Pilgujärvi Volcanic Formation and partly to the Ansemjoki and Mennel formations of the South Pechenga Group (Fig. 4.33).

Given the lack of reliable radiometric ages and great uncertainties in geological relationships of different South Pechenga Group units, depositional environments can only be tentative. Cratonic sediments, such as found in the lower part of the North Pechenga Group, are absent, whereas deep-water sediments, partially deposited from turbiditic currents, prevail. The mafic to ultramafic volcanism seems to have taken place mainly or solely subaqueously. In contrast, calc-alkaline, andesitic to rhyolitic volcanic rocks (Kaplya formation) record mostly subaerial, explosive volcanism. From the geochemistry of the volcanic rocks, it is evident that magmas were derived from different sources and erupted in different geotectonic settings (Melezhik and Sturt 1994a; Melezhik et al. 1994b). The picrites of the Mennel formation are geochemically similar to Hawaiian picrites (cf. Rhodes 1996), suggesting a within-plate origin for these rocks. On the other hand, the intermediate to felsic volcanic rocks show an obvious subduction-related geochemical affinity in their chemical composition (Melezhik and Sturt 1994a). The South Pechenga Group thus seems to represent a collage of lithological units that were generated in both extensional and compressional geotectonic settings and were later emplaced tectonically together (Melezhik and Sturt 1994a; Melezhik et al. 1994b). The generation and emplacement of the plutons have been linked to island arc magmatism at c. 1.94–1.95 Ga and a later collisional event at c. 1.90–1.91 Ga (Melezhik and Sturt 1994a; Daly et al. 2006). However, Sharkov and Smol'kin (1997) suggested that the South Pechenga calc-alkaline volcanism was associated with a continent-continent collision event between the Kola and Karelian cratons and related development of a subduction zone.

The above-presented, concise description of the South Pechenga Group lithostratigraphy and depositional interpretations should be considered as provisional. Although the presence of a major tectonic break, the Poritash fault zone, between the North and South Pechenga Zones is generally accepted, its position with respect to the Poritash dacites, for example, varies on different maps and its geotectonic significance has been interpreted in different ways. As was already briefly acknowledged above, different researchers describe, subdivide and combined these diverse lithological units in contrasting models. Formations have been defined differently and as have correlations with the North Pechenga Group and even with older Archaean complexes (Zagorodny et al. 1964; Gilyarova 1967; Polyak 1968; Predovsky et al. 1974; Kravtsov 1984; Melezhik et al. 1994a, b; Smol'kin and Skuf'in 1995). Resolution of these various models must await obtaining robust depositional and formational ages of the various rock units.

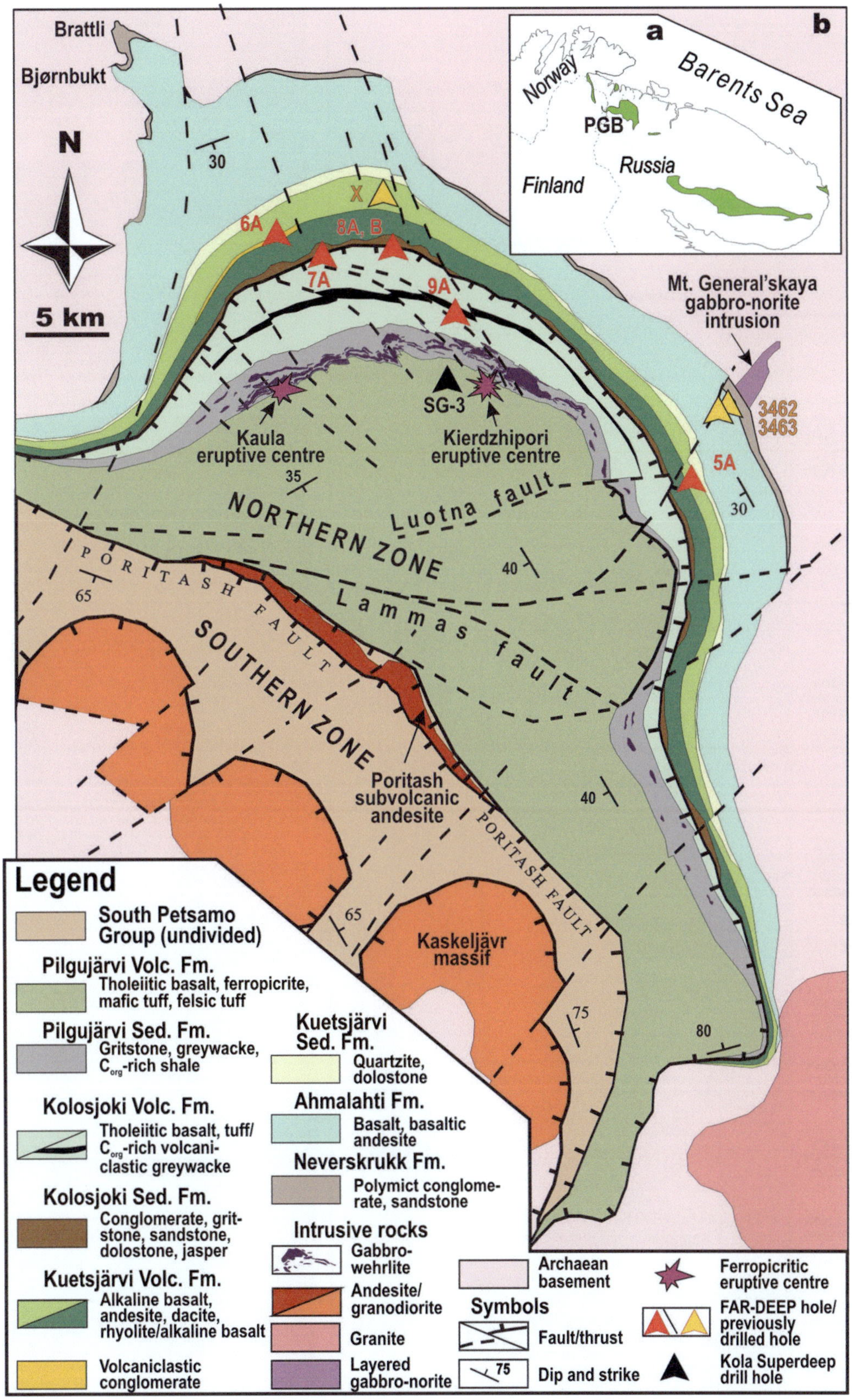

Fig. 4.15 (**a**) Geographic location of the Pechenga Greenstone Belt (PGB). (**b**) Simplified geological map of the Pechenga Greenstone Belt (Modified from Zagorodny et al. 1964). Note that contacts of the Pilgujärvi Sedimentary Formation are structurally compromised in places

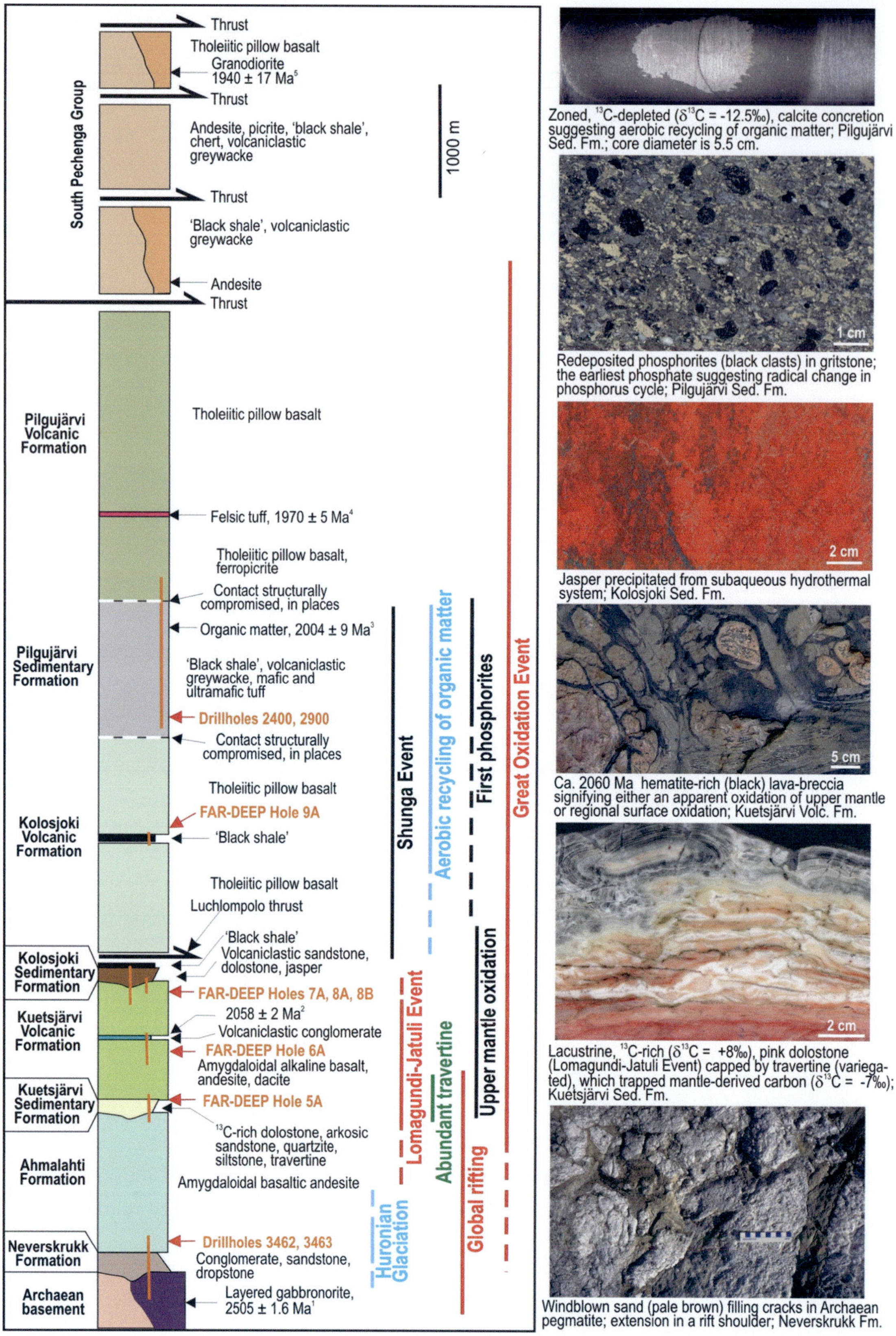

Fig. 4.16 Simplified lithological column of the Pechenga Greenstone Belt, positions of FAR-DEEP and other relevant drillholes, and examples of photos of sedimentary and volcanic rocks from Pechenga Belt formations. Also shown is how the evolution of the Pechenga Greenstone Belt is related to global palaeoenvironmental events. Superscripts denote radiometric ages from (1) Amelin et al. (1995), (2) Melezhik et al. (2007), (3) Hannah et al. (2004), (4) Hanski (1992) and (5) Vetrin et al. (2008)

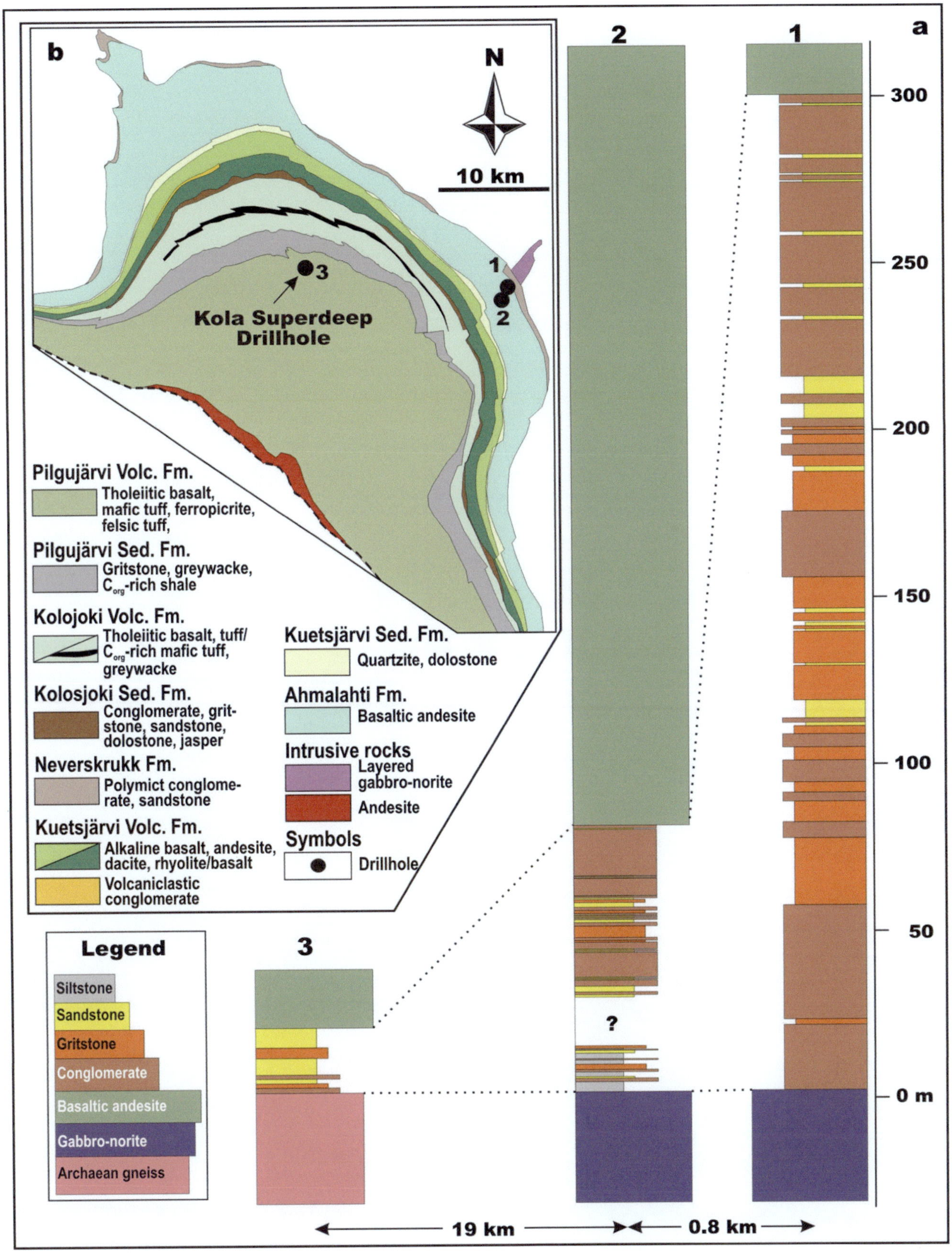

Fig. 4.17 (**a**) Lithological columns of the Neverskrukk Formation, illustrating facies and thickness variations. (**b**) Geological sketch showing the positions of the measured sections

Fig. 4.18 Images of sedimentary rocks of the Neverskrukk Formations. (**a**) Surface of fractured Archaean pegmatite at Brattli; fractures filled with *pale brown* sandstone. (**b**) Disintegrated Archaean granite (regolith) at the base of the Neverskrukk Formation east of Mt. General'skaya. (**c**) Cross-section at Brattli in the *lower part* of the formation composed of polymict clast-supported conglomerate; note that some boulders and pebbles in the *upper part* of the section show unidirectional imbrication

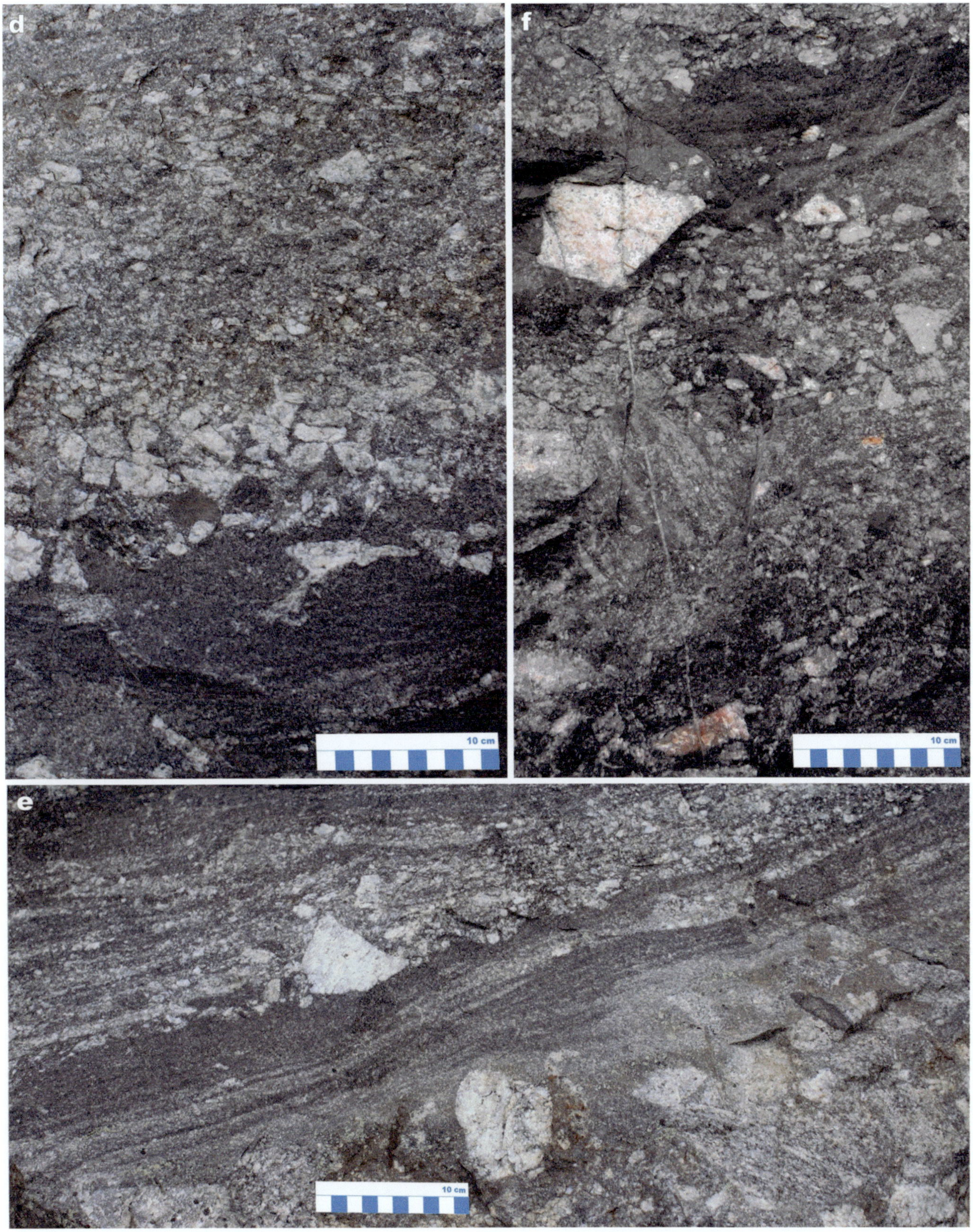

Fig. 4.18 (continued) (**d**) Cross-section at Brattli in the *middle part* of the formation composed of *dark grey* sandstone passing upward into small-fragment breccia grading into coarse-grained sandstone with scattered, larger, angular clasts. (**e, f**) Interbedded polymict conglomerate, bedded sandstone and gritstone with oversized, unsorted, angular clasts; a road-cut section between Brattli and Bjørnbukt (Fig. 4.14)

Fig. 4.18 (continued) (**g**) Sawn drillcore of gritty conglomerate with polymict, unsorted clasts supported by greywacke matrix in the middle part of the formation. (**h**) Drillcore of polymict, clast-supported conglomerate with well-rounded and unsorted clasts in the *middle part* of the formation. (**i**) Sawn drillcore of polymict sandstone with scattered oversized clasts in the *upper part* of the formation. (**j**) Sawn drillcore of thickly-bedded greywacke deposited on top of gritty sandstone in the upper part of the formation. Core diameter in (**g–j**) is 5 cm (Photographs by Victor Melezhik. All samples are from a section at Mt. General'skaya (Fig. 4.14))

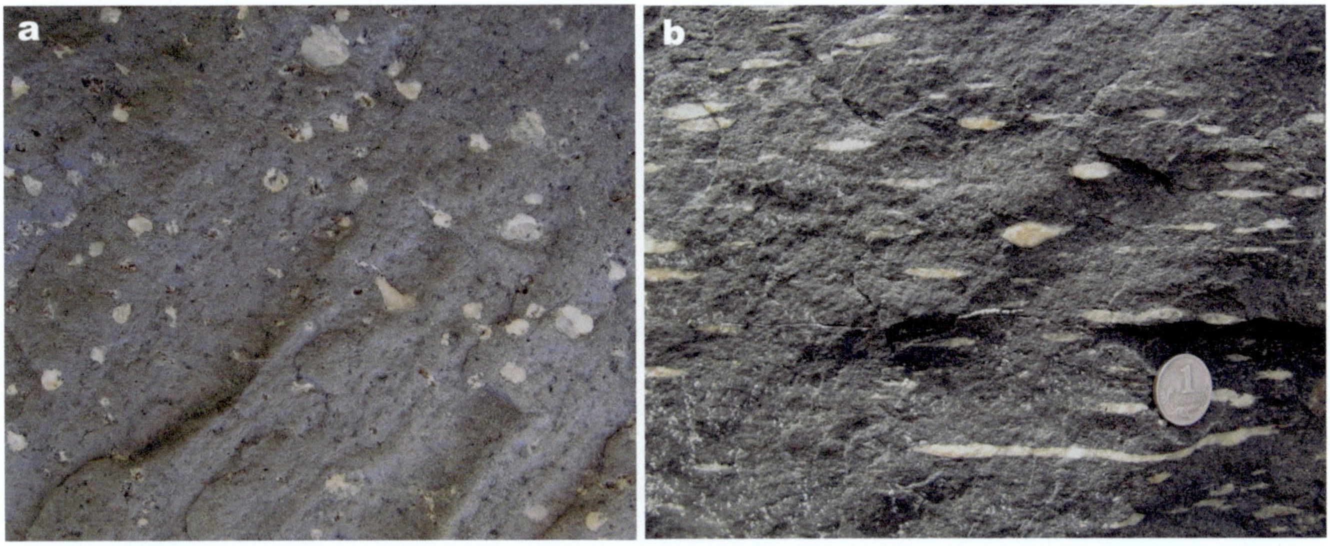

Fig. 4.19 Amygdaloidal basaltic andesite of the Ahmalahti Formation. Photograph (**a**) taken by Victor Melezhik west of the Mt. General'skaya area, photograph (**b**) by Aivo Lepland south of Bjørnbukt (Fig. 4.15)

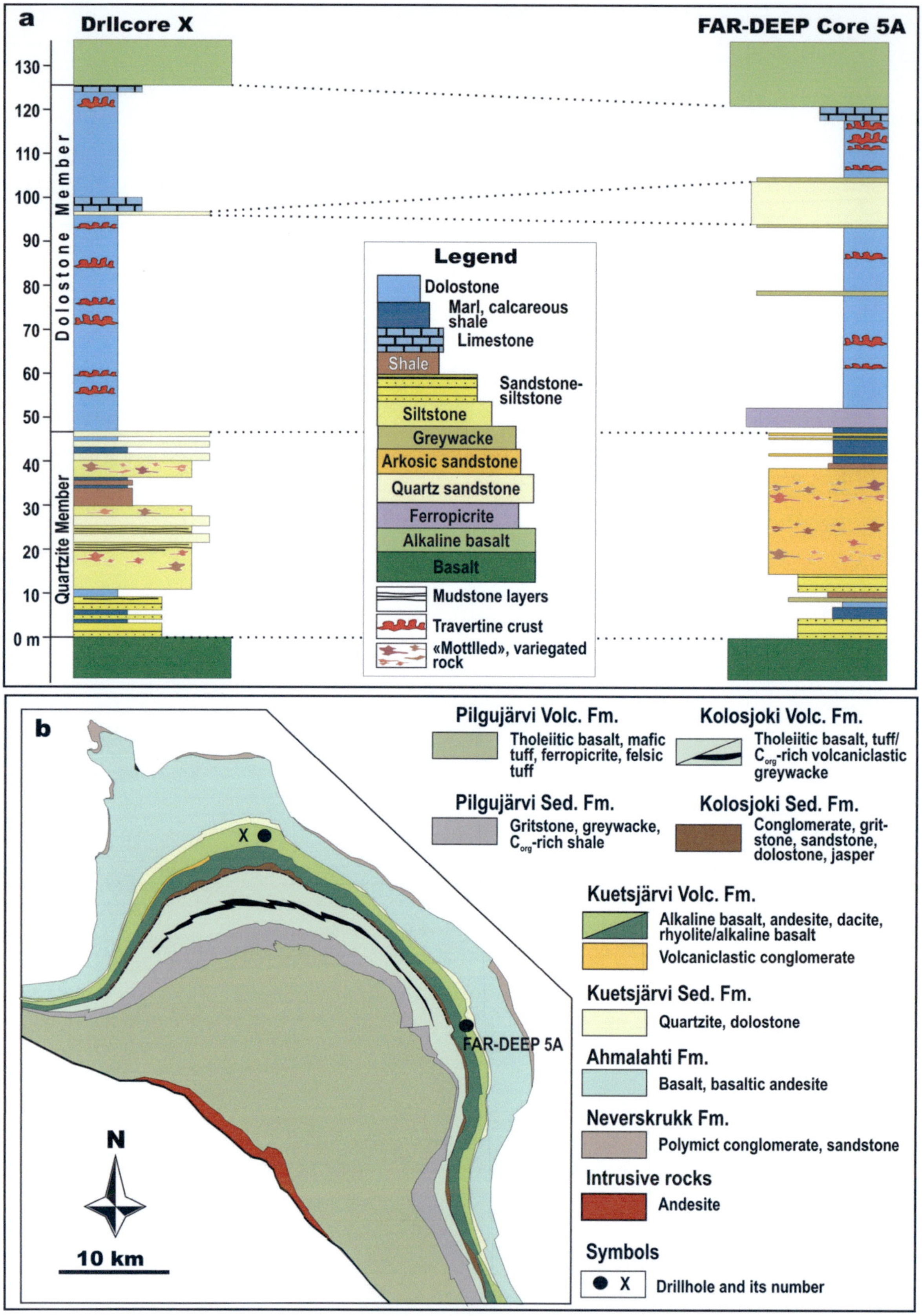

Fig. 4.20 (**a**) Drillcore-based logs illustrating facies variations in the Kuetsjärvi Sedimentary Formation over a distance of 25 km strike-length. (**b**) Geological sketch showing the positions of the measured sections. Drillcore X log is modified from Melezhik and Fallick (2005); FAR-DEEP Core 5A log is simplified from FAR-DEEP database

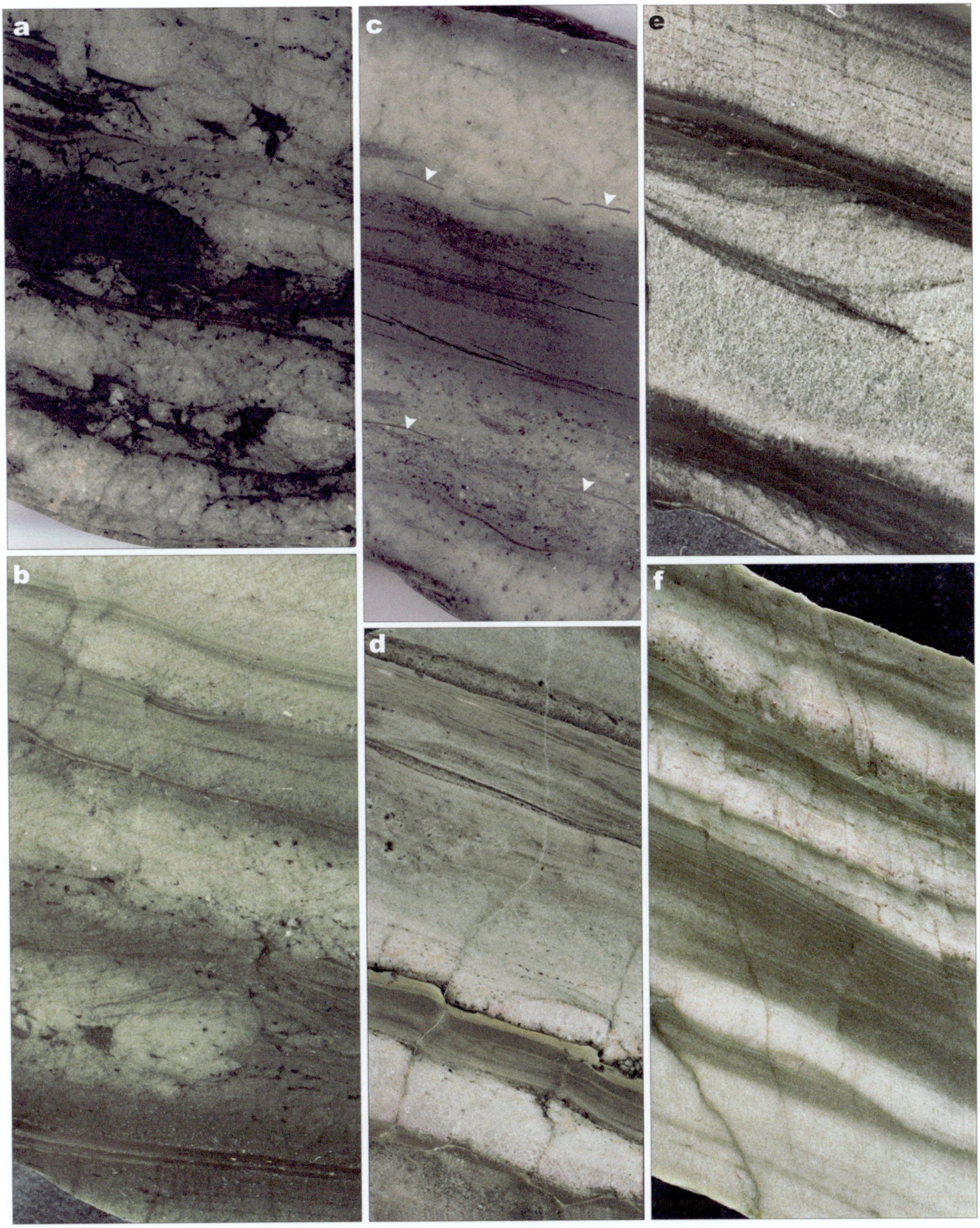

Fig. 4.21 Images of sedimentary rocks of the Kuetsjärvi Sedimentary Formations. (**a**) Soft-sediment loading structures along bases of light-coloured sandstone beds. (**b**) Graded sandstone-to-mudstone beds displaying flat- to wavy-parallel and ripple cross-lamination; note loading and small-scale scouring along base of sandstones. (**c**) Sandstone bed with mud rip-up intraclasts (*arrowed*) passing upward into dark grey siltstone with low-angle cross-lamination, which is, in turn, erosionally overlain by sandstone bed with siltstone intraclasts in its basal part. (**d**) Variably developed Bouma sequences composed of sharp-based, graded sandstone beds, faintly parallel-laminated sandstone, and ripple cross-laminated and wavy- to low-angle laminated mudstone (dark coloured layers). (**e**) Bouma sequence exhibiting uneven spacing and containing graded-bedded sandstone, cross-bedded silty sandstone and planar-laminated, dark-coloured mudstone; note that the lowermost thin sandy layer has been cleaved in a direction opposite that of ripple cross-lamination. (**f**) Thinly interbedded, fine sandstone (light coloured) and siltstone (dark coloured) exhibiting planar to low-angle ripple cross-lamination; grading can be seen in the upper two sandstone beds. Photographs were taken from drillcore X. Core diameter is 5 cm

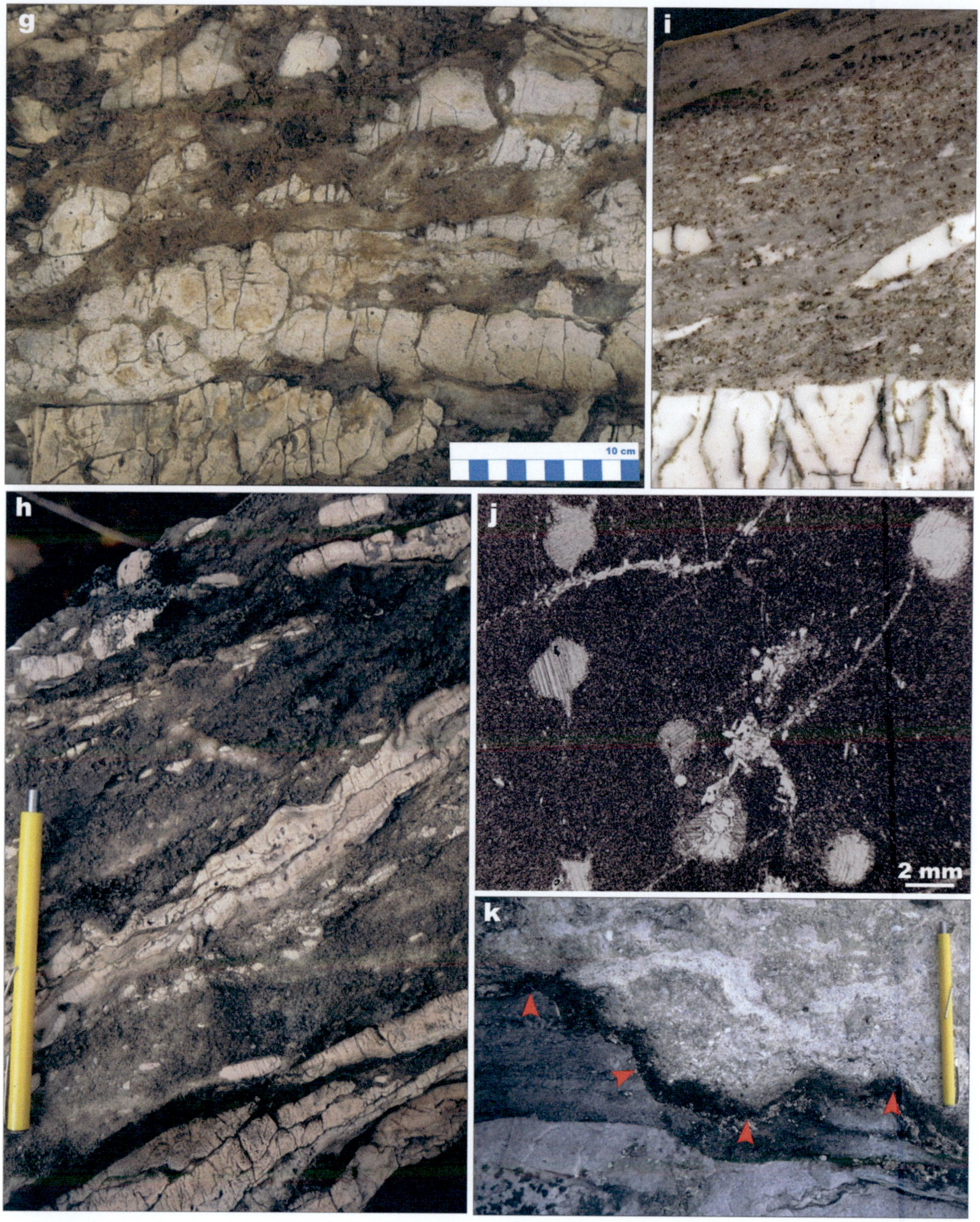

Fig. 4.21 (continued) (**g**) Intensely desiccated and dismembered dolostone beds (*light yellow*) encased in a muddy sandstone matrix; exposure located 7 km west of drillhole X (Fig. 4.15). (**h**) Thin, irregular, undulose-bedded dolostones alternating with poorly-sorted, matrix-supported, intraclastic (clasts are dolomicrites) sedimentary breccias; exposure located 8.6 km west of drillhole X; pencil length 15 cm. (**i**) Drillcore showing a dolomicrite bed (*bottom*) sharply overlain by sedimentary breccia composed of dispersed, imbricated, tabular, dolomicrite clasts in a muddy sandstone matrix. (**j**) Photomicrograph in transmitted, non-polarised light of probable dolomite-pseudomorphed evaporite nodules showing spheroidal and castellated margins. (**k**) Small channel (marked by *red arrows*) filled with poorly sorted dolarenite, and cutting into the underlying bed of dark purple, haematite-rich, arkosic siltstone and pale purple arkosic sandstone; exposure located 8.2 km west of drillhole X; pencil length 15 cm. Photographs (**i, j**) taken from drillcore X. Core diameter is 5 cm

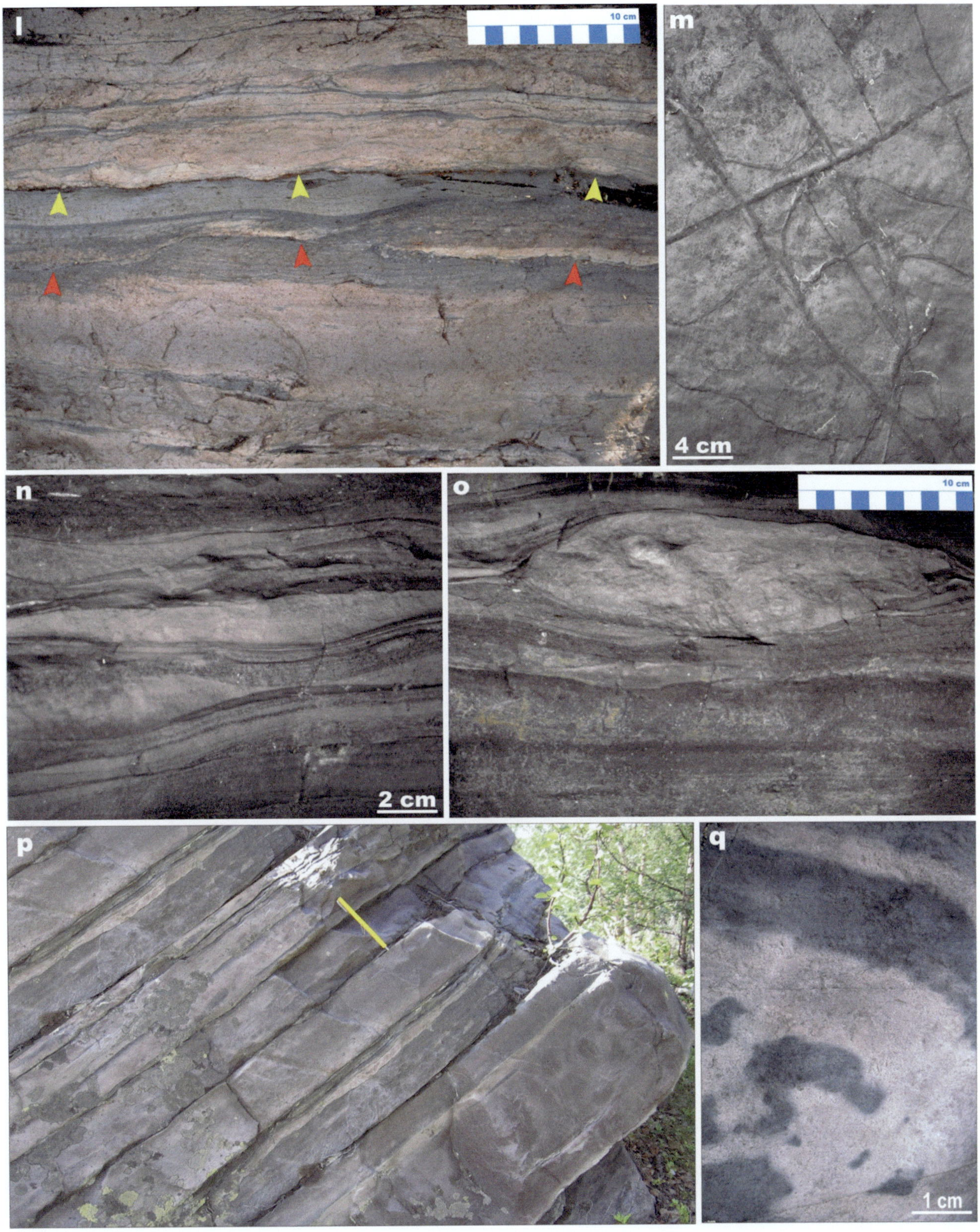

Fig. 4.21 (continued) (**l**) *Pale brown* siltstone bed with finer intraclasts (darker coloured) grades upward into wavy- to flat-parallel laminated mudstone, which is sharply truncated (erosion surface marked by *red arrows*) by thin-based sandstone bed that passes rapidly upward into wavy-laminated mudstone; the latter erosionally overlain (basal scour marked by *yellow arrows*) by another thin sandstone that grades upward into wavy-parallel laminated and wispy-bedded siltstone-mudstone. (**m**) Desiccation cracks in mudstone. (**n**) Lenticular, wavy and linsen-bedded (starved ripples), fine sandstone and siltstone. (**o**) Slumped horizon in an interbedded, fine sandstone-siltstone interval. Note how slump appears to be draped and possibly onlapped by (left-hand-side of slump) overlying layers; pencil length 15 cm. (**p**) Thickly bedded arkosic sandstone with thin sericite shale interbeds; pencil length 14 cm. (**q**) Soft-deformation in arkosic sandstone expressed by bent, haematite-rich layers. All exposures are located 7 km east of drillhole X (for location, see Fig. 4.15)

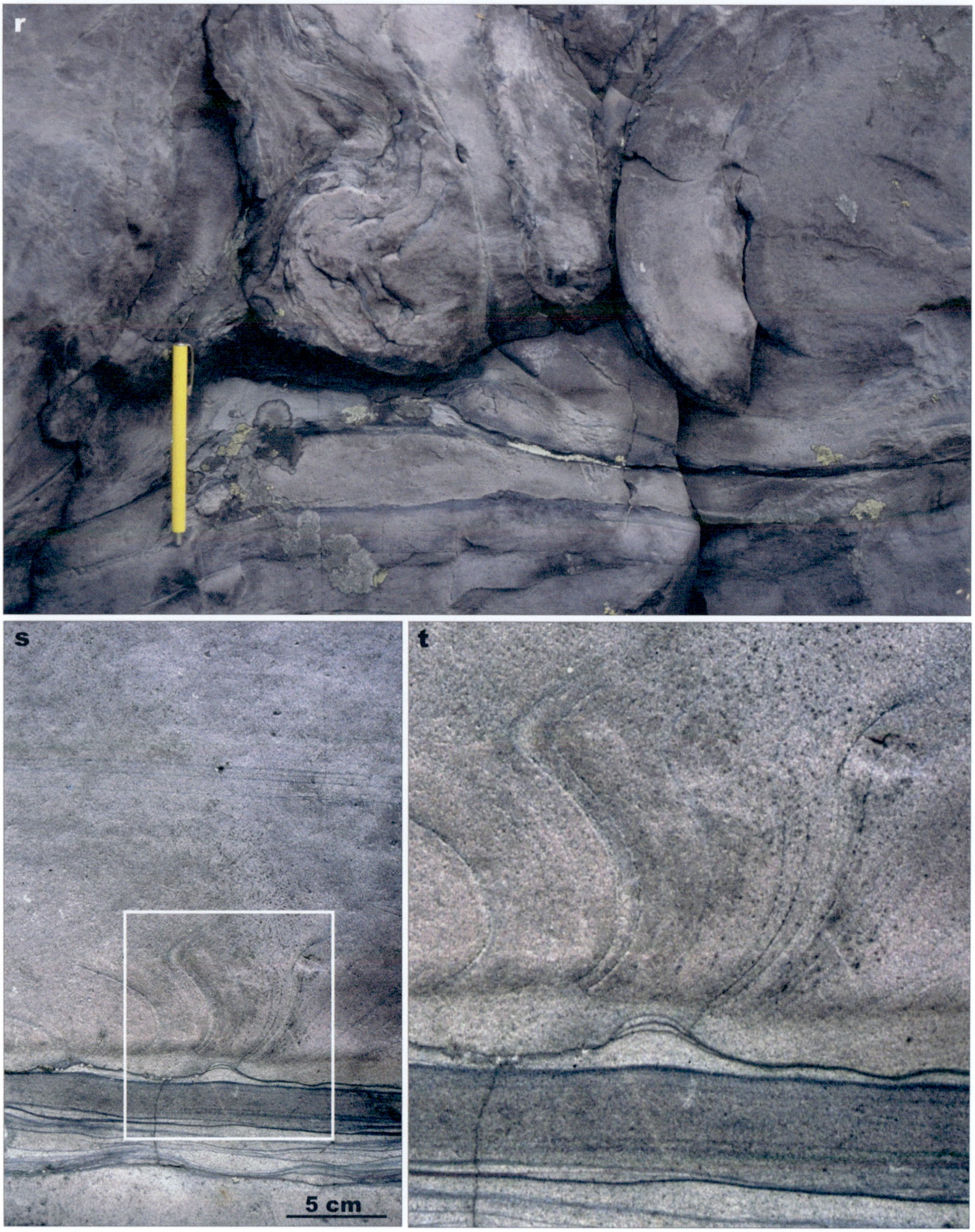

Fig. 4.21 (continued) (**r**) 'Ball-and-pillow' structure in arkosic sandstones. Sandstone beds are separated by thin, haematite-rich mudstone beds (note upward injection of mudstone into and between the 'ball-and-pillow' structure); pencil length 15 cm. (**s**) Injection (fluid escape?) structure in otherwise massive sandstone bed overlying thinly interbedded siltstones and fine sandstones separated by filamentous wisps to more continuous partings of mudstone. (**t**) Close-up view of fluid escape(?) structure. Exposures located 7 km east of drillhole X (for location, see Fig. 4.15)

Fig. 4.21 (continued) (**u**) Stacked planar cross-bed co-sets in quartzitic sandstone. (**v**) Stacked trough cross-bed co-sets in quartzitic sandstone. (**w**) Finely laminated, haematite-rich mudstone-siltstone interbeds. The lighter-coloured layer was caused by infiltration of catagenetic reducing fluids. (**x**) Haematite-rich siltstone with fine, varve-like lamination. (**y**) 'Mottled' siltstone in which the darker-coloured layers are primary sedimentary laminae whereas the lighter-coloured layers and blebs are due to reduction by migrating catagenetic fluids. Photographs (**u, v**) were taken from exposures located 7 km east of drillhole X (for location see Fig. 4.15) and (**w–y**) from drillcore X. Core diameter is 5 cm

Fig. 4.21 (continued) (**z**) Finely laminated, variegated siltstone. (**aa**) Stacked cross-bed co-sets in dolomite-cemented arkosic sandstone. (**ab**) Two beds of massive dolomicrite (*bottom*) sharply overlain by cross-laminated, allochemical dolostone that grades into dark-coloured, dolomite-cemented quartzite, which is, in turn, overlain by two other allochemical dolostone layers sharply capped by light-coloured, faintly laminated dolomicrite (*top*). (**ac**) *Pale yellow*, faintly laminated, micritic dolostone sharply capped by travertine crust composed of clusters of semi-spheroidal travertine mounds exhibiting radial fabrics and irregularly developed siliceous sinters (dark coloured). (**ad**) Allochemical sandy dolostone (light-coloured) overlain by quartzitic sandstone composed of well-rounded and sorted quartz grains in a dolomite matrix. Photographs were taken from drillcore X. Core diameter is 5 cm

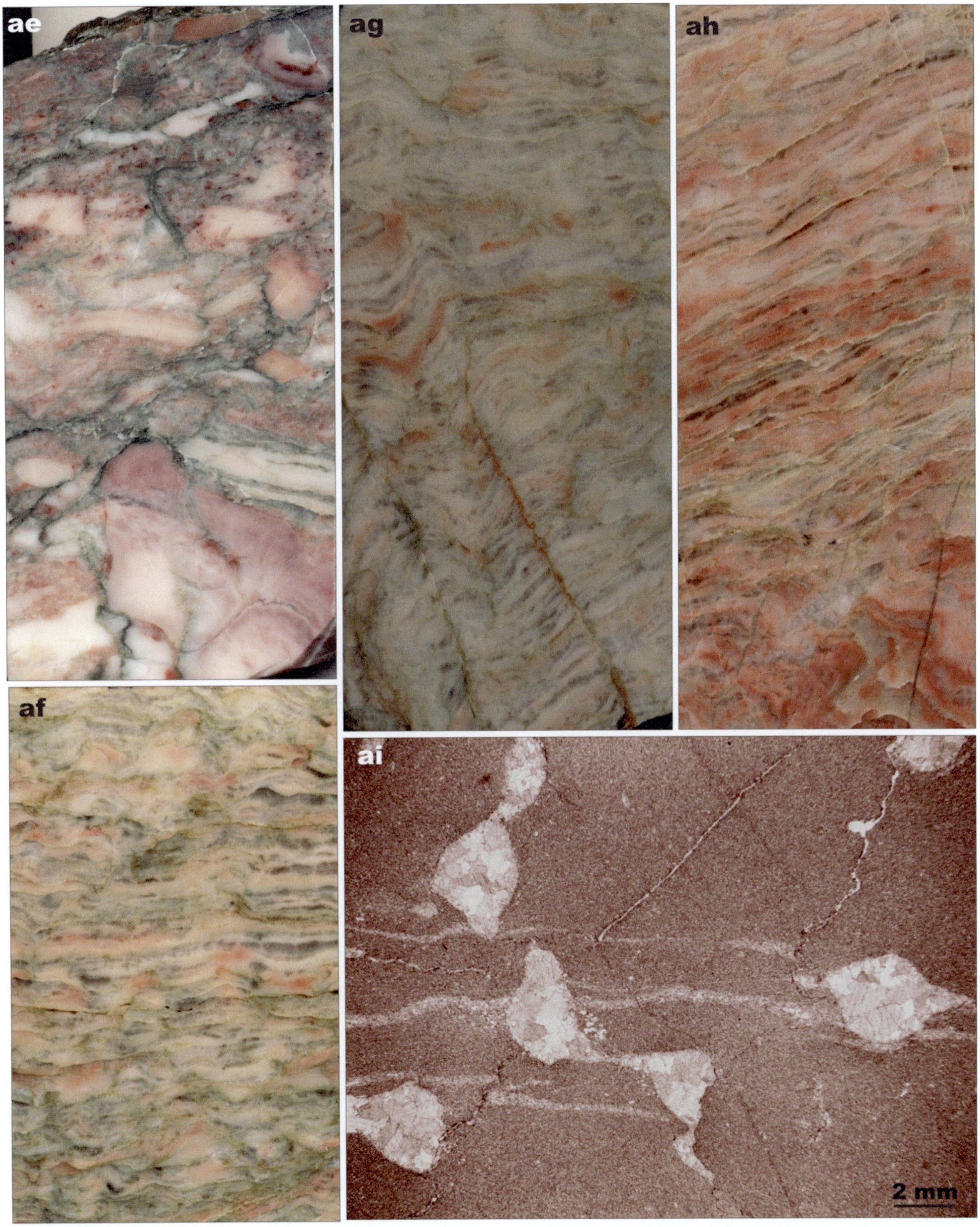

Fig. 4.21 (continued) (**ae**) Clast-supported dolorudite; clasts are variegated dolomicrite in a dark mudstone matrix. (**af, ag**) Vertical section through mottled, flat-laminated stromatolite layers with flat, curly, and blister laminae separated by thin, irregular partings of silica sinters (*pale grey* colours). (**ah**) Flat-laminated stromatolitic dolostone deformed by tepee structure (*bottom*). (**ai**) Plane-polarised light photomicrograph of probable sulphate micronodules pseudomorphed by dolospar in a dolomicritic groundmass. Photographs were taken from drillcore X. Core diameter is 5 cm

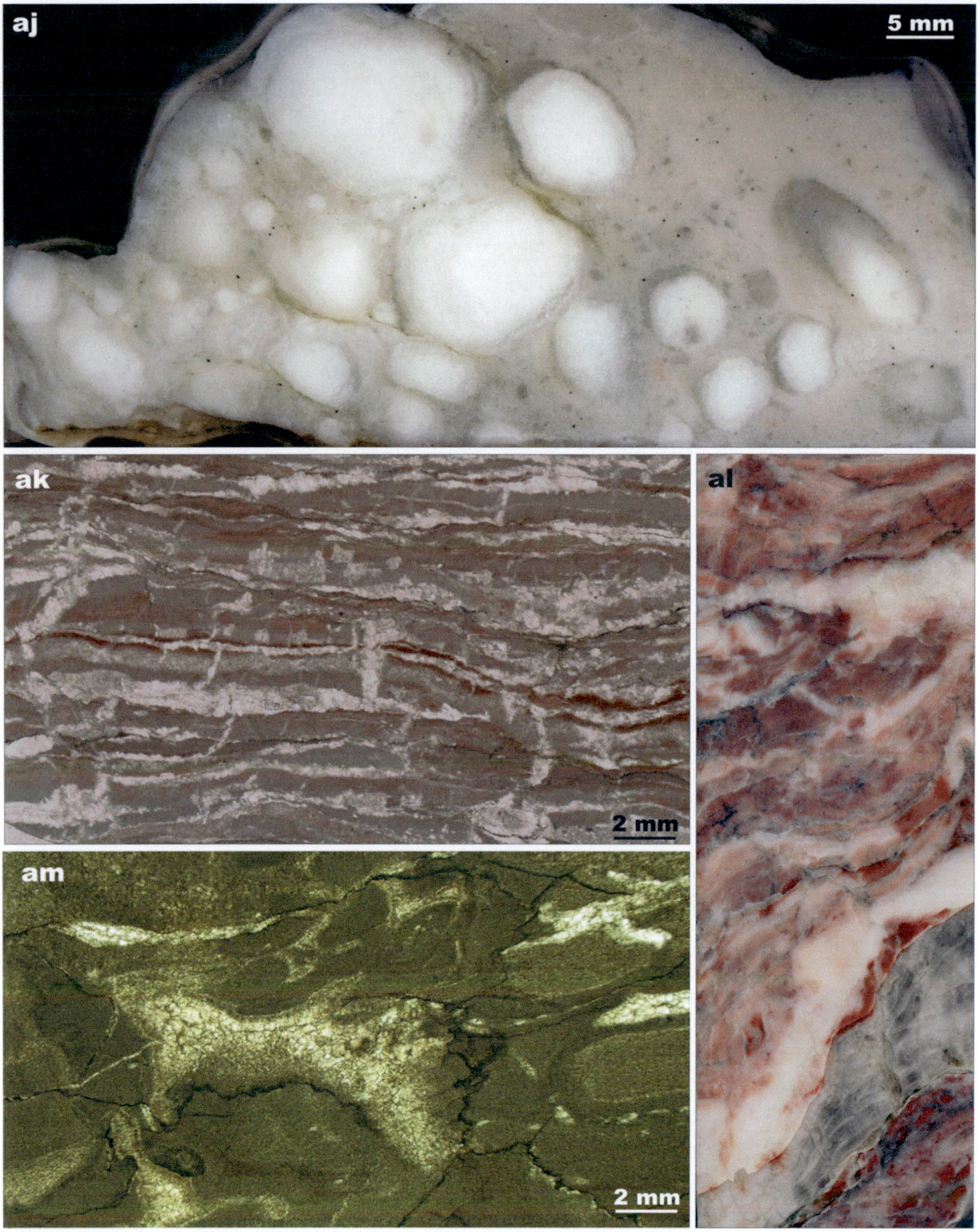

Fig. 4.21 (continued) (**aj**) Probable sulphate nodules replaced by dolospar in a dolarenite matrix. (**ak**) Photomicrograph in transmitted, non-polarised light showing a cross-section view of intensively cracked, stromatolitic dolostone with mudcracks (vertical) and sheetcracks (horizontal) filled with dolosparite and microcrystalline quartz. (**al**) Vertical section through iron-stained dolocrete with tepee-related deformation features and travertine-filled cracks (*pale grey*). (**am**) Brecciated, microbial (?) dolostone with abundant fenestrae filled with sparry dolomite. Photograph (**aj**) was taken from a quarry located near FAR-DEEP Hole 5A (Fig. 4.15) and (**ak–am**) from drillhole X. Core diameter is 5 cm

Fig. 4.21 (continued) (**an**) Polished slab exhibiting crudely bedded, allochemical dolostone; uniform, sub-spherical dolomite particles in the lower part may represent recrystallised oolites, whereas clasts in the upper bed range in size, shape and composition, and represent fragments of dolomicrite, dolocrete (*pink*) and dolomite-cemented sandstone. Photograph was taken from a quarry located near FAR-DEEP Hole 5A

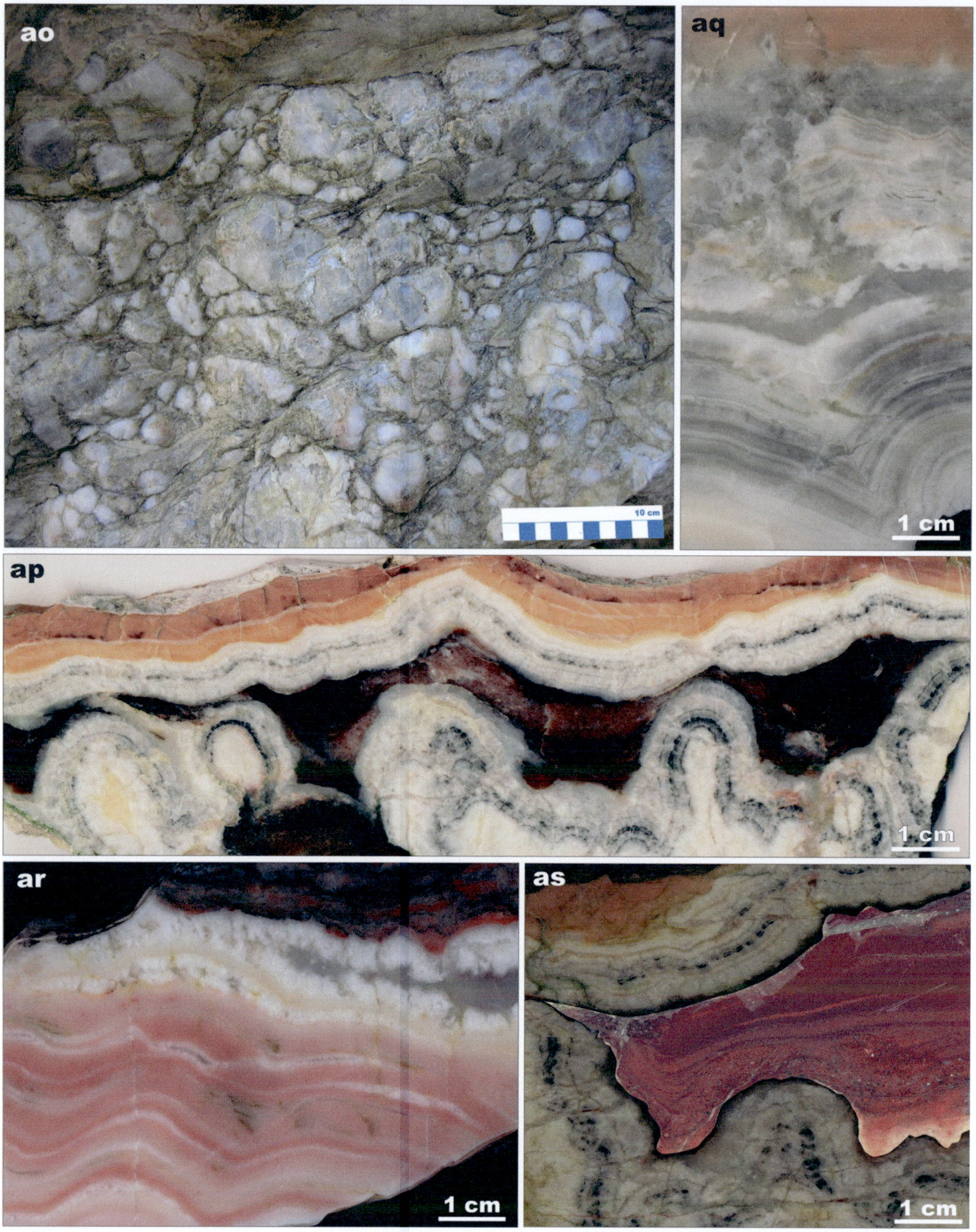

Fig. 4.21 (continued) (**ao**) Quarry exposure of travertine crust on the surface of a bedding plane exhibiting clusters of small spheroidal structures. (**ap**) Cross-section slab of dolomite travertine mounds buried under red, clayey sandstone and capped by a travertine crust. (**aq**) Cross-section slab of dolomite travertine exhibiting light and dark banding (lower part) capped by a disrupted (desiccated) and silicified travertine crust followed by beige dolomicrite. (**ar**) Cross-section slab of dolomite travertine exhibiting white and pale pink banding (*lower part*) overlain by white vuggy travertine (vug is filled with pale grey, finely-crystalline quartz) and capped by dark-coloured, haematite-rich dolocrete. (**as**) Space in dolomite travertine filled with pink, heamatite-rich quartz sandstone grading into laminated siltstone. Photographs were taken from a quarry located near FAR-DEEP Hole 5A (Fig. 4.15)

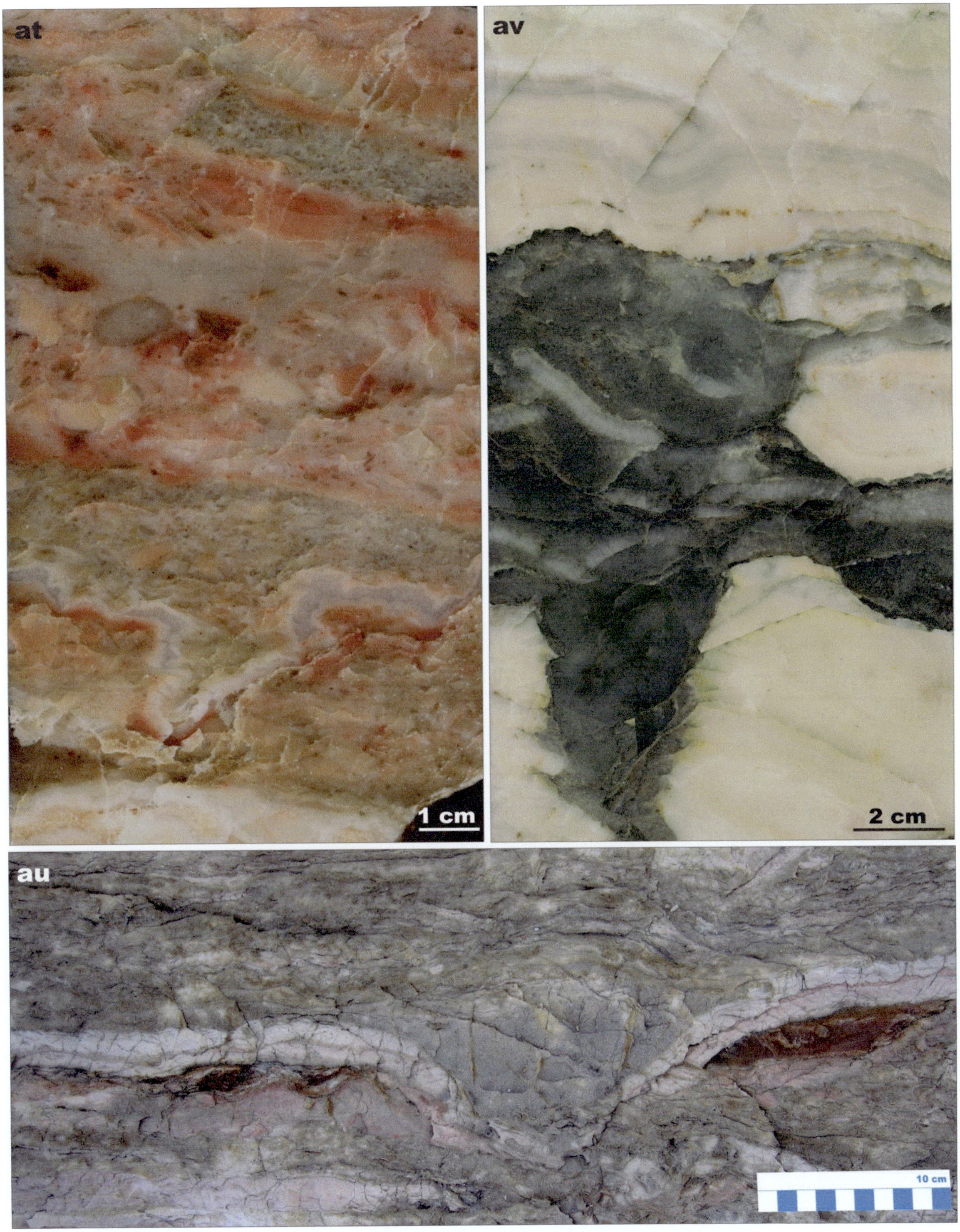

Fig. 4.21 (continued) (**at**) Polished slab showing the vertical section through a dissolution cavity of surface origin in allochemical, sandy, micritic dolostone; cavity walls marked by reddening and veneered by silica sinter and filled by sandy to fine-pebble dolostone conglomerate. (**au**) Eroded surface of a dolarenite bed covered with lenses of brown silcrete and a thin crust of light-coloured travertine. (**av**) Polished slab illustrating the vertical section through an epikarstic cavity in travertine filled with quartz sandstone containing fragments of travertine crust. Photographs were taken by Victor Melezhik from a quarry located near FAR-DEEP Hole 5A

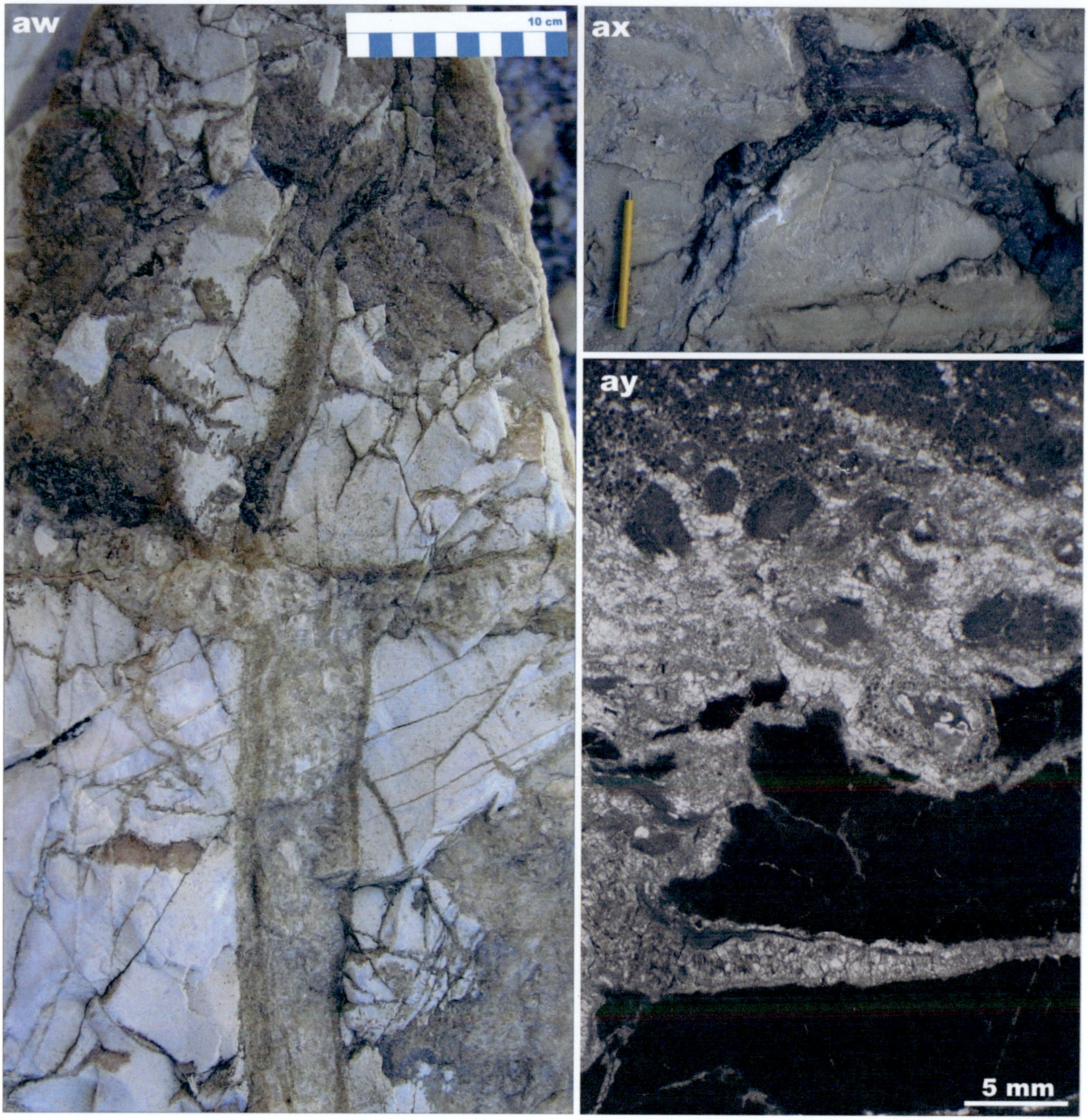

Fig. 4.21 (continued) (**aw**) Cross-section view of intensely brecciated dolostone (*white*) with syndepositional cracks filled with sandstone, gritstone and angular dolostone fragments. (**ax**) Bedding plane exhibiting syndepositional cracks filled with arkosic sandstone. (**ay**) Photomicrograph in transmitted, non-polarised light, showing the cross-section through a dolocrete profile composed of a silicified, nodular zone (*upper part*) developed on the dissolved surface of fractured, iron-stained, non-laminated dolomicrite (*dark brown*). Photographs were taken by Victor Melezhik from a quarry located near FAR-DEEP Hole 5A. Photographs (**a–o, r, s, w–ai, al, at**) are reproduced from Melezhik and Fallick (2005) by permission of the Royal Society of Edinburgh; photograph (**ak**) is reproduced from Melezhik et al. (2005) with permission of Elsevier. Other photographs by Victor Melezhik

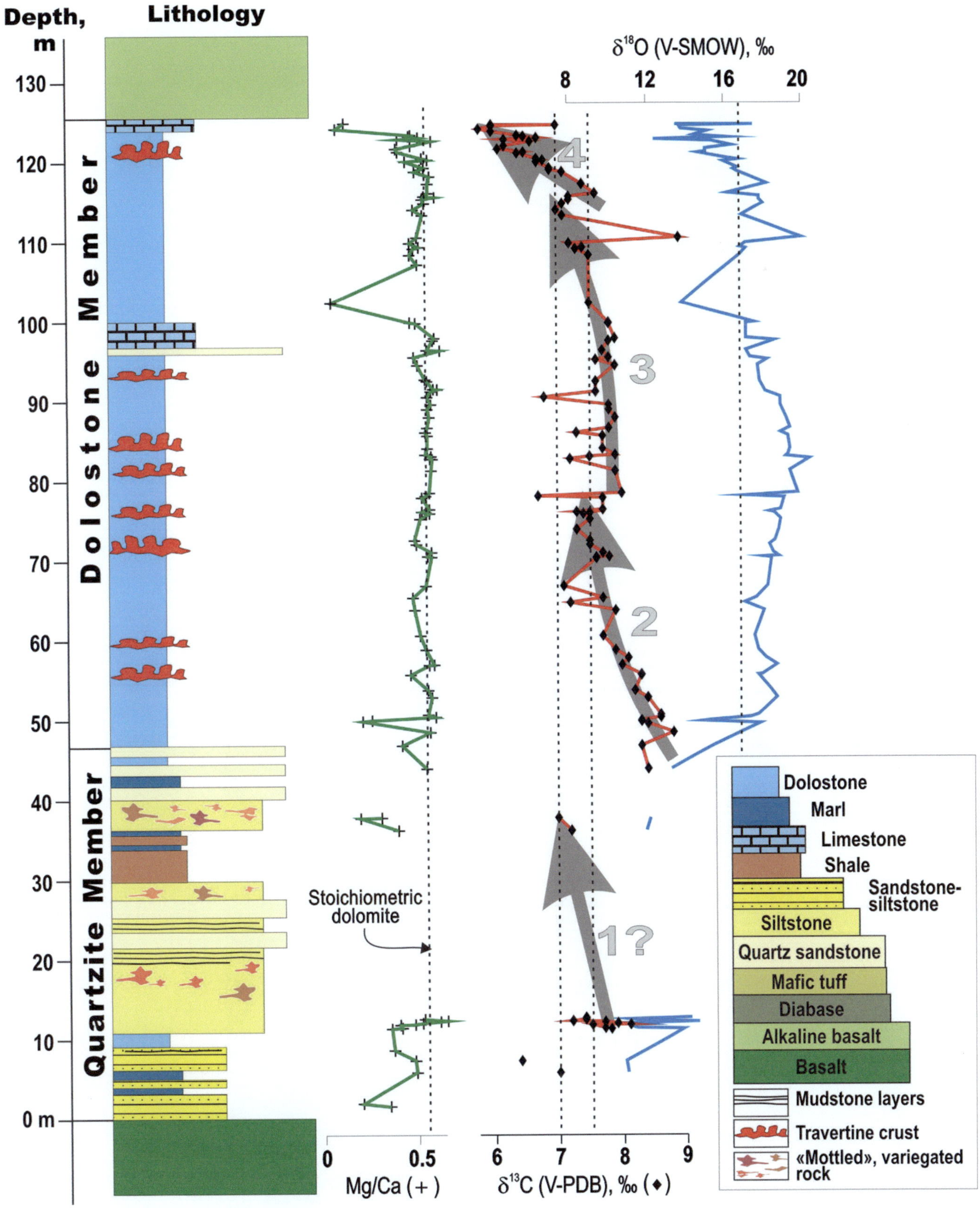

Fig. 4.22 Lithofacies and Mg/Ca, carbon and oxygen isotopic compositions in a section of the carbonate rocks of the Kuetsjärvi Sedimentary Formation; large, *pale grey* numbers against *arrowhead lines* indicate short term, stratigraphic variations of $\delta^{13}C$ (Modified from Melezhik and Fallick 2005)

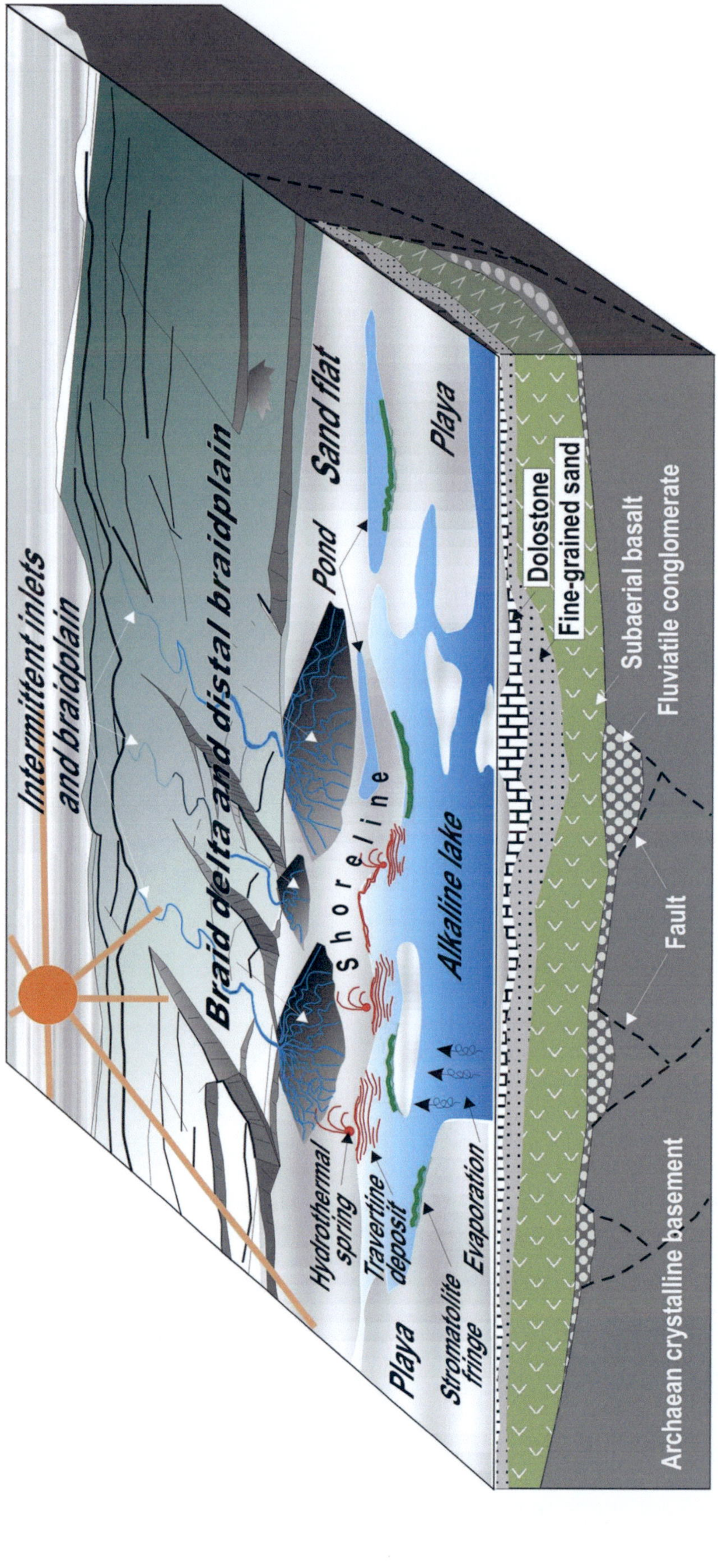

Fig. 4.23 Reconstruction of the Kuetsjärvi rift-related depositional system (Modified from Melezhik and Fallick 2005)

Fig. 4.24 Selected images of volcanic and sedimentary rocks of the Kuetsjärvi Volcanic Formations. (**a**) Amygdaloidal basalt from the base of the formation; core diameter is 5 cm. (**b**) Contact between two basaltic lava flows; the lower lava flow has an amygdale-rich (*pink* adularia) *top* while the *upper* lava flow has fewer, subvertical pipe amygdales. (**c**) Plan-view of haematite-rich, felsic lava with fragments of fluidal lava showing soft-deformation

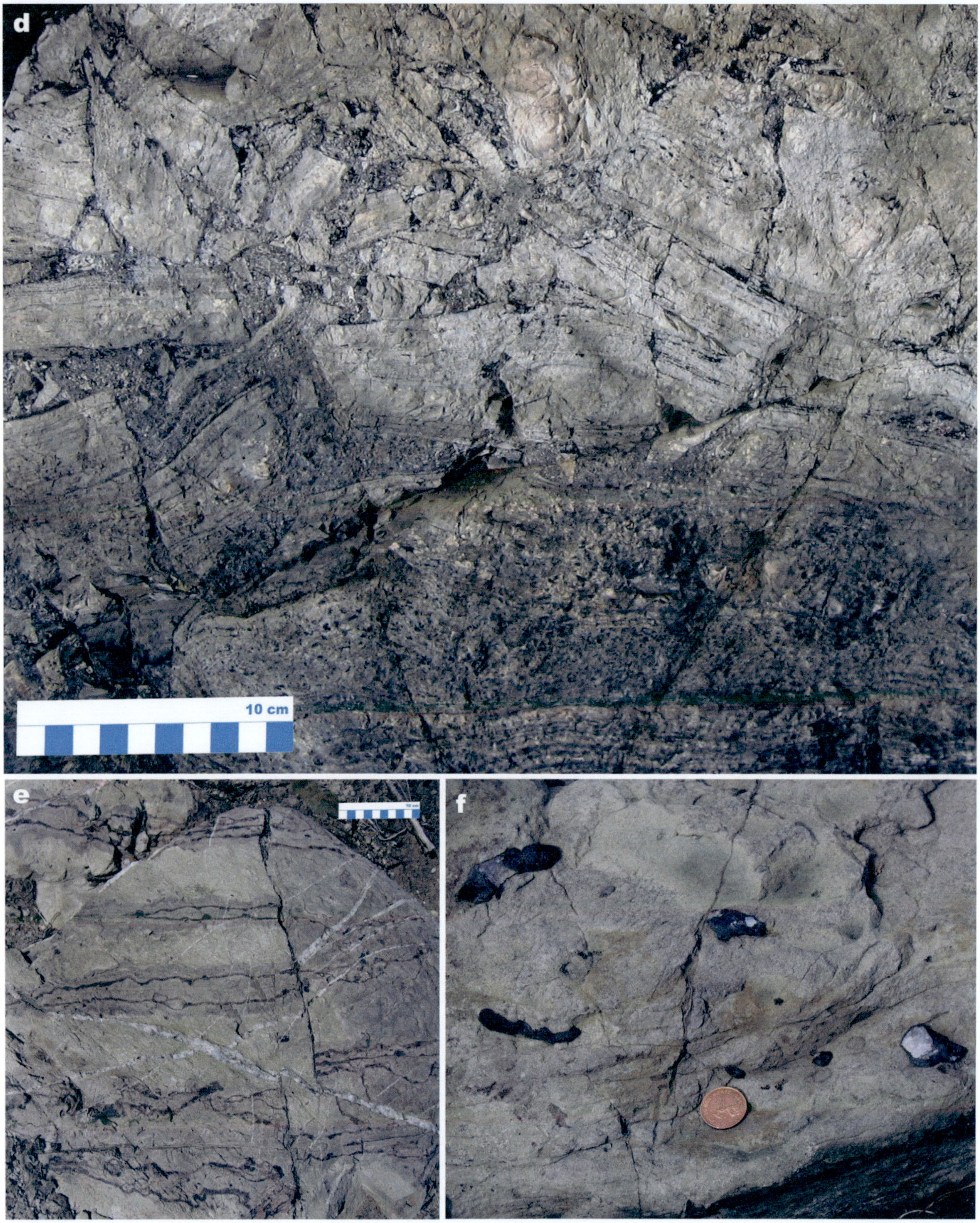

Fig. 4.24 (continued) (**d**) Cross-section view of felsic lava breccia composed of angular fragments embedded in fluidal lava. (**e**) Cross-section view through a series of thin dacitic lava flows; note thin haematite bands developed on either side of each flow contact. (**f**) Fluidal dacitic lava with *black* haematite amygdales; coin diameter is 1.5 cm

Fig. 4.24 (continued) (**g**) Plan view of lava breccia composed of fragments of basaltic lava (*dark grey*) embedded in *pale grey* dacitic lava; glove is c. 15 cm in length

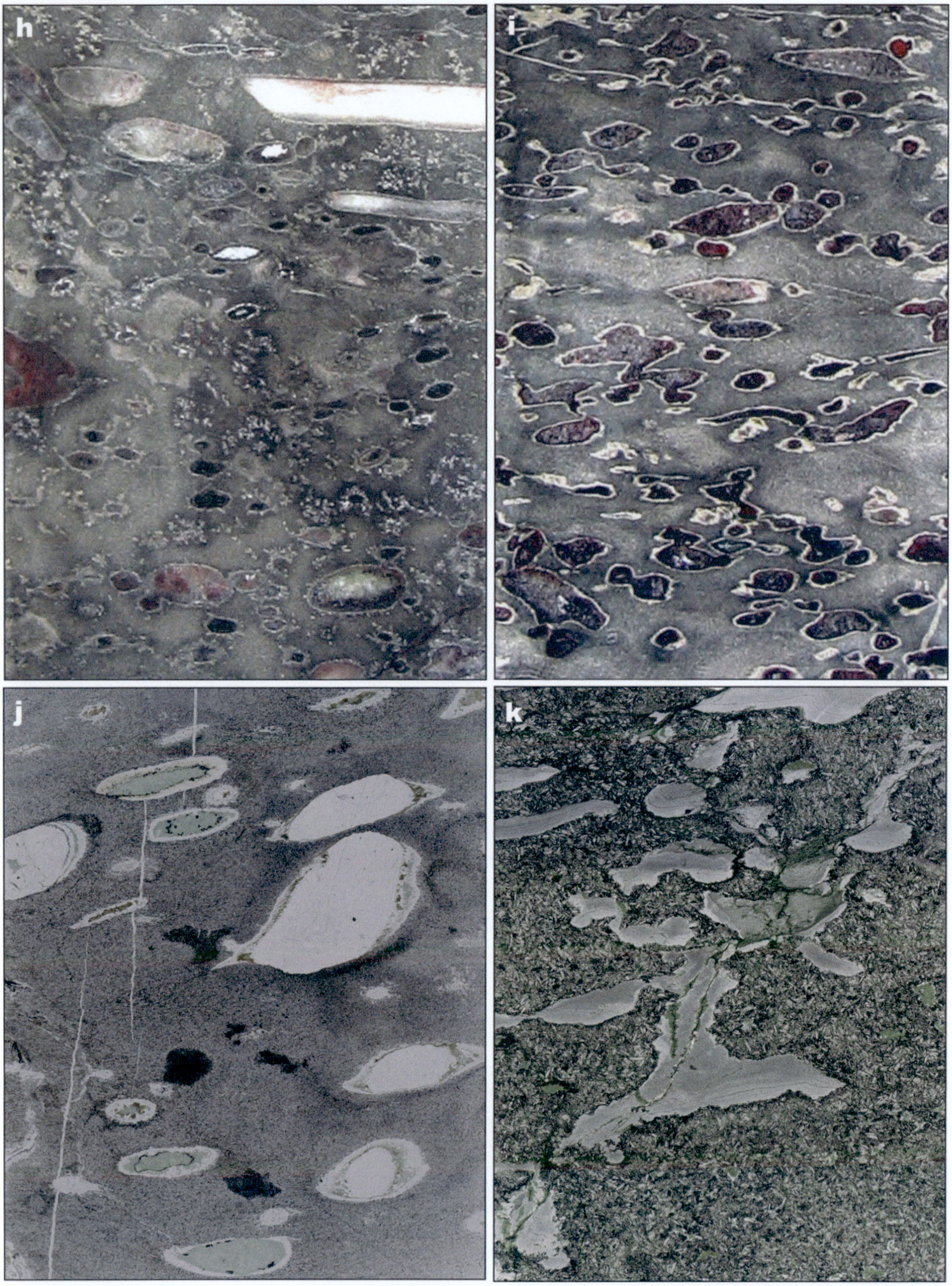

Fig. 4.24 (continued) (**h**) Rhyodacitic lava with abundant amygdales composed of white quartz, *pink* and *dark red* jasper, and dark green chlorite. (**i**) Abundant vesicles filled with *dark-red* jasper rimmed by white quartz in rhyodacitic lava. (**j**) Scanned thin section of rhyodacitic lava containing abundant ovoidal vesicles filled with white quartz, *pale green* chlorite, *brown* biotite and *black* haematite. (**k**) Scanned thin section of amygdaloidal alkali basalt, showing irregular-shaped, zoned vesicles filled with with microcrystalline quartz and chlorite. Core diameter in (**h**) and (**i**) is 5 cm. Width of images (**j**) and (**k**) is 20 cm

Fig. 4.24 (continued) (**l**) Volcaniclastic sandstone-siltstone couplets; note normal grading in sandstone and parallel lamination in siltstone units. (**m**) Cross-section through three beds of volcaniclastic conglomerates; note that although the bed boundaries are well defined (*arrowed*), beds themselves do not exhibit any bedding; compass length is 12 cm. (**n–o**) Volcaniclastic conglomerate consisting of rounded clasts of mafic, andesitic and dacitic lavas; note haematised rims in several clasts of dacitic lava. Core diameter in (**l, n**) is 5 cm

Fig. 4.24 (continued) (**p**) Volcaniclastic conglomerate composed of large, mainly rounded fragments of dacitic lava supported by *black*, haematite-magnetite-rich matrix. (**q**) Black, haematite-magnetite-rich shale overlain by basaltic pillow lava incorporating a large volume of fragments from the underlying shale

Fig. 4.24 (continued) (**r**) Basaltic pillow lava from a single, 10-m-thick flow unit; width of outcrop is 12 m. (**s**) Columnar jointing in a basaltic lava flow overlying the pillow basalt shown in photographs (**q**) and (**r**); width of outcrop is 15 m

Fig. 4.24 (continued) (**t**) Basaltic lava with a large amygdale composed of red adularia, white calcite, green epidote, violet axinite and black haematite-magnetite; core diameter is 5 cm. (**u**) Basaltic lava with a large amygdale filled with violet axinite rimmed with *green* epidote. (**v**) Basalt with vesicles partially filled with quartz showing concentric banding. (**w**) Basaltic lava breccia containing gas tubes filled with *green* epidote and *red* adularia. (**x**) Basaltic lava flow with quartz-filled amygdales which are partially elongated along the flow direction

Fig. 4.24 (continued) (**y**) Dacitic lava breccia with a *black*, amygdaloidal top enriched in oxidised iron oxides. (**z**) Amygdaloidal dacitic lava flow with *dark brown* brecciated top enriched in oxidised iron oxides. (**aa**) Lava breccia with a *green* epidote vein and *black* haematite-magnetite cement. (**ab**) Dacitic lava flow with gas tubes and haematite amygdales (*black*). (**ac**) Basaltic lava flow with the top composed mainly of quartz-filled vesicles

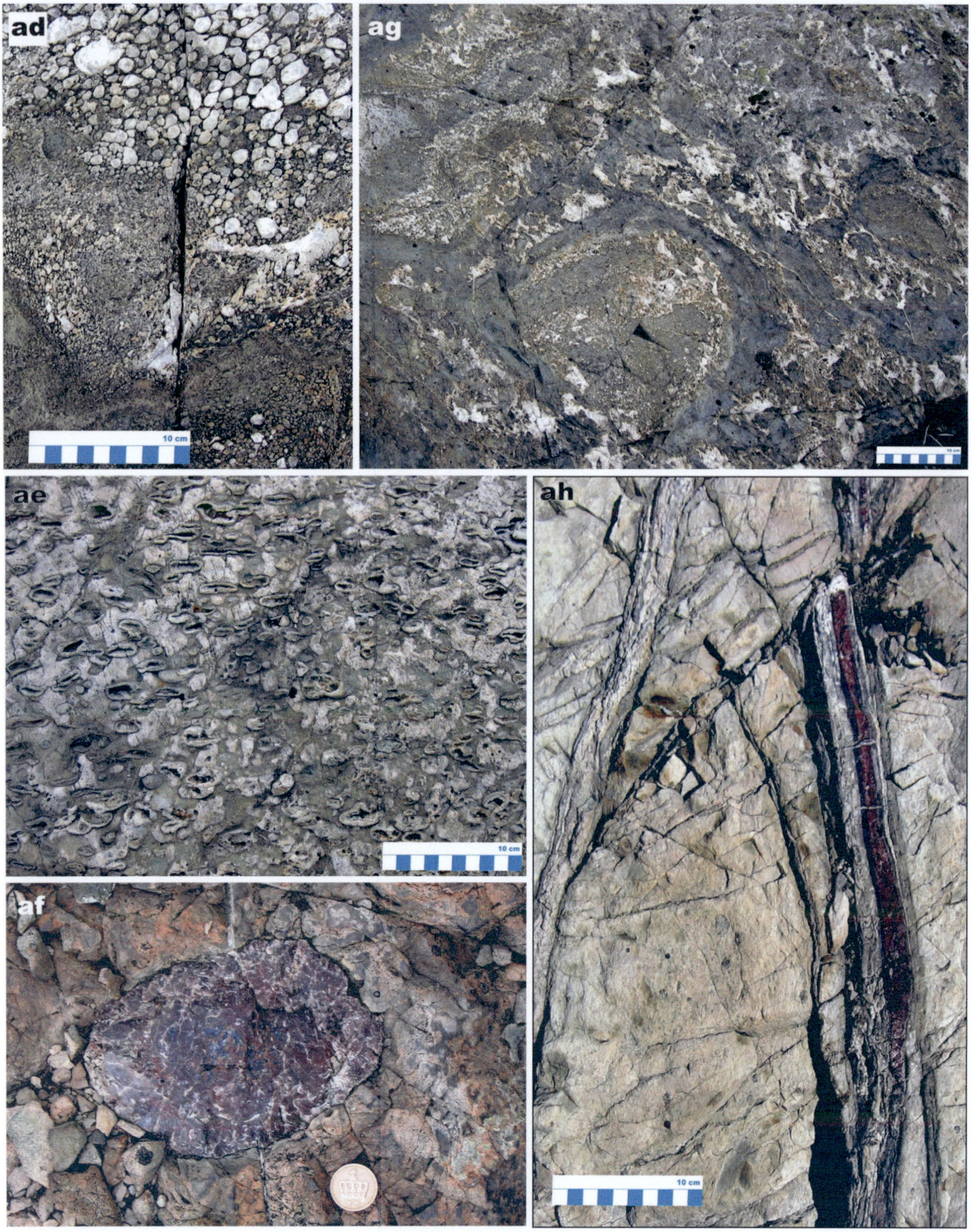

Fig. 4.24 (continued) (**ad**) Basaltic lava flow with a top rich in quartz-filled vesicles. (**ae**) Basaltic lava with vesicles partially filled with quartz, chlorite and calcite. (**af**) A large jasper amygdale in rhyodacite. (**ag**) Incipient pillows in basaltic lava. (**ah**) *Dark red* jasper vein in basaltic lava

Fig. 4.24 (continued) (**ai**) A large quartz amygdale in basaltic lava. (**aj**) A vesicle filled with *red* adularia, white calcite and violet axinite in basaltic lava. (**ak**) Lava beccia containing a rounded clast of weathered-out dolostone (*dark brown*); coin diameter is 1.5 cm. (**al**) Lava breccia immediately below the overlying Kolosjoki Formation; rounded dacite fragments are surrounded by *black*, haematite- and magnetite-rich material. Core diameter in (**ai**) and (**aj**) is 5 cm (Photographs (**a**), (**h–k**), (**l–n**), (**t**) and (**ai–aj**) are taken from FAR-DEEP Core 5A. Photograph (**m**) courtesy of Kauko Laajoki; photographs (**b, j** and **k**) by Eero Hanski; the rest of the photographs by Victor Melezhik)

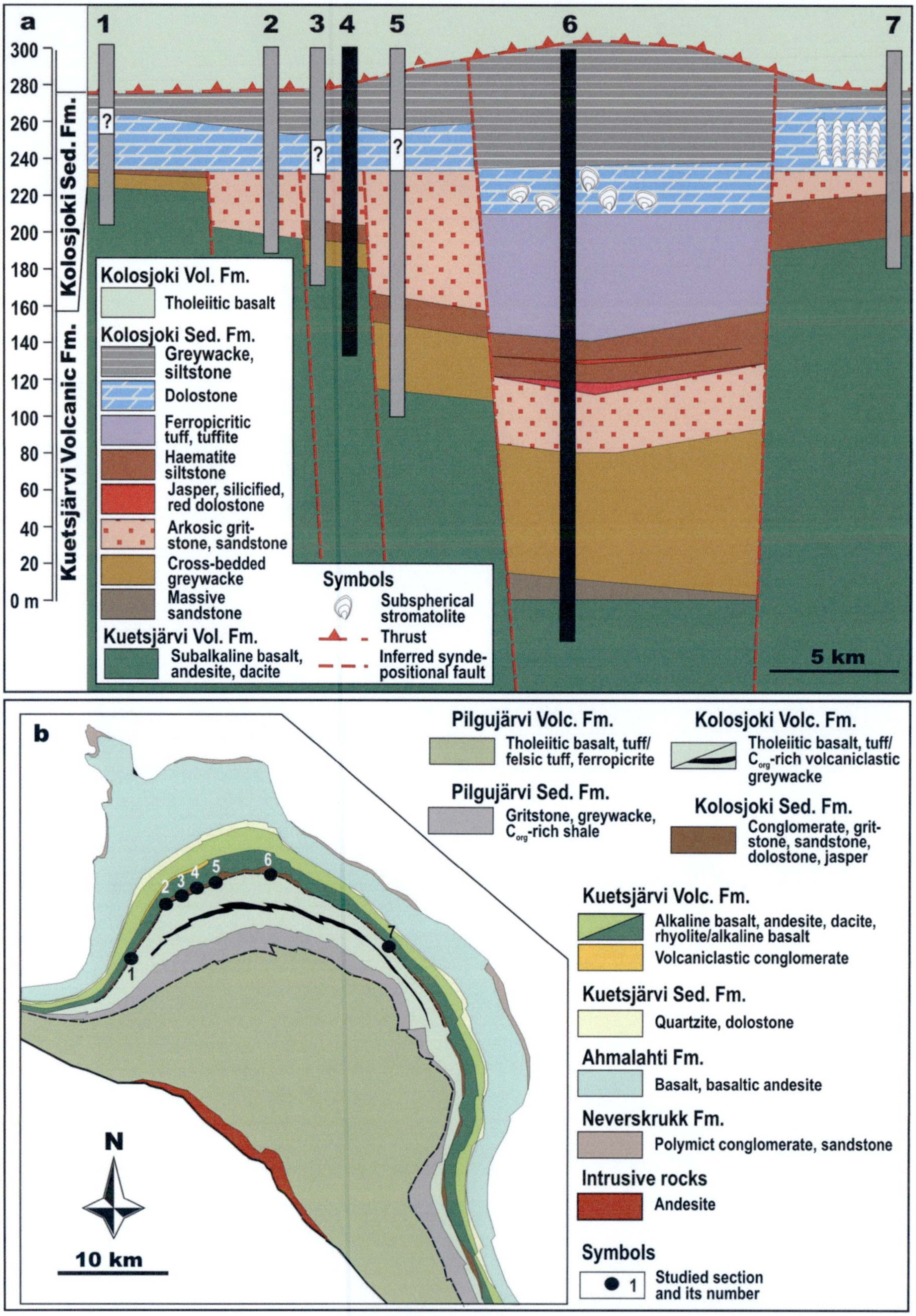

Fig. 4.25 (**a**) Outcrop (*grey bars*)- and drillcore (*black bars*)-based logs illustrating facies variations in the Kolosjoki Sedimentary Formation over a distance of 30 km strike-length. (**b**) Geological sketch showing the positions of the measured sections. Sections 1, 5 and 7 adopted from Predovsky et al. (1974). Section 4 is based on log of exploration drillhole 2812 (V. Melezhik, unpubl. data). Section 6 is based on simplified log of FAR-DEEP Holes 8A and 8B

Fig. 4.26 Selected images of sedimentary rocks of the Kolosjoki Sedimentary Formation. (**a**) Basal Kolosjoki greywacke resting with erosional contact on oxidised lava breccia of the Kuetsjärvi Volcanic Formation. (**b**) Blastomylonite affected by extensional cracks from the upper contact of the Kolosjoki Sedimentary Formation

Fig. 4.26 (continued) (c) Volcaniclastic conglomerate that occurs as a series of small channels within massive greywackes at the base of the Kolosjoki Sedimentary Formation; clasts are fluidal dacitic lava (two large pale grey fragments), red jasper amygdales, and various volcanic rocks derived from the underlying Kuetsjärvi Volcanic Formation. (d) Cross-section view of fine-grained volcaniclastic greywacke with trough cross-bedding. (e) Core image of fine-grained, cross-bedded, volcaniclastic greywacke with dark laminae enriched in haematite; core is 4 cm in diameter

Fig. 4.26 (continued) (**f**) Bedding-parallel view of trough cross-bedded, volcaniclastic greywacke with scattered fragments of black mud rip-up clasts. (**g**) Cross-section view of dark grey, parallel-bedded siltstone and pale grey, tabular cross-bedded, volcaniclastic greywacke; note steeply-dipping cross-beds; cross-beds in the lower unit are overturned; siltstone unit in the middle is effected by slumping. (**h**) Cross-section view of a ball-and-pillow structure expressed by soft-sediment deformed black mudstone in greywacke. Hammer head in (**f**) and (**g**) is 13 cm long

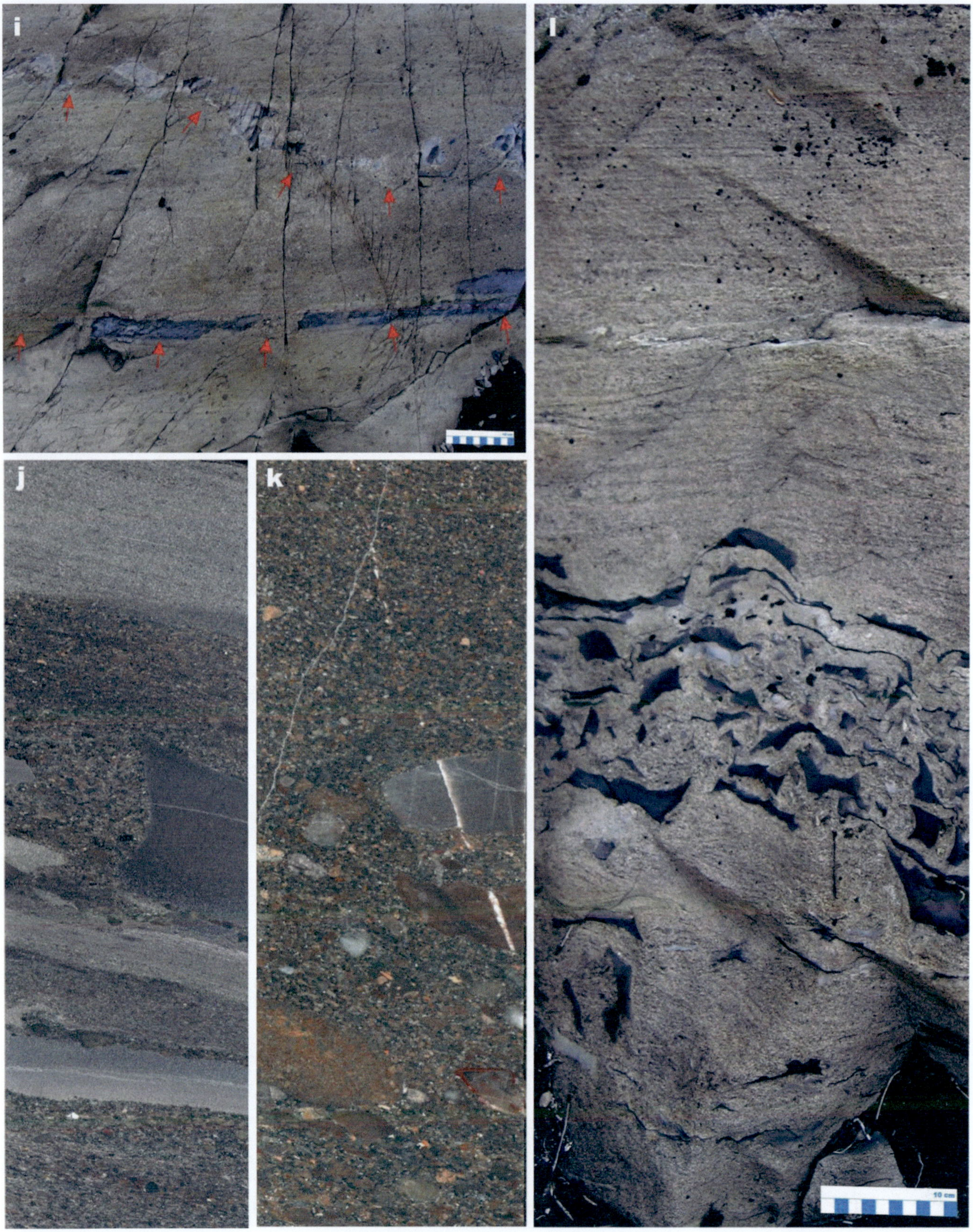

Fig. 4.26 (continued) (**i**) Erosion surfaces (*arrowed*) at the base of cross-bedded cosets; note that the basal parts of cosets contain large, platy fragments of black, haematite-rich mudstone. (**j**) Volcaniclastic sandstone-siltstone; the thickest sandstone bed exhibits normal grading with angular fragments of lava in its lower part. (**k**) Coarse-grained, volcaniclastic sandstone with scattered fragments of dacitic and basaltic lava. (**l**) Cross-section view of desiccated and dismembered, black, haematite-rich mudstone chips in cross-bedded, volcaniclastic greywacke overlain by parallel-bedded, volcaniclastic greywacke. Core diameter in (**j**) and (**k**) is 4 cm

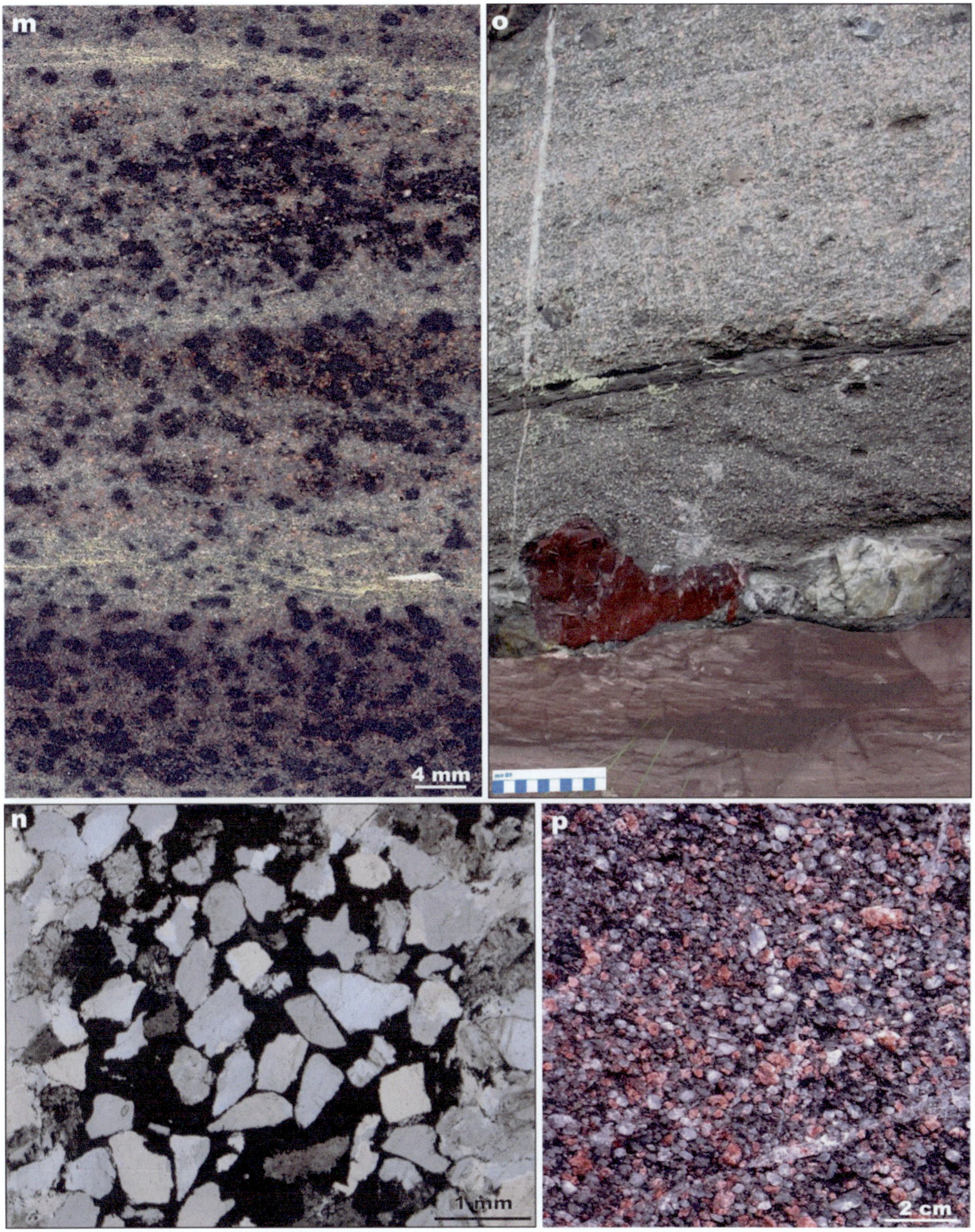

Fig. 4.26 (continued) (**m**) Spotted, cross-bedded greywacke; black spots are due to patchy cementation by haematite. (**n**) Photomicrograph in transmitted, non-polarised light of patchy haematite cement occurring in spotty greywacke shown in (**m**). (**o**) Dark brown, haematite-rich mudstone overlain by arkosic sandstone bed containing large, reworked quartz (*white*) and jasper (*red*) amygdales at the base, and thin mudstone drape on top; this is followed by pale grey, massive arkosic sandstone. (**p**) Close-up view of arkosic sandstone overlying dark brown mudstone in (**o**)

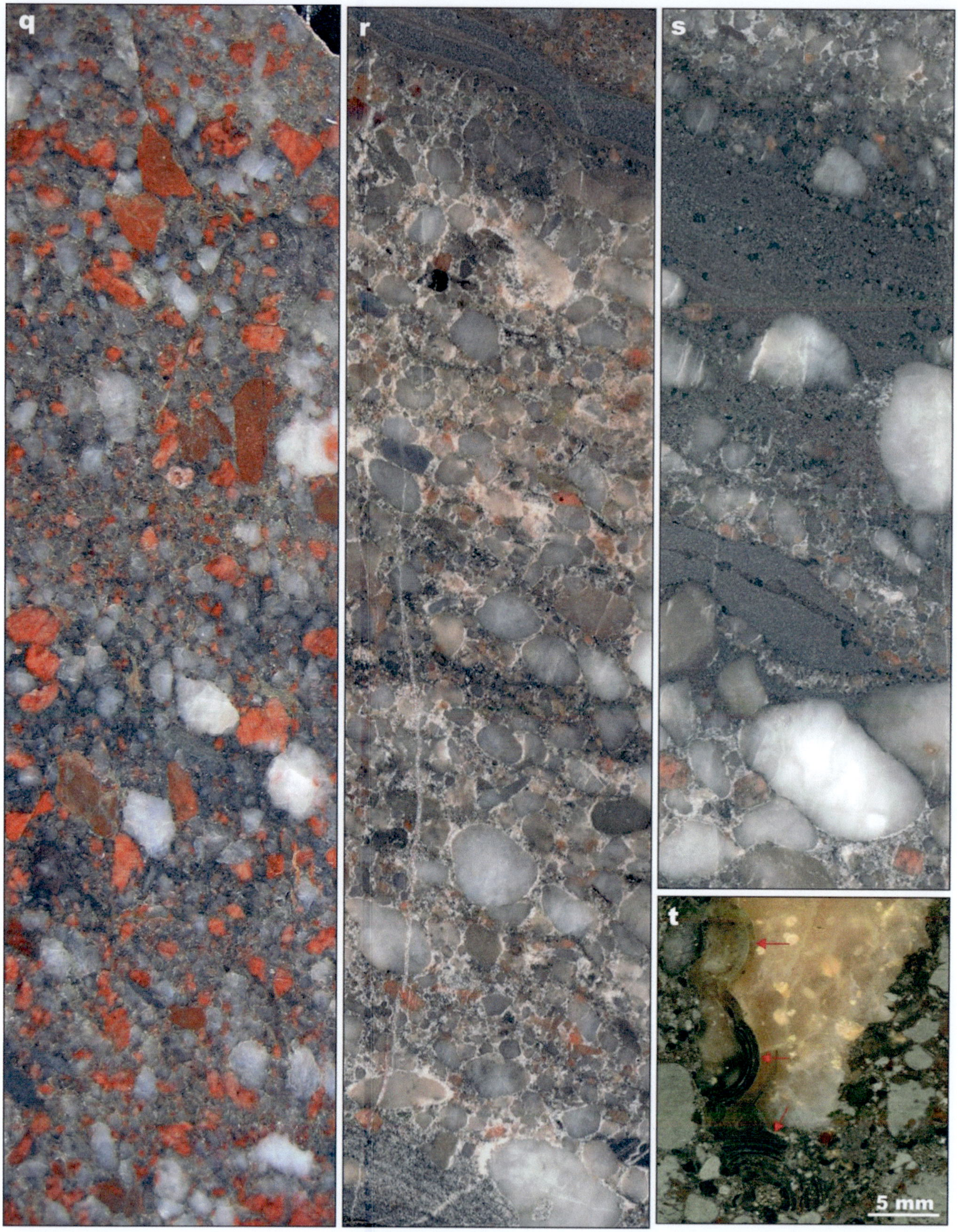

Fig. 4.26 (continued) (**q**) Massive arkosic gritstone composed of clasts of microcline (*red*), dacitic lava (light brown) and quartz (*white*, *pale grey*). (**r**) Fine-pebble conglomerate draped by haematite-rich sandstone; well-rounded and sorted clasts are mainly volcanic rocks and quartz amygdales reworked from the Kuetsjarvi Volcanic Formation rocks. (**s**) Volcaniclastic conglomerate-gritstone-sandstone with reworked, white quartz amygdales (**t**) Close-up view of reworked, *pale yellow* quartz amygdale with concentric banding along its left margin (*arrowed*); bright grey, grounded clasts are haematite. Core diameter in (**q**) is 5 cm and in (**r–s**) 4 cm

Fig. 4.26 (continued) (**u**) Rounded and angular haematite clasts at the base of a sandstone bed. (**v**) Fine-pebble, haematite conglomerate grading into haematite sandstone, which is overlain by arkosic sandstone composed of rounded quartz fragments (*black*) and haematite particles (*bright grey*); the *grey yellow* clast is a reworked quartz amygdale. (**w**) Beds of clastic haematite in sandstone; pen length is 12 cm

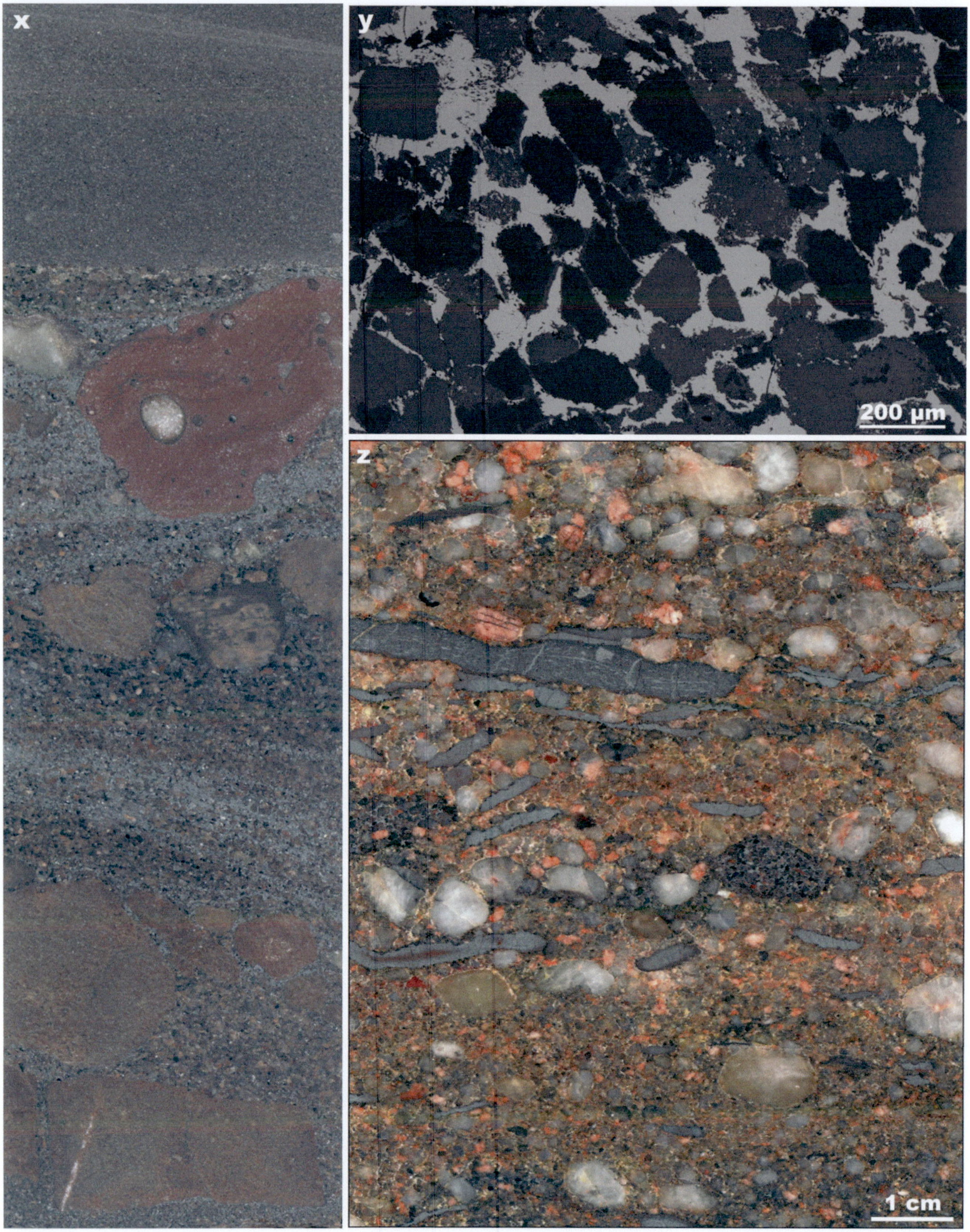

Fig. 4.26 (continued) (**x**) Cross-bedded sandstone with scattered fragments of fluidal and amygdaloidal dacitic lava; bright grey particles in the sandstone are haematite grains. (**y**) Back-scattered electron image of arkosic sandstone; quartz (*black*) and microcline (*pale grey*) are cemented by haematite (*bright*). (**z**) Volcaniclastic conglomerate-gritstone containing platy fragments of haematite "ore" (*grey*), reworked quartz amygdales (*white*, *pale yellow*), microcline and altered dacitic lava (*red*, *pale brown*)

Fig. 4.26 (continued) (**aa**) Bedding surface of a haematite bed exhibiting an "elephant-skin" texture. (**ab**) Scanned polished section showing sandstone passing into parallel-laminated, haematitic siltstone; rounded clasts in the sandstone are quartz (*grey*) and microcline (*red*). (**ac**) Sandstone grading into *brown* siltstone with a rippled surface covered by parallel-laminated, haematitic siltstone

Fig. 4.26 (continued) (**ad**) Back-scattered electron image of graded sandstone with clay-haematite clasts; note that many haematite platy fragments (*bright*) are interlaminated with clay laminae (*black*)

Fig. 4.26 (continued) (**ae**) Greywacke (*bottom*) overlain by chert breccia passing into jasperised dolostone; core diameter is 5 cm. (**af**) Polished slab of jasper with a colloform structure. (**ag**) Parallel-bedded, haematite-rich siltstone. Core diameter in (**ae**) and (**af**) is 5 cm

Fig. 4.26 (continued) (**ah**) A series of jasper sandstone grading into brown siltstone intersected by jasper veinlets. (**ai**) Tuff composed of unsorted clasts of intensively altered ferropicritic lava in psammitic ferropicritic tuff matrix; core diameter is 5 cm. (**aj**) Cross-section view of grey, cross-bedded greywacke interbedded with dolostone breccias, in which angular clasts are mainly dolostone (*pale yellow*) and chert (*white*); the grewacke contains clasts of chert (*white*, *pale pink*) and dolostone (*pale yellow*)

Fig. 4.26 (continued) (**ak**) Polished slab of breccia composed of dolostone (*pale yellow* and *pink*) and chert (*pale grey*) clasts in a greywacke matrix. (**al**) Pink dolostone with wavy bedding; core diameter is 5 cm. (**am**) Dolostone (*pale yellow*) interbedded with chert (*bright*). Photograph (**o**) by Eero Hanski, the rest by Victor Melezhik. Cores (**e**), (**j**), (**k**), (**r**), (**s**), (**u**) and (**x**) are from drillhole 2812 (Sect. 4 in Fig. 4.24); cores (**q**) and (**ae**) and (**ag–ai**) are from FAR-DEEP Hole 8B

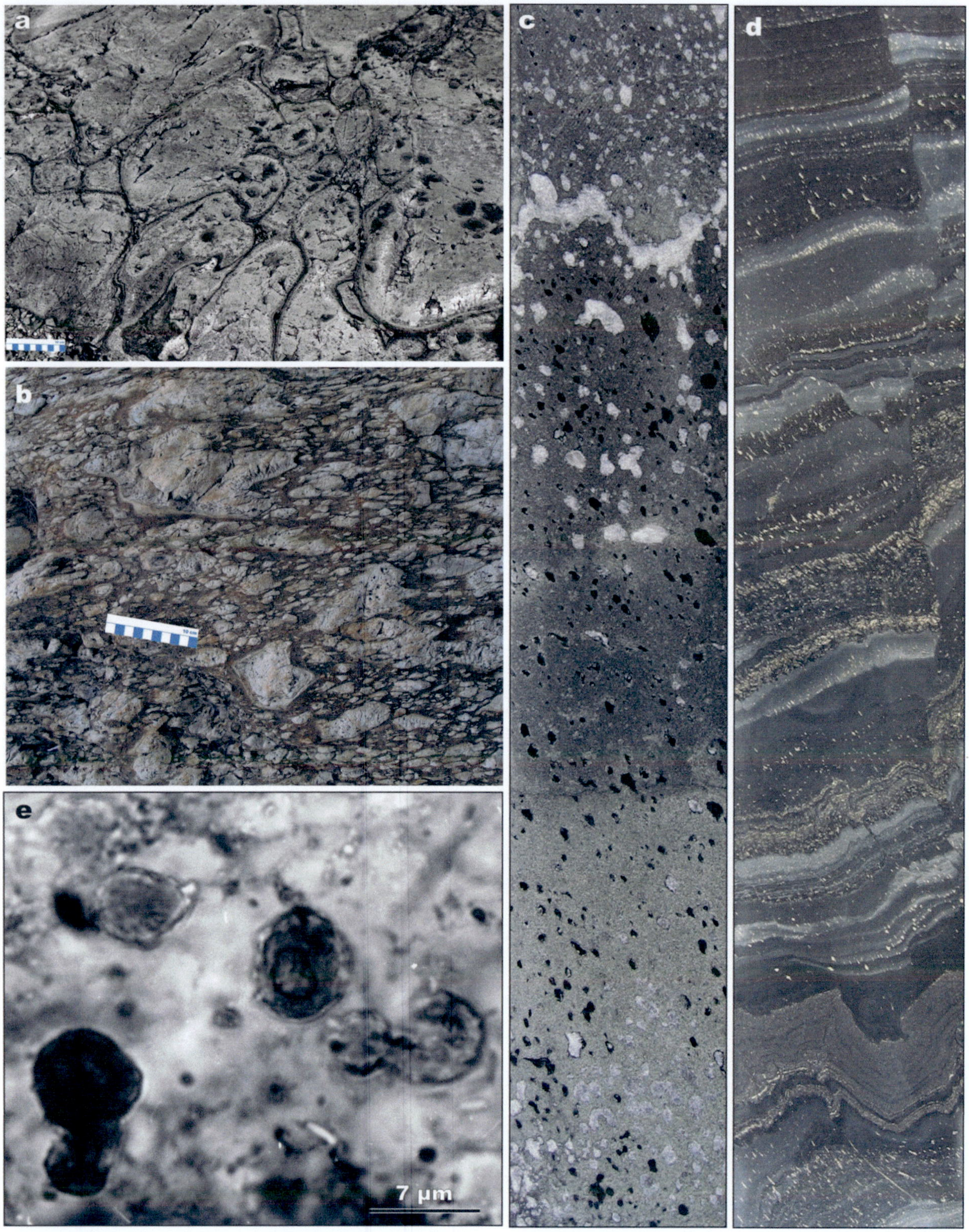

Fig. 4.27 Images of rocks of the Kolosjoki Volcanic Formation. (**a**) Pillowed tholeiitic basalt forming an essential part of the formation. (**b**) Lava breccia of tholeiitic basalt; some clasts exhibit chilled margins. (**c**) Two amygdaloidal, basaltic lava flows separated by an uneven boundary emphasised by formation of white calcite; the *upper part* of the *lower* flow represents a chilled margin and exhibits a darker colour; amygdales in both flows are filled by calcite (*white*) and chlorite (dark-coloured). (**d**) Laminated, sulphide-rich, volcaniclastic greywacke and mafic tuff comprising a thick unit in the middle part of c. 2,000-m-thick pile of pillowed tholeiitic basalts; some beds show normal grading and were apparently deposited from turbidity currents; primary bedding has been modified by post-depositional folding and faulting. (**e**) Chert with dark-coloured, zoned, spherical objects interpreted as coccoidal microfossils. (**c**) and (**d**) are from FAR-DEEP Core 9A; core diameter is 5 cm (Photograph (**b**) by Eero Hanski, the rest by Victor Melezhik)

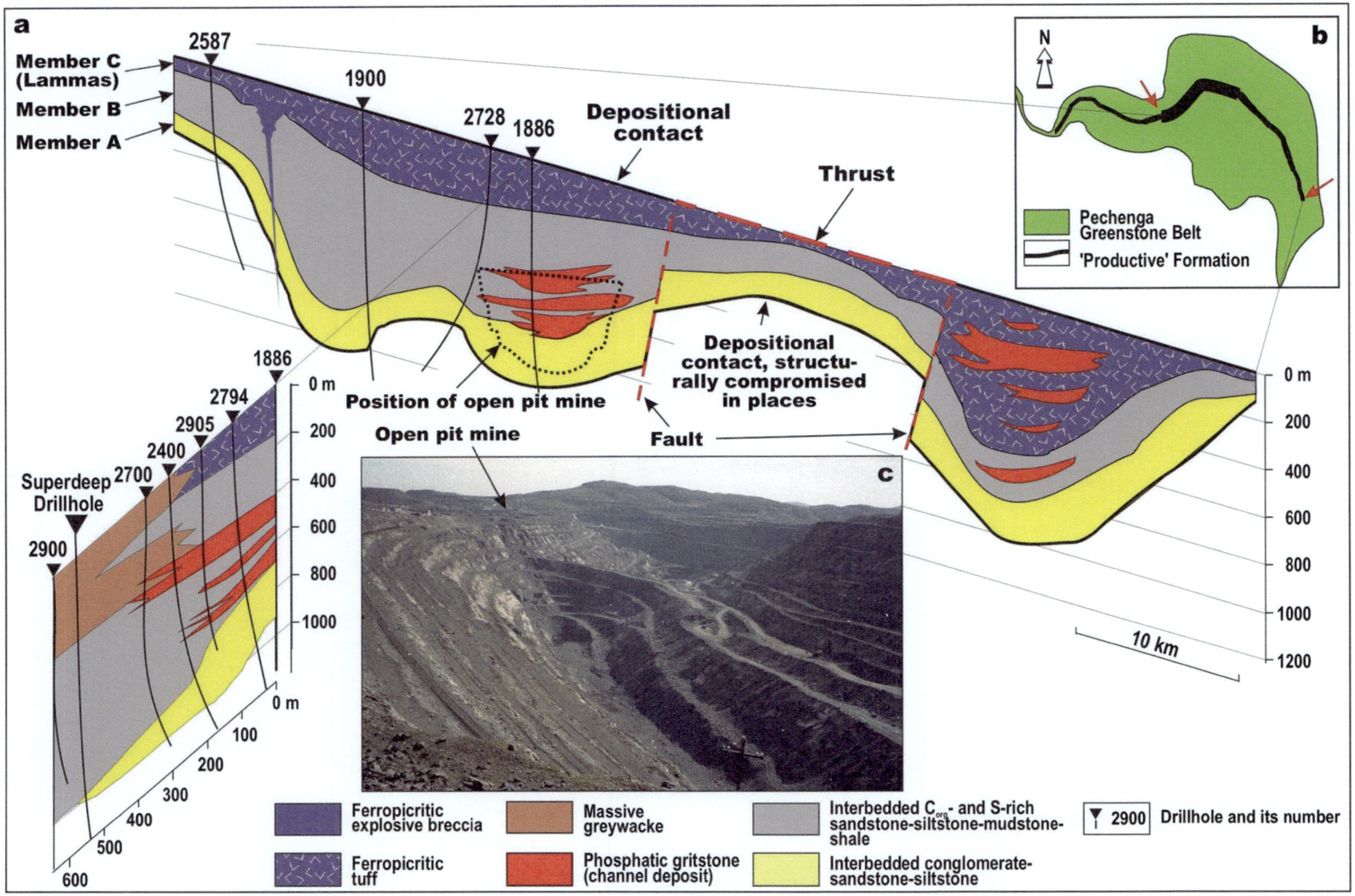

Fig. 4.28 Lithostratigraphy of the Pilgujärvi Sedimentary Formation. (**a**) Two-dimensional section of the formation (Modified after Melezhik et al. 1998). (**b**) Position of the section within the Pechenga Greenstone Belt. (**c**) Panoramic view of the Zdanov open pit Ni-Cu mine exposing an essential part of Members A and B and strata containing gritstones with phosphorite clasts. Note that contacts of the Pilgujärvi Sedimentary Formation are structurally compromised in places

Fig. 4.29 Selected images of sedimentary and volcanic rocks of the Pilgujärvi Sedimentary Formation. *Member A:* (**a**) Interbedded, dark-coloured siltstone-mudstone followed by a thick, graded, arkosic sandstone bed overlain by siltstone-sandstone with lenticular bedding. (**b**) Sandstone-siltstone rhythmites showing thinning- and fining-upward sequences. (**c**) Sandstone-siltstone rhythmites showing thickening- and coarsening-upward sequences

Fig. 4.29 (continued) (**d**) A series of cross-bedded sandstone beds with erosive and graded bases capped by black mudstone. (**e**) Cross-bedded sandstone with erosive base in laminated sandstone-siltstone. (**f**) Bedded sandstone mudstone at the base followed by a series of parallel-bedded, graded sandstone beds. (**g**) Dismembered and soft-sediment-deformed black mudstones beds in massive greywacke. All core diameters are 6 cm

Fig. 4.29 (continued) *Member B:* (**h**) Sandstone-siltstone rhythmites with several layers of massive pyrrhotite (*pale yellow*). (**i**) Conglomerate composed of rounded, unsorted clasts of black, massive mudstone supported by greywacke matrix. (**j**) Cross-bedded sandstone in sandstone-siltstone rhythmites. (**k**) Sandstone grading into mudstone showing rhythmic bedding; note that mudstone units are affected by penetrative cleavage; several layers are sulphidised (*brown*). (**l**) Siltstone-mudstone rhythmites with several sulphidised layers (*bright*). Core diameter in (**i**) and (**j**) is 6 cm

Fig. 4.29 (continued) (**m**) Thick sandstone-mudstone rhythmites forming a thinning-upward cycle; note that sandstone beds are much thicker and have internal division, and both sandstone and mudstone beds exhibit fine, horizontal lamination. (**n**) Sandstone-mudstone rhythmites; sandstone units are either massive or faintly graded. Core diameter in (**n**) is 6 cm

Fig. 4.29 (continued) (**o**) Intensively sulphidised, parallel-laminated mudstone. (**p**) Sulphidised, parallel-laminated, tectonically folded mudstone erosively overlain by cross-bedded sandstone followed by parallel-laminated sandstone-siltstone-mudstone. Core diameter in (**p**) is 6 cm

Fig. 4.29 (continued) (**q, r, s**) Thinly-laminated sandstone-siltstone with rare mudstone units; some light-coloured sandstone beds have an erosive base. (**t**) Rhythmically- and parallel-bedded siltstone-mudstone; siltstone exhibits faint grading. Core diameter is 6 cm

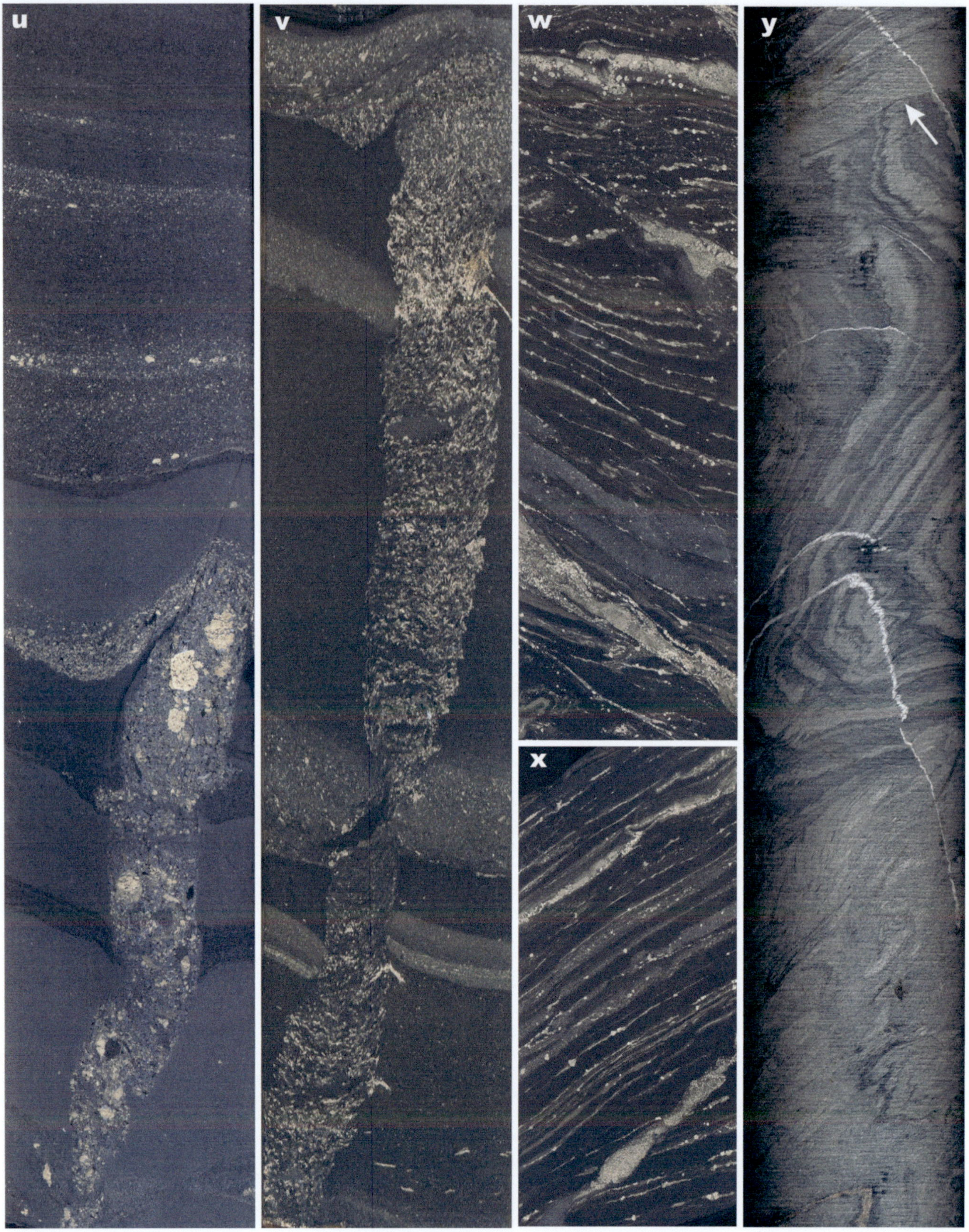

Fig. 4.29 (continued) (**u, v**) Sedimentary dykes injected upwards through thickly bedded sandstone-mudstone rhythmites; dykes are composed of unsorted mudstone-sandstone clasts and fragments of pyrite concretions and pyrite layers. (**w, x**) Sedimentary dykes largely composed of pyrite ("pyrite dykes") injected at oblique angles to the bedding through black, C_{org}- and pyrite-rich siltstone-mudstone. (**y**) Syn-sedimentary, soft-sediment deformed sandstone-mudstone rhythmites; note that the folded unit is erosively overlain by a non-deformed sandstone bed (arrowed). Core diameter is 6 cm

Fig. 4.29 (continued) (**z**) Interbedded, massive greywacke and calcite-cemented breccia capped by black mudstone bed on top; fragments in breccia beds are mainly angular, soft-sediment-deformed, black mudstone and greywacke. (**aa**) Rhythmic interbedding of massive sandstone and breccia composed of dismembered and soft-sediment-deformed mudstone and greywacke cemented by calcite. (**ab**) Breccia with thin greywacke interlayers; clasts are soft-sediment-deformed mudstone and greywacke cemented by calcite. (**ac**) Clast-supported breccia composed of soft-sediment-deformed mudstone and greywacke clasts with thin, softly-deformed greywacke interlayers. Core diameter is 6 cm

Fig. 4.29 (continued) (**ad–ag**) Dismembered and soft-sediment-deformed mudstone beds and layers embedded in massive greywacke forming an essential part of Member B. Core diameter is 6 cm

Fig. 4.29 (continued) (**ah**) Massive gritstone composed of rounded clasts of phosphorites (*black*), pyrite and pyrrhotite (*pale yellow*); pale grey and grey clasts are shale, greywacke and limestone

Fig. 4.29 (continued) (**ai**) Gritstone beds with clasts of phosphorites and sulphides (*arrowed*) in horizontally bedded sandstone-mudstone; all sandstone beds are pale brown due to a high content of iron sulphides. (**aj**) Close-up view of phosphatic gritstone (at the base) overlain by horizontally laminated sandstone-mudstone and massive sulphide layers (*pale brown*). (**ak**) Close-up view of phosphatic gritstone composed of black, rounded clasts of phosphorites and rounded and angular clasts of pyrite and pyrrhotite (*bright*)

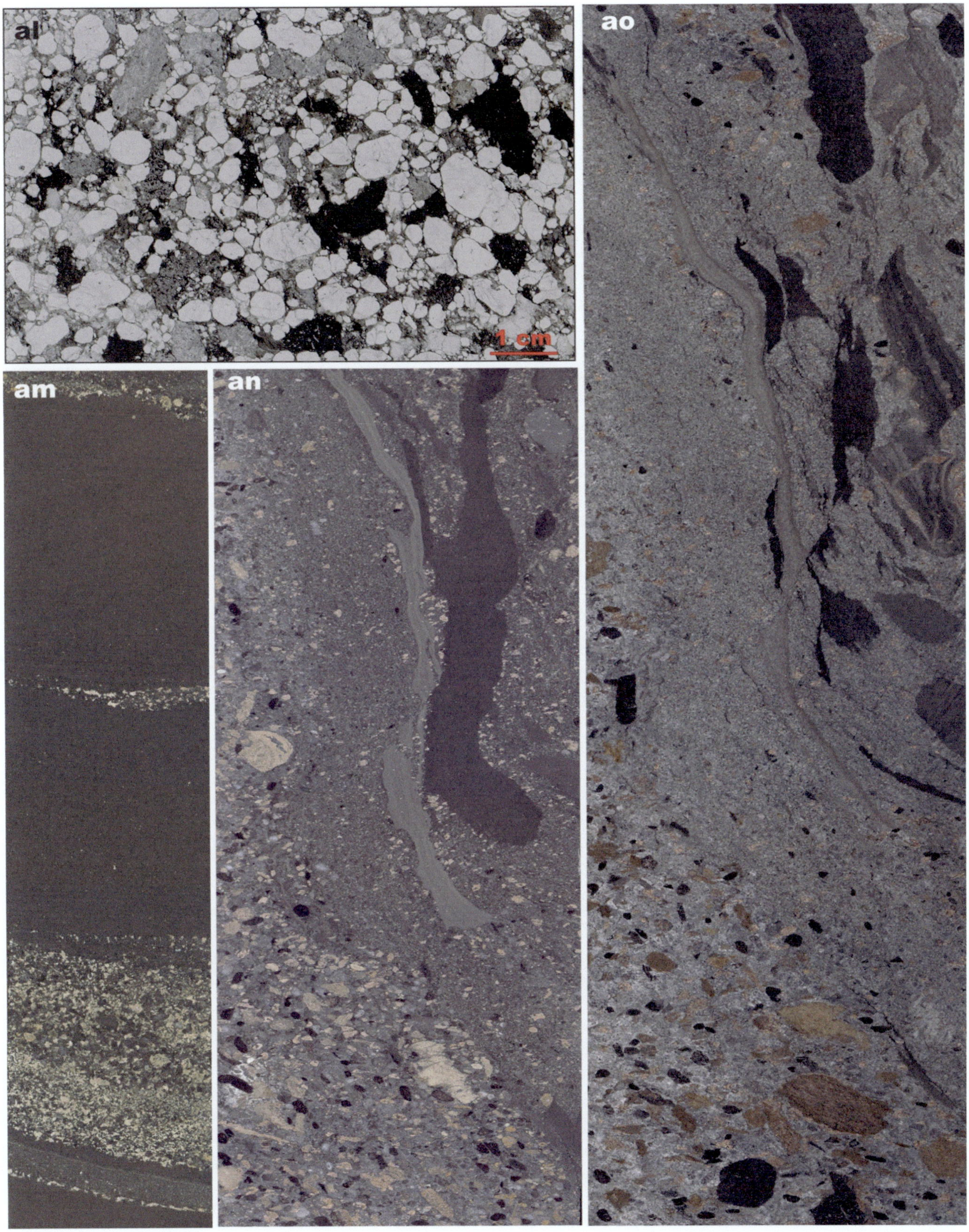

Fig. 4.29 (continued) (**al**) Scanned thin section of phosphatic sandstone composed of rounded clasts of quartz (*white*), iron sulphides and phosphorites (*black*), limestones (*grey*) and shales (*dark grey*). (**am**) Beds and lenses of phosphatic sandstone (*bright*) interbedded with black, massive mudstone. (**an, ao**) Phosphatic gritstone grading into greywacke with soft-sediment-deformed fragments of shale (*grey*) and mudstone (*black*); black, rounded clasts in gritstone are phosphorite and bright and pale brown ones represent iron sulphides. Core diameter in (**am–ao**) is 6 cm

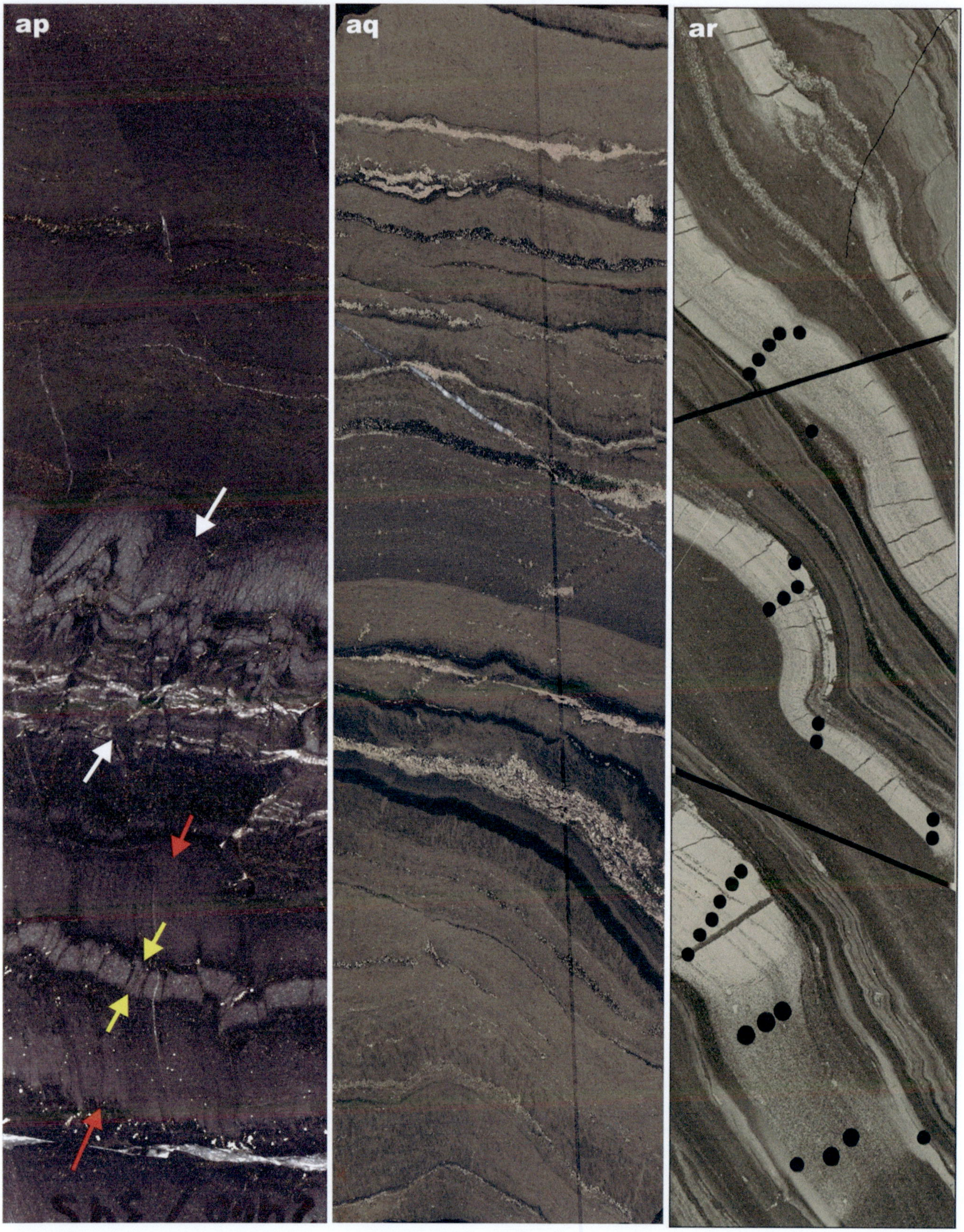

Fig. 4.29 (continued) (**ap**) C_{org}-bearing, low-^{13}C, diagenetic limestone (*arrowed*) in C_{org}-rich mudstone; note that the red-white-arrowed bed is soft-sediment-deformed; whereas the red- and yellow-arrowed bed is zoned with a brighter internal layer (*yellow-arrowed*). (**aq**) Pyrite-rich layers (*brownish*) with thin, black, C_{org}-rich mudstone interlayers and cross-cutting pyrite veinlets (*bright*). (**ar**) Rhythmic interbedding of massive pyrite layers and laminated, pyrite- and C_{org}-rich siltstone-mudstone. Core diameter in (**ap–ar**) is 6 cm

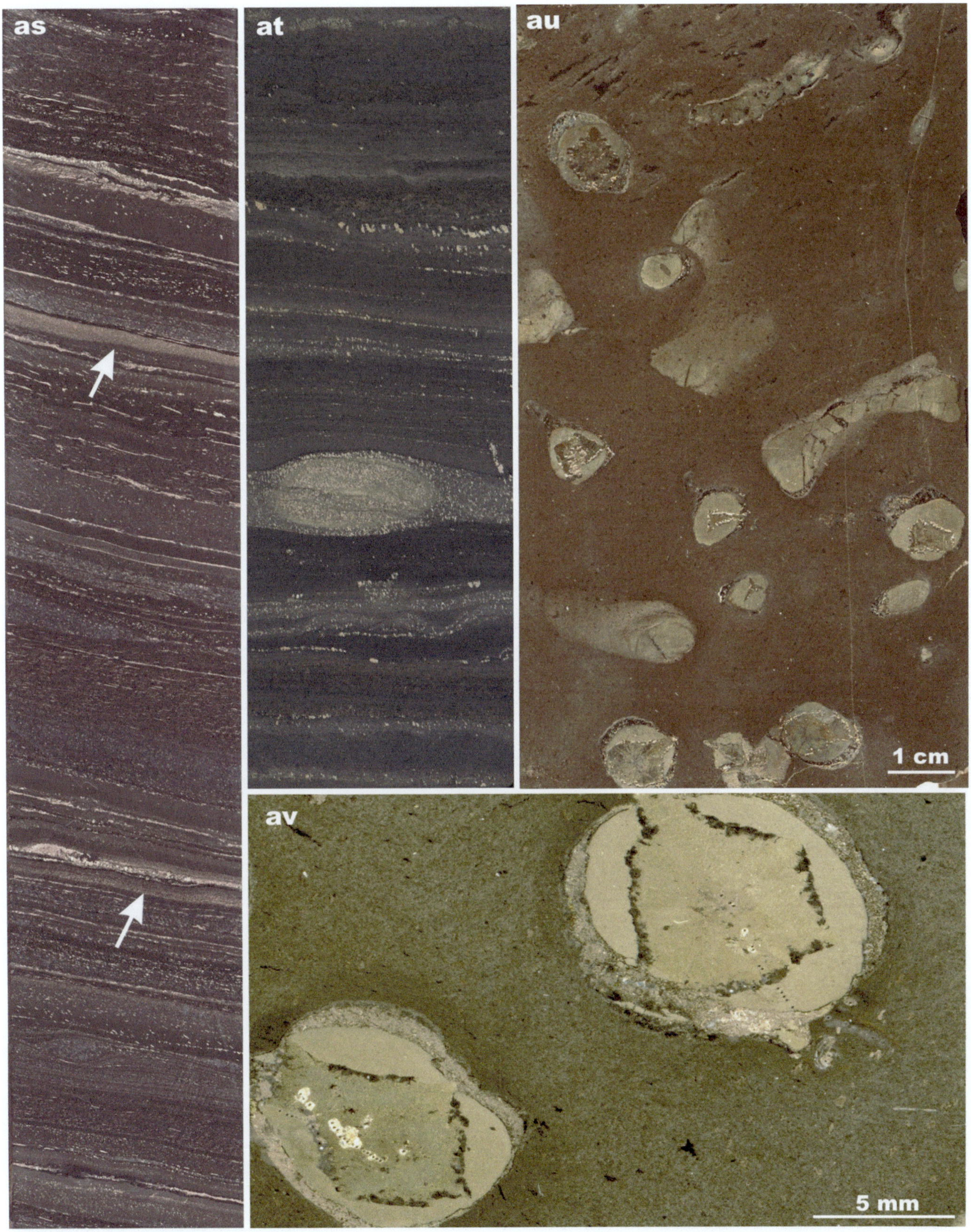

Fig. 4.29 (continued) (**as**) Thinly laminated mudstone with layers of massive pyrrhotite; the thickest layers are arrowed. (**at**) Black, C_{org}-rich, bedded mudstone with an early diagenetic pyrite concretion retaining cross-lamination of the host greywacke bed. (**au**) Zoned pyrite concretions in a massive, C_{org}-rich mudstone. (**av**) Close-up view of zoned pyrite concretions with metamorphic pyrrhotite overgrowth; note the vague septarian core surrounded by thin, black, discontinued, C_{org}-rich band. Drillcore diameter in (**as, at**) is 6 cm

Fig. 4.29 (continued) *Member C:* (**aw**) Crudely bedded, ferropicritic tuff overlain by a black mudstone bed with a soft-sediment-deformed and dismembered pyrite layer followed by a thin bed of ferropicritic tuff and thinly laminated, C_{org}-rich mudstone

Fig. 4.29 (continued) (**ax, ay**) Breccia containing angular, unsorted clasts of various sedimentary rocks including large clasts of rhythmically bedded sandstone-mudstone (**ax**) containing cracked pyrite concretions (**ay**). (**az**) Massive greywacke beds containing scattered clasts and having indistinct boundaries (*arrowed*). Core diameter in (**ax-az**) is 6 cm

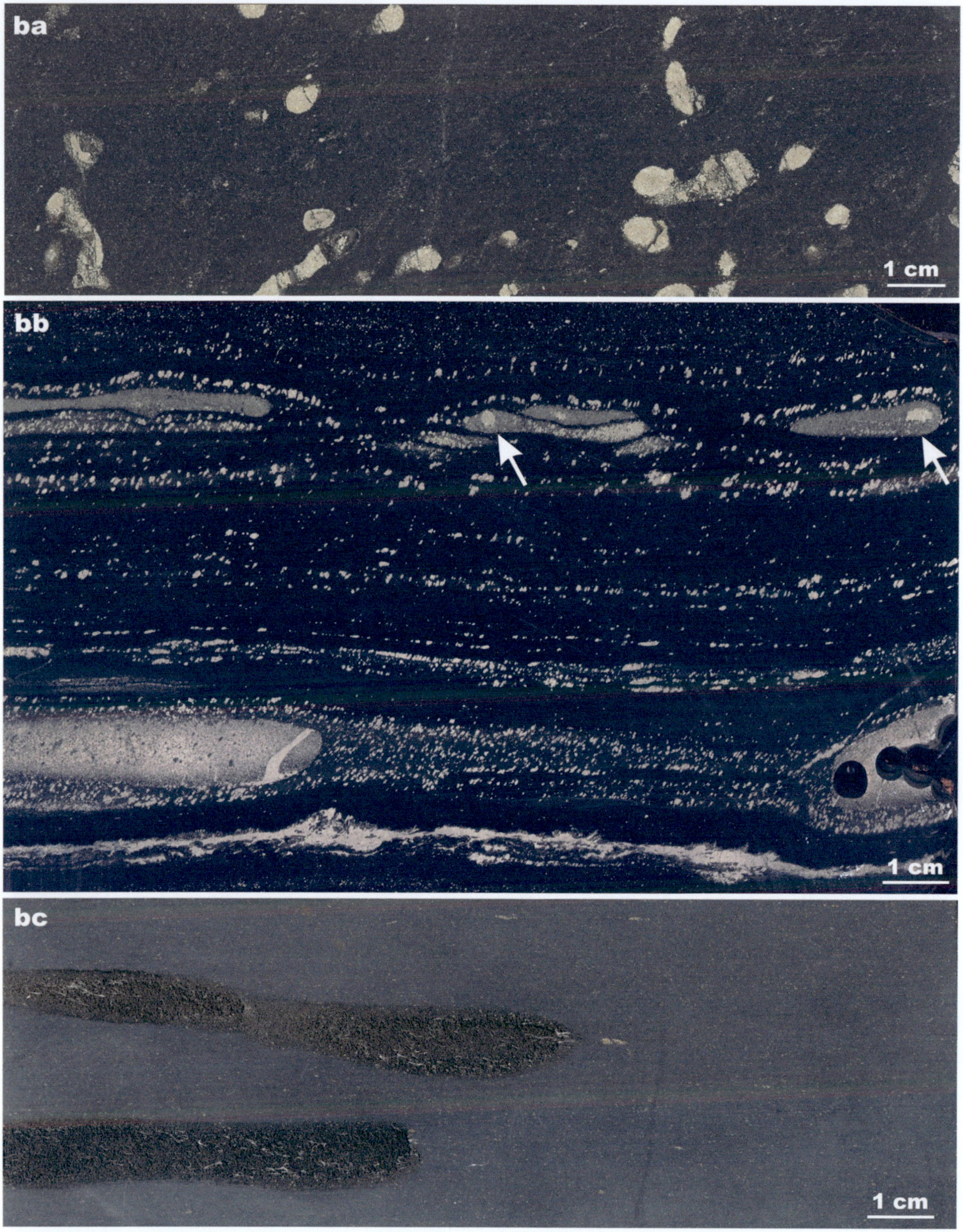

Fig. 4.29 (continued) (**ba**) Pyrite concretions in massive greywacke (**bb**) Pyrrhotite concretions and disseminated pyrrhotite in bedded mudstone; note that the arrowed pyrrhotite concretions have a pyrite core (*bright*). (**bc**) Dark grey, C_{org}-bearing, lensoidal calcite concretions in massive greywacke

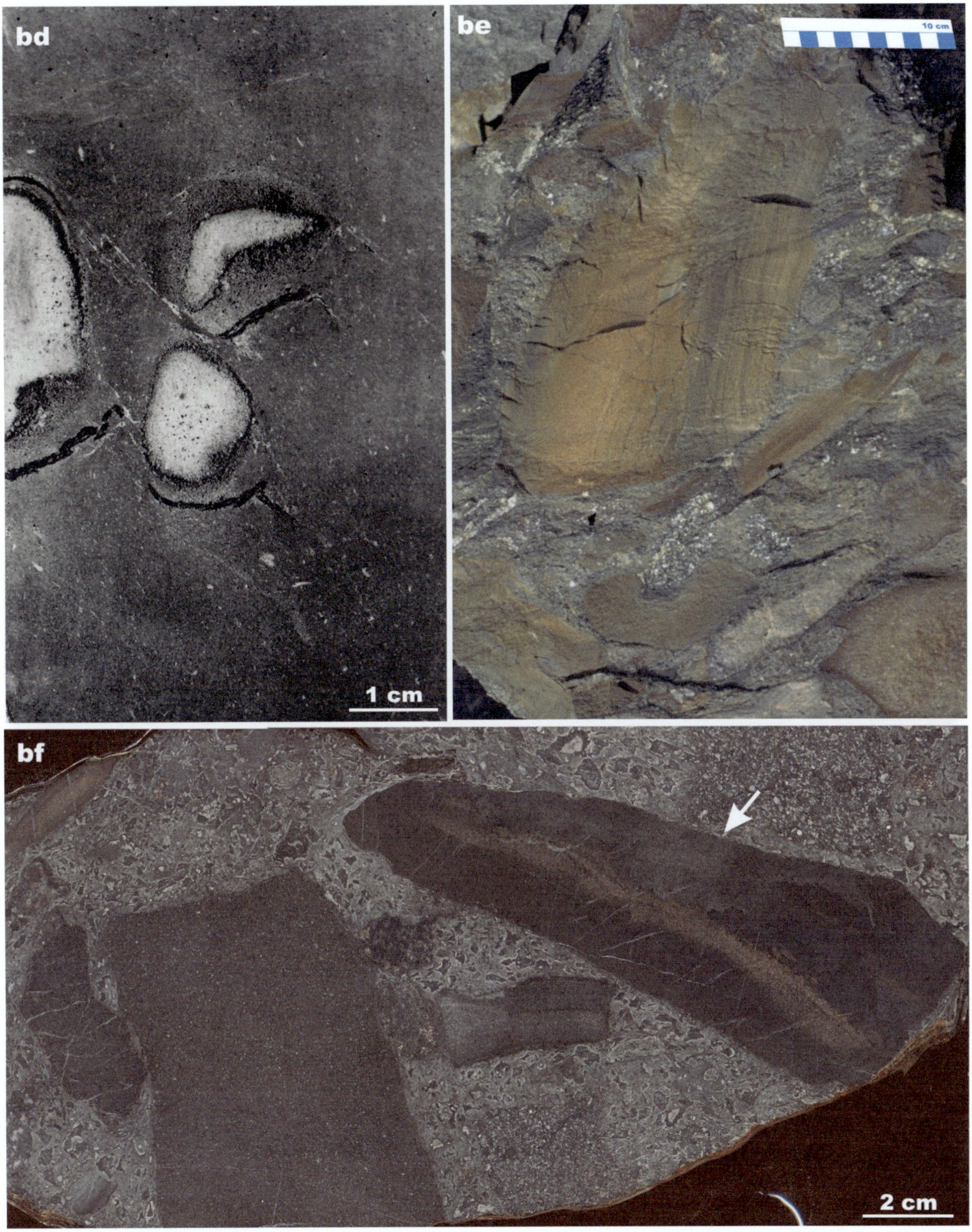

Fig. 4.29 (continued) (**bd**) Spherical calcite concretions (*white*) in massive greywacke. (**be**) Fragment-supported eruptive breccia consisting of clasts of various sedimentary and volcanic rocks in a ferropicritic tuff matrix. (**bf**) Eruptive breccia composed of angular, phosphorus-rich limestone clasts supported by a ferropicritic tuff matrix

Fig. 4.29 (continued) (**bg**) Fragment-supported eruptive breccia consisting of clasts of various sedimentary and volcanic rocks in a ferropicritic tuff matrix (Photographs (**a–f, gt, v–aq, as–ba, bc–bg**) taken by Victor Melezhik. Photographs (**ar** and **bb**) are reproduced from Melezhik et al. (1998) with permision from Elsevier)

Fig. 4.30 Images of the main rock types of the Pilgujärvi Volcanic Formation. (**a**) Pillowed tholeiitic basalt, the rock type, which constitutes the main portion of the formation; hammer head is 14 cm long. (**b**) Interpillow hyaloclastic breccia. (**c**) Basaltic lava breccia. This, together with massive lavas, are also important lithologies of the formation. (**d**) Ferropicritic lava with contraction joints; such a lava type occurs as thin flows between tholeiitic basalts in the lower part of the formation; compass length is 12 cm. (**e**) Globular ferropicritic lava

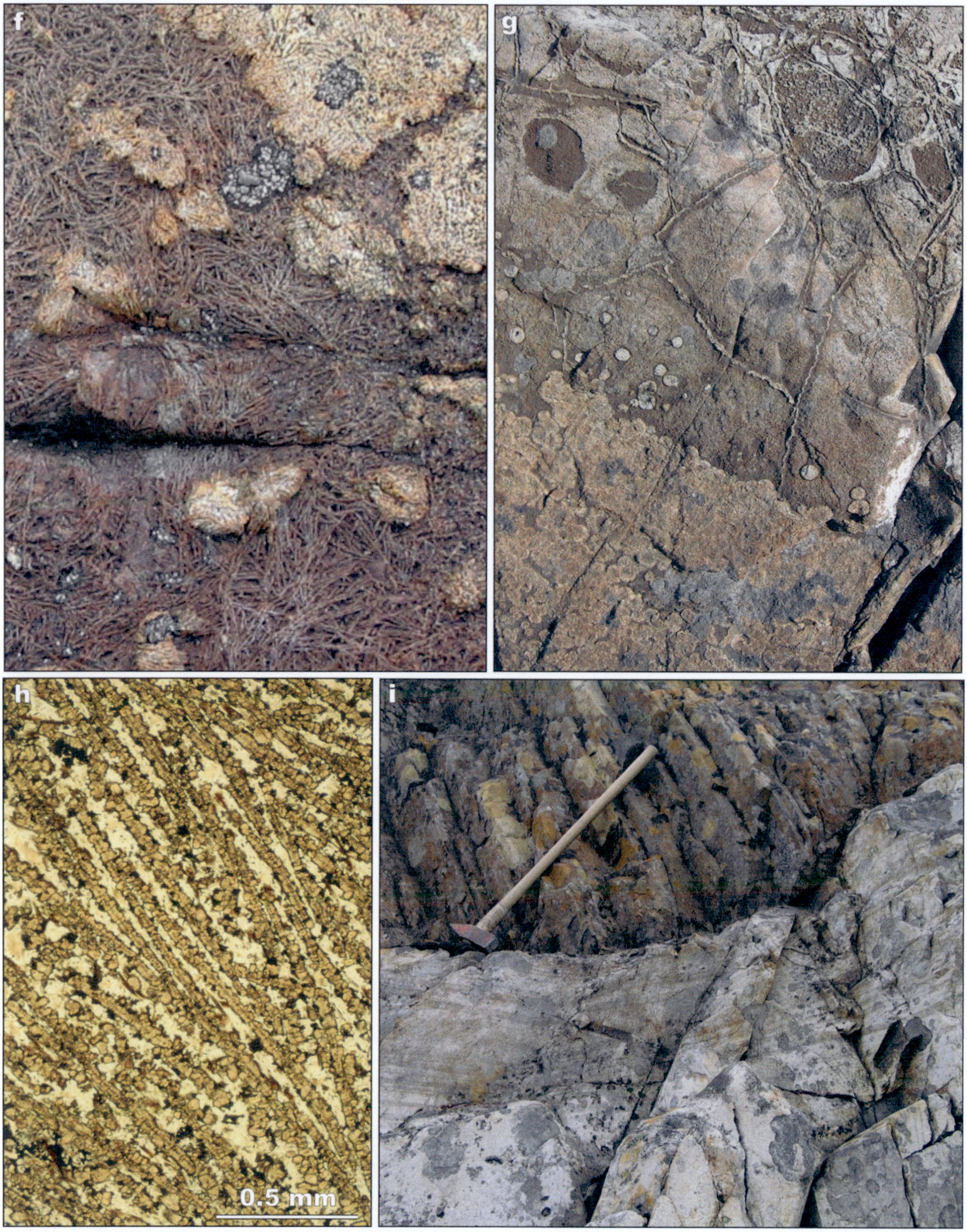

Fig. 4.30 (continued) (**f**) Globular ferropicritic lava with spinifex texture. (**g**) Ferropicritic dyke exhibiting a globular structure and contraction joints. (**h**) Photomicrograph in plane-polarised light of a pyroxene spinifex texture from the *upper part* of a layered ferropicritic lava flow; width of field is 1.8 mm. (**i**) Felsic tuff overlain by tholeiitic basalts with columnar jointing. Stratigraphic position of (**g** and **i**) images shown in Fig. 4.32

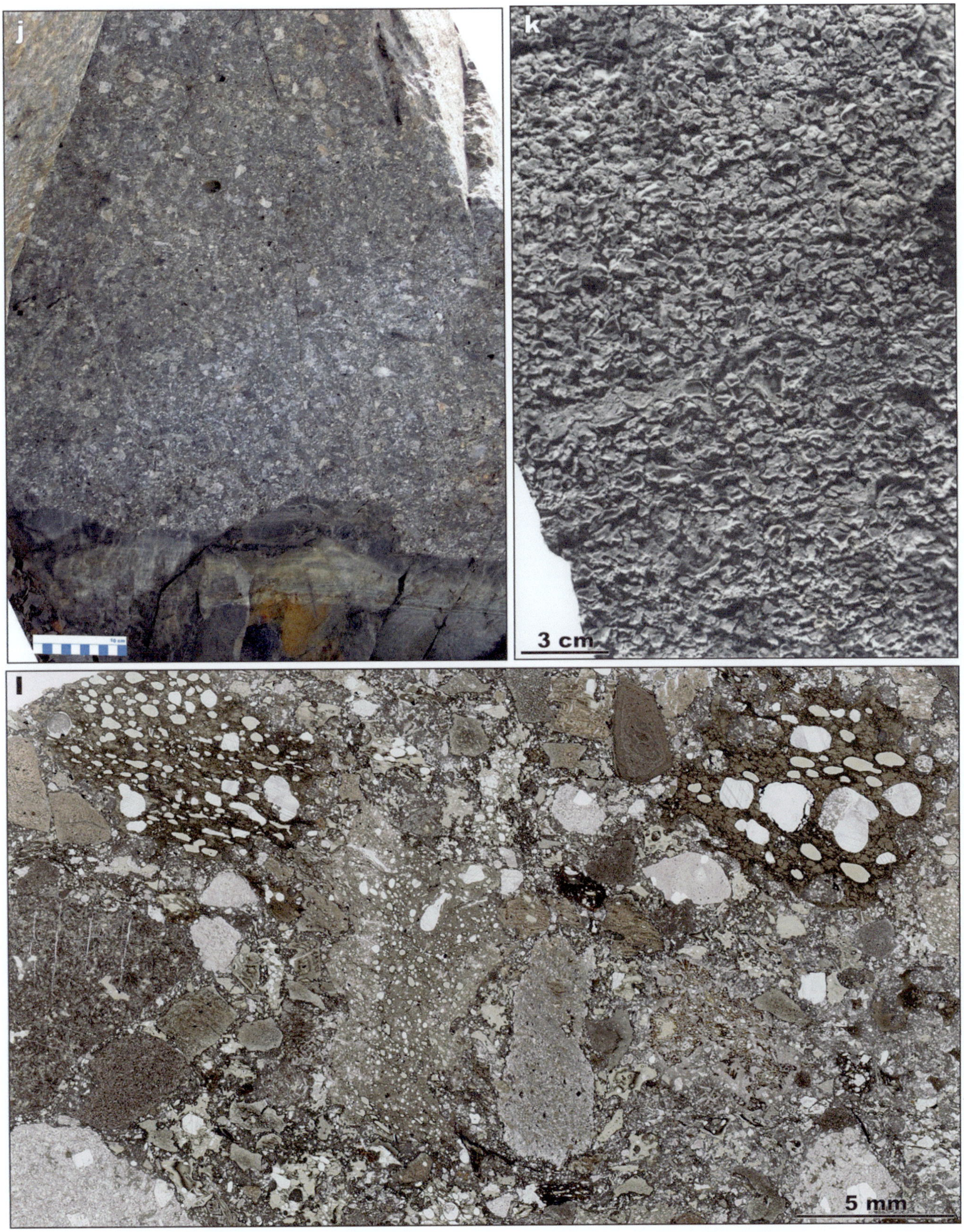

Fig. 4.30 (continued) (**j**) Laminated, glassy, felsic tuff overlain erosively by non-bedded, fragmental, felsic tuff containing clasts of various volcanic rocks. (**k**) Massive, rhyolitic lapilli tuff; note that the lapilli display well-developed chilled margins (bright-coloured rims with a positive relief). (**l**) Scanned thin section of fragmental tuff containing clasts of various "ferro-series" volcanic rocks ranging in composition from evolved ferropicrites and ferrobasalts to ferroandesites, ferrodacites and ferrorhyolites (e.g. Melezhik et al. 1994c). Stratigraphic position of the images shown in Fig. 4.32

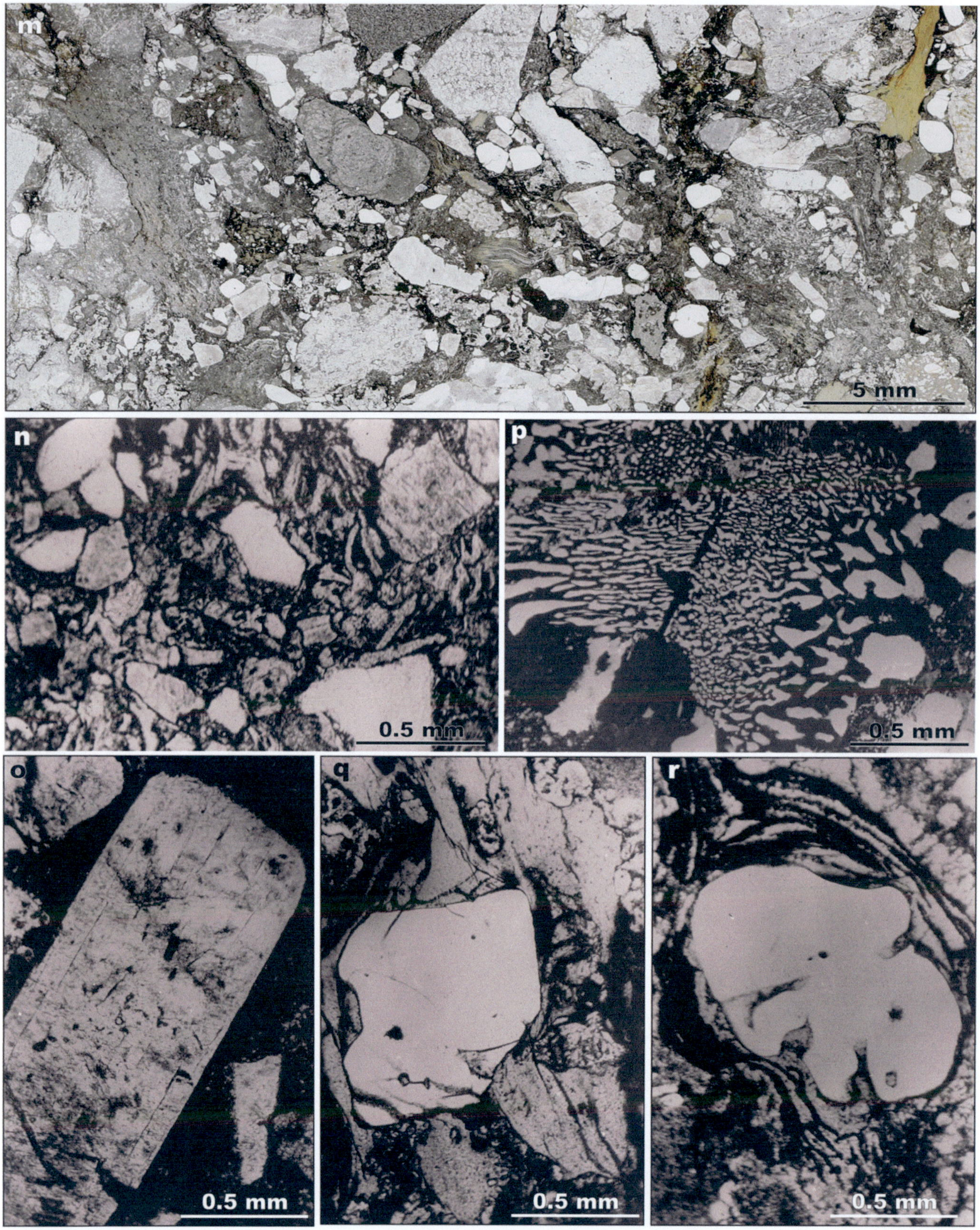

Fig. 4.30 (continued) (**m**) Scanned thin section of fragmental tuff containing mainly rhyolite clasts and fragments of quartz and K-feldspar. (**n**) Rhyolitic crystalloclastic tuff composed of quartz and K-feldspar crystalloclasts. (**o**) K-feldspar crystal in a chloritised glassy matrix. (**p**) Microgranophyre fragment in litho-crystalloclastic felsic tuff. (**q**) Partially-resorbed, early-formed quartz phenocryst in felsic tuff. (**r**) Resorbed, early-formed quartz phenocryst surrounded by a secondary melt. Photomicrographs (**n–r**) were taken in cross-polarised light. Stratigraphic position of (**m**) image shown in Fig. 4.32

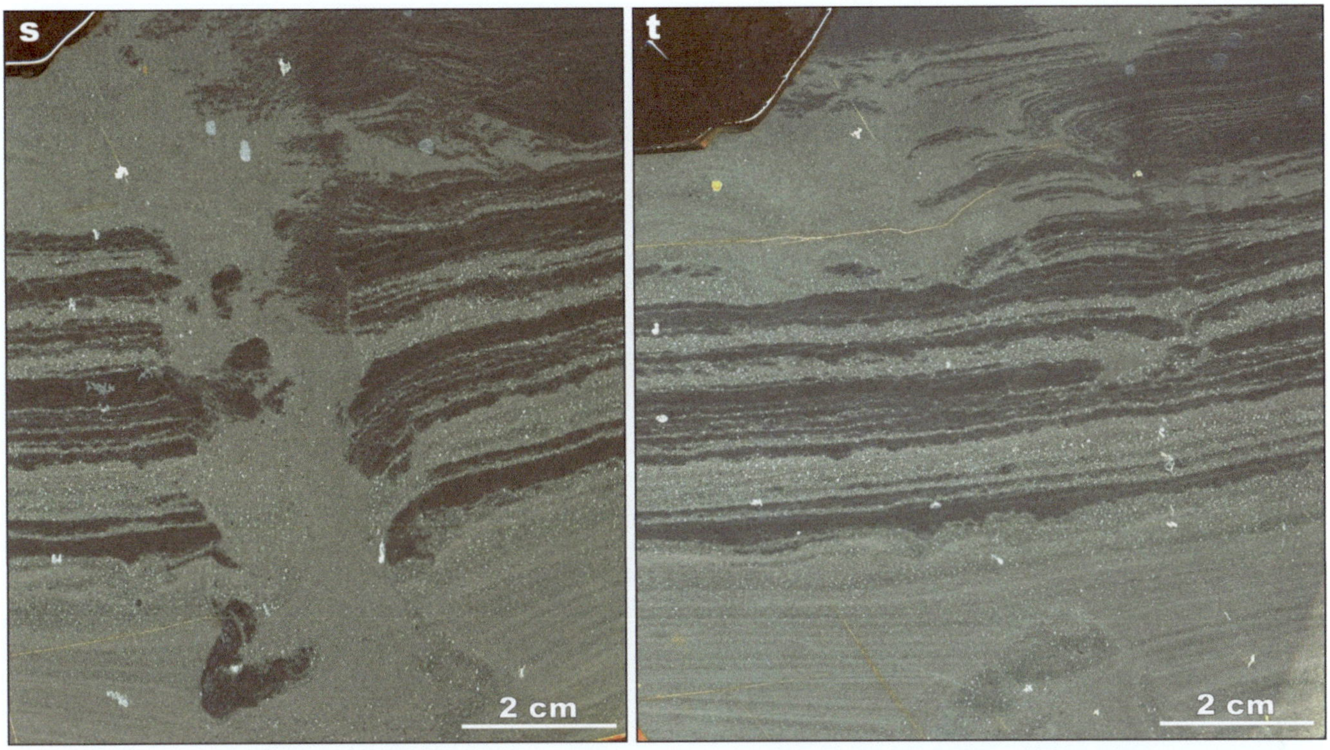

Fig. 4.30 (continued) (**s**) Bedded, microcrystalline felsic tuff cut by a sedimentary dyke filled with similar tuff material. (**t**) Slumping in bedded, microcrystalline felsic tuff. Stratigraphic position of the images shown in Fig. 4.32 (Photographs (**d, j**) courtesy of Kauko Laajoki, photographs (**a–f, h–i, l–m**) by Eero Hanski, photographs (**g, k, n–t**) by Victor Melezhik)

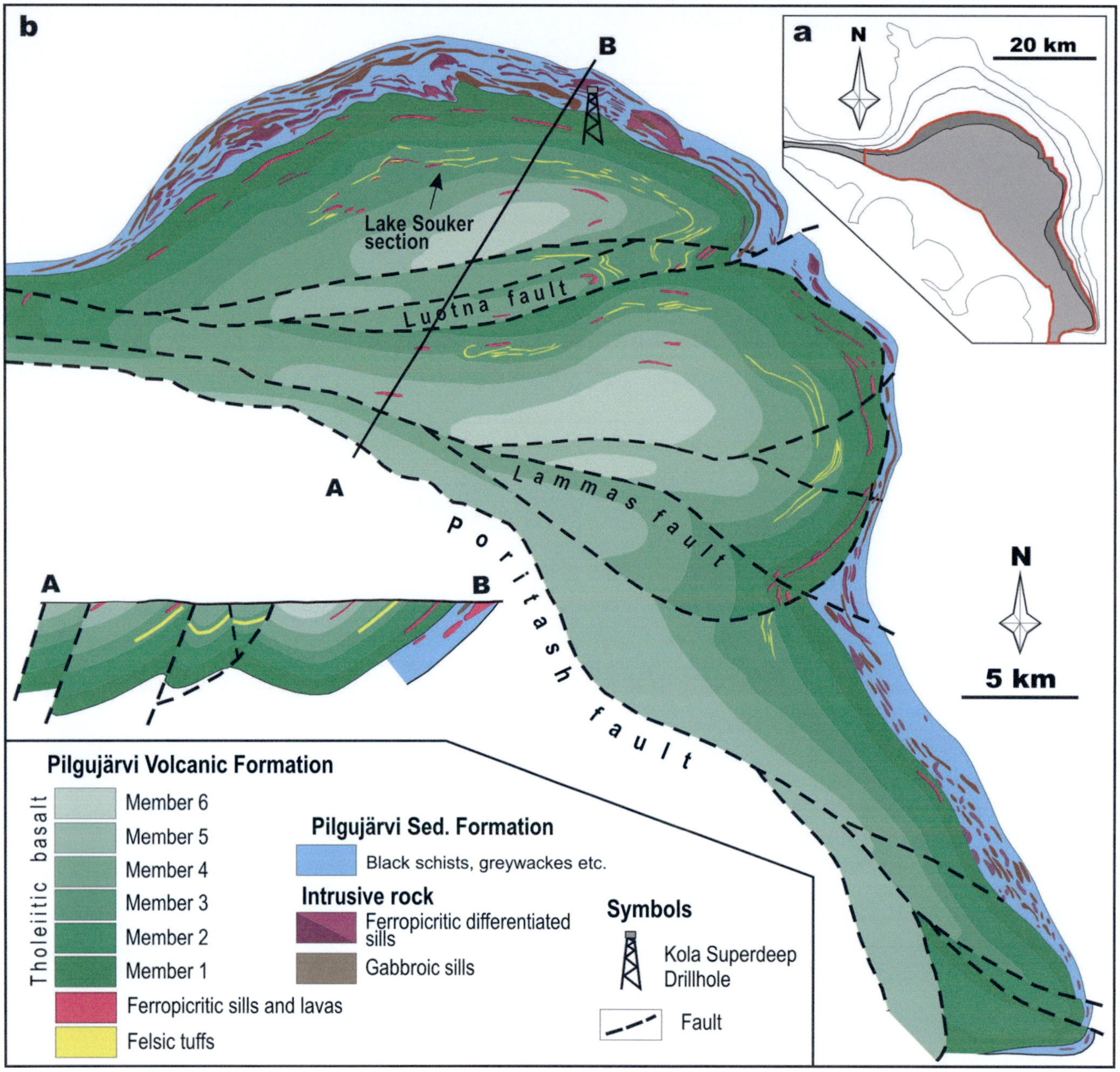

Fig. 4.31 A schematic geological map of the Pilgujärvi Volcanic Formation. (**a**) Outline of the Pechenga Greenstone Belt; area confined by red contour is enlarged in (**b**), which illustrates a tentative subdivision of the Pilgujärvi Volcanic Formation into several units (Modified from Rusanov 1981)

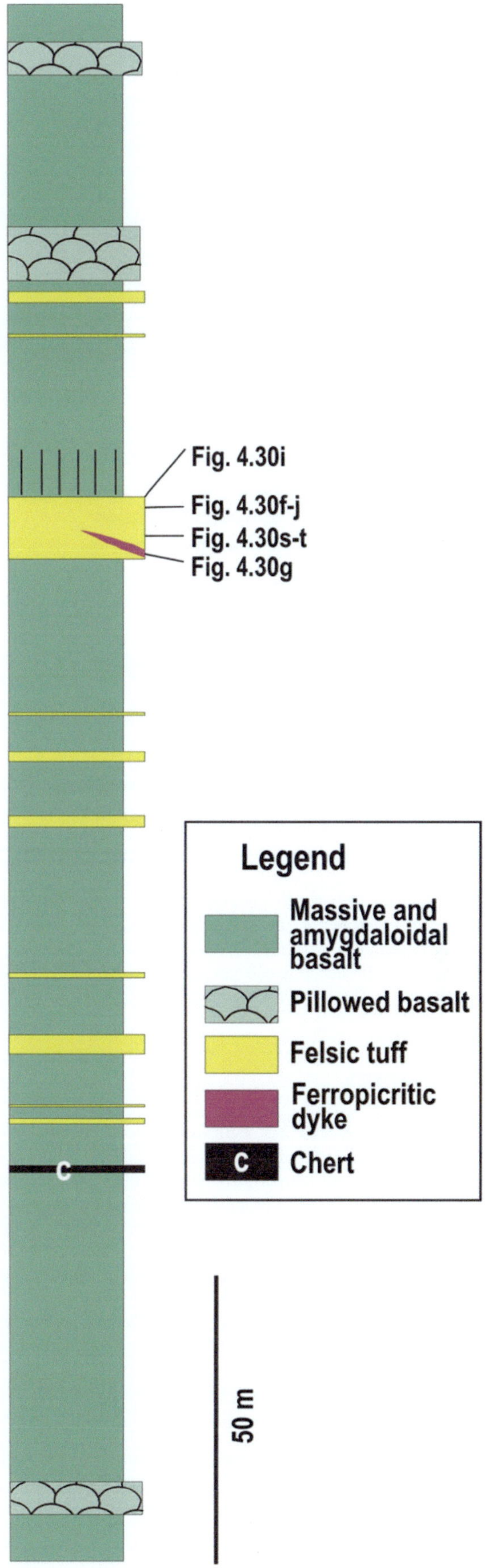

Fig. 4.32 Lithological section through middle part of the Pilgujärvi Volcanic Formation (near Lake Souker; for location see Fig. 4.31) showing stratigraphic distribution of felsic tuffs and structural variations in tholeiitic basalts

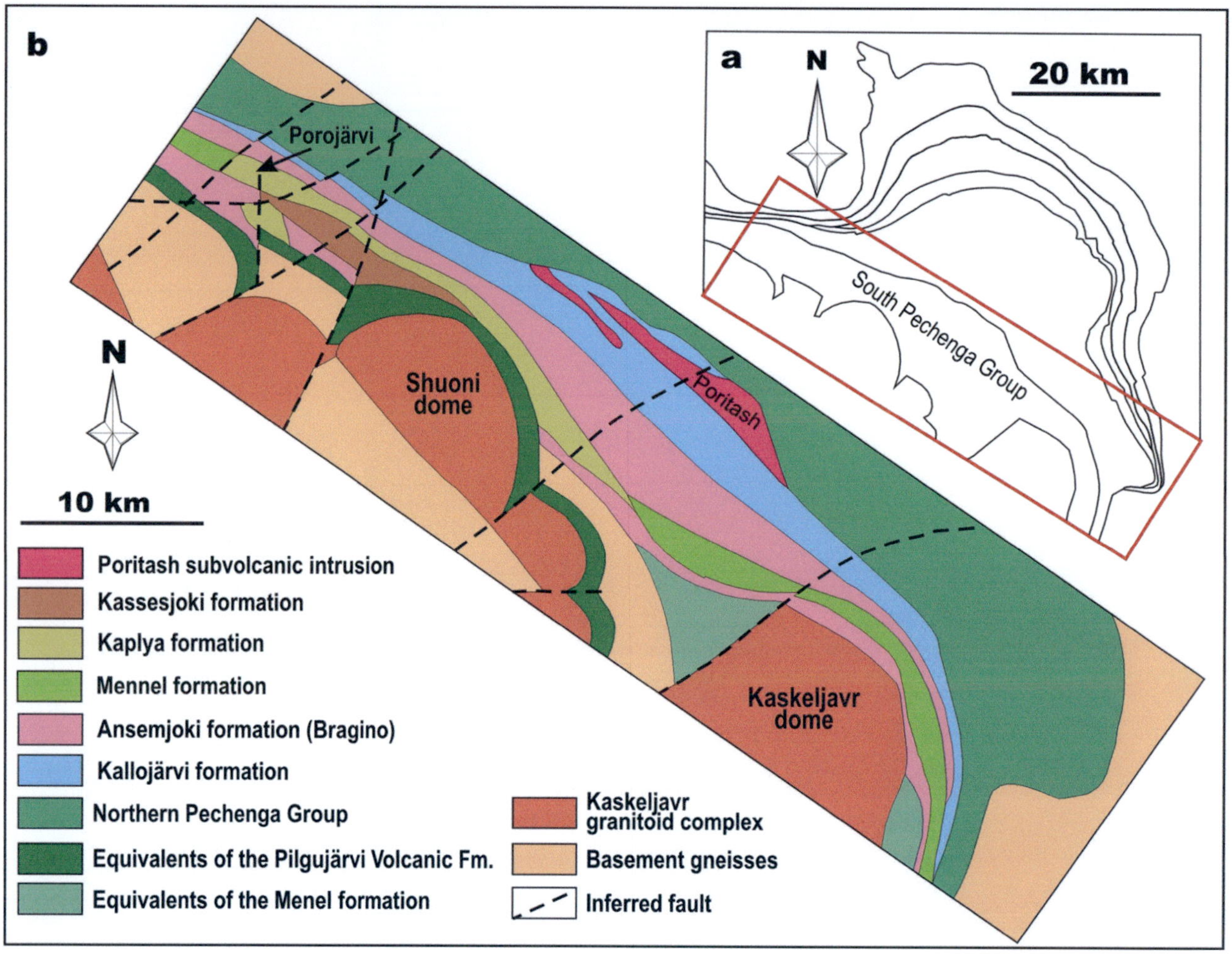

Fig. 4.33 A schematic geological map of the Southern Pechenga Zone. (**a**) Outline of the Pechenga Greenstone Belt; red rectangle defines the area enlarged in (**b**), which illustrates a tentative subdivision of the the South Pechenga Group into smaller, informally defined units; note that boundaries between the units are not specified and may represent a combination of primary stratigraphic contacts and thrust faults (Modified from Skuf'in and Theart 2005)

References

Akhmedov AM (1972a) Ferrous metasedimentary rocks of the Pechenga complex, and their genesis. In: Bel'kov IV (ed) Materials on geology and metallogeny of the Kola Peninsula, 1st edn. Kola Science Centre, Apatity, pp 125–131 (in Russian)

Akhmedov AM (1972b) Haematitic oolites of the Pechenga complex. In: Bel'kov IV (ed) Materials on mineralogy of the Kola Peninsula, 8th edn. Kola Science Centre, Apatity, pp 41–55 (in Russian)

Akhmedov AM, Chekushin VA (1977) Copper-bearing Precambrian volcano-sedimentary formations of the Kola Peninsula. Sov Geol 8:88–97 (in Russian)

Akhmedov AM, Krupenik VA (1990) Turbidity conditions of sedimentation and pyrite formation in the Early Proterozoic Pechenga Basin. Sov Geol 11:51–60

Amelin YuV, Heaman LM, Semenov VS (1995) U-Pb geochronology of layered mafic intrusions in the eastern Baltic Shield: implications for the timing and duration of Palaeoproterozoic continental rifting. Precambrian Res 75:31–46

Balashov YuA (1996) Paleoproterozoic geochronology of the Pechenga-Varzuga structure, Kola Peninsula. Petrology 4:1–22

Barnes S-J, Melezhik VA, Sokolov SV (2001) The composition and mode of formation of the Pechenga nickel deposits, Kola Peninsula, northwestern Russia. Can Mineral 39:447–471

Bayanova T, Ludden J, Mitrofanov F (2009) Timing and duration of Palaeoproterozoic events producing ore-bearing layered intrusions of the Baltic Shield: metallogenic, petrological and geodynamic implications. In: Reddy SM, Mazumder R, Evans DAD, Collins AS (eds) Palaeoproterozoic supercontinents and global evolution, Geol Soc, Lond, Spec Pub 323: 165–198

Bekasova NB (1985) Pechenga palaeogeography in early Pilgujärvi time of the early Proterozoic. Lithol Min Resour 1:127–137 (in Russian)

Bekasova NB, Dudkin OB (1981) Composition and nature of early Precambrian Pechenga concretionary phosphorites (Kola Peninsula). Lithol Min Resour 6:107–113 (in Russian)

Borisov AE (1990) Volcanism and native copper mineralisation in the Palaeoproterozoic of the Kola Peninsula. Kola Science Centre, Apatity, p 70 (in Russian)

Borisov AE (1995) A structural pattern of the Matert suite within the Luotn fault zone. In: Mitrofanov FP, Smol'kin VF (eds)

Magmatism, sedimentogenes and geodynamics of the Pechenga Riftogenic structure. Kola Science Centre, Apatity, pp 95–100 (in Russian)

Borisov AE, Smol'kin VF (1992) On the genesis of high-Si rocks of the fourth volcanogenic unit of the Pechenga structure. Proc USSR Acad Sci, Geol Ser 7:66–78 (in Russian)

Brügmann GE, Hanski EJ, Naldrett AJ, Smolkin VF (2000) Sulfide segregation in ferropicrites from the Pechenga Complex, Kola Peninsula, Russia. J Petrol 41:1721–1742

Daly JS, Balagansky VV, Timmerman MJ, Whitehouse MJ (2006) The Lapland-Kola orogen: Palaeoproterozoic collision and accretion of the northern Fennoscandian lithosphere. In: Gee DG, Stephenson RA (eds) Eur Lithos Dynam, Geol Soc Memoir 32:579–598

Fallick AE, Melezhik VA, Simonson B (2008) The ancient anoxic biosphere was not as we know it. In: Dobretsov N, Kolchanov N, Rozanov A, Zavarzin G (eds) Biosphere origin and evolution. Springer, New York, pp 169–188

Fedotov ZhA (1985) Evolution of Precambrian volcanism of the eastern part of the Pechenga/Varzuga Belt. Kola Science Centre, Apatity, p 105 (in Russian)

Fiorentini ML, Beresford SW, Deloule E, Hanski E, Stone WE, Pearson NJ (2008) The role of mantle-derived volatiles in the petrogenesis of Palaeoproterozoic ferropicrites in the Pechenga Greenstone Belt, northwestern Russia: insights from in-situ microbeam and nanobeam analysis of hydromagmatic amphibole. Earth Planet Sci Lett 268:2–14

Fliegel D, Wirth R, Simonetti A, Furnes H, Staudigel H, Hanski E, Muehlenbachs K (2010) Septate-tubular textures in 2.0 Ga pillow lavas from the Pechenga Greenstone Belt, a nano-spectroscopic approach to investigate their biogenicity. Geobiology 8:372–390

Gilyarova MA (1967) Stratigraphy and structure of Pechenga. Leningrad University, Leningrad, p 95 (in Russian)

Gorbatsevich FF, Ikorsky SV, Zharikov AV (2010) Structure and permeability of deep-seated rocks in the Kola Superdeep Borehole section (SG-3). Acta Geodynam Geomater 7:145–152

Gorbunov GI (1968) Geology and genesis of the Pechenga Copper-Nickel deposits. Nedra, Moscow, p 352 (in Russian)

Gorbunov GI, Yakovlev YuN, Goncharov YuV, Gorelov VA, Telnov VA (1985) The nickel areas of the Kola Peninsula. In: Papunen H, Gorbunov GI (eds) Nickel-Copper deposits of the Baltic Shield and Scandinavian Caledonides, Geol Surv Finl Bull 333: 41–109

Gorbunov GI, Korchagin AU, Mednikov AI (1989) On the presence of ore in Pechenga picrites. Trans USSR Acad Sci 304(5):1205–1208

Green AH, Melezhik VA (1999) Geology of the Pechenga ore deposits – a review with comments on ore forming processes. In: Keays RR, Lesher CM, Lightfoot PC, Farrow CEG (eds) Dynamic processes in magmatic ore deposits and their application in mineral exploration, short course notes 13. Geological Association of Canada, Sudbury/Ontrario/Canada, pp 287–328

Grinenko LN, Smol'kin VF (1991) Isotopic composition and content of sulfur in the ferropicrites and gabbro-wehrlites of the Pechenga zone. Geochemistry 9:1250–1261 (in Russian)

Grinenko LN, Grinenko VA, Lyakhnitskaya IV (1967) Sulfur isotope composition of sulfides in the Ni-Cu deposits of the Kola Peninsula. Geol Ore Deposits 4:3–17 (in Russian)

Hannah JL, Stein HJ, Zimmermman A, Yang G, Markey RJ, Melezhik VA (2006) Precise 2004 ± 9 Ma Re-Os age for Pechenga black shale: comparison of sulfides and organic material. In: 16th annual V.M. Goldschmidt conference, 27 Aug–1 Sept 2006, Melbourne, Geochim Cosmochim Acta 70: A228

Hanski EJ (1992) Petrology of the Pechenga ferropicrites and cogenetic, Ni-bearing gabbro-wehrlite intrusions, Kola Peninsula, Russia. Geol Surv Finl Bull 367:192

Hanski E (1993) Globular ferropicritic rocks at Pechenga, Kola Peninsula: immiscibility versus alteration. Lithos 29:197–216

Hanski EJ, Smolkin VF (1995) Iron- and LREE-enriched mantle source for early Proterozoic intraplate magmatism as exemplified by the Pechenga ferropicrites, Kola Peninsula, Russia. Lithos 34:107–125

Hanski EJ, Huhma H, Smol'kin VF, Vaasjoki M (1990) The age of the ferropicritic volcanics and comagmatic intrusions at Pechenga, Kola Peninsula, USSR. Bull Geol Soc Finl 62:123–133

Hanski E, Luo Zh-Y, Oduro H, Walker RJ (2011) The Pechenga Ni-Cu sulfide deposits, NW Russia: a review with new constraints from the feeder dikes. Rev Econ Geol 17:145–162

Hickey R, Frey FA (1982) Geochemical characteristics of boninite series volcanics: implication for their sources. Geochim Cosmochim Acta 46:2099–2115

Hiscott RN, Pickering KT, Bouma AH, Hand BM, Kneller BC, Postma G, Soh W (1997) Basin-floor fans in the North Sea: sequence stratigraphic models vs. sedimentary facies: discussion. Am Assoc Petroleum Geol Bull 81:662–665

ISSC (International Subcommision on Stratigraphic Classification of IUGS Comission on Stratigraphy) (1976), In: Hedberg HD (ed) International stratigraphic guide: a guide to stratigraphic classification, terminology and procedure, Willey, New York, p 200

Ivanova LV, Chapina OS, Melezhik VA (1988) Discovery of coccoidal microfossils in Early Precambrian metamorphosed cherts. Commun USSR Acad Sci 303:210–211 (in Russian)

Jones JG (1969) Pillow lavas as depth indicators. Am J Sci 267:181–195

Jones AP, Mutanen T, Tuisku P, Hanski E, Price GD (2003) The Pechenga structure, Russia: giant Ni-Cu mineralisation related to large meteorite impact? Fermor Flagship meeting. World class mineral deposits and Earth evolution, 19–21 Aug 2003, Cardiff University and The National Museum of Wales, Abstracts, Trans Inst Min Metal: Sect B: Appl Earth Sci 112: B149–B150

Karhu JA (1993) Palaeoproterozoic evolution of the carbon isotope ratios of sedimentary carbonates in the Fennoscandian Shield. Geol Surv Finl Bull 371:1–87

Kazansky VI (1992) Deep structure and metallogeny of Early Proterozoic mobile belts in the light of superdeep drilling in Russia. Precambrian Res 58:289–303

Kozlova NE, Balagansky VV (1995) Structural evolution of the Southern zone rocks. In: Mitrofanov FP, Smol'kin VF (eds) Magmatism, sedimentogenes and geodynamics of the Pechenga Palaeoriftogenic structure. Kola Science Centre, Apatity, pp 151–163 (in Russian)

Kozlovsky YeA (ed) (1984) The superdeep well of the Kola Peninsula. Springer, Berlin, p 558

Kravtsov NA (1984) Sections of the Southern-Pechenga series. In: Zagorodny VG (ed) Geology and history of formation of the Precambrian Structures of the Kola Peninsula. Kola Science Centre, Apatity, pp 28–32 (in Russian)

Kremenetsky AA, Ovchinnikov LN (1986) Geochemistry of deep-seated rocks. Nauka (Science), Moscow, p 262 (in Russian)

Lowe DR (1982) Sediment gravity flows: II. Depositional models with special reference to the deposits of high-density turbidity currents. J Sediment Petrol 52:279–297

Lybtsov VV (1979) Stromatolites of the Palaeoproterozoic Pechenga Complex, Kola Peninsula. Trans USSR Acad Sci 247:419–421 (in Russian)

Lybtsov VV, Akhmedov AM, Predovsky AA (1971) Volcanism and some palaeontological and litho-geochemical pecularities of Pechenga carbonate rocks. In: Sidorenko AB (ed) Problems of Precambrian sedimentary geology, vol 3. Science (Nauka), Moscow, pp 170–176 (in Russian)

Melezhik VA (1992) Early Proterozoic sedimentary and rock-forming basins of the Baltic Shield. Nauka (Science), St. Petersburg, p 258 (in Russian)

Melezhik VA (1996) The North Transfennoscandian Greenstone Belt: and overview. ICDP Project 336 symposium, Rovaniemi, 21–23

Aug 1996, University of Turku, Finland, Program and Abstracts, pp 23–24
Melezhik VA, Fallick AE (1996) A widespread positive $\delta^{13}C_{carb}$ anomaly at around 2.33–2.06 Ga on the Fennoscandian Shield: a paradox? Terra Nova 8:141–157
Melezhik VA, Fallick AE (2001) Palaeoproterozoic travertines of volcanic affiliation from a ^{13}C-rich rift lake environment. Chem Geol 173:293–312
Melezhik VA, Fallick AE (2005) The Palaeoproterozoic, rift-related, shallow-water, ^{13}C-rich, lacustrine carbonates, NW Russia – Part I: sedimentology and major element geochemistry. Trans R Soc Edinburgh: Earth Sci 95:393–421
Melezhik VA, Predovsky AA (1982) Geochemistry of early Proterozoic lithogenesis. Nauka (Science), St. Petersburg, p 208 (in Russian)
Melezhik VA, Sturt BA (1994a) General geology and evolutionary history of the early Proterozoic Polmak-Pasvik-Pechenga-Imandra/Varzuga-Ust'Ponoy Greenstone Belt in the north-eastern Baltic Shield. Earth-Sci Rev 36:205–241
Melezhik VA, Sturt BA (1994b) The Paleoproterozoic (2.5–1.7 Ga) Midcontinent rift system of the northeastern Fennoscaandian Shield *versus* The early Proterozoic Pechenga-Varzuga Belt: a case of back-arck spreading, discussion. Can J Earth Sci 35:720–725
Melezhik VA, Sturt BA, Mokrousov VA, Ramsay DM, Nilsson L-P, Balashov YuA (1994a) The early Proterozoic Pasvik-Pechenga Greenstone Belt: 1:200,000 geological map, stratigraphic correlation and revision of stratigraphic nomeclature. Norges Geologiske Undersøkelse, Spec Pub 7:81–91
Melezhik VA, Hudson-Edwards KA, Skuf'in PK, Nilsson LP (1994b) Pechenga area, Russia – Part 1: geological setting and comparison with Pasvik, Norway. Trans Instit Min Metall: Sect B: Appl Earth Sci 103:B129–B145
Melezhik VA, Hudson-Edwards KA, Green AH, Grinenko LN (1994c) Pechenga area, Russia – Part 2: nickel-copper deposits and related rocks. Transaction of Institutions of Mining and Metallurgy: Section B: Applied Earth Sciences 103:B146–B161
Melezhik VA, Grinenko LN, Fallick AE (1998) 2000 Ma sulphide concretions from the 'Productive' formation of the Pechenga Greenstone Belt, NW Russia: genetic history based on morphological and isotopic evidence. Chem Geol 148:61–94
Melezhik VA, Fallick AE, Medvedev PV, Makarikhin VV (1999) Extreme $^{13}C_{carb}$ enrichment in ca. 2.0 Ga magnesite-stromatolite-dolomite-'red beds' association in a global context: a case for the world-wide signal enhanced by a local environment. Earth-Sci Rev 48:71–120
Melezhik VA, Fallick AE, Smirnov YuP, Yakovlev YuN (2003) Fractionation of carbon and oxygen isotopes in ^{13}C-rich Palaeoproterozoic dolostones in the transition from medium-grade to high-grade greenschist facies: a case study from the Kola Superdeep Drillhole. J Geol Soc Lond 160:71–82
Melezhik VA, Fallick AE, Grillo SM (2004) Subaerial exposure surfaces in a Palaeoproterozoic ^{13}C-rich dolostone sequence from the Pechenga Greenstone Belt: palaeoenvironmental and isotopic implications for the 2330–2060 Ma global isotope excursion of $^{13}C/^{12}C$. Precambrian Res 133:75–103
Melezhik VA, Fallick AE, Rychanchik DV, Kuznetsov AB (2005) Palaeoproterozoic evaporites in Fennoscandia: implications for seawater sulphate, $\delta^{13}C$ excursions and the rise of atmospheric oxygen. Terra Nova 17:141–148
Melezhik VA, Huhma H, Condon DJ, Fallick AE, Whitehouse MJ (2007) Temporal constraints on the Paleoproterozoic Lomagundi-Jatuli carbon isotopic event. Geology 35:655–658
Melezhik VA, Fallick AE, Filippov MM, Lepland A, Rychanchik DV, Deines JE, Medvedev PV, Romashkin AE, Strauss H (2009) Petroleum surface oil seeps from Palaeoproterozoic petrified giant oilfield. Terra Nova 21:119–126
Mints MV (1993) Palaeotectonic reconstructions of the early Precambrian of the eastern part of the Baltic Shield. Tectonics 1:39–56 (in Russian)
Mints MV (1996) The sedimentary-volcanic Pechenga-Imandra-Varzuga belt. In: Leonov YuG (ed) The early Precambrian of the Northeastern Baltic Shield: paleogeodynamics, crustal structure and evolution. Scientific World, Moscow, pp 77–112 (in Russian)
Mints MV (1998) Regional structures and paleogeodynamic models. In: Orlov VP, Laverov NP (eds) Kola superdeep, research results and experience. MF Tekhnoneft'gaz, Moscow, pp 146–156 (in Russian)
Mints MV, Glaznev VN, Raevsky AB (1995) Three-dimensional geological model of the geological structure of the Earth's upper crust, the Kola superdeep borehole and adjacent areas of the Kola Peninsula. Geotectonics 6:457–475
Mints MV, Glasnev VN, Konilov AN, Kunina NM, Nikitichev AP, Raevsky AB, Sedikh YuN, Stupak VM, Fonarev VI (1996) The early Precambrian of the Northeastern Baltic Shield: paleogeodynamics, crustal structure and evolution. Scientific World, Moscow, p 287 (in Russian)
Mirskaya DD, Zagorodny VG (1966) Quartz porphyry and porphyry tuffs in the upper part of the Pechenga volcano-sedimentary series. In: Ivanov AM (ed) Ancient sedimentary-volcanic and metamorphic complexes in the Kola Peninsula. Nauka (Science), Leningrad, pp 37–42
Mitrofanov FP, Smol'kin VF (eds) (1995) Magmatism, sedimentogenesis and geodynamics of the Pechenga Palaeoriftogenic structure. Kola Science Centre, Apatity, p 258 (in Russian)
Mitrofanov FP, Skuf'in PK, Bayanova TB, Smirnov YuP, Levkovich NV (2001) An intrusive body of rhyodacite in the Kola Superdeep Drillhole section of the early Proterozoic Pechenga complex. Trans Acad Sci 380(4):540–544
Negrutsa VZ (1984) Early Proterozoic stages of evolution of the Eastern Baltic Shield. Nedra, Leningrad, p 270 (in Russian)
Negrutsa VZ (1995) Evolution of exogenic processes in the Pechenga paleobasin. In: Mitrofanov FP, Smol'kin VF (eds) Magmatism, sedimentogenesis and geodynamics of the Pechenga Palaeorift. Kola Science Centre, Apatity, pp 101–123 (in Russian)
Orlov VP, Laverov NP (eds) (1998) Kola superdeep, research results and experience. MF Tekhnoneft'gaz, Moscow, p 260 (in Russian)
Petrov VP, Voloshina ZM (1995) Regional metamorphism of the Pechenga area rocks. In: Mitrofanov FP, Smol'kin VF (eds) Magmatism, sedimentogenesis and geodynamics of the Pechenga Palaeorift. Kola Science Centre, Apatity, pp 164–182 (in Russian)
Polyak EA (1968) Geological build-up of the Pechenga structural zone, In: Geology and deep structure of the eastern part of the Baltic Shield, Nauka (Science), Leningrad, pp 111–119 (in Russian)
Predovsky AA, Akhmedov AM, Kazeblin PL, Voinov AS (1970) Haematite quartzite of the second sedimentary unit of the Pechenga complex and some peculiarities of middle Proterozoic sedimentation. In: Bel'kov IV (ed) Iron-silica formations of the Kola Peninsula. Nauka (Science), Leningrad, pp 73–80 (in Russian)
Predovsky AA, Fedotov ZhA, Akhmedov AM (1974) Geochemistry of the Pechenga Complex. Nauka (Science), Leningrad, p 139 (in Russian)
Predovsky AA, Melezhik VA, Bolotov VA, Fedotov ZhA, Basalaev AA, Kozlov NE, Ivanov AA, Zhangurov AA, Skuf'in PK, Lyubtsov VV (1987) Volcanism and sedimentology of the Precambrian in the North-Eastern part of the Baltic Shield. Nauka (Science), Leningrad, p 185 (in Russian)
Rhodes JM (1996) Geochemical stratigraphy of lava flows sampled by the Hawaii Scientific Drilling Project. J Geophys Res 101: 11729–11746

Roman'ko AE, Shilov VN, Efremova LB, Savichev AT (1991) REE geochemistry and geodynamics of the Pechenga Zone in the Kola Peninsula, Trans (Doklady) USSR Acad Sci Geochemist 321(5):1075–1079 (in Russian)

Rusanov MS (1981) Tholeiite-komatiite formation of the Pechenga complex. Sov Geol 2:98–112 (in Russian)

Sedykh YuN, Dodin DA, Stupak VM, Nikitichev AP (1998) New data on the structure and ore potential of deep sections of the Pechenga synclinorium. Doklady Earth Sci 359:220–223

Shanmugam G (1996) High-density turbidity currents: are they sandy debris flows? J Sediment Res 66:2–10

Sharkov EV, Smol'kin VF (1997) The early Proterozoic Pechenga-Varzuga Belt: a case of back-arc spreading. Precambrian Res 82:135–151

Sharov NV, Rispolozhensky YuA, Karaev NA, Polyakova VA, Ronin AL, Lizinsky MD, Isanina EV, Yepinatieva AM, Galdin NE (1998) Characteristics of seismic boundaries and heterogeneities from on-land and downhole observesations. In: Orlov VP, Laverov NP (eds) Kola Superdeep. Scientific results and research experience, TECHNONEFTEGAZ, Moscow, pp 159–167 (in Russian)

Skuf'in PK (1990) On the paleoignimbrites and kagusites in the volcanic section of the Pechenga Series (Kola Peninsula). News High Edu Inst, Geol Expl 11:26–33 (in Russian)

Skuf'in PK (1995a) On the picritic volcanism in the South Pechenga structural zone (Kola Peninsula). Bull Mosc Soc Natural Geol Sect 70(6):28–38

Skuf'in PK (1995b) Geology of the Poritash volcanic centre: new results. Norges Geologiske Undersøkelse, Spec Pub 7:111–116

Skuf'in PK, Bayanova TB (2006) Early Proterozoic central-type volcano in the Pechenga Structure and its relation to the ore-bearing gabbro-wehrlite complex of the Kola Peninsula. Petrology 14:609–627

Skuf'in PK, Theart HFJ (2005) Geochemical and tectono-magmatic evolution of the volcano-sedimentary rocks of Pechenga and other greenstone fragments of the Kola Greenstone Belt. Precambrian Res 141:1–48

Skuf'in PK, Yakovlev YuN (2007) Geological position and geochemistry of volcanic rocks of the Majärvi, Pirttijärvi and Orshoaivi suites from the Kola Superdeep Hole and near-surface zones. Proc Murmansk State Tech Univ 10(2):173–197 (in Russian)

Skuf'in PK, Pushkarev YuD, Kravchenko MP (1986) Mugearite-trachyte volcanic formation of the Pechenga volcano-tectonic depression. Proc USSR Acad Sci, Geol Ser 1:18–29

Skuf'in PK, Gavrilenko BV, Fedotov ZhA (1988) Porojärvi volcanic cupola-like structure in the southern Pechenga zone (Kola Peninsula), Bull Mosc Soc Natural Geol Sect 63(5):98–107

Skuf'in PK, Bayanova TB, Levkovich NV (1999) Lamprophyres in the Early Proterozoic volcanic complex of the South Pechenga structure, Kola Peninsula. Petrology 7(3):299–315 (in Russian)

Skuf'in PK, Bayanova TB, Mitrofanov FP, Apanasevich EA, Levkovich NV (2000) The absolute age of granitoids from the Shuoniyavri pluton in the southern framework of the Pechenga structure, the Kola Peninsula. Doklady Earth Sci 370:114–117

Smol'kin VF (1992) Early Precambrian Komatiitic and Picritic Magmatism of the Baltic Shield. Nauka (Science), St. Petersburg, p 278 (in Russian)

Smol'kin VF (1995) Early Proterozoic. In: Mitrofanov FP (ed) Geology of the Kola Peninsula (Baltic Shield). Kola Science Centre, Russian Academy of Sciences, Apatity, pp 38–75

Smol'kin VF (1997) The Paleoproterozoic (2.5–1.7 Ga) Midcontinent rift system of the northeastern Fennoscandian Shield. Can J Earth Sci 34:426–443

Smol'kin VF, Skuf'in PK (1995) Volcanic association of Northern and Southern Zones (composition, evolution and genesis). In: Mitrofanov FP, Smol'kin VF (eds) Magmatism, sedimentogenes and geodynamics of the Pechenga Riftogenic structure. Kola Science Centre, Apatity, pp 37–81 (in Russian)

Smol'kin VF, Borisov AE, Marakushev AA (1987) Evidence for differentiation by silicate liquid immiscibility in the Pechenga picritic basalts. Trans (Doklady) USSR Academy Sci Geochemist 294:669–673 (in Russian)

Smol'kin VF, Skuf'in PK, Mokrousov VA (1995) Stratigraphic position, geochemistry and genesis of volcanic associations of the Early Proterozoic Pechenga area. Norges Geologiske Undersøkelse, Spec Pub 7:50–75

Smol'kin VF, Bayanova TB, Fedotov ZhA (2003) Ore-bearing mafic-ultramafic rocks of the Pechenga-Allarechka area, Kola region: isotopic dating. In: Kosakov IK, Kotov AB (eds) Proceedings of the II Russian conference on isotope geochemistry. IGGP RAS, St. Petersburg, pp 467–470 (in Russian)

Sturt BA, Melezhik VA, Ramsay DM (1994) Early Proterozoic regolith at Pasvik, NE Norway: palaeotectonical implications for the Baltic Shield. Terra Nova 6:618–633

Suslova SN (1966) Tufogenic conglomerates at Lake Porojärvi (Pechenga region). In: Ivanov AM (ed) Ancient sedimentary-volcanic and metamorphic complexes in the Kola Peninsula. Nauka (Science), Leningrad, pp 43–52

Tel'nov VA (ed) (1986) Geological Map of the Pechenga-Allarechka Region, 1:200,000. Ministry of Geology of the Soviet Union, Moscow

Väyrynen H (1938) Petrologie des Nickelerzfeldes Kaulatunturi-Kammikivitunturi in Petsamo. Bulletin de la Commission Geólogique Finlande 116:1–198

Vetrin VR, Turkina OM, Rodionov NV (2008) U-Pb age and genesis of granitoids in the southern framing of the Pechenga structure, Baltic Shield. Trans Russ Acad Sci 418(6):806–810

Walker RJ, Morgan JW, Hanski EJ, Smolkin VF (1997) Re-Os systematics of early Proterozoic ferropicrites, Pechenga Complex, Russia: evidence for ancient ^{187}Os enriched plumes. Geochim Cosmochim Acta 61:3145–3160

Zagorodny VG, Mirskaya DD (1967) On ultramafic volcanic rocks of the fourth volcanogenic unit of the Pechenga series. In: Bel'kov IV (ed) Mafic and ultramafic rocks of the Kola Peninsula. Nauka (Science), Leningrad, pp 3–10 (in Russian)

Zagorodny VG, Mirskaya DD, Suslova SN (1964) Geology of the Pechenga sedimentary-volcanogenic series. Nauka (Science), Leningrad, p 218 (in Russian)

Zak SI, Makarov VN, Kochnev-Pervukhov VI, Proskuryakov VV, Zaskind ES, Batashev EV, Kolesnikov GP (1982) Geology, magmatism and ore formation in the Pechenga ore belt. Nedra, Leningrad, p 112 (in Russian)

4.3 The Onega Basin

V.A. Melezhik, P.V. Medvedev, and S.A. Svetov

4.3.1 Introduction

The Onega Basin represents one of the largest fragments of continental margin preserved on the eastern part of the Fennoscandian Shield (Fig. 4.34). It formed initially as a Palaeoproterozoic rift basin and contains a c. 2440–1980 Ma, low metamorphic-grade, volcano-sedimentary succession that has served as a stratotype area for the establishment of a formal lithostratigraphic subdivision of the Palaeoproterozoic of the former Soviet Union (e.g. Semikhatov et al. 1991). The Onega Basin succession is widely known for its ^{13}C-rich sedimentary carbonates (Galimov et al. 1968; Schidlowski et al. 1975), which record the greatest perturbation of the global carbon cycle in Earth history, the Lomagundi-Jatuli Event (e.g. Melezhik et al. 2005a). Another remarkable feature is the earliest known petrified giant oilfield (e.g. Melezhik et al. 2009). A recent discovery of thick halite and massive anhydrite beds of Palaeoproterozoic age (Morozov et al. 2010) makes the Onega Basin unique for this time period.

The history of geological investigation of the Onega area dates back to the seventeenth century when early prospecting for iron ores and small-scale mining operations served local iron smelters (Rozhkov 1895). However, the main regional geological features of the area were established only in the 1950s during the 1:200,000 mapping and prospecting for sedimentary Fe-ores (Perevozchikova 1957; Kratz 1963), and were summarised in the 37th volume of "Geology of the USSR" (Yakovleva and Gilyarova 1960). This important compilation was subsequently supplemented by intensive academic research in the fields of stratigraphy and sedimentology (Sokolov 1963), stratigraphy, tectonics and petrology (Kharitonov 1966; Gilyarova 1972), and the findings were summarised by Mikhailjuk (1988). Mapping was complemented by considerable contributions from various research programmes performed by academic institutions and synthesised in a series of books devoted to the general geology, stratigraphy and sedimentology (Sokolov 1972a), mineral deposits (Chernov 1975), volcanology (Svetov 1972) and tectonics (Sokolov 1973). Extensive exploration drilling for uranium and shungite (C_{org}-rich rocks) helped to define the stratigraphic framework (Sokolov 1987a).

More recent investigations were devoted to tectonics (Systra 1991), geochemistry of the volcanic rocks (Golubev and Svetov 1983; Svetov et al. 2004) and sedimentology (Golubev et al. 1984; Melezhik et al. 2000, 2001). Metasedimentary rocks have been the subject of numerous sedimentological (Sokolov et al. 1970; Heiskanen 1990; Ojakangas et al. 2001; Negrutsa 1984; Akhmedov et al. 2004) and palaeontological (Makarikhin and Kononova 1983; Makarikhin 1992) investigations and intensive research has been carried out on determining the carbon, oxygen and strontium isotope compositions of carbonate rocks (Yudovich et al. 1991; Akhmedov et al. 1993; Heiskanen and Rychanchik 1999; Melezhik et al. 1999a; Gorokhov et al. 1998; Kuznetsov et al. 2010). Galdobina (1982), Golubev et al. (1984), Filippov (1994, 2002, 2004) and Melezhik et al. (1999b, 2004) have studied the stratigraphy and genesis of the organic-carbon-rich rocks.

The modern concept of the deep crustal structure on the Onega Basin has been recently presented in a summary book that synthesised geological and geophysical data and the material obtained from drilling exploration over last 50 years (Glushanin et al. 2011).

Despite this extensive multidisciplinary research in the Onega Basin, there are very few geochronological data constraining the depositional age of the sedimentary and volcanic rocks (Bibikova et al. 1990; Puchtel et al. 1992, 1998, 1999; Philippov et al. 2007; Ovchinnikova et al. 2007). As well as being limited in number, the available

V.A. Melezhik (✉)
Geological Survey of Norway, Postboks 6315 Sluppen, NO-7491 Trondheim, Norway

Centre for Geobiology, University of Bergen, Allegaten 41, N-5007 Bergen, Norway
e-mail: victor.melezhik@ngu.no

V.A. Melezhik et al. (eds.), *Reading the Archive of Earth's Oxygenation*, Volume 1: The Palaeoproterozoic of Fennoscandia as Context for the Fennoscandian Arctic Russia - Drilling Early Earth Project,
DOI 10.1007/978-3-642-29682-6_9, © Springer-Verlag Berlin Heidelberg 2013

dates are imprecise, and hence the deposition of the Onega Basin volcano-sedimentary succession remains radiometrically poorly constrained.

4.3.2 General Geological Features

The Onega Basin is located in the southeastern part of the Archaean Karelian craton of the Fennoscandian Shield. It is known in the Russian literature as the Onega structure and, in recently published accounts (e.g. Kulikov et al. 2011a), it consists of a North-Onega synclinorium and a South-Onega trough. The former includes a series of early Palaeoproterozoic volcano-sedimentary formations, whereas the latter is filled with sedimentary rocks whose deposition postdated the 1890–1790 Ma Svecofennian orogeny. As the scope and main focus of this contribution is the early Palaeoproterozoic volcano-sedimentary rocks, the late Palaeoproterozoic succession of the South-Onega trough will not be considered further. South of Onega Lake, the Palaeoproterozoic succession is unconformably overlain and covered by the Palaeozoic sedimentary rocks of the Russian platform (Fig. 4.34).

The early Palaeoproterozoic volcano-sedimentary succession underwent deformation and greenschist-facies metamorphism during the 1890–1790 Ma Svecofennian orogeny. The succession was deformed into a system of roughly parallel, northwest-southeast trending folds that overall form a synclinorium (Kharitonov 1966). The individual folds are characterised by a shallow plunge to the southeast with the synclines having wavelengths of 6–12 km whereas the anticlines are much narrower, and in places culminate to form vertical "diapirs" (Leonov et al. 2011). Where well-exposed, such "diapirs" appear as linear structures bound by vertical faults and composed of irregularly brecciated rocks. The structural discrepancy between large-scale open synclines and steep anticlines can be explained by a progression from early gently folding to later intense shortening accommodated by brittle, subvertical northwest-southeast trending fault zones; a number of these run parallel to fold axes and structurally complicate the synclinorium.

4.3.3 Stratigraphic Subdivision and Age

The Palaeoproterozoic volcano-sedimentary succession of the Onega Basin was traditionally subdivided into six lithostratigraphic units, from oldest to youngest: the Sumian, Sariolian, Jatulian, Ludicovian, Livian and Vepsian super-horizons (e.g. Sokolov 1987a). The Livian Super-Horizon has been recently renamed the Kalevian (e.g. Kulikov et al. 2011a). The Vepsian Super-Horizon (1800–1600 Ma) is not considered in this contribution (Fig. 4.35).

The lowermost Sumian Super-Horizon rests unconformably on Archaean granites, gneisses and amphibolites. However, the relationship between the Sumian and overlying Sariolian Super-Horizon has long been controversial (Sokolov 1987a). Heiskanen et al. (1977) used composition as a criterion for subdividing the pre-Jatulian volcanic rocks and assigned felsic volcanic rocks to the Sumian Super-Horizon (their dacite-liparite complex) whereas intermediate-mafic volcanic rocks were assumed to represent the Sariolian Super-Horizon (their andesite-basalt complex).

Later workers (e.g. Kulikov et al. 1978; Sokolov 1987a) suggested that all pre-Jatulian volcanic rocks be collectively known as Sumi-Sariolian. However, as discussed in Chaps. 3.2 and 3.4, the Sariolian-age rocks in the Kola Peninsula and Finland are younger than 2.44 Ga, and their deposition took place after the gabbro-norite layered intrusions and associated Sumian-age rocks were uplifted and deeply eroded, so their amalgamation is not a useful solution. Since direct, precise dating of neither Sariolian nor Sumian rocks exists, and field evidence on the relationship between the pre-Jatulian volcanic rocks and Sumian-age intrusive rocks in the Onega Basin is not available, the confusion is apparently unavoidable and will last.

Despite all the uncertainties and differences in the opinions on how to distinguish between the Sariolian and Sumian rocks in the Onega Basin, Svetov et al. (2004) tentatively suggested recently that the mafic to intermediate volcanic rocks in the northwestern part of the Onega Basin (the andesite-basalt complex of Heiskanen et al. 1977) be assigned to the Sumian Super-Horizon. In the current contribution, we have adopted this approach.

The remaining four super-horizons, namely Sariolian, Jatulian, Ludicovian, and Kalevian, are separated from each other by unconformities and palaeoweathered surfaces (e.g. Sokolov 1987a) that can be readily identified in the field. It was inferred that these four lithostratigraphic units span the time interval of 2500–1800 Myr (e.g. Semikhatov et al. 1991; Kulikov et al. 2011a); however, their exact depositional ages and boundaries in the Onega Basin remain to be determined. Currently, there is only one imprecise Pb-Pb age of 2090 ± 70 Ma obtained from dolomite of the Tulomozero Formation in the Jatulian Super-Horizon (Ovchinnikova et al. 2007). Two gabbro sills, one within the Jatulian Super-Horizon and the other within the Ludicovian Super-Horizon, have been dated by U-Pb (zircon) and Sm-Nd and Pb-Pb techniques at 1983.4 ± 6.5 and c. 1980 Ma, respectively (Philippov et al. 2007; Puchtel et al. 1992, 1998).

The five super-horizons have been subdivided further into a series of suites (e.g. Kulikov et al. 2011a) (Figs. 4.35 and 4.36). Although in this contribution the original lithostratigraphic names are generally utilised, the Russian term "suite" and "subsuite" are substituted with "formation" and "member", respectively (e.g. Melezhik et al. 1999a); these are discussed in detail below.

The Sumian Super-Horizon

In the Onega Basin, the Sumian Super-Horizon occurs as a series of remnants infolded into Archaean basement structures in the northern and northwestern parts of the basin (Figs. 4.34 and 4.37). The Sumian rocks rest unconformably on various metamorphic complexes of the Archaean basement (Heiskanen 1987a) and have been subdivided recently into the Glubokozero and Kumsa formations (Golubev et al. 2011; Fig. 4.36); cumulative stratigraphic thicknesses of the sedimentary and volcanic rocks are c. 50 m and c. 1,300 m, respectively.

The Glubokozero Formation

The Glubokozero Formation (Korosov and Nazarova 1989; Korosov 1991) occurs in the Kumsa (Fig. 4.38a) and Lake Paljeozero areas. The formation rests with angular unconformity on variably eroded Archaean granites, tonalites and gabbros (Fig. 4.37) (Svetov 2009). The formation is 150 m thick and subdivided into a Lower (sedimentary) and an Upper (volcanic) member (Fig. 4.38a).

The Lower member is a 20-m-thick unit. In places, it rests on Archaean granite, which shows signs of palaeoweathering (removal of silica and accumulation of Al_2O_3 and K_2O) that can extend as much as 10 m downward from the unconformity surface (Korosov and Nazarova 2011). This palaeoweathered interval is transitional through sericite schists with a relict granite structure, expressed by quartz crystals embedded in the sericite mass, upward into a several-m-thick, massive sandstone composed of variably weathered fragments of the underlying granite. The sandstone contains thin intervals of quartz-sericite schists. This, in turn, passes upward into alternating grey, sericite-cemented arkosic and quartzitic sandstones that range in thickness from 2 to 4 m (Fig. 4.38a). They contain minor intervals of gritstone and polymict pebble conglomerate and are mainly massive or show indistinct parallel layering.

The upper part of the sedimentary member is composed of a c. 2-m-thick unit of pale grey, medium-grained, quartz sandstone interbedded with grey, fine-grained, arkosic sandstones. The sandstones display well-developed, planar cross-bedding. Magnetite-enriched (up to 30 wt.% magnetite) layers, up to 10 cm thick, were described from a single exposure (Korosov and Nazarova 2011).

The Upper member is a c. 130-m-thick volcanogenic unit consisting of a series of basaltic andesite lava flows (Fig. 4.38a) that vary in thickness from 12 to 35 m. Lower parts of the flows are composed of plagioporphyritic basaltic andesites with rare amphibole porphyroblasts; the middle parts comprise variolitic and amygdaloidal rocks and tops are either "foamy" or fragmented basaltic andesites. Pillow lavas are rare (Fig. 4.38a), with the pillow size ranging between 0.5 and 2 m and interpillow-space filled with an amphibole-chlorite mass.

The volcanic rocks of the Glubokozero Formation are basaltic andesite and andesites, which contain 2.2–3.4 wt.% MgO and 54.3–57.0 wt.% SiO_2. The rocks are characterised by an elevated concentration of Cr (26–70 ppm) and Ni (40–120 ppm), and have a low content of HREE. The rocks fall into the field of basaltic andesite-andesite series with a normal alkalinity (3–7 wt.% $Na_2O + K_2O$). They are depleted in high-field strength elements, such as Nb and Ti, relative to large-ion lithophile elements.

Sedimentological and lithological features of sedimentary and volcanic rocks are indicative of deposition in an intraplate rift under subaerial environments.

The Kumsa Formation

The Kumsa Formation occurs mainly in the Kumsa area near the town of Medvezhegorsk (Fig. 4.37), and partly in the Lake Paljeozero area (Fig. 4.34). It is c. 1,200 m thick and has been subdivided into a Lower (sedimentary) and an Upper (volcanic) member (Svetov and Golubev 1972; Korosov 1991).

The Lower member is composed mainly of chemically mature, clastic rocks, mostly quartz sandstones with subordinate greywackes and arkosic sandstones (Fig. 4.38b). The member varies greatly in thickness. It reaches c. 45 m in the Kumsa area, where it conformably overlies the Glubokozero basaltic andesites. Elsewhere, it is either absent or, as in the in the Semch and Koikary areas (Fig. 4.39, for location see Fig. 4.34), it consists of thin layers of eluvial breccia, resting directly on Archaean basement rocks (Fig. 4.40a).

At Kumsa, eight descrete units have been described, with the lowermost one (Unit 1, c. 3 m thick, Fig. 4.38b) being composed of fine-grained greywacke with lenses of polymict sandstones displaying flaser bedding and then a c. 1.5-m-thick bed (Unit 2) of massive polymict sandstone. Greywackes interbedded with arkosic sandstones (Unit 3) occur above these units and, at the top, are arkosic sandstones that are c. 9-m-thick with parallel lamination (Unit 4).

The unit above the arkosic sandstones is comprised mainly of quartz sandstones (Fig. 4.40b) and is more than 20 m thick. At the base, the quartz sandstones are beige in colour and graded, cross-bedded and interbedded with thin layers of greywackes (Unit 5). Above these rocks is a rhythmically interbedded grey and grey-pink, haematite-bearing quartz sandstone (Unit 6). This is overlain by massive quartz sandstones (Unit 7; c. 5 m thick) followed by a c. 1.5-m-thick interval of carbonate-cemented sandstones interbedded with grey quartz sandstones (Unit 8). The sedimentary member ends with a c. 5-m-thick interval (Unit 9) of fine-grained, greenish greywackes (Fig. 4.38b).

The sedimentary member contains two populations of detrital zircons. The first consists of well-rounded grains yielding U-Pb ages between 2831 and 2883 Ma whereas the second suite is less rounded with ages of 2726–2788 Ma (Berezhnaya et al. 2003). The nearest tonalite intrusion in the Kumsa area has an age of 2875 ± 4 Ma (Berezhnaya et al. 2003) and thus can be considered as one of the possible sources for the zircons. Igneous rocks providing the second zircon population are yet to be identified.

The Upper member is a c. 1,200-m-thick pile of lava flows. They were erupted either on the underlying sedimentary member (Kumsa, Semch and Koikary areas; Figs. 4.34 and 4.39) or directly on Archaean basement. The most complete succession (1,150 m thick) was documented at Kumsa (Fig. 4.39) where it comprises 35 lava flows (Korosov and Nazarova 1989). In the Semch and Elmus areas, the succession thins to 560 m and is composed of 21 and 20 lava flows, respectively (Eliseev and Robonen 1983). In the Koikary area, only 18 flows are documented with a total thickness of 630 m. The representative rock images, characterising volcanic rocks from these areas, are presented in Fig. 4.40b–m.

At Kumsa, individual lava flows range in thickness from 9 to 45 m and are composed of massive, porphyritic, amygdaloidal (Fig. 4.40b–e), variolitic, fluidal and, less commonly, pillowed basaltic andesites. The lava flows were grouped into seven piles (Svetov et al. 2011) and four marker horizons have been identified. Flow 3 (150 m thick) at the base of the second volcanic pile defines the first marker horizon (Fig. 4.39). It is composed of fine-grained basaltic andesite at the base, passing upward gradually into a coarser-grained variety (gabbro-dolerite), and then again into fine-grained basaltic andesite with 2- to 3-mm-sized varioles filled with pink albite. Higher in the lava flow, the varioles appear as 1- to 8-cm-wide, lensoidal clusters giving rise to a banded and patchy appearance. The flow top is composed of amygdaloidal basaltic andesite.

The second marker horizon occurs at the upper part of the second volcanic pile. It is 65 m thick and consists of Flows 5–8 that are mostly variolitic basaltic andesites, some of which contain abundant pink-albite-filled vesicles (up to 15 cm in size) and veins (Flow 8 does not contain albite amygdales). The third marker horizon includes Flows 17–21, totalling 105 m in thickness, and is part of the fourth volcanic pile (Fig. 4.39). It is composed of plagioporphyritic basaltic andesites with ubiquitous and randomly distributed pink-albite amygdales (Fig. 4.40d). The fourth marker horizon is in the sixth volcanic pile (Fig. 4.39), which includes Flows 30–33, all composed of variolitic basaltic andesites with rare pink-albite amygdales.

Other lava flows in all areas exhibit similar structural organisation; the lower part is composed of fine-grained basaltic andesite passing upward into coarser-grained varieties with a gabbroic texture, whereas flow tops are commonly vesicular (Fig. 4.40d, e) or fragmented. The contacts between lava flows are well defined showing irregular topography. Tuff beds (0.5–4.5 m thick) occur between lava flows but are only characteristic of the upper part of the volcanic member. However, at Semch, pyroclastic rocks are more abundant. Here, agglomerates comprise a 50-m-thick lens, together with welded tuffs, massive, amygdaloidal and variolitic basaltic andesites; these are interpreted as the remnant of a volcanic cone (Svetov et al. 2004).

Most of the volcanic rocks of the Kumsa Formation are basaltic andesites and andesites, which are rich in MgO (4.5–10.3 wt.%) and SiO_2 (53.2–58.2 wt.%). The rocks contain high Cr (33–890 ppm) and Ni (73–259 ppm) concentrations and are low in HREE abundances (0.24 ppm Tm, 1.40 ppm Yb, 0.20 ppm Lu). The rocks fall into the field of basaltic andesite-andesite series with normal alkalinity (3–6 wt.% $Na_2O + K_2O$); a few exceptions have up to 7 wt.% $Na_2O + K_2O$ (Svetov et al. 2011). The rocks are depleted in high-field strength elements, such as Nb and Ti, relative to large-ion lithophile elements. Svetov et al. (2004) classified the Kumsa Formation rocks as bajaites; the terms used for a variety of boninite with c. 8 wt.% MgO, c. 56 wt.% SiO_2 and characterised by high contents of Sr (>1,000 ppm) and high K/Rb ratios (>1,000). Alternatively, these rocks are crustally contaminated komatiitic basalts (for details see Chap. 3.4). In fact, (Puchtel and Humayun 2001) provided an in-depth analysis of the komatiite versus boninite identification problem and argued against the boninitic nature of the Vetreny lavas on the basis of the PGE and chalcophile element behavior. This also applies to the Kumsa Formation volcanic rocks.

The geochemical pattern of volcanic rocks, their petrological features and the sedimentological characteristics of the Kumsa Formation point collectively to a depositional setting associated with an active continental margin. The basal greywackes of the sedimentary member with flaser bedding are indicative of a tidal flat environment, hence a marine basin. The rest of the sedimentary succession is composed of predominantly chemically and physically mature quartz sandstones, which are suggestive of a coastal marine depositional setting allowing multiple reworking of clastic material. Abundant amygdaloidal and rare pillow lavas comprising the volcanic member suggest that volcanism occurred in a rather shallow-marine, perhaps even non-marine, setting.

The Sariolian Super-Horizon

Eskola (1925) was the first to report polymict, immature conglomerates and associated sandstones (Fig. 4.41) in the Onega region, which he termed Sariolian. He demonstrated that they are older than the Jatulian rocks, rest unconformably on older gneisses, and suggested that their origin was linked to an ancient glacial event (Eskola 1919). In the Onega Basin, the Sariolian Super-Horizon is represented by the Paljeozero Formation (Korosov et al. 2011).

The Paljeozero Formation

The Paljeozero Formation (Korosov 1989) occurs in a series of discontinuously developed, graben-like structures, remnants of which are now infolded into the Archaean basement in the western and northern parts of the Onega Basin (Fig. 4.34). The formation overlies unconformably the Archaean basement and Sumian-age sedimentary and volcanic rocks. The formation has a greatly varying thickness ranging from a few to as much as 450 m. As suggested by Eskola (1919), the considerable thickness variation was controlled by uneven palaeotopography. This was later confirmed by Heiskanen et al. (1977) who documented several prominent features of Sariolian palaeorelief such as vertical cliffs, horsts and grabens at Krasnaja Rechka and Svjatnavolok (for locations see Fig. 4.34).

The Paljeozero Formation has considerable lithological heterogeneity (Korosov et al. 2011). Although it is composed mainly of polymict conglomerates with subordinate interbeds and lenses of sandstones and shales (Fig. 4.41, 4.42), the relative proportion of the different lithologies exhibits marked lateral variation. A type section was identified in the 1960s by Kratz (1963) on the western side of Lake Paljeozero (Fig. 4.34a). He described a three-member section comprising mainly polymict conglomerates containing variably rounded, unsorted boulders and pebbles of granites and gneisses (Fig. 4.41a). The conglomerates grade irregularly upward through a 400-m-thick section into pebble-sized varieties, which are overlain by a c. 40-m-thick unit of massive, fine-grained sandstones composed of fragments of quartz and plagioclase in a chlorite matrix. The sandstones contain lenses and pockets of polymict conglomerates and solitary pebbles (lonestones) of basalts and granites. The formation ends with a c. 20-m-thick, fine-grained, chlorite schist (Fig. 4.41a) showing a varve-type structure.

Several other sections have been described from the Koikary, Semch, Elmus and Krasnaja Rechka areas (for location see Fig. 4.39a), all exhibiting a variable thickness and lithological composition. The thickest (>500 m) conglomerate unit is exposed at Svjatnavolok (for location see Fig. 4.34). West of Medvezhegorsk, the base of the Paljeozero conglomerates cuts down through the Kumsa and Glubokozero formations, sealing an intensely faulted and eroded ancient surface (Fig. 4.37). Clastic material commonly shows a local source. For example, at Elmus the conglomerate overlies unconformably the Kumsa Formation and Archaean rocks (Fig. 4.41b). In the first case, variably rounded, unsorted boulders and pebbles consist of plagioporphyrites and variolitic basaltic andesites derived from the underlying Kumsa Formation and in the latter the blocks and boulders consist of Archaean gneiss, granite, tonalite, diorite, silicite and dacite. Matrix composition in the conglomerate ranges from arkosic sandstone and gritstone to greywacke, and in places contains sulphides.

Despite greatly variable thickness and lithological heterogeneity, an attempt was made to construct a composite section (Korosov et al. 2011), which includes six members (Fig. 4.41c). Member 1 is a basal unit (1–2 m thick) commonly resting on a palaeoweathered surface developed upon Archaean and older Palaeoproterozoic sedimentary and volcanic rocks. The basal unit is composed of eluvial breccias (Fig. 4.42a) passing upward into monomict conglomerates containing locally derived, angular and poorly sorted clasts. Member 2 (1.5–2 m thick) has a transitional contact with the underlying rocks and is composed of polymict conglomerates with variably rounded but unsorted fragments derived from the immediate basement (Fig. 4.42b–h). The conglomerates are overlain by a 0.5- to 1.5-m-thick bed of massive polymict sandstones of Member 3, followed by Member 4, which is the thickest of all (80–300 m). It is composed of boulder to pebble, polymict conglomerates grading upward into small-pebble conglomerate with concomitant increase in the degree of clast roundness and sorting. The conglomerates contain lenses of polymict sandstones (Fig. 4.42i), the amount of which increase in the uppermost part of the member. The conglomerates are crudely stratified, whereas the sandstones are massive or show crude, indistinct, discontinuous bedding, contain lonestones, and have transitional contacts with the host conglomerates (Fig. 4.42j).

Member 5 is up to 100 m thick and is termed a transitional unit of rhythmically interbedded conglomerates, sandstones and siltstones. Bed thicknesses range between 0.3 and 1.5 m. Beds in the lower and middle parts of the member have conglomeratic bases, whereas the rest of the section is composed of sandstone-siltstone rhythms. Member 6 ends the section of the Paljeozero Formation. It is a c. 100-m-thick succession of finely laminated chlorite schists resembling varves and which contain lonestones and probable dropstones (Korosov et al. 2011).

Apart from an apparent glaciogenic origin, other typical features of the Paljeozero Formation rocks include their immature composition. Heiskanen (1987b) and Ojakangas et al. (2001) assigned their accumulation to graben-like structures. Svetov and Kulikov (2011) categorised them as molasse deposits accumulated in pull-apart rift basins.

The Jatulian Super-Horizon

The Jatulian-age rocks have the widest distribution of any of the older Palaeoproterozoic rocks in the Onega Basin (Fig. 4.34). In contrast to the Sariolian Super-Horizon, the Jatulian rocks were deposited on a peneplained surface, and variably overlie older Archaean, Sumian or Sariolian rocks, in places with prominent angular unconformities (Fig. 4.37). The Jatulian rocks record a time of strong chemical weathering as evidenced by widespread regoliths (e.g. Sokolov and Heiskanen 1966) and the common occurrence of quartz-pebble conglomerates, and quartz arenites (Negrutsa 1984; Sokolov 1987b). Chemical weathering beneath the Jatulian deposits appears to be a widespread phenomenon not only in the Onega basin but over the entire Fennoscandian Shield (for details see Chap. 7.9.2).

The Jatulian sedimentary-volcanic successions attracted attention of many geologists and had been a target of multidisciplinary research over several decades starting from the 1950s (Sokolov 1955). There were several attempts to subdivide the Jatulaian Super-Horizon into smaller lithological units. Kratz (1963) and Sokolov et al. (1970) recognised three transgressive-regressive series, and consequently subdivided the Jatulian Super-Horizon into the Lower, Middle and Upper series. Later, the Lower series was further subdivided into the Majmjärvi and Jangozero horizons, and the Middle series was renamed the Medvezhgora Horizon, whereas the Upper series was termed the Tulomozero Horizon (e.g. Sokolov 1987b). In contrast, Kharitonov (1966) and Negrutsa (1984) suggested two series within the Jatulian succession. Satzuk et al. (1988) considered them as horizons and termed the lower one as "Segozero" and the upper one as "Onega" (Fig. 4.35). The former comprises predominantly clastic sedimentary and mafic volcanic rocks, whereas the latter is composed mainly of carbonate rocks and subordinate clastic sedimentary rocks.

The stratotype section of the entire Jatulian Super-Horizon was originally defined at, and described from, Segozero (Timofeev 1935; Kharitonov 1938; Negrutsa 1963, 1967; Sokolov et al. 1966, 1970; Sokolov 1972a, b); a lake located northwest of the Onega Basin per se (Fig. 4.34). The stratotype section at Segozero is more than 500 m in thickness and comprises clastic sedimentary rocks exhibiting exceptionally well-preserved and diverse sedimentary structures. Terrestrial and shallow-marine clastic sedimentary rocks display planar unidirectional and trough cross-bedding, numerous rippled and desiccated surfaces, and spectacular red beds (Fig. 4.43). The early Jatulian lithofacies at Segozero represents a succession deposited in an epeiric sea associated with the marine transgression following the post-Huronian deglaciation (see Chaps. 3.3 and 7.2). Unfortunately, the type-section was flooded when Lake Segozero was dammed in the 1970s.

A parastratotype section was later defined from, and described in, the same area, though located on a group of islands in Lake Segozero. Lithostratigraphic correlation of the parastratotype section with other sections, including those located in the Onega Basin, is given in Sokolov (1987b) and several other publications. However, the lack of precise age constraints renders equivocal validity of various proposed lithological subdivisions. In order to minimise confusion and uncertainties, the lithostratigraphic overview presented below is restricted to description of well studied and defined successions at the rank of formations. In the Onega Basin, the Jatulian Super-Horizon comprises the Jangozero, Medvezhegorsk and Tulomozero formations. It currently remains unresolved as to how these formations correlate chronostratigraphically to other Jatulian-age lithostratigraphic subdivisions outside the Onega Basin.

The Jangozero Formation

The boundaries of the Jangozero Formation were established during a 1:50,000 mapping programme (Sivaev et al. 1982). The complete formation was intersected by several drillholes in the western part of the Onega Basin (Fig. 4.44) where it has a thickness of c. 30 m; further south it thins to 13 m (Sokolov 1963). The formation was subdivided into lower and upper units termed in the following text as members. The lithological description presented below is based on drillcore 17 (Fig. 4.44; Sivaev et al. 1982).

The Lower member is a c. 30-m-thick, terrigeneous succession comprising three distinct units (Fig. 4.44, drillhole 17): a lower 12-m-thick unit of sericite-cemented, quartz-feldspar sandstone (interpreted as an in situ palaeoweathering crust) having a diffuse, transitional contact with the underlying Archaean granite and gneiss (Sivaev et al. 1982); a middle unit that is c. 6 m thick and consists of quartz-feldspar gritstone with rare quartz pebbles capped by a c. 1-m-thick quartz-pebble conglomerate; and an upper unit that is c. 10 m thick and composed of quartz-sericite-chlorite siltstone with thin beds of arkosic sandstones at the base and top (Fig. 4.44), with the latter containing siltstone clasts.

The Upper member of the Jangozero Formation is represented in the drilled area by a single, 11- to 17-m-thick, massive tholeiitic basalt with an amygdaloidal top. Further to the north, near Medvezhegorsk (for location see Fig. 4.34), the member comprises four flows of amygdaloidal tholeiitic basalts totalling 52 m, in thickness, though all rapidly wedge out laterally (Golubev 2011).

Sokolov (1987b) suggested deposition of the Jangozero sedimentary rocks in terrestrial environments and recognised lacustrine and alluvial facies. The absence of pillow lavas, the presence of widespread massive and amygdaloidal tholeiitic basalts with crustal geochemical features (Golubev 2011), modest thicknesses and rapid lateral thickness variations are suggestive of subaerial eruptions. These combined features suggest an intraplate rift setting.

The Medvezhegorsk/Koikary Formation

The Medvezhegorsk Formation was originally defined by Negrutsa (1963) and included two basaltic lava flows separated by a sedimentary unit. Later, the lower volcanic unit was assigned to the Jangozero Formation (Sokolov 1984). The most complete section of the formation was described outside the Onega Basin, north of Segozero (Sokolov et al. 1970). In the western part of the Onega Basin, a lithostratigraphic unit correlative to the Medvezhegorsk Formation was defined as the Koikary Formation. The formation was intersected by several drillholes during a 1:50,000 mapping programme and described by Sivaev et al. (1982) who subdivided the formation into a lower and an upper unit termed here as members. A description of the lithological section based on drillcore 41 (Fig. 4.44) is briefly given below.

The Lower member conformably overlies the Jangozero Formation basalts. It has a thickness of c. 20 m and comprises six distinct beds. Bed 1 (c. 7 m thick) has at its base thin intervals (up to 5 cm) of small-pebble, quartz conglomerate interbedded with magnetite-bearing quartz gritstone containing scattered plagioclase fragments. These lithologies pass into a fining-upward succession consisting of rhythmically interbedded coarse- and medium-grained gritstones.

Bed 2 (1 m thick) comprises a matrix-supported conglomerate with ellipsoidal pebbles of translucent quartz. The matrix is a haematite- and magnetite-bearing arkosic gritstone; this also forms thin, discrete layers in the conglomerate.

Bed 3 (5 m thick) is a rhythmically bedded succession of matrix-supported, quartz-pebble conglomerate that fine upward into arkosic gritstone to coarse-grained arkosic sandstone. Outcrops show trough cross-bedding, planar, unidirectional cross-bedding and rippled bedding surfaces (Fig. 4.45a–c). All lithologies contain finely dispersed, clastic magnetite and haematite. The iron oxides also occur in a concentrated form as 1- to 3-cm-thick, discrete layers containing up to 60 wt.% magnetite and haematite.

Bed 4 (7 m thick) is a rhythmically bedded succession of coarse-grained through fine-grained, clayey, arkosic sandstones to clayey siltstones. The sandstones forming the base of Bed 4 are partially calcite-cemented.

Bed 5 (1.5 m thick) is represented by calcite-cemented quartz sandstone with angular, platy clasts of dark brown, clayey siltstones; these are apparently chips derived from ripped-up desiccated layers, as suggested by neighbouring outcrops (Fig. 4.45d).

Bed 6 (3 m thick) is a volcaniclastic unit consisting of dark brown and brown-red tuffaceous siltstones with layers of black, tuffaceous sandstones and thin intervals (5 cm thick) of amygdaloidal basaltic tuffs. All rocks contain dispersed haematite, and some intervals are cemented with calcite. Bed 6 represents a unit transitional into the Upper (volcanic) member.

The Upper member is only 10 m thick in the drillhole 41 section (Fig. 4.44) and is composed of amygdaloidal, tholeiitic basalts. The thickness of the volcanic member shows a considerable variation, from less than 3 m (Fig. 4.44, drillhole 17) to 250 m near Medvezhegorsk and further east. The thickest sections contain up to 9 lava flows separated by thin intervals of ash beds and cherts. The volcanic rocks occur as massive, amygdaloidal, fluidal and pahoehoe lava flows; some flows have a fragmented top. Columnar jointing is common, whereas pillows are rare (Fig. 4.45e–j). The volcanic rocks are tholeiitic basalts with crustal geochemical affinities and a high degree of iron oxidation (Golubev 2011). Many flows exhibit reddening caused by haematisation. Golubev (2011) classified the Medvezhegorsk mafic volcanic rocks as intraplate flood-basalts erupted predominantly in subaerial environments. This is in coherence with sedimentological features of the Lower member, which were interpreted by Sokolov (1987b) to represent fluvial systems.

The Tulomozero Formation

The Tulomozero Formation was defined by Perevozchikova (1957) and was first studied in detail by Molotkova and Anistchenkova in 1964 (cited in Negrutsa 1984). The formation has been a subject of intensive sedimentological (Heiskanen 1975; Akhmedov and Krupenik 1995; Sokolov 1987b; Sokolov et al. 1970; Satzuk et al. 1988; Akhmedov et al. 1993; Lavrov 1999; Melezhik et al. 2000, 2001; Brasier et al. 2011), geochemical (Yudovich et al. 1991; Tikhomirova and Makarikhin 1993; Gorokhov et al. 1998; Heiskanen and Rychanchik 1999; Melezhik et al. 1999a; Akhmedov et al. 2004; Kuznetsov et al. 2010; Reuschel et al. 2012) and palaeontological research (Butin 1959, 1966; Vologdin 1966; Makarikhin and Kononova 1983; Makarikhin 1992, 2011). Well-exposed sections are not readily available, hence the lithological composition, thickness and facies variations of the formation across the basin are not well understood. Currently available lithostratigraphic descriptions are based on research of surface outcrops supported by logs of a series of drillholes that were made for the purpose of establishing the local lithostratigraphy. Deposition of the Tulomozero Formation has been constrained by the Pb-Pb age on dolomite of 2090 ± 70 Ma (Ovchinnikova et al. 2007), and a gabbro sill (Koikary-Svjatnavolok) intruded into the Tulomozero rocks has been dated at 1983.4 ± 6.5 Ma (Philippov et al. 2007).

No agreed upon subdivision of the Tulomozero Formation exists. Molotkova and Anistchenkova (cited in Negrutsa

1984) divided the Tulomozero Formation into lower, middle and upper units. Sokolov (1963) renamed the units as "piles"; the lower unit became the Conglomerate-Sandstone pile and the middle and upper units were divided into six piles and three piles, respectively (Fig. 4.46). Akhmedov et al. (1993) used the terms Lower, Middle and Upper sub-suites, and subdivided the two youngest sub-suites into several "piles" (Fig. 4.46). For example, their Middle sub-suite piles are, from oldest to youngest: (1) Limestone-Dolostone-Mudstone, (2) Phosphate-Dolostone, (3) Magnesite-Dolostone, (4) Stromatolite-Dolostone, (5) Siltstone, (6) Pisolite-Dolostone and (7) Haematite-Sandstone (note that these names do not correspond to the lithologically defined piles of Sokolov 1963). However, subsequent work showed that lithologies were variable in their development, e.g. limestones and phosphorus-containing rocks are absent in drillholes 5177 and 4699 (Fig. 4.47) but thin magnesite beds occur throughout the lower half of the formation and are most abundant in Pisolite-Dolostone pile (Melezhik et al. 2000). Similarly, stromatolite beds can be abundant (drillholes 7 and 9, Fig. 4.48; for drillhole locations see Fig. 4.47) or rare (FAR-DEEP Hole 10B, see Chap. 6.3.1). Thus, it became apparent that the specifically defined lithological names were difficult to recognise in practice. Consequently, Melezhik et al. (1999a, 2000) defined members, designating these with letters (starting with "A") (Fig. 4.46), some of which consisted of several 'piles' (e.g. Piles 2, 3 and 4 comprise Member B because these are indistinguishable in some of the drilled sections, Fig. 4.47b). In the latest compilation, Medvedev et al. (2011) adopted the lithological subdivision and names proposed by Sokolov (1963) and suggested a subdivision based on stromatolite morphologies (Fig. 4.46).

In this contribution, the lithostratigraphic subdivisions of the Tulomozero Formation are a combination of the sub-suites of Akhmedov et al. (1993) and the members of Melezhik et al. (1999a, 2000). Description of the units is based on drillcores 5177, 4699, 7, 9 and 41, and sedimentological research in the vicinities of the drilling sites (Akhmedov and Krupenik 1995; Lavrov 1999; Melezhik et al. 1999a, 2000). Drillholes 5177 and 4699 are in the eastern part of the basin (Fig. 4.47a) and overlap, intersecting c. 500 m of section. Drillholes 7 and 9 in the western part of the basin (Fig. 4.47a) overlap and intersected c. 700 m of section. The two pairs of drillholes penetrated the entire formation except its lowermost part, which was intersected by drillcore 41 (Fig. 4.44b). Photographs in Fig. 4.48 show sedimentological features of the main rocks types of the Tulomozero Formation in the drillcores and from natural outcrops.

The Lower sub-suite is as much as 20 m thick but is highly variable and wedges out locally. In drillcore 41, its base is a 4-m-thick unit composed of interbedded haematite-bearing pink quartz sandstone and quartz conglomerate resting on the Medvezhegorsk Formation basalt (Fig. 4.44b). The lowermost bed of the quartz conglomerate contains fragments of the underlying basalt. The upper part of the sub-suite is c. 5 m of interbedded quartz gritstone grading into sandstone.

The Middle sub-suite is recorded in drillcores 5177/4699 and consists of Members A-E of Melezhik et al. (1999a, 2000). Member A (Fig. 4.47b) begins with a c. 30-m-thick unit of variegated, sparry and crystalline dolostones with flat-laminated and markedly divergent, small columnar stromatolites. Flat-laminated stromatolitic layers are often multicoloured and syngenetically brecciated. The dolostone unit contains in its lower part a c. 5-m-thick magnesite bed and an interval of dissolution-collapse breccia (Fig. 4.47b). The magnesite is white, grey, yellowish, fine- to coarse-grained and structureless (for details see Melezhik et al. 2001). The dissolution-collapse dolostone breccias appear as poorly cemented fragments of brown, pink and white dolostones embedded in a dark brown dolomite-sericite-chlorite material enriched in iron oxide.

The dolostone unit with magnesite and breccias is followed by five sandstone-siltstone-dolostone cycles. The thickness of individual cycles ranges between 30 and 70 m (Fig. 4.47b). The lowermost one starts with reddish, ripple-marked and cross-stratified, dolomite-cemented, quartz sandstones (c. 8 m thick). The sandstone unit has an erosional and channelised base and passes upward into a thicker (c. 20 m thick) unit composed of variegated sandstone-mudstone variably with flaser, wavy and lenticular bedding. Some beds display herringbone cross-bedding. The uppermost portion of the first cycle is 10 m thick and comprises variegated, sparry and crystalline dolostones with highly cracked and syngenetically brecciated, flat-laminated stromatolitic layers. The second, third and fourth cycles have similar lithologies but exhibit variable thicknesses and lack the basal sandstone member. The fourth has, in addition, a dissolution-collapse breccia on top. The fifth cycle is incomplete and comprises only a variegated, heterolithic sandstone-mudstone unit with lenticular bedding.

A characteristic feature of Member A is abundant evaporitic growth of sulphates and their partial or total dissolution (Fig. 4.48a–k). These are exceptionally well preserved in drillcores 7 and 9 and include dolomite- and calcite-replaced bedded and nodular Ca-sulphates (Fig. 4.48a), dolomite-pseudomorphed gypsum crystals and their clusters (Fig. 4.48b–d, g, h), former chicken-wire anhydrite (Fig. 4.48f), diagenetic growths of sulphates (Fig. 4.48i, j) and dissolution-collapse breccias (Fig. 4.48e, k).

Member B has a thickness and lithological composition differing in the 4699 and 5177 sections (Fig. 4.47b). In both cores, thin magnesite beds mark the base of the member. In the 4699 section, the member is composed of a c. 60-m-thick dolostone bed with c. 20 m of dissolution breccias on top. In contrast, in the 5177 section, the member comprises two sandstone-dolostone layers, 20 and 70 m in thickness.

In outcrop, magnesite layers are associated with individual and coalesced former sulphate nodules formed in clayey dolostone passing downwards into bedded dolarenite. One nodular bed has a mud-draped, irregular surface and is overlain by magnesium-rich dolarenite (Akhmedov et al. 1993; Melezhik et al. 2005b). Elsewhere in the Onega Basin, the lower part of the member has been described (Akhmedov et al. 1993) as containing the phosphorus-bearing mineral, wagnerite ($(Mg, Fe)_2PO_4F$) in the form of a concretion, and abundant silcretes.

Dolostones in both sections are represented by stromatolitic mat lithofacies and rare red dolorudites (Fig. 4.48l, m, p). Outcrops exhibit beds of desiccated microbial laminates, columnar stromatolites and a variety of biostromal and biohermal stromatolites (Fig. 4.48n–o). In places, stromatolitic beds are synsedimentary deformed into tepee structures (Akhmedov et al. 1993). Columnar stromatolites (*Nuclephyton, Carelozoon, Sundosia*; Makarikhin and Kononova 1983) and oncolites are abundant and a characteristic feature of the member.

The sandstones, which occur at the base of the drillhole 5177 section, comprise the lower parts of two sandstone-dolostone cycles. The sandstone beds have erosional bases, are grey, medium-grained, dolomite-cemented, ripple-marked, cross-stratified or structureless. The dolostones are sparry and micritic, structureless or crudely stratified, and contain cauliflower and castellated quartz aggregates replacing calcium sulphates (Melezhik et al. 2000). Similar pseudomorphs containing abundant gas and fluid inclusions are present in the drillhole 4699 section (Fig. 4.48q, r).

The breccia occurring on top of the member in the 4699 section is composed of dolomite-cemented fragments of brown, pink and white dolostones (Fig. 4.48s, t). The breccia contains small pockets of dark brown, dolomite-sericite-chlorite-haematite material.

Member C is c. 90 m thick and contains four complete sandstone/siltstone-dolostone cycles, 20 to 30 m in thickness; these are lithologically similar to those and described in Member A (Fig. 4.47b). Figure 4.48u–ae illustrate some sedimentological features of Member C; photographs are mainly from core 7 whose chemostratigraphic correlation with cores 5177 and 4699 is shown in Fig. 4.49. The lower part of the cycles (10–30 m in thickness) are composed of beige, pale pink and purple, fine-grained, dolomite-rich sandstones and siltstones. Some beds show grading (Fig. 4.48v), but primary structure is generally expressed by lenticuls and lenses of siltstone and sandstone, and by cross-bedded sandstone layers and lenses (ripples) in a silty material; some silty layers are desiccated (Fig. 4.48w). Mud-cracks, tepee structures and pseudomorphs after gypsum have been observed (Akhmedov et al. 1993). The upper part of the cycles (5–10 m in thickness) consists of red, crystalline, siliceous dolostone and pink massive and bedded dolomarls with thin intervals of wavy and lensoidal lamination. Some intervals show desiccation and contain small dolomite-pseudomorphed crystals of gypsum (Fig. 4.48ac–ae). Nearby outcrops and other core material show that Member C contains sporadic oncolites and abundant beds with columnar stromatolites (*Colleniella, Omachtenia, Carelozoon*) forming laterally continuous biostromes (Akhmedov et al. 1993). Figure 4.48u, x–ab illustrate some stromatolite morphologies.

Member D is 60 m thick, with its base defined by a c. 2-m-thick magnesite bed (Fig. 4.47b). The member consists of crystalline and stromatolitic dolostones, a bed of heterolithic sandstone-siltstone in the middle part of the section, and magnesite beds and dissolution breccias in its upper part (Fig. 4.48af–aj). Dolostones in nearby outcrops contain spaced, solitary, large stromatolite columns (*Colonnella*) and abundant oncolites. Magnesite occurs as two 2-m-thick beds as well as an uppermost one 15 m thick between two intervals of dissolution breccias. Magnesite occurs only in core 5177 and was not documented in cores 7 and 9. It appears as white, grey or yellow, mainly structureless rock having micritic, crystalline or coarsely crystalline stucture (Fig. 4.48ah). Finely laminated micritic and stromatolitic magnesite is rare (for details see Melezhik et al. 2001).

The dissolution-collapse breccias appear as poorly-cemented fragments of brown, pink and white dolostones and red, finely-laminated mudstones, either encased in insoluble residue or cemented by white calcite (Fig. 4.48af). The bed of sandstone-siltstone consists of 1–4 cm thick sandstone-siltstone couplets. Many sandstone layers show herringbone cross-bedding. Brown mudstone beds occurring in the upper part of the member contain abundant dolomite- and quartz-pseudomorphed gypsum crystals (Fig. 4.48ai, aj).

Member E is c. 120 m thick and is characterised by intercalated brown, dark grey and bright red, haematite-stained heterolithic quartz sandstones and subordinate siltstones. There also are three discrete dolostone beds (Fig. 4.47b). The sandstones are chemically mature, dolomite-cemented and have tabular cross-bedding, and some beds contain pebbles of quartz-haematite rock. The sandstones form 5- to 70-cm-thick layers separated by thinner siltstone intervals that are mud-cracked and display fine laminae defining low-relief hummocks and swells with amplitudes of less than 1 cm and wavelengths of 1–3 cm. Previous workers (e.g. Akhmedov et al. 1993) have reported on the presence of casts of halite cubes and a 1- to 2-m-thick bed of clastic haematite ore associated with the siltstone facies. The dolostones of Member E are white, grey, beige, pink and crystalline rocks, and, in most cases, are structureless or indistinctly parallel-laminated. Rare shale laminae and flat-laminated stromatolites have been observed.

The Upper sub-suite begins with Member F, whose base is marked by a c. 50-m-thick flow of amygdaloidal basalt that was documented in several drillholes and outcrops across the Onega Basin (Satzuk et al. 1988); hence it was considered as a marker horizon (e.g. Krupenik et al. 2011a; Krupenik and Sveshnikova 2011). The basalt is overlain by two sandstone/mudstone-dolostone cycles, 10 and 20 m in thickness. The lower unit in both cycles is composed of variegated, dolomite-cemented sandstones and siltstones with flaser and lenticular bedding, which commonly show disruption from desiccation and growth of evaporitic minerals (Fig. 4.48ak, al). The dolostones comprising the upper part of the cycles are crystalline, though there are intervals retaining stromatolite layers (Fig. 4.48am). In a section intersected by drillhole 7, the member contains calcite-pseudomorphed halite crystals in red mudstone (Fig. 4.48an).

Member G consists of c. 30-m-thick of white, fine-grained dolostones, which are devoid of siliciclastic intercalations and exhibit indistinct wavy lamination (Fig. 4.48ao). Corresponding outcrops show several intervals characterised by in situ brecciation and formation of stone rosettes (Fig. 4.48ap, aq). Akhmedov et al. (1993) reported an oncolite bed (30 m in thickness) and a cupola-like bioherm in the upper part of the member.

Member H comprises the top of the Tulomozero Formation. It is less than 20 m thick and composed of flat-laminated and columnar stromatolitic dolostones in its lower part (Fig. 4.48ar) and red or brown, fine-grained, thinly laminated, micritic, allochemical dolarenites in the uppermost part of the member (Fig. 4.48as). Allochems are represented exclusively by dolomitic oolites. Dolostones contain relicts of primary porosity (i.e., birdseyes) and vugs filled with cryptocrystalline quartz.

The Tulomozero Formation has been central in a number of studies investigating the global carbon cycle during the Palaeoproterozoic. Since the work of Galimov et al. (1968) and Schidlowski et al. (1975), who discovered considerable enrichment of ^{13}C in rocks considered to be equivalent to the Tulomozero, the Tulomozero carbonate succession has been the focus of numerous detailed isotopic studies (Yudovich et al. 1991; Akhmedov et al. 1993; Karhu 1993; Tikhomirova and Makarikhin 1993; Melezhik and Fallick 1996; Heiskanen and Rychanchik 1999; Melezhik et al. 1999a, 2005b). A well-established, drillcore-based $\delta^{13}C_{carb}$ profile shows an overall stratigraphic trend from +18‰ at the base to +6‰ at the top of the succession (Fig. 4.49; for details see Chap. 7.3). Some workers have stressed local factors as potentially enhancing a global seawater $\delta^{13}C$ signal (e.g. Heiskanen and Rychanchik 1999; Melezhik et al. 1999a, 2005b) and Akhmedov et al. (2004) emphasised complexity of the depositional setting of the Tulomozero Formation, but details remain too sketchy for a robust understanding of the possible influence of depositional environments on the geochemical/isotopic trends. Hopefully, though, a recent c. 3,537-m-deep parametric drillhole (Morozov et al. 2010) that intersected the Tulomozero Formation at a depth of 2,900–2,100 m combined with FAR-DEEP 10A, 10B and 11A drillholes, may yield greater insight into nature of the facies distribution across the Tulomozero basin (an area of 60 × 100 km; Fig. 4.47a), and the influence of local factors on the isotopic signals.

Figure 4.50 illustrates the lithofacies variation documented in FAR-DEEP drillholes and the various difficulties regarding stratigraphic correlations (for details see Chaps. 6.3.1 and 6.3.2). If the basalt at the base of the Lower sub-suite (Member F, in Fig. 4.47b) is taken as a marker horizon (as suggested by Krupenik et al. 2011a; Krupenik and Sveshnikova 2011), then the section in 11A from above the basalt and below the overlying Zaonega Formation (i.e., the Upper sub-suite) differs markedly from that in 10B (Fig. 4.50). These sections are less than 10 km apart (Fig. 4.47a) yet the 11A section is entirely "dolomitic", with several thin beds of dissolution breccia at the base and several thick beds of carbonate conglomerate in the middle, whereas the section in 10B is siliciclastic at the base, shaly in the middle with several thick intervals of collapse breccias and is dominantly "calcitic" in the top. Moreover, neither lithologies of the 10B section nor the lithologies of the 11A section can be readily correlated with the type section intersected by drillholes 5177/4699 (Fig. 4.51).

In the case of drillcore 10A, it represents a continuation of the 10B section below the basalt (see a tentative correlation in Fig. 4.50), hence belonging to the upper part of the Middle sub-suite. Unfortunately, this core was drilled nearly parallel to dip (for detailed description see Chap. 6.3.1), but, nevertheless, it contains three magnesite beds and a collapse breccia that might be correlative to Member D in the 5177/4699 section (Fig. 4.51). If correct, then the interbedded sandstone-shale-dolostone in 10A that occur above the magnesite interval could correspond to Member E in the 5177/4699 section. However, the strata below the magnesite interval is composed largely of marls, which are not present in the 5177/4699 section below Member D. Hence, it is evident that there is considerable lithofacies variation across the basin.

The 3,537-m-deep parametric drillhole (Morozov et al. 2010; Krupenik et al. 2011a) made in the southern part of the basin (Fig. 4.47a) represents additional remarkable evidence of facies variations in the Tulomozero basin (Fig. 4.51). The section begins with a 294-m-thick interval of red and grey halite with fragments of anhydrite and magnesite (2,944–2,750 m) resting directly on 2711 ±17 Ma Archaean granite (Krupenik et al. 2011a). The halite bed is followed by a 345-m-thick (2,750–2,405 m) succession composed of rhythmically interbedded siltstone-magnesite-anhydrite. The cycles are 10–16 m in thickness and comprised of variegated, laminated siltstone (lower unit), laminated and massive, fine-grained magnesite (middle unit) and white, massive, coarsely-crystalline anhydrite-magnesite rocks (upper unit). Anhydrite-magnesite rocks also occur as individual units several tens of metres thick in the middle of the succession (Fig. 4.51) and are injected by numerous anhydrite-magnesite veins. Both the halite and the cyclically bedded siltstone-magnesite-anhydrite succession were termed, the Halite and Ulita strata, respectively (Krupenik et al. 2011a). Neither the halite nor the massive anhydrite-magnesite rocks were previously documented within the Tulomozero Formation in other parts of the basin. Sulphur isotopic composition ($\delta^{34}S$) measured from massive anhydrite ranges between +4.8 and +6.1‰ (Krupenik et al. 2011b). In contrast, $\delta^{34}S$ obtained from carbonate-associated sulphates and from relicts of anhydrite and barite in quartz-pseudomorphed sulphate nodules and dissolution breccias in the FAR-DEEP cores range between 9.0 ± 1.1‰ and 11.0 ±3.1 ‰ (Reuschel et al. 2012).

A succession overlying the Ulita strata has a thickness of 290 m (2,405–2,115 m) and was subdivided into two parts. The lower part (2,405–2,333 m) is composed of interbedded dolomarls, dolostones and magnesite-anhydrite rocks. The upper part, above 2,333 m, comprises variegated dolomarls, which are rhythmically interbedded with white and pale pink dolostones. The cycles have a thickness ranging between 0.2 and 0.5 m, though the dolostone occurs in part as 1- to 2-m-thick, discrete beds (Krupenik et al. 2011a). Many of the dolostone interbeds are fragmented and appear to be dissolution collapse breccias with white, blocky, dolomite cement. At a depth of 2,264–2,205 m, the dolomarls contain abundant dolomite-replaced anhydrite nodules and gypsum crystals. This section was correlated with the Siltstone, Pisolite-Dolomite and Haematite-Sandstone piles (Krupenik and Sveshnikova 2011), which correspond to Members C–E in the 5177/4699 section (Fig. 4.51). Beds of stromatolitic and nodular dolostone (former anhydrite nodules) occuring at a depth of 2,272–2,264 m were correlated with the Stromatolite-Dolomite pile of Akhmedov and Krupenik (1995), which corresponds to the upper part of Member B in the 5177/4699 section (Fig. 4.51).

The interbedded dolomarls and dolostones are capped by mafic lava (the marker horizon; Krupenik et al. 2011a; Krupenik and Sveshnikova 2011) at a depth 2,156–2,146 m, and then followed by a c. 30-m-thick succession of dolostones and marls (2,146–2,115 m). Within this interval, a 14- m-thick (2,129–2,115 m) interval of red, indistinctly laminated dolostones and dolomite-cemented sandstone defines the uppermost part of the succession. Krupenik and Sveshnikova (2011) correlated these rocks with the Redbed-Dolostone pile of Akhmedov et al. (1993), which correspond to Member H (Fig. 4.51).

Figure 4.51 summarises the lithological comparison/contrast of the Tulomozero Formation revealed by various drilling programmes in different parts of the Onega Basin. The striking feature is a spatial lithological heterogeneity apparently caused by a much higher degree of isolation and evaporation in the southern part of the Onega Basin. Other details of depositional variability across the basin are yet to emerge, in particular thickness variations and the presence/absence of internal erosional surfaces. The succession occurring between the basalt marker-horizon and the overlying Zaonega Formation (the Upper sub-suite) shows a significant thickness variation from 30 m in the southern part of the basin (Onega parametric hole, Fig. 4.51) to 60–80 m in the eastern and western parts (drillholes 7 and 5177) to almost 200 m in the northern part (FAR-DEEP Hole 11A). This implies that the southern section either was eroded or had reduced accommodation. An alternative involves a tectonically enhanced (folded) thickness of the Upper sub-suite and its tectonic contact with the Zaonega formations in the 11A section (for details see Chap. 6.3.2).

Interpretation of core material from drillholes 5177, 4699, 7 and 9, and sedimentological features obtained from available and studied outcrops suggest a complex combination of terrestrial and marine depositional environments involved in the accumulation of the Tulomozero Formation (Sokolov 1963; Sokolov et al. 1970; Akhmedov and Krupenik 1995; Lavrov 1999: Melezhik et al. 2000; Akhmedov et al. 2004; Brasier et al. 2011). The lower part of the formation (the Lower sub-suite and Member A of the Middle sub-suite) contains abundant evidence for sulphate deposition, and as a whole, was apparently accumulated in lower-energy intertidal, playa lake and coastal sabkha environments influenced by braided fluvial systems (Melezhik et al. 2000). In the southern part of the basin, the Onega parametric hole data suggest evaporitic settings with a high degree of basinal restriction leading to deposition of halite. However, accumulation of several hundreds of metres of massive anhydrite-magnesite rocks implies a significant and continuous influx of seawater, hence the basin was connected to the sea. The succession of the middle part of the formation (Members B, C and lower D), containing abundant stromatolite lithofacies, bears many features of intertidal and partially subtidal depositional environments (Melezhik et al. 2000). Stromatolitic strata in the western part of the basin (area around drillholes 7 and 9; Fig. 4.47a) have been interpreted as reefogenic biohermal stromatolite facies (Satzuk and Kononova 1975; Akhmedov and Krupenik 1995; Lavrov 1999; Akhmedov et al. 2004). Thus, this part of the Tulomozero succession may represent a carbonate platform (see Chap. 3.3). The upper part of Member D and most of Member E in the western and eastern parts of the basin (drillholes 5177/4699 and 7/9) are marked by widespread desiccation features and halite cube casts; all are indicative of a playa setting (Melezhik et al. 2000). A shallow-water depositional environment has been suggested for the upper part of the succession in the same area

(Melezhik et al. 2000), whereas an upper tidal and supratidal sabkha has been inferred for the northern part of the basin (FAR-DEEP Hole 10B; Brasier et al. 2011).

The Ludicovian Super-Horizon

The Ludicovian Super-Horizon was originally defined by Sokolov and Galdobina (1982) to include a thick succession of grey and black, C_{org}-rich, sedimentary rocks and mafic igneous rocks overlying red and variegated Tulomozero dolostones. Later, the super-horizon was subdivided into two lithostratigraphic units: the lower, termed the Zaonega Horizon, and the upper, termed the Suisari Horizon. Within the Onega Basin, these correspond to the Zaonega and Suisari formations, respectively (Galdobina 1987a).

The Zaonega Formation

The Zaonega Formation has the greatest areal distribution of all formations in the Onega Basin (Fig. 4.34). It was originally defined by Perevozchikova (1957) who subdivided it into two sub-formations: a lower carbonate-argillite and an upper volcano-sedimentary, which contains voluminous C_{org}-bearing rocks. The Zaonega Formation represents a stratotype section of the Zaonega Horizon (Rychanchik 2011a). Its lower contact was documented in various places, but interpretation of its nature remains controversial (e.g. Negrutsa 1984; Galdobina 1987a). In a summary provided by Galdobina (1987a), variable relationships were described within the Onega Basin. In its southwestern part, the Zaonega Formation lies on white and pink Tulomozero dolostones conformably but with an erosional surface implying a hiatus in the geological record. In the northwestern part, at Belaja Gorka (for location see Fig. 4.34), red, laminated marls of the Zaonega Formation show a sharp contact on underlying red, iron-hydroxide-stained, brecciated dolostone (which rests on a grey and pink dolostone having an irregular karstified surface) that was interpreted as "terra rossa" (a 'residue' of red clay and other non-soluble material after weathering of carbonate rocks under oxidising conditions). If this is not due to modern weathering, then it would imply uplift, erosion and weathering of the Tulomozero succession, and a significant non-depositional break prior to the deposition of the Zaonega Formation. At Sundozero (Fig. 4.34), the contact between the two formations is marked by a bed composed of dolostone conglomerate and breccia with fragments of stromatolites (e.g. Sokolov 1963). Similarly, in other places of the northwestern part of the basin, Negrutsa (1984) noted an as much as 32-m-thick interval of red dolostone conglomerate, breccias, and dolomite-cemented sandstone and gritstone at the base of the Zaonega Formation. Plisov and Sapozhnikova (cited in Negrutsa 1984) interpreted these rocks as eluvial facies that signify a non-depositional break and erosion of the Tulomozero succession. In contrast, south of Sundozero, other workers interpreted the contact between the two formations as a transitional passage from the Tulomozero red dolomarls to grey-green dolomarls and siltstones of the Zaonega Formation (Medvedev et al. 2011). As a result, the nature of the contact, whether conformable or defining a variety of unconformities, renmains unresolved.

FAR-DEEP Hole 11A intersected the base of the formation in the northern part of the basin (Fig. 4.34). The contact is sharp and shows red staining by iron-oxides. Tulomozero marine, stromatolitic dolostones beneath the contact dip steeply. Above the contact the Zaonega Formation comprises flat-lying, red, parallel-laminated and cross-bedded, non-marine sandstones that pass up-section into grey-green, laminated and graded sandstone-siltstone-mudstone (see Chap. 6.3.2). Hence the contact between the two formations could define an angular unconformity. The latter interpretation implies a period of deformation, uplift and erosion of the pre-Zaonega rocks. This is far-reaching implications for the tectonostratigraphic evolution of the Karelian craton and it requires additional research before robust conclusions are drawing.

There are two views on the lithostratigraphy of the Zaonega Formation (for references see Rychanchik 2011a). Polekhovsky and Golubev (1989) subdivided the formation into three sub-suites, starting from the oldest: mudstone-carbonate-schist (records a transgressive cycle); volcanogenic-sedimentary; and (3) sedimentary volcanogenic (igneous rocks comprise >50 % of the section). In contrast, Galdobina (1987a) suggested a subdivision into Lower and Upper sub-suites. The former composed of lower and upper "piles" of sedimentary rocks (renamed as Members A and B by Melezhik et al. (1999b)) and latter comprised of three piles (Members A, B and C of Melezhik et al. (1999b)) of sedimentary and volcanic rocks with three spatially separated types of section: volcanogenic (southwestern part of the basin); sedimentary-volcanogenic (central part); and (3) volcanogenic-sedimentary (north-eastern part of the basin). Local geologists favour the "two-sub-suite" subdivision (Rychanchik 2011a).

The Lower sub-suite is a c. 200-m-thick succession composed mainly of clastic, horizontally bedded, cross-bedded and laminated rocks (Galdobina 1987a). Member A of the sub-suite comprises 100–130 m of rhythmically interbedded feldspar-quartz sandstones and siltstones with a variable content of carbonate and mica. Rare lenses of dolomitic conglomerate occur at the base. Light grey, thin, dolomite layers intercalated with siltstone occur locally and increase up-section. Many beds are graded and contain small-scale slumps and slump-scars (Fig. 4.52a–c). The rocks contain disseminated pyrite. Grey and dark-grey, graded, carbonaceous, quartz-sericite, quartz-sericite-chlorite and quartz-sericite-chlorite-carbonate shales with disseminated sulphides (Fig. 4.52d–e) overlie those rocks. The graded pattern is emphasised by the distribution of organic matter and sulphides that enrich the upper portions of the graded beds. The rocks of Member A have been affected by secondary (epigenetic) oxidation resulting in variegated colouration and decomposition of pyrite.

Member B has a thickness of 60–90 m and consists of fine-grained sandstone, siltstone and shale. The middle part contains a unit of thinly laminated, marly rhythmites. The rhythmites have conchoidal fractures and a variegated colour ranging from dark grey and grey through grey-green to grey-pink. Shale components differ in dolomite content and each consecutive rhythm exhibits a slightly variable colour. The thickness of individual rhythmites ranges between c. 1 mm to less than 1 cm and closely resembles that of varves (Fig. 4.52f–h). A specific appearance of these rocks, termed locally as "krivoserite" (the term was introduced in the nineteenth century by Schmidt), makes them a marker bed that has been documented in many drillholes and traced across the Onega Basin (Rychanchik 2011a). The upper portion of the member is formed by grey and dark-grey, fine-grained, horizontally bedded, quartz-feldspar sandstone, siltstone and quartz-sericite-chlorite shale. The rocks of the member are typically variegated in colour due to postdepositional (epigenetic) infiltration of several pulses of oxidising and chemically reducing fluids (Fig. 4.52i–j).

The Onega parametric drillcore data (Krupenik et al. 2011b) show that the carbonate rocks of the Lower sub-suite exhibit a positive carbon-isotopic excursion from +5‰ to +10‰ over 100 m and then decline to 0‰ over another 100 m. Hence, the double isotopic excursion, if primary in origin, may record the final termination of the Lomagundi-Jatuli isotopic event (for details see Chap. 7.3).

The Upper sub-suite is 600–650 m thick and differs from the Lower sub-suite in that it contains several intervals of C_{org}-rich sedimentary rocks interbedded with abundant lava flows and intruded by gabbro sills. Although three types of section have been recognised in the Onega Basin (volcanogenic, sedimentary-volcanogenic, and volcanogenic-sedimentary), their individual descriptions are not available in accessible literature and a "summary" section is presented below based on previously and recently published data (Galdobina 1987a; Rychanchik 2011a). Detailed bed-by-bed description and photo-documentation of individual sections based on FAR-DEEP material are presented in Chaps. 6.3.3 and 6.3.4.

Member A is a c. 200-m-thick unit (Galdobina 1987a) and consists mostly of cyclically interbedded greywackes and mudstones with subordinate calcareous greywackes, sandy limestones, and minor intraformational, matrix-supported conglomerates and breccias (Fig. 4.52k–w). The succession of interbedded greywacke and mudstone exhibits Bouma sequences, hence implying deposition from turbidity currents. Clastic rocks have enhanced abundances of Na_2O and were assigned to "sodium-series" rocks (Filippov et al. 1994). Scattered carbonate beds and lenses (mainly dolomitic) of sedimentary origin as well as diagenetic concretions and concretionary layers (Fig. 4.52l) occur throughout the member. Extensional cracks in greywackes are filled with calcite and a pyrobitumen-rich material (Fig. 4.52r, t). The member contains three 12- to 35-m-thick intervals of C_{org}-rich rocks. Igneous rocks are represented by amygdaloidal-textured mafic lava flows (Fig. 4.52x, y) and gabbro sills. Both sills and lava flows are associated with the formation of peperite (Biske et al. 2004; Poleshchuk 2011).

Member B is c. 300 m thick (Galdobina 1987a) and composed of greywacke, black C_{org}-bearing siltstone and mudstone. Similar to Member A, the siliciclastic rocks show rhythmic bedding and contain Bouma sequences (Fig. 4.52z–ac). Clastic rocks have $K_2O/Na_2O > 1$ and hence were assigned to "potassium-series" rocks (Filippov et al. 1994). Abundant dark grey dolostones, limestones and dark grey cherts (locally termed "lydites") are present in the upper part of the section, and distinguish Member B from other lithological units of the Zaonega Formation. Both the carbonate rocks and the cherts occur as lenses (or diagenetic nodules) and appear mostly as massive rocks (Fig. 4.52ad–ah). Melezhik et al. (1999b) considered the massive, lensoidal dolostones and limestones as diagenetic megaconcretions. Matrix-supported, polymict, intraformational conglomerate and breccias are abundant (Fig. 4.52ai), with some cemented by pyrobitumen-rich material (Fig. 4.52aj).

Basaltic, amygdaloidal and pillow lava flows (Fig. 4.52ak–am) and associated peperites are present, but only in the sedimentary-volcanogenic and volcanogenic-sedimentary types of section. The member contains six intervals of C_{org}-rich rocks ranging in thickness from 5 to 120 m. Among the C_{org}-rich rocks (>20 wt.% total organic carbon, TOC) Rychanchik et al. (2011) recognised three lithological/genetic types termed "maksovite" (up to 45 wt. % TOC), "shungite" (up to 80 wt.% TOC) and anthraxolite (up to 98 wt.% TOC) (Fig. 4.52an–aq); for detailed characteristics of the C_{org}-rich rocks see Chap. 7.6. The association of dolomite-lydite-"shungite" (C_{org}-rich rocks) of Member B can potentially be used as a marker horizon for correlation of Zaonega sections elsewhere in the Onega Basin.

Member C is a 200- to 300-m-thick unit. In places, its base is marked by a bed of tuffogenic conglomerate that passes laterally into tuffogenic sandstone; clasts are derived from Members A and B (Galdobina 1987a). Member C comprises siltstones, tuffite, quartz-biotite-chlorite and biotite-albite-quartz schists with beds of black, C_{org}-bearing dolostones.

Galdobina (1987a) emphasised that abundant mafic lavas are an essential component of the middle and upper parts of the Zaonega Formation. Thicknesses of individual lava flows and their packages vary across the Onega Basin (Golubev and Kulikov 2011). However, despite numerous drillholes, which intersected different stratigraphic intervals of the Zaonega Formation, there are no published bed-by-bed descriptions that characterise individual volcanogenic (south-western part of the basin), sedimentary-volcanogenic (central part), and volcanogenic-sedimentary (north-eastern part of the basin) types of successions. Galdobina (1987a) stated that in the western part of the basin (apparently within the volcanogenic type of succession) a series of lava flows (8–10 in total) comprises a 200- to 300-m-thick package, which would be the thickest known in the Zaonega Formation. On Lelikov island (for location see Fig. 4.34) mafic rocks dominate the succession and three volcanic packages have been observed. Each starts with a coarse-grained plagioclase basalt gradually passing into a fine-grained basalt. The coarse-grained plagioclase basalts can be easily confused with a gabbro when sampled/studied from isolated exposures. Golubev and Kulikov (2011) recently provided an up-to-date compilation on the Zaonega volcanic rocks and described several volcanic piles from western, central and northeastern parts of the basin. However, in this compilation the stratigraphic positions of the described sections remained unspecified, and it is unclear how much of the formational thicknesses are occupied by volcanic rocks in various locations.

The recent Onega parametric drillhole (southwest Onega Basin, volcanogenic succession) and FAR-DEEP drillholes (northeast Onega Basin, volcanogenic-sedimentary succession) allow a quantitative comparison/contrast of the lithological composition of the Upper sub-suite deposited in different settings. In the volcanogenic-sedimentary succession (FAR-DEEP Holes 12A, 12B and 13A), mafic lavas, tuffs and gabbro constitute 30–35 vol.% of the succession (Fig. 4.53) and the sedimentary part contains a significant amount of calcareous greywackes and marls, as well as bedded and massive limestone and dolostones. Black cherts (lydites) are also a characteristic feature of the upper part of the volcanogenic-sedimentary succession (Fig. 4.53; for details see Chaps. 6.3.3 and 6.3.4). In contrast, 70 vol.% of the volcanogenic succession intersected by the Onega drillhole is composed of igneous rocks (Fig. 4.53). The sedimentary rocks contain a significant proportion of the volcaniclastic material, whereas carbonate rocks are largely absent apart from diagenetic dolomite concretions ($\delta^{13}C$ ranging between −19 and −1‰ with one outlier at +9‰; Krupenik et al. 2011b).

Galdobina (1987a) emphasised that all the rocks of the Upper sub-suite are rich in sulphides occurring in the form of fine disseminations, massive and laminated beds, diagenetic concretions and cross-cutting veins. The sulphur isotopic composition of these sulphides varies between −22 and +31‰ (Akhmedov and Efremov 1989; Shatsky 1990). A somewhat similar range (−14 to +26‰) has been reported from the Onega parametric drillhole, though this covers the entire thickness of the Zaonega Formation (Krupenik et al. 2011b). This drillhole data reveal a three-modal $\delta^{34}S$ distribution with maxima at −12‰, +8‰ and +20‰, and suggest a complicated stratigraphic trend from c. +12‰ in the lower part, with a rapid increase to +20‰ in the middle part followed by an abrupt decrease to c. +7‰ in the upper part of the section (Krupenik et al. 2011b). The greatest enrichment in ^{34}S is associated with the thickest, continuous interval of C_{org}-rich rocks, containing massive varieties with C_{org} >30 wt.%. This interval is also characterised by a dense, fine dissemination (including framboidal pyrite) of pyrite (up to 50 vol.%), abundant pyrite and pyrrhotite concretions (>20 cm in diameter), beds of massive pyrrhotite (1–4 cm thick), and diagenetic dolomite nodules (Krupenik et al. 2011a). Krupenik et al. (2011b) reported no correlation between the sulphur isotopic values, types of sulphide occurrences and host rock lithology. In contrast, Melezhik et al. (1999b) noted that rocks with C_{org} >20 wt.% are most enriched in ^{34}S. Although a large overall spread of $\delta^{34}S$ values (53‰) and the presence of negative values are indicative of biological sulphate reduction (Shatsky 1990; Melezhik et al. 1999b), the overall stratigraphic change of $\delta^{34}S$ remains understudied and unexplained.

The carbon isotopic composition of organic matter obtained from the Onega parametric drillcore (Krupenik et al. 2011b) shows a negative excursion from −16 % in the Lower sub-suite to −25 % through the lower part of the Upper sub-suite. This is followed by a rapid decline to −41 % in the middle part of the Upper sub-suite, and then a positive shift to −30 % in the uppermost part of the section. The overall stratigraphic trend confirms previous observations (Filippov and Golubev 1994; Melezhik et al. 1999b). Different interpretations were advanced for such large fluctuation of $\delta^{13}C_{org}$, including influxes of mantle carbon (Galdobina et al. 1986), organic carbon cycling through methanogenesis and methanotrophy (Yudovich et al. 1991), oil generation and subsequent thermal maturation and metamorphic alteration of organic matter (Melezhik et al. 1999b). More petrographic and isotopic work is required for understanding the isotopic composition of primary biomass and processes involved in postdepositional recycling and thermal maturation of the organic matter.

Currently, there is no self-consistent model explaining enhanced accumulation and preservation of a large amount of organic matter during the deposition of the Zaonega Formation. Previously proposed depositional settings include a shallow-water basin for the Lower sub-suite followed by a brackish-water lagoonal environment for the Upper sub-suite. The brackish water conditions were inferred based on a biased sampling of rocks with low S/C ratios (Melezhik et al. 1999b). This is not supported by new data obtained through the FAR-DEEP and Onega drillcore material. Abundant sulphides occurring throughout the Zaonega Formation have sulphur isotope values which are consistent with bacterial sulphate reduction and suggest marine, sulphate-bearing water. Voluminous mafic volcanism associated with the accumulation of sedimentary rocks favours an active continental margin setting. Other features, such as seafloor hydrothermal activity and enhanced preservation of organic matter are suggestive of rift conditions. Hence, the overall data can be reconciled with a rift basin developed on an active continental margin (for details see Chaps. 6.3.3 and 6.3.4).

The Suisari Formation

The Suisari Formation was first described by Timofeev (1935). The formation occurs mainly in the western part of the Onega Basin within an area of c. 2,000 km^2 (Kulikov et al. 2011b). A bed of volcaniclastic, polymict conglomerates, gritstones and rare tuffs mark the base of the formation and rests either with a low-angle discordance, or sub-concordantly, on the Zaonega Formation basalts (Kulikov et al. 1999). In the absence of the basal volcaniclastic bed, the lower boundary is defined by the first appearance of a high-MgO lava flow. Both the upper and lower boundary of the original formation defined by Timofeev were revised and changed by excluding a pile of basaltic andesites at the base (included into the Zaonega Formation), and by excluding a sedimentary unit on the top (now recognised as the Kondopoga Formation) (Kulikov et al. 1999). The Suisari formation is composed essentially of mafic-ultramafic volcanic rocks, which are not known in any other lithostratigraphic unit of the Onega Basin (Kulikov et al. 2011b). Sedimentary rocks are rare but pyroclastic products are widespread. A depositional age is constrained by Sm-Nd mineral and Pb-Pb whole-rock isochron ages of 1975 $\pm$ 24 and 1980 $\pm$ 57 Ma obtained from a gabbro-peridotite sill and associated lavas (Puchtel et al. 1992, 1998).

The stratotype section of the Suisari Formation is based on a core log obtained from drillhole 5 (Fig. 4.54). Here, the drilled succession was subdivided into five piles (renamed here as members). The section has an incomplete thickness of 420 m (total thickness is c. 700 m) of which 105 m constitute 18 beds of tuffs and tuffites (25 % of the section), and 43 lava flows form the remaining 315 m. Pyroclastic beds range in thickness from 0.3 to 15.5 m, whereas individual lava flows show thicknesses from 0.8 to 29.5 m (Kulikov et al. 2011b). The section description is detailed below and is based on a review of data reported by Kulikov et al. (2011b).

Member 1 is a 12-m-thick unit resting on interbedded greywacke and siltstone of the Zaonega Formation (drilled thickness is 5.7 m). It consists of interbedded psammitic to pelitic, lithocrystalloclastic, basaltic tuffs with subordinate layers containing angular clasts and rounded pebbles of basalt and dolerite. The pelitic tuffs are parallel-bedded, whereas the psammitic ones show unidirectional cross-bedding. Outside the type-section area, in the southern part of the Onega Basin, Member 1 reaches a thickness of more than 70 m and is represented by massive, amygdaloidal, pillowed and variolitic lavas forming a series of flows (Fig. 4.55). Their thicknesses range between 3 and 20 m, and some flows are separated by agglomerates (Svetov 2008). Volcanic rocks of Member 1 have a low MgO content (6–9 wt.%).

Member 2 has a thickness of 75 m. At its base is a 5.3-m-thick picrobasaltic flow (Flow 1) with a fragmented top. The rest of the pile consists of interbedded tuffites and psammitic, and aleuritic and pelitic lithocrystalloclastic basaltic tuffs. The tuffites and tuffs have parallel bedding, and rare unidirectional cross-bedding. Volcanic rocks of Member 2 have a high MgO content showing a lateral variation from 8–13 wt.% to 15–25 wt.%.

Member 3 is 100 m thick. The base comprises a fine-grained basalt flow (Flow 2; 7.9 m thick), which is followed by parallel-bedded basaltic tuffs (4.6 m thick). The upper part is represented by 12 lava flows (Flows 3–14) composed of fine-grained basalts with amygdaloidal and fragmented tops. The basalts have a MgO content ranging between 5 and 8 wt.%, hence the lowest among the units of the Suisari Formation (Kulikov et al. 2011b).

Member 4 has a thickness of 198.9 m. Its base comprises four lava flows (Flows 15–18, 28 m) of fine-grained, massive, high-Mg basalts with fragmented tops. These are overlain by a 3.5-m-thick unit of rhythmically bedded, psammitic and aleuropelitic high-Mg basaltic tuffs. Beds of psammitic tuffs are 1- to 3 cm thick, whereas beds of the aleuropelitic tuffs are much thicker ranging between 30 and 60 cm. The tuffs exhibit parallel bedding. The overlying unit is 58.5 m thick and contains five picritic lava flows (Flows 19–23) separated by thin beds of lithocrystalloclastic tuffs. The lower part of the flows is composed of aphanitic picrite, whereas the top is amygdaloidal and fragmental. The upper part of the member is represented by 13 lava flows (Flows 24–36) of massive picrite with fragmental tops totalling 118.9 m in thickness. The igneous rocks of Member 4 have the highest MgO content (10–24 wt.%) of all members and these rocks in particular are what makes the Suisari

Formation distinctly different from other Palaeoproterozoic volcanic rocks of the Onega Basin.

Member 5 forms the upper part of the drilled section; it has an incomplete thickness of 29 m because it is overlain by Quaternary moraine. The member consists of seven lava flows (Flows 37–43) with a single tuff bed. The flows are composed of fine-grained basalts with fragmental tops.

In the Onega Basin as a whole, the Suisari volcanic succession is intruded by numerous dykes, sills and necks; the latter were not confirmed by drilling and remain a matter for debate. Dykes of basaltic andesites, basalts, high-Mg basalts and picrobasalts are commonly 1–2 m in thickness, tens to hundreds of metres in length and commonly trend NE (5°–20°; three exceptions having an EW direction). Picrite-dolerite sills, more than 10 in number, have variable thicknesses with the thickest reaching 100 m. Geochemical and petrographic data indicate that the picrite-dolerite sills are coeval and comagmatic with the c. 1980 Ma volcanic rocks of Members 2 and 4 of the Suisari Formation (Puchtel et al. 1995; Kulikov et al. 2011b).

Puchtel et al. (1999) considered the Suisari volcanic rocks to represent one of the oldest continental flood basalt provinces. They also reported that the age data and the petrological, geochemical and isotopic features suggest that the Suisari lavas originated from the same mantle plume as the Pilgujärvi volcanic rocks in the Pechenga Greenstone Belt, though showing a minimal crustal contamination. The presence of pillow lavas indicates subaqueous eruption, whereas abundant pyroclastic products (25 % in the type section) suggest that subaerial explosions also played an important role. Most of the pyroclastic products exhibit parallel bedding, hence were accumulated in a deep-water basin. The presence of rhythmically bedded tuffs is indicative of redeposition of pyroclastic material by turbidity currents. Unidirectional cross-bedding in tuffs from the lower part of the succession suggests that pyroclastic material was resedimented by currents. Presently, no further sedimentological details are available to better discriminate a depositional setting.

The Kalevian Super-Horizon

There are many unresolved issues regarding the lithostratigraphy of the Kalevian Super-Horizon in the Onega Basin. Several formations were defined during regional mapping and drilling projects, but no agreement exists as to how they correlate or how to distinguish their lower and upper boundaries, and their depositional ages are not constrained (a complete overview is presented in Medvedev (2011)). Kajryak (1973) defined and described the Besovets Series, which he subdivided into the Padoss (c. 700 m thick) and Shuja (c. 500 m thick) suites north of Petrozavodsk, and the Kodopoga (c. 600 m thick) and Vashozero (c. 170 m thick) suites in the environs of Kondopoga. In the central part of the Onega Basin, Bondar (1972) recognised the Munozero Suite, which is supposedly correlative to the Vashozero Suite. In contrast, Galdobina (1987b) correlated the lower part of the Besovets Series with the Suisari Formation. Considering current uncertainties involved in intrabasinal correlation of various suites, their relationships with, and distinction from, both older and younger lithostratigraphic subdivisions, our characterisation of the Kalevian Super-Horizon is focused on the Kondopoga and Vashozeo suites, which were studied in much greater detail and whose stratigraphic position is less problematic.

The Kondopoga Formation

The Kondopoga Formation occurs within the central part of the Onega Basin in a series of NW-SE-trending synforms (Fig. 4.34). The description of the type section is based on cores from two drillholes. Drillhole #1 (525 m deep) and drillhole #2 (467.8 m deep) made in 1991 (Fig. 4.56) intersected the entire thickness of the formation including both the lower and upper contacts (Lavrov 1999). These data were supplemented by outcrop exposures of the Kondopoga Formation and the geological relationship with underlying and overlying strata.

In drillhole #2 (Fig. 4.56), the Kondopoga succession lies directly on the Suisari lavas with a thin basal bed of eluvial breccia (or polymict, volcaniclastic conglomerate) containing clasts of Suisari lavas in a sandstone matrix (Lavrov 1999). In the vicinity of drillhole #2, the Kondopoga volcaniclastic sandstones sit with angular discordance (c. 20°–35°) on Suisari lavas (Kajryak 1973). In areas where the Suisari Formation is absent, the Kondopoga Formation or its correlatives (e.g. the Pados Formation) rests directly on the Zaonega Formation (Lavrov 1999), or even the Tulomozero Formation (Akhmedov et al. 1990).

The Kondopoga Formation is composed of immature siliciclastic rocks showing different scales of rhythmic bedding. Although the succession includes rocks ranging from small-pebble conglomerate to mudstone, it is rather monotonous and dominated by greywackes (Fig. 4.57). In a recently published review (Rychanchik 2011b), the formation was subdivided into two sub-suites, and each sub-suite was further subdivided into three piles (renamed here as members, Fig. 4.56). In the following text, the lithofacies comprising these six members are illustrated by core images (drillholes #1 and #2) supplemented by photographs in the Nigozero aggregate quarry (Fig. 4.58).

Member 1 (60 m thick) comprises conglomerate, gritstone, greywacke, siltstone and mudstone appearing as rhythmically bedded successions or Bouma sequences (Fig. 4.58cl–cn). The thickness of individual Bouma sequences ranges from a few millimetres to a few metres and decreases irregularly up section. The sequences are typically comprised of Bouma T_a, T_b and T_d. Bouma T_a are either massive or graded conglomerate to greywacke in the lower part of the member, and can be as much as 2 m thick. The upper part of the member is thinner bedded and finer grained, from coarse- to fine-grained greywacke with thicknesses between a few millimetres to 10 cm. T_b are composed of parallel-laminated, fine-grained greywacke or siltstone and are commonly less than 1 cm thick. T_d are represented by laminated mudstone, only a few millimetres thick. T_c is rare but where present consists of cross-bedded sandstone-siltstone.

Member 2 (Figs. 4.56 and 4.58bi–ck) is one of the thickest in the formation (c. 150 m thick) and is composed mainly (c. 60 % of the section) of numerous, several metre-thick beds of massive greywackes having a thin, graded basal part (Bouma T_a). The upper part of the member comprises 1 to 3 cm thick beds of massive siltstone (Bouma T_a) and parallel-laminated mudstone (Bouma T_d). Here, there are also a couple of 5- to 10-cm-thick rhythms composed of near-complete Bouma sequences from graded and massive to parallel-laminated then cross-laminated and finely parallel-laminated siltstone-mudstone.

The massive sandstone lithofacies contains rare and randomly distributed intraformational clasts of greywacke, siltstone and resedimented ankerite concretions (Fig. 4.58bi, bq, cd–cf). In the Nigozero quarry, these sandstones are seen to contain 2- to 3-cm-thick interlayers of dark-coloured siltstones as Bouma T_d components (Fig. 4.58bk, ch) and some beds have 1-m-thick intervals with a much greater concentration of intraformational clasts; these are mainly grey and dark grey siltstone and redeposited carbonate concretions and their fragments (Fig. 4.58cg, ci–ck).

The middle part of Member 2 is fine-grained, laminated greywacke-siltstone with cm-thick layers, lenses and small concretions of ankerite and/or siderite (Fig. 4.58bs, by). In the Nigozero quarry, ankerite and/or siderite beds and lenses represent early-diagenetic cementation and/or partial or complete replacement of permeable sandy layers by carbonate phases (Fig. 4.58bt, bu, ca–cc). Considerable differential compaction within and outside concretions suggests their early diagenetic origin (Fig. 4.58bu). Hence, diagenetic replacement can preserve the original shape of sandy bedforms including parallel-laminated and graded beds, ripples and slumps (Fig. 4.58ca–cc) and, in some instances, ripples replaced by ankerite on the bedding surfaces occur as either aligned or randomly oriented "sausages" (Fig. 4.58bv–bx).

The uppermost beds of the massive sandstone lithofacies of Member 2 (drillhole #2, 120–122 m interval, Fig. 4.56) contain rare, mm-sized, angular clasts of pyrobitumen (Fig. 4.58bi–bj). Similar lithofacies occur in the Nigozero quarry (Fig. 4.58bm) but there the abundances of pyrobitumen clasts can be much greater and range from small, angular fragments to large, flattened, pancake-like inclusions (Fig. 4.58bn–bp). The latter have either sharp or diffuse boundaries and even show partial injection into their host rock (Melezhik et al. 2009). Clasts and inclusions of pyrobitumen were identified as former petroleum (Mishunina 1979) and were interpreted by Melezhik et al. (2009) as redeposited surface and subaqueous oil-seeps derived from the underlying tectonically-compromised Zaonega oil reservoirs.

Member 3 (c. 50 m thick) is composed of rhythmically bedded greywacke/feldspathic sandstone, siltstone and mudstone (Fig. 4.58y–s, ac–bh). Similar to Member 1, the rhythmically bedded succession is composed of Bouma sequences whose thicknesses vary from a few millimetres (in the middle and upper parts) to a couple of metres (in the lower part). In drillhole #2, the base of the member comprises a 1-m-thick, matrix-supported conglomerate followed by a c. 2-m-thick bed of massive sandstone. The conglomerate has an erosional base and consists of unsorted, angular and variably rounded fragments of laminated and bedded greywacke-siltstone-mudstone (up to 4 cm in size), and redeposited ankerite/siderite concretions (Fig. 4.58be, bf). Similar conglomerate or breccia beds occur in the Nigozero quarry. Here, a few closely spaced, c. 2 m-thick and 10 m-wide lenses of a matrix-supported, intraformational, polymict breccia and conglomerate are associated with massive greywacke and encased within overall rhythmically bedded rocks. Up to 60-cm-large clasts of greywacke, siltstones and mudstones and redeposited carbonate nodules and their fragments show a limited degree of sorting and a variable roundness; all are embedded in a greywacke matrix (Fig. 4.58bg, bh). Sedimentological features are consistent with massflow deposition in a channelised subaqueous environment (Melezhik et al. 2009).

The lithofacies overlying the intraformational conglomerate are represented in cores #1 and #2 by a rhythmically bedded succession comprising 0.5- to 15-cm-thick Bouma T_a, (massive and graded sandstone), T_b (parallel-laminated siltstone) and T_d (parallel-laminated mudstone) beds (Fig. 4.58q, w, y, z, ac–ag). A corresponding, and apparently contemporaneous, succession exposed in the Nigozero quarry shows Bouma sequences similar in thickness and composition (Fig. 4.58r, s, aa, ag, ao). Both the cores and quarry exposures provide evidence for syndepositional folding, faulting and slumping (Fig. 4.58w, x, ah–an), as well as tilting and subaqueous erosion (Fig. 4.58p). Loading structures are also common (Fig. 4.58al, ap), whereas sedimentary dykes are rare (Fig. 4.58o). Resedimented mud-chips are present in the upper part of the section (Fig. 4.58t, u). Siltstones and mudstones contain iron- and copper-sulphides; when the latter are oxidised, the leached copper reacts to form malachite and produces bright green staining on bedding and cleavage surfaces (Fig. 4.58v).

In the quarry, the rhythmically bedded lithofacies of Member 3 contain abundant beds and lenses of ankerite/siderite-cemented sandstone (Fig. 4.58aq–ay) and commonly display sedimentary/diagenetic boudinage caused by loading (Fig. 4.58av, ay). The boudinage forms complex structures including polygonal and ridged patterns in overlying and underlying siltstones/mudstones (Fig. 4.58ba–bc), and recessive polygonal relief on either side of the sandstone beds (Fig. 4.58az, bd). Such polygonal patterns were previously interpreted either as desiccation cracks or even as stromatolites (Makarikhin and Kononova 1983).

Member 4 (c. 200 m thick) is the thickest lithostratigraphic unit of the Kondopoga Formation (Fig. 4.56). It is a rhythmically bedded succession with 70 % of its thickness comprised of siltstones (Rychanchik 2011b). The lower part is marked by grey-bluish, siltstone-mudstone varve-like rhythmites with approximately equal thicknesses of siltstone and mudstone beds ranging in thickness between 0.2 and 2 cm. Bed contacts are typically straight and sharp, although some siltstone units have a gently erosive base. These distinctive rocks define a marker interval that occurs in the lower part (490–457 m) of core #1 (Fig. 4.58h, j, l) and the uppermost part (27–63 m) of core #2 (Fig. 4.58ab); the interval enables correlation of these two holes that are c. 30 km apart (Fig. 4.56). Such rhythmites have not been observed in the Nigozero quarry, which may indicate that the quarry section is located stratigraphically below Members 4–6. The marker interval has a few thicker interbeds of mudstone containing siltstone intraclasts (Fig. 4.58i), and a c. 0.5-m-thick bed of massive sandstone with mudstone intraclasts (Fig. 4.58k). Above this interval are rhythmites composed mainly of dark grey, flat laminated siltstone-mudstone. The thickness of siltstone-mudstone couplets ranges between 0.2 and 5 cm with the siltstone beds dominanting and often displaying normal grading. Thus, these rocks are readily interpreted as T_a and T_d beds of a Bouma sequence. In rare cases, mostly in the lower part of the member, T_a beds are composed of graded sandstone (Fig. 4.58g).

In several intervals of core #1 (i.e., 423 m and 385.4 m), there is a complex system of randomly oriented calcite veinlets (stockwork), commonly sourced from sandy beds and expanding upward over a distance of 5 to 10 cm (Fig. 4.58a, b). Similar features, but not necessarily contemporaneous with the core intervals, have been documented in the Nigozero quarry. When observed on the bedding plane, such veinlets are projected as a pattern resembling desiccation cracks; however, three-dimensional observations (Fig. 25c–f) suggest that such systems of randomly oriented veinlets resemble "molar tooth" structures. The latter commonly occur in carbonate substrates (Shields 2002 and references therein), though here they are in siliciclastic rocks. We tentatively interpret the observed structures as earthquake-induced extensional cracks developed in partially solidified sediments.

Member 5 (c. 50 m thick) is distinguished by: (*1*) thick massive, feldspathic sandstone beds in the middle part; (*2*) dominance of parallel-bedded sandstone; (*3*) scarcity of mudstone beds; and (*4*) absence of molar-tooth structure. The contact with member 4 is, in practice, difficult to place.

Member 6 (c. 30 m) is composed mainly of parallel-bedded, feldspathic sandstones with thin and rare interlayers of siltstone. The sandstones appear to be either massive or indistinctly bedded in the upper part of the member.

The Kondopoga Formation rocks contain redeposited organic matter (Gorlov and Filippov 1987). The C_{org} content is <1 wt.%, and the rocks are generally devoid of sulphides (Filippov and Golubev 1994), except minor pyrite and chalcopyrite in the upper part of Member 3. Abundant ankerite/siderite concretions with low $\delta^{13}C$ (at around -15‰; Fallick et al. 2008) suggest that they incorporated bicarbonate derived from oxidation of organic matter. The iron fixation in diagenetic carbonates and the scarcity of sulphides indicate that the organic carbon was not cycled through bacterial sulphide reduction. Assuming oxic conditions, such geochemical features have been interpreted as consistent with a fresh-water, sulphate-free, lacustrine depositional system (Melezhik et al. 2009).

The most common lithologies of the Kondopoga Formation are greywackes, feldspathic sandstones and siltstones. Generally thin Bouma sequences and fine-grained clastic material (Members 1–4) are indicative of deposition from turbidity currents. Intraformational conglomerates/breccias and massive sandstone lithofacies located in Members 2 and 3 may represent channelised massflow deposition (Melezhik et al. 2009). Sedimentological characteristics of Members 5 and 6 of parallel-bedded sandstones, and siltstone with rare and thin mudstone interlayers are suggestive of deposition on a marine shelf, but the lack of tidal features and the absence of sulphides suggest that a nearshore lacustrine setting with deposition from river-fed underflows is a more probable setting.

The Vashozero Formation

The contact between the Kondopoga and Vashozero formations was defined on geochemical criteria; the former marked by sodium-series rocks and the latter potassium-series rocks (e.g. Rychanchik 2011c). Drillhole #1 intersected the formational boundary but no clear lithological expression of such a geochemical difference could be distinguished. In general, the lower part of the Vashozero Formation (c. 70 m thick) is composed of greywackes interbedded with siltstones, mudstones, clayey cherts and rare beds of limestone. The greywackes contain several intervals of intraformational breccias (Rychanchik 2011b).

The overlying succession (c. 22 m thick) comprises polymict, pebbly sandstones with lenses of pebble conglomerates, and the so-called “limestone-gritstone-oncolite” marker horizon (Akhmedov and Krupenik 1991). In core #1, this part of the section (c. 139–110 m) apparently includes massive sandstone fining upward into siltstones with numerous, 2- to 5-cm-thick beds of “fenestral” shales that, in places, are calcareous (Fig. 4.59a).

This is followed by a succession (c. 47 m thick) of thinly interbedded polymict sandstones, siltstones and oncolite dolostones, some of which have microstromatolites. The oncolite dolomite interval is 15–20 cm thick and was defined by Akhmedov and Krupenik (1991) as an important marker bed termed “dolostone-sandstone-oncolite”. In core #1, the corresponding succession (c. 110–50 m) apparently includes fine-grained sandstone-siltstone lithofacies with lenticular and flaser bedding and ripple cross-lamination. Normally graded sandstones with massive mudstone tops, thin beds of matrix-supported conglomerates and cross-bedded silty sandstones (Fig. 4.59b) are present in the upper part of the unit (c. 63–51 m).

The formation ends with a succession (c. 30 m thick) of interbedded sandstones, laminated cherts and chert conglomerates and breccias (Rychanchik 2011b). In core #1, this interval includes parallel-bedded sandstones and siltstones. Many sandstone beds are graded from coarse- to fine-grained, and some contains rare mudstone chips. Sandstone and siltstone beds have variable thickness, from a few millimetres to as much as 15 cm.

The depositional environment of the Vashozero Formation remains to be studied. However, a series of natural outcrops of the Shuja Formation (apparently correlative to the middle part of the Vashozero Formation) expose distinct lithofacies, including a unit composed of sandstone-siltstone with lenticular and flaser bedding and ripple cross-lamination (Fig. 4.59c). The unit also contains a series of 10- to 20-cm-thick and several-metre-long lenses of cross-bedded sandstones (Fig. 4.59d), as well as numerous small-scale sandstone dykes (Fig. 4.59e). Bedding surfaces exhibit desiccated mud layers (Fig. 4.59f) and flat, asymmetrical sand-ripples formed on silty-muddy substrate (Fig. 4.59g).

Lenticular and flaser bedding are commonly formed in intertidal and supratidal zones (Reineck and Wunderlich 1968). In such environments, the cross-bedded sandstone lenses may represent tidal channels. The sand-dykes indicate over-pressurised, non-cemented sand beds, and these can be attributed to seismic disturbance and liquefaction of high water-content sediments (Audemard and de Santis 1991; Peterson 1997; Ettensohn et al. 2002). The existence of desiccated surfaces suggests phases of emergence. Taking all indicators together, the most probable depositional environment is a shallow-water, high-energy supratidal zone on a clastic marine shelf, whereas the association of coarse-grained and graded sandstone beds, parallel-bedded sandstone and parallel-laminated siltstone in the upper part of the formation may be indicative of a lacustrine or non-marine setting.

The Vashozero Formation is the youngest lithostratigrapic unit in the Onega Basin per se. Although there are several younger lithostratigraphic units south and southwest of the Onega Basin (e.g. Petrozavodsk and Shoksha formations), they comprise a separate basin the Southern Onega Trough, which is outside the scope of this contribution.

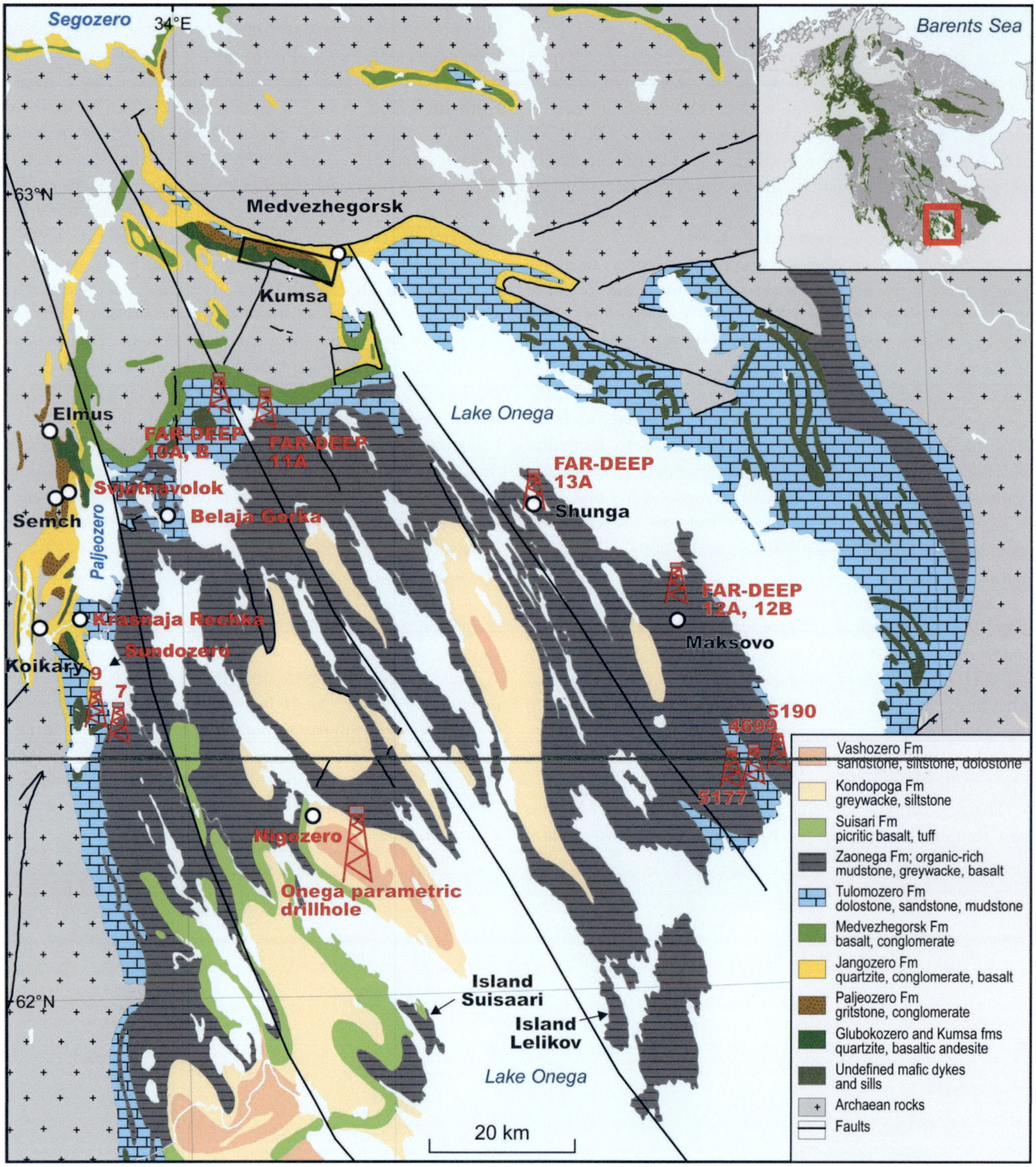

Fig. 4.34 Simplified geological map of the Onega Basin (Based on Koistinen et al. (2001) with modification by Aivo Lepland)

<table>
<tr><th>Complex</th><th>Super-Horizon (Super Group*)</th><th>Horizon (Group*)</th><th>Suite (Formation*)</th></tr>
<tr><td rowspan="12">Karelian (Karelides)</td><td rowspan="2">Vepsian</td><td>Shoksha</td><td>Shoksha</td></tr>
<tr><td>Petrozavodsk</td><td>Petrozavodsk</td></tr>
<tr><td rowspan="2">Kalevian</td><td rowspan="2"></td><td>Vashozero/Shuja</td></tr>
<tr><td>Kondopoga/Padass</td></tr>
<tr><td rowspan="2">Ludicovian</td><td>Suisari</td><td>Suisari</td></tr>
<tr><td>Zaonega</td><td>Zaonega</td></tr>
<tr><td rowspan="3">Jatulian</td><td>Onega</td><td>Tulomozero</td></tr>
<tr><td rowspan="2">Segozero</td><td>Medvezhegorsk</td></tr>
<tr><td>Jangzero</td></tr>
<tr><td>Sariolian</td><td></td><td>Paljeozero</td></tr>
<tr><td rowspan="2">Sumian</td><td rowspan="2"></td><td>Kumsa</td></tr>
<tr><td>Glubokozero</td></tr>
</table>

Fig. 4.35 Lithostratigraphic subdivision of the Palaeoproterozoic rocks of the Onega Basin (Based on Kulikov et al. 2011a). Asterisks denote stratigraphic terms used in the current contribution

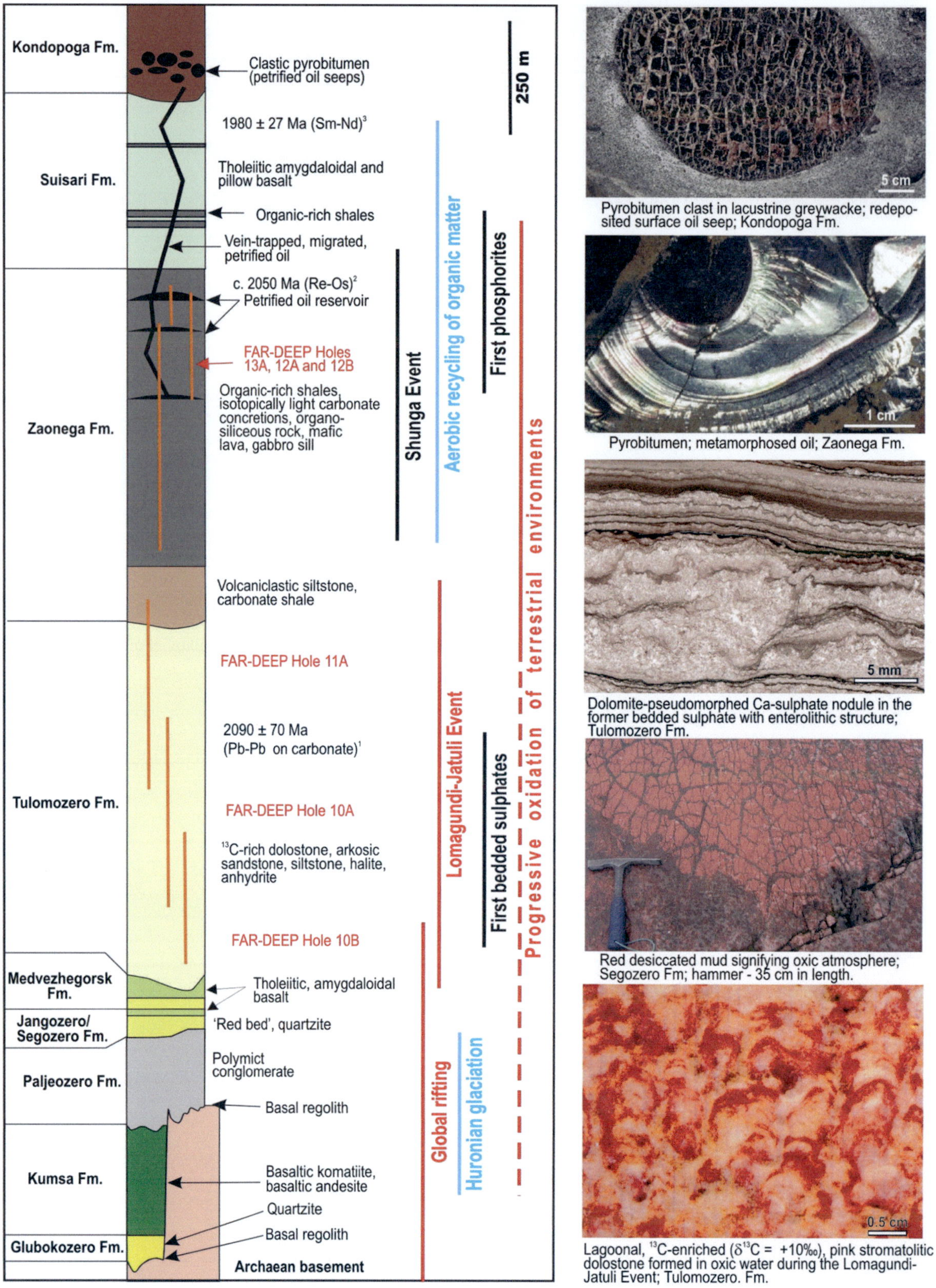

Fig. 4.36 Lithostratigraphic column of the Onega Basin, positions of FAR-DEEP drillholes, and examples of photos of sedimentary rocks from Onega Basin formations. Also shown is how the evolution of the Onega Basin is related to global palaeoenvironmental events. Superscripts denote radiometric ages from (1) Ovchinnikova et al. (2007), (2) Hannah et al. (2008) and (3) Puchtel et al. (1992, 1998)

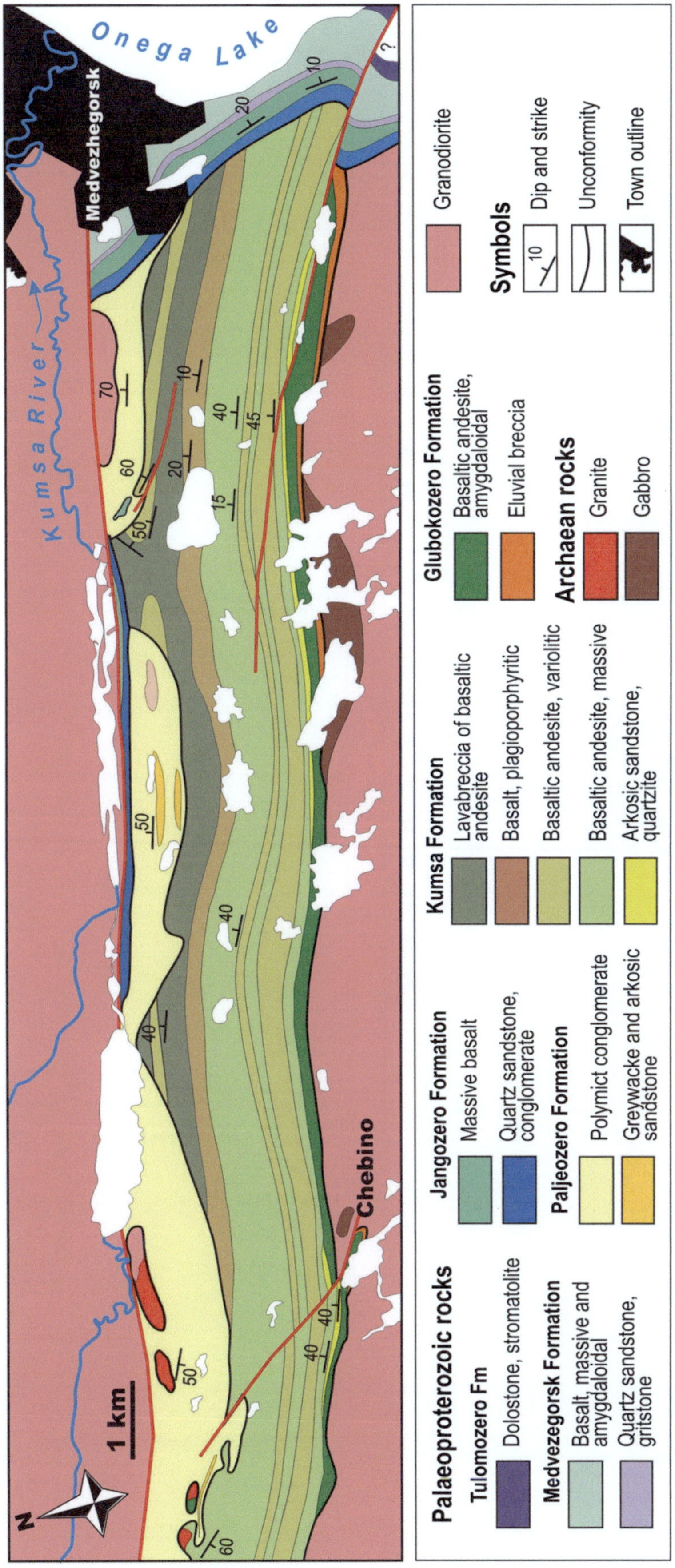

Fig. 4.37 Simplified geological map of the Kumsa area (Modified from Korosov and Nazarova 2011). Map position in the Onega Basin area is shown in Fig. 4.34

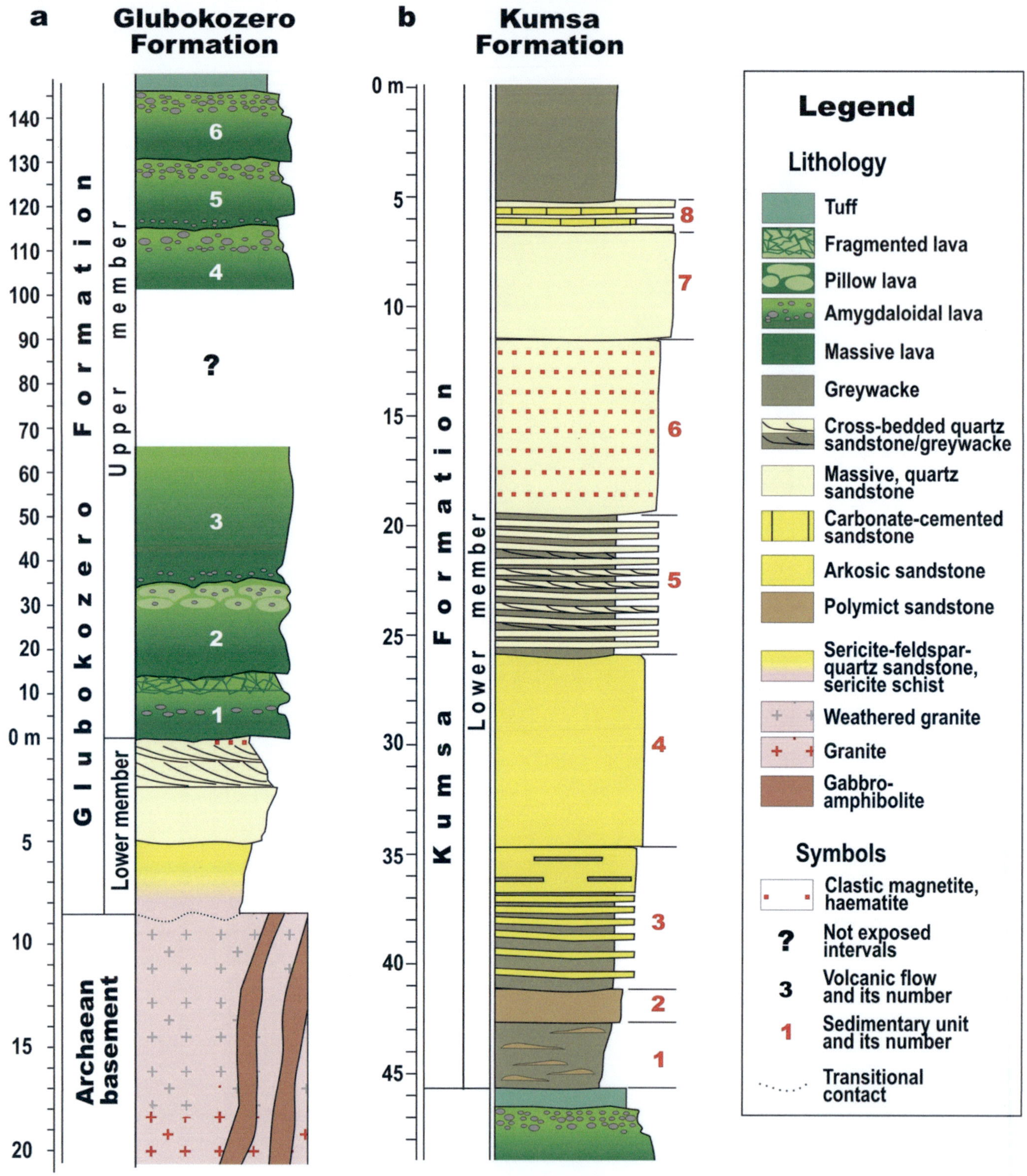

Fig. 4.38 The Glubokozero and Kumsa formations in the Kumsa area. (**a**) Lithological column of the type section of the Glubokozero Formation is based on the description by Korosov and Nazarova (2011). (**b**) Lithological column of the Lower member of the Kumsa Formation from Kumsa; the column is based on the description by Svetov et al. (2011). Location of the Kumsa area is shown by rectangle near Medvezhegorsk on Fig. 4.34

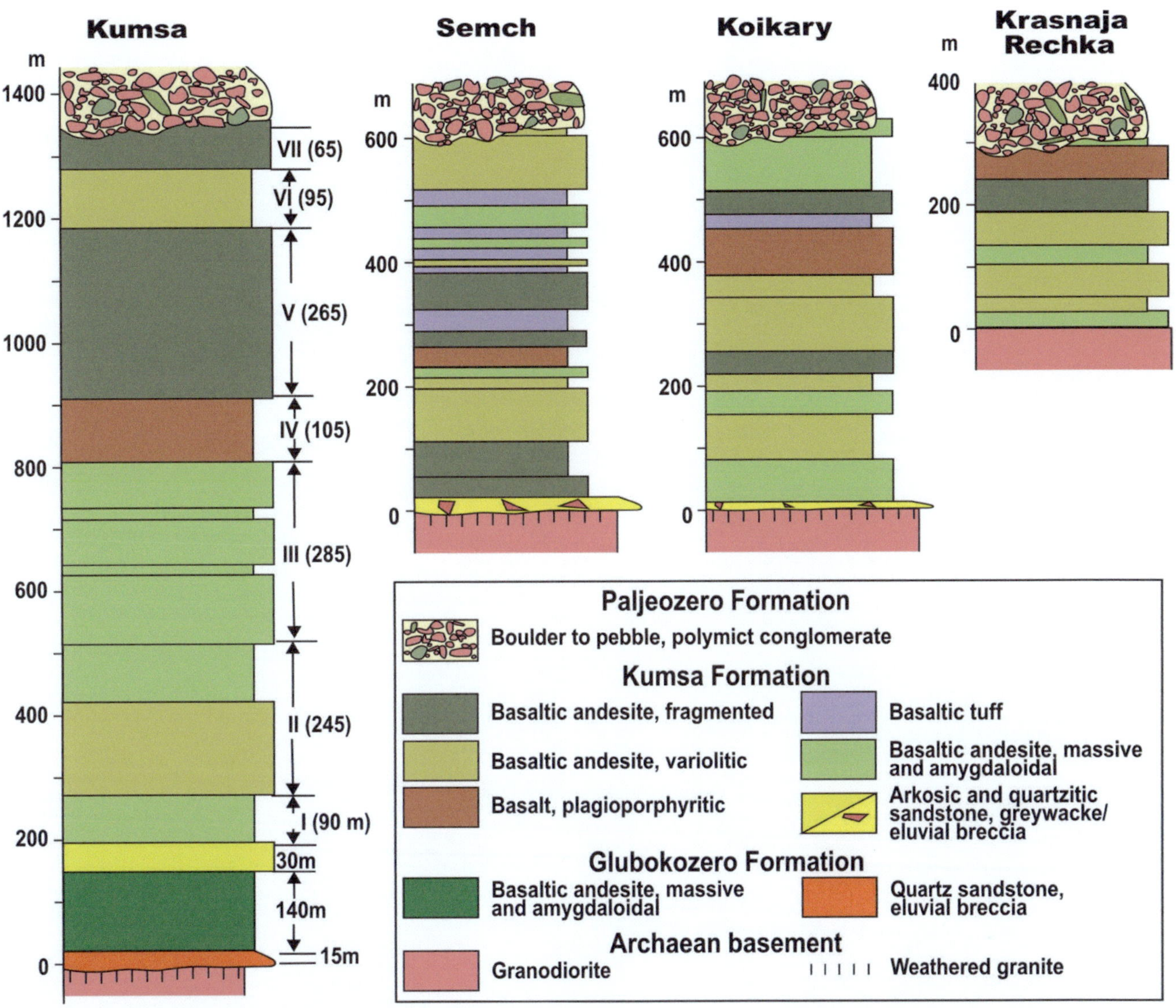

Fig. 4.39 Lithological columns of the Upper member of the Kumsa Formation from various locations; modified from Svetov et al. (2011). I–VII – volcanic piles and thicknesses

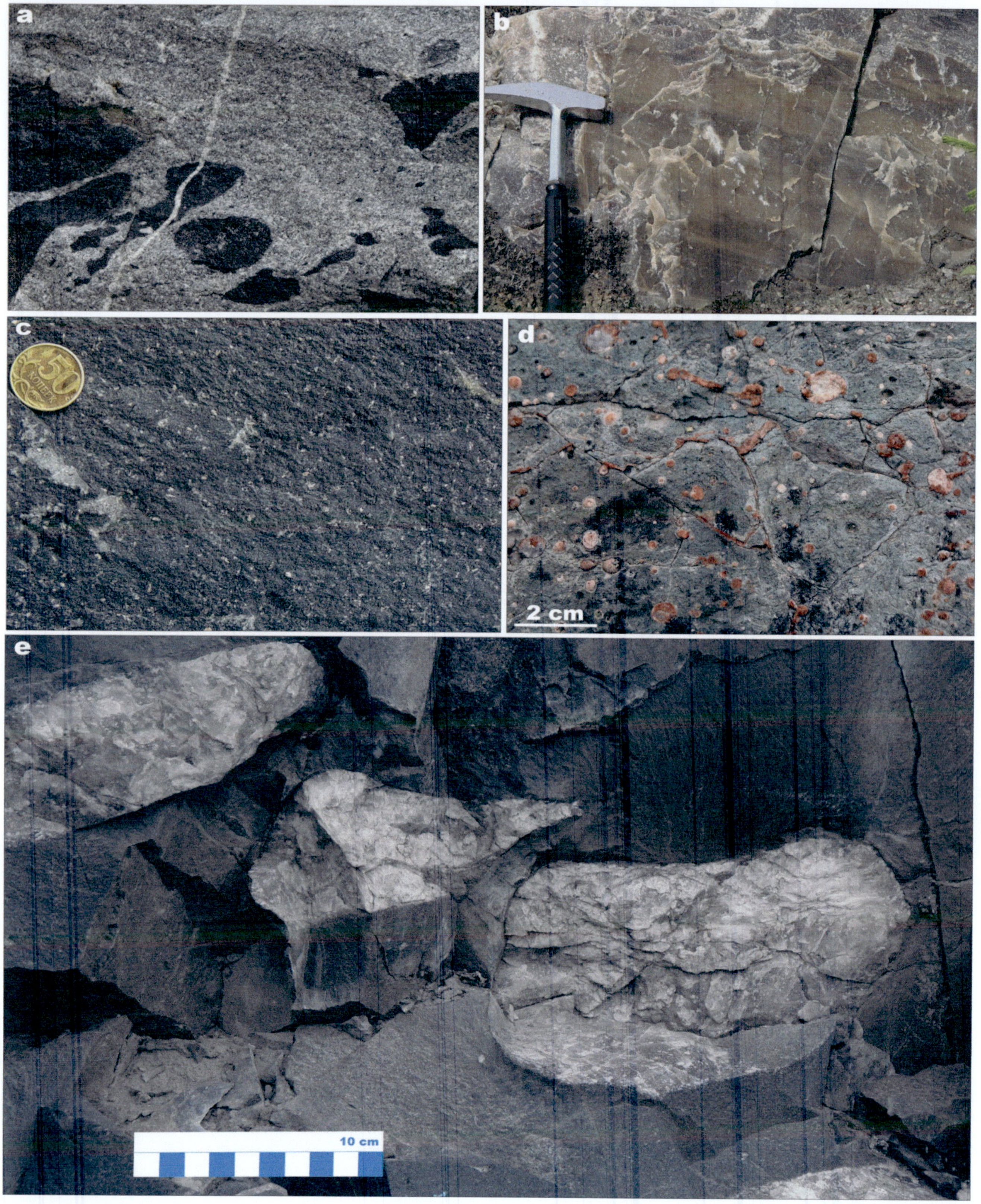

Fig. 4.40 Photographs of selected rock types from the Kumsa Formation and its basement. (**a**) Archaean tonalite with amphibolite xenoliths; base of the Kumsa Formation; width of the photo is c. 1 m. *Kumsa Formation, Lower member*: (**b**) Quartz sandstone with relicts of vague parallel bedding; hammer head is 15 cm. *Kumsa Formation, Upper member, Kumsa area:* (**c**) Massive, plagiophyric basaltic andesite; coin diameter is 2 cm. (**d**) Vesicular basaltic andesite from the middle part of a lava flow; the vesicles are filled with red albite. (**e**) Basaltic andesite lava flow with large quartz amygdales

Fig. 4.40 (continued) *Kumsa Formation, Upper member, Semch area:* (**f**) Amygdaloidal basaltic andesite from the top of a lava flow. (**g**) Vesicular basaltic andesite from the *top* of a lava flow; vesicles are filled by quartz and albite. (**h**) Foamy top of a lava flow of basaltic andesite. (**i**) Contraction joints in basaltic andesite lava flow. *Kumsa Formation, Upper member, Elmus area:* (**j**) Basaltic pillow lava

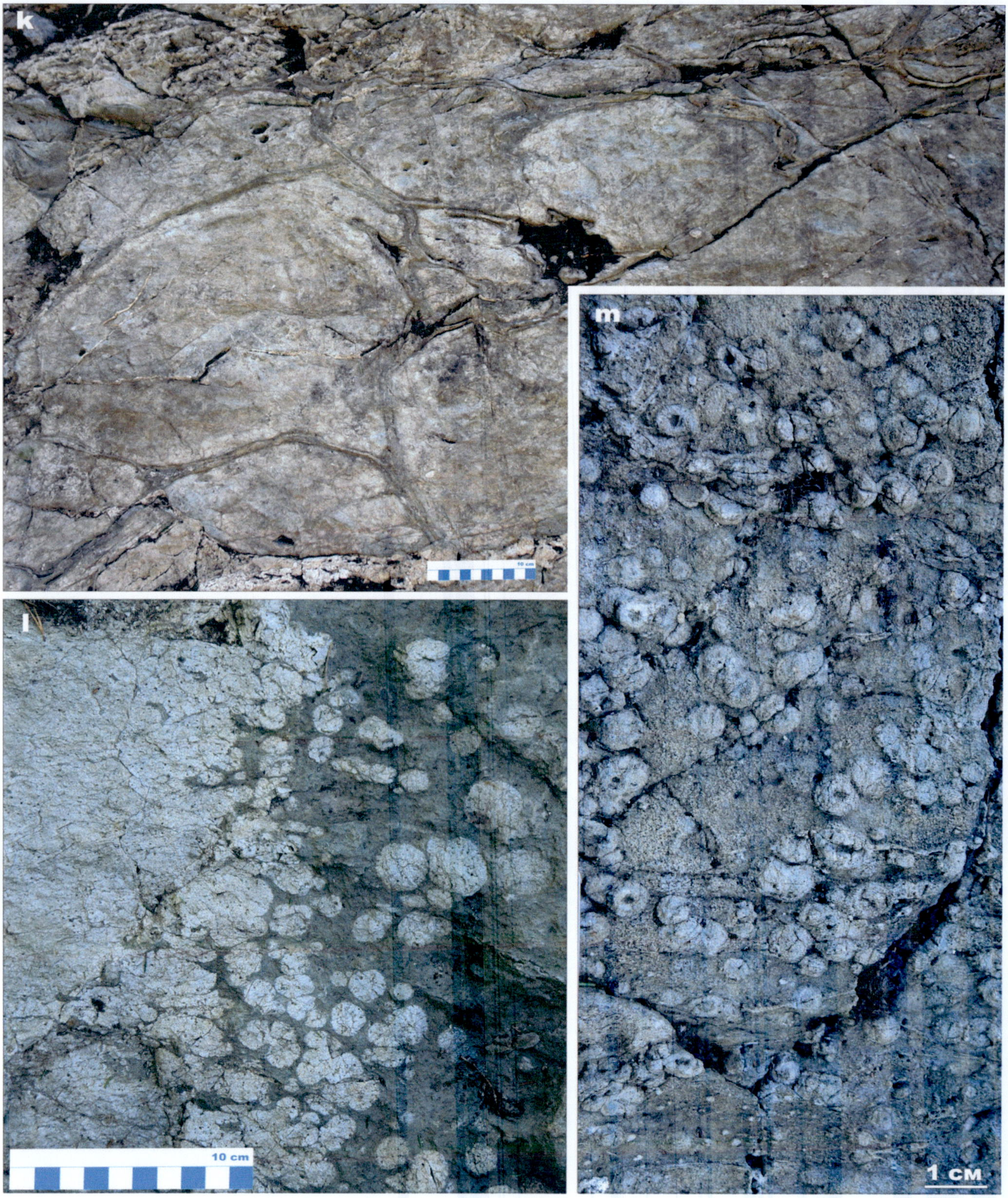

Fig. 4.40 (continued) (**k**) Pillow lava of basaltic andesite; note that some pillows have a well-developed sagging tail. (**l**) Segregation of basaltic andesite (*bright*) occurring as a lens and varioles within basaltic lava (*dark coloured*). (**m**) Variolitic basaltic andesite from the *top* of a lava flow. Photographs by Sergey Svetov

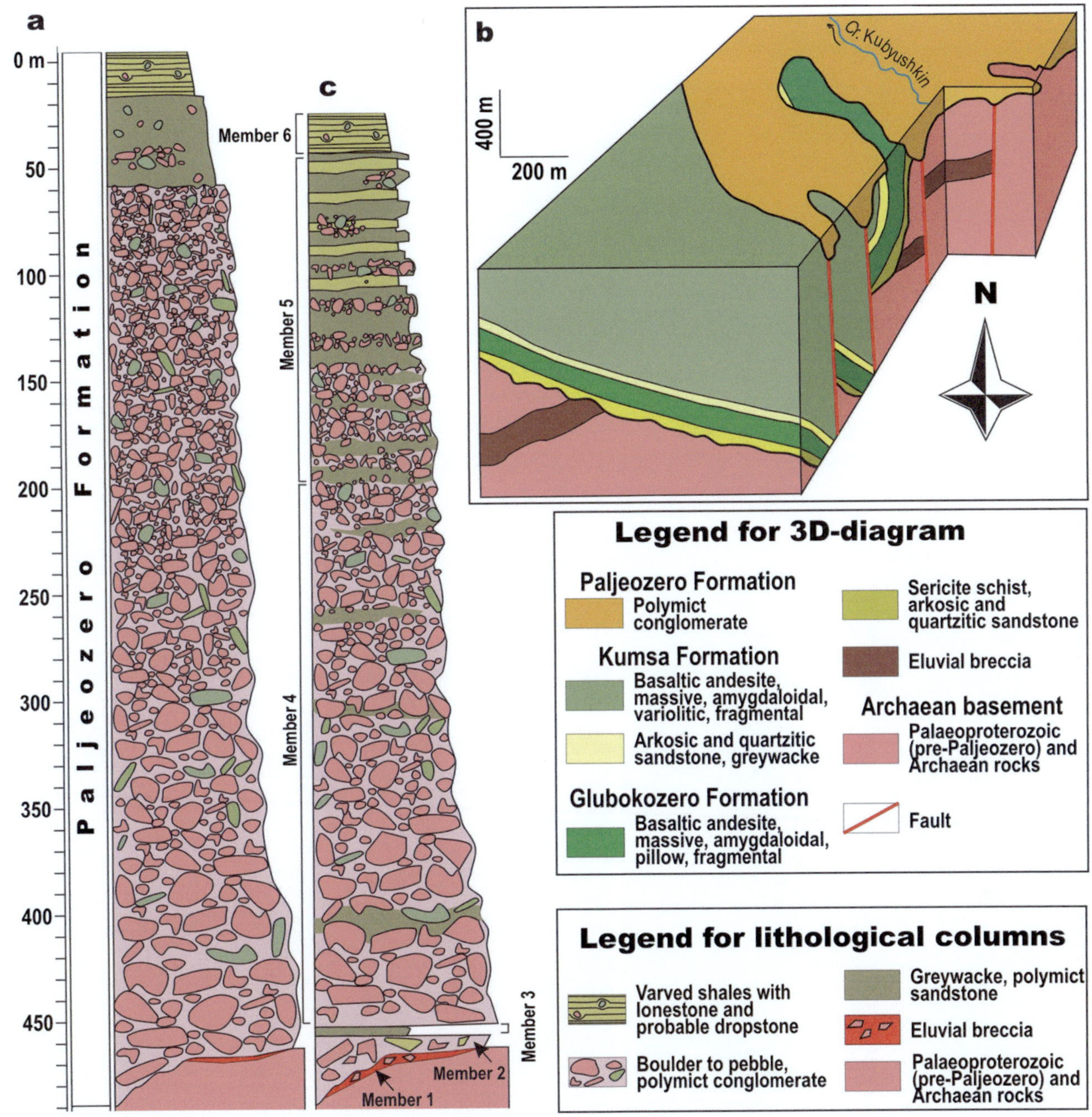

Fig. 4.41 Paljeozero Formation in the Onega Basin (**a**) Lithological section from the type locality is based on the description by Kratz (1963). (**b**) Fence diagram illustrating the stratigraphic relationship between Sumian (Glubokozero and Kumsa formations) and Sariolian (Paljeozero Formation) successions; these appear to be separated by an unconformity formed after a deformation episode followed by deep erosion (simplified after Korosov et al. 2011). (**c**) A composite lithological section of the Paljeozero Formation; the section is based on the description by Korosov et al. (2011)

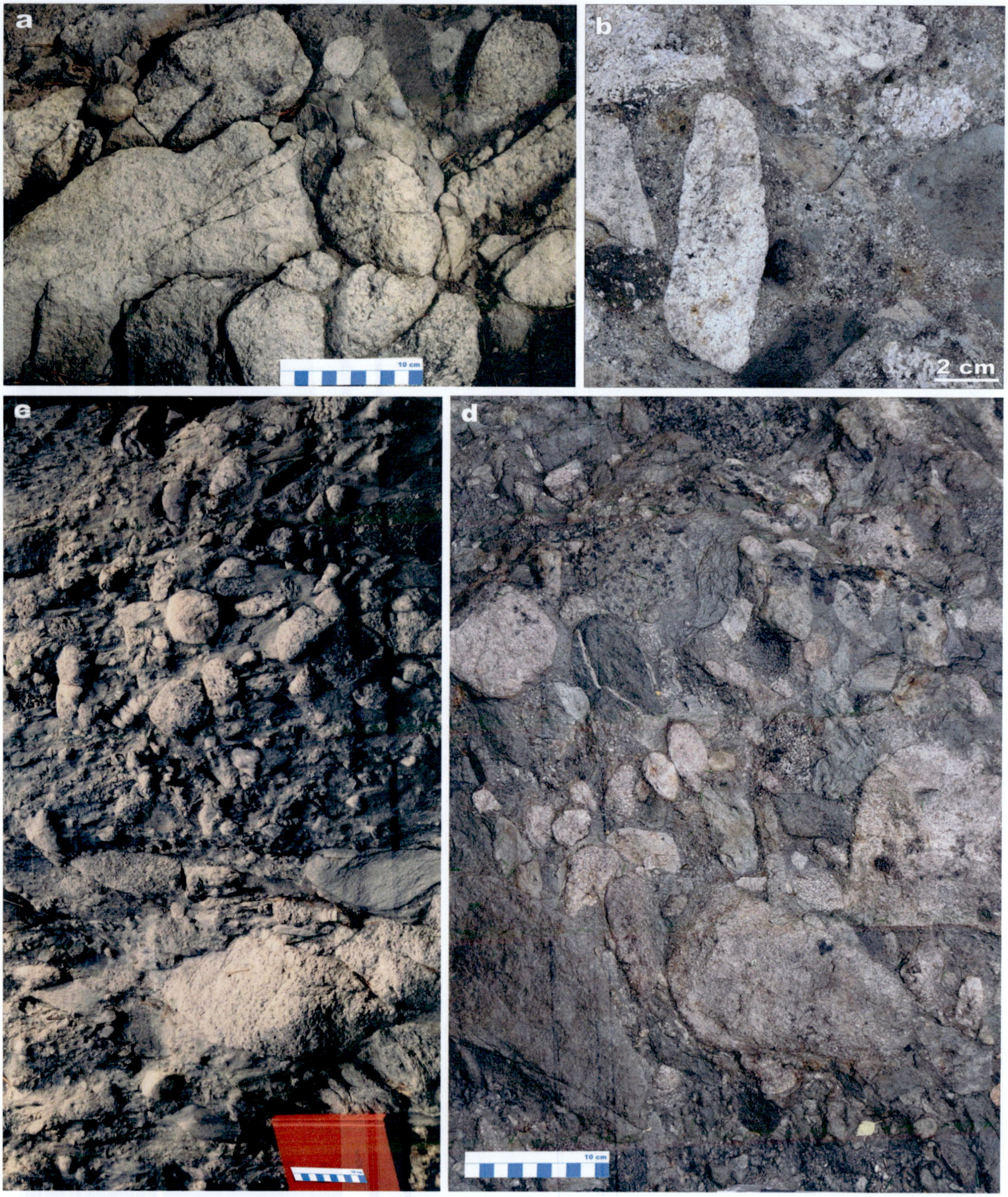

Fig. 4.42 Photographs of selected rock types from the Paljeozero Formation. (**a**) Boulder conglomerate overlying partially disintegrated Archaean granodiorite (eluvial breccia). (**b**) Polymict conglomerate composed of unsorted fragments of granite (*bright*) and mafic volcanic rocks (*dark coloured*). (**c**) Polymict conglomerate composed of mainly unsorted and variably rounded fragments of Archaean granodiorite embedded in greywacke matrix. (**d**) Crudely stratified, polymict conglomerate composed of unsorted and rounded fragments of Archaean granodiorite (*bright*), amphibolite (*dark coloured*) and various schists

Fig. 4.42 (continued) (**e**) Polymict conglomerate containing unsorted boulders and pebbles of Archaean granite (*bright*), mafic volcanic rock (*dark coloured*) and gabbro (*arrowed*) embedded into greywacke matrix. (**f**) Polymict, clast-supported conglomerate containing poorly sorted and variably rounded fragments of rocks derived from the immediate Archaean basement. (**g**) Polished slab of polymict conglomerate containing fragments of Archaean granodiorite (*bright*) as well as basalt (*dark coloured*) and amygdaloidal basaltic andesite (*arrowed*) derived from the underlying Kumsa Formation

Fig. 4.42 (continued) (**h**) Polished slab of polymict gritstone to small-pebble conglomerate containing fragments of basaltic andesite and basalt derived from the underlying Kumsa Formation; note the large granite pebble (*arrowed*) apparently piercing layering. (**i**) Graded bed of a greywacke with a lonestone (*arrowed*) overlain by small-pebble, polymict conglomerate. (**j**) Small-pebble, poorly sorted, polymict conglomerate passing upward into crudely stratified greywacke with scattered fragments of granite, amphibolite and basalt. Photographs by Victor Melezhik (**a**, **c**) and Sergey Svetov (**b**, **d–f**, **i**, **j**) Photographs (**g, h**) courtesy of Aivo Lepland

Fig. 4.43 Photographs of selected rock types representing part of the original type section of the Segozero Formation at Segozero, and illustrating classical Palaeoproterozoic "red beds". The illustrated part of the section may represent either an upper part of the Segozero or a lower part of the Onega horizons. (**a**) Cross-section view of massive gritstone overlain by *pale pink* sandstone with lenses of *bright pink*, clayey siltstone; hammer head for scale here and in the following photographs is 17 cm long. (**b**) Interbedded *pale pink*, coarse-grained sandstone and *bright pink*, clayey siltstone; note that some sandstone beds are channelised (*arrowed*). (**c**) Grey, arkosic sandstone; the lower bed exhibits planar, steep, cross bedding

Fig. 4.43 (continued) (d) Cross-section view of stacked planar and trough cross-beds in coarse-grained arkosic sandstone. (e) Plan view of trough cross-bedded and parallel-bedded, coarse-grained, arkosic sandstone; note scattered mudstone rip-ups. (f) Trough cross-bedded, coarse-grained, arkosic sandstone; note that some cross beds are overturned

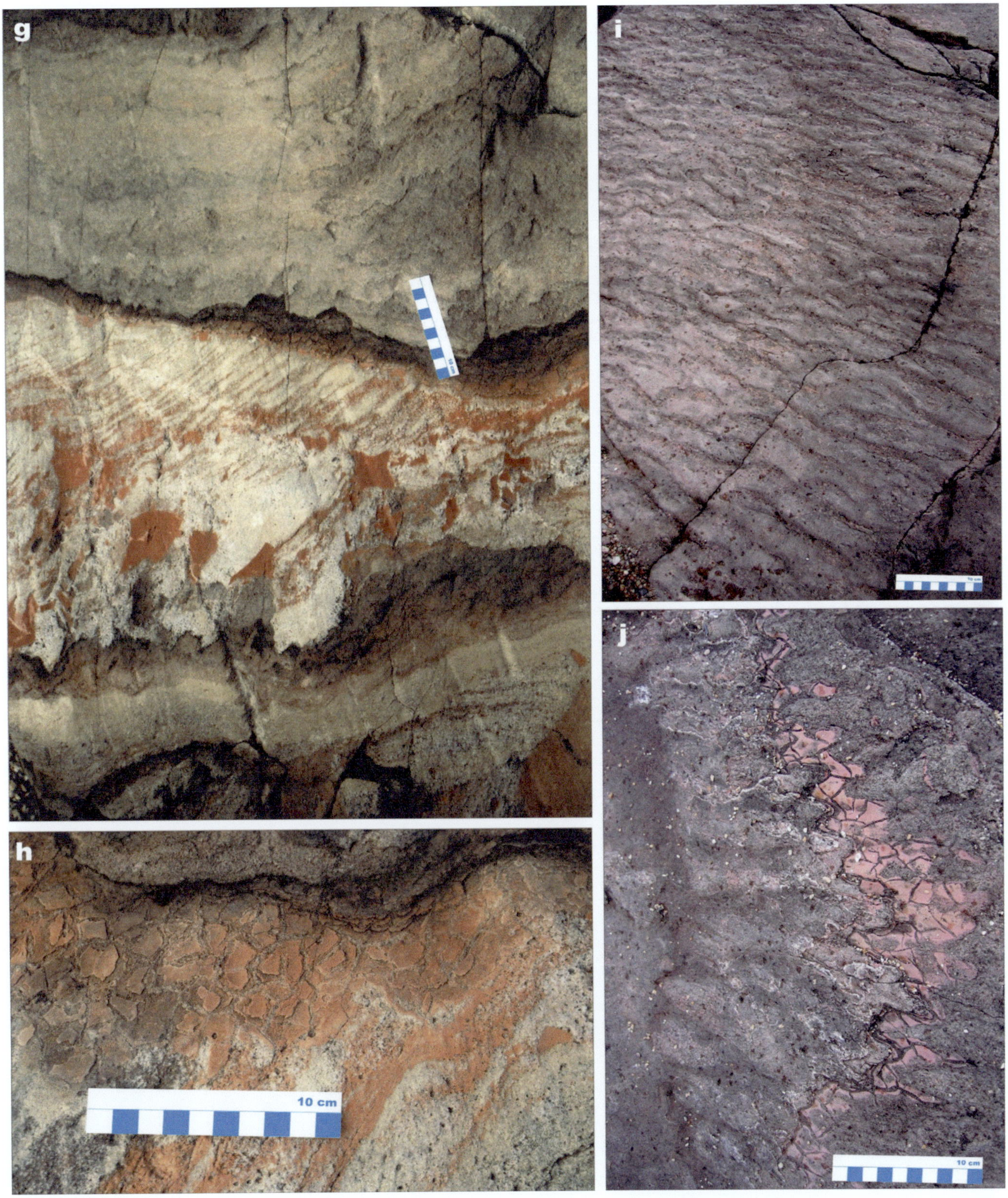

Fig. 4.43 (continued) (**g**) White, cross-bedded, sandstone with chips of *red* mudstone, note that both contacts of the cross-bedded unit show uneven (erosional) surfaces. (**h**) *Pale grey*, coarse-grained sandstone and desiccated, *red* mudstone erosively overlain by clayey sandstone. (**i**) Sandstone bed with asymmetric ripples. (**j**) Rippled sandstone with desiccated and partially dismembered, *red* mudstone layer

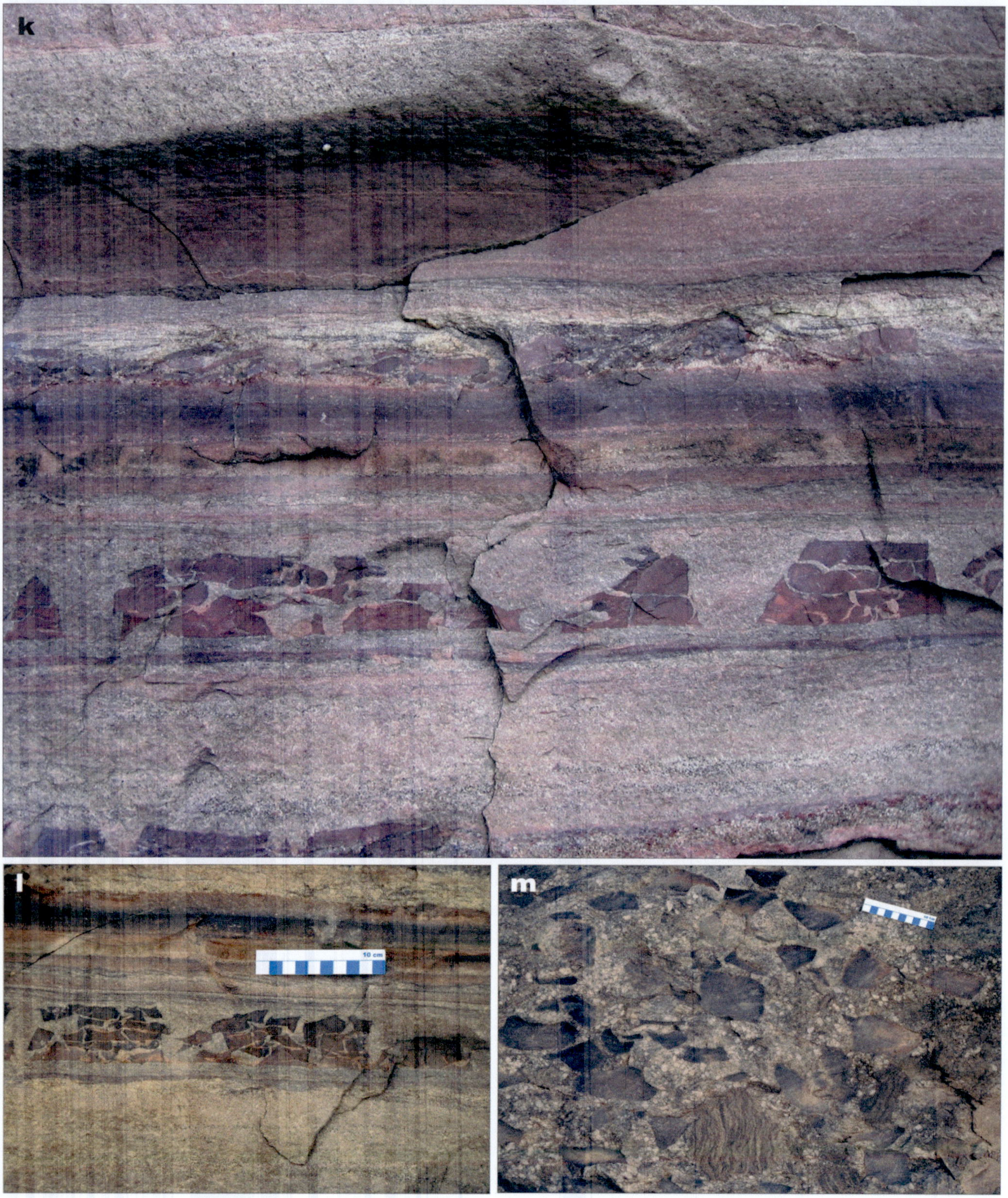

Fig. 4.43 (continued) (**k, l**) Cross-section view of parallel-bedded, coarse-grained, arkosic sandstone with a series of mudstone beds affected by desiccation; note that both bedding-parallel (sheet cracks) and vertical cracks affected the mudstone beds. (**m**) Chips of mudstones on the surface of a sandstone bed

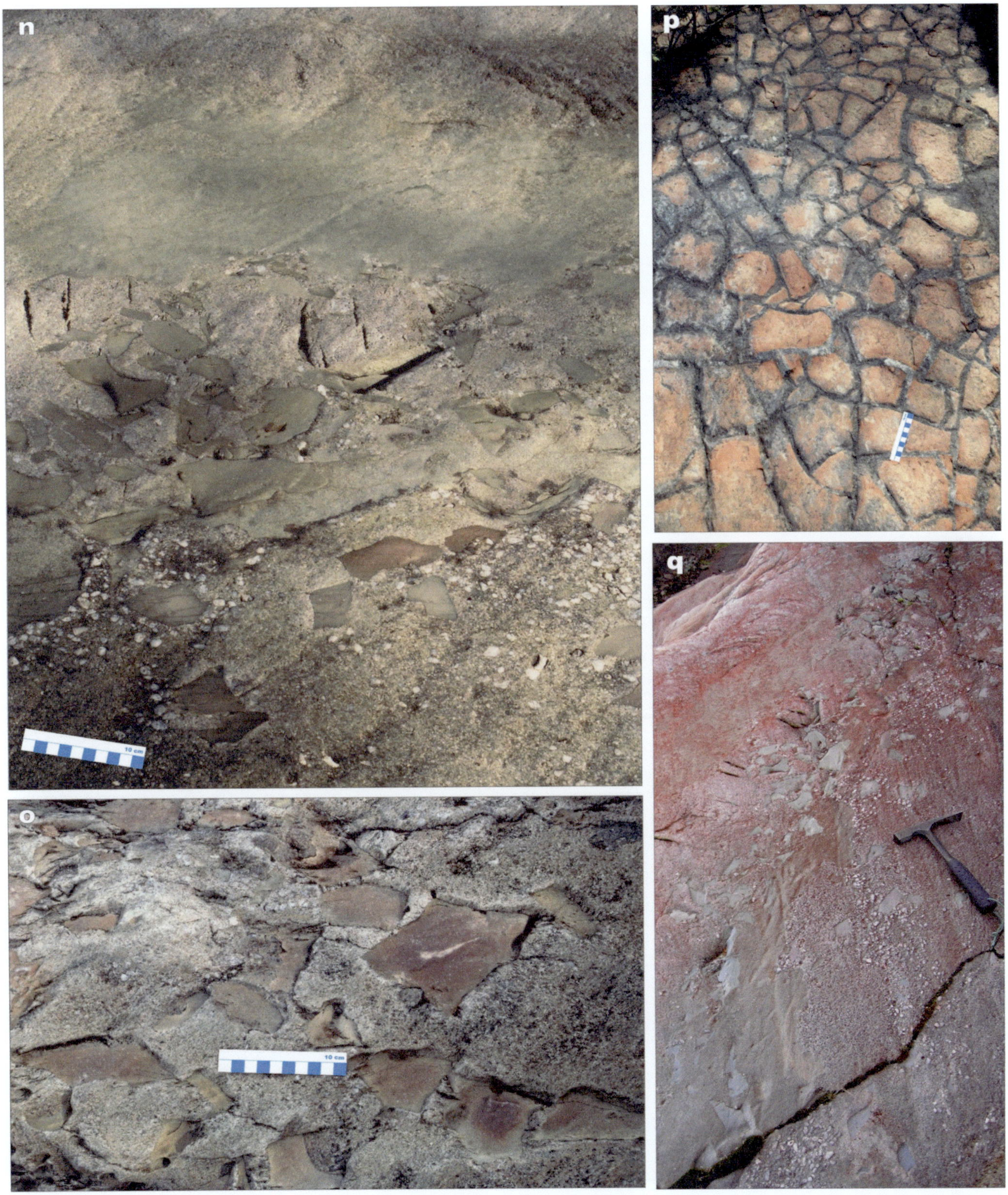

Fig. 4.43 (continued) (**n**) Gritty sandstone with scattered chips of mudstone overlain by massive sandstone. (**o**) Intensely desiccated and dismembered mudstone layer on top of gritty sandstone. (**p**) Polygonal cracks in *red* mudstone bed ("red bed"). (**q**) Bedding surface view of massive sandstone with pockets (channels) of gritty sandstone with scattered chips of grey mudstone

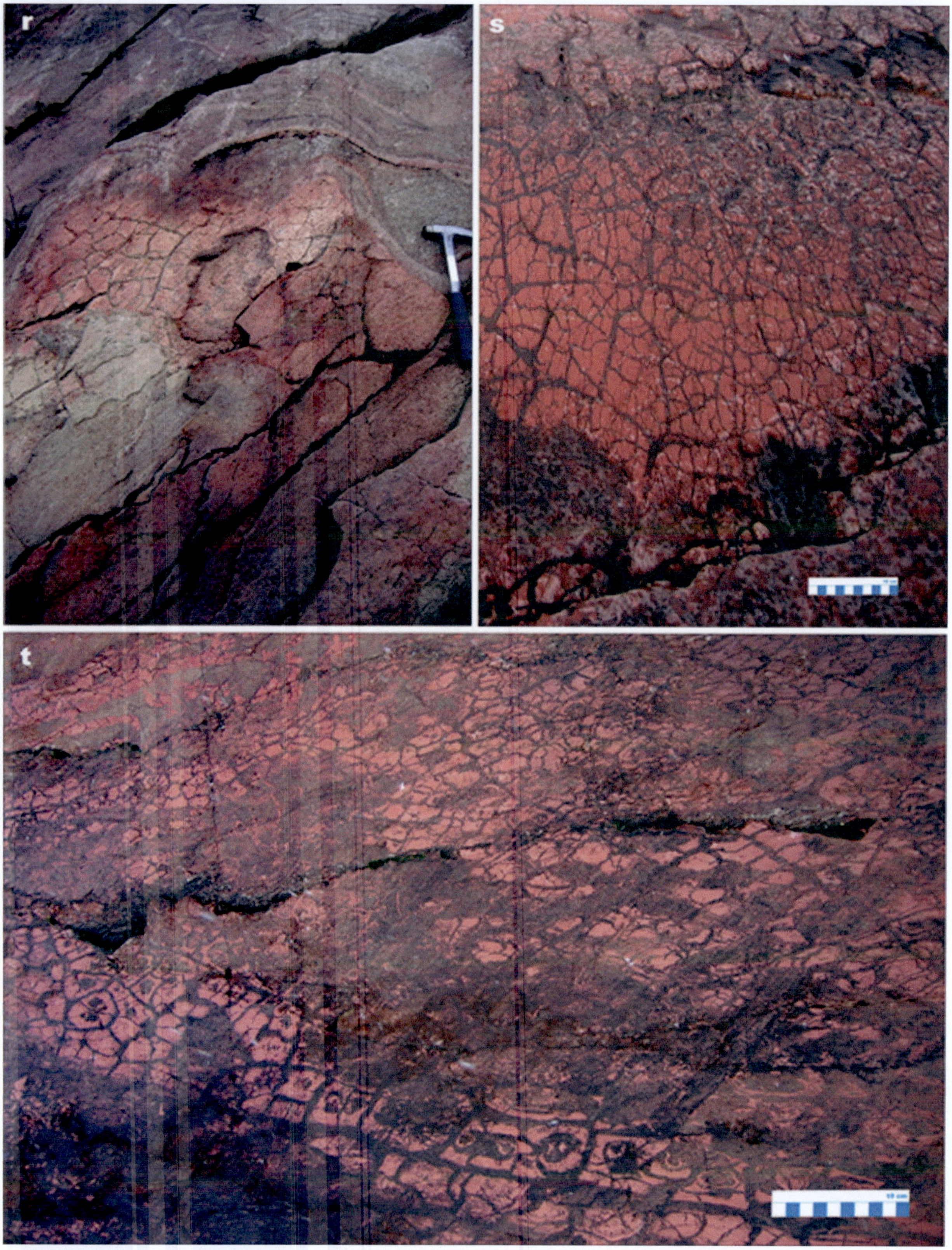

Fig. 4.43 (continued) (**r**) Oblique surface to bedding exhibiting partially desiccated, variegated, clayey sandstone ("red bed") overlain by channelised sandstone unit. (**s**) Two generations of desiccation cracks in a thick, *red* mudstone bed ("red bed"). (**t**) Bedding-parallel view of *red* mudstone bed ("red bed") showing intensive desiccation (Photographs by Victor Melezhik)

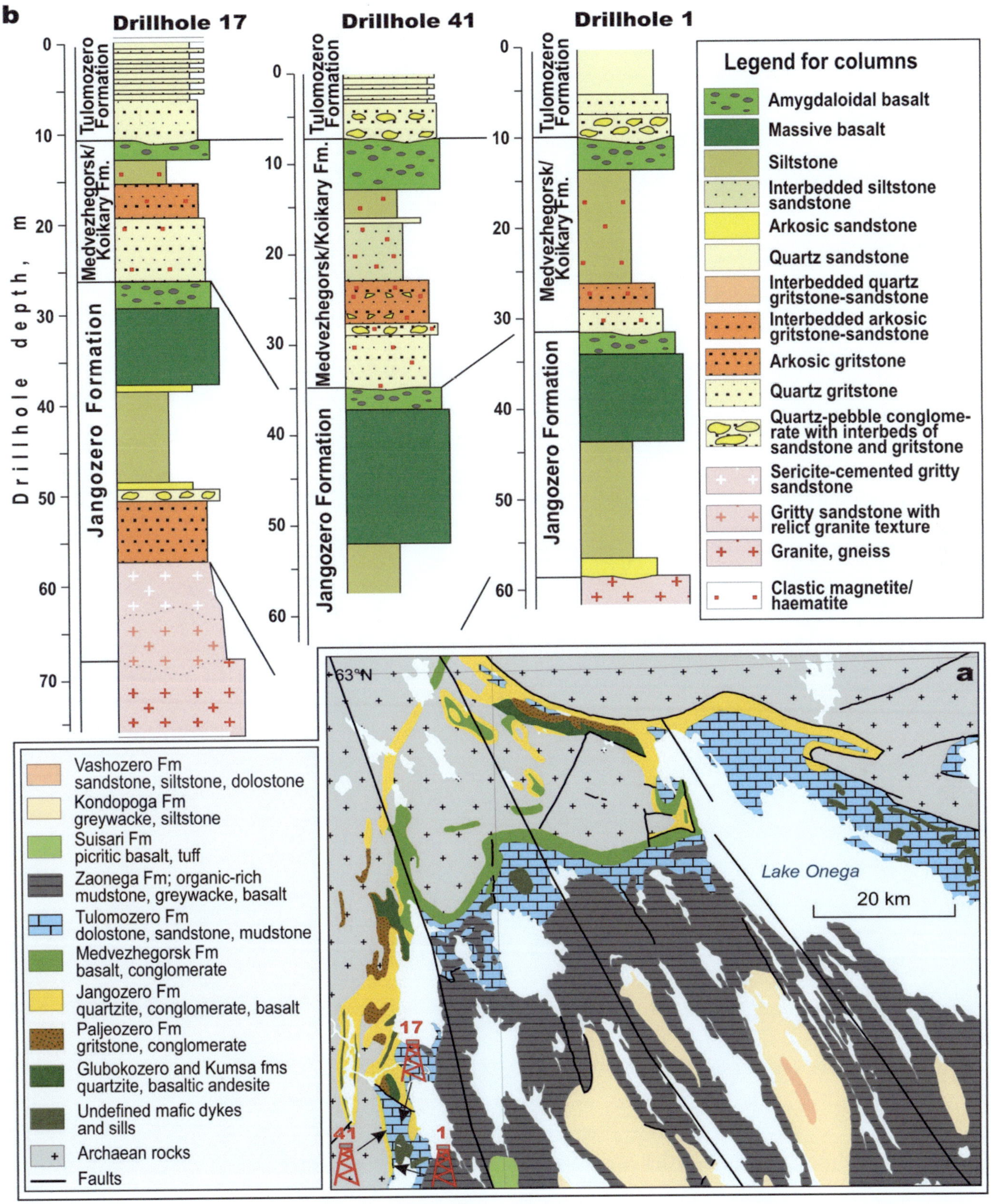

Fig. 4.44 Lithostratigraphic columns of the Jangozero and Medvezhegorsk formations based on drilled sections. (**a**) Geological map showing position of the drillholes that intersected the Jangozero and Medvezhegorsk formations. (**b**) Lithostratigraphic columns of the Jangozero and Medvezhegorsk formations based on description of cores presented in Sivaev et al. (1982; cores 14 and 41) and Gromova et al. (1954; core 1)

Fig. 4.45 Photographs of selected rock types representing the Medvezhegorsk Formation. *Lower (sedimentary) member*: (**a, b**) Cross-section through quartz sandstone with planar, unidirectional cross bedding (*arrowed*) and multiple sets of small-scale, trough cross-beds. (**c**) Bedding surface of quartz sandstone with bifurcating ripples

Fig. 4.45 (continued) (**d**) Bedding surface of quartz sandstone with desiccated mudstone layer. *Upper (volcanic) member*: (**e, f**) Fragmented top of a mafic lava flow (lava breccias)

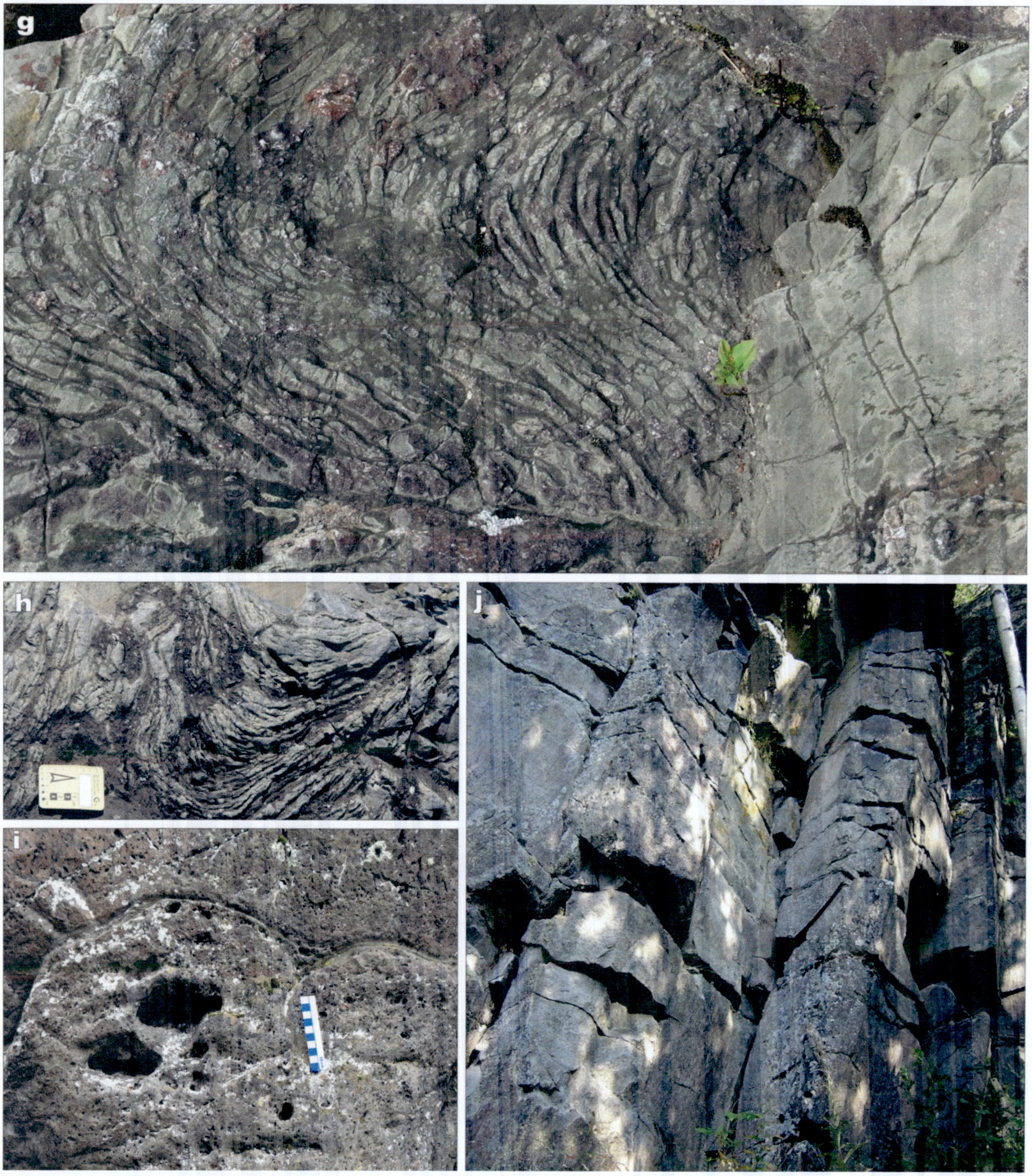

Fig. 4.45 (continued) (**g, h**) Two examples of mafic lava flow with a ropy surface (pahoehoe lava); width of the photograph (**g**) is c 1.2 m, tag length in (**h**) is 12 cm. (**i**) Mafic pillow lava; note thin, chilled-margins in all pillows and large vesicles in the center of the larger pillow. (**j**) Columnar joints in mafic lava; width of the photograph (**j**) is c 5 m (Photographs by Victor Melezhik (**a, c, d**), Pavel Medvedev (**b, j**) and Sergey Svetov (**e-i**))

Akhmedov et al., 1993			Medvedev et al. (2011)		Melezhik et al. (1999a) and current contribution	
Suite	Sub-suite	Pile	Beds with *Lithophyta*	Pile (Sokolov, 1963)	Member	Formation
Tulomozero	Upper	Redbed-Dolostone	*Calevia ruokanensis*	Redbed-Dolostone	H	Tulomozero
		Oncolite-Dolostone	*Butinella*	Dolostone	G	
		Dolostone-Mudstone		Dolostone-Sandstone	F	
	Middle		*Omachtenia kintsiensis*	Volcanogenic		
		Haematite-Sandstone		Haematite-Sandstone-Mudstone	E	
		Pisolite-Dolostone			D	
		Siltstone			C	
		Stromatolite-Dolostone	*Sundosia*	Chert-Dolostone	B	
		Magnesite-Dolostone	*Nuclephyton*	Dolostone-Sandstone-Mudstone		
		Phosphate-Dolostone	*Lukanoa*	Sandstone-Dolostone		
		Limestone-Dolostone-Mudstone		Limestone-Breccia-Mudstone	A	
	Lower	Undivided		Conglomerate-Sandstone	Undivided	

Fig. 4.46 Various lithostratigraphic subdivisions of the Tulomozero Formation in the Onega Basin

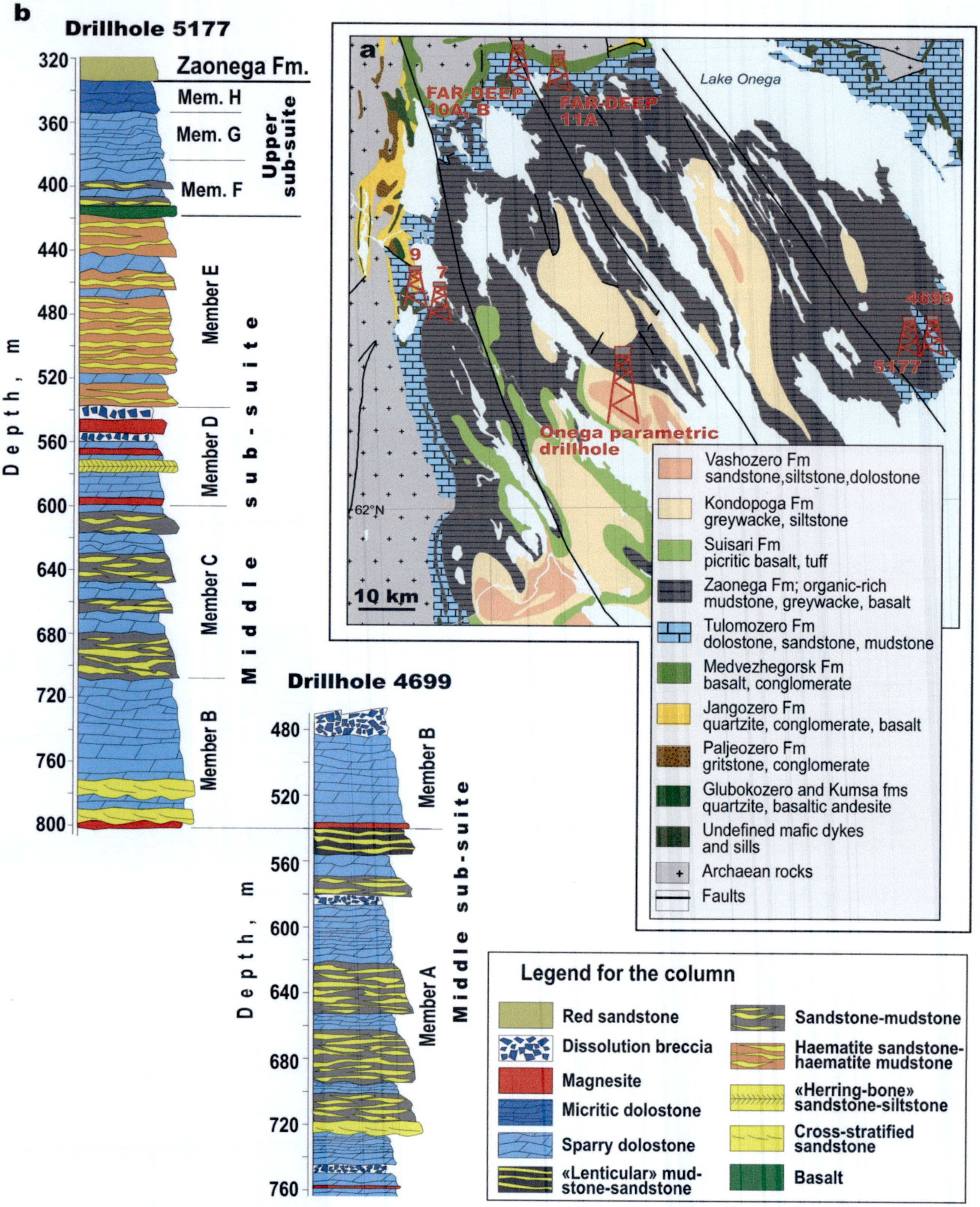

Fig. 4.47 Tulomozero Formation in the Onega Basin (**a**) Geological map of the Onega Basin with locations of studied drillholes. (**b**) Lithological column of the Tulomozero Formation based on drillholes 5177 and 4699; the lithological column is modified from Melezhik et al. (1999a)

Fig. 4.48 Photographs illustrating main sedimentological features of selected rock types of the Tulomozero Formation (core diameter is 4 cm here and in all following photographs unless specified otherwise; numbers correspond to drillhole number/depth in metres). *Member A:* (**a**) *Red*, laminated siltstone with nodules of former Ca-sulphates pseudomorphed by calcite; note small-scale enterolithic folds in the upper part of the core. (**b**, **c**, **d**) Dolomite-pseudomorphed gypsum crystals and clusters in brown, black and variegated mudstone. (**e**) Dissolution, collapse breccia composed of fragments of *brown*, *grey* and *pink* mudstone cemented by white calcite

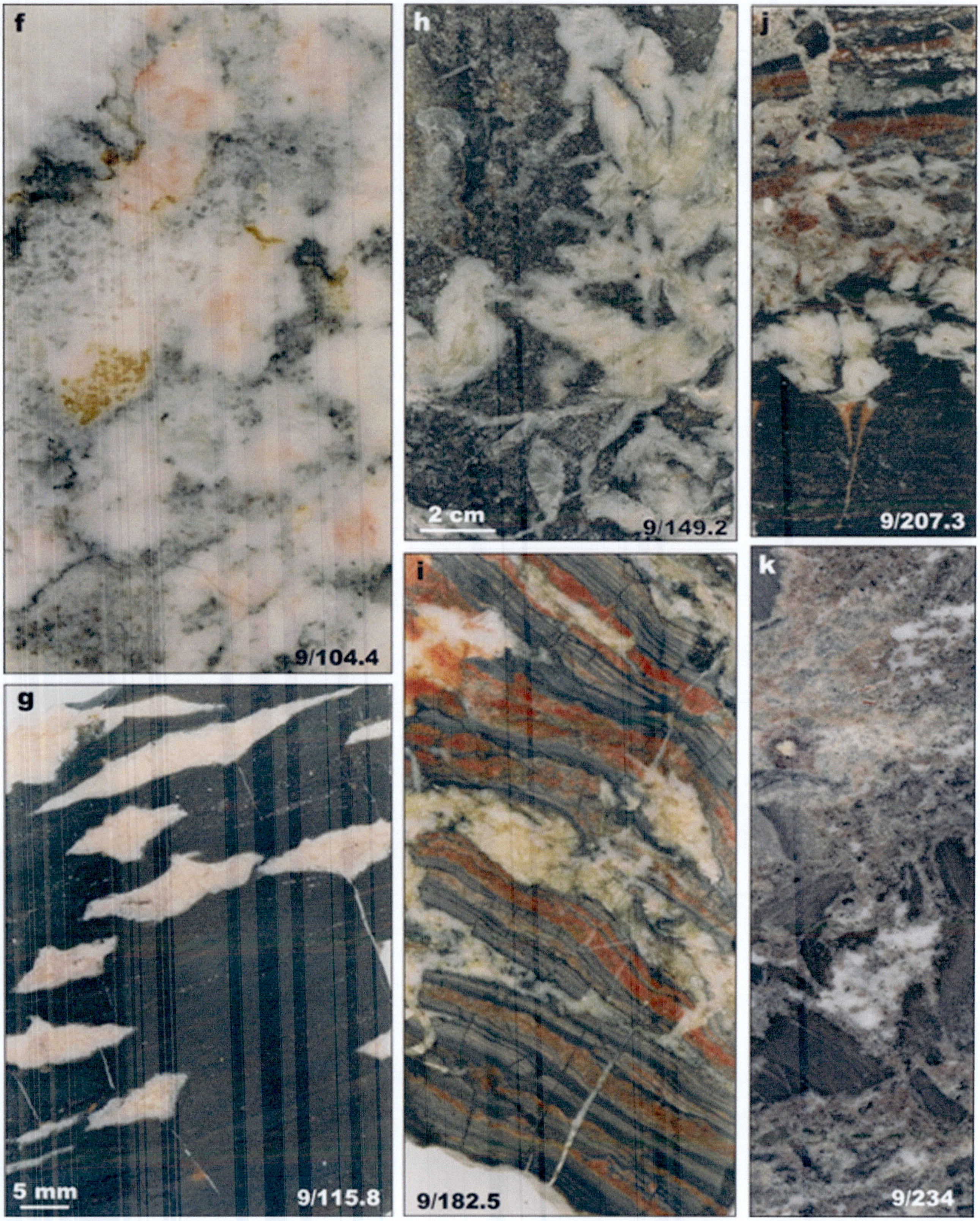

Fig. 4.48 (continued) (**f**) Chicken-wire anhydrite pseudomorphed by white dolomite. (**g**) Scanned thin section of *dark brown* mudstone with flat crystals of dolomite-pseudomorphed gypsum. (**h**) Scanned thin section of dolomite-pseudomorphed clusters of gypsum crystals in dolomarl. (**i, j**) Variegated mudstones with dolomite-pseudomorphed Ca-sulphates disrupting the lamination. (**k**) Dissolution, collapse breccia composed of mudstone fragments cemented by white calcite

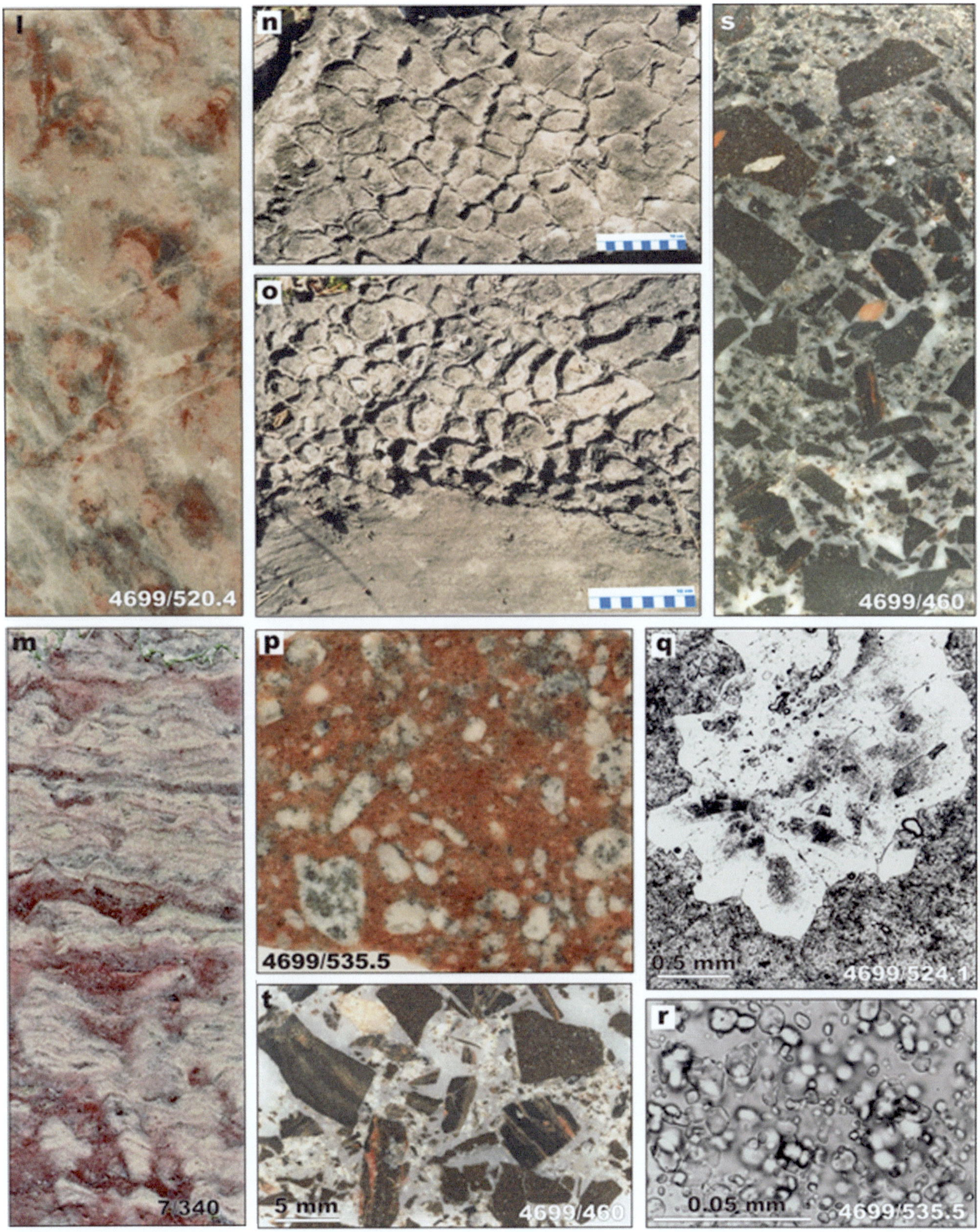

Fig. 4.48 (continued) *Member B:* (**l**) Recrystallised dolostone with relicts of stromatolites. (**m**) Flat-laminated, gently domed, small columnar stromatolites in *red* dolarenite. (**n, o**) Bedding-parallel surface of stromatolitic dolostone; columnar stromatolies form recessive relief with respect to intercolumnar, silicified dolarenite infill; outcrop at Sundozero (for location see Fig. 4.34). (**p**) Matrix-supported dolorudite. (**q**) Photomicrograph in transmitted, non-polarised light of silica-pseudomorphed gypsum rosette with ubiquitous gas/fluid micro-inclusions (dark coloured clusters) in dolomite-cemented sandstone. (**r**) Photomicrograph in transmitted, non-polarised light of ubiquitous gas/fluid inclusions in silica-pseudomorphed gypsum rosette. (**s, t**) Dissolution, collapse breccia composed of fragments of brown mudstone cemented by white calcite

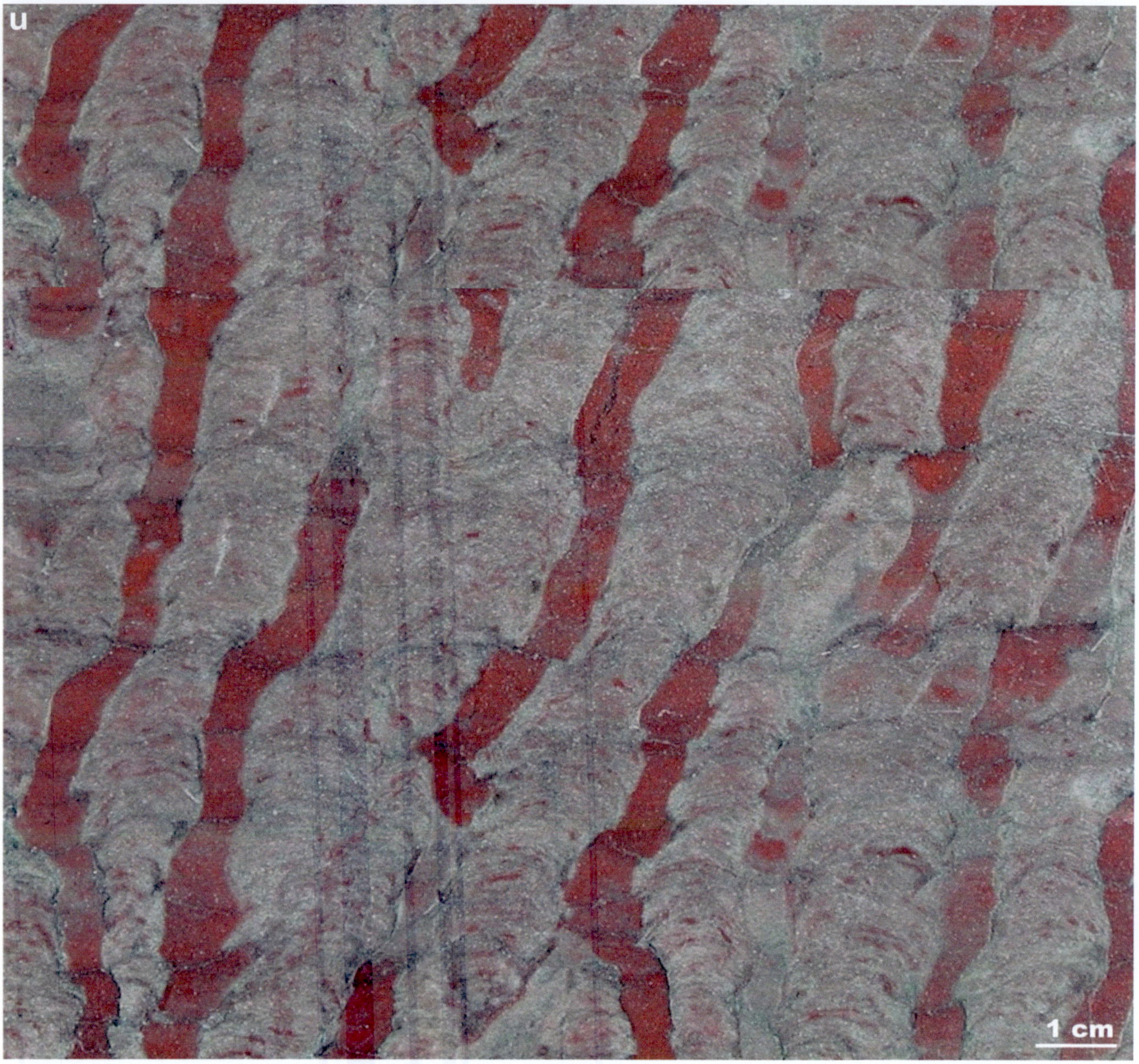

Fig. 4.48 (continued) *Member C:* (**u**) Polished slab of columnar stromatolite with *red* dolarenite filling inter-columnar areas

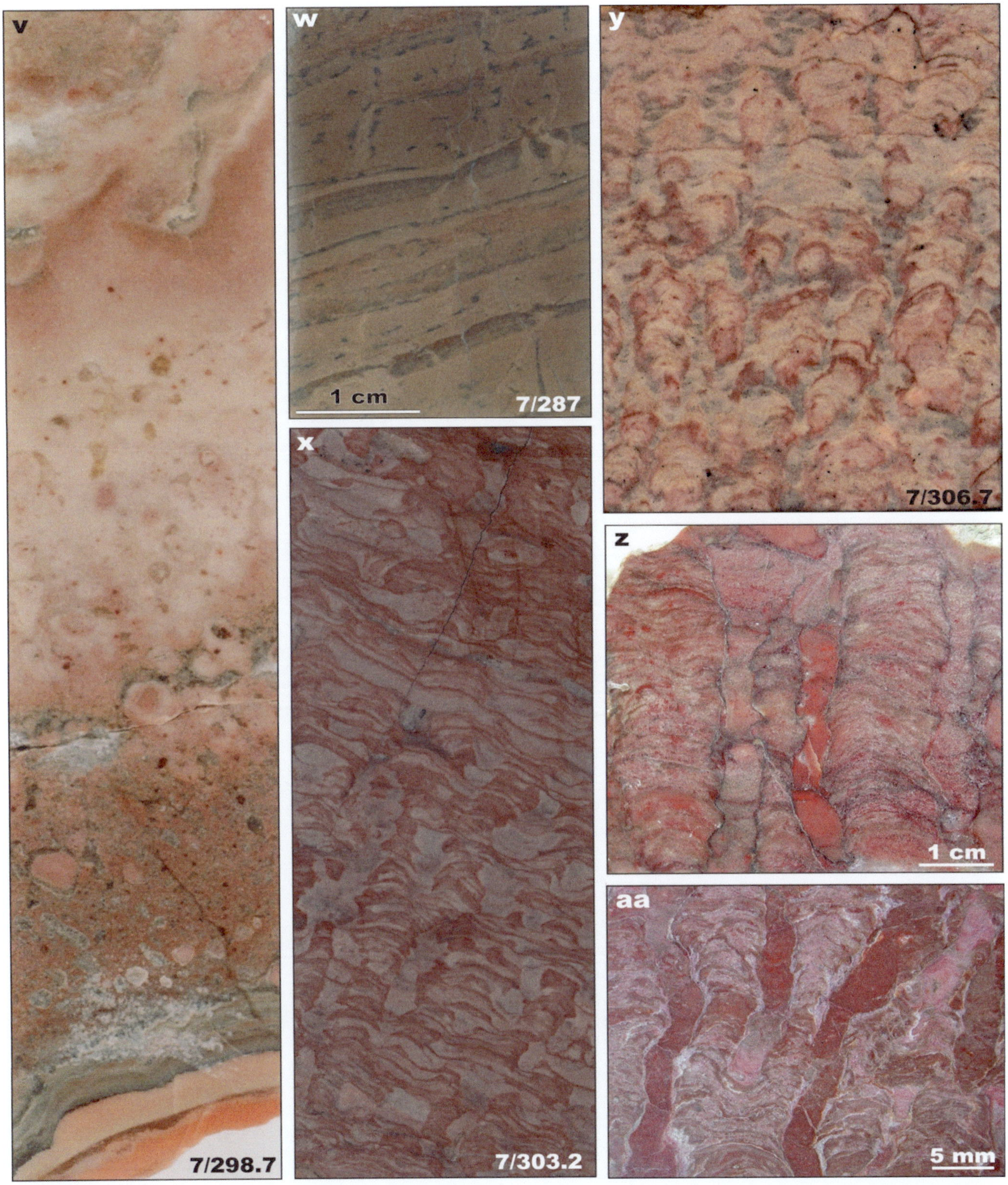

Fig. 4.48 (continued) (**v**) Thin, rippled layers of *pale pink* dolomicrite overlain by *pink* dolorudite grading upward into *pale pink* dolarenite. (**w**) Scanned thin section of pale brown dolomarl with desiccated cracks. (**x–aa**) Variable morphologies of columnar stromatolites as seen in core (**x**, **y**) and polished slabs from hand specimens (**z**, **aa**)

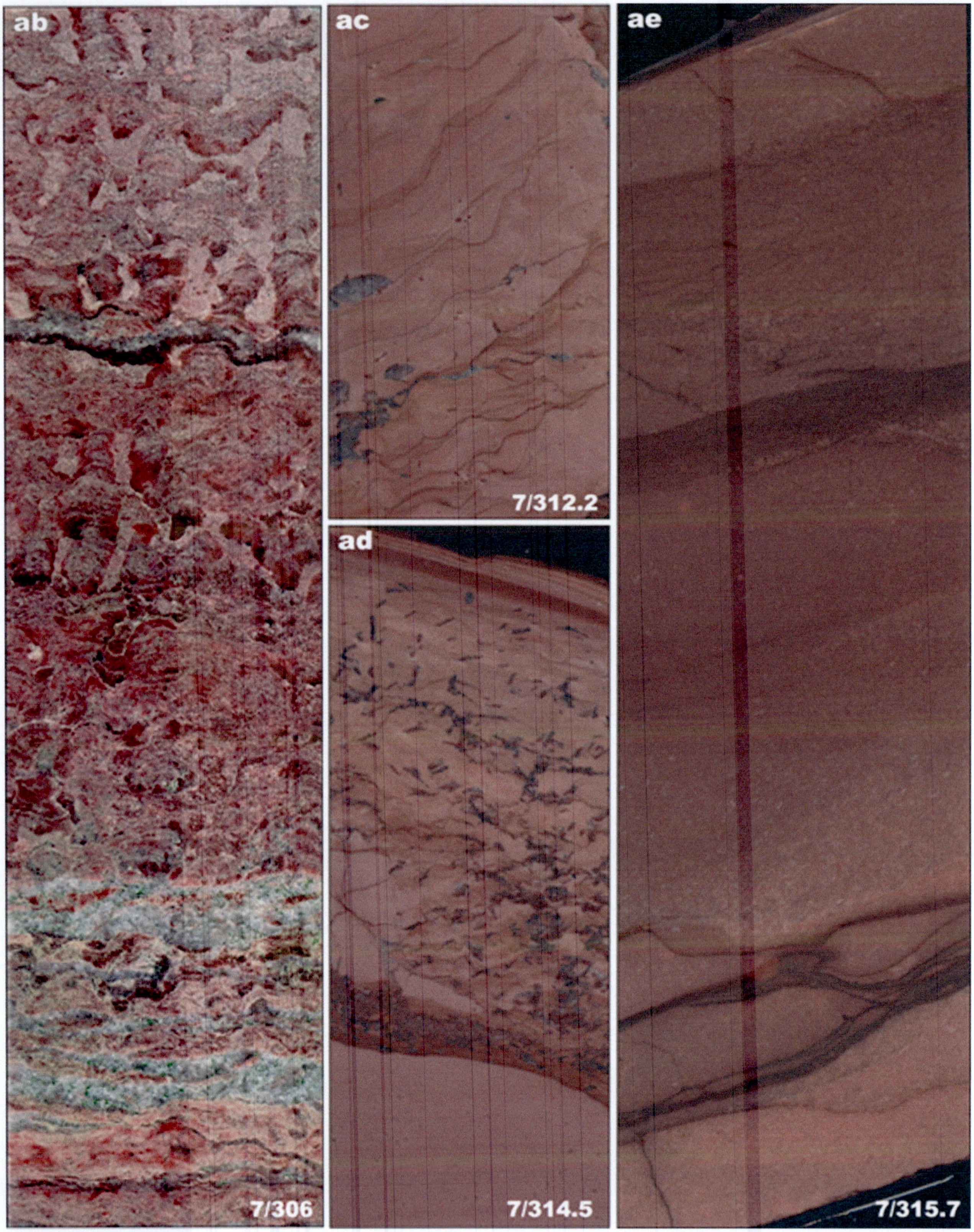

Fig. 4.48 (continued) (**ab**) Alternating beds of flat-laminated and small columnar stromatolites. (**ac**) *Pink*, rippled fine-grained dolarenite with thin, darker mud drapes. (**ad**) *Pink*, massive, fine-grained dolomicrite overlain by dolarenite with silica-pseudomorphed gypsum crystals. (**ae**) *Dark pink*, fine-grained dolarenite with flaser bedding in the lower part defined by *dark pink* dolarenite ripples and brown dolomarl; thicker overlying bed shows vague grading and is draped by darker dolomarl

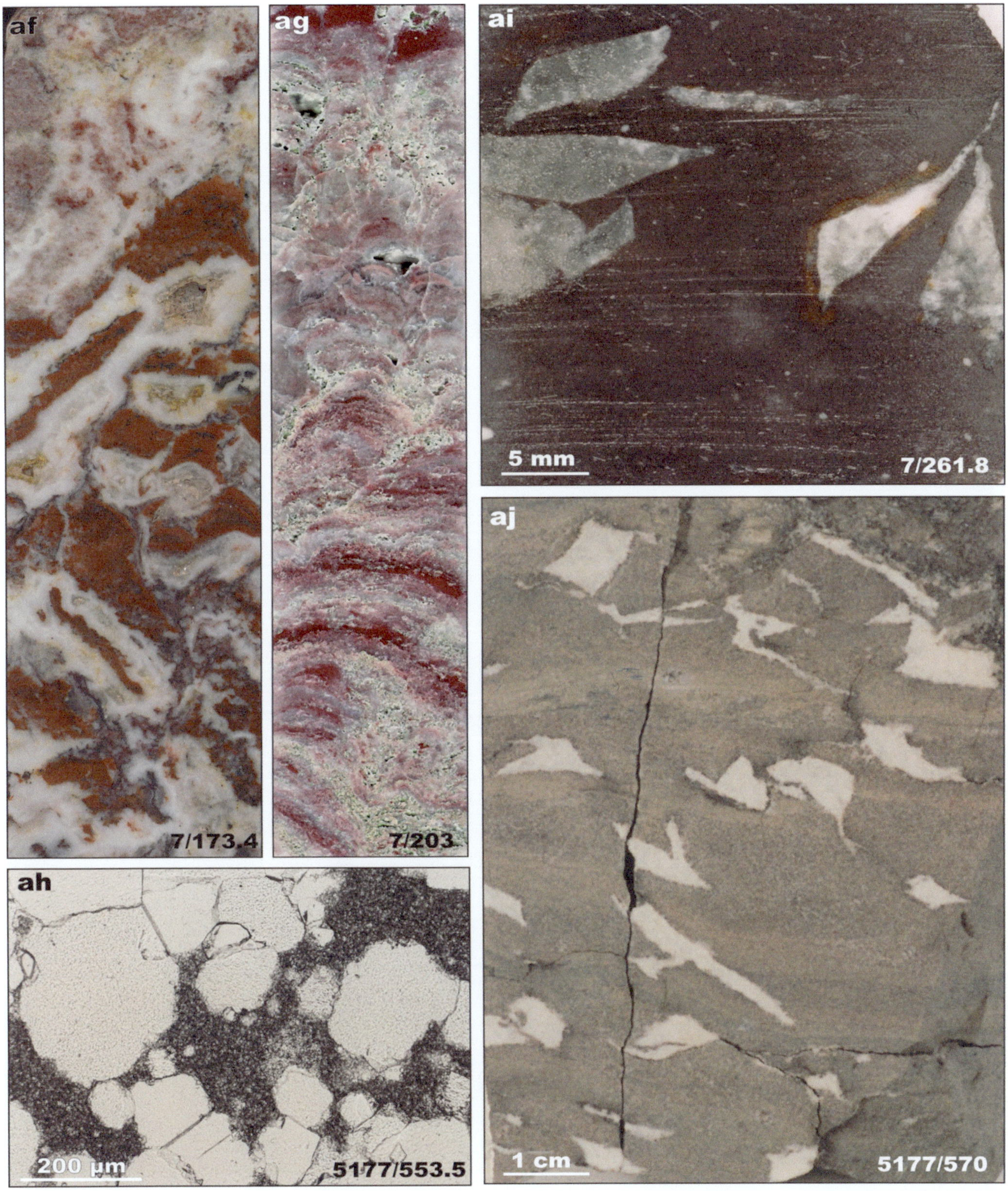

Fig. 4.48 (continued) *Member D:* (**af**) Dissolution brecccia composed of *red* mudstone fragments cemented with white dolomite. (**ag**) Porous stromatolitic dolostone. (**ah**) Photomicrograph in transmitted non-polarised light showing large magnesite crystals (*bright*) grown in an earlier, fine-grained magnesite mass (*dark grey*). (**ai**) Scanned polished section with quartz-pseudomorphed gypsum crystals in brown mudstone. (**aj**) Dolomite-pseudomorphed gypsum crystals in brown mudstone; note that many crystals show swallow-tail twinning

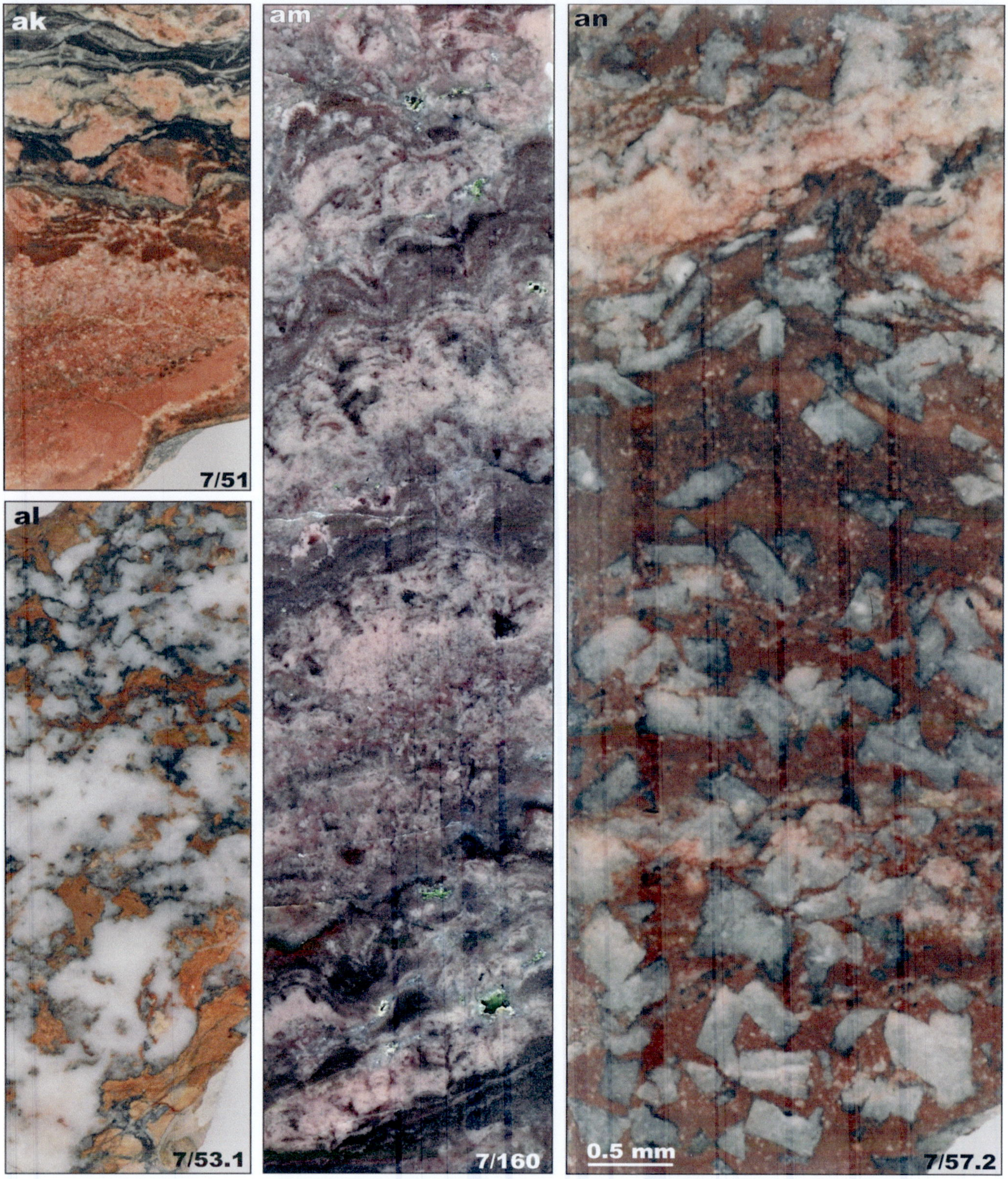

Fig. 4.48 (continued) *Member F*: (**ak**) *Red* dolarenite overlain by dark mudstone with lamination distorted by evaporitic growth of apparent sulphates, which were later replaced by *pale pink* dolomite. (**al**) Dolomite-replaced chicken-wire anhydrite with relicts of *red* mudstones dismembered by growing sulphates. (**am**) Recrystallised, porous, clayey, stromatolitic dolostone. (**an**) Polished section of *red* mudstone with dolomite-pseudomorphed crystals of halite

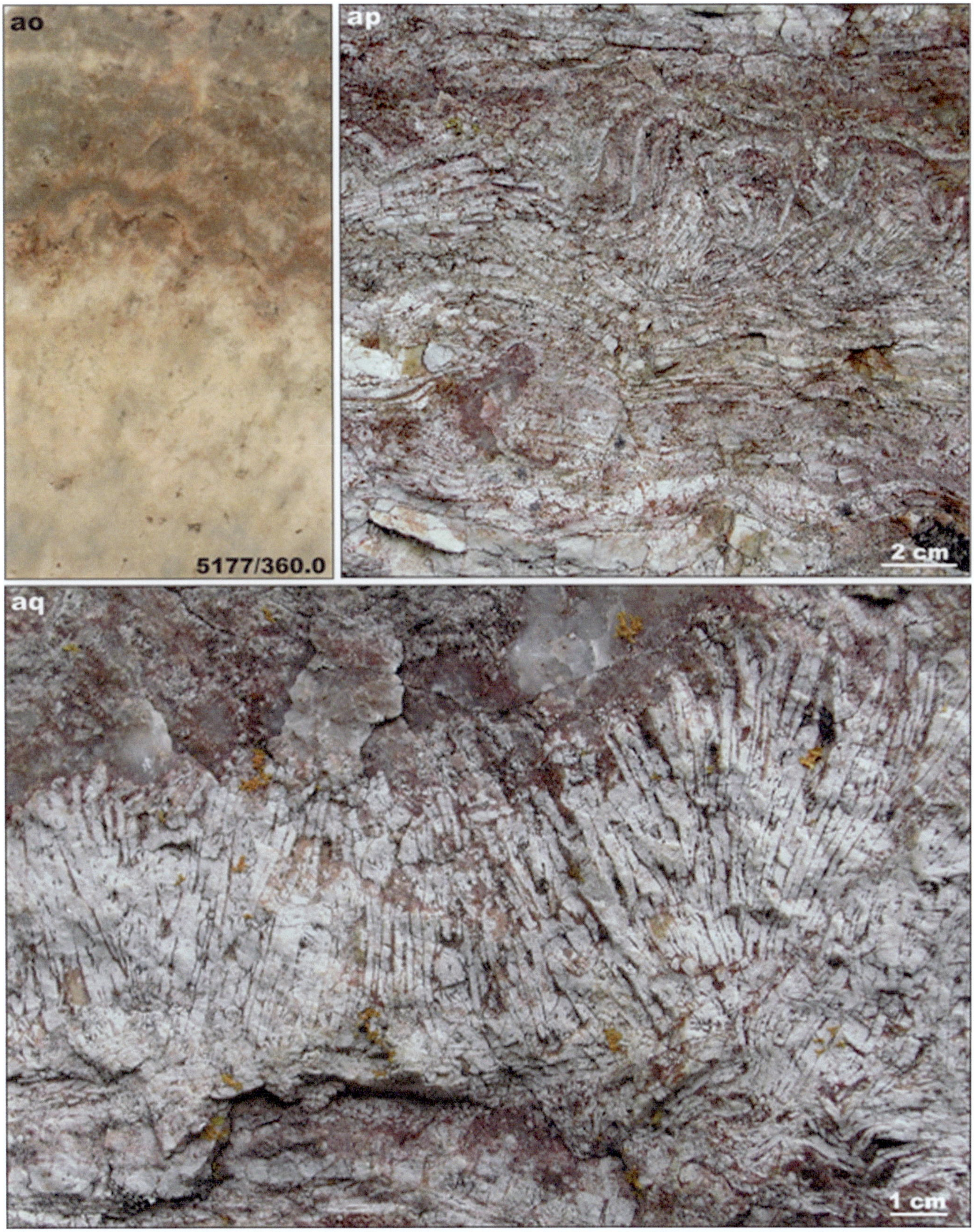

Fig. 4.48 (continued) *Member G:* (**ao**) Beige, massive dolostone capped by *pale brown* stromatolite layer passing upward into indistinctly bedded dolostone. (**ap**) Ribbonite dolostone overlain by a bed with stone-rosettes; note syndepositional deformation of "ribbonite". (**aq**) Stone rosettes on an irregularly eroded surface of a dolostone bed

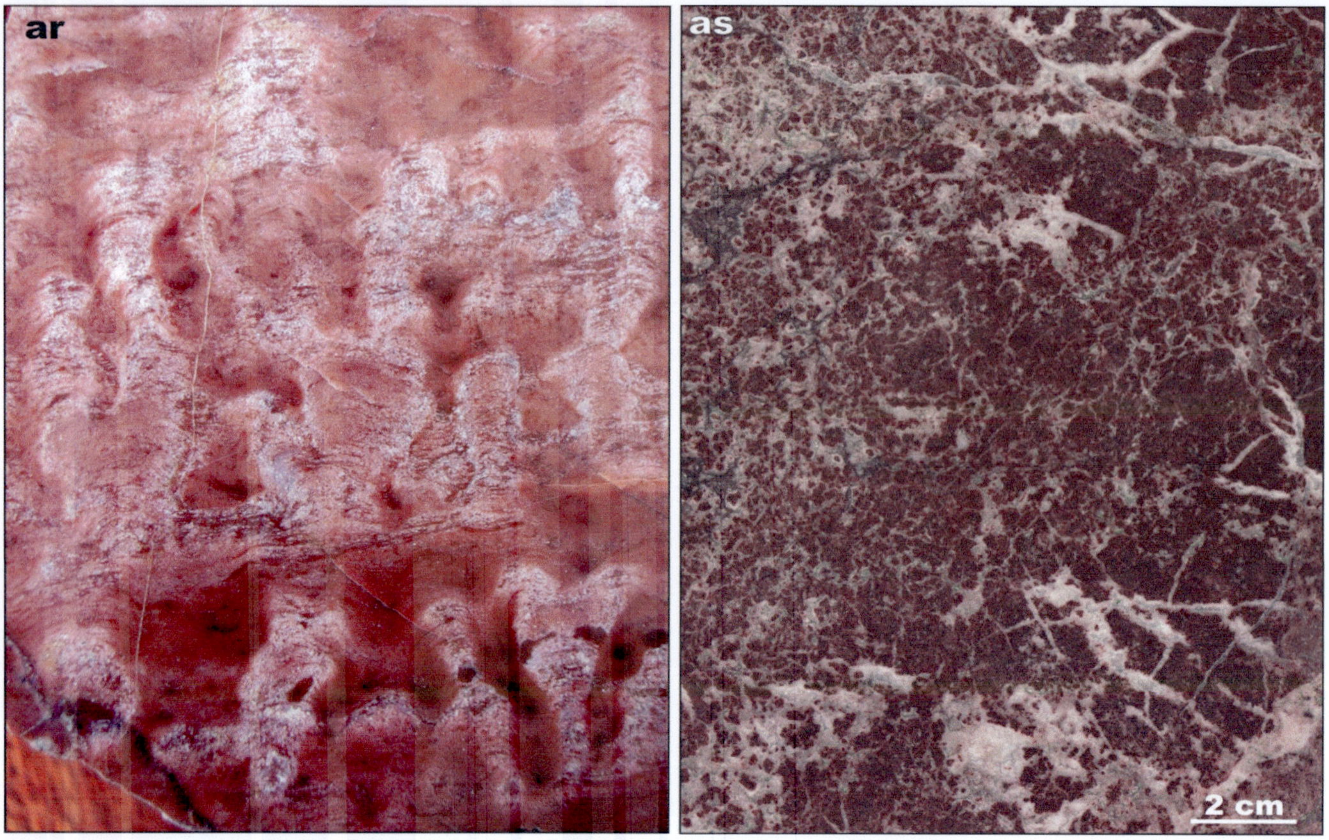

Fig. 4.48 (continued) *Member G:* (**ar**) Markedly divergent columnar stromatolites in *red* dolarenite. (**as**) Recrystallised and jointed oolitic (oncolitic?) dolostone; the joints are filled with white dolomite (Photographs by Victor Melezhik (**a, d–f, i–l, t, s, v–x, ac–af, ah, ai, ak–al, an**), Vladimir Makarikhin (**u, z, ap, aq**) and Pavel Medvedev (**m, ab, ag, am, ar, as**). Photographs (**b, c, g, h, n–p, q, r, y, aj, a**) are reproduced from Melezhik et al. (1999a, 2000, 2005b) with permission from Elsevier and Blackwell Publishing Ltd.)

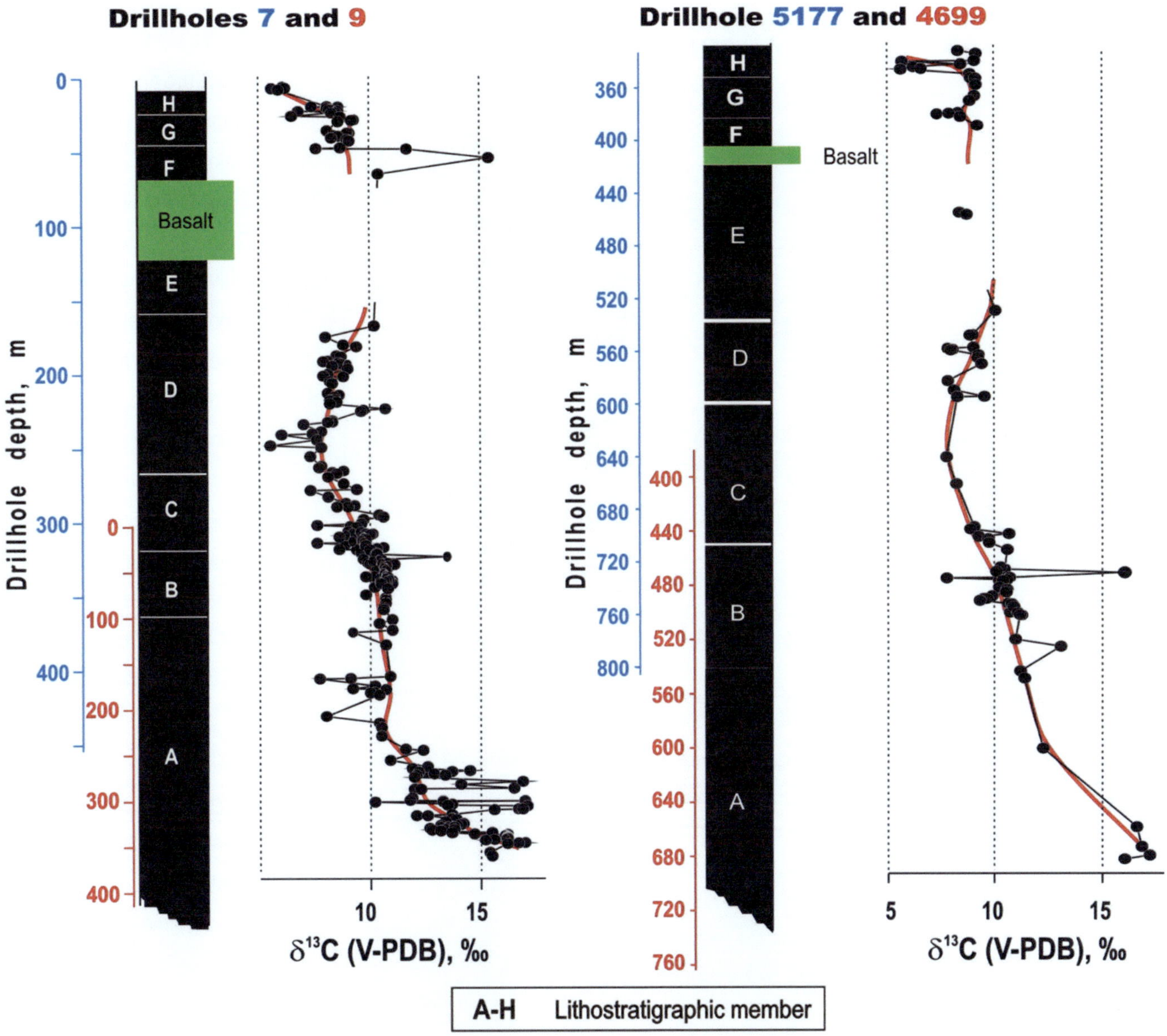

Fig. 4.49 Correlation of two sections of the Tulomozero Formation intersected by overlapping drillholes. The drilled sections are separated by a distance of c. 80 km, and correlated based on $\delta^{13}C_{carb}$ isotopic trends (Modified from Melezhik et al. 2005b). Drillhole locations are shown in Fig. 4.47

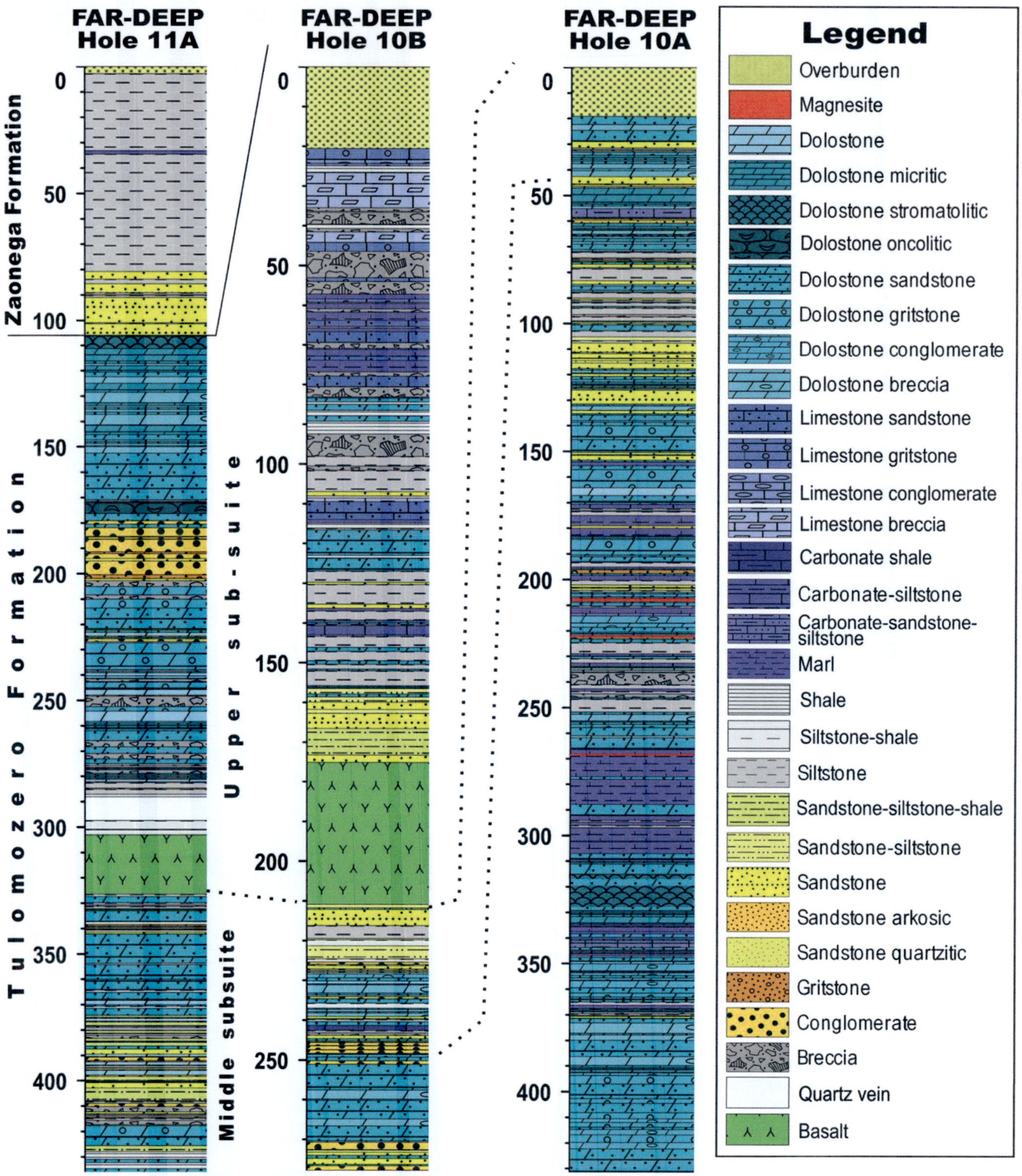

Fig. 4.50 Lithological columns of the Tulomozero Formation based on FAR-DEEP cores 10A, 10B and 11A (for location see Fig. 4.47) showing the gross lithological compositions of the drilled secton; for more detailed descriptions see Chaps. 6.3.1 and 6.3.2

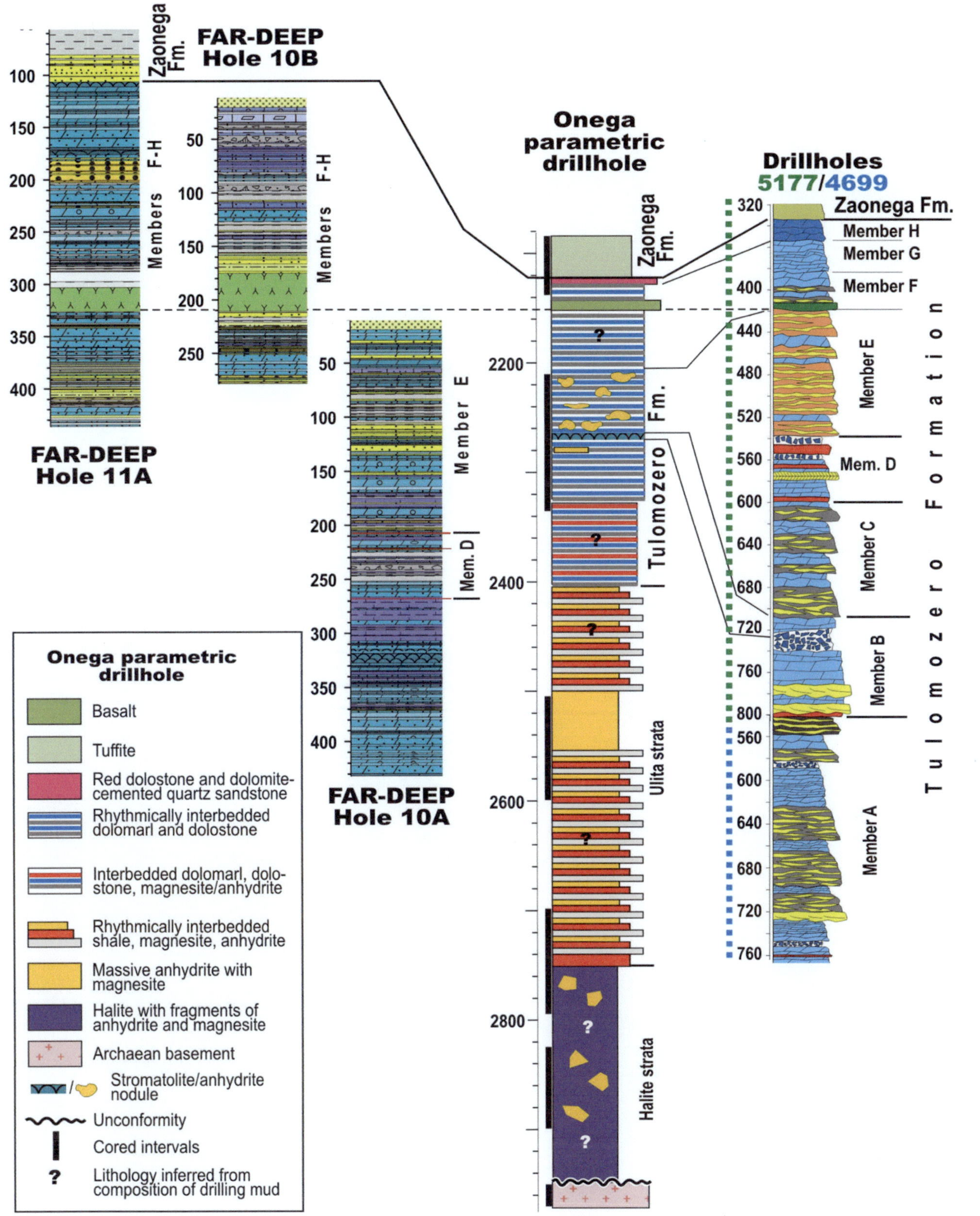

Fig. 4.51 Tulomozero sections intersected by drillholes in different parts of the Onega Basin; note variations in lithologies from core-to-core (for location see Fig. 4.47). The Onega parametric drillhole section is modified after Morozov et al. (2010) and Krupenik et al. (2011a)

Fig. 4.52 Photographs illustrating main sedimentological and petrological features of selected rock types of the Zaonega Formation. In all photographs, unless otherwise marked, the scale is the core width of 5 cm. Numbers denote drillhole number/depth in metres. *Lower sub-suite, Member A*: (**a**) Variegated, coarse-grained, cross-bedded sandstone. (**b**) *Red*, cross-bedded sandstone with bright, chemically reduced spots. (**c**) Variegated, graded beds of greywacke; note that some beds have erosional bases and syndepositional faulting. (**d**) *Pale green*, parallel-laminated greywacke grading into massive siltstone. (**e**) A series of graded greywacke beds with a small-scale loading structure at the base

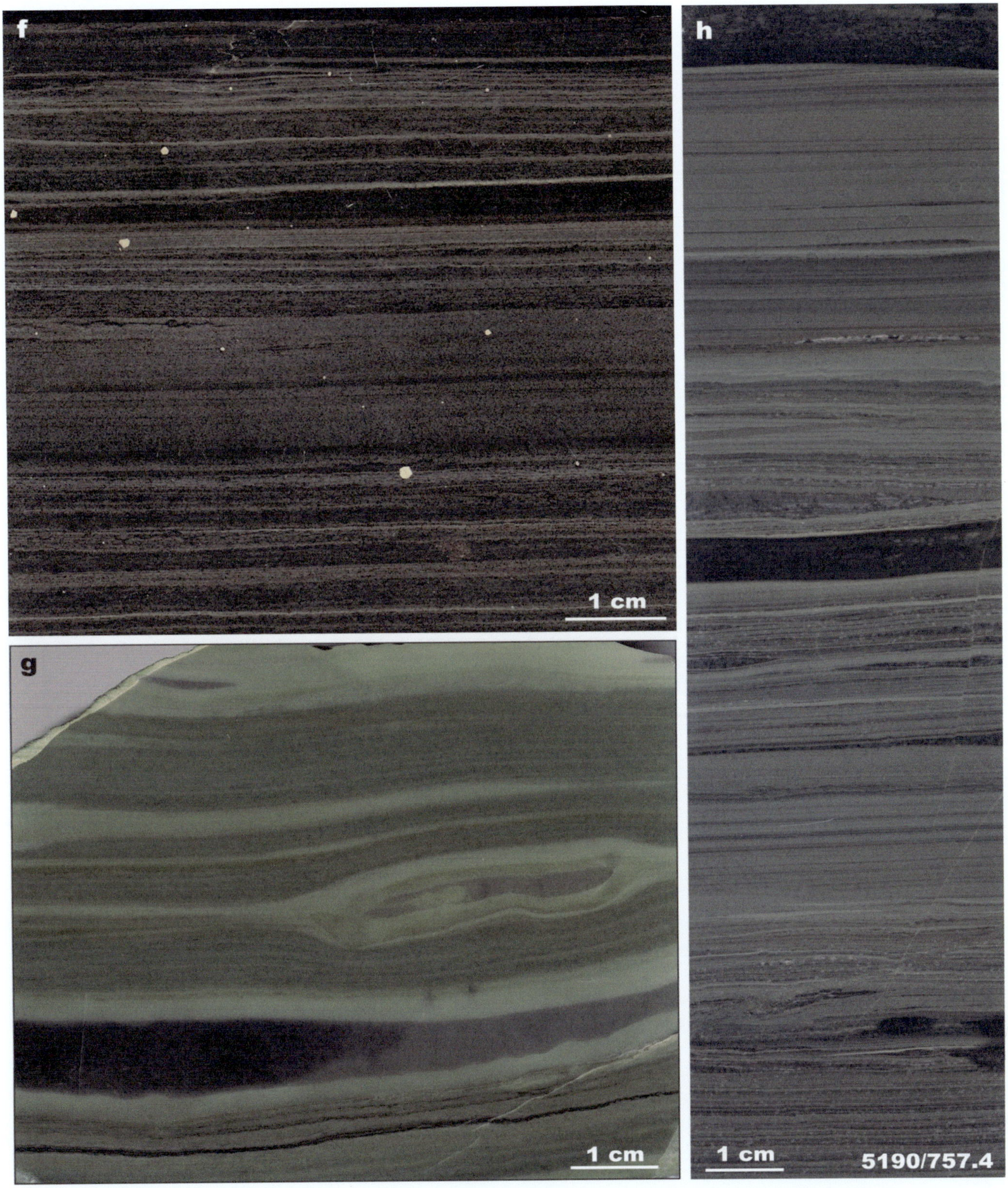

Fig. 4.52 (continued) *Lower sub-suite, Member B*: (**f**) Polished slab of parallel-laminated, calcareous siltstone locally termed "krivozerite" after a type locality at Krivozero; sample from the type locality. (**g**) Polished slab of krivozerite; variegated, parallel-bedded siltstone with siliceous concretion and mudstone layer (*dark coloured*); type locality at Krivozero. (**h**) Krivozerite; parallel-laminated and low-angle, cross-laminated greywacke with mudstone layers (*dark coloured*)

Fig. 4.52 (continued) (**i, j**) Krivozerite; variegated siltstone; note that postdepositional (katagenetic/epigenetic) oxidation has resulted in considerable obliteration of well-pronounced parallel and cross lamination. *Upper sub-suite, Member A*: (**k**) C_{org}-bearing, parallel-bedded greywacke with white silica-rich laminated interval (*arrowed*) and sulphide veins (*yellow*)

Fig. 4.52 (continued) *Upper sub-suite, Member A*: (**k**) C_{org}-bearing, parallel-bedded greywacke with white silica-rich laminated interval (*arrowed*) and sulphide veins (*yellow*). *Upper sub-suite, Member A*: (**l**) Greywacke with thin layers of C_{org}-rich mudstone (*black*) and concretionary calcite layer (arrowed). (**m**) Thick intervals of *pale grey*, massive grewacke with thin mudstone drapes (*black*). (**n, o**) Pale grey greywacke interbedded with dark grey siltstone and black mudstone layers. (**p**) Parallel-laminated, *dark grey* siltstone and *pale grey* greywacke passing upward into greywacke with thin interlyers of *dark grey* siltstone. (**q**) Parallel-laminated siltstone and greywacke

Fig. 4.52 (continued) (**r**) C_{org}-bearing, parallel-laminated siltstone with thin pyrobitumen-rich veins (*yellow arrowed*) and small sulphide nodules (*red arrowed*). (**s**) Rhythmically bedded greywacke-siltstone. (**t**) Parallel-laminated siltstone with pyrobitumen-rich vein (*yellow arrowed*) displaced along bedding-parallel shear-band (*red arrowed*). (**u**) Intraformation conglomerate composed of soft-sediment deformed fragments of greywacke (*pale grey*) in siltstone matrix (*dark coloured*)

Fig. 4.52 (continued) (**v**) Matrix-supported, syndepositional breccia gradually passing upward into laminated greywacke; note that greywacke matrix is structureless and clasts are angular. (**w**) Matrix-supported, intraformational conglomerate composed of rounded clasts of greywacke embedded in massive silty mudstone. (**x**) Amygdaloidal mafic lava flow with faintly fragmented top overlain by dark grey siltstone. (**y**) Sulphidised, C_{org}-rich, bedded siltstone eroded by the overlying mafic lava flow. All cores are from Hole

Fig. 4.52 (continued) *Upper sub-suite, Member B*: (**z**, **aa**, **ab**) Rhythmically bedded greywacke (*pale grey* and *grey*) and C_{org}-rich siltstone-mudstone (*black*). (**ac**) Parallel-bedded, C_{org}-rich siltstone-mudstone with sulphide concretions and concretionary layers (*dark yellow*). (**ad**) Pale grey, massive dolostone intervals separated by thin, C_{org}-rich shale layers and cross-cut by ptygmatic, pyrobitumen-rich veinlets. (**ae**) Large calcite concretion (*bright*) in C_{org}-rich siltstone; depth of 310.4 m

Fig. 4.52 (continued) (**af**) Massive dolostone interval in contact with C_{org}-rich siltstone; note that the siltstone is partially dolomitised at the contact with the dolostone (*arrowed*). (**ag**) Rhythmically bedded, C_{org}-rich siltstone-mudstone with chert nodules (*arrowed*). (**ah**) Parallel-laminated, C_{org}-rich siltstone with a chert nodule (*red arrowed*) and chert layer (*yellow arrowed*); note differential compaction of the siltstone around the chert nodule. (**ai**) Matrix-supported, polymict, intraformational conglomerate containing greywacke, siltstone and marl clasts with some showing soft-sediment deformation (*arrowed*)

Fig. 4.52 (continued) (**aj**) In situ brecciated dolostone cemented with pyrobitumen-rich material. (**ak**) Vesicular, mafic lava flow with small-scale, incipient pillows; note that the lava is intensely sulphidised and sulphide-filled vesicles are zoned. (**al**) Pillows with thin, chilled margins (*dark coloured*). (**am**) Sulphide-fringed pyrobitumen vein in mafic lava

Fig. 4.52 (continued) (**an**) Massive, organosiliceous rock locally termed maksovite. (**ao**) Maksovite with incipient polygonal shrinkage cracks (columnar joints in three-dimensional space) cemented with sulphides; photograph was taken from a quarry at Maksovo (for location see Fig. 4.34). (**ap**) Natural exposure of maksovite breccia cemented with quartz. (**aq**) Vertical wall at the entrance of the Shunga adit exhibiting C_{org}-rich rock resembling anthracite; for location of Shunga see Fig. 4.34. (**ar**) Pyrobitumen (former petroleum) from a c. 10-cm-thick layer in the Shunga adit (Photographs by Victor Melezhik (**a–e**, **h–aq**) and Pavel Medvedev (**f**, **g**))

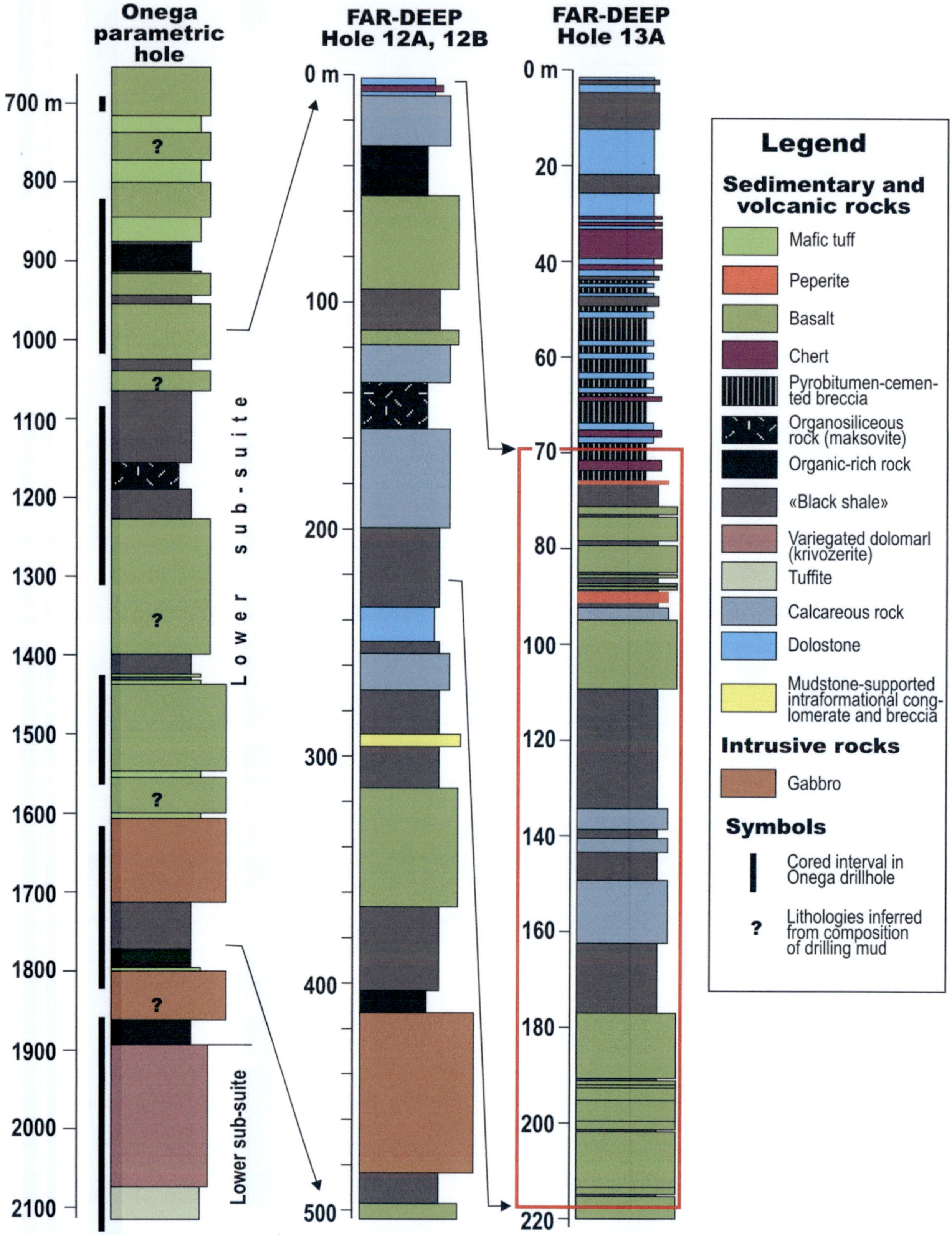

Fig. 4.53 Lithological sections of the Zaonega Formation intersected by the Onega parametric and FAR-DEEP drillholes in different parts of the Onega Basin (for drillhole location see Fig. 4.34). The section of the Onega parametric drillhole is modified after Morozov et al. (2010) and Krupenik et al. (2011a). The sections of FAR-DEEP drillholes are simplified from Chaps. 6.3.3 and 6.3.4

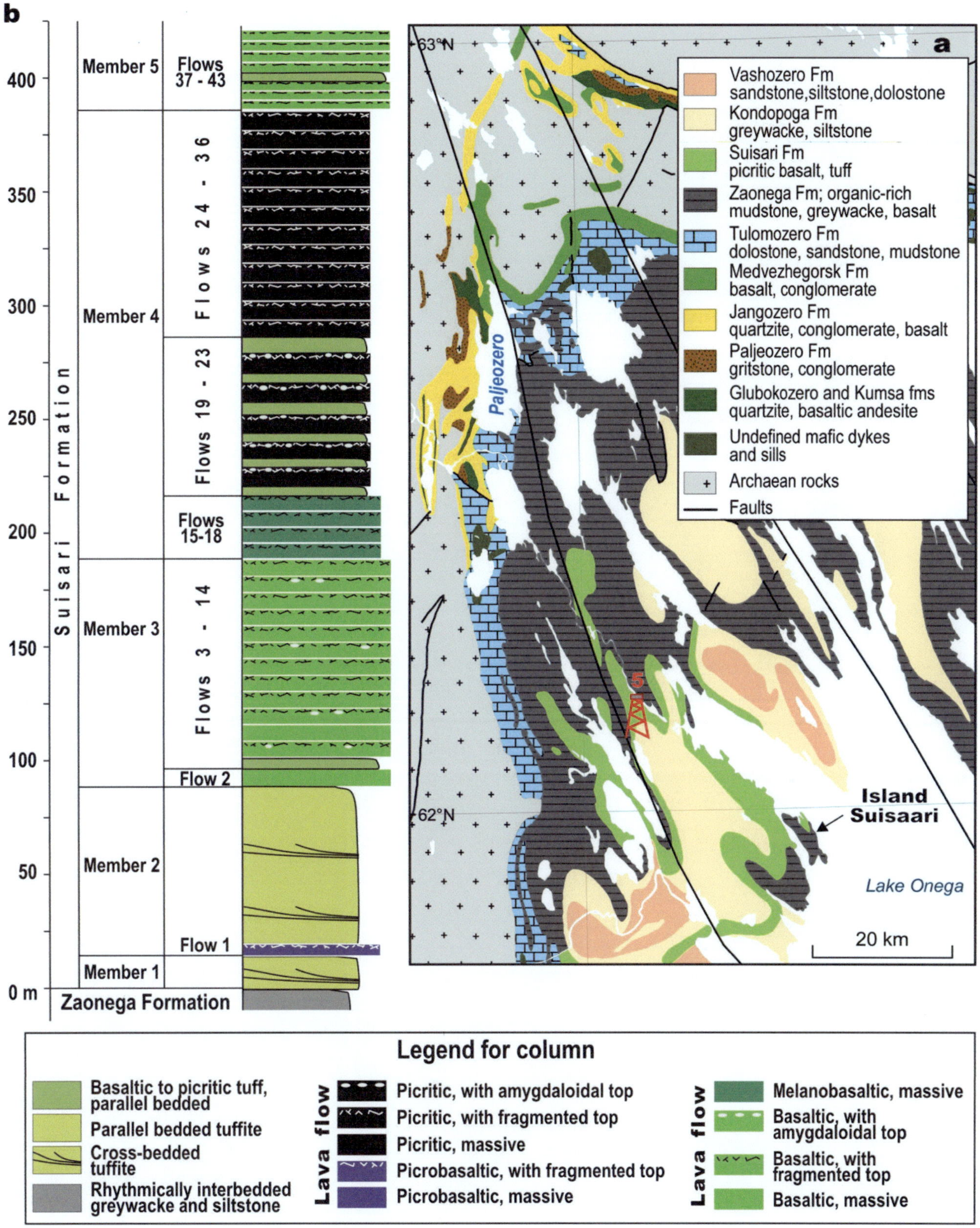

Fig. 4.54 The Suisari Formation in the Onega Basin. (**a**) Location of drillhole # 5 intersecting the Suisari Formation at the type locality. (**b**) Lithological section of the Suisari Formation based on description of drillcore # 5 by Kulikov et al. (2011b)

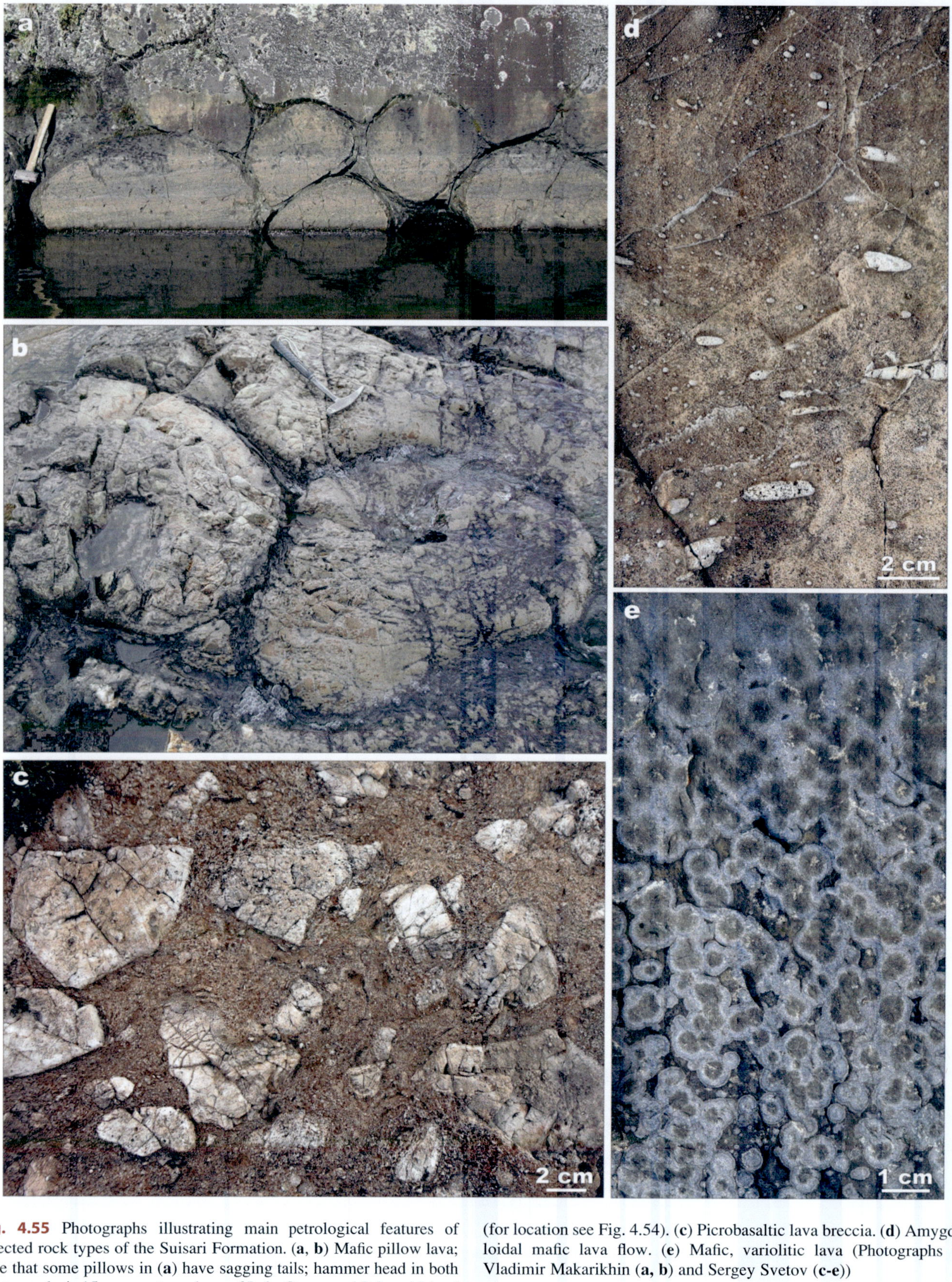

Fig. 4.55 Photographs illustrating main petrological features of selected rock types of the Suisari Formation. (**a**, **b**) Mafic pillow lava; note that some pillows in (**a**) have sagging tails; hammer head in both photographs is 15 cm; western shore of Lake Onega and Suisaari Island (for location see Fig. 4.54). (**c**) Picrobasaltic lava breccia. (**d**) Amygdaloidal mafic lava flow. (**e**) Mafic, variolitic lava (Photographs by Vladimir Makarikhin (**a**, **b**) and Sergey Svetov (**c**-**e**))

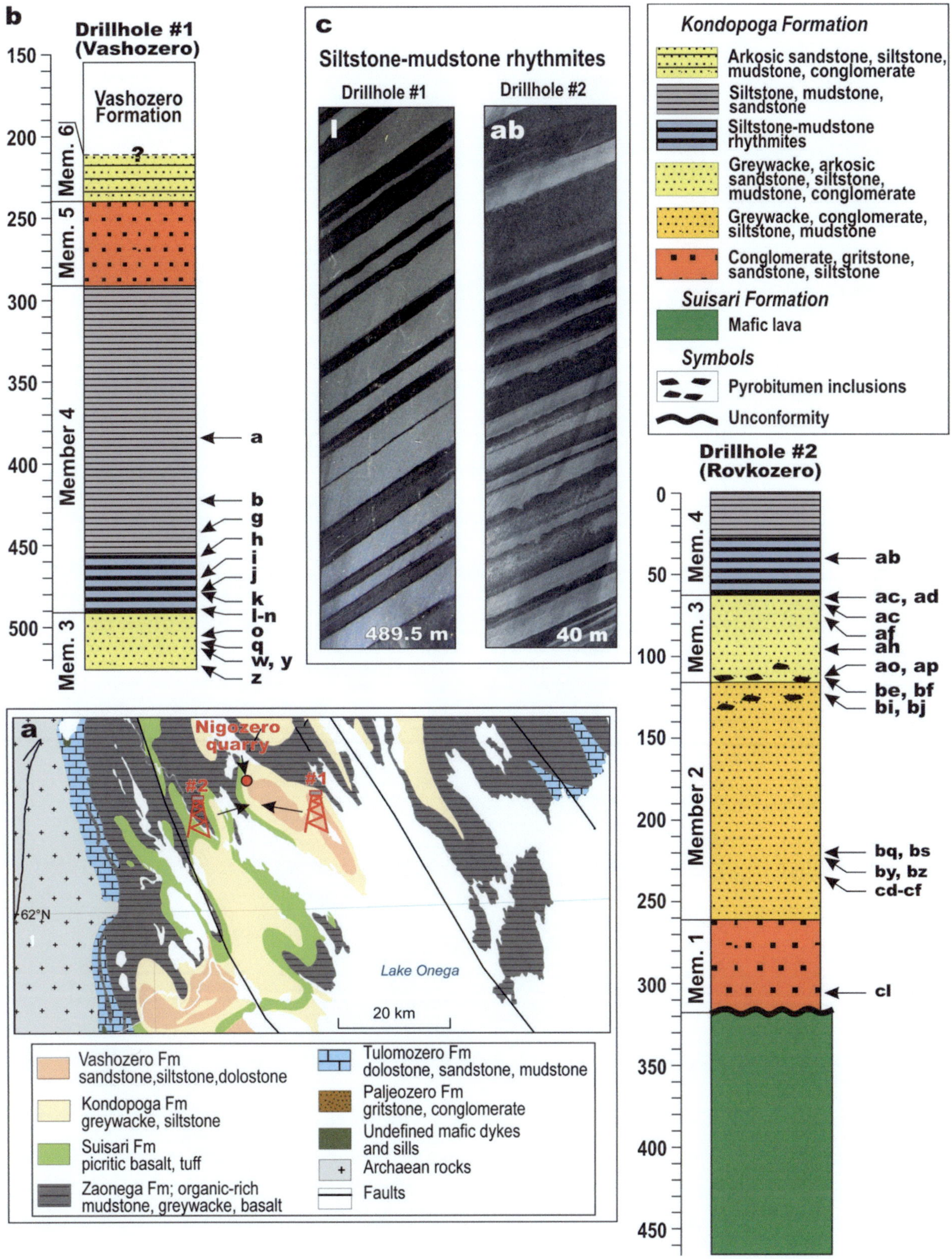

Fig. 4.56 The Kondopoga Formation in the Onega Basin. (**a**) Location of drillholes #1 and #2 intersecting the Kondopoga Formation at the type locality. (**b**) Lithological sections of the Kondopoga Formation based on drillcore #1 and #2, and the description by Rychanchik (2011b); vertical scale is in metres. (**c**) Rhythmites, which represent a marker horizon, were used to correlate sections intersected by drillholes #1 and #2

Fig. 4.57 Photographs illustrating the dominant lithology of the Kondopoga Formation. (**a**) A vertical quarry wall with rhythmically bedded greywacke-siltstone representing distal turbiditic facies; note numerous, pale brown bands of diagenetically formed, ankeritic concretionary layers; width of the photograph is 2.5 m

Fig. 4.57 (continued) (**b**) Rhythmically bedded greywacke overlain by thick beds of coarse-grained massive sandstone; note that many beds have an uneven base resulting from erosion or loading (Photographs by Victor Melezhik)

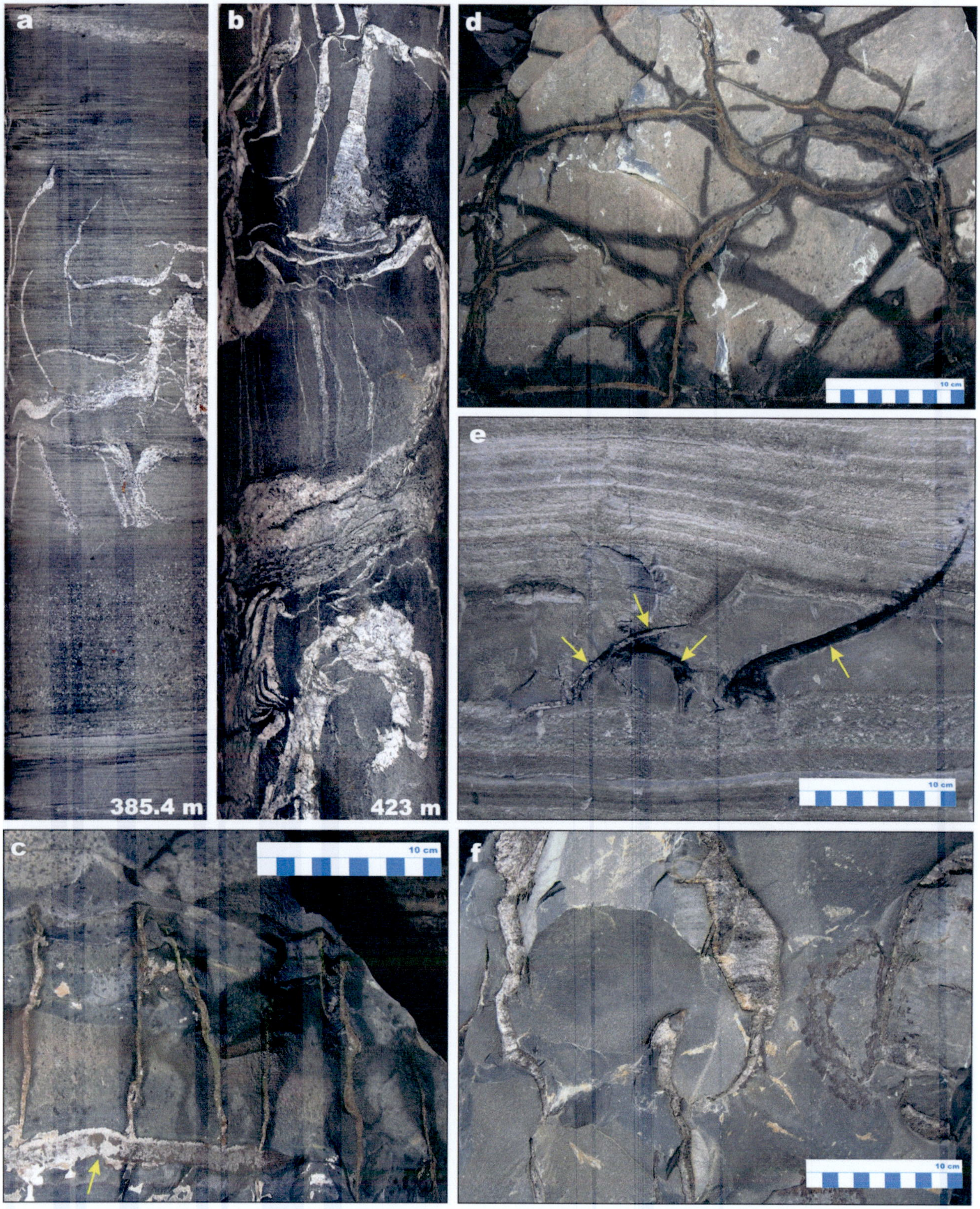

Fig. 4.58 Photographs illustrating main sedimentological features of the selected rock types of the Kondopoga Formation. In all photographs, unless otherwise marked, the scale is a core width of 4 cm. Numbers denote drillhole number/depth in metres. *Drillhole #1 and Nigozero quarry, Member 4*: (**a, b**) Drillcore of massive siltstone with irregular, calcite-cemented cracks resembling a "molar tooth" structure. (**c**) Cross-section view of a greywacke bed with sub-vertical, calcite-cemented cracks, which originated from *pale brown* layer. (**d**) Bedding-parallel view of the same bed as in figure (**c**) showing horizontal projection of sub-vertical cracks. (**e**) Cross-section view of a parallel-bedded greywacke-siltstone bed with thick, massive sandstone, which shows irregular cracks expanding from the layer beneath. (**f**) Bedding-parallel view of the same bed as in figure (**e**) showing horizontal projection of sub-vertical cracks

Fig. 4.58 (continued) (**g**) Interbedded greywacke-siltstone composed of T_a and T_d Bouma units; note that graded greywacke beds have erosional bases (*arrowed*). (**h**) Siltstone-mudstone rhythmite; a specific, varve-like appearance and bluish-greenish colour makes them a prominent marker in the succession. (**i**) Dark grey mudstone with layers of greenish, fine-grained siltstone, which also occurs as angular fragments. (**j**) Siltstone-mudstone rhythmite; note that siltstone beds have erosive bases (*arrowed*)

Fig. 4.58 (continued) (**k**) Massive greywacke bed with scattered, rounded clasts of black mudstone. (**l**) Bluish, siltstone-mudstone, varve-like rhythmite from the marker bed. (**m**) Siltstone-mudstone rhythmite with thick massive greywacke beds (*arrowed*). (**n**) Interbedded mudstone (*dark grey*) and siltstone (*pale grey*) with carbonate-cemented sandstone (*arrowed*)

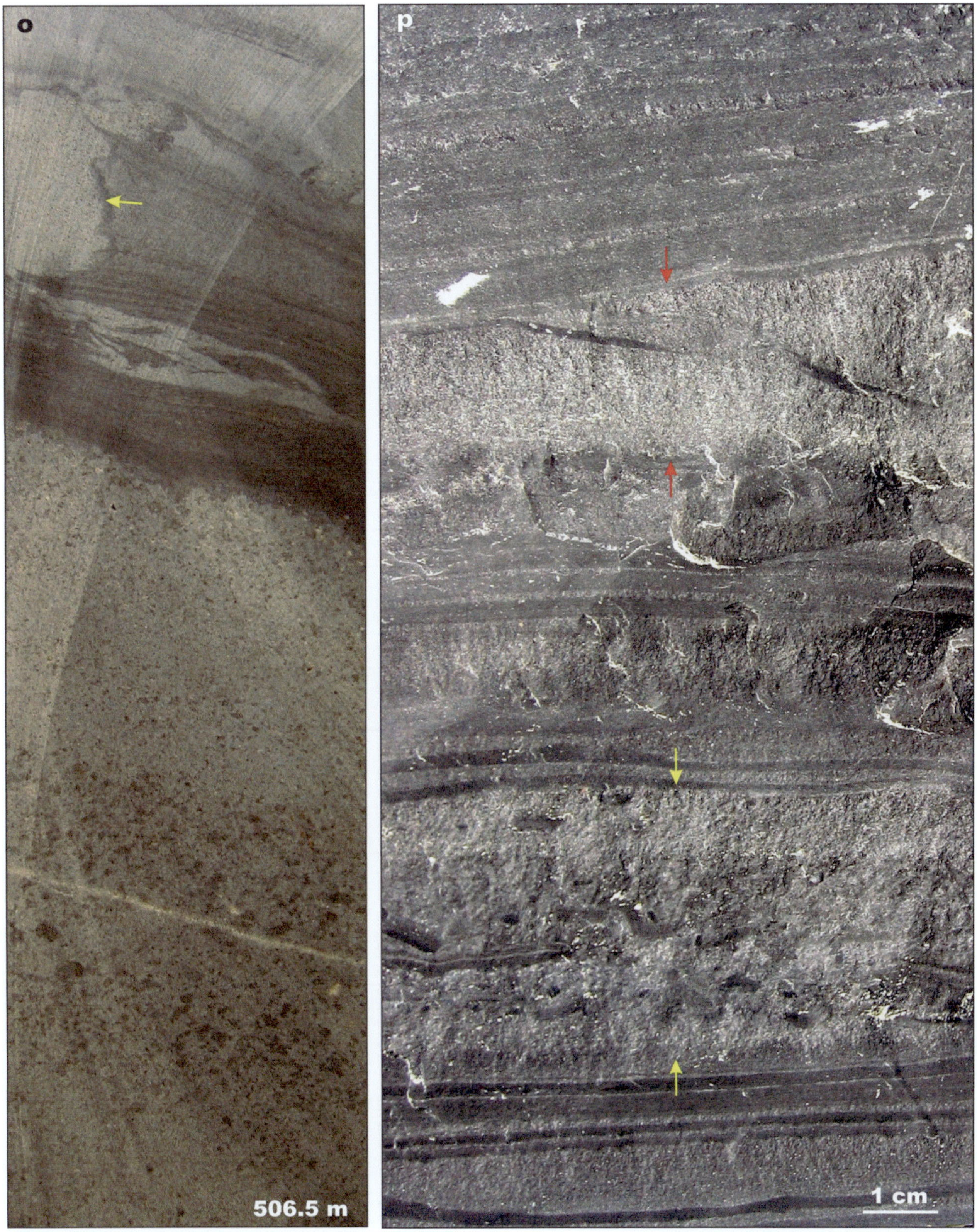

Fig. 4.58 (continued) *Drillhole #1 and Nigozero quarry, Member 3*: (**o**) Greywacke bed overlain by laminated siltstone cut by sandstone sedimentary dyke (*arrowed*). (**p**) A block from the Nigozero quarry showing rhythmically bedded greywacke-siltstone with a bed of matrix-supported, intraformational conglomerate (*yellow arrowed*) and a bed of coarse-grained sandstone (*red arrowed*) with imbricated platy fragments of black mudstone; note that both beds have an erosional base and the conglomerate grades upward into coarse-grained sandstone with scattered mudstone, also, the bedding above the *red-arrowed* sandstone is at an oblique angle to the underlying bedding, hence suggesting syndepositional tilting/rotation

Fig. 4.58 (continued) (**q**) Bouma sequences (T_a and T_d) in rhythmically bedded greywacke-siltstone; T_a unit is represented by a graded bed of greywacke (*pale grey*), whereas T_d unit is represented by laminated, dark-coloured siltstone. (**r**) Rhythmically interbedded thinly laminated siltstone-mudstone with a silicate concretion at the base of a massive greywacke bed; Nigozero quarry. (**s**) Thinly interbedded siltstone-mudstone; note that some sandstone layers are completely or partially (along contacts) replaced by ankerite with a *brown* or *yellow* colour (*arrowed*)

Fig. 4.58 (continued) (**t**, **u**) Bedding surface of a greywacke bed covered with angular, platy fragments of clayey siltstone; Nigozero quarry. (**v**) A vertical wall of Nigozero quarry covered by a film of malachite

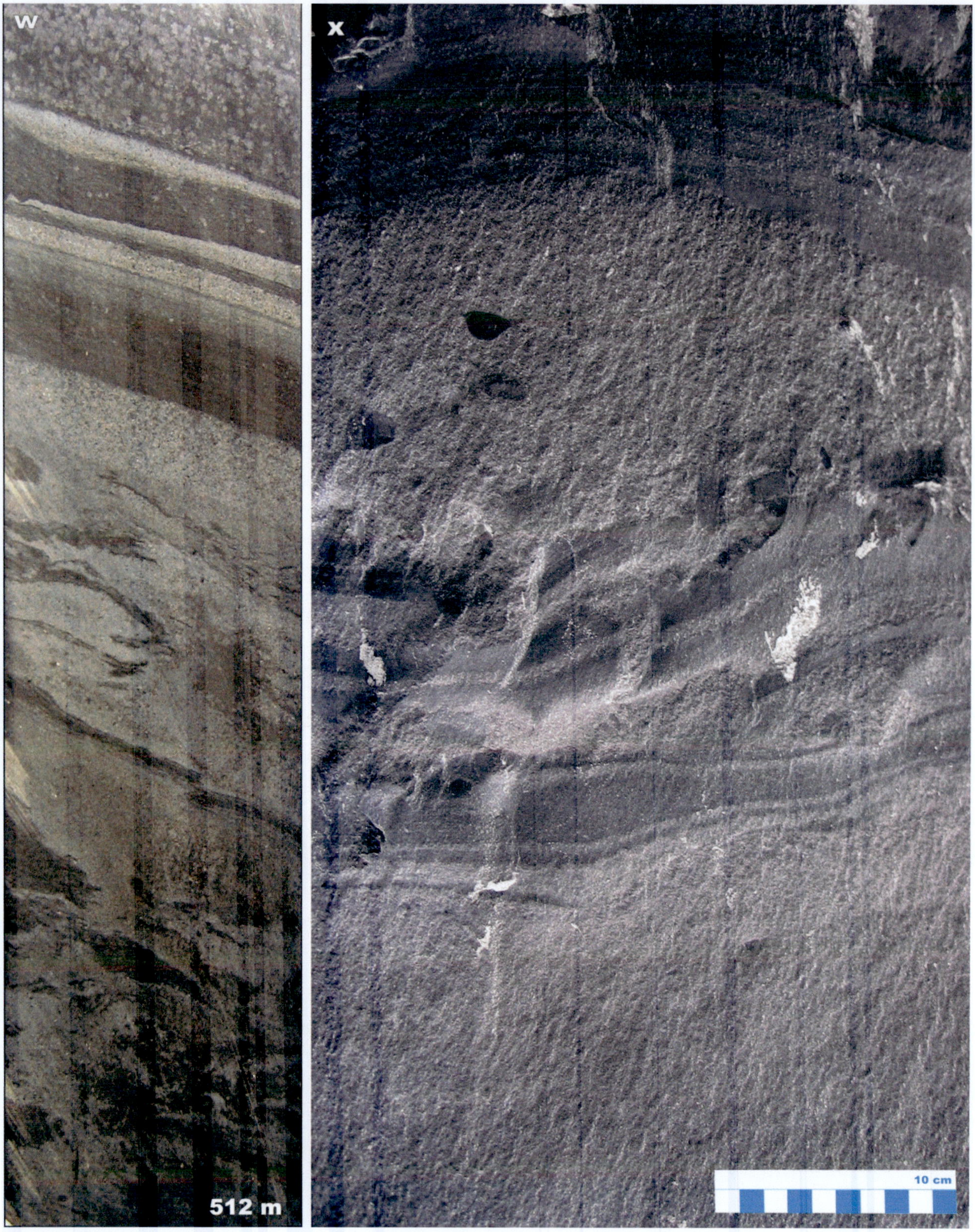

Fig. 4.58 (continued) (**w**) Intraformational breccias that represent mass-flow deposit composed of soft-sediment deformed, partially or entirely dismembered layers of greywacke and mudstone (dark coloured), overlain by parallel-bedded greywacke and mudstone (*dark coloured*). (**x**) Massive greywacke bed with a bedded greywacke-siltstone top overlain erosively by a greywacke bed (with scattered mudstone intraclasts) passing upward into dark-coloured siltstone-mudstone; note that siltstone layers at the base of the second sandstone bed are partially dismembered and produce clasts incorporated into the sandstone; Nigozero quarry

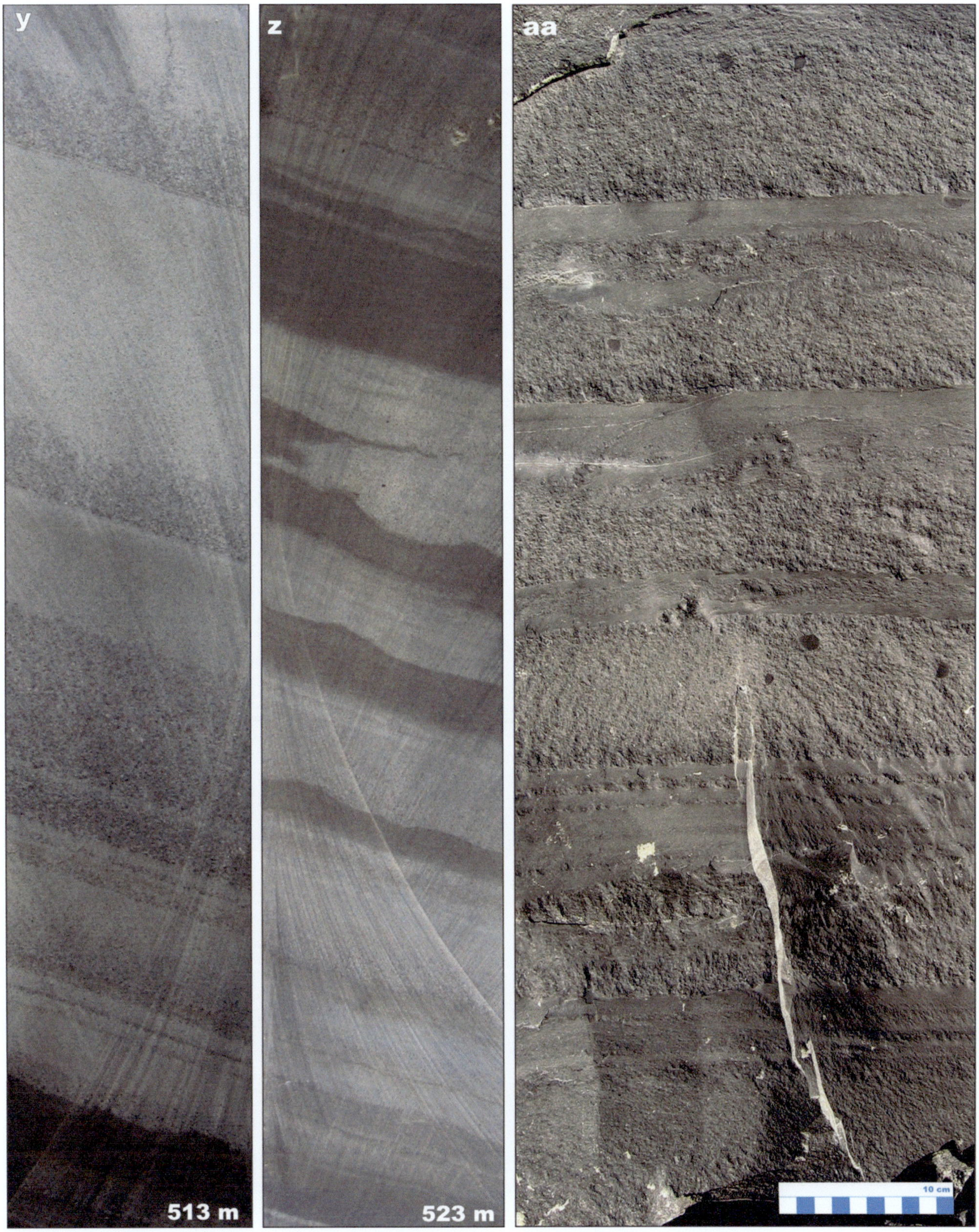

Fig. 4.58 (continued) (**y**) Thickly bedded greywacke-siltstone. (**z**) Rhythmically interbedded greywacke and mudstone (dark-coloured); note that the greywacke beds have an erosional base. (**aa**) Bouma sequence represented by T_a (graded greywacke beds) and T_d (*dark grey*, laminated siltstone) units in a rhythmically bedded greywacke-siltstone succession; Nigozero quarry

Fig. 4.58 (continued) *Drillhole #2 and Nigozero quarry, Member 4:* (**ab**) Siltstone-mudstone rhythmite with a varve-like appearance and bluish-greenish colour from the marker horizon. *Drillhole #2 and Nigozero quarry, Member 3*: (**ac, ad**) Rhythmically bedded greywacke and siltstone, mainly parallel-bedded with a cross-bedded bed (T_b Bouma unit) (**ad**) (*arrowed*). (**ae**) Grey, parallel-bedded siltstone overlain by a greywacke bed with a coarse-grained base (*yellow arrowed*); the greywacke grades into thicker mudstone (*black*) followed by two thick mudstone beds with a thin sandstone base (*red arrowed*)

Fig. 4.58 (continued) (**af**) Rhythmically interbedded graded greywacke (T_a Bouma unit) and thinner, laminated siltstone-mudstone (T_d Bouma unit). (**ag**) Rhythmically bedded greywacke-siltstone eroded by overlying, massive greywacke bed with a large intraclast at the base; Nigozero quarry

Fig. 4.58 (continued) (**ah**) Slumped greywacke-siltstone beds separated by an undisturbed bed with parallel lamination. (**ai**) Soft-sediment deformation and partial dismembering of bedded greywacke-siltstone within a massive greywacke bed. (**aj**) Dismembered mudstone bed deformed into a tight, slump fold within a massive greywacke bed. (**ak**) Tight slump folds in rhythmically bedded greywacke-siltstone. (**al**) Bedding surface with slump folds; all photos taken in Nigozero quarry

Fig. 4.58 (continued) (**am**) Rhythmically bedded greywacke-siltstone succession affected by erosion and associated syndepositional faulting; the eroded surface is covered by a slumped greywacke unit with large angular intraformational clasts. (**an**) Two blocks of soft-sediment deformed, rhythmically bedded greywacke-siltstone; note the pyrobitumen inclusion in the upper block; both photos taken in Nigozero quarry

Fig. 4.58 (continued) (**ao**) Rhythmically bedded, graded greywacke-siltstone-mudstone; note formation of incipient flame structures (*arrowed*). (**ap**) Loading structure on top of a mudstone bed. (**aq**) Selective cementation of sandstone beds by ankerite (*brown*). (**ar**) Close-up view of ankerite-cemented sandstone beds. (**as**) *Brown*, former sandstone beds, in which a considerable amount of siliciclastic material was replaced by ankerite; all photos taken in Nigozero quarry

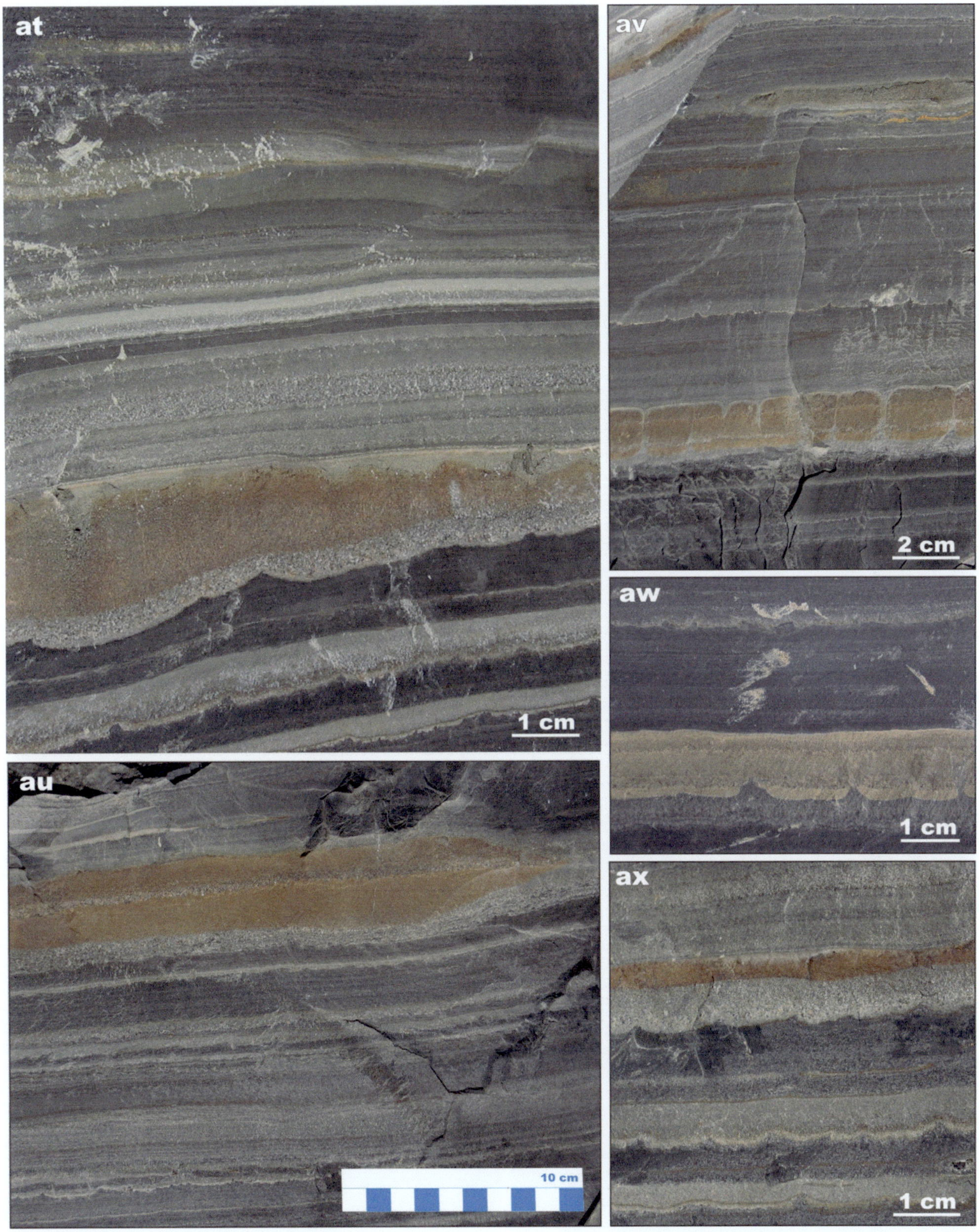

Fig. 4.58 (continued) (**at**) Rhythmically bedded greywacke-siltstone-mudstone with ankerite-cemented, graded sandstone bed (*brown*); note that the coarse-grained part of the bed was not cemented. (**au**) Ankerite-cemented sandstone bed; note that only part of the bed experienced cementation resulting in a lensoidal (concretionary) appearance; differential compaction within and outside the concretion suggests that the cementation occurred during diagenesis. (**av**) Two sandstone beds affected by ankerite cementation in a rhythmically bedded succession; only part of the upper bed was cemented resulting in the formation of a lensoidal concretion with a significantly greater compaction outside the concretion; the lower bed experienced more complete cementation resulting in the formation of a concretionary layer, which was affected by partial sedimentary boudinage. (**aw**) Ankerite-cemented sandstone bed, whose lower part was affected by incipient sedimentary boudinage. (**ax**) Several graded sandstone beds showing a variable degree of cementation by ankerite. Photographs (**at**–**ax**) taken in Nigozero quarry

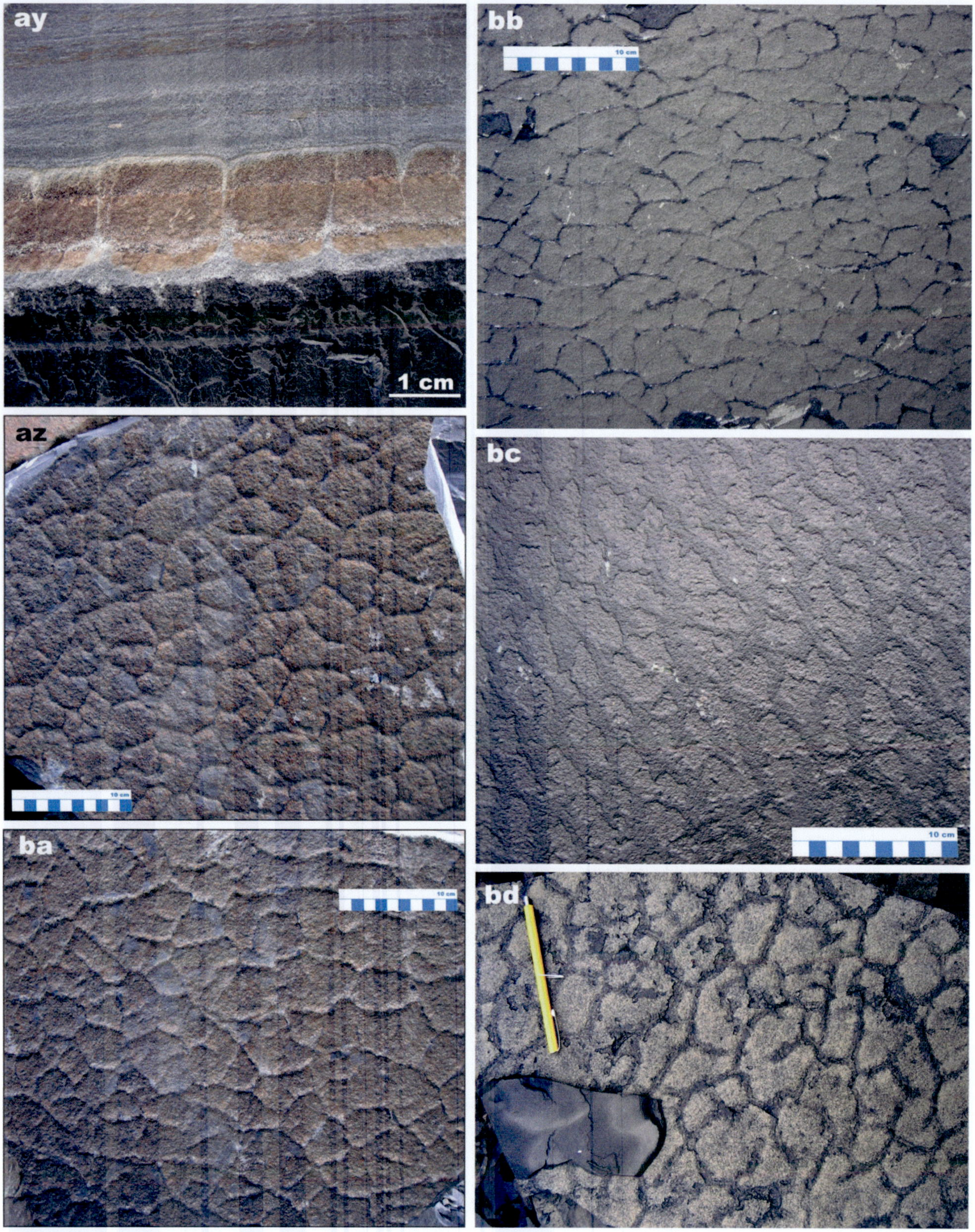

Fig. 4.58 (continued) (**ay**) Ankerite-cemented sandstone bed (concretionary layer) affected by incipient sedimentary boudinage resulting in cracking and injection of overlying and underlying siltstone-mudstone into the cracks. (**az**, **ba**) Sedimentary boudinage creates a polygonal pattern either with recessive relief (on either side of the sandstone bed surface; **az**) or with positive relief (on siltstone-mudstone bedding surfaces; **ba**). (**bb**–**bd**) Bedding surfaces exhibiting variable polygonal patterns associated with sedimentary boudinage

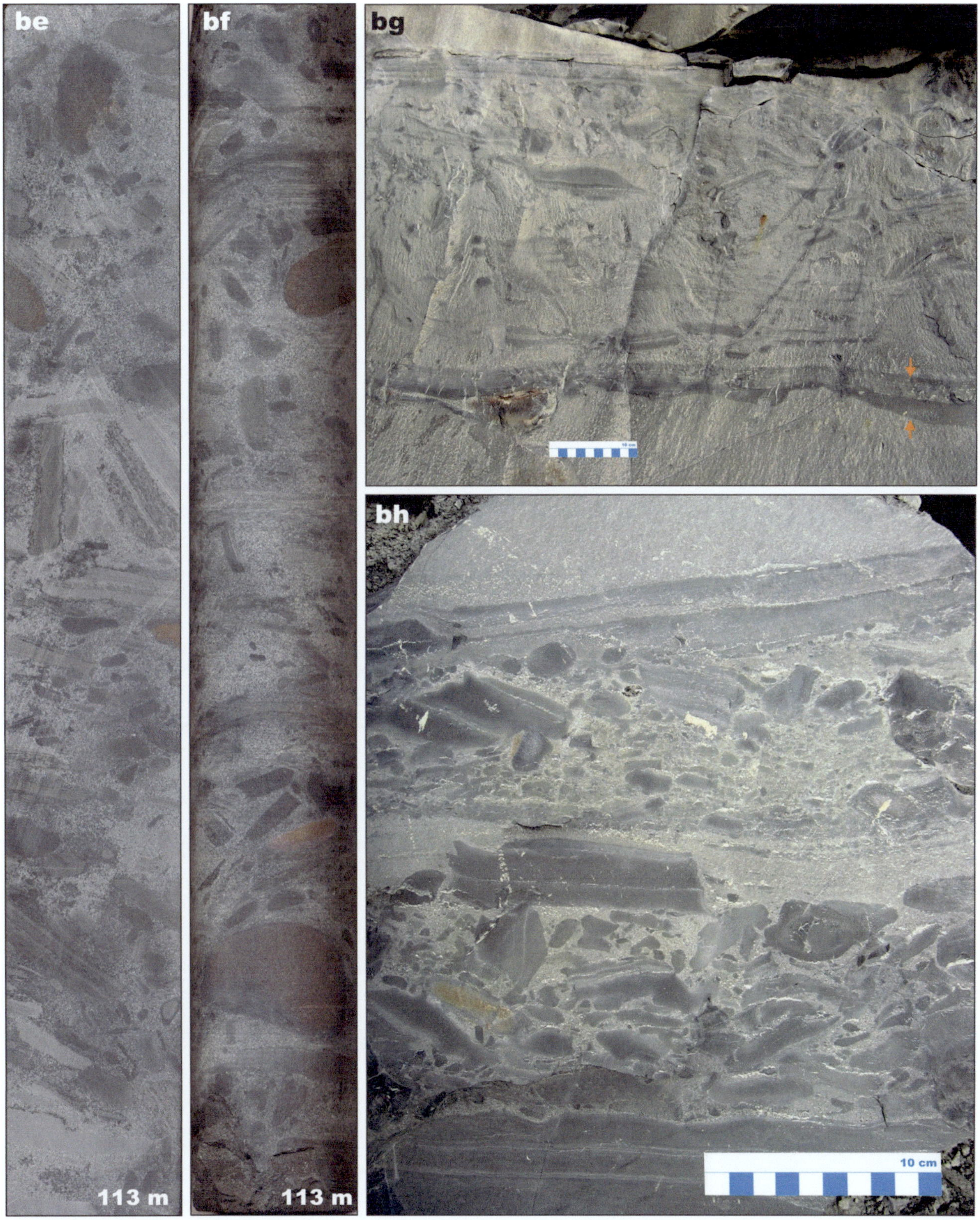

Fig. 4.58 (continued) (**be**, **bf**) Matrix-supported breccias (mass-flow deposit) composed of chaotically distributed angular and variably rounded intraformational fragments of greywacke-siltstone-mudstone and ankerite concretions or ankerite-cemented sandstone (*brown*). (**bg**) Two mass-flow beds in Nigozero quarry; the thinner, lower interval (*arrowed*) exhibits a highly erosive base on the underlying massive sandstone, whereas the thicker upper interval has a straight base; both beds are composed of angular fragments and clasts of greywacke-siltstone-mudstone in a coarse-grained, structureless greywacke matrix. (**bh**) Two mass-flow beds in Nigozero quarry both with erosive bases and containing angular greywacke-siltstone-mudstone clasts in a structureless, coarse-grained, greywacke matrix; note the slight angular discordance of the topmost beds to those below implying tilting or rotation

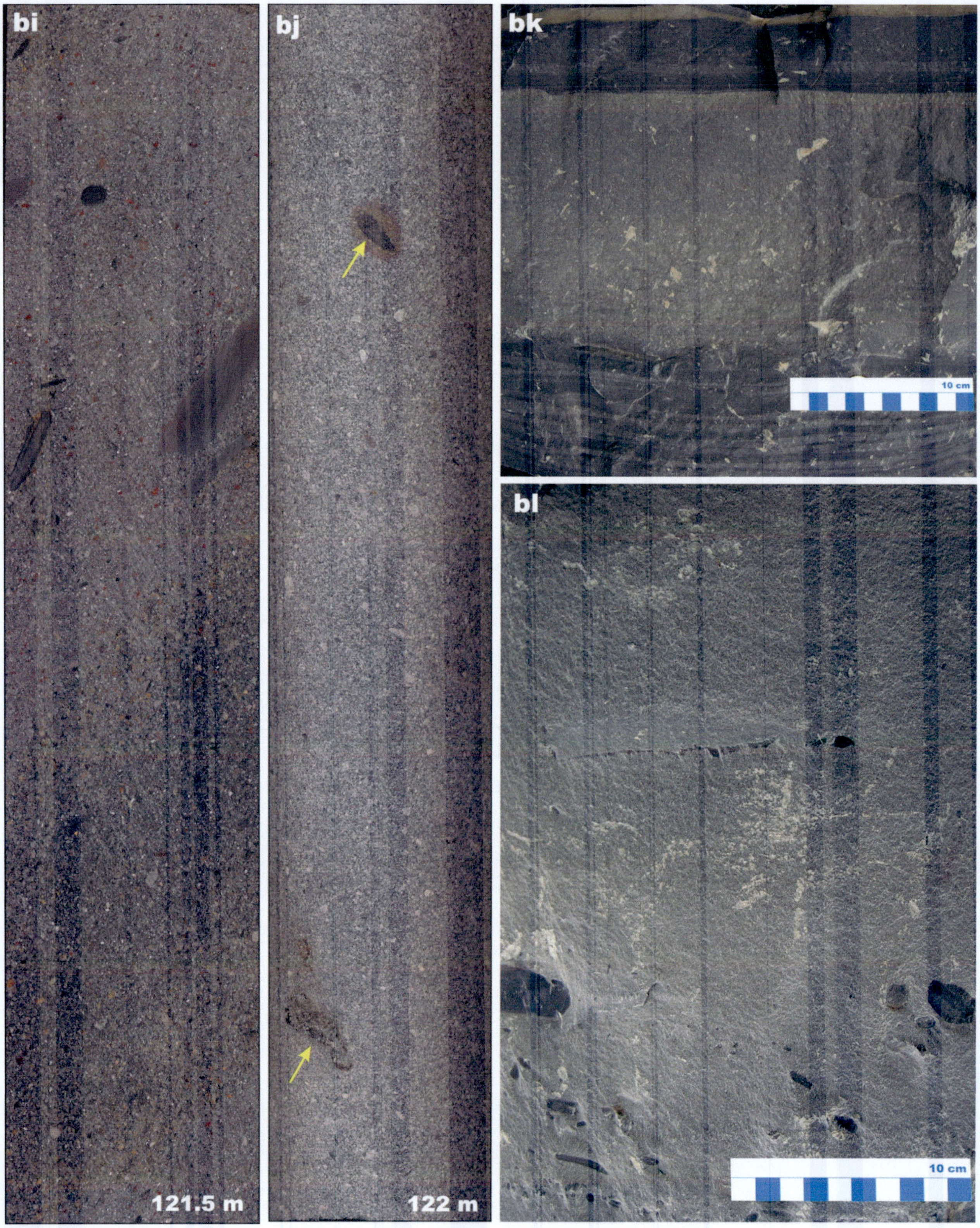

Fig. 4.58 (continued) *Drillhole #2 and Nigozero quarry*, Member 2: (**bi**, **bj**) Massive, coarse-grained greywacke with scattered clasts of grey and black siltstone and mudstone, and inclusions of pyrobitumen (*arrowed*). (**bk**) Rhythmically bedded, dark grey greywacke-siltstone overlain by massive greywacke in turn capped sharply by faintly bedded, dark grey siltstone with an ankerite-cemented bed at the top; Nigozero quarry. (**bl**) Bed of massive, coarse-grained greywacke that fines and becomes darker upward; scattered clasts are intraformational mudstone and greywacke; Nigozero quarry

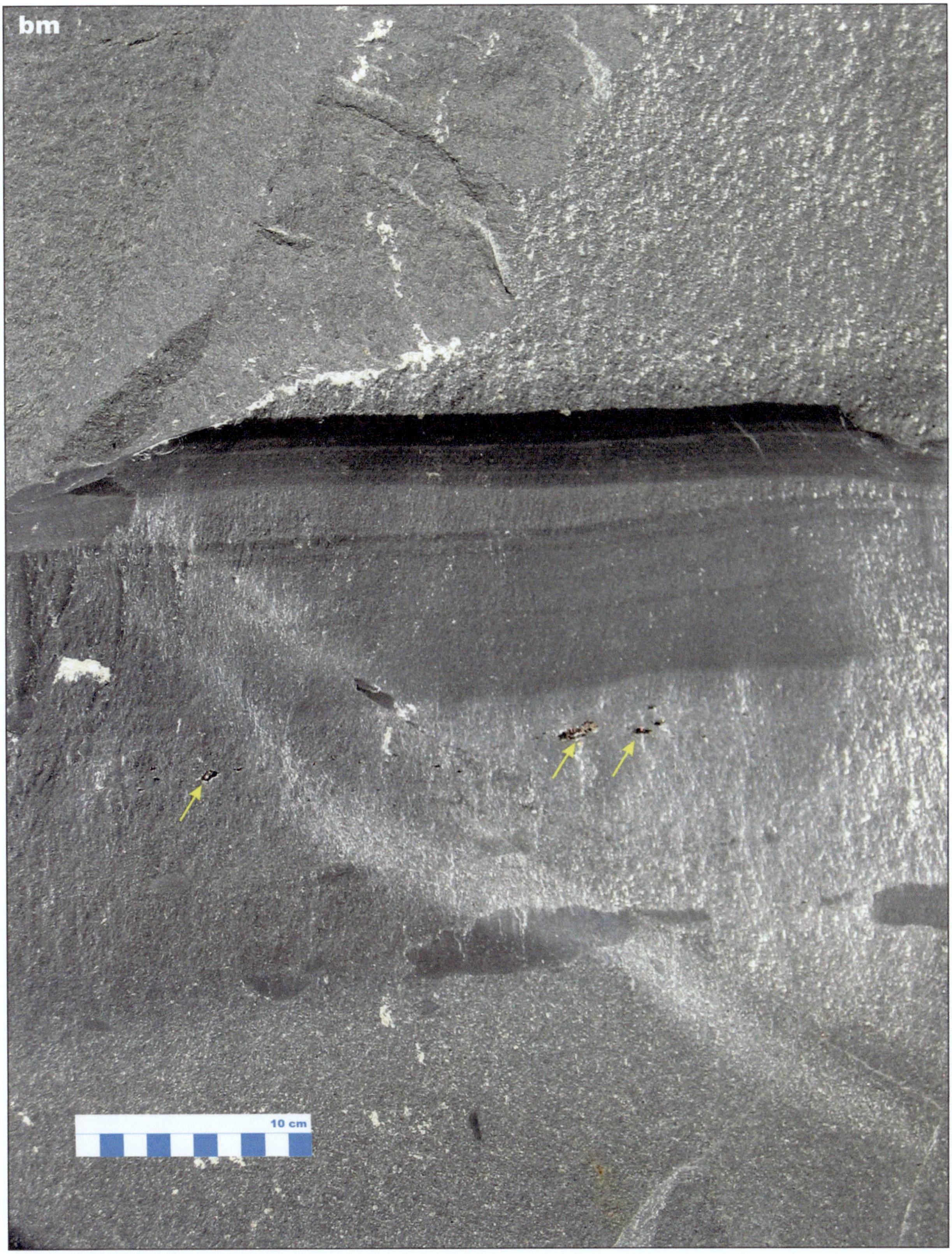

Fig. 4.58 (continued) (**bm**) Coarse-grained sandstone bed with scattered clasts of pyrobitumen (*arrowed*) and intraformational fragments of siltstone; the massive bed grades into bedded siltstone, which is overlain by the next graded bed of coarse-grained sandstone; Nigozero quarry

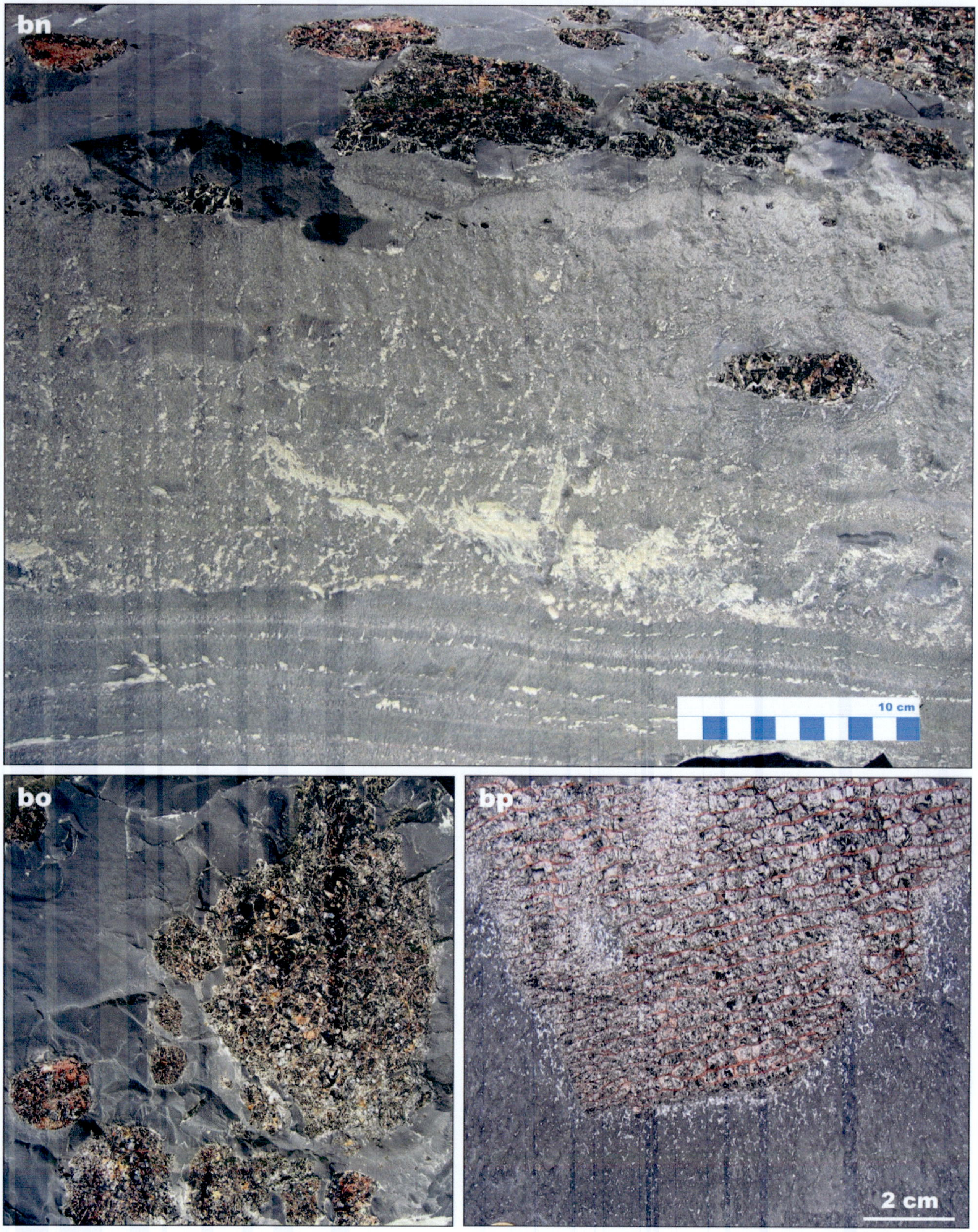

Fig. 4.58 (continued) (**bn**) Pyrobitumen inclusions in a mass-flow deposit. (**bo**) Bedding surface with a cluster of flattened inclusions of pyrobitumen. (**bp**) Bedding surface with a pancake-like pyrobitumen inclusion having a diffuse boundary with hosting greywacke; shrinkage cracks are filled with quartz; all photos taken in Nigozero quarry

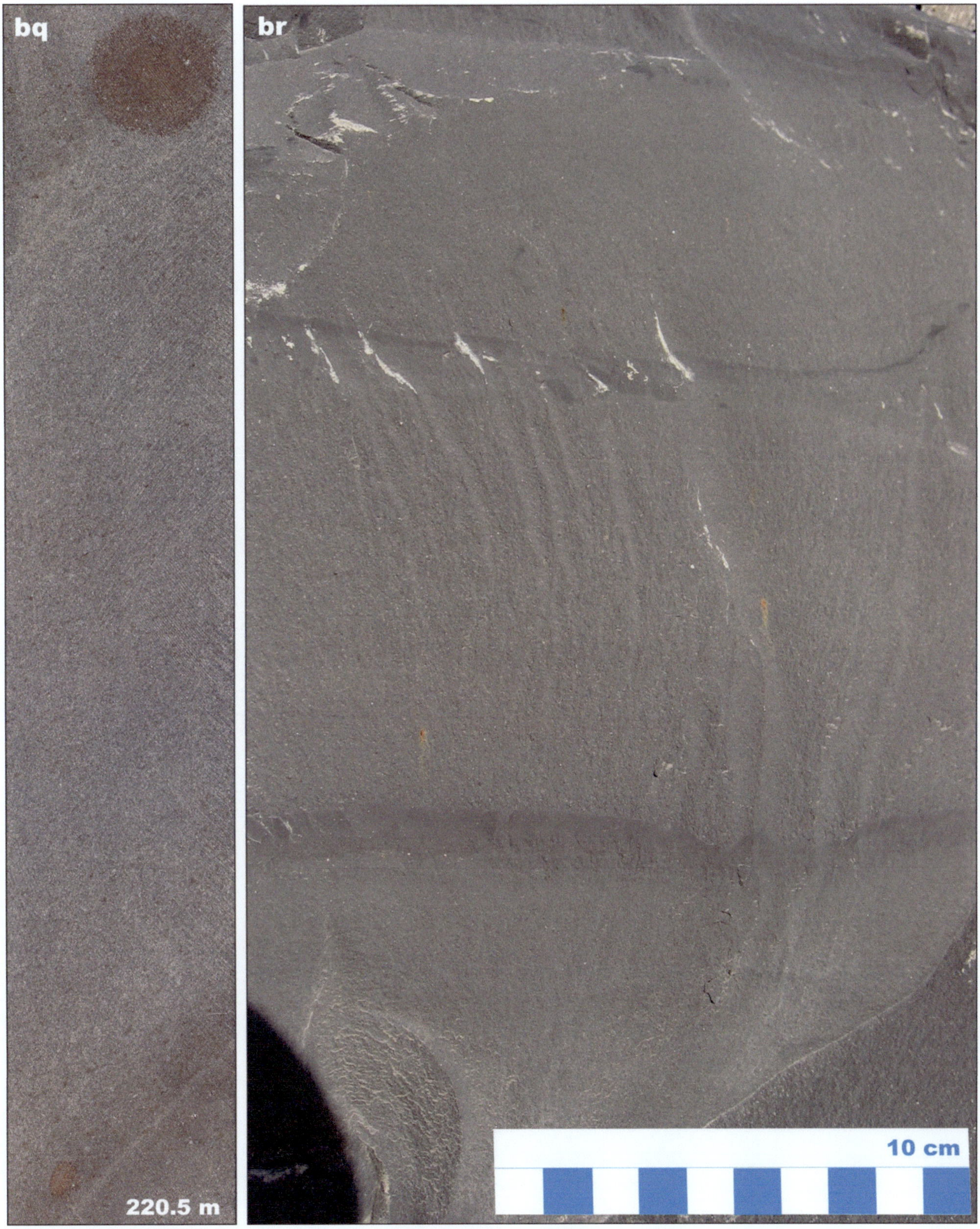

Fig. 4.58 (continued) (**bq**) Massive sandstone with brown spots, which commonly developed around pyrobitumen inclusions (see Fig. **bj**). (**br**) Beds of graded, massive sandstone (T_a Bouma unit) with a thin top of dark grey siltstone (T_d Bouma unit) partially eroded by the overlying sandstone bed; Nigozero quarry

Fig. 4.58 (continued) Laminated siltstone with ankeritic concretions (*brown*). (**bt**) Concretionary layers and lenses of ankerite (arrowed) in rhythmically bedded greywacke-siltstone. (**bu**) Details of an ankeritic concretion demonstrating its formation by partial cementation of a sandstone bed; differential compaction around the concretion indicates an early diagenetic origin; Nigozero quarry

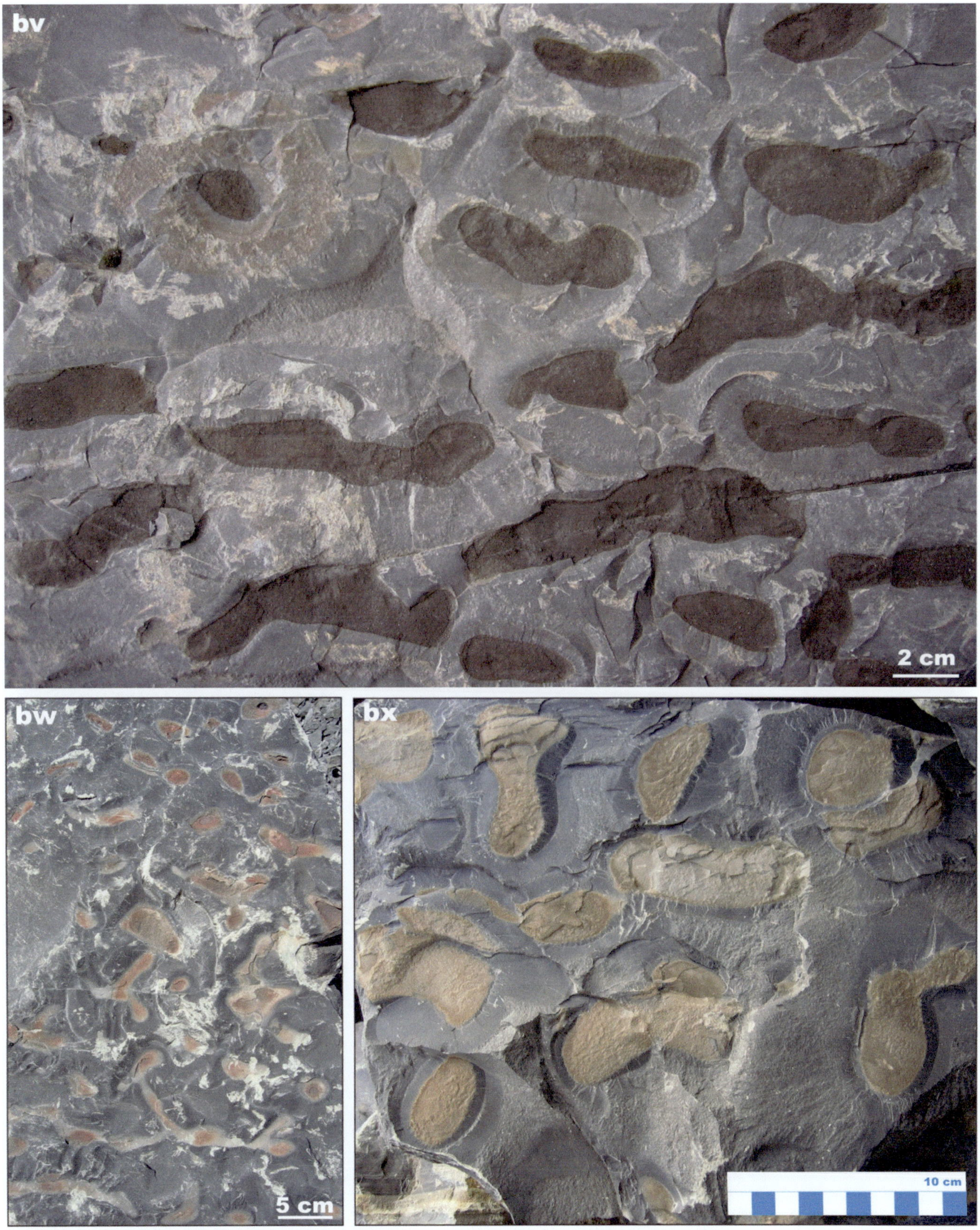

Fig. 4.58 (continued) (**bv**–**bx**) Bedding-parallel views showing clusters of ankeritic concretions, which in each case were formed by selective cementation of a single sand bed; Nigozero quarry

Fig. 4.58 (continued) (**by**, **bz**) Ankeritic layers in parallel-bedded siltstone-mudstone. (**ca**) Ankeritic concretionary layer in a sandstone bed preserving primary lamination. (**cb**) Sand ripples cemented by ankerite. (**cc**) Slump structure with subsequent and selective cementation of the sandy part of the slump by ankerite (*brown*); all photos taken in Nigozero quarry

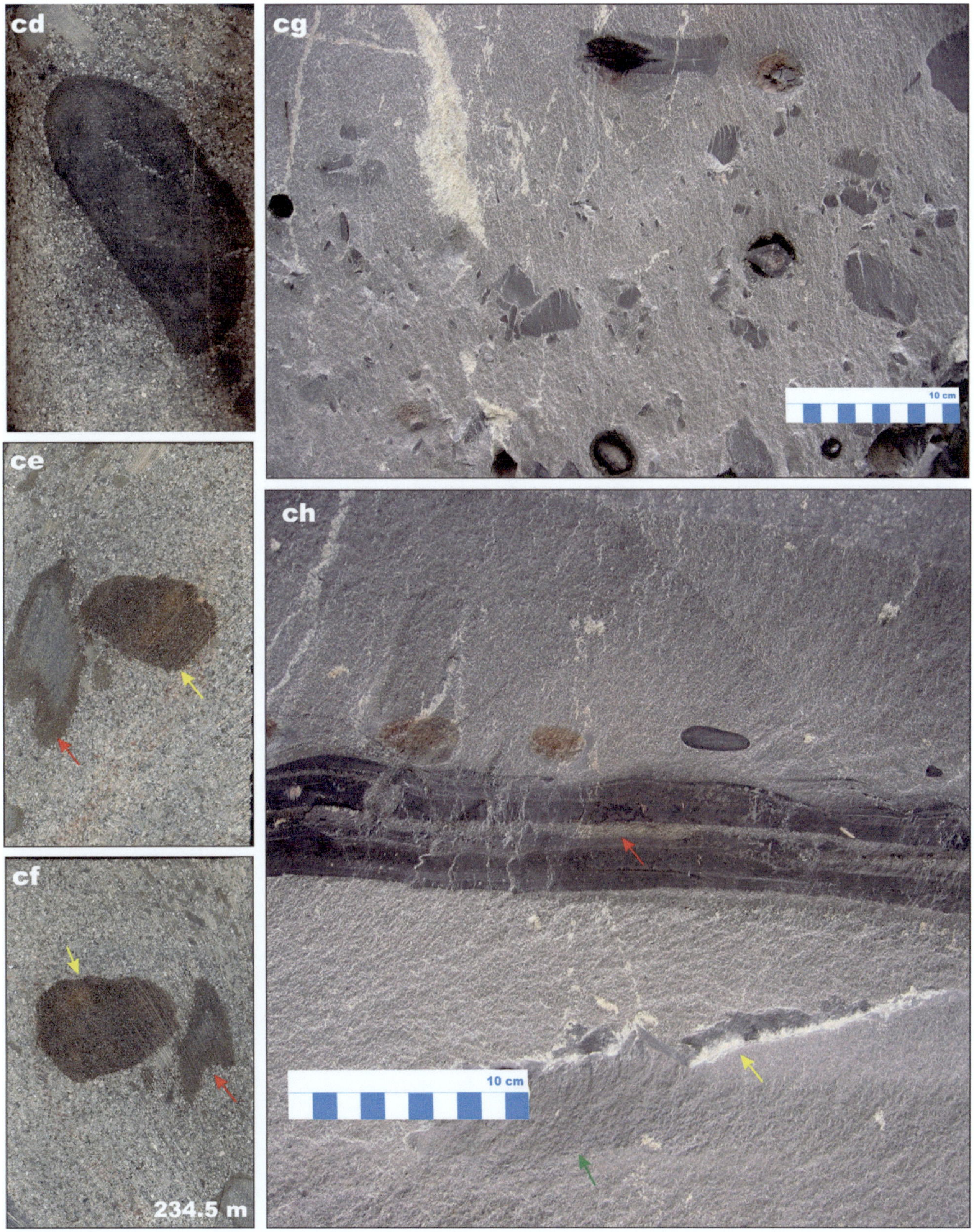

Fig. 4.58 (continued) (**cd**–**cf**) Coarse-grained, massive sandstone with rounded clasts of mudstone (*black*), ankerite concretion (*yellow arrowed*) and variably rounded sandstone with ankerite-cemented margin (*red arrowed*). All core images are from the same depth interval. (**cg**) Massive coarse-grained sandstone (mass-flow deposit) with scattered fragments of grey siltstone and zoned carbonate concretions (**ch**) Three sandstone-siltstone cycles: the lower massive sandstone bed has a laminated siltstone top (*dark grey*) and contains a fragmented layer of siltstone (*yellow arrowed*) and a large, rounded clast of sandstone (*green arrowed*) with a diffuse boundary; an overlying thin sandstone layer (*red arrowed*) with a thicker, laminated siltstone top (*dark grey*); and then a, thicker, massive sandstone bed with fragments of dark mudstone and ankeritic concretions at the base; note that the bases of the sandstone beds are erosive; Nigozero quarry

Fig. 4.58 (continued) (**ci**, **cj**) Mass-flow deposits composed of irregularly distributed fragments of grey siltstone and pale brown dolostone; Nigozero quarry. (**ck**) *Pale yellow*, massive sandstone containing two layers of intraformational conglomerate-breccia with clasts of siltstone; sandstone rests with a sharp contact on dark grey siltstone; Nigozero quarry

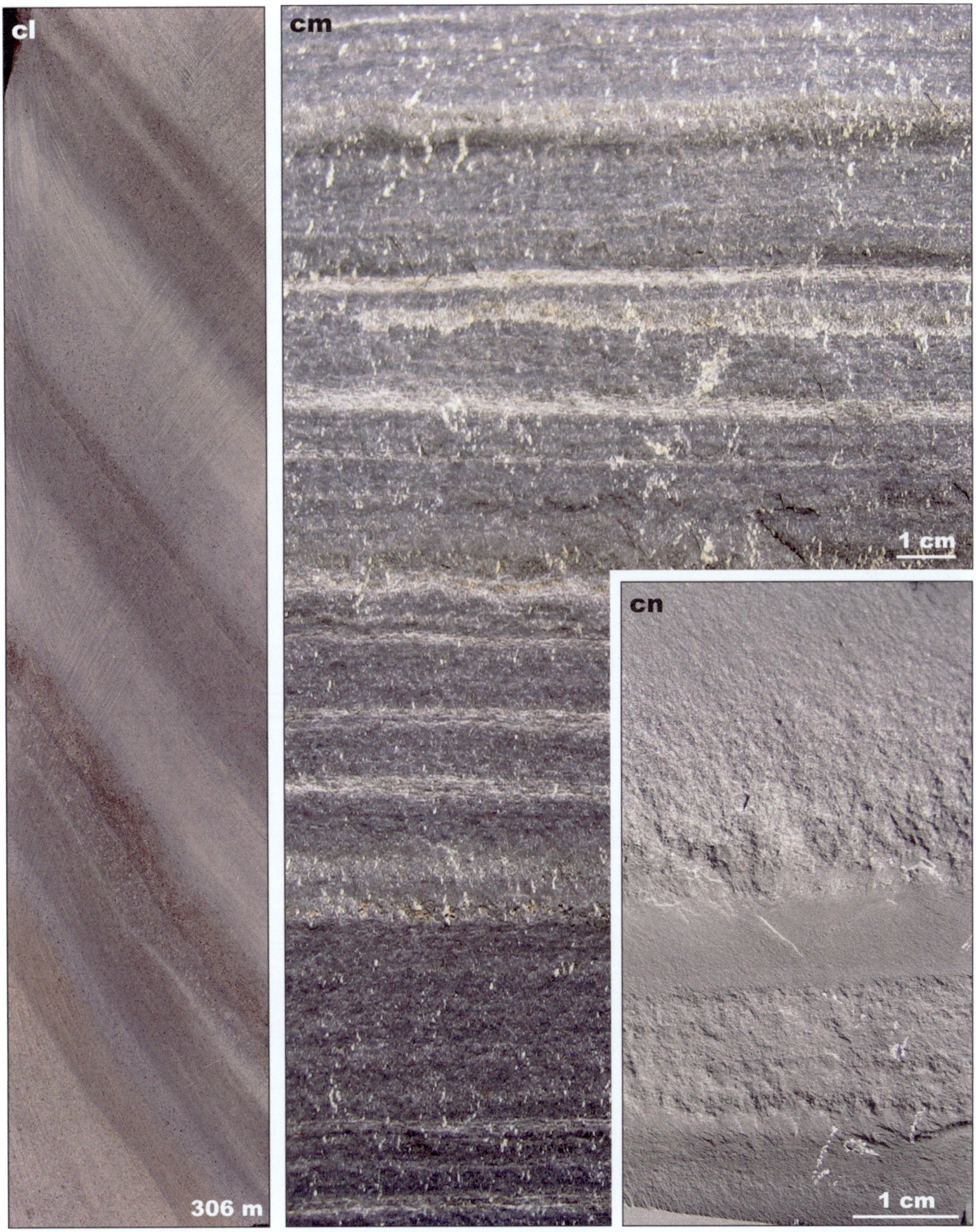

Fig. 4.58 (continued) *Drillhole #2, Member 1*: (**cl**) Parallel-laminated, rhythmically bedded siltstone-mudstone. (**cm, cn**) Rhythmically bedded, coarse- and medium-grained sandstone; Nigozero quarry (All photographs by Victor Melezhik)

Fig. 4.59 Photographs illustrating some sedimentological features of the selected rock types of the Vashozero Formation based on drillhole #1 and natural outcrops. (**a**) Drillcore of calcareous shale overlain by "fenestral" mudstone; core diameter is 4 cm. (**b**) Drillcore of grey, cross-bedded, silty sandstone overlain by massive sandstone with rounded fragments of black mudstone; core diameter is 4 cm. (**c**) Sandstone-siltstone with lenticular and flaser bedding and ripple cross-lamination. (**d**) A lens of white, cross-bedded sandstone within lenticular- and flaser-bedded sandstone-siltstone. (**e**) Sandstone-siltstone with lenticular and flaser bedding injected by sandstone dykes (arrowed). (**f**) Bedding surfaces of desiccated mud layer. (**g**) Flat, asymmetrical sand-ripples formed on silty/muddy substrate. Photographs (**c–g**) represent the Shuja Formation and were taken from a series of outcrops exposed along Shuja river (Photographs by Victor Melezhik)

References

Akhmedov AM, Efremov AM (1989) Isotopic composition of sulphur in sulphide concretions from carbonaceous formations of the Baltic Shield. In: Zaritsky PV (ed) Concretions of the Precambrian. Nauka (Science), Leningrad, pp 121–129 (in Russian)

Akhmedov AM, Krupenik VA (1991) Defining marker horizons in the reference sections of the Tulomozero and Suisari suites in the Onega structure. Report #803, VSEGEI, Leningrad, p 139 (in Russian)

Akhmedov AM, Krupenik VA (1995) Lithologo-geochemical investigation of the reference sections of Early Proterozoic, carbonate-siliciclastic complexes. In: Methodological recommendations, VSEGEI, St. Petersburg, p 63 (in Russian)

Akhmedov AM, Gushchin VS, Savitsky AB, Melnikov EK, Kondakov SN, Gordienko VI (1990) The Karelides of Southern Karelia. Sov Geol 1:96–105

Akhmedov AM, Krupenik VA, Makarikhin VV, Medvedev PV (1993) Carbon isotope composition of carbonates in Early Proterozoic sedimentary basins. Published report of the Institute of Geology, Institute of Geology, Karelian Research Centre of RAS, Petrozavodsk, p 56 (in Russian)

Akhmedov AM, Panova EG, Krupenik VA, Sveshnikova KY (2004) Early Proterozoic and Devonian arid palaeobasins from a joint zone of the Baltic Shield and Russian Platform. Trans St Petersburg Soc Nat 2(86):1–140 (in Russian)

Audemard FA, de Santis F (1991) Survey of liquefaction structures induced by recent moderate earthquakes. Bull Int Assoc Eng Geol 44:5–16

Berezhnaya NG, Myskova TA, Arestova NA (2003) Protolith ages of Sumian quartzites from the Kumsa structure, In: Isotope geochemistry applied to problems of the geodynamic and ore genesis. Proceedings of the II Russian conference on isotope geochronology, Centre of Informational Culture, St. Petersburg, pp 63–65 (in Russian)

Bibikova EV, Kirnozova TI, Lazarev YuI, Makarov VA, Nikolaev AA (1990) U–Pb isotopic age of the Karelian Vepsian. Trans USSR Acad Sci 310:189–191

Biske NC, Romashkin AE, Rychanchik DV (2004) Proterozoic peperite-structures of Lebestchina. In: Golubev AI, Slabunov AI (eds) Geology and mineral deposits, Proceedings of the Institute of Geology, Karelian Research Centre of RAS, Petrozavodsk, 7: 193–199 (in Russian)

Bondar LF (1972) On geology of the Upper Jatuli-Suisari. The Munozero area, In: Sokolov VA (ed) Problems of geology of the middle Proterozoic of Karelia, Publ. "Karelia", Petrozavodsk, pp 144–152 (in Russian)

Brasier AT, Fallick AE, Prave AR, Melezhik VA, Lepland A, Scientists FAR-DEEP (2011) Coastal sabkha dolomites and calcitised sulphates preserving the Lomagundi-Jatuli carbon isotope signal. Precambrian Res 189:193–211

Butin PV (1959) Fossil *Cynophyceae* in the Proterozoic formations of Southern Karelia. Proc Karelian Kola Sci Branch 12:47–51 (in Russian)

Butin PV (1966) Proterozoic fossil algae of Karelia. In: Vologdin AG (ed) Remnants of organisms and problematics of the Proterozoic formations in Karelia. Karelian Book, Petrozavodsk, pp 34–64 (in Russian)

Chernov VM (ed) (1975) Geology and mineral deposits of Karelia. Karelia, Petrozavodsk, p 141 (in Russian)

Eliseev MA, Robonen VI (1983) On structural relationship between Sumian and Sariolian at Elmus. In: Heiskanen KI (ed) Geology and stratigraphy of Precambrian formations in Karelia. Institute of Geology, Karelian Branch of Academy of the USSR Academy of Sciences, Petrozavodsk, pp 11–13 (in Russian)

Eskola PE (1919) Hufvuddragen av Onega-Karelens geologi. Meddelande från Geologiska Föreningen i Helsingfors 1917–1918:13–18

Eskola P (1925) On the petrology of Eastern Fennoscandia. 1. The mineral development of basic rocks in the Karelian formations. Fennia 45(19):93

Ettensohn FR, Rast N, Brett CE (eds) (2002) Ancient seismites. Geol Soc Am, Spec Pap 359: 190

Fallick AE, Melezhik VA, Simonson B (2008) The ancient anoxic biosphere was not as we know it. In: Dobretsov N, Kolchanov N, Rozanov A, Zavarzin G (eds) Biosphere origin and evolution. Springer, New York, pp 169–188

Filippov MM (ed) (1994) The organic matter of Karelian Shungite rocks (Genesis, evolution and the methods of study). Institute of Geology, Karelian Research Centre of RAS, Petrozavodsk, p 208 (in Russian)

Filippov MM (2002) Shungite rocks of the Zaonega structure. Institute of Geology, Karelian Research Centre of RAS, Petrozavodsk, p 280 (in Russian)

Filippov MM (2004) Shungite rocks of Karelia. Institute of Geology, Karelian Research Centre of RAS, Petrozavodsk, p 488 (in Russian)

Filippov MM, Golubev AI (1994) Carbon isotope composition of shungite rocks. In: Filippov MM (ed) The organic matter of Karelian Shungite rocks (Genesis, evolution and the methods of study). Institute of Geology, Karelian Research Centre of RAS, Petrozavodsk, pp 32–43 (in Russian)

Filippov MM, Golubev AI, Romashkin AE, Rychanchik DV (1994) Mineral constituent of shungite-bearing rocks: primary composition, sources and relation to shungite matter. In: Filippov MM (ed) The organic matter of Karelian Shungite rocks (Genesis, evolution and the methods of study). Kola Science Centre, Petrozavodsk, pp 78–93 (in Russian)

Galdobina LP (1982) The geology of Proterozoic Shungite-Bearing volcano-sedimentary formations of Karelia. Institute of Geology. Karelian Research Centre of the USSR Academy of Sciences, Petrozavodsk, p 204, (in Russian)

Galdobina LP (1987a) The Ludicovian Super-Horizon. In: Sokolov VA (ed) Geology of Karelia. Nauka (Science), Leningrad, pp 59–67 (in Russian)

Galdobina LP (1987b) The Livvian Super-Horizon. In: Sokolov VA (ed) Geology of Karelia. Nauka (Science), Leningrad, pp 71–74 (in Russian)

Galdobina LP, Kalinin YuK, Kuprjakov SV (1986) Endogenic origin of Proterozoic shungite rocks in Karelia. In: Abstract, the 2nd all-union conference on carbon geochemistry, Institute of Geochemistry and Analytical Chemistry, Moscow, pp 79–81 (in Russian)

Galimov IM, Kuznetsova NG, Prokhorov VS (1968) The problem of the composition of the Earth's ancient atmosphere in connection with results of isotopic analyses of carbon from Precambrian carbonates. Geochemistry 11:1376–1381 (in Russian)

Gilyarova MA (1972) Precambrian stratigraphy and structures of Karelia and Kola Peninsula. Nauka (Science), Leningrad, p 218 (in Russian)

Glushanin LV, Sharov NV, Shchptsov VV (eds) (2011) Palaeoproterozoic Onega Structure (Geology, tectonics, deep structure and mineralogeny). Karelian Research Centre, Petrozavodsk, p 431 (in Russian)

Golubev AI (2011) The Jatulian dolerite-basalt complex. In: Glushanin LV, Sharov NV, Shchiptsov VV (eds) The Onega Palaeoproterozoic structure (Geology, tectonics, deep structure and minerageny). Institute of Geology, Karelian Research Centre of RAS, Petrozavodsk, pp 61–63 (in Russian)

Golubev AI, Kulikov VS (2011) The Zaonega dolerite-basalt complex. In: Glushanin LV, Sharov NV, Shchiptsov VV (eds) The Onega Palaeoproterozoic structure (Geology, tectonics, deep structure and minerageny). Institute of Geology, Karelian Research Centre of RAS, Petrozavodsk, pp 87–91 (in Russian)

Golubev AI, Svetov AP (1983) Geochemistry of Basalts of platform volcanism in Karelia. Karelia, Petrozavodsk, p 192 (in Russian)

Golubev AI, Akhmedov AM, Galdobina LP (1984) Geochemistry of lower Proterozoic Black Shale complexes of the Karelo-Kola Region. Nauka (Science), Leningrad, p 192 (in Russian)

Golubev AI, Svetov SA, Korosov VI, Svetova AI, Lavrov MM, Trofimov NN, Nazarova TN (2011) The Sumi. In: Glushanin LV, Sharov NV, Shchiptsov VV (eds) The Onega Palaeoproterozoic Structure (Geology, tectonics, deep structure and minerageny). Institute of Geology, Karelian Research Centre of RAS, Petrozavodsk, pp 32–52 (in Russian)

Gorlov VI, Filippov MM (1987) On genesis of the shungite carbon of Lower Proterozoic Suisari suite rocks of Karelia. In: Golod MI (ed) Methods and results of the geophysical investigation of Precambrian rocks of the Baltic Shield. Institute of Geology, Karelian Research Centre of RAS, Petrozavodsk, pp 105–122 (in Russian)

Gorokhov IM, Kuznetsov AB, Melezhik VA, Konstantinova GV, Melnikov NN (1998) Sr isotope composition of upper Jatulian dolostones of the Tulomozero Formation, south-eastern Karelia. Trans Russ Acad Sci, Geochem 360:533–536 (in Russian)

Gromova ZT, Yuzhakova VV, Zak CI (1954) Report on structural and exploration drilling for Haematite in the Kondopoga, Petrov and Medvezhgora areas of the KFSSR, and on Exploration Work for Cupper in the Medvezhgora area of the KFSSR (Tivdiy, Medvezhgora and Pjalm Geological Parties, 1952–1953), CSTGU, KKGRE, Petrozavodsk, p 363 (in Russian)

Hannah JL, Stein HJ, Zimmerman A, Yang G, Melezhik VA, Filippov MM, Turgeon SC, Creaser RA (2008) Re-Os geochronology of a 2.05 Ga fossil oil field near Shunga, Karelia, In: Abstract, the 33rd international geological congress, Oslo

Heiskanen KI (1975) A dynamic system of the Jatulian sedimentation in Central Katelia. In: Proceedings of the institute of geology, Karelian Branch of Academy of the USSR Academy of Sciences, Issue 28, Nauka (Science), Leningrad, p 104 (in Russian)

Heiskanen KI (1987a) The Sumian Super-Horizon. In: Sokolov VA (ed) Geology of Karelia. Nauka (Science), Leningrad, pp 40–44 (in Russian)

Heiskanen KI (1987b) The Sariolian Super-Horizon. In: Sokolov VA (ed) Geology of Karelia. Nauka (Science), Leningrad, pp 44–51 (in Russian)

Heiskanen KI (1990) Palaeogeography of the Baltic Shield in Karelian time. Institute of Geology. Karelian Research Centre of RAS, Petrozavodsk, p 124 (in Russian)

Heiskanen KI, Rychanchik DV (1999) The Jatulian, Early Proterozoic, carbonates with anomalously heavy carbon of the Baltic Shield. Stratigr Geol Corr 7(6):14–19 (in Russian)

Heiskanen KI, Golubev AI, Bondar LF (1977) Orogenic volcanism of Karelia. Nauka (Science), Leningrad, p 216 (in Russian)

Kajryak AI (1973) The Besovets series in the Onega structure. Nedra, Leningrad, p 176 (in Russian)

Karhu JA (1993) Palaeoproterozoic evolution of the carbon isotope ratios of sedimentary carbonates in the Fennoscandian Shield. Geol Surv Finl Bull 371:1–87

Kharitonov LYa (1938) New data on stratigraphy and tectonics of the Karelian formation from the Onego-Segozero watershed. Trans Leningrad Geol Trust 17:1–52 (in Russian)

Kharitonov LYa (1966) Structure and stratigraphy of the Karelides of the Eastern Part of the Baltic Shield. Nedra, Moscow, p 360 (in Russian)

Koistinen T, Stephens MB, Bogatchev V, Nordgulen Ø, Wennesström M, Korhonen J (comps.) (2001) Geological map of the Fennoscandian Shield, Scale 1:2 000 000, Espoo/Trondheim/Uppsala/Moscow

Korosov VI (1989) The Paljeozero suite. In: Shurkin KA (ed) Stratigraphic dictionary of the USSR, The lower Precambrian. Nauka (Science), Leningrad, p 215

Korosov VI (1991) The Geology of the Pre-Jatulian Proterozoic of the Eastern Part of the Baltic Shield (Sumian, Sariolian). Institute of Geology, Karelian Research Centre of RAS, Petrozavodsk, p 118 (in Russian)

Korosov VI, Nazarova TN (1989) General section of the sedimentary-volcanogenic formations and their relationship with Sariolian conglomerates in the Kumsa structure. In: Sokolov VA (ed) Problems of lower Proterozoic Stratigraphy of Karelia. Institute of Geology, Karelian Research Centre of RAS, Petrozavodsk, pp 39–47 (in Russian)

Korosov VI, Nazarova TN (2011) The Glubokozero Formation. In: Glushanin LV, Sharov NV, Shchiptsov VV (eds) The Onega Palaeoproterozoic structure (Geology, tectonics, deep structure and minerageny). Institute of Geology, Karelian Research Centre of RAS, Petrozavodsk, pp 32–34 (in Russian)

Korosov VI, Robonen VI, Nazarova TN (2011) Sariolian. In: Glushanin LV, Sharov NV, Shchiptsov VV (eds) The Onega Palaeoproterozoic structure (Geology, tectonics, deep structure and minerageny). Institute of Geology, Karelian Research Centre of RAS, Petrozavodsk, pp 50–52 (in Russian)

Kratz KO (1963) The geology of the Karelides of Karelia. Proc Lab Geol Precambrian 16:205 (in Russian)

Krupenik VA, Sveshnikova KY (2011) Correlation of the Onega parametric hole with the reference sections of the Onega Structure. In: Glushanin LV, Sharov NV, Shchiptsov VV (eds) The Onega Palaeoproterozoic structure (Geology, tectonics, deep structure and minerageny). Institute of Geology, Karelian Research Centre of RAS, Petrozavodsk, pp 190–195 (in Russian)

Krupenik VA, Akhmedov AM, Sveshnikova KYu (2011a) Structure of the section of the Onega structure based on the Onega parametric drillhole. In: Glushanin LV, Sharov NV, Shchiptsov VV (eds) The Onega Palaeoproterozoic structure (Geology, tectonics, deep structure and minerageny). Institute of Geology, Karelian Research Centre of RAS, Petrozavodsk, pp 172–189 (in Russian)

Krupenik VA, Akhmedov AM, Sveshnikova KYu (2011b) Isotopic composition of carbon, oxygen and sulphur in the Ludicovian and Jatulian rocks. In: Glushanin LV, Sharov NV, Shchiptsov VV (eds) The Onega Palaeoproterozoic Structure (Geology, Tectonics, Deep Structure and Minerageny). Institute of Geology, Karelian Research Centre of RAS, Petrozavodsk, pp 250–255 (in Russian)

Kulikov VS, Svetov AP, Golubev AI (1978) Main features of petrochemical evolution of the volcanism. In: Sokolov VA (ed) Volcanic structures of the Proterozoic in Karelia. Nauka (Science), Leningrad, pp 33–42 (in Russian)

Kulikov VS, Kulikova VV, Lavrov BS, Puchtel IS, Pisarevsky SA, Sokolov SYa (1999) Palaeoproterozoic Suisarian Picrite-Basalt complex of Karelia. Institute of Geology, Karelian Research Centre of RAS, Petrozavodsk, p 96 (in Russian)

Kulikov VS, Medvedev PV, Golubev AI (2011a) Summary geological map of the Onega Structure. In: Glushanin LV, Sharov NV, Shchiptsov VV (eds) The Onega Palaeoproterozoic structure (Geology, tectonics, deep structure and minerageny). Institute of Geology, Karelian Research Centre of RAS, Petrozavodsk, pp 19–23 (in Russian)

Kulikov VS, Svetov AP, Golubev AI (2011b) The Suisari Horizon (Suisari picrite-basalt complex). In: Glushanin LV, Sharov NV, Shchiptsov VV (eds) The Onega Palaeoproterozoic structure (Geology, tectonics, deep structure and minerageny). Institute of Geology, Karelian Research Centre of RAS, Petrozavodsk, pp 91–101 (in Russian)

Kuznetsov AB, Melezhik VA, Gorokhov IM, Melnikov NN, Konstantinova GV, Kutyavin EP, Turchenko TL (2010) Sr isotopic composition of Paleoproterozoic ^{13}C-rich carbonate rocks: the Tulomozero formation, SE Fennoscandian Shield. Precambrian Res 182:30–312

Lavrov BS (1999) Report on compilation of the lower Proterozoic reference stratigraphic sections within the Onega structure. Karelian Geological Expedition, Petrozavodsk, p 670 (in Russian)

Leonov MG, Kulikov VS, Zykov DC, Kolodyazhny SYu, Poleshchuk AV (2011) Tectonics. In: Glushanin LV, Sharov NV, Shchiptsov VV (eds) The Onega Palaeoproterozoic structure (Geology, tectonics, deep structure and minerageny). Institute of Geology, Karelian Research Centre of RAS, Petrozavodsk, pp 127–170 (in Russian)

Makarikhin VV (1992) Lower Precambrian stromatolite associations of Karelia. In: Schidlowski M, Golubic S, Kimberley MM, McKirdy DM, Trudinger PA (eds) Early organic evolution. Springer, Berlin/Heidelberg/New York, pp 463–467

Makarikhin VV (2011) Palaeontological study of Palaeoproterozoic rocks in Karelia, In: Geology of Karelia from the Archaean to the Present. In: Proceedings of the all-Russian conference convened to celebrate the 50th anniversary of the foundation of the Institute of Geology, Karelian Research Centre, RAS, Petrozavodsk, 24–26 May 2011, Karelian Research Centre, Petrozavodsk, pp 71–79 (in Russian with English abstract)

Makarikhin VV, Kononova GM (1983) Early Proterozoic Phytolithes of Karelia. Nauka (Science), Leningrad, p 180 (in Russian)

Medvedev PV (2011) Parastratotype of the Kalevian Super-Horizon in the Onega structure. In: Glushanin LV, Sharov NV, Shchiptsov VV (eds) The Onega Palaeoproterozoic structure (Geology, tectonics, deep structure and minerageny). Institute of Geology, Karelian Research Centre of RAS, Petrozavodsk, pp 101–105 (in Russian)

Medvedev PV, Makarikhin VV, Golubev AI, Rychanchik DV, Trofimov NN (2011) The Jatuli. In: Glushanin LV, Sharov NV, Shchiptsov VV (eds) The Onega Palaeoproterozoic structure (Geology, tectonics, deep structure and minerageny). Institute of Geology, Karelian Research Centre of RAS, Petrozavodsk, pp 52–67 (in Russian)

Melezhik VA, Fallick AE (1996) A widespread positive $\delta^{13}C_{carb}$ anomaly at around 2.33–2.06 Ga on the Fennoscandian Shield: a paradox? Terra Nova 8:141–157

Melezhik VA, Fallick AE, Filippov MM, Larsen O (1999a) Karelian shungite – an indication of 2000 Ma-year-old metamorphosed oil-shale and generation of petroleum: geology, lithology and geochemistry. Earth-Sci Rev 47:1–40

Melezhik VA, Fallick AE, Medvedev PV, Makarikhin VV (1999b) Extreme $^{13}C_{carb}$ enrichment in ca. 2.0 Ga magnesite-stromatolite-dolomite-'red beds' association in a global context: a case for the world-wide signal enhanced by a local environment. Earth-Sci Rev 48:71–120

Melezhik VA, Fallick AE, Medvedev PV, Makarikhin VV (2000) Palaeoproterozoic magnesite–stromatolite–dolostone–'red bed' association, Russian Karelia: palaeoenvironmental constraints on the 2.0 Ga positive carbon isotope shift. Norsk Geologisk Tidsskrift 80:163–186

Melezhik VA, Fallick AE, Medvedev PV, Makarikhin V (2001) Palaeoproterozoic magnesite: lithological and isotopic evidence for playa/sabkha environments. Sedimentology 48:379–397

Melezhik VA, Filippov MM, Romashkin AE (2004) A giant Palaeoproterozoic deposit of shungite in NW Russia: genesis and practical applications. Ore Geol Rev 24:135–154

Melezhik VA, Fallick AE, Hanski E, Kump L, Lepland A, Prave A, Strauss H (2005a) Emergence of the aerobic biosphere during the Archean-Proterozoic transition: challenges of future research. Geol Soc Am Today 15:4–11

Melezhik VA, Fallick AE, Rychanchik DV, Kuznetsov AB (2005b) Palaeoproterozoic evaporites in Fennoscandia: implications for seawater sulphate, $\delta^{13}C$ excursions and the rise of atmospheric oxygen. Terra Nova 17:141–148

Melezhik VA, Fallick AE, Filippov MM, Lepland A, Rychanchik DV, Deines YE, Medvedev PV, Romashkin AE, Strauss H (2009) Petroleum surface oil seeps from Palaeoproterozoic petrified giant oilfield. Terra Nova 21:119–126

Mikhailjuk EM (1988) Report on the topic A-12-148, Years 1984–1987, Book 5, United 1:200,000 map of the Onega structure, Petrozavodsk (in Russian)

Mishunina ZA (1979) Petroleum genesis in carbonate formations. In: Nalivkin VD, Ivanchuk PK, Dvali MF (eds) Problems in the geology of oil and gas. Nedra, Leningrad, pp 139–143 (in Russian)

Morozov AF, Hakhaev BN, Petrov OV, Gorbachev VI, Tarkhanov GB, Tsvetkov LD, Erinchek YuM, Akhmedov AM, Krupenik VA, Sveshnikova KYu (2010) Rock-salts in Palaeoproterozoic strata of the Onega depression of Karelia (based on data from the Onega parametric drillhole). Trans Acad Sci 435(2):230–233 (in Russian)

Negrutsa VZ (1963) Experience in the sedimentological study of Proterozoic (Jatulian) formations of Central Karelia. Sov Geol 7:52–76 (in Russian)

Negrutsa VZ (1967) Experience in making palaeogeographic maps of the Karelian Proterozoic on the basis of structural-facial study of the type structures of the Karelides. In: Bel'kov IV (ed) Problems of research of the Precambrian geology. Nauka (Science), Leningrad, pp 181–188 (in Russian)

Negrutsa VZ (1984) Early Proterozoic stages of evolution of the Eastern Baltic Shield. Nedra, Leningrad, p 270 (in Russian)

Ojakangas RW, Marmo JS, Heiskanen KI (2001) Basin evolution of the Paleoproterozoic Karelian Supergroup of the Fennoscandian (Baltic) Shield. Sediment Geol 141–142:255–285

Ovchinnikova GV, Kusnetzov AB, Melezhik VA, Gorokhov IM, Vasil'eva IM, Gorokhovsky BM (2007) Pb-Pb age of Jatulian carbonate rocks: the Tulomozero Formation in south-eastern Karelia. Stratigr Geol Correlat 4:20–33 (in Russian)

Perevozchikova VV (1957) Proterozoic geology of Karelia. Mater Geol Miner Resour Northwest USSR 1:25–51 (in Russian)

Peterson CD (1997) Coseismic paleoliquefaction evidence in the central Cascadia margin, USA, Oregon. Geology 59:51–74

Philippov NB, Trofimov NN, Golubev AI, Sergeev SA, Huhma H (2007) New geochronological data on the Koikary-Svjatnavolok and Pudozhgora gabbro-dolerite intrusives. In: Golubev VI, Shchiptsov VV (eds) Geol Miner Resour Karelia 10: 49–68 (in Russian with English abstract)

Polekhovsky YuS, Golubev AI (1989) The Ludicovian Horizon of the Onega depression. In: Svetov AP (ed) Problems of the Palaeoproterozoic Stratigraphy of Karelia. Institute of Geology, Karelian Branch of the Academy of Sciences of the USSR, Petrozavodsk, pp 106–117 (in Russian)

Poleshchuk AV (2011) Sill genesis in the Paleoproterozoic tectonic evolution of the Onega Trough, Baltic Shield. Trans Acad Sci Geol 439(3):365–369 (in Russian)

Puchtel IS, Humayun M (2001) Platinum group element fractionation in a komatiitic basalt lava lake. Geochim Cosmochim Acta 65:2979–2993

Puchtel IS, Zhuravlev DZ, Ashikhmina NA, Kulikov VS, Kulikova VV (1992) Sm-Nd age of the Suisarian suite on the Baltic Shield. Trans Russ Acad Sci 326:706–711 (in Russian)

Puchtel IS, Bogatikov OA, Kulikov VS, Kulikova VV, Zhuravlev DZ (1995) The role of crustal and mantle sources in the petrogenesis of continental magmatism: Evidence from isotope and geochemical study of the early Proterozoic picrites from the Onega plateau, Baltic Shield. Petrology 3:397–419 (in Russian)

Puchtel IS, Arndt NT, Hofmann AW, Haase KM, Kröner A, Kulikov VS, Kulikova VV, Garbe-Schönberg C-D, Nemchin AA (1998) Petrology of mafic lavas within the Onega plateau, Central Karelia: evidence for the 2.0 Ga plume-related continental crustal growth in the Baltic Shield. Contrib Miner Petrol 130:134–153

Puchtel IS, Brügmann GE, Hofmann AW (1999) Precise Re-Os mineral isochron and Pb-Nd-Os isotope systematics of a mafic-ultramafic sill in the 2.0 Ga Onega plateau (Baltic Shield). Earth Planet Sci Lett 170:447–461

Reineck H, Wunderlich F (1968) Classification and origin of flaser and lenticular bedding. Sedimentology 11:99–104

Reuschel M, Melezhik VA, Whitehouse MJ, Lepland A, Fallick AE, Strauss H (2012) Isotopic evidence for a sizeable seawater sulfate reservoir at 2.1 Ga. Precambrian Res 192–195:78–88

Rozhkov VI (1895) Mining activity in the Olonets district. Olonets Govern Prov News 71 (in Russian)

Rychanchik DV (2011a) The Zaonega Horizon. In: Glushanin LV, Sharov NV, Shchiptsov VV (eds) The Onega Palaeoproterozoic structure (Geology, tectonics, deep structure and minerageny). Institute of Geology, Karelian Research Centre of RAS, Petrozavodsk, pp 67–69 (in Russian)

Rychanchik DV (2011b) The Kondopoga Suite. In: Glushanin LV, Sharov NV, Shchiptsov VV (eds) The Onega Palaeoproterozoic structure (Geology, tectonics, deep structure and minerageny). Institute of Geology, Karelian Research Centre of RAS, Petrozavodsk, pp 105–109 (in Russian)

Rychanchik DV (2011c) Petrogeochemical and geochemical characteristic of the Kondopoga and Vashozero suites. In: Glushanin LV, Sharov NV, Shchiptsov VV (eds) The Onega Palaeoproterozoic structure (Geology, tectonics, deep structure and minerageny). Institute of Geology, Karelian Research Centre of RAS, Petrozavodsk, pp 110–114 (in Russian)

Rychanchik DV, Filippov MM, Romashkin AE (2011) Shungite-bearing rocks of the Upper sub-suite. In: Glushanin LV, Sharov NV, Shchiptsov VV (eds) The Onega Palaeoproterozoic structure (Geology, tectonics, deep structure and minerageny). Institute of Geology, Karelian Research Centre of RAS, Petrozavodsk, pp 67–69 (in Russian)

Satzuk YuI, Kononova GM (1975) Reefogenic complex from the upper algal horizon of the Jatuli in the Onega basin. In: Chernov VM (ed) Geology and mineral resources of Karelia. Karelia, Pertozavodsk, pp 110–116 (in Russian)

Satzuk YuI, Makarikhin VA, Medvedev PV (1988) Jatulian geology of the Onega-Segozero Watershed. Nauka (Science), Leningrad, p 96 (in Russian)

Schidlowski M, Eichmann R, Junge CE (1975) Precambrian sedimentary carbonates: carbon and oxygen isotope geochemistry and implications for the terrestrial oxygen budget. Precambrian Res 2:1–69

Semikhatov MA, Shurkin KA, Aksenov EM (1991) A new stratigraphic scale for the Precambrian of the USSR. Proc USSR Acad Sci, Geol Ser 4:3–16 (in Russian)

Shatsky GV (1990) Isotope composition of sulphides from the Zazhogino shungite deposit. Lithol Min Res 1:20–28 (in Russian)

Shields GA (2002) 'Molar-tooth microspar': a chemical explanation for its disappearance 750 Ma. Terra Nova 14:108–113

Sivaev VV, Goroshko AF, Gorbatyuk LV (1982) Report on a 1:50,000 Geological Mapping and Investigation of the North-Western Limb of the Onega Depression and its Surroundings Conducted in 1978–1982, CSPGO, KKGRE, Petrozavodsk, p 424 (in Russian)

Sokolov VA (1955) On the geology of Upper Karelian formations of the northern Onega area. Proc Karelo-Finnish Branch USSR Acad Sci 3:17–27 (in Russian)

Sokolov VA (1963) Geology and lithology of middle Proterozoic Carbonate rocks of Karelia. Nauka (Science), Moscow-Leningrad, p 196 (in Russian)

Sokolov VA (ed) (1972a) Problems of geology of the middle Proterozoic of Karelia. Karelia, Petrozavodsk, p 187 (in Russian)

Sokolov VA (1972b) Type sections of the Jatulian deposits. Probl Mid Proterozoic Geol Karelia 16:124–128 (in Russian)

Sokolov VA (ed) (1973) Stages of tectonic evolution of Precambrian of Karelia. Nauka (Science), Leningrad, p 175 (in Russian)

Sokolov VA (ed) (1984) Precambrian stratigraphy of KASSR. Karelian Branch of the USSR Academy of Sciences, Petrozavodsk, p 115 (in Russian)

Sokolov VA (ed) (1987a) Geology of Karelia. Nauka (Science), Leningrad, p 231 (in Russian)

Sokolov VA (1987b) The Jatulian Super-Horizon. In: Sokolov VA (ed) Geology of Karelia. Nauka (Science), Leningrad, pp 51–59 (in Russian)

Sokolov VA, Galdobina (1982) The Ludicovi – a new stratigraphic subdivision of the lower Proterozoic in Karelia. Trans Russ Acad Sci 267(1):187–190 (in Russian)

Sokolov VA, Heiskanen KI (1966) Geological and geochemical charcteristics of Proterozoic (Jatulian) weathering crusts. In: Sidorenko AV (ed) Problems of sedimentary geology of the Precambrian, vol 1. Nauka (Science), Moscow, pp 176–185 (in Russian)

Sokolov VA, Galdobina LP, Ryleev AB, Catzuk YuI, Heiskanen KI (1966) On lithology and palaeogeography of the Jatulian formations of the Segozero-Elmozero area. In: Borisov PA, Borovikov PP, Bogachev AI, Voenushkin SF, Kratz KO, Lebedev VI, Ozhinskiy IS, Pokrovskiy SD, Porotova GS, Robonen VI, Sokolov VA (eds) Questions on geology and spatial regularities of the distributions of mineral recourses of Karelia. Karelian Book, Petrozavodsk, pp 168–186 (in Russian)

Sokolov VA, Galdobina LP, Ryleev AB, Satzuk YuI, Heiskanen KI (1970) Geology, sedimentology and palaeogeography of the Jatuli, Central Karelia. In: Proceedings of the institute of geology, Karelian Branch of Academy of the USSR Academy of Sciences, Issue 6, Karelia, Petrozavodsk, p 367 (in Russian)

Svetov AP (1972) Jatulian palaeovolcanology of Central Karelia. In: Proceedings of the institute of geology, Karelian Branch of the USSR Academy of Sciences, Issue 11, Nauka (Science), Leningrad, p 115 (in Russian)

Svetov SA (2008) Liquation differentiation in basalt systems (on example of Suisari variolites from the Jalguba Ridge). In: Golubev AI, Slabunov AI (ed) Geol Miner Resour Karelia 11: 120–134 (in Russian)

Svetov SA (2009) The oldest Adakites of the Fennoscandian Shield. Institute of Geology, Karelian Research Centre of RAS, Petrozavodsk, p 115 (in Russian)

Svetov AP, Golubev AI (1972) The structure of Lower Proterozoic volcanogenic formation from the Onega area (Karelia). Trans USSR Acad Sci 206:1429–1432 (in Russian)

Svetov SA, Kulikov VS (2011) Petrological-geodynamic model. In: Glushanin LV, Sharov NV, Shchiptsov VV (eds) The Onega Palaeoproterozoic structure (Geology, tectonics, deep structure and minerageny). Institute of Geology, Karelian Research Centre of RAS, Petrozavodsk, pp 378–383 (in Russian)

Svetov SA, Golubev AI, Svetova AI (2004) Geochemistry of Sumian basaltic andesites of Central Karelia. Geochem Int 42:630–640

Svetov SA, Korosov VI, Svetova AI, Nazarova TN (2011) The Kumsa Suite. In: Glushanin LV, Sharov NV, Shchiptsov VV (eds) The Onega Palaeoproterozoic structure (Geology, tectonics, deep structure and minerageny). Institute of Geology, Karelian Research Centre of RAS, Petrozavodsk, pp 34–39 (in Russian)

Systra YuI (1991) Tectonics of the Karelian region. Nauka (Science), St. Petersburg, p 176 (in Russian)

Tikhomirova M, Makarikhin VV (1993) Possible reasons for the $\delta^{13}C$ anomaly of Lower Proterozoic sedimentary carbonates. Terra Res 5:244–248

Timofeev VM (1935) Petrography of Karelia. USSR Academy of Sciences, Moscow-Leningrad, p 256 (in Russian)

Vologdin AG (1966) Relicts of Proterozoic algae from Karelia. In: Vologdin AG (ed) Remnants of organisms and Problematics of the Proterozoic formation in Karelia. Karelian Book Publisher, Petrozavodsk, pp 65–95 (in Russian)

Yakovleva VV, Gilyarova MA (1960) The Middle Proterozoic. The north-western Onega area, In: Kratz KO (ed) Geology of the USSR. Karelian ASSR, vol 37, pp 138–152 (in Russian)

Yudovich YE, Makarikhin VV, Medvedev PV, Sukhanov NV (1991) Carbon isotope anomalies in carbonates of the Karelian Complex. Geochem Int 28:56–62

MIX
Papier aus verantwortungsvollen Quellen
Paper from responsible sources
FSC® C105338

If you have any concerns about our products,
you can contact us on
ProductSafety@springernature.com

In case Publisher is established outside the EU,
the EU authorized representative is:
Springer Nature Customer Service Center GmbH
Europaplatz 3, 69115 Heidelberg, Germany

Printed by Libri Plureos GmbH
in Hamburg, Germany